Published for
OXFORD INTERNATIONAL
AQA EXAMINATIONS

International A Level
CHEMISTRY

AS and
A LEVEL

Ted Lister
Janet Renshaw

OXFORD
UNIVERSITY PRESS

OXFORD
UNIVERSITY PRESS

Great Clarendon Street, Oxford, OX2 6DP, United Kingdom

Oxford University Press is a department of the University of Oxford. It furthers the University's objective of excellence in research, scholarship, and education by publishing worldwide. Oxford is a registered trade mark of Oxford University Press in the UK and in certain other countries

First published in 2016

British Library Cataloguing in Publication Data
Data available

978-0-19-837602-6

10 9 8

Paper used in the production of this book is a natural, recyclable product made from wood grown in sustainable forests.
The manufacturing process conforms to the environmental regulations of the country of origin.

Printed in United Kingdom

Acknowledgements
The publishers would like to thank the following for permissions to use their photographs:

Cover: Francisco Chanes/Custom Medical Stock Photo/Science Photo

p17: John Mclean/Science Photo Library; **p19**: Jpl-Caltech/Msss/NASA; **p32**: Martyn Chillmaid; **p36**: Hulton-Deutsch Collection/Corbis; **p40**: Martyn Chillmaid; **p70**: Nicolas/iStockphoto; **p81**: Martyn F. Chillmaid/Science Photo Library; **p89**: Martyn F. Chillmaid; **p111t**: Shutterstock; **p111b**: Shutterstock; **p117**: Science Photo Library; **p120**: Science Photo Library; **p147**: Science Photo Library; **p160**: Andrew Fletcher/Shutterstock; **p163t**: Green Gate Publishing Services, Tonbridge, Kent; **p163b**: Russ Munn/Agstockusa/Science Photo Library; **p165**: Sciencephotos/Alamy; **p168**: Shutterstock; **p171**: Andrew Lambert Photography/Science Photo Library; **p190t**: Molekuul. Be/Shutterstock; **p198t**: Martyn F. Chillmaid; **p198b**: Martyn F. Chillmaid/Science Photo Library; **p202**: Green Gate Publishing Services; **p204**: E.R.Degginger/Science Photo Library; **p205t**: Kzenon/Shutterstock;**p205b**: Science Photo Library; **p207**: Shutterstock; **p212**: Martyn F. Chillmaid/Science Photo Library; **p223**: Martyn F. Chillmaid; **p230**: Corel; **p231**: Martyn F. Chillmaid/Science Photo Library; **p232**: iStockphoto; **p239**: Martyn F. Chillmaid/Science Photo Library; **p244**: Andrew Lambert Photography/Science Photo Library; **p245**: Andrew Lambert Photography/Science Photo Library; **p249**: Corel; **p250**: Colin Cuthbert/Newcastle University/ Science Photo Library; **p271**: Shutterstock; **p322**: Colin Cuthbert/Science Photo Library; **p328**: Public Health England/Science Photo Library; **p353t - p444**: Shutterstock; **p456**: MeePoohyaPhoto/Shutterstock; **p461**: Shutterstock; **p465**: dt03mbb/iStockphoto; **p503**: Shutterstock; **p521**: Science Photo Library.

Although we have made every effort to trace and contact all copyright holders before publication this has not been possible in all cases. If notified, the publisher will rectify any errors or omissions at the earliest opportunity.

Links to third party websites are provided by Oxford in good faith and for information only. Oxford disclaims any responsibility for the materials contained in any third party website referenced in this work.

AQA material is reproduced by permission of AQA.
With special thanks to Colin Chambers for his advice and guidance, and Emma Gadsden for her editing and management of the project.

Contents

Answers to the Practice Questions and Section Questions are available at
www.oxfordsecondary.com/oxfordaqaexams-alevel-chemistry

AS/A Level course structure

This book has been written to support students studying for Oxford AQA International AS and A Level Chemistry. The sections covered are shown in the contents list, which shows you the page numbers for the main topics within each section. There is also an index at the back to help you find what you are looking for. If you are studying for AS Chemistry, you will only need to know the content in the blue box for the AS exams.

AS exam

Year 1 content

1 Physical chemistry 1

2 Inorganic chemistry 1

3 Organic chemistry 1

Year 2 content

4 Physical chemistry 2

5 Inorganic chemistry 2

6 Organic chemistry 2

A level exam

A Level exams will cover content from Year 1 and Year 2 and will be at a higher demand than the AS exams. You will also carry out practical activities throughout your course.

How to use this book

Learning objectives

- At the beginning of each topic, there is a list of learning objectives.
- These are matched to the specification and allow you to monitor your progress.
- A specification reference is also included.
  Specification reference: 3.1.1

This book contains many different features. Each feature is designed to foster and stimulate your interest in chemistry, as well as supporting and developing the skills you will need for your examinations.

Terms that it is important you are able to define and understand are highlighted in **bold orange text**. You can look these words up in the glossary.

Sometimes a word appears in **bold**. These are words that are useful to know but are not used on the specification. They therefore do not have to be learnt for examination purposes.

Synoptic link

These highlight how the sections relate to each other. Linking different areas of chemistry together is important, and you will need to be able to do this.

There are also links to the mathematical skills on the specification. More detail can be found in the maths section.

Application features

These features contain important and interesting applications of chemistry in order to emphasise how scientists and engineers have used their scientific knowledge and understanding to develop new applications and technologies. There are also application features to develop your maths skills, and to develop your practical skills.

Extension features

These features contain material that is beyond the specification designed to stretch and provide you with a broader knowledge and understanding and lead the way into the types of thinking and areas you might study in further education. As such, neither the detail nor the depth of questioning will be required for the examinations. But this book is about more than getting through the examinations.

1 Extension and application features have questions that link the material with concepts that are covered in the specification. Short answers are inverted at the bottom of the feature, whilst longer answers where appropriate can be found in the answers section at the back of the book.

Study tips

Study tips contain prompts to help you with your revision. They can also support the development of your practical skills.

Hint

Hint features give other information or ways of thinking about a concept to support your understanding. They can also relate to practical or mathematical skills.

Summary questions

1 These are short questions that test your understanding of the topic and allow you to apply the knowledge and skills you have acquired. The questions are ramped in order of difficulty.

Introduction at the opening of each section summarises what you need to know.

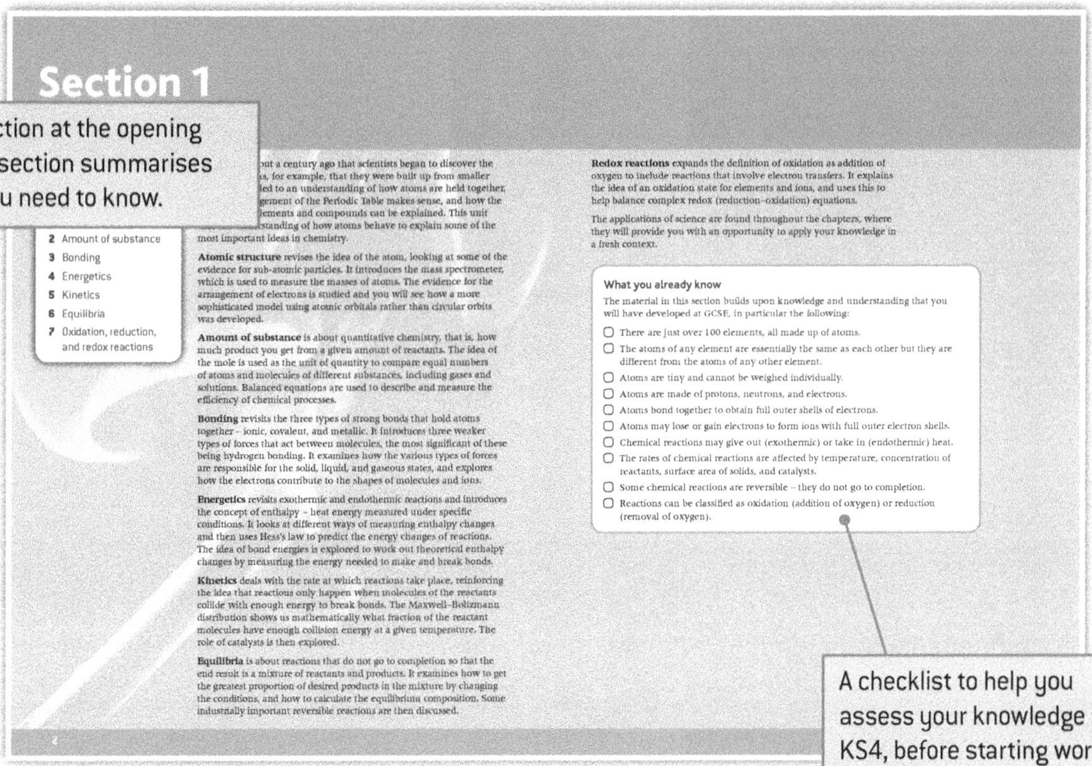

Section 1

...out a century ago that scientists began to discover the ...s, for example, that they were built up from smaller ...led to an understanding of how atoms are held together, ...gement of the Periodic Table makes sense, and how the ...lements and compounds can be explained. This unit ...standing of how atoms behave to explain some of the most important ideas in chemistry.

2 Amount of substance
3 Bonding
4 Energetics
5 Kinetics
6 Equilibria
7 Oxidation, reduction, and redox reactions

Atomic structure revises the idea of the atom, looking at some of the evidence for sub-atomic particles. It introduces the mass spectrometer, which is used to measure the masses of atoms. The evidence for the arrangement of electrons is studied and you will see how a more sophisticated model using atomic orbitals rather than circular orbits was developed.

Amount of substance is about quantitative chemistry, that is, how much product you get from a given amount of reactants. The idea of the mole is used as the unit of quantity to compare equal numbers of atoms and molecules of different substances, including gases and solutions. Balanced equations are used to describe and measure the efficiency of chemical processes.

Bonding revisits the three types of strong bonds that hold atoms together – ionic, covalent, and metallic. It introduces three weaker types of forces that act between molecules, the most significant of these being hydrogen bonding. It examines how the various types of forces are responsible for the solid, liquid, and gaseous states, and explores how the electrons contribute to the shapes of molecules and ions.

Energetics revisits exothermic and endothermic reactions and introduces the concept of enthalpy – heat energy measured under specific conditions. It looks at different ways of measuring enthalpy changes and then uses Hess's law to predict the energy changes of reactions. The idea of bond energies is explored to work out theoretical enthalpy changes by measuring the energy needed to make and break bonds.

Kinetics deals with the rate at which reactions take place, reinforcing the idea that reactions only happen when molecules of the reactants collide with enough energy to break bonds. The Maxwell–Boltzmann distribution shows us mathematically what fraction of the reactant molecules have enough collision energy at a given temperature. The role of catalysts is then explored.

Equilibria is about reactions that do not go to completion so that the end result is a mixture of reactants and products. It examines how to get the greatest proportion of desired products in the mixture by changing the conditions, and how to calculate the equilibrium composition. Some industrially important reversible reactions are then discussed.

Redox reactions expands the definition of oxidation as addition of oxygen to include reactions that involve electron transfers. It explains the idea of an oxidation state for elements and ions, and uses this to help balance complex redox (reduction–oxidation) equations.

The applications of science are found throughout the chapters, where they will provide you with an opportunity to apply your knowledge in a fresh context.

What you already know

The material in this section builds upon knowledge and understanding that you will have developed at GCSE, in particular the following:

- ☐ There are just over 100 elements, all made up of atoms.
- ☐ The atoms of any element are essentially the same as each other but they are different from the atoms of any other element.
- ☐ Atoms are tiny and cannot be weighed individually.
- ☐ Atoms are made of protons, neutrons, and electrons.
- ☐ Atoms bond together to obtain full outer shells of electrons.
- ☐ Atoms may lose or gain electrons to form ions with full outer electron shells.
- ☐ Chemical reactions may give out (exothermic) or take in (endothermic) heat.
- ☐ The rates of chemical reactions are affected by temperature, concentration of reactants, surface area of solids, and catalysts.
- ☐ Some chemical reactions are reversible – they do not go to completion.
- ☐ Reactions can be classified as oxidation (addition of oxygen) or reduction (removal of oxygen).

2

A checklist to help you assess your knowledge from KS4, before starting work on the section.

Visual summaries of each section show how some of the key concepts of that section interlink.

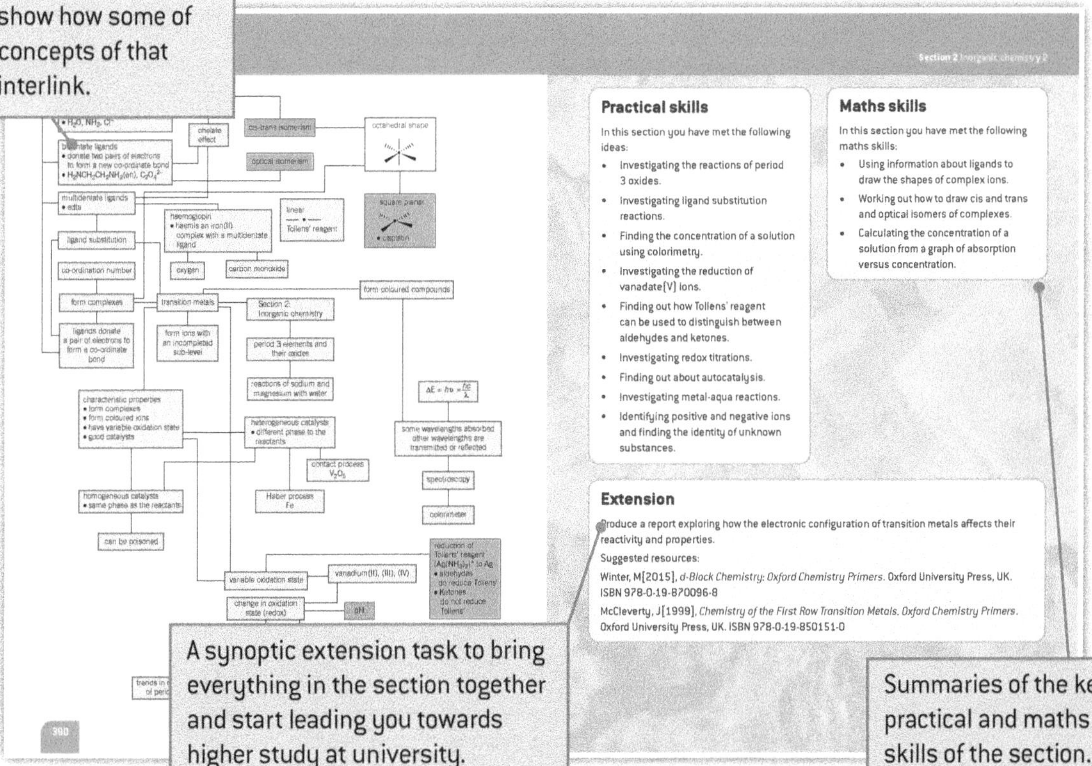

Practical skills

In this section you have met the following ideas:

- Investigating the reactions of period 3 oxides.
- Investigating ligand substitution reactions.
- Finding the concentration of a solution using colorimetry.
- Investigating the reduction of vanadate(V) ions.
- Finding out how Tollens' reagent can be used to distinguish between aldehydes and ketones.
- Investigating redox titrations.
- Finding out about autocatalysis.
- Investigating metal-aqua reactions.
- Identifying positive and negative ions and finding the identity of unknown substances.

Maths skills

In this section you have met the following maths skills:

- Using information about ligands to draw the shapes of complex ions.
- Working out how to draw cis and trans and optical isomers of complexes.
- Calculating the concentration of a solution from a graph of absorption versus concentration.

Extension

Produce a report exploring how the electronic configuration of transition metals affects their reactivity and properties.

Suggested resources:

Winter, M[2015], *d-Block Chemistry: Oxford Chemistry Primers*. Oxford University Press, UK. ISBN 978-0-19-870096-8

McCleverty, J[1999], *Chemistry of the First Row Transition Metals. Oxford Chemistry Primers*. Oxford University Press, UK. ISBN 978-0-19-850151-0

390

A synoptic extension task to bring everything in the section together and start leading you towards higher study at university.

Summaries of the key practical and maths skills of the section.

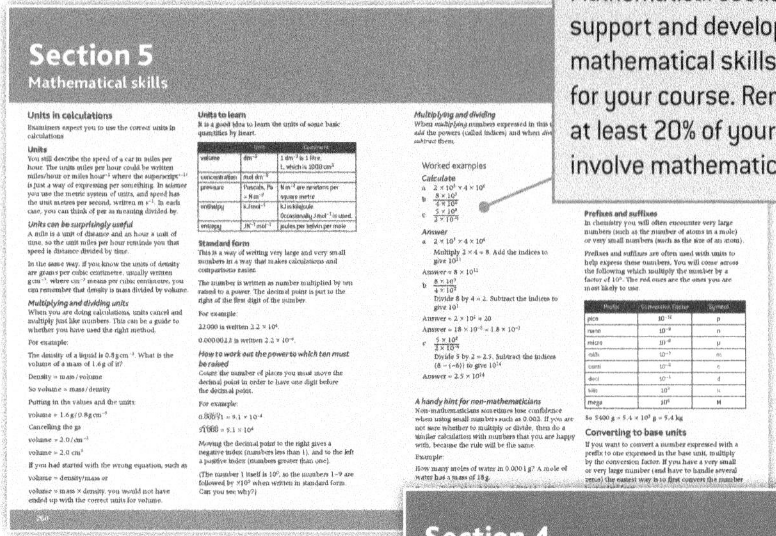

Mathematical section to support and develop your mathematical skills required for your course. Remember, at least 20% of your exam will involve mathematical skills.

Practical skills section with questions for each suggested practical on the specification. Remember, at least 15% of your exam will be based on practical skills.

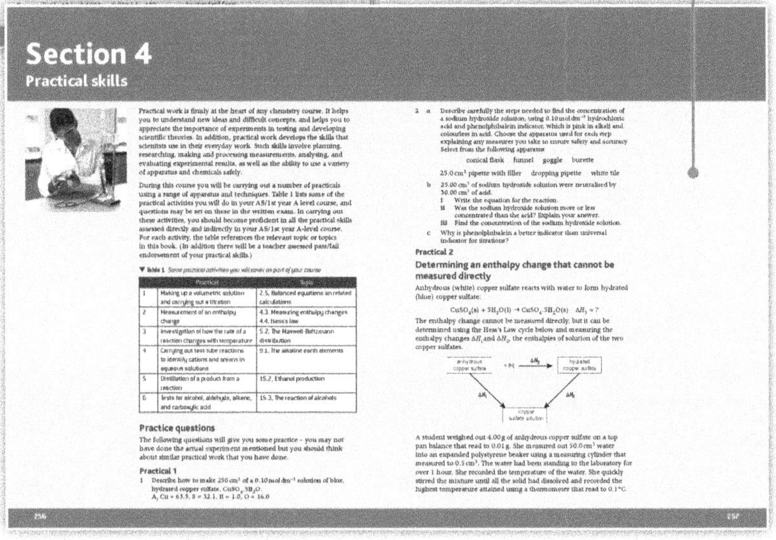

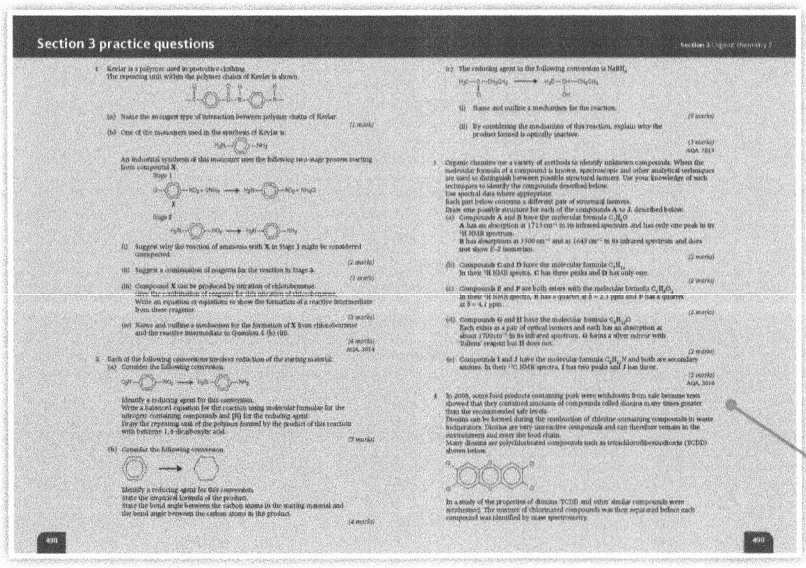

Practice questions at the end of each chapter and each section, including questions that cover practical and maths skills. There are also additional practice questions at the end of the book.

Section 1

AS Physical chemistry 1

It was only about a century ago that scientists began to discover the nature of atoms, for example, that they were built up from smaller particles. This led to an understanding of how atoms are held together, why the arrangement of the Periodic Table makes sense, and how the properties of elements and compounds can be explained. This unit uses the understanding of how atoms behave to explain some of the most important ideas in chemistry.

Atomic structure revises the idea of the atom, looking at some of the evidence for sub-atomic particles. It introduces the mass spectrometer, which is used to measure the masses of atoms. The evidence for the arrangement of electrons is studied and you will see how a more sophisticated model using atomic orbitals rather than circular orbits was developed.

Amount of substance is about quantitative chemistry, that is, how much product you get from a given amount of reactants. The idea of the mole is used as the unit of quantity to compare equal numbers of atoms and molecules of different substances, including gases and solutions. Balanced equations are used to describe and measure the efficiency of chemical processes.

Bonding revisits the three types of strong bonds that hold atoms together – ionic, covalent, and metallic. It introduces three weaker types of forces that act between molecules, the most significant of these being hydrogen bonding. It examines how the various types of forces are responsible for the solid, liquid, and gaseous states, and explores how the electrons contribute to the shapes of molecules and ions.

Energetics revisits exothermic and endothermic reactions and introduces the concept of enthalpy – heat energy measured under specific conditions. It looks at different ways of measuring enthalpy changes and then uses Hess's law to predict the energy changes of reactions. The idea of bond energies is explored to work out theoretical enthalpy changes by measuring the energy needed to make and break bonds.

Kinetics deals with the rate at which reactions take place, reinforcing the idea that reactions only happen when molecules of the reactants collide with enough energy to break bonds. The Maxwell–Boltzmann distribution shows us mathematically what fraction of the reactant molecules have enough collision energy at a given temperature. The role of catalysts is then explored.

Equilibria is about reactions that do not go to completion so that the end result is a mixture of reactants and products. It examines how to get the greatest proportion of desired products in the mixture by changing the conditions, and how to calculate the equilibrium composition. Some industrially important reversible reactions are then discussed.

Chapters in this section

Oxidation, reduction, and redox reactions expands the definition of oxidation as addition of oxygen to include reactions that involve electron transfers. It explains the idea of an oxidation state for elements and ions, and uses this to help balance complex redox (reduction–oxidation) equations.

The applications of science are found throughout the chapters, where they will provide you with an opportunity to apply your knowledge in a fresh context.

What you already know

The material in this section builds upon knowledge and understanding that you will have developed at GCSE, in particular the following:

- ☐ There are just over 100 elements, all made up of atoms.
- ☐ The atoms of any element are essentially the same as each other but they are different from the atoms of any other element.
- ☐ Atoms are tiny and cannot be weighed individually.
- ☐ Atoms are made of protons, neutrons, and electrons.
- ☐ Atoms bond together to obtain full outer shells of electrons.
- ☐ Atoms may lose or gain electrons to form ions with full outer electron shells.
- ☐ Chemical reactions may give out (exothermic) or take in (endothermic) heat.
- ☐ The rates of chemical reactions are affected by temperature, concentration of reactants, surface area of solids, and catalysts.
- ☐ Some chemical reactions are reversible – they do not go to completion.
- ☐ Reactions can be classified as oxidation (addition of oxygen) or reduction (removal of oxygen).

1 Atomic structure
1.1 Fundamental particles

Developing ideas of the atom

The Greek philosophers had a model in which matter was made up of a single continuous substance that produced the four elements – earth, fire, water, and air. The idea that matter was made of individual atoms was not taken seriously for another 2000 years. During this time alchemists built up a lot of evidence about how substances behave and combine. Their aim was to change other metals into gold. Here are a few of the steps that led to our present model.

1661 Robert Boyle proposed that there were some substances that could not be made simpler. These were the chemical elements, as we now know them.

1803 John Dalton suggested that elements were composed of indivisible atoms. All the atoms of a particular element had the same mass and atoms of different elements had different masses. Atoms could not be broken down.

1896 Henri Becquerel discovered radioactivity. This showed that particles could come from inside the atom. Therefore the atom was not indivisible. The following year, J J Thomson discovered the electron. This was the first sub-atomic particle to be discovered. He showed that electrons were negatively charged and electrons from all elements were the same.

As electrons had a negative charge, there had to be some source of positive charge inside the atom too. Also, as electrons were much lighter than whole atoms, there had to be something to account for the rest of the mass of the atom. Thompson suggested that the electrons were located within the atom in circular arrays, like plums in a pudding of positive charge, see Figure 1.

1911 Ernest Rutherford and his team found that most of the mass and all the positive charge of the atom was in a tiny central nucleus.

So, for many years, it has been known that atoms themselves are made up of smaller particles, called sub-atomic particles. The complete picture is still being built up in 'atom smashers' such as the one at CERN, near Geneva.

Learning objectives:

- → State the relative masses of protons, neutrons, and electrons.
- → State the relative charges of protons, neutrons, and electrons.
- → Explain how these particles are arranged in an atom.

Specification reference: 3.1.1

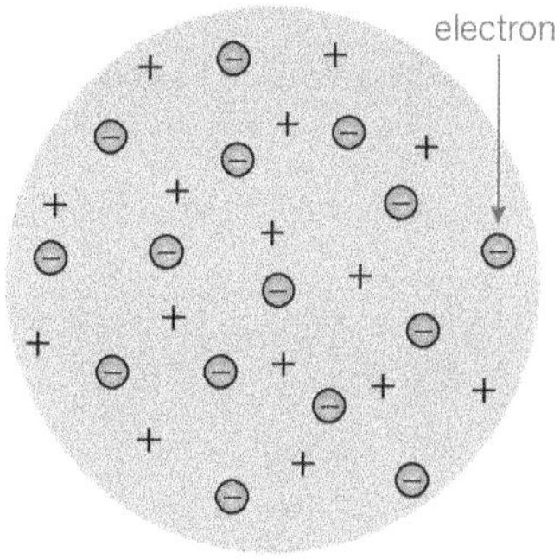

▲ **Figure 1** *The plum pudding model of the atom – electrons located in circular arrays within a sphere of positive charge*

The sub-atomic particles

Atoms are made of three fundamental particles – **protons**, **neutrons**, and **electrons**.

The protons and neutrons form the **nucleus**, in the centre of the atom.

- Protons and neutrons are sometimes called **nucleons** because they are found in the nucleus.
- The electrons surround the nucleus.

The properties of the sub-atomic particles are shown in Table 1.

▲ **Figure 2** *Atoms can only be seen indirectly. This photograph of xenon atoms was taken by an instrument called a scanning tunnelling electron microscope*

Maths link

The masses and charges of sub-atomic particles are very small so they are expressed in standard form. Refer to Section 8, Mathematical skills, if you are not sure about this notation.

▼ **Table 1** *The properties of the sub-atomic particles*

Property	Proton p	Neutron n	Electron e
Mass / kg	1.673×10^{-27}	1.675×10^{-27}	0.911×10^{-30} (very nearly 0)
Charge / C	$+1.602 \times 10^{-19}$	0	-1.602×10^{-19}
Position	in the nucleus	in the nucleus	around the nucleus

These numbers are extremely small. In practice, *relative* values for mass and charge are used. The relative charge on a proton is taken to be +1, so the charge on an electron is −1. Neutrons have no charge, see Table 2.

Study tip

You must remember the relative masses and charges of a proton, neutron, and an electron, as given in Table 2.

▼ **Table 2** *The relative masses and charges of the sub-atomic particles*

	Proton p	Neutron n	Electron e
Relative mass	1	1	$\frac{1}{1840}$
Relative charge	+1	0	−1

In a neutral atom, the number of electrons must be the same as the number of protons because their charge is equal in size and opposite in sign.

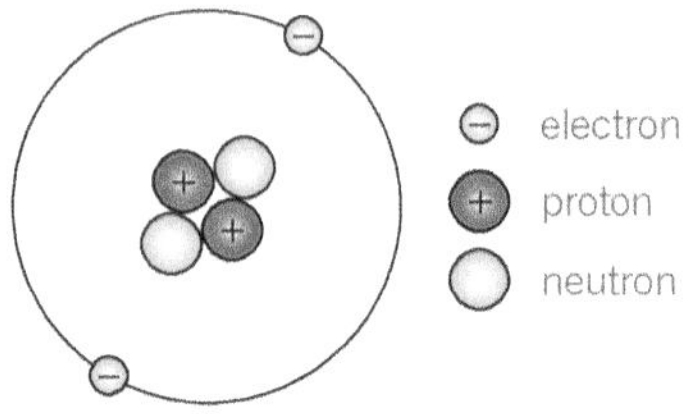

▲ **Figure 3** *The sub-atomic particles in a helium atom (not to scale)*

The arrangement of the sub-atomic particles

The sub-atomic particles (protons, neutrons, and electrons) are arranged in the atom as shown in Figure 3.

The protons and neutrons are in the centre of the atom, held together by a force called the **strong nuclear force**. This is much stronger than the **electrostatic forces** of attraction that hold electrons and protons together in the atom, so it overcomes the repulsion between the protons in the nucleus. It acts only over very short distances, that is, within the nucleus.

The nucleus is surrounded by electrons. Electrons are found in a series of shells, sometimes referred to as orbits or levels, which get further and further away from the nucleus. This is a simplified picture that will develop in Topic 1.5.

Extension

The diameter of the nucleus of a hydrogen atom is about 2×10^{-15} m, while the diameter of the atom itself is about 1×10^{-10} m, about 50 000 times larger. This means that if the nucleus were the size of a fly, the whole atom would be roughly the size of a cathedral.

St Paul's Cathedral is roughly 200 m long. Estimate the length of a fly and, without using a calculator, check that the analogy is realistic.

Summary questions

1 **a** Identify which of the following – protons, neutrons, or electrons:

- **i** are nucleons
- **ii** have the same relative mass
- **iii** have opposite charges
- **iv** have no charge
- **v** are found outside the nucleus

b Explain why we assume that there are the same number of protons and electrons in an atom.

1.2 Mass number, atomic number, and isotopes

Mass number and atomic number

Atomic number *Z*

As you have seen in Topic 1.1, atoms consist of a tiny nucleus made up of protons and neutrons that is surrounded by electrons. The number of protons in the nucleus is called the atomic number or the **proton number** *Z*.

The number of electrons in the atom is equal to the proton number, so atoms are electrically neutral. The number of electrons in the outer shell of an atom determines the chemical properties of an element (how it reacts) and what sort of element it is. The atomic number defines the chemical identity of an element.

atomic number (proton number) Z = number of protons

All atoms of the same element have the same atomic number. Atoms of different elements have different atomic numbers.

Mass number *A*

The total number of protons plus neutrons in the nucleus (the total number of nucleons) is called the mass number *A*. It is the nucleons that are responsible for almost all of the mass of an atom because electrons weigh virtually nothing.

mass number A = number of protons + number of neutrons

Isotopes

Every single atom of any particular element has the same number of protons in its nucleus and therefore the same number of electrons. But the number of neutrons may vary.

- Atoms with the same number of protons but different numbers of neutrons are called **isotopes**.
- Different isotopes of the same element react chemically in exactly the same way as they have the same electron configuration.
- Atoms of different isotopes of the same element vary in mass number because of the different number of neutrons in their nuclei.

All atoms of the element carbon, for example, have atomic number 6. That is what makes them carbon rather than any other element. However, carbon has three isotopes with mass numbers 12, 13, and 14 respectively (Table 1). All three isotopes will react in the same way, for example, burning in oxygen to form carbon dioxide.

Isotopes are often written like this: ${}^{13}_{6}C$. The superscript 13 is the mass number of the isotope, and the subscript 6 the atomic number.

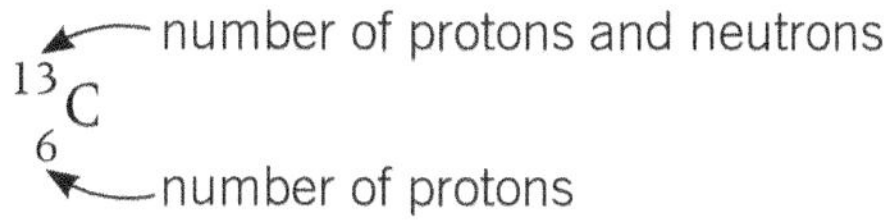

Learning objectives:

→ Define the terms mass number, atomic number, and isotope.

→ Explain why isotopes of the same element have identical chemical properties.

Specification reference: 3.1.1

Study tip

The mass number of an isotope must always be bigger than the atomic number (except in ${}^{1}_{1}H$). Typically it is around twice as big.

▼ **Table 1** *Isotopes of carbon*

Name of isotope	carbon-12	carbon-13	carbon-14
Symbol	$^{12}_{6}C$	$^{13}_{6}C$	$^{14}_{6}C$
Number of protons	6	6	6
Number of neutrons	6	7	8
Abundance	98.89%	1.11%	trace

Summary questions

1 Isotopes are usually identified by the name of the element and the mass number of the isotope, as in carbon-13. However, isotopes of hydrogen have their own names. Hydrogen-2 is often called **deuterium**, and hydrogen-3 is called **tritium**. However, both these isotopes behave chemically just like the most common isotope, hydrogen-1. State how many protons, neutrons, and electrons the atoms of the following have.

a deuterium

b tritium

2 $^{31}_{15}W$, $^{14}_{7}X$, $^{16}_{8}Y$, $^{15}_{7}Z$

Identify which of these atoms (not their real symbols) is a pair of isotopes.

3 For each element in question **2**, state:

a the number of protons

b the mass number

c the number of neutrons

Carbon dating

Isotopes of an element have different numbers of neutrons in their nuclei and most elements have some isotopes. Sometimes these isotopes are unstable and the nucleus of the atom itself breaks down giving off bits of the nucleus or energetic rays. This is the cause of radioactivity. Radioactive isotopes have many uses. Each radioactive isotope decays at a rate measured by its half life. This is the time taken for half of its radioactivity to decay.

One well-known radioactive isotope is carbon-14. It has a half life of 5730 years and is produced by cosmic-ray activity in the atmosphere. It is used to date organic matter. Radiocarbon dating can find the age of carbon-based material up to 60 000 years old, though it is most accurate for materials up to 2000 years old.

There is always a tiny fixed proportion of carbon-14 in all living matter. All living matter takes in and gives out carbon in the form of food and carbon dioxide, respectively. As a result, the level of carbon-14 stays the same. Once the living material dies, this stops happening. The radioactive carbon breaks down and the level of radioactivity slowly falls. So, knowing the half life of carbon-14, scientists work backwards. They work out how long it has taken for the level of radioactivity to fall from what it is in a living organism to what it is in the sample. So, a sample with half the level of radioactivity expected in a living organism would have been dead for 5730 years, while one with a quarter of the expected level would have been dead for twice as long.

The radioactivity in a wooden bowl was found to be $\frac{1}{8}$ of that found in a sample of living wood.

1 How old is the wood from the bowl?

2 Does this tell us the age of the bowl? Explain your answer.

Carbon-14

Radiocarbon dating was introduced in 1949 by the American Willard Libby who won the Nobel Prize for the technique. Carbon-14 is produced in the atmosphere by a nuclear reaction in which a neutron (from a cosmic ray) hits a nitrogen atom and ejects a proton:

$$^{14}_{7}N + ^{1}_{0}n \rightarrow ^{14}_{6}C + ^{1}_{1}p$$

If the half life of ^{14}C is taken to be 6000 years, 24 000 years is four half lives so the remaining radioactivity will be

$$\frac{1}{2} \times \frac{1}{2} \times \frac{1}{2} \times \frac{1}{2} = \frac{1}{16}$$

of the original activity.

Suggest why 60 000 years is the practical limit for ^{14}C dating.

1.3 The mass spectrometer

The mass spectrometer

The mass spectrometer is the most useful instrument for the accurate determination of **relative atomic masses** A_r. Relative atomic masses are measured on a scale on which the mass of an atom of ^{12}C is defined as *exactly* 12. No other isotope has a relative atomic mass that is exactly a whole number. This is because neither the proton nor the neutron has a mass of exactly 1.

$$\text{relative atomic mass } A_r = \frac{\text{average mass of 1 atom of an element}}{\frac{1}{12}\text{ mass of 1 atom of } ^{12}\text{C}}$$

$$\text{relative molecular mass } M_r = \frac{\text{average mass of a molecule}}{\frac{1}{12}\text{ mass of 1 atom of } ^{12}\text{C}}$$

The mass spectrometer determines the mass of separate atoms (or molecules). Mass spectrometers are an essential part of a chemist's toolkit of equipment. For example, they are used by forensic scientists to help identify substances such as illegal drugs.

There are several types of mass spectrometer, but all work on the principle of forming ions from the sample and then separating the ions according to the ratio of their charge to their mass. The type described here is called an electro spray ionisation time of flight (TOF) instrument. The layout of this type of mass spectrometer is shown in Figure 2.

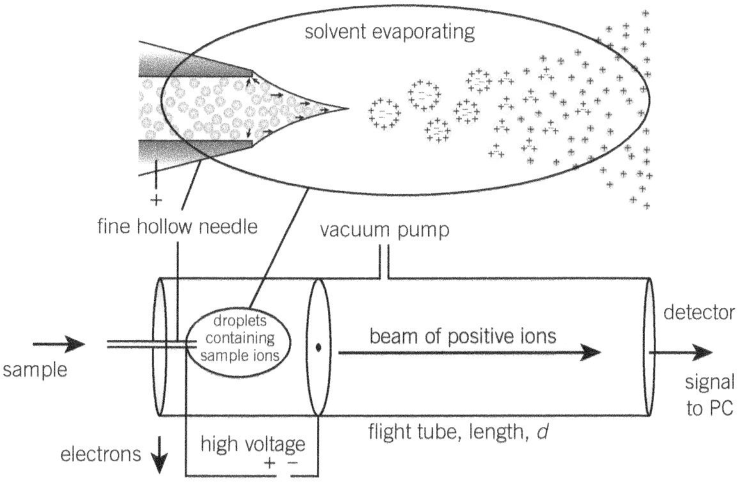

▲ **Figure 2** *The layout of an electron spray ionisation time of flight mass spectrometer*

What happens in a time of flight mass spectrometer?

In outline, the substance(s) in the sample are converted to positive ions, accelerated to high speeds (which depend on their mass to charge ratio), and arrive at a detector. The steps are described in more detail below.

- **Vacuum** The whole apparatus is kept under a high vacuum to prevent the ions that are produced colliding with molecules from the air.
- **Ionisation** The sample to be investigated is dissolved in a volatile solvent and forced through a fine hollow needle that is connected

Learning objectives:

→ Explain how a mass spectrometer works and what it measures.

Specification reference: 3.1.1

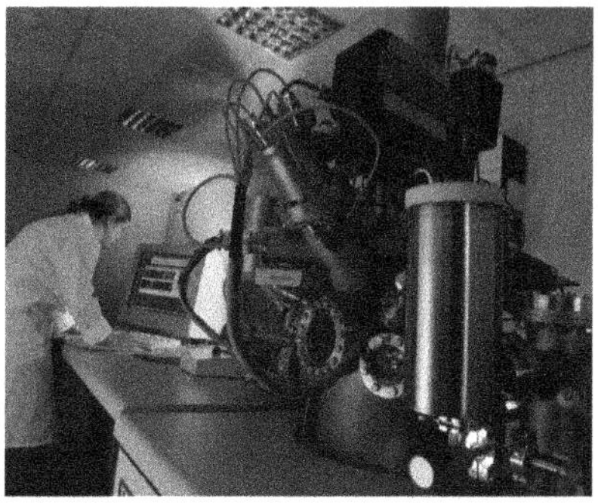

▲ **Figure 1** *A modern mass spectrometer*

Synoptic link

Mass spectrometry can also be used to measure relative molecular masses and much more, as you will see in Topic 16.2, Mass spectrometry.

Synoptic link

You will find out more about relative atomic mass in Topic 2.1, Relative atomic and molecular masses, the Avogadro constant, and the mole.

The relationship between mass and time of flight of an ion

The kinetic energy of the ions in the flight tube is given by

$KE = \frac{1}{2}mv^2$ and velocity, $v = \frac{d}{t}$, where d is the length of the flight tube.

So $KE = \frac{1}{2}m\left(\frac{d^2}{t^2}\right)$, so $t^2 = m\left(\frac{d^2}{2KE}\right)$

Since KE and d are constant, $m \propto t^2$

to the positive terminal of a high voltage supply. This produces tiny positively charged droplets which have lost electrons to the positive charge of the supply. The solvent evaporates from the droplets into the vacuum and the droplets get smaller and smaller until they may contain no more than a single positively charged ion of mass m.

- **Acceleration** The positive ions are attracted towards a negatively charged plate and accelerate towards it. Lighter ions and more highly charged ions achieve a higher speed.
- **Ion drift** The ions, all of which have the same kinetic energy, KE, pass through a hole in the negatively charged plate, forming a beam and travel along a tube, called the flight tube, to a detector with velocity, v.
- **Detection** When ions with the same charge arrive at the detector, the lighter ones are first as they have higher velocities. The flight times are recorded. The positive ions pick up an electron from the detector, which causes a current to flow.
- **Data analysis** The signal from the detector is passed to a computer which generates a mass spectrum like those in Figures 3 and 4.

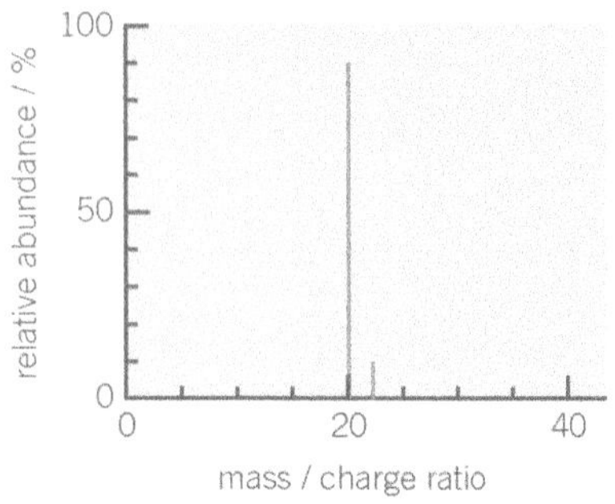

▲ **Figure 3** *The mass spectrum of neon. Even though the relative atomic mass is 20.2, there is no peak at 20.2 because no neon atoms actually have this mass*

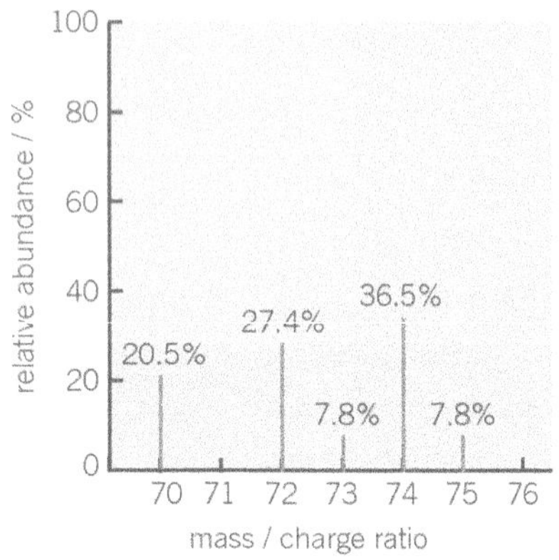

▲ **Figure 4** *The mass spectrum of germanium (the percentage abundance of each peak is given)*

Mass spectra of elements

The mass spectrometer can be used to identify the different isotopes that make up an element. It detects individual ions, so different isotopes are detected separately because they have different masses. This is how the data for the neon, germanium, and chlorine isotopes in Figures 3, 4, and 5 were obtained. The peak height gives the relative abundance of each isotope and the horizontal scale gives the mass to charge ratio, m/z, which, for a singly charged ion is numerically the same as the mass number A.

Mass spectrometers can measure relative atomic masses to five decimal places of an atomic mass unit – this is called high resolution mass spectrometry. However, most work is done to one decimal point – this is called low resolution mass spectrometry.

Low resolution mass spectrometry

The low resolution mass spectrum of neon is shown in Figure 3. This shows that neon has two isotopes, with mass numbers 20 and 22, and abundances to the nearest whole number of 90% and 10%, respectively. From this we can say that neon has an average relative atomic mass of:

$$\frac{(90 \times 20) + (10 \times 22)}{100} = 20.2$$

When calculating the relative atomic mass of an element, you must take account of the relative abundances of the isotopes. The relative atomic mass of neon is not 21 because there are far more atoms of the lighter isotope.

Another example is the mass spectrum of the element germanium, which is shown in Figure 4.

Isotopes of chlorine

Chlorine has two isotopes. They are ${}^{35}_{17}Cl$, with a mass number of 35, and ${}^{37}_{17}Cl$, with a mass number of 37. They occur in the ratio of almost exactly 3 : 1.

${}^{35}Cl$ ${}^{35}Cl$ ${}^{35}Cl$	${}^{37}Cl$
three of these	to every one of this

So there are 75% ^{35}Cl and 25% ^{37}Cl atoms in naturally occurring chlorine gas (see Figure 5).

The average mass of these is 35.5, as shown below.

Mass of 100 atoms = $(35 \times 75) + (37 \times 25) = 3550$

Average mass = $\frac{3550}{100} = 35.5$

This explains why the relative atomic mass of chlorine is *approximately* 35.5.

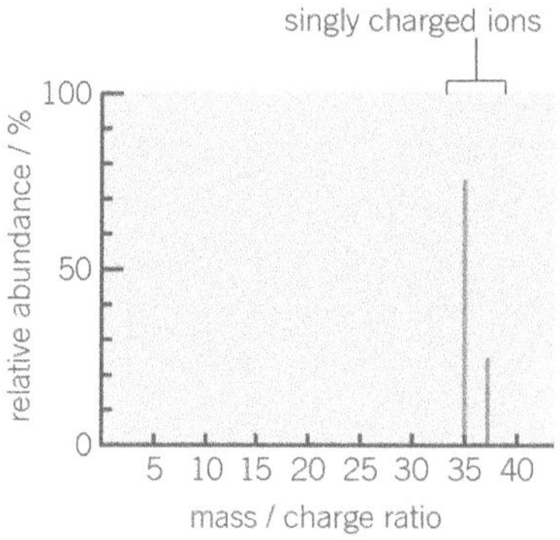

▲ **Figure 5** *The mass spectrum of chlorine*

Identifying elements

All elements have a characteristic pattern that shows the relative abundances of their isotopes. This can be used to help identify any particular element. Chlorine, for example, shows two peaks at mass 35 and mass 37. The peak of mass 35 is three times the height of the peak of mass 37 because there are three times as many ^{35}Cl atoms in chlorine.

The spectrum will also show peaks caused by ionised Cl_2 molecules. These are called molecular ions. There will be three of these:

- at *m/z* 70, due to $^{35}Cl^{35}Cl$
- at *m/z* 72, due to $^{35}Cl^{37}Cl$
- at *m/z* 74, due to $^{37}Cl^{37}Cl$

High resolution mass spectrometers can measure the masses of atoms to several decimal places. This allows us to identify elements by the exact masses of their atoms that (apart from carbon-12 whose relative atomic mass is exactly 12) are not exactly whole numbers.

What will be the relative abundances of the three Cl_2^+ ions of *m/z* 70, 72, and 74 respectively? The relative abundances of the atoms are $^{35}Cl : ^{37}Cl = 3 : 1$. i.e., $\frac{3}{4} : \frac{1}{4}$

Study tip

Relative atomic masses are weighted averages of the mass numbers of the isotopes of the element, taking account of both the masses and their abundances, relative to the ^{12}C isotope, which is exactly 12. Chlorine has isotopes of mass number 35 and 37 but the relative atomic mass of chlorine is *not* 36, it is 35.5.

Mass spectrometers in space

Space probes such as the Mars Rover Curiosity carry mass spectrometers. They are used to identify the elements in rock samples. The Huygens spacecraft that landed on Titan, one of the moons of Saturn, in January 2005 carried a mass spectrometer used to identify and measure the amounts of the gases in Titan's atmosphere. After landing, it also analysed vaporised samples of the surface.

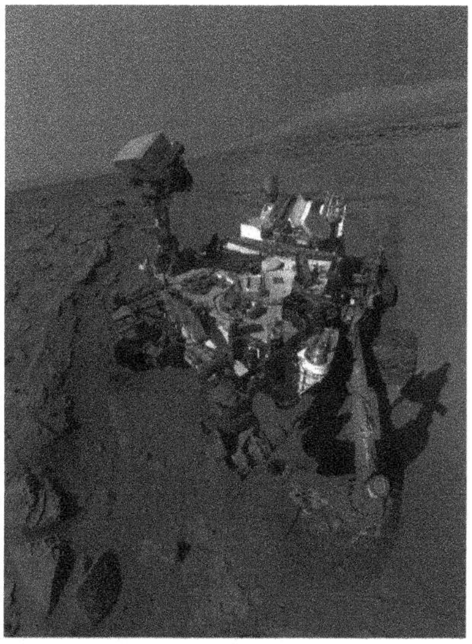

▲ **Figure 6** *The Mars Rover Curiosity carries a mass spectrometer to look for compounds of carbon that may suggest that there was once life on Mars.*

Mini-mass spectrometer

The latest development in mass spectrometry is a unit small enough to carry as a back pack. The unit, including rechargeable batteries, weighs 10 kg, light enough to be carried by scene of crime officers looking for drugs, explosives, or chemical weapons. Other uses include investigating chemical spills.

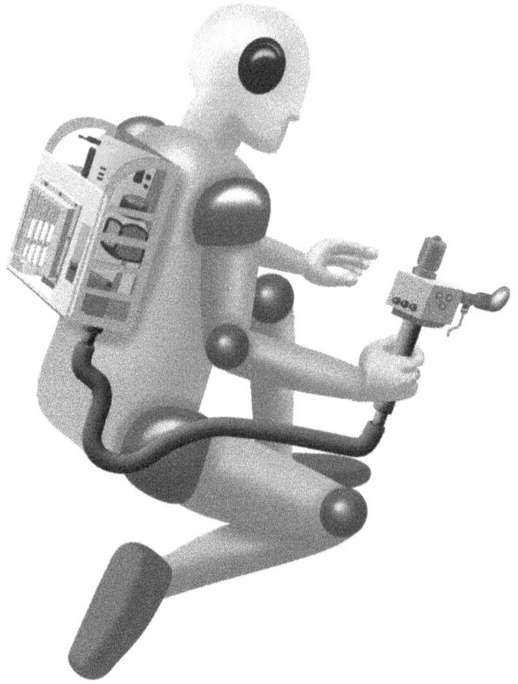

▲ **Figure 7** *The mini-mass spectrometer*

The mass spectrometer includes software to match spectra of samples investigated with a library of spectra and so identify them. The instrument can be used by operators with little or no chemical knowledge.

Summary questions

1. Explain why the ions formed in a mass spectrometer have a positive charge.
2. Explain what causes the ions to accelerate through the mass spectrometer.
3. Describe what forms the ions into a beam.
4. State which ions will arrive at the detector first.
5. Use the information about germanium in Figure 4 to calculate its relative atomic mass.
6. Figure 8 shows the mass spectrum of copper. Calculate the relative atomic mass of copper.

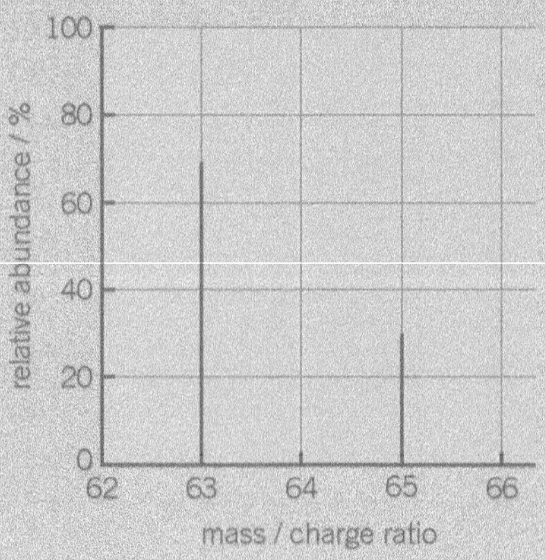

▲ **Figure 8** *The mass spectrum of copper*

1.4 The arrangement of the electrons

The atom and electrons

During the early years of the twentieth century, physicists made great strides in understanding the structure of the atom. These are some of the landmarks.

1913 Niels Bohr put forward the idea that the atom consisted of a tiny positive nucleus orbited by negatively-charged electrons to form an atom like a tiny solar system. The electrons orbited in shells of fixed size and the movement of electrons from one shell to the next explained how atoms absorbed and gave out light. This was the beginning of what is called quantum theory.

1926 Erwin Schrödinger, a mathematical physicist, worked out an equation that used the idea that electrons had some of the properties of waves as well as those of particles. This led to a theory called **quantum mechanics**, which can be used to predict the behaviour of sub-atomic particles.

1932 James Chadwick discovered the neutron.

At the same time, chemists were developing their ideas about how electrons allowed atoms to bond together. One important contributor was the American, Gilbert Lewis. He put forward the ideas that:

- the inertness of the noble gases was related to their having full outer shells of electrons
- ions were formed by atoms losing or gaining electrons to attain full outer shells
- atoms could also bond by sharing electrons to form full outer shells.

Lewis' theories are the basis of modern ideas of chemical bonding, and explain the formulae of many simple compounds using the idea that atoms tend to gain the stable electronic structure of the nearest noble gas.

Evolving ideas

Early theories model the electron as a minute solid particle. Later theories suggest you can also think of electrons as smeared out clouds of charge, so you can never say exactly where an electron is at any moment. You can merely state the probability that it can be found in a particular volume of space that has a particular shape. However, chemists still use different models of the atom for different purposes.

- Dalton's model can still be used to explain the geometries of crystals.
- Bohr's model can be used for a simple model of ionic and covalent bonding.
- The charge cloud idea is used for a more sophisticated explanation of bonding and the shapes of molecules.
- The simple model of electrons orbiting in shells is useful for many purposes, particularly for working out bonding between atoms.

You will be familiar with the electron diagrams in this section from GCSE. They lead on to the more sophisticated models of electron structure described in Topic 1.5. However, they can still be useful, for example, in predicting and explaining the formulae of simple compounds and the shapes of molecules.

Learning objectives:

- → Describe how electrons are arranged in an atom.
- → Recognise that the electron can behave as a particle, a wave, or a cloud of charge.
- → Describe how the structure of an atom developed from Dalton to Schrödinger.

Specification reference: 3.1.1

Synoptic link

This topic revises your knowledge of electron arrangements from GCSE. This will be useful when you study Topic 1.5, More about electron arrangements in atoms.

Quantum theory in practice

Quantum theory makes predictions that seem to contradict our everyday experience, such as the fact that an electron can pass through two different holes at once! However, it is an extremely successful theory and underlies electronic gadgets such as computers, mobile phones, and DVD players.

Study tip

You must remember the number of electrons in each shell.

Electron shells

The first shell, which is closest to the nucleus, fills first, then the second, and so on. The number of electrons in each shell = $2n^2$, where n is the number of the shell, so:

- the first shell holds up to two electrons
- the second shell holds up to eight electrons
- the third shell holds up to 18 electrons.

Electron diagrams

If you know the number of protons in an atom, you also know the number of electrons it has. This is because the atom is neutral. You can therefore draw an electron diagram for any element. For example, carbon has six electrons. The four electrons in the outer shell are usually drawn spaced out around the atom (Figure 1).

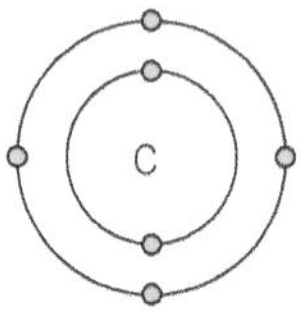

carbon (2,4)

▲ **Figure 1** *Electron diagram of carbon*

Sulfur has 16 electrons. It has six electrons in its outer shell. It helps when drawing bonding diagrams to space out the first four (as in carbon), and then add the next two electrons to form pairs (Figure 2).

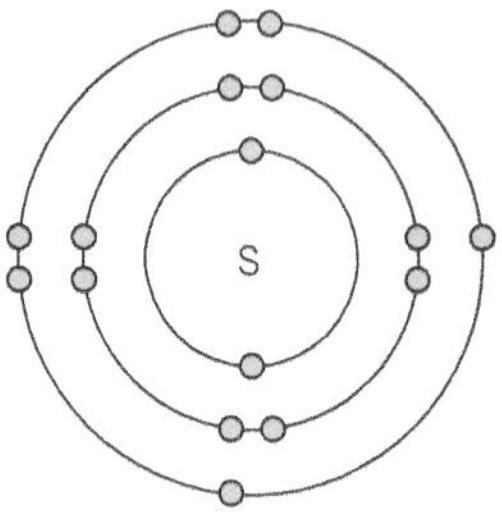

sulfur (2,8,6)

▲ **Figure 2** *Electron diagram of sulfur*

You can also draw electron diagrams of ions, as long as you know the number of electrons. For example, a sodium *atom*, Na, has 11 electrons, but its *ion* has 10, so it has a positive charge, Na^+ (Figure 3).

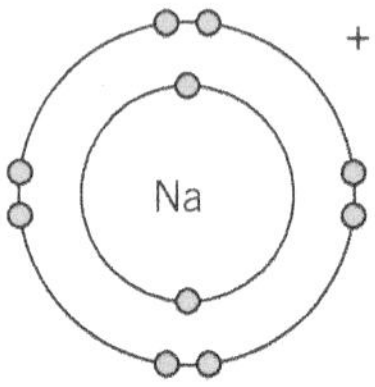

Na^+ sodium ion
11 protons, 10 electrons
(2,8)

▲ **Figure 3** *Electron diagram of a sodium ion*

An oxygen *atom* has eight electrons, but its *ion* has 10, so it has a negative charge, O^{2-} (Figure 4).

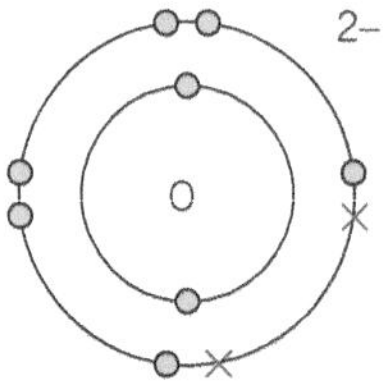

O^{2-} oxygen ion
8 protons,
10 electrons (2,8)

▲ **Figure 4** *Electron diagram of an oxygen ion*

You can write electron diagrams in shorthand:

- write the number of electrons in each shell, starting with the inner shell and working outwards
- separate each number by a comma.

For carbon you write 2,4; for sulfur 2,8,6; for Na^+ 2,8.

Summary questions

1 Draw the electron arrangement diagrams of atoms that have the following numbers of electrons:

a 3 **b** 9 **c** 14

2 State, in shorthand, the electron arrangements of atoms with:

a 4 electrons **b** 13 electrons **c** 18 electrons

3 Identify which of the following are atoms, positive ions, or negative ions. Give the size of the charge on each ion, including its sign. Use the Periodic Table to identify the elements A–E.

	Number of protons	Number of electrons
A	12	10
B	2	2
C	17	18
D	10	10
E	3	2

1.5 More about electron arrangements in atoms

As you have seen in Topic 1.4, in a simple model of the atom the electrons are thought of as being arranged in shells around the nucleus. The shells can hold increasing numbers of electrons as they get further from the nucleus – the pattern is 2, 8, 18, and so on.

Learning objectives:

→ Illustrate how the electron configurations of atoms and ions are written in terms of s, p, and d electrons.

Specification reference: 3.1.1

Energy levels

Electrons in different shells have differing amounts of energy. They can therefore be represented on an energy level diagram. The shells represent energy levels and they are labelled 1, 2, 3, and so on (Figure 1). Each main shell can hold up to a maximum number of electrons given by the formula $2n^2$, where n is the number of the main shell. So, you can have two electrons in the first main shell, eight in the next, 18 in the next, and so on.

Apart from the first shell, these main energy levels are divided into sub-shells, called s, p, d, and f, which have slightly different energies (Figure 2). Shell 2 has an s-sub-shell and a p-sub-shell. Shell 3 an s-sub-shell, a p-sub-shell, and a d-sub-shell.

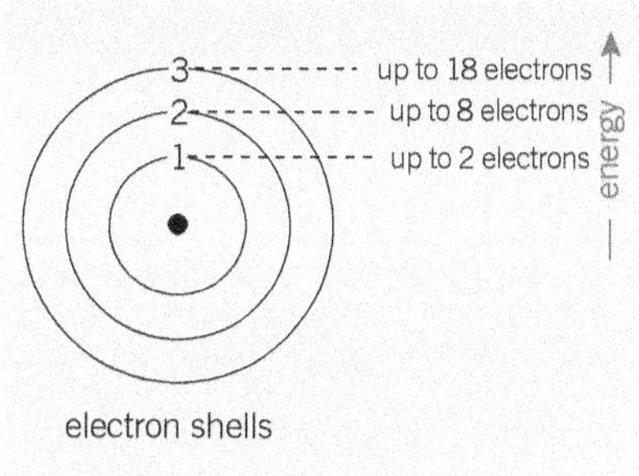

▲ **Figure 1** *Electron shells*

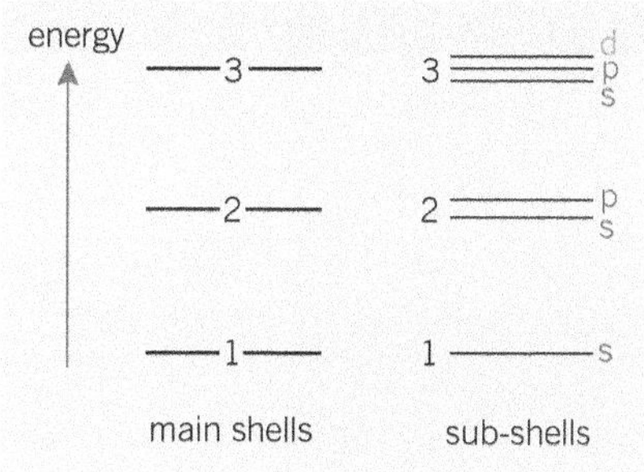

▲ **Figure 2** *Main shells and sub-shells*

Quantum mechanics

For a more complete description of the electrons in atoms a theory called quantum mechanics is used, which was developed during the 1920s. This describes the atom mathematically with an equation (the Schrödinger equation). The solutions to this equation give the *probability* of finding an electron in a given *volume* of space called an atomic orbital.

Atomic orbitals

The electron is no longer considered to be a particle but a cloud of negative charge. An electron fills a volume in space called its **atomic orbital**. The concept of the main shells and the sub-shells is then included in the following way.

- Different atomic orbitals have different energies. Each orbital has a number that tells us the main energy shell that it corresponds to: 1, 2, 3, and so on.
- The atomic orbitals of each main shell have different shapes, which in turn have slightly different energies. These are the sub-shells. They are described by the letters s, p, d, and f.

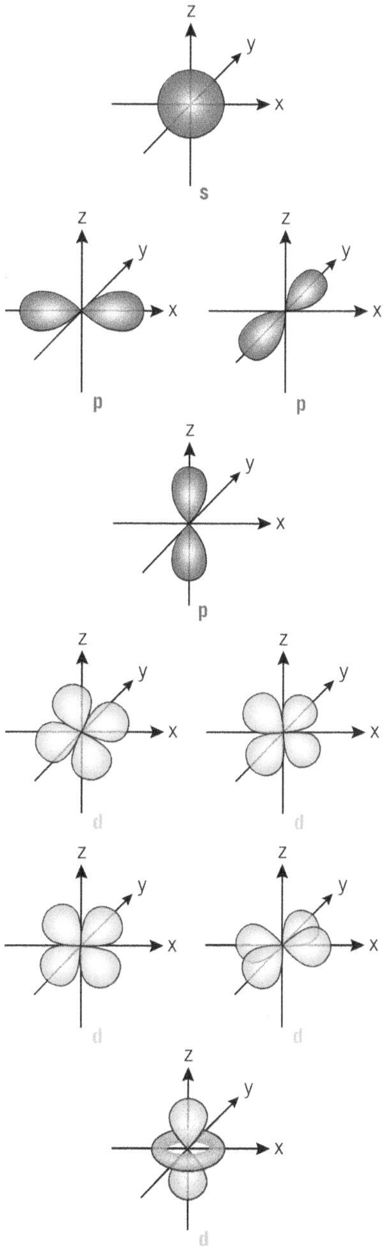

▲ **Figure 3** *The shapes of s-, p-, and d-orbitals*

Study tip

You should know how many s-, p-, and d-orbitals there are in each main shell. However, you only need to know the shapes of the s- and p-orbitals.

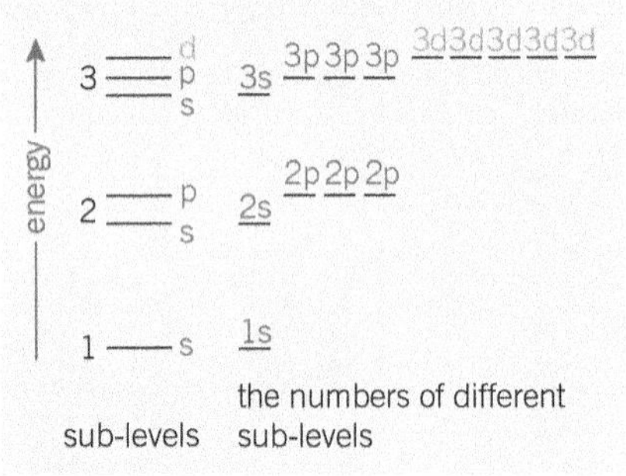

▲ **Figure 4** *The subdivisions of orbitals*

The shapes of the s-, p-, and d-orbitals are shown in Figure 3. The shapes of f-orbitals are even more complicated.

- These shapes represent a volume of space in which there is a 95% probability of finding an electron and they influence the shapes of molecules.
- The first main shell consists of a single s-orbital. The second main shell has a single s-orbital and three p-orbitals of a slightly higher energy, the third main shell has a single s-orbital, three p-orbitals of slightly higher energy, and five d-orbitals of slightly higher energy still, and so on, see Figure 4.
- Any single atomic orbital can hold a maximum of two electrons.
- s-orbitals can hold up to two electrons.
- p-orbitals can hold up to two electrons each, but always come in groups of three of the same energy, to give a total of up to six electrons in the p-sub-shell.
- d-orbitals can hold up to two electrons each, but come in groups of five of the same energy to give a total of up to 10 electrons in the d-sub-shell.

Table 1 summarises the number of electrons in the different shells and sub-shells.

▼ **Table 1** *The number of electrons in the different levels and sub-levels*

Main energy level (shell)	1	2		3			4			
sub-shell(s)	s	s	p	s	p	d	s	p	d	f
number of orbitals in sub-shell	1 (2 electrons)	1 ($2e^-$)	3 ($6e^-$)	1 ($2e^-$)	3 ($6e^-$)	5 ($10e^-$)	1 ($2e^-$)	3 ($6e^-$)	5 ($10e^-$)	7 ($14e^-$)
total number of electrons in main shell	2	8		18			32			

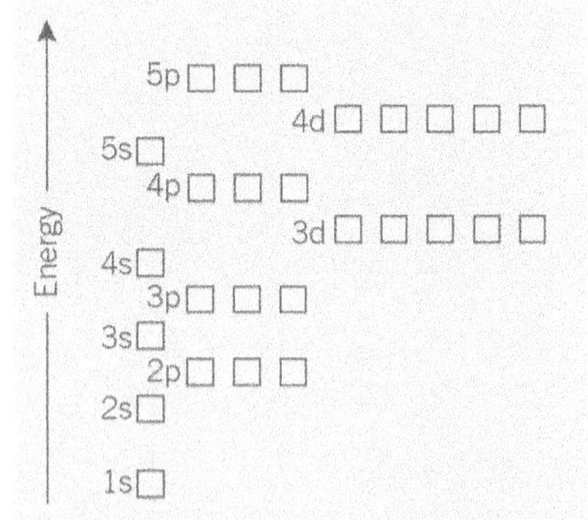

▲ **Figure 5** *The energies of the first few atomic orbitals*

The energy level diagram in Figure 5 shows the energies of the orbitals for the first few elements of the Periodic Table. Notice that the first main shell has only an s-orbital. The second main shell has an s- and p-sub-shell, and the p-sub-shell is composed of three p-orbitals of equal energy. The third main shell has an s-, p-, and d-sub-shell, and the d-sub-shell is composed of five atomic orbitals of equal energy.

- Each 'box' in Figure 5 represents an orbital of the appropriate shape that can hold up to two electrons.
- Notice that 4s is actually of slightly lower energy than 3d for neutral atoms, though this can change when ions are formed.

Spin

Electrons also have the property called **spin**.

- Two electrons in the same orbital must have opposite spins.
- The electrons are usually represented by arrows pointing up or down to show the different directions of spin.

Study tip

Although we use the term spin, the electrons are not actually spinning.

Putting electrons into atomic orbitals

Remember that the label of an atomic orbital tells us about the energy (and shape) of an electron cloud. For example, the atomic orbital 3s means the main shell is 3 and the sub-level (and therefore the shape) is spherical.

There are three rules for allocating electrons to atomic orbitals:

1. Atomic orbitals of lower energy are filled first – so the lower main shell is filled first and, within this shell, sub-shells of lower energy are filled first.
2. Atomic orbitals of the same energy fill singly before pairing starts. This is because electrons repel each other.
3. No atomic orbital can hold more than two electrons.

The electron diagrams for the elements hydrogen to sodium are shown in Figure 6.

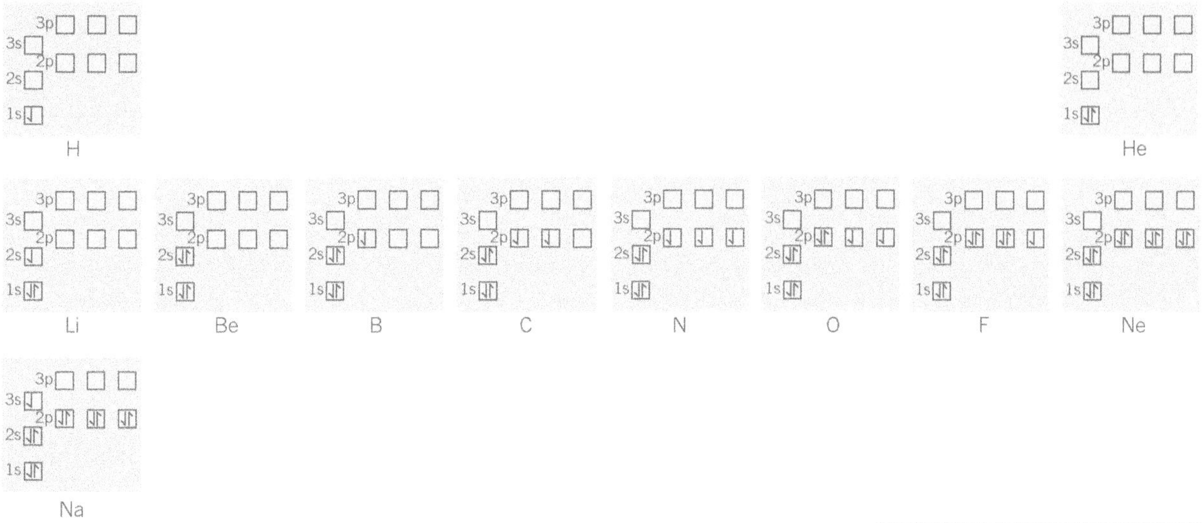

▲ **Figure 6** *The electron arrangements for the elements hydrogen to sodium – note how they obey the rule above*

Writing electronic structures

A shorthand way of writing electronic structures is as follows, for example, for sodium which has 11 electrons:

$1s^2$	$2s^2\ 2p^6$	$3s^1$
2	8	1

Note how this matches the simpler 2,8,1 you used at GCSE.

Calcium, with 20 electrons would be:

$1s^2$ $2s^2\ 2p^6$ $3s^2\ 3p^6$ $4s^2$ which matches 2,8,8,2

Notice how the 4s orbital is filled before the 3d orbital because it is of lower energy.

After calcium, electrons begin to fill the 3d orbitals, so vanadium with 23 electrons is: $1s^2\ 2s^2\ 2p^6\ 3s^2\ 3p^6\ 3d^3\ 4s^2$

Krypton with 36 electrons is: $1s^2\ 2s^2\ 2p^6\ 3s^2\ 3p^6\ 3d^{10}\ 4s^2\ 4p^6$

Sometimes it simplifies things to use the previous noble gas symbol. So the electron arrangement of calcium, Ca, could be written [Ar] $4s^2$ as a shorthand for [$1s^2\ 2s^2\ 2p^6\ 3s^2\ 3p^6$] $4s^2$ because $1s^2\ 2s^2\ 2p^6\ 3s^2\ 3p^6$ is the electron arrangement of argon.

You can use the same notation for ions. So a sodium ion, Na^+, would have the electron arrangement $1s^2\ 2s^2\ 2p^6$, one less than a sodium atom, $1s^2\ 2s^2\ 2p^6\ 3s^1$.

Study tip

Practise working out the shorthand electronic structure of all the elements at least up to krypton (atomic number 36).

Synoptic link

You will learn how electron arrangements affect the properties of the transition metals in Topic 23.1, The general properties of transition metals.

Summary questions

1. **a** Give the full electron arrangement for phosphorus.
   b Give the electron arrangement for phosphorus using an inert gas symbol as a shorthand.
2. **a** Give the full electron arrangements of:
   i Ca^{2+} and **ii** F^-
   b Give their electron arrangements using an inert gas symbol as a shorthand.

1.6 Electron arrangements and ionisation energy

Learning objectives:

- → State the definition of ionisation energy.
- → Describe the trend in ionisation energies a) down a group and b) across a period in terms of electron configurations.
- → Explain how trends in ionisation energies provide evidence for the existence of electron shells and sub-shells.

Specification reference: 3.1.1

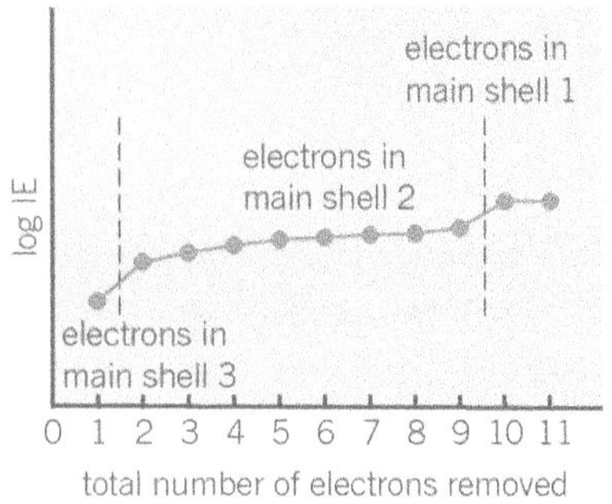

▲ **Figure 1** *The successive ionisation energies of sodium against number of electrons removed. Note that the log of the ionisation energy has been plotted in order to fit the large range of values on the scale*

The patterns in first ionisation energies across a period provide evidence for electron sub-shells.

Ionisation energy

Electrons can be removed from atoms and the energy it takes to remove them can be measured. This is called **ionisation energy** because as the electrons are removed, the atoms become positive ions.

- Ionisation energy is the energy required to remove a mole of electrons from a mole of atoms in the gaseous state, and is measured in kJ mol^{-1}.
- Ionisation energy has the abbreviation IE.

Removing the electrons one by one

You can measure the energies required to remove the electrons one by one from an atom, starting from the outer electrons and working inwards.

- The first electron needs the least energy to remove it because it is being removed from a neutral atom. This is the first IE.
- The second electron needs more energy than the first because it is being removed from a +1 ion. This is the second IE.
- The third electron needs even more energy to remove it because it is being removed from a +2 ion. This is the third IE.
- The fourth needs yet more, and so on.

These are called **successive ionisation energies**.

For example, sodium:

$Na(g)$	$\rightarrow Na^{+}(g) + e^{-}$	first IE	$= +496$ kJ mol^{-1}
$Na^{+}(g)$	$\rightarrow Na^{2+}(g) + e^{-}$	second IE	$= +4563$ kJ mol^{-1}
$Na^{2+}(g)$	$\rightarrow Na^{3+}(g) + e^{-}$	third IE	$= +6913$ kJ mol^{-1}

and so on, see Table 1.

▼ **Table 1** *Successive ionisation energies of sodium*

Electron removed	1st	2nd	3rd	4th	5th	6th	7th	8th	9th	10th	11th
Ionisation energy / kJ mol^{-1}	496	4563	6913	9544	13 352	16 611	20 115	25 491	28 934	141 367	159 079

Study tip

The shape of the graph in Figure 1 has to be thought about carefully. The first electron removed is in the outer main shell and the 10th and 11th electrons removed are in the innermost main shell.

Notice that the second IE is *not* the energy change for

$$Na(g) \rightarrow Na^{2+}(g) + 2e^{-}$$

The energy for this process would be (first IE + second IE).

If you plot a graph of the values shown in Table 1 you get Figure 1.

Notice that one electron is relatively easy to remove, then comes a group of eight that are more difficult to remove, and finally two that are very difficult to remove.

This suggests that sodium has:

- *one* electron furthest away from the positive nucleus (easy to remove)
- *eight* electrons nearer in to the nucleus (harder to remove)
- *two* electrons very close to the nucleus (very difficult to remove because they are nearest to the positive charge of the nucleus).

This tells you about the number of electrons in each main shell or orbit: 2,8,1. The eight electrons in shell 2 are in fact sub-divided into two further groups that correspond to the $2s^2$, $2p^6$ electrons in the second main shell, but this is not visible on the scale of Figure 1.

You can find the number of electrons in each main shell of *any* element by looking at the jumps in successive ionisation energies.

Trends in ionisation energies across a period in the Periodic Table

The trends in first ionisation energies moving across a period in the Periodic Table can also give information about the energies of electrons in main shells and sub-shells. Ionisation energies generally increase across a period because the nuclear charge is increasing and this makes it more difficult to remove an electron.

The data for Period 3 are shown in Table 2.

▼ **Table 2** *The first ionisation energies of the elements in Period 3 in kJ mol^{-1}*

Na	Mg	Al	Si	P	S	Cl	Ar
496	738	578	789	1012	1000	1251	1521

nuclear charge increasing →

Plotting a graph of these values shows that the increase is not regular (Figure 2). In going from magnesium ($1s^2$, $2s^2$, $2p^6$, $3s^2$) to aluminium ($1s^2$, $2s^2$, $2p^6$, $3s^2$, $3p^1$), the ionisation energy actually goes down, despite the increase in nuclear charge. This is because the outer electron in aluminium is in a 3p orbital which is of a slightly higher energy than the 3s orbital. It therefore needs less energy to remove it, see Figure 3.

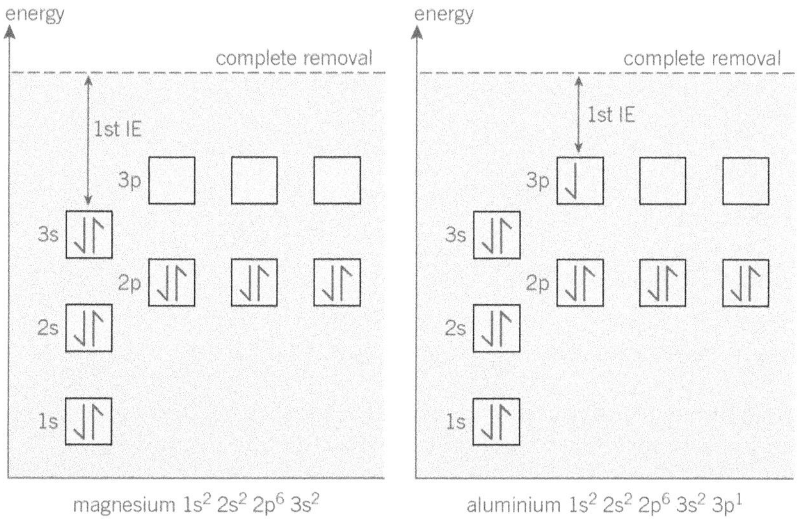

▲ **Figure 3** *The first ionisation energy of aluminium is less than that of magnesium*

Study tip

The energy change for the formation of a negative ion is called the **electron affinity**. The term ionisation energy is used *only* for the formation of *positive* ions.

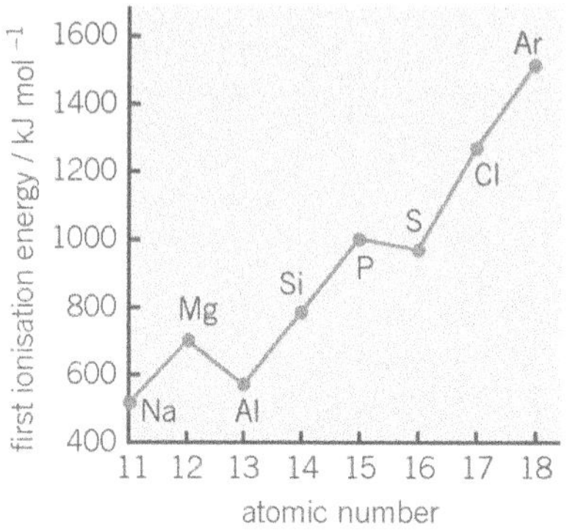

▲ **Figure 2** *Trends in first ionisation energies across Period 3*

In Figure 2, notice the small drop between phosphorus ($1s^2$, $2s^2$, $2p^6$, $3s^2$, $3p^3$) and sulfur ($1s^2$, $2s^2$, $2p^6$, $3s^2$, $3p^4$). In phosphorus, each of the three 3p orbitals contains just one electron, while in sulfur, one of the 3p orbitals must contain two electrons. The repulsion between these paired electrons makes it easier to remove one of them, despite the increase in nuclear charge, see Figure 4.

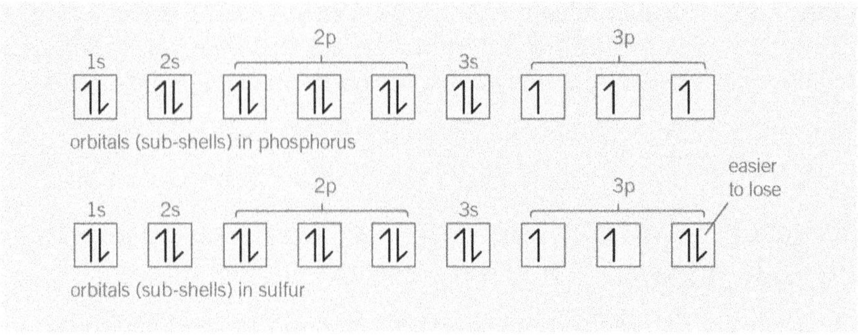

▲ **Figure 4** *Electron arrangements of phosphorus and sulfur*

Both these cases, which go against the expected trend, are evidence that confirms the existence of s- and p-sub-shells. These were predicted by quantum theory and the Schrödinger equation.

Trends in ionisation energies down a group in the Periodic Table

Figure 5 shows that there is a general decrease in first ionisation energy going down Group 2, and the same pattern is seen in other groups. This is because the outer electron is in a main shell that gets further from the nucleus in each case.

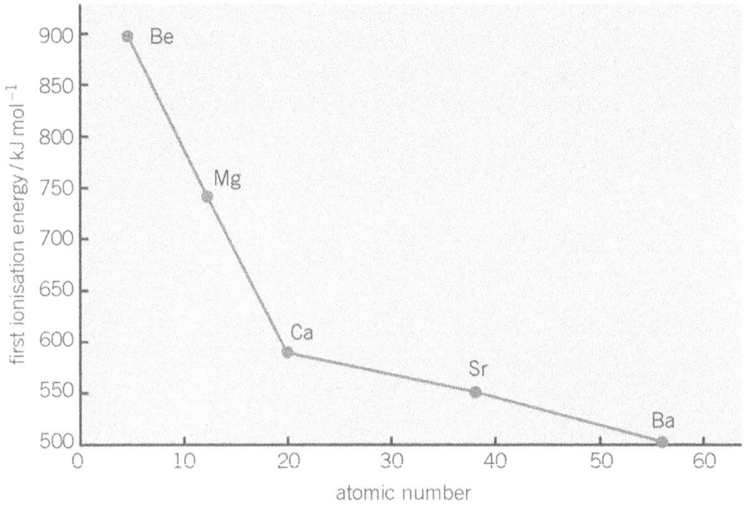

▲ **Figure 5** *The first ionisation energies of the elements of Group 2*

Going down a group, the nuclear charge increases. At first sight you might expect that this would make it *more* difficult to remove an electron. However, the actual positive charge 'felt' by an electron in the outer shell is less than the full nuclear charge. This is because of the effect of the inner electrons shielding the nuclear charge.

Summary questions

1 State why the second ionisation energy of any atom is larger than the first ionisation energy.

2 Sketch a graph similar to Figure 1 of the successive ionisation energies of aluminium (electron arrangement 2,8,3).

3 An element X has the following values (in kJ mol^{-1}) for successive ionisation energies: 1093, 2359, 4627, 6229, 37 838, 47 285.

  a Identify which group in the Periodic Table it is in.

  b Explain your answer to **a**.

Practice questions

1 The diagram to the right shows the first ionisation energies of some Period 3 elements.

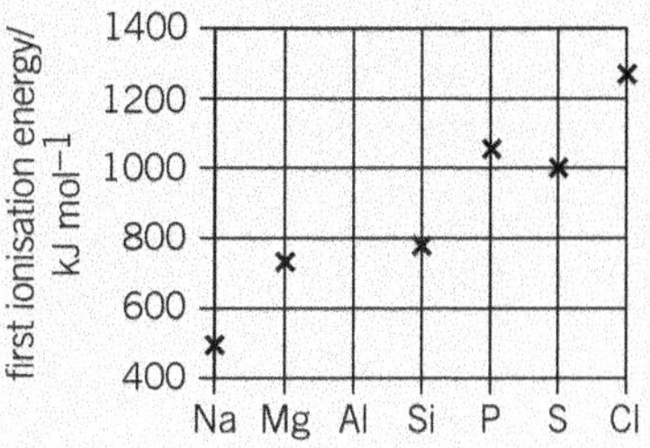

(a) Draw a cross on the diagram to show the first ionisation energy of aluminium.

(1 mark)

(b) Write an equation to show the process that occurs when the first ionisation energy of aluminium is measured.

(2 marks)

(c) State which of the first, second, or third ionisations of aluminium would produce an ion with the electron configuration $1s^2\ 2s^2\ 2p^6\ 3s^1$.

(1 mark)

(d) Explain why the value of the first ionisation energy of sulfur is less than the value of the first ionisation energy of phosphorus.

(2 marks)

(e) Identify the element in Period 2 that has the highest first ionisation energy and give its electron configuration.

(2 marks)

(f) State the trend in first ionisation energies in Group 2 from beryllium to barium. Explain your answer in terms of a suitable model of atomic structure.

(3 marks)

AQA, 2010

2 **(a)** One isotope of sodium has a relative mass of 23.

(i) Define, in terms of the fundamental particles present, the meaning of the term *isotopes*.

(ii) Explain why isotopes of the same element have the same chemical properties.

(3 marks)

(b) Give the electronic configuration, showing all sub-shells, for a sodium atom.

(1 mark)

(c) An atom has half as many protons as an atom of ^{28}Si and also has six fewer neutrons than an atom of ^{28}Si. Give the symbol, including the mass number and the atomic number, of this atom.

(2 marks)

AQA, 2004

3 The values of the first ionisation energies of neon, sodium, and magnesium are 2080, 494, and 736 kJ mol^{-1}, respectively.

(a) Explain the meaning of the term *first ionisation energy* of an atom.

(2 marks)

(b) Write an equation using state symbols to illustrate the process occurring when the **second** ionisation energy of magnesium is measured.

(2 marks)

(c) Explain why the value of the first ionisation energy of magnesium is higher than that of sodium.

(2 marks)

(d) Explain why the value of the first ionisation energy of neon is higher than that of sodium.

(2 marks)

AQA, 2004

4 A sample of iron from a meteorite was found to contain the isotopes ^{54}Fe, ^{56}Fe, and ^{57}Fe.

(a) The relative abundances of these isotopes can be determined using a mass spectrometer. In the mass spectrometer, the sample is first vaporised and then ionised.

(i) State what is meant by the term *isotopes*.

(ii) Explain how, in a mass spectrometer, ions are detected and how their abundance is measured.

(5 marks)

(b) (i) Define the term *relative atomic mass* of an element.

(ii) The relative abundances of the isotopes in this sample of iron were found to be as follows.

m/z	54	56	57
Relative abundance %	5.80	91.60	2.60

Use the data above to calculate the relative atomic mass of iron in this sample. Give your answer to the appropriate number of significant figures.

(2 marks)

AQA, 2005

5 The diagram shows the layout of a time of flight mass spectrometer.

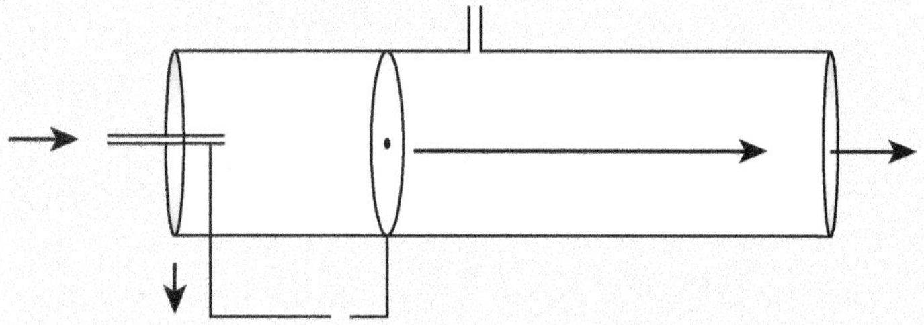

(a) Explain how positive ions are formed from the sample.

(1 mark)

(b) Explain why the instrument is kept under vacuum.

(1 mark)

(c) Explain how the ions are accelerated and separated by mass in the instrument.

(3 marks)

(d) Explain how an electric current is produced when an ion arrives at the detector.

(1 mark)

(e) The low resolution mass spectrum of magnesium shows three peaks.

Mass / charge	Relative abundance / %
24	79.0%
25	10.0%
26	11.0%

(i) Give the numbers of protons and neutrons in the nuclei of each isotope.

(1 mark)

(ii) Calculate the relative atomic mass of a sample of magnesium. Give your answer to the appropriate number of significant figures.

(2 marks)

Answers to the Practice Questions and Section Questions are available at
www.oxfordsecondary.com/oxfordaqaexams-alevel-chemistry

2 Amount of substance

2.1 Relative atomic and molecular masses, the Avogadro constant, and the mole

Relative atomic mass A_r

The actual mass in grams of any atom or molecule is too tiny to find by weighing. Instead, the masses of atoms are compared and *relative* masses are used.

This was done in the past by defining the relative atomic mass of hydrogen, the lightest element, as 1. The average mass of an atom of oxygen (for example) is 16 times heavier, to the nearest whole number, so oxygen has a relative atomic mass of 16. Scientists now use the isotope carbon-12 as the baseline for relative atomic mass, because the mass spectrometer has allowed us to measure the masses of individual isotopes extremely accurately. One-twelfth of the relative atomic mass of carbon-12 is given a value of *exactly* 1. The carbon-12 standard (defined below) is now accepted by all chemists throughout the world.

The relative atomic mass A_r is the weighted average mass of an atom of an element, taking into account its naturally occurring isotopes, relative to $\frac{1}{12}$ the relative atomic mass of an atom of carbon-12.

$$\text{relative atomic mass } A_r = \frac{\text{average mass of one atom of an element}}{\frac{1}{12}\text{ mass of one atom of } ^{12}\text{C}}$$

$$= \frac{\text{average mass of one atom of an element} \times 12}{\text{mass of one atom of } ^{12}\text{C}}$$

Relative molecular mass M_r

Molecules can be handled in the same way, by comparing the mass of a molecule with that of an atom of carbon-12.

The relative molecular mass, M_r, of a molecule is the mass of that molecule compared to $\frac{1}{12}$ the relative atomic mass of an atom of carbon-12.

$$\text{relative molecular mass } M_r = \frac{\text{average mass of one molecule}}{\frac{1}{12}\text{ mass of one atom of } ^{12}\text{C}}$$

$$= \frac{\text{average mass of one molecule} \times 12}{\text{mass of one atom of } ^{12}\text{C}}$$

You find the **relative molecular mass** by adding up the relative atomic masses of all the atoms present in the molecule, and you find this from the formula.

Learning objectives:

- → State the definition of relative atomic mass.
- → State the definition of relative molecular mass.
- → State the meaning of the Avogadro constant.
- → State what the same numbers of moles of different substances have in common.
- → Calculate the number of moles present in a given mass of an element or compound.

Specification reference: 3.1.2

Study tip

The weighted average mass must be used to allow for the presence of isotopes, using their percentage abundances in calculations.

Study tip

It would be useful to learn the exact definitions of A_r and M_r.

Study tip

In practice, the scale based on $^{12}C = 12$ exactly is virtually the same as the scale based on hydrogen = 1. This is because, on this scale, A_r for hydrogen = 1.000 7.

▼ **Table 1** *Examples of relative molecular mass*

Molecule	Formula	A_r of atoms	M_r
water	H_2O	$(2 \times 1.0) + 16.0$	18.0
carbon dioxide	CO_2	$12.0 + (2 \times 16.0)$	44.0
methane	CH_4	$12.0 + (4 \times 1.0)$	16.0

Relative formula mass

The term **relative formula mass** is used for ionic compounds because they don't exist as molecules. However, this has the same symbol M_r.

▼ **Table 2** *Some examples of the relative formula masses of ionic compounds*

Ionic compound	Formula	A_r of atoms	M_r
calcium fluoride	CaF_2	$40.1 + (2 \times 19.0)$	78.1
sodium sulfate	Na_2SO_4	$(2 \times 23.0) + 32.1 + (4 \times 16.0)$	142.1
magnesium nitrate	$Mg(NO_3)_2$	$24.3 + (2 \times (14.0 + (16.0 \times 3))$	148.3

▲ **Figure 1** *Large numbers of coins or bank notes are counted by weighing them*

The Avogadro constant and the mole

One atom of any element is too small to see with an optical microscope and impossible to weigh individually. So, to count atoms, chemists must weigh large numbers of them. This is how cashiers count money in a bank (Figure 1).

Working to the nearest whole number, a helium atom ($A_r = 4$) is four times heavier than an atom of hydrogen. A lithium atom ($A_r = 7$) is seven times heavier than an atom of hydrogen. To get the same number of atoms in a sample of helium or lithium, as the number of atoms in 1 g of hydrogen, you must take 4 g of helium or 7 g of lithium.

In fact, if you weigh out the relative atomic mass of *any* element, this amount will also contain this same number of atoms.

The same logic applies to molecules. Water H_2O, has a relative molecular mass M_r of 18. So, one molecule of water is 18 times heavier than one atom of hydrogen. Therefore, 18 g of water contain the same number of *molecules* as there are *atoms* in 1 g of hydrogen. A molecule of carbon dioxide is 44 times heavier than an atom of hydrogen, so 44 g of carbon dioxide contain this same number of molecules.

If you weigh out the relative or formula mass M_r of a compound in grams you have *this same number* of **entities**.

The Avogadro constant

The actual number of atoms in 1 g of hydrogen atoms is unimaginably huge:

602 200 000 000 000 000 000 000 usually written 6.022×10^{23}.

The difference between this scale, based on H = 1 and the scale used today based on ^{12}C, is negligible, for most purposes.

The Avogadro constant or Avogadro number is the number of atoms in 12 g of carbon-12.

The mole

The amount of substance that contains 6.022×10^{23} particles is called a **mole**.

Study tip

The Avogadro constant is the same as the number of *atoms* in 1 g of hydrogen H_2, not the number of hydrogen *molecules*.

Study tip

Entity is a general word for a particle. It can refer to an atom, molecule, ion, electron, or the simplest formula unit of a giant ionic structure, such as sodium chloride, NaCl.

Study tip

You can also use the term *molar mass*, which is the mass per mole of a substance. It has units kg mol^{-1} or g mol^{-1}. The molar mass in g mol^{-1} is the same numerically as M_r.

The relative atomic mass of any element in grams contains one mole of atoms. The relative molecular mass (or relative formula mass) of a substance in grams contains one mole of entities. You can also have a mole of ions or electrons.

It is easy to confuse moles of *atoms* and moles of *molecules*, so always give the formula when working out the mass of a mole of entities. For example, 10 moles of hydrogen could mean 10 moles of hydrogen atoms or 10 moles of hydrogen molecules, H_2, which contains twice the number of atoms. Using the mole, you can compare the *numbers* of different particles that take part in chemical reactions.

▼ **Table 3** *Examples of moles*

Entities	Formula	Relative mass	Mass of a mole / g = molar mass
oxygen atoms	O	16.0	16.0
oxygen molecules	O_2	32.0	32.0
sodium ions	Na^+	23.0	23.0
sodium fluoride	NaF	42.0	42.0

Number of moles

If you want to find out how many moles are present in a particular mass of a substance you need to know the substance's formula. From the formula you can then work out the mass of one mole of the substance.

You use:

$$\text{number of moles } n = \frac{\text{mass } m \text{ (g)}}{\text{mass of 1 mole } M \text{ (g)}}$$

Worked example: Finding the number of moles

How many moles are there in 0.53 g of sodium carbonate, Na_2CO_3?

A_r Na = 23.0, A_r C = 12.0, A_r O =16.0,
so M_r of Na_2CO_3 = (23.0 × 2) + 12.0 + (16.0 × 3) = 106.0,
so 1 mole of calcium carbonate has a mass of 106.0g.

Number of moles = $\frac{0.53}{106.0}$ = 0.0050 mol

Worked example: Finding the number of atoms

You have 3.94g of gold, Au, and 2.70g of aluminium, Al. Which contains the greater number of atoms? (A_r Au = 197.0, A_r Al = 27.0)

Number of moles of gold atoms = $\frac{3.94}{197.0}$ = 0.020 mol

Number of moles of aluminium atoms = $\frac{2.70}{27.0}$ = 0.100 mol

There are more atoms of aluminium.

Summary questions

1 Calculate the M_r for each of the following compounds.

a CH_4 **b** Na_2CO_3
c $Mg(OH)_2$ **d** $(NH_4)_2SO_4$

Use these values for the relative atomic masses (A_r): C = 12.0, H =1.0, Na = 23.0, O = 16.0, Mg = 24.3, N = 14.0, S = 32.1

2 Imagine an atomic seesaw with an oxygen atom on one side. Find six combinations of other atoms that would make the seesaw balance. For example, one nitrogen atom and two hydrogen atoms would balance the seesaw. Use values of A_r to the nearest whole number.

3 Calculate the number of moles in the given masses of the following entities.

a 32.0 g CH_4
b 5.30 g Na_2CO_3
c 5.83 g $Mg(OH)_2$

4 Identify which contains the fewest molecules: 0.5 g of hydrogen H_2, 4.0 g of oxygen O_2, or 11.0 g of carbon dioxide CO_2

5 Identify the quantity in Question **3** that contains the greatest number of atoms.

2.2 Moles in solutions

Learning objectives:

- → Calculate the number of moles of a substance from the volume of a solution and its concentration.

Specification reference: 3.1.2

Solutions

A solution consists of a solvent with a solute dissolved in it, (Figure 1).

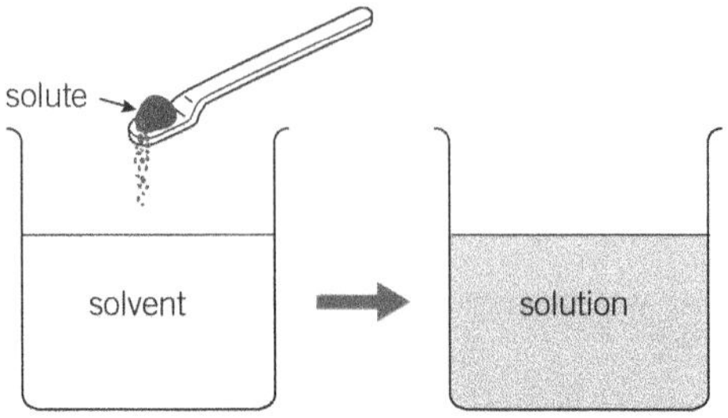

▲ **Figure 1** *A solution contains a solute and a solvent*

Study tip

To get a solution with a concentration of 1 mol dm^{-3} you have to add the solvent to the solute until you have 1 dm^3 of solution. You *do not* add 1 mol of solute to 1 dm^3 of solvent. This would give more than 1 dm^3 of solution.

The units of concentration

The concentration of a solution tells us how much solute is present in a known volume of solution.

Concentrations of solutions are measured in mol dm^{-3}. 1 mol dm^{-3} means there is 1 mole of solute per cubic decimetre of solution; 2 mol dm^{-3} means there are 2 moles of solute per cubic decimetre of solution, and so on.

Study tip

It is important to state units in the answers to numerical questions.

Worked example: Finding the concentration in mol dm^{-3}

1.17 g of sodium chloride was dissolved in water to make 500 cm^3 of solution. What is the concentration of the solution in mol dm^{-3}? A_r Na = 23.0, A_r Cl = 35.5

The mass of 1 mole of sodium chloride, NaCl, is 23.0 + 35.5 = 58.5 g.

$$\text{number of moles } n = \frac{\text{mass } m \text{ (g)}}{\text{mass of 1 mole } M \text{ (g)}}$$

So 1.17 g of NaCl contains $\frac{1.17}{58.5}$ mol = 0.020 mol to 3 s.f.

This is dissolved in 500 cm^3, so 1000 cm^3 (1 dm^3) would contain 0.040 mol of NaCl. This means that the concentration of the solution is 0.040 mol dm^{-3}.

The general way of finding a concentration is to remember the relationship:

$$\text{concentration } c \text{ (mol dm}^{-3}\text{)} = \frac{\text{number of moles } n}{\text{volume } V \text{ (dm}^3\text{)}}$$

Substituting into this gives:

$$\text{concentration} = \frac{0.020}{0.500} = 0.040 \text{ mol dm}^{-3}$$

Study tip

1 decimetre = 10 cm, so one cubic decimetre, 1 dm^3, is 10 cm × 10 cm × 10 cm = 1000 cm^3. This is the same as 1 litre (1 l or 1 L). If you are not confident about conversion factors and writing units, see Section 8, Mathematical skills.

The small negative sign in mol dm^{-3} means per and is sometimes written as a slash, mol/dm^3.

The number of moles in a given volume of solution

You often have to work out how many moles are present in a particular volume of a solution of known concentration. The general formula for the number of moles in a solution of concentration c (mol dm^{-3}) and volume V (cm^3) is:

$$\text{number of moles in solution, } n = \frac{\text{concentration } c\ (\text{mol dm}^{-3}) \times \text{volume } V\ (\text{cm}^3)}{1000}$$

Here is an example of how you reach this formula in steps.

Worked example: Moles in a solution

How many moles are present in 24.70 cm^3 of a solution of concentration 0.100 mol dm^{-3}?

From the definition,

1000 cm^3 of a solution of 1.00 mol dm^{-3} contains 1 mol

So 1000 cm^3 of a solution of 0.100 mol dm^{-3} contains 0.100 mol

So 1.0 cm^3 of a solution of 0.100 mol dm^{-3} contains $\frac{0.10}{1000}$ mol = 0.000 10 mol

So 24.7 cm^3 of a solution of 0.10 mol dm^{-3} contains 24.7 × 0.000 10 = 0.002 47 mol

Using the formula gives the same answer:

$$n = \frac{c \times V}{1000}$$

$$= \frac{0.10 \times 24.7}{1000} = 0.00247\ \text{mol}$$ i.e., 0.0025 to 4 significant figures

Summary questions

1 Calculate the concentration in mol dm^{-3} of the following.

- a 0.500 mol acid in 500 cm^3 of solution
- b 0.250 mol acid in 2000 cm^3 of solution
- c 0.200 mol solute in 20 cm^3 of solution

2 Calculate how many moles of solute there are in the following.

- a 20.0 cm^3 of a 0.100 mol dm^{-3} solution
- b 50.0 cm^3 of a 0.500 mol dm^{-3} solution
- c 25.0 cm^3 of a 2.00 mol dm^{-3} solution

3 0.234 g of sodium chloride was dissolved in water to make 250 cm^3 of solution.

- a State the M_r for NaCl.
  A_r Na = 23.0, A_r Cl = 35.5
- b Calculate how many moles of NaCl is in 0.234 g.
- c Calculate the concentration in mol dm^{-3}.

Error in measurements

Every measurement has an inherent uncertainty (also known as error). In general, the uncertainty in a single measurement from an instrument is *half the value of the smallest division*. The uncertainty of a measurement may also be expressed by ± sign at the end. For example, the mass of an electron is given as $9.109\,382\,91 \times 10^{-31}$ kg $\pm\ 0.000\,000\,40 \times 10^{-31}$ kg, that is, it is between 9.109 383 31 and $9.109\,382\,51 \times 10^{-31}$ kg

For example, a 100 cm^3 measuring cylinder has 1 cm^3 as its smallest division so the measuring error can be taken as 0.5 cm^3. So if you measure 50 cm^3, the percentage error is $\left(\frac{0.5}{50}\right) \times 100\% = 1\%$

What is the percentage error if you use a 100 cm^3 measuring cylinder to measure

- a 10 cm^3
- b 100 cm^3

a 5% b 0.5%

2.3 The ideal gas equation

Learning objectives:

- → State the ideal gas equation.
- → Describe how it is used to calculate the number of moles of a gas at a given volume, temperature, and pressure.

Specification reference: 3.1.2

▲ **Figure 1** *The German airship Hindenburg held about 210 000 m^3 of hydrogen gas*

The Hindenburg airship (Figure 1) was originally designed in the 1930s to use helium as its lifting gas, rather than hydrogen, but the only source of large volumes of helium was the USA and they refused to sell it to Germany because of Hitler's aggressive policies. The airship was therefore made to use hydrogen. It held about 210 000 m^3 of hydrogen gas, but this volume varied with temperature and pressure.

The volume of a given mass of any gas is not fixed. It changes with pressure and temperature. However, there are a number of simple relationships for a given mass of gas that connect the pressure, temperature, and volume of a gas.

Boyle's law

The product of pressure and volume is a constant as long as the temperature remains constant.

$$\text{pressure } P \times \text{volume } V = \text{constant}$$

Charles' law

The volume is proportional to the temperature as long as the pressure remains constant.

$$\text{volume } V \propto \text{temperature } T \quad \text{and} \quad \frac{\text{volume } V}{\text{temperature } T} = \text{constant}$$

Gay-Lussac's law (also called the constant volume law)

The pressure is proportional to the temperature as long as the volume remains constant.

$$\text{pressure } P \propto \text{temperature } T \quad \text{and} \quad \frac{\text{pressure } P}{\text{temperature } T} = \text{constant}$$

Combining these relationships gives us the equation:

$$\frac{\text{pressure } P \times \text{volume } V}{\text{temperature } T} = \text{constant for a fixed mass of gas}$$

The ideal gas equation

In one mole of gas, the constant is given the symbol R and is called the gas constant. For n moles of gas:

$$\underset{P\ (\text{Pa})}{\text{pressure}} \times \underset{V\ (\text{m}^3)}{\text{volume}} = \underset{\text{moles } n}{\text{number of}} \times \underset{R\ (\text{J K}^{-1}\ \text{mol}^{-1})}{\text{gas constant}} \times \underset{T\ (\text{K})}{\text{temperature}}$$

$$PV = nRT$$

The value of R is 8.31 J K^{-1} mol^{-1}.

This is the ideal gas equation. No gases obey it exactly, but at room temperature and pressure it holds quite well for many gases. It is often useful to imagine a gas which obeys the equation perfectly – an ideal gas.

Notes on units

When using the ideal gas equation, consistent units must be used. If you want to calculate n, the number of moles:

P must be in Pa (N m^{-2}) T must be in K

V must be in m^3 R must be in J K^{-1} mol^{-1}

Maths link

If you are not sure about proportionality and changing the subject of an equation, see Section 8, Mathematical skills.

Study tip

The units used here are part of the Système Internationale (SI) of units. This is a system of units for measurements used by scientists throughout the world. The basic units used by chemists are: metre m, second s, Kelvin K, and kilogram kg.

Using the ideal gas equation

Using the ideal gas equation, you can calculate the volume of one mole of gas at any temperature and pressure. Since none of the terms in the equation refers to a particular gas, this volume will be the same for any gas.

This may seem very unlikely at first sight, but it is the space between the gas molecules that accounts for the volume of a gas. Even the largest gas particle is extremely small compared with the space in between the particles.

Rearranging the ideal gas equation to find a volume gives:

$$V = \frac{nRT}{P}$$

The worked example tells you that the volume of a mole of *any* gas at room temperature and pressure is approximately 24 000 cm^3 (24 dm^3). For example, one mole of sulfur dioxide gas, SO_2 (mass 64.1 g) has the same volume as one mole of hydrogen gas, H_2 (mass 2.0 g).

In a similar way, pressure can be found using $P = \frac{nRT}{V}$

Finding the number of moles *n* of a gas

If you rearrange the equation $PV = nRT$ so that n is on the left-hand side, you get:

$$n = \frac{PV}{RT}$$

If T, P, and V are known, then you can find n.

Worked example: Volume from the ideal gas equation

If temperature = 20.0 °C (293.0 K), pressure = 100 000 Pa, and $n = 1$ for one mole of gas

$$V = \frac{8.31\ \text{JK}^{-1}\text{mol}^{-1} \times 293\ \text{K}}{100\,000\ \text{Pa}}$$
$$= 0.0243\ \text{m}^3$$
$$= 0.0243 \times 10^6\ \text{cm}^3$$
$$= 24\,300\ \text{cm}^3$$

Worked example: Finding the number of moles

How many moles of hydrogen molecules are present in a volume of 100 cm^3 at a temperature of 20.0 °C and a pressure of 100 kPa? $R = 8.31\ \text{JK}^{-1}\text{mol}^{-1}$

First, convert to the base units:

P must be in Pa, and 100 kPa = 100 000 Pa

V must be in m^3, and 100 cm^3 = $100 \times 10^{-6}\ m^3$

T must be in K, and 20 °C = 293 K (add 273 to the temperature in °C)

Substituting into the ideal gas equation:

$$n = \frac{PV}{RT}$$
$$= \frac{100\,000 \times 100 \times 10^{-6}}{8.31 \times 293}$$
$$= 0.00411\ \text{moles}$$

Study tip

Remember to convert to SI units and to cancel the units.

Study tip

To convert °C to K add 273.

Finding the relative molecular mass of a gas

If you know the number of moles present in a given mass of gas, you can find the mass of one mole of gas and this tells us the relative molecular mass.

Study tip

Using 24 000 cm^3 as the volume of a mole of any gas is not precise, and it is always necessary to apply the ideal gas equation in calculations.

Study tip

In reporting a measurement, you should include the best value (e.g., the average) and an estimate of its uncertainty. One common practice is to round off the experimental result so that it contains the digits known with certainty *plus* the first *uncertain* one. The total number of digits is the number of significant figures used.

Study tip

In the equation $PV = nRT$, the units *must* be P in Pa (not kPa), V in m^3, and T in K.

Summary questions

1 a Calculate approximately how many moles of H_2 molecules were contained in the Hindenburg airship at 298 K.

b The original design used helium. State how many moles of helium atoms it would have contained.

2 a Calculate the volume of 2 moles of a gas if the temperature is 30 °C, and the pressure is 100 000 Pa.

b Calculate the pressure of 0.5 moles of a gas if the volume is 11 000 cm^3, and the temperature is 25 °C.

3 Calculate how many moles of hydrogen molecules are present in a volume of 48 000 cm^3, at 100 000 Pa and 25 °C.

4 State how many moles of carbon dioxide molecules would be present in Question **3**. Explain your answer.

Finding the relative molecular mass of lighter fuel

The apparatus used to find the relative molecular mass of lighter fuel is shown in Figure 2.

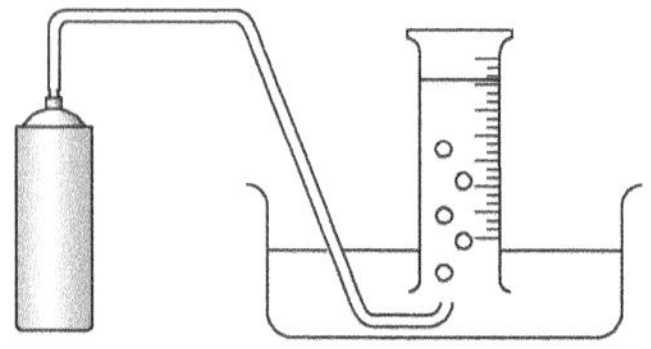

▲ **Figure 2** *Measuring the relative molecular mass of lighter fuel*

The lighter fuel canister was weighed.

1000 cm^3 of gas was dispensed into the measuring cylinder, until the levels of the water inside and outside the measuring cylinder were the same, so that the pressure of the collected gas was the same as atmospheric pressure.

The canister was reweighed.

Atmospheric pressure and temperature were noted.

The results were

loss of mass of the can = 2.29 g

temperature = 14 °C = 287 K

atmospheric pressure = 100 000 Pa

Volume of gas = 1000 cm^3 = 1000×10^{-6} m^3

$$n = \frac{PV}{RT}$$

$$= \frac{100\,000 \times 1000 \times 10^{-6}}{8.31 \times 287}$$

$$= 0.042 \text{ mol}$$

0.042 mol has a mass of 2.29 g

So, 1 mol has a mass of $\frac{2.29}{0.042 \text{ g}} = 54.5$ g

So, $M_r = 54.5$

2.4 Empirical and molecular formulae

The empirical formula

The **empirical formula** is the formula that represents the simplest whole number ratio of the atoms of each element present in a compound. For example, the empirical formula of carbon dioxide, CO_2, tells us that for every carbon atom there are two oxygen atoms.

To find an empirical formula:

1. Find the masses of each of the elements present in a compound (by experiment).
2. Work out the number of moles of atoms of each element.

$$\text{number of moles} = \frac{\text{mass of element}}{\text{mass of 1 mol of element}}$$

3. Convert the number of moles of each element into a whole number ratio.

Worked example: Finding empirical formula of calcium carbonate

10.01 g of a white solid contains 4.01 g of calcium, 1.20 g of carbon, and 4.80 g of oxygen. What is its empirical formula?

(A_r Ca = 40.1, A_r C = 12.0, A_r O = 16.0)

Step 1 Find the masses of each element.

Mass of calcium = 4.01 g

Mass of carbon = 1.20 g

Mass of oxygen = 4.80 g

Step 2 Find the number of moles of atoms of each element.

A_r Ca = 40.1

Number of moles of calcium = $\frac{4.01}{40.1}$ = 0.10 mol

A_r C = 12.0

Number of moles of carbon = $\frac{1.2}{12.0}$ = 0.10 mol

A_r O = 16

Number of moles of oxygen = $\frac{4.8}{16.0}$ = 0.30 mol

Step 3 Find the simplest ratio.

	calcium	carbon	oxygen
Ratio in moles of	0.10	0.10	0.30
So the simplest whole number ratio is:	1	1	3

The formula is therefore $CaCO_3$.

Learning objectives:

→ State the definitions of empirical formula and molecular formula.

→ Calculate the empirical formula from the masses or percentage masses of the elements present in a compound.

→ Calculate the additional information needed to work out a molecular formula from an empirical formula.

Specification reference: 3.1.2

Study tip

The mass of 1 mole in grams is the same as the relative atomic mass of the element.

Worked example: Finding the empirical formula of copper oxide

0.795 g of black copper oxide is reduced to 0.635 g of copper when heated in a stream of hydrogen (Figure 1). What is the formula of copper oxide? A_r Cu = 63.5, A_r O = 16.0

Step 1 Find the masses of each element.

Mass of copper = 0.635 g

Started with 0.795 g of copper oxide and 0.635 g of copper were left, so:

Mass of oxygen = 0.795 − 0.635 = 0.160 g

Step 2 Find the number of moles of atoms of each element.

A_r Cu = 63.5

Number of moles of copper $= \frac{0.635}{63.5} = 0.01$

A_r O = 16.0

Number of moles of oxygen $= \frac{0.16}{16.0} = 0.01$

Step 3 Find the simplest ratio.

The ratio of moles of copper to moles of oxygen is:

copper : oxygen

0.01 : 0.01

So the simplest whole number ratio is 1 : 1

The simplest formula of black copper oxide is therefore one Cu to one O, CuO. You may find it easier to make a table.

	Copper Cu	Oxygen O
mass of element	0.635 g	0.160 g
A_r of element	63.5	16.0
number of moles = $\frac{\text{mass of element}}{A_r}$	$\frac{0.635}{63.5} = 0.01$	$\frac{0.160}{16.0} = 0.01$
ratio of elements	1	1

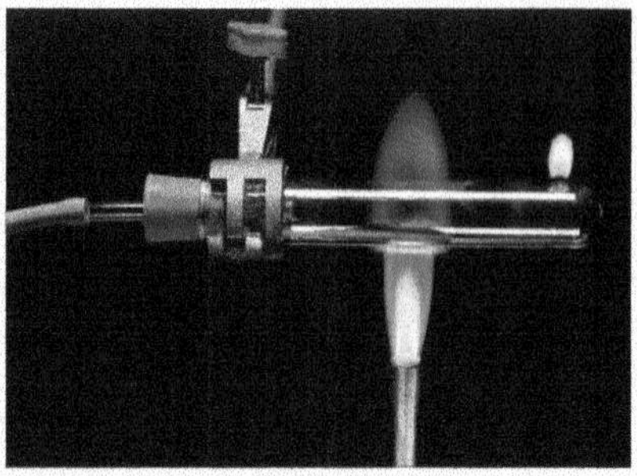

▲ **Figure 1** *Finding the empirical formula of copper oxide*

Finding empirical formula of copper oxide

In this experiment explain why:

1. there is a flame at the end of the tube
2. this flame goes green
3. droplets of water form near the end of the tube
4. the flame at the end of the tube is kept alight until the apparatus is cool.

Erroneous results

One student carried out the experiment to find the formula of black copper oxide with the following results:
0.735 g of the oxide was reduced to 0.635 g after reduction.

1. Confirm that these results lead to a ratio of 0.01 mol, copper to 0.006 25 mol of oxygen, which is incorrect.
2. Suggest what the student might have done wrong to lead to this apparently low value for the amount of oxygen.

Finding the simplest ratio of elements

Sometimes you will end up with ratios of moles of atoms of elements that are not easy to convert to whole numbers. If you divide each number by the smallest number you will end up with whole numbers (or ratios you can recognise more easily). Here is an example.

Worked example: Empirical formula

Compound X contains 50.2 g sulfur and 50.0 g oxygen. What is its empirical formula? A_r S = 32.1, A_r O = 16.0

Step 1 Find the number of moles of atoms of each element.

A_r S = 32.1

Number of moles of sulfur $= \frac{50.2}{32.1} = 1.564$

A_r O =16

Number of moles of oxygen $= \frac{50.0}{16.0} = 3.125$

Step 2 Find the simplest ratio.

Ratio of sulfur : oxygen : 1.564 : 3.125

Now divide each of the numbers by the smaller number.

Ratio of sulfur : oxygen

$$\frac{1.564}{1.564} : \frac{3.125}{1.564} = 1{:}2$$

The empirical formula is therefore SO_2. Sometimes you may end up with a ratio of moles of atoms, such as 1:1.5. In these cases you must find a whole number ratio, in this case 2:3.

Finding the molecular formula

The **molecular formula** gives the actual number of atoms of each element in one molecule of the compound. (It applies only to substances that exist as molecules.)

The empirical formula is not always the same as the molecular formula. There may be several units of the empirical formula in the molecular formula.

For example, ethane (molecular formula C_2H_6) would have an empirical formula of CH_3.

To find the number of units of the empirical formula in the molecular formula, divide the relative molecular mass by the relative mass of the empirical formula.

For example, ethene is found to have a relative molecular mass of 28.0 but its empirical formula, CH_2, has a relative mass of 14.0.

$$\frac{\text{Relative molecular mass of ethene}}{\text{Relative mass of empirical formula of ethene}} = \frac{28.0}{14.0} = 2$$

So there must be two units of the empirical formula in the molecule of ethene. So ethene is $(CH_2)_2$ or C_2H_4.

Another oxide of copper

There is another oxide of copper, which is red. In a reduction experiment similar to that for finding the formula of black copper oxide, 1.43 g of red copper oxide was reduced with a stream of hydrogen and 1.27 g of copper were formed. Use the same steps as for black copper oxide to find the formula of the red oxide.

1. Find the masses of each element.
2. Find the number of moles of atoms of each element.
3. Find the simplest ratio.

Synoptic link

Having more than one oxide (and other compounds) with different formulae is a typical property of transition metals, see Topic 23.1, The general properties of transition metals.

Study tip

- When calculating empirical formulae from percentages, check that all the percentages of the compositions by mass add up to 100%. (Don't forget any oxygen that may be present.)
- Remember to use relative atomic masses from the Periodic Table, *not* the atomic number.

Synoptic link

Once we know the formula of a compound we can use techniques such as infrared spectroscopy and mass spectrometry to help work out its structure, see Chapter 16, Organic analysis.

Combustion analysis

Organic compounds are based on carbon and hydrogen. One method of finding empirical formulae of new compounds is called combustion analysis (Figure 2). It is used routinely in the pharmaceutical industry. It involves burning the unknown compound in excess oxygen and measuring the amounts of water, carbon dioxide, and other oxides that are produced. The gases are carried through the instrument by a stream of helium.

The basic method measures carbon, hydrogen, sulfur, and nitrogen. It is assumed that oxygen makes up the difference after the other four elements have been measured. Once the sample has been weighed and placed in the instrument, the process is automatic and controlled by computer.

The sample is burnt completely in a stream of oxygen. The final combustion products are water, carbon dioxide, and sulfur dioxide. The instrument measures the amounts of these by infrared absorption. They are removed from the gas stream leaving the unreacted nitrogen which is measured by thermal conductivity. The measurements are used to calculate the masses of each gas present and hence the masses of hydrogen, sulfur, carbon, and nitrogen in the original sample. Oxygen is found by difference.

Traditionally, the amounts of water and carbon dioxide were measured by absorbing them in suitable chemicals and measuring the increase in mass of the absorbents. This is how the composition data for the worked example below were measured. The molecular formula can then be found, if the relative molecular mass has been found using a mass spectrometer.

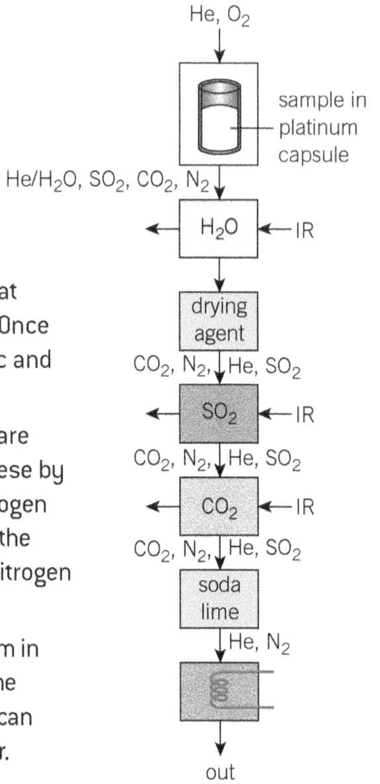

▲ **Figure 2** *Combustion analysis*

Soda lime is a mixture containing mostly calcium hydroxide, $Ca(OH)_2$. Construct a balanced symbol equation for the reaction of calcium hydroxide with carbon dioxide.

$Ca(OH)_2 + CO_2 \rightarrow CaCO_3 + H_2O$

Worked example: Molecular formula

An organic compound containing only carbon, hydrogen, and oxygen was found to have 52.17% carbon and 13.04% hydrogen. What is its molecular formula if $M_r = 46.0$?

100.00 g of this compound would contain 52.17 g carbon, 13.04 g hydrogen and (the rest) 34.79 g oxygen.

Step 1 Find the empirical formula.

	Carbon	Hydrogen	Oxygen
mass of element/g	52.17	13.04	34.79
A_r of element	12.0	1.0	16.0
number of moles = $\frac{\text{mass of element}}{A_r}$	$\frac{52.17}{12.0} = 4.348$	$\frac{13.04}{1.0} = 13.04$	$\frac{34.79}{16.0} = 2.174$
divide through by the smallest	$\frac{4.348}{2.174} = 2$	$\frac{13.04}{2.174} = 6$	$\frac{2.174}{2.174} = 1$
ratio of elements	2	6	1

So the empirical formula is C_2H_6O.

Step 2 Find M_r of the empirical formula.

$(2 \times 12.0) + (6 \times 1.0) + (1 \times 16.0) = 46.0$

C H O

So, the molecular formula is the same as the empirical formula, C_2H_6O.

Worked example: Molecular formula by combustion analysis

0.53 g of a compound X containing only carbon, hydrogen, and oxygen, gave 1.32 g of carbon dioxide and 0.54 g of water on complete combustion in oxygen. What is its empirical formula? What is its molecular formula if its relative molecular mass is 58.0?

To calculate the empirical formula:

carbon 1.32 g of CO_2 ($M_r = 44.0$) is $\frac{1.32}{44.0} = 0.03$ mol CO_2

As each mole of CO_2 has 1 mole of C, the sample contained 0.03 mol of C atoms.

hydrogen 0.54 g of H_2O ($M_r = 18.0$) is $\frac{0.54}{18.0} = 0.03$ mol H_2O

As each mole of H_2O has 2 moles of H, the sample contained 0.06 mol of H atoms.

oxygen 0.03 mol of carbon atoms ($A_r = 12.0$) has a mass of 0.36 g

0.06 mol of hydrogen atoms ($A_r = 1.0$) has a mass of 0.06 g

Total mass of carbon and hydrogen is 0.42 g

The rest (0.58 – 0.42) must be oxygen, so the sample contained 0.16 g of oxygen.

0.16 g of oxygen ($A_r = 16.0$) is $\frac{0.16}{16.0} = 0.01$ mol oxygen atoms

So the sample contains 0.03 mol C, 0.06 mol H, and 0.01 mol O

	C	H	O
Dividing by the smallest number 0.06 gives the ratio:			
so the **empirical formula** is C_3H_6O.	3	6	1

M_r of this unit is 58, so the molecular formula is also C_3H_6O.

Summary questions

1 Calculate the empirical formula of each of the following compounds? (You could try to name them too.)

a A liquid containing 2.0 g of hydrogen, 32.1 g sulfur, and 64.0 g oxygen.

b A white solid containing 4.0 g calcium, 3.2 g oxygen, and 0.2 g hydrogen.

c A white solid containing 0.243 g magnesium and 0.710 g chlorine.

2 3.888 g magnesium ribbon was burnt completely in air and 6.448 g of magnesium oxide was produced.

a Calculate how many moles of magnesium and oxygen are present in 6.448 g of magnesium oxide.

b State the empirical formula of magnesium oxide.

3 State the empirical formula of each of the following molecules.

a cyclohexane, C_6H_{12} b dichloroethene, $C_2H_2Cl_2$ c benzene, C_6H_6

4 M_r for ethane-1,2-diol is 62.0. It is composed of carbon, hydrogen, and oxygen in the ratio by moles of 1 : 3 : 1. Identify its molecular formula.

5 An organic compound containing only carbon, hydrogen, and oxygen was found to have 62.07% carbon and 10.33% hydrogen. Identify the molecular formula if $M_r = 58.0$.

6 A sample of benzene of mass 7.8 g contains 7.2 g of carbon and 0.6 g of hydrogen. If M_r is 78.0, identify:

a the empirical formula b the molecular formula.

2.5 Balanced equations and related calculations

Learning objectives:

- ➔ Demonstrate how an equation can be balanced if the reactants and products are known.
- ➔ Calculate the amount of a product using experimental data and a balanced equation.

Specification reference: 3.1.2

Equations represent what happens when chemical reactions take place. They are based on experimental evidence. The starting materials are reactants. After these have reacted you end up with products.

reactants → products

Word equations only give the names of the reactants and products, for example:

hydrogen + oxygen → water

Once the idea of atoms had been established, chemists realised that atoms react together in simple whole number ratios. For example, two hydrogen molecules react with one oxygen molecule to give two water molecules.

2 hydrogen molecules +		1 oxygen molecule →		2 water molecules
2	:	1	:	2

The ratio in which the reactants react and the products are produced, in simple whole numbers, is called the **stoichiometry** of the reaction.

You can build up a stoichiometric relationship from experimental data by working out the number of moles that react together. This leads us to a balanced symbol equation.

Balanced symbol equations

Balanced symbol equations use the formulae of reactants and products. There are the same number of atoms of each element on both sides of the arrow. (This is because atoms are never created or destroyed in chemical reactions.) Balanced equations tell us about the amounts of substances that react together and are produced.

Study tip

Learn these four state symbols.

State symbols can also be added. These are letters, in brackets, which can be added to the formulae in equations to say what state the reactants and products are in – (s) means solid, (l) means liquid, (g) means gas, and (aq) means aqueous solution (dissolved in water).

Writing balanced equations

When aluminium burns in oxygen it forms solid aluminium oxide. You can build up a balanced symbol equation from this and the formulae of the reactants and product – Al, O_2, and Al_2O_3.

1 Write the word equation

aluminium + oxygen → aluminium oxide

2 Write in the correct formulae

$$Al + O_2 \rightarrow Al_2O_3$$

This is not balanced because:

- there is one aluminium atom on the reactants side (left-hand side) but two on the products side (right-hand side)
- there are two oxygen atoms on the reactants side (left-hand side) but three on the products side (right-hand side).

3 To get two aluminium atoms on the left-hand side put a 2 in front of the Al:

$$2Al + O_2 \rightarrow Al_2O_3$$

Now the aluminium is correct but not the oxygen.

4 If you multiply the oxygen on the left-hand side by 3, and the aluminium oxide by 2, you have six O on each side:

$$2Al + 3O_2 \rightarrow 2Al_2O_3$$

5 Now you return to the aluminium. You need four Al on the left-hand side:

$$4Al + 3O_2 \rightarrow 2Al_2O_3$$

The equation is balanced because there are the same numbers of atoms of each element on both sides of the equation.

The numbers in front of the formulae (4, 3, and 2) are called **coefficients**.

6 You can add state symbols.

The equation tells you the numbers of moles of each of the substances that are involved. From this you can work out the masses that will react together: (using Al = 27.0, O = 16.0)

$4Al(s)$	+	$3O_2(g)$	$\rightarrow$	$2Al_2O_3(s)$
4 moles		3 moles		2 moles
108.0 g		96.0 g		204.0 g

The total mass is the same on both sides of the equation. This is another good way of checking whether the equation is balanced.

Hint

Since we know that 1 mole of *any* gas at room conditions has a volume of approximately 24 dm^3 we can see that the volume of oxygen required is approximately 3 x 24 dm^3, i.e., 72 dm^3

Ionic equations

In some reactions you can simplify the equation by considering the ions present. Sometimes there are ions that do not take part in the overall reaction. For example, when any acid reacts with an alkali in solution, you end up with a salt (also in solution) and water. Look at the reaction between hydrochloric acid and sodium hydroxide:

$HCl(aq)$	+	$NaOH(aq)$	$\rightarrow$	$NaCl(aq)$	+	$H_2O(l)$
hydrochloric acid	+	sodium hydroxide	$\rightarrow$	sodium chloride	+	water

The ions present are:

$HCl(aq)$	$H^+(aq)$ and $Cl^-(aq)$
$NaOH(aq)$	$Na^+(aq)$ and $OH^-(aq)$
$NaCl(aq)$	$Na^+(aq)$ and $Cl^-(aq)$

If you write the equation using these ions and then strike out the ions that appear on each side we have:

$$H^+(aq) + \cancel{Cl^-(aq)} + \cancel{Na^+(aq)} + OH^-(aq) \rightarrow \cancel{Na^+(aq)} + \cancel{Cl^-(aq)} + H_2O(l)$$

Overall, the equation is

$$H^+(aq) + OH^-(aq) \rightarrow H_2O(l)$$

$Na^+(aq)$ and $Cl^-(aq)$ are called **spectator ions** – they do not take part in the reaction.

Study tip

The charges balance as well as the elements. On the left +1 and −1 (no overall charge) and no charge on the right.

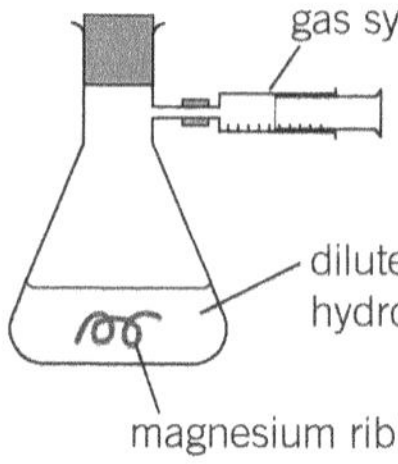

▲ **Figure 1** *Apparatus for collecting hydrogen gas*

Synoptic link

See Practical 1 on page 521.

Whenever an acid reacts with an alkali, the overall reaction will be the same as the one above.

Useful tips for balancing equations

- You *must* use the correct formulae – you cannot change them to make the equation balance.
- You can only change the numbers of atoms by putting a number, called a coefficient, in front of formulae.
- The coefficient in front of the symbol tells you how many moles of that substance are reacting.
- It often takes more than one step to balance an equation, but too many steps suggests that you may have an incorrect formula.
- When dealing with ionic equations the total of the charges on each side must also be the same.

Working out amounts

You can use a balanced symbol equation to work out how much product is produced from a reaction. For example, the reaction between magnesium and hydrochloric acid produces hydrogen gas, as shown in Figure 1.

Worked example: Calculating the mass of product

How much magnesium chloride is produced by 0.120 g of magnesium ribbon and excess hydrochloric acid? A_r Mg = 24.3, A_r H = 1.0, A_r Cl = 35.5

(The word 'excess' means there is more than enough acid to react with all the magnesium.)

Step 1 Write the correct formulae equation.

$$Mg(s) + HCl(aq) \rightarrow MgCl_2(aq) + H_2(g)$$

magnesium + hydrochloric acid → magnesium chloride + hydrogen

Step 2 Balance the equation. The number of Mg atoms is correct. There are two Cl atoms and two H atoms on the right-hand side so you need to add a 2 in front of the HCl.

$$Mg(s) + 2HCl(aq) \rightarrow MgCl_2(aq) + H_2(g)$$

Now find the numbers of moles that react.

$$Mg(s) + 2HCl(aq) \rightarrow MgCl_2(aq) + H_2(g)$$

$$1\ \text{mol} \quad 2\ \text{mol} \rightarrow 1\ \text{mol} \quad 1\ \text{mol}$$

1 mol of Mg has a mass of 24.3 g because its $A_r = 24.3$.

So, 0.12 g of Mg is $\frac{0.12}{24.3} = 0.0049$ mol.

From the equation, you can see that one mole of magnesium reacts to give one mole of magnesium chloride. Therefore, 0.0049 mol of magnesium produces 0.0049 mol of magnesium chloride.

M_r $MgCl_2 = 24.3 + (2 \times 35.5) = 95.3$

So the mass of $MgCl_2 = 0.0049 \times 95.3 = 0.47$ g to 2 s.f.

Finding concentrations using titrations

Titrations can be used to find the concentration of a solution, for example, an alkali by reacting an acid with the alkali using a suitable indicator.

You need to know the concentration of the acid and the equation for the reaction between the acid and alkali.

The apparatus is shown in Figure 2.

The steps in a titration are:

1. Fill a burette with the acid of known concentration.
2. Accurately measure an amount of the alkali using a calibrated pipette and pipette filler.
3. Add the alkali to a conical flask with a few drops of a suitable indicator.
4. Run in acid from the burette until the colour just changes, showing that the solution in the conical flask is now neutral.
5. Repeat the procedure, adding the acid dropwise as you approach the end point, until two values of the volume of acid used at neutralisation are the same, within experimental error.

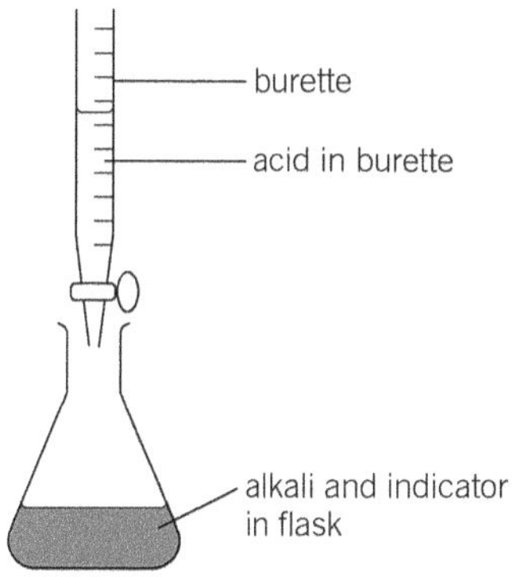

▲ **Figure 2** *Apparatus for a titration*

Worked example: Finding concentration

25.00 cm^3 of a solution of sodium hydroxide, NaOH, of unknown concentration was neutralised by 22.65 cm^3 of a 0.100 mol dm^{-3} solution of hydrochloric acid, HCl. What is the concentration of the alkali?

First write a balanced symbol equation and then the numbers of moles that react:

NaOH(aq) +	HCl(aq) →	NaCl(aq) +	$H_2O(l)$
sodium hydroxide	hydrochloric acid	sodium chloride	water
1 mol	1 mol	1 mol	1 mol

1 mol of sodium hydroxide reacts with 1 mol of hydrochloric acid.

$$\text{number of moles of HCl} = \frac{c \times V}{1000} = \frac{22.65 \times 0.100}{1000}$$

From the equation, there must be an equal number of moles of sodium hydroxide and hydrochloric acid for neutralisation:

number of moles of NaOH = number of moles of HCl

So you must have $\frac{22.65 \times 0.100}{1000}$ mol of NaOH in the 25.00 cm^3 of sodium hydroxide solution.

The concentration of a solution is the number of moles in 1000 cm^3.

Therefore the concentration of the alkali

$$= \frac{22.65 \times 0.100}{1000} \times \frac{1000}{25.00} \text{ mol dm}^{-3} = 0.0906 \text{ mol dm}^{-3}$$

A note on significant figures

22.65 and 25.00 both have 4 s.f. but 0.100 has only 3 s.f. So we can quote the answer to 3 s.f only. So rounding up the final digit gives the concentration of the alkali as 0.091 mol dm^{-3} to 3 s.f.

Summary questions

1 Balance the following equations.

a $Mg + O_2 \rightarrow MgO$

b $Ca(OH)_2 + HCl \rightarrow CaCl_2 + H_2O$

c $Na_2O + HNO_3 \rightarrow NaNO_3 + H_2O$

2 State the concentration of hydrochloric acid if 20.0 cm^3 is neutralised by 25.0 cm^3 of sodium hydroxide of concentration 0.200 mol dm^{-3}.

3 In the reaction

$Mg(s) + 2HCl(aq) \rightarrow MgCl_2(aq) + H_2$

2.60 g of magnesium was added to 100 cm^3 of 1.00 mol dm^{-3} hydrochloric acid.

a State if there is any magnesium left when the reaction finished. Explain your answer.

b Calculate the volume of hydrogen produced at 25 °C and 100 kPa.

4 a Write the balanced equation for the reaction between sulfuric acid and sodium hydroxide

i in full

ii in terms of ions.

b Identify the spectator ions in this reaction.

2.6 Balanced equations, atom economies, and percentage yields

Learning objectives:

- → Describe the atom economy of a chemical reaction.
- → State how an equation is used to calculate an atom economy.
- → Describe the percentage yield of a chemical reaction.
- → Calculate percentage yields.

Specification reference: 3.1.2

Once you know the balanced equation for a chemical reaction, you can calculate the theoretical amount that you should be able to make of any of the products. Most chemical reactions produce two (or more) products but often only one of them is required. This means that some of the products will be wasted. In a world of scarce resources, this is obviously not a good idea. One technique that chemists use to assess a given process is to determine the percentage atom economy.

Atom economy

The **atom economy** of a reaction is found directly from the balanced equation. It is theoretical rather than practical. It is defined as:

$$\text{\% atom economy} = \frac{\textbf{mass of desired product}}{\textbf{total mass of reactants}} \times 100$$

You can see what atom economy means by considering the following real reaction.

Chlorine, Cl_2, reacts with sodium hydroxide, NaOH, to form sodium chloride, NaCl, water, H_2O, and sodium chlorate, NaOCl. Sodium chlorate is used as household bleach – this is the useful product.

From the equation you can work out the mass of each reactant and product involved.

2NaOH	+	Cl_2	→	NaCl	+	H_2O	+	NaOCl
2 mol		1 mol	→	1 mol		1 mol		1 mol
80.0 g		71.0 g	→	58.5 g		18.0 g		74.5 g
Total	151.0 g			Total		151.0 g		

$$\text{\% Atom economy} = \frac{\text{mass of desired product}}{\text{total mass of reactants}} \times 100$$
$$= \frac{74.5}{151} \times 100$$
$$= 49.3\%$$

So only 49.3% of the starting materials are included in the desired product, the rest is wasted.

It may be easier to see what has happened if you colour the atoms involved. Those coloured in green are included in the final product and those in red are wasted – one atom of sodium, one of chlorine, two of hydrogen, and one of oxygen.

$$\mathbf{NaOH + NaOH + ClCl \rightarrow NaCl + H_2O + NaOCl}$$

Another example is the reaction where ethanol breaks down to ethene (the product wanted) and water (which is wasted).

C_2H_5OH	→	$CH_2{=}CH_2$	+	H_2O
46.0 g	→	28.0 g		18.0 g

$$\text{\% Atom economy} = \frac{28.0}{46.0} \times 100 = 60.9\%$$

Some reactions, in theory at least, have no wasted atoms.

For example, ethene reacts with bromine to form 1,2-dibromoethane

$$\begin{array}{cccc} CH_2{=}CH_2 & + \; Br_2 & \rightarrow & CH_2BrCH_2Br \\ 28.0\,g & 160.0\,g & \rightarrow & 188.0\,g \\ & \textbf{Total 188.0 g} & & \textbf{Total 188.0 g} \end{array}$$

$$\%\text{ Atom economy} = \frac{188.0}{(28.0 + 160.0)} \times 100 = 100\%$$

Atom economy – a dangerous fuel

Hydrogen can be made by passing steam over heated coal, which is largely carbon.

$$C(s) + 2H_2O(g) \rightarrow 2H_2(g) + CO_2(g)$$

$$12.0 + (2 \times 18.0) \rightarrow (2 \times 2.0) + 44.0$$

As the only useful product is hydrogen, the atom economy of this reaction is $\left(\frac{4.0}{48.0}\right) \times 100\% = 8.3\%$ – not a very efficient reaction! The reason that it is so inefficient is that all of the carbon is discarded as useless carbon dioxide.

However, under different conditions a mixture of hydrogen and carbon monoxide can be formed (this was called water gas or town gas).

$$C(s) + H_2O(g) \rightarrow H_2(g) + CO(g)$$

Both hydrogen and carbon monoxide are useful fuels, so nothing is discarded and the atom economy is 100%. You can check this with a calculation if you like.

Carbon monoxide is highly toxic. However, almost incredibly to modern eyes, town gas was supplied as a fuel to homes in the days before the country converted to natural gas (methane, CH_4) from the North Sea.

Even methane is not without its problems. When it burns in a poor supply of oxygen, carbon monoxide is formed and this can happen in gas fires in poorly-ventilated rooms. This has sometimes happened in student flats, for example, where windows and doors have been sealed to reduce draughts and cut energy bills resulting in a lack of oxygen for the gas fire. Landlords are now recommended to fit a carbon monoxide alarm.

Write a balanced formula equation for the formation of carbon monoxide by the combustion of methane in a limited supply of oxygen.

$2CH_4 + 3O_2 \rightarrow 2CO + 4H_2O$

Atom economies

There are clear advantages for industry and society to develop chemical processes with high atom economies. A good example is the manufacture of the over-the-counter painkiller and anti-inflammatory drug ibuprofen. The original manufacturing process had an atom economy of only 44%, but a newly-developed process has improved this to 77%.

The percentage yield of a chemical reaction

The yield of a reaction is different from the atom economy.

- The atom economy tells us *in theory* how many atoms *must* be wasted in a reaction.
- The **yield** tells us about the practical efficiency of the process, how much is lost by:

  a the *practical* process of obtaining a product and

  b as a result of reactions that do not go to completion.

As you have seen, once you know the balanced symbol equation for a chemical reaction, you can calculate the amount of any product that you should be able to get from given amounts of starting materials if the reaction goes to completion. For example:

$2KI(aq)$	$+\ Pb(NO_3)_2(aq)$	$\rightarrow$	$PbI_2(s)$	$+$	$2KNO_3(aq)$
potassium iodide	lead nitrate	$\rightarrow$	lead iodide		potassium nitrate
2 mol	1 mol		1 mol		2 mol
332 g	331 g		461 g		202 g

So starting from 3.32 g $\left(\frac{2}{100}\text{mol}\right)$ of potassium iodide in solution and adding 3.31 g $\left(\frac{1}{100}\text{mol}\right)$ of lead nitrate in aqueous solution should produce 4.61 g $\left(\frac{1}{100}\text{mol}\right)$ of a precipitate of lead iodide which can be filtered off and dried.

However, this is in theory only. When you pour one solution into another, some droplets will be left in the beaker. When you remove the precipitate from the filter paper, some will be left on the paper. This sort of problem means that in practice you never get as much product as the equation predicts. Much of the skill of the chemist, both in the laboratory and in industry, lies in minimising these sorts of losses.

$$\text{The yield of a chemical reaction} = \frac{\text{the number of moles of a specified product}}{\text{theoretical maximum number of moles of the product}} \times 100\%$$

It can equally well be defined as:

$$\frac{\text{the number of grams of a specified product obtained in a reaction}}{\text{theoretical maximum number of grams of the product}} \times 100\%$$

If you had obtained 4.00 g of lead iodide in the above reaction, the yield would have been:

$$\frac{4.00}{4.61} \times 100\% = 86.8\%$$

A further problem arises with reactions that are reversible and do not go to completion. This is not uncommon. One example is the Haber process in which ammonia is made from hydrogen and nitrogen. Here is it impossible to get a yield of 100% even with the best practical skills. However, chemists can improve the yield by changing the conditions.

√x Percentage yields

Yields of multi-step reactions can be surprisingly low because the overall yield is the yield of each step multiplied together. So a four step reaction in which each step had an 80% yield would be

80% × 80% × 80% × 80% = 41%

What would be the overall yield of a three step process if the yield of each separate step were 80%, 60%, and 75% respectively?

36%

Summary questions

1 Lime (calcium oxide, CaO) is made by heating limestone (calcium carbonate, $CaCO_3$) to drive off carbon dioxide gas, CO_2.

$CaCO_3 \rightarrow CaO + CO_2$

Calculate the atom economy of the reaction.

2 Sodium sulfate can be made from sulfuric acid and sodium hydroxide.

$H_2SO_4 + 2NaOH \rightarrow Na_2SO_4 + H_2O$

If sodium sulfate is the required product, calculate the atom economy of the reaction.

3 Ethanol, C_2H_6O, can be made by reacting ethene, C_2H_4, with water, H_2O.

$C_2H_4 + H_2O \rightarrow C_2H_6O$

Without doing a calculation, state the atom economy of the reaction. Explain your answer.

4 Consider the reaction

$CaCO_3 \rightarrow CO_2 + CaO$

a Calculate the theoretical maximum number of moles of calcium oxide, CaO, that can be obtained from 1 mole of calcium carbonate, $CaCO_3$.

b Starting from 10 g calcium carbonate, calculate the theoretical maximum number of grams of calcium oxide that can be obtained.

c If 3.6 g of calcium oxide was obtained, calculate the yield of the reaction.

Practice questions

1 Potassium nitrate, KNO_3, decomposes on strong heating, forming oxygen and solid **Y** as the only products.

(a) A 1.00 g sample of KNO_3 ($M_r = 101.1$) was heated strongly until fully decomposed into **Y**.

(i) Calculate the number of moles of KNO_3 in the 1.00 g sample.

(ii) At 298 K and 100 kPa, the oxygen gas produced in this decomposition occupied a volume of $1.22 \times 10^{-4}\ m^3$.
State the ideal gas equation and use it to calculate the number of moles of oxygen produced in this decomposition.
(The gas constant $R = 8.31\ JK^{-1}mol^{-1}$)

(5 marks)

(b) Compound **Y** contains 45.9% of potassium and 16.5% of nitrogen by mass, the remainder being oxygen.

(i) State what is meant by the term *empirical formula*.

(ii) Use the data above to calculate the empirical formula of **Y**.

(4 marks)

(c) Deduce an equation for the decomposition of KNO_3 into **Y** and oxygen.

(1 mark)

AQA, 2006

2 Ammonia is used to make nitric acid, HNO_3, by the Ostwald process.
Three reactions occur in this process.

Reaction 1 $4NH_3(g) + 5O_2(g) \rightarrow 4NO(g) + 6H_2O(g)$

Reaction 2 $2NO(g) + O_2(g) \rightarrow 2NO_2(g)$

Reaction 3 $3NO_2(g) + H_2O(I) \rightarrow 2HNO_3(aq) + NO(g)$

(a) In one production run, the gases formed in Reaction 1 occupied a total volume of $4.31\ m^3$ at 25 °C and 100 kPa.
Calculate the amount, in moles, of NO produced.
Give your answer to the appropriate number of significant figures.
(The gas constant $R = 8.31\ JK^{-1}mol^{-1}$)

(4 marks)

(b) In another production run, 3.00 kg of ammonia gas were used in Reaction 1 and all of the NO gas produced was used to make NO_2 gas in Reaction 2.
Calculate the mass of NO_2 formed from 3.00 kg of ammonia in Reaction 2 assuming an 80.0% yield.
Give your answer in kilograms.

(5 marks)

(c) Consider Reaction 3 in this process.

$3NO_2(g) + H_2O(l) \rightarrow 2HNO_3(aq) + NO(g)$

Calculate the concentration of nitric acid produced when 0.543 mol of NO_2 is reacted with water and the solution is made up to $250\ cm^3$.

(2 marks)

(d) Suggest why a leak of NO_2 gas from the Ostwald process will cause atmospheric pollution.

(1 mark)

(e) Give one reason why excess air is used in the Ostwald process.

(1 mark)

(f) Ammonia reacts with nitric acid as shown in this equation.

$NH_3 + HNO_3 \rightarrow NH_4NO_3$

Deduce the type of reaction occurring.

(1 mark)

AQA, 2013

3 Zinc forms many different salts including zinc sulfate, zinc chloride, and zinc fluoride.

(a) People who have a zinc deficiency can take hydrated zinc sulfate, $ZnSO_4.xH_2O$, as a dietary supplement.
A student heated 4.38 g of hydrated zinc sulfate and obtained 2.46 g of anhydrous zinc sulfate.
Use these data to calculate the value of the integer x in $ZnSO_4.xH_2O$.
Show your working.

(3 marks)

(b) Zinc chloride can be prepared in the laboratory by the reaction between zinc oxide and hydrochloric acid.
The equation for the reaction is:

$$ZnO + 2HCl \rightarrow ZnCl_2 + H_2O$$

A 0.0830 mol sample of pure zinc oxide was added to 100 cm^3 of 1.20 $mol\,dm^{-3}$ hydrochloric acid.
Calculate the maximum mass of anhydrous zinc chloride that could be obtained from the products of this reaction. Give your answer to the appropriate number of significant figures.

(4 marks)

(c) Zinc chloride can also be prepared in the laboratory by the reaction between zinc and hydrogen chloride gas.

$$Zn + 2HCl \rightarrow ZnCl_2 + H_2$$

An impure sample of zinc powder with a mass of 5.68 g was reacted with hydrogen chloride gas until the reaction was complete. The zinc chloride produced had a mass of 10.7 g.
Calculate the percentage purity of the zinc metal. Give your answer to 3 significant figures.

(4 marks)

AQA, 2013

4 In this question give all your answers to the appropriate number of significant figures.
Magnesium nitrate decomposes on heating to form magnesium oxide, nitrogen dioxide, and oxygen as shown in the following equation.

$$2Mg(NO_3)_2(s) \rightarrow 2MgO(s) + 4NO_2(g) + O_2(g)$$

(a) Thermal decomposition of a sample of magnesium nitrate produced 0.741 g of magnesium oxide.

(i) Calculate the amount, in moles, of MgO in 0.741 g of magnesium oxide.

(2 marks)

(ii) Calculate the total amount, in moles, of gas produced from this sample of magnesium nitrate.

(1 mark)

(b) In another experiment, a different sample of magnesium nitrate decomposed to produce 0.402 mol of gas. Calculate the volume, in dm^3, that this gas would occupy at 333 K and 1.00×10^5 Pa.
(The gas constant $R = 8.31\ J\,K^{-1}\,mol^{-1}$)

(3 marks)

(c) A 0.0152 mol sample of magnesium oxide, produced from the decomposition of magnesium nitrate, was reacted with hydrochloric acid.

$$MgO + 2HCl \rightarrow MgCl_2 + H_2O$$

This 0.0152 mol sample of magnesium oxide required 32.4 cm^3 of hydrochloric acid for complete reaction. Use this information to calculate the concentration, in $mol\,dm^{-3}$, of the hydrochloric acid.

(2 marks)

AQA, 2010

Answers to the Practice Questions and Section Questions are available at
www.oxfordsecondary.com/oxfordaqaexams-alevel-chemistry

3 Bonding

3.1 The nature of ionic bonding

Why do chemical bonds form?

The bonds between atoms always involve their outer electrons.

- Noble gases have full outer main shells of electrons (Figure 1) and are very unreactive.
- When atoms bond together they share or transfer electrons to achieve a more stable electron arrangement, often a full outer main shell of electrons, like the noble gases.
- There are three types of strong chemical bonds – **ionic**, **covalent**, and **metallic**.

Learning objectives:

→ State how ions form and why they attract each other.

→ State the properties of ionically bonded compounds.

→ Describe the structure of ionically bonded compounds.

Specification reference: 3.1.3

Ionic bonding

Metals have one, two, or three electrons in their outer shells, so the easiest way for them to attain the electron structure of a noble gas is to lose their outer electrons. Non-metals have spaces in their outer shells, so that the easiest way for them to attain the electron structure of a noble gas is to gain electrons.

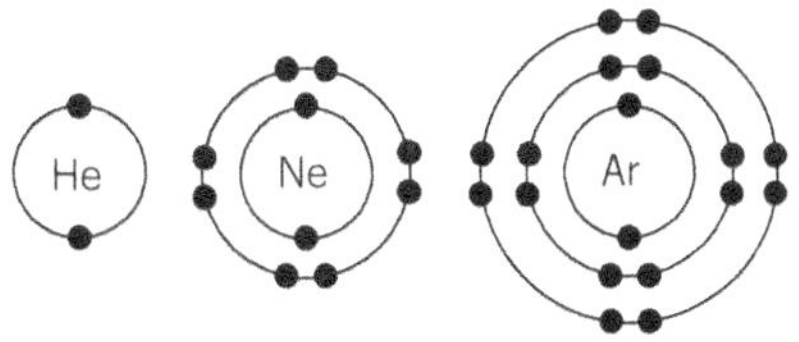

▲ **Figure 1** *Noble gases*

- Ionic bonding occurs between metals and non-metals.
- Electrons are transferred from metal atoms to non-metal atoms.
- Positive and negative ions are formed.

Sodium chloride (Figure 2) has ionic bonding.

- Sodium, Na, has 11 electrons (and 11 protons). The electron arrangement is $1s^2\ 2s^2\ 2p^6\ 3s^1$.
- Chlorine, Cl, has 17 electrons (and 17 protons). The electron arrangement is $1s^2\ 2s^2\ 2p^6\ 3s^2\ 3p^5$.
- An electron is transferred. The single outer electron of the sodium atom moves into the outer shell of the chlorine atom.
- Each outer main shell is now full.

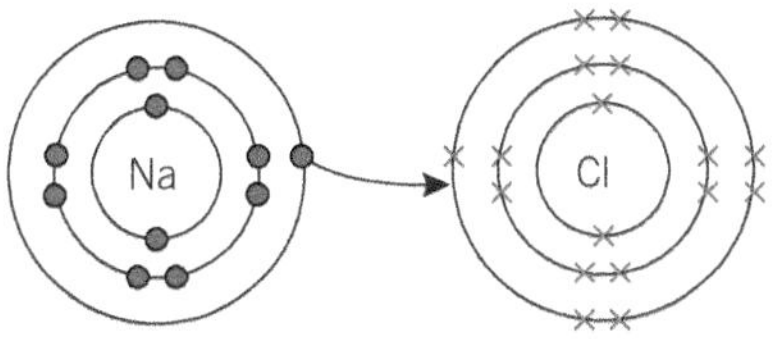

▲ **Figure 2** *A dot-and-cross diagram to show the transfer of the $3s^1$ electron from the sodium atom to the 3p orbital on a chlorine atom. Remember that electrons are all identical whether shown by a dot or a cross*

Noble gas compounds

The noble gases do form a few compounds although they are mostly unstable. The first, Xe PtF_6, was made in 1961 by Neil Bartlett, pictured below.

There are as yet no compounds of helium or neon. Xenon has the largest number of known compounds. In most of them xenon forms a positive ion by losing an electron.

Suggest why it is easier for xenon to form a positive ion than for helium or neon.

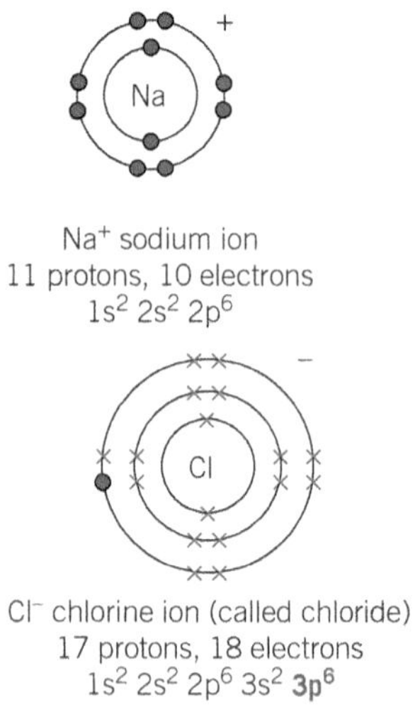

▲ **Figure 3** *The ions that result from electron transfer*

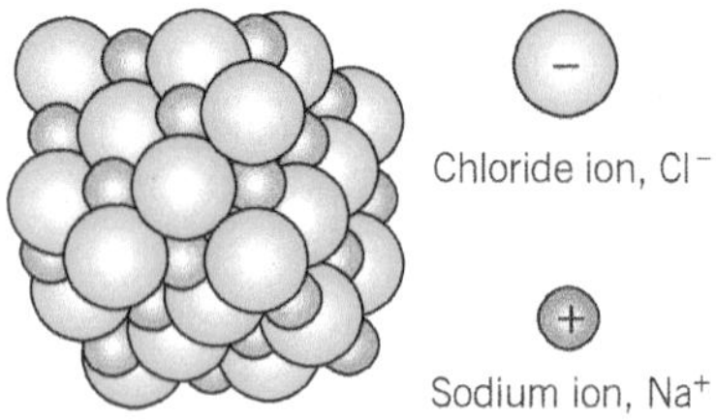

▲ **Figure 4** *The sodium chloride structure. This is an example of a giant ionic structure. The strong bonding extends throughout the compound and because of this it will be difficult to melt.*

Study tip

Dot-and-cross diagrams can help you to understand the principles of bonding and to predict the shapes of molecules – see Topic 3.5, The shapes of molecules and ions.

- Both sodium and chlorine now have a noble gas electron arrangement. Sodium has the neon noble gas arrangement whereas chlorine has the argon noble gas arrangement (compare the ions in Figure 3 with the noble gas atoms in Figure 1).

The two charged particles that result from the transfer of an electron are called **ions**.

- The sodium ion is positively charged because it has *lost* a negative electron.
- The chloride ion is negatively charged because it has *gained* a negative electron.
- The two ions are attracted to each other and to other oppositely charged ions in the sodium chloride compound by **electrostatic forces**.

Therefore ionic bonding is the result of electrostatic attraction between oppositely charged ions. The attraction extends throughout the compound. Every positive ion attracts every negative ion and vice versa. Ionic compounds always exist in a structure called a **lattice**. Figure 4 shows the three-dimensional lattice for sodium chloride with its singly charged ions.

The formula of sodium chloride is NaCl because for every one sodium ion there is one chloride ion.

Example: magnesium oxide

Magnesium, Mg, has 12 electrons. The electron arrangement is $1s^2\ 2s^2\ 2p^6\ 3s^2$.

Oxygen, O, has eight electrons. The electron arrangement is $1s^2\ 2s^2\ 2p^4$.

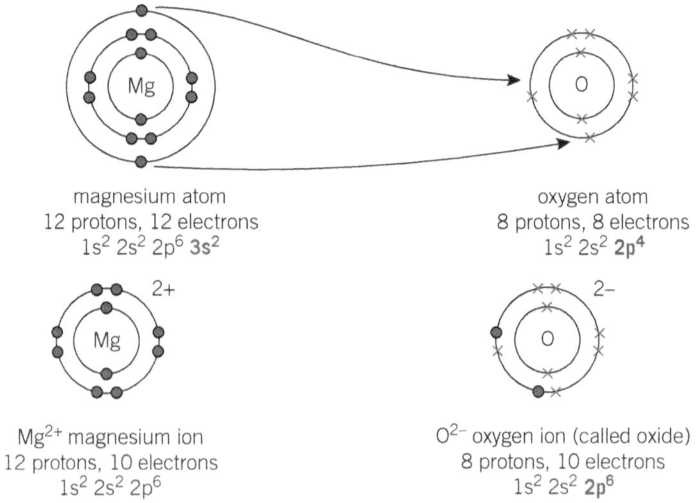

▲ **Figure 5** *Ionic bonding in magnesium oxide, MgO*

This time, two electrons are transferred from the 3s orbitals on each magnesium atom. Each oxygen atom receives two electrons into its 2p orbital.

- The magnesium ion, Mg^{2+}, is positively charged because it has lost two negative electrons.
- The oxide ion, O^{2-}, is negatively charged because it has gained two negative electrons.
- The formula of magnesium oxide is MgO.

The formulae of ionic compounds

Group 1 metals form 1^+ ions, and Group 2 metals form 2^+ ions. Group 7 non-metals form 1^- ions and Group 6 non-metals form 2^- ions. Since all compounds are neutral, we can predict the formulae of simple ionic compounds. For example, calcium fluoride is Ca^{2+} and $2F^-$, ie CaF_2, lithium oxide is $2Li^+$ and O^{2-}, ie Li_2O. The same idea applies to compound ions, see the Study tip. For example, sodium sulfate is $2Na^+$ and SO_4^{2-}, ie Na_2SO_4 and ammonium hydroxide is NH_4^+ and OH^-, ie NH_4OH.

Properties of ionically bonded compounds

Ionic compounds are always solids at room temperature. They have giant structures and therefore high melting temperatures. This is because in order to melt an ionic compound, energy must be supplied to break up the lattice of ions.

Ionic compounds conduct electricity when molten or dissolved in water (aqueous) but not when solid. This is because the ions that carry the current are free to move in the liquid state but are not free in the solid state (Figure 6).

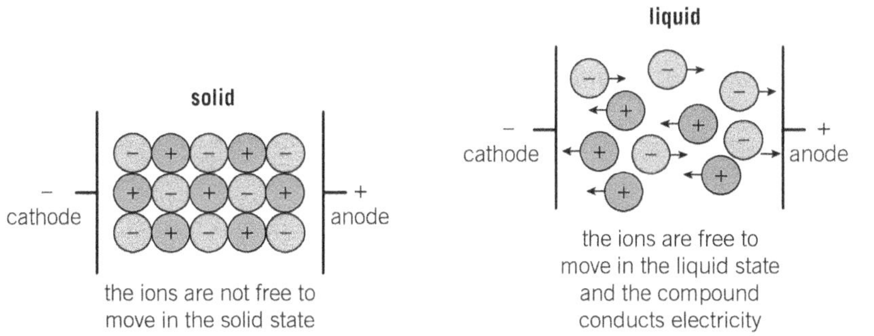
solid
–
cathode
+
anode
the ions are not free to move in the solid state
liquid
–
cathode
+
anode
the ions are free to move in the liquid state and the compound conducts electricity

▲ **Figure 6** *Ionic liquids conduct electricity, ionic solids do not*

Ionic compounds are *brittle* and shatter easily when given a sharp blow. This is because they form a lattice of alternating positive and negative ions, see Figure 7. A blow in the direction shown may move the ions and produce contact between ions with like charges.

Summary questions

1. Identify which of the following are ionic compounds and explain why.
   a CO **b** KF **c** CaO **d** HF
2. Explain why ionic compounds have high melting temperatures.
3. Describe the conditions where ionic compounds conduct electricity.
4. Draw dot-and-cross diagrams to show the formation of the following ions. Include the electronic configuration of the atoms and ions involved.
   a the ions being formed when magnesium and fluorine react
   b the ions being formed when sodium and oxygen react.
5. Give the formulae of the compounds formed in question **4**.
6. Look at the electron arrangements of the Mg^{2+} and O^{2-} ions. State the noble gas they correspond to.

Study tip

Some ions consist of groups of covalently bonded atoms which have an overall charge. You will need to remember the formulae and charges of the following:

sulfate, SO_4^{2-}; hydroxide, OH^-; nitrate, NO_3^-, carbonate CO_3^{2-}, ammonium NH_4^+.

When working out formulae of compounds these behave exactly like ions with a single atom.

Hint

A current of electricity is a flow of charge. In metals, negative electrons move. In ionic compounds, charged ions move.

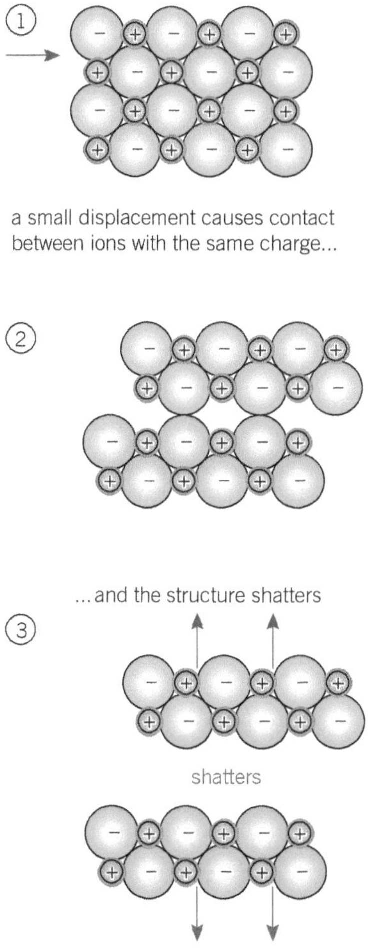
1
a small displacement causes contact between ions with the same charge...
2
3
...and the structure shatters
shatters

▲ **Figure 7** *The brittleness of ionic compounds*

3.2 Covalent bonding

Learning objectives:

→ Describe a covalent bond.

→ Describe a co-ordinate bond.

→ Describe the properties of covalently bonded molecules.

Specification reference: 3.1.3

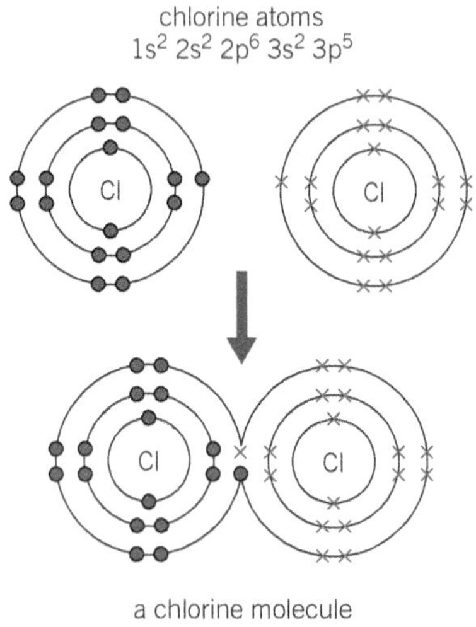

▲ **Figure 1** *Formation of a chlorine molecule – the two atoms share a 3p electron from each atom*

Non-metal atoms need to *receive* electrons to fill the spaces in their outer shells.

- A covalent bond forms between a pair of non-metal atoms.
- The atoms *share* some of their outer electrons so that each atom has a stable noble gas arrangement.
- A covalent bond is a shared pair of electrons.

Forming molecules by covalent bonding

A small group of covalently bonded atoms is called a molecule. For example, chlorine exists as a gas that is made of molecules, Cl_2, see Figure 1.

Chlorine has 17 electrons and an electron arrangement $1s^2\ 2s^2\ 2p^6\ 3s^2\ 3p^5$. Two chlorine atoms make a chlorine molecule:

- The two atoms share one pair of electrons.
- Each atom now has a stable noble gas arrangement.
- The formula is Cl_2.
- Molecules are neutral because no electrons have been transferred from one atom to another.

You can represent one pair of shared electrons in a covalent bond by a line, Cl—Cl.

Example: methane

Methane gas is a covalently bonded compound of carbon and hydrogen. Carbon, C, has six electrons with electron arrangement $1s^2\ 2s^2\ 2p^2$ and hydrogen, H, has just one electron $1s^1$.

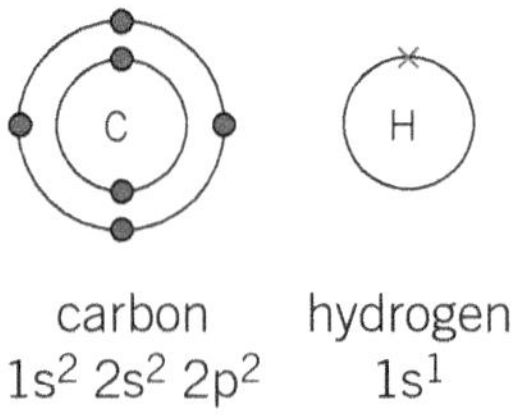

In order for carbon to attain a stable noble gas arrangement, there are four hydrogen atoms to every carbon atom.

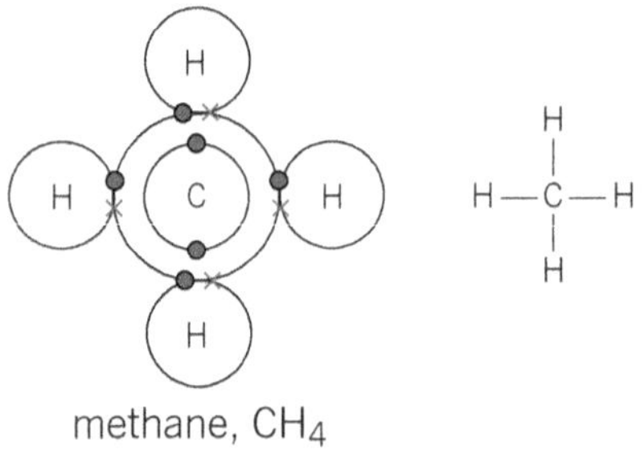

The formula of methane is CH_4. The four 2p electrons from carbon and the $1s^1$ electron from the four hydrogen atoms are shared.

Hint

Another way of picturing covalent bonds is to think of electron orbitals on each atom merging to form a molecular orbital that holds the shared electrons.

Hint

The hydrogen has a filled outer shell with only two electrons ($1s^2$). It fills the first shell to get the structure of the noble gas helium. The carbon atoms have an electron arrangement $1s^2\ 2s^2\ 2p^6$.

How does sharing electrons hold atoms together?

Atoms with covalent bonds are held together by the electrostatic attraction between the nuclei and the shared electrons. This takes place within the molecule. The simplest example is hydrogen. The hydrogen molecule consists of two protons held together by a pair of electrons. The electrostatic forces are shown in Figure 2. The attractive forces are in black and the repulsive forces in red. These forces just balance when the nuclei are a particular distance apart.

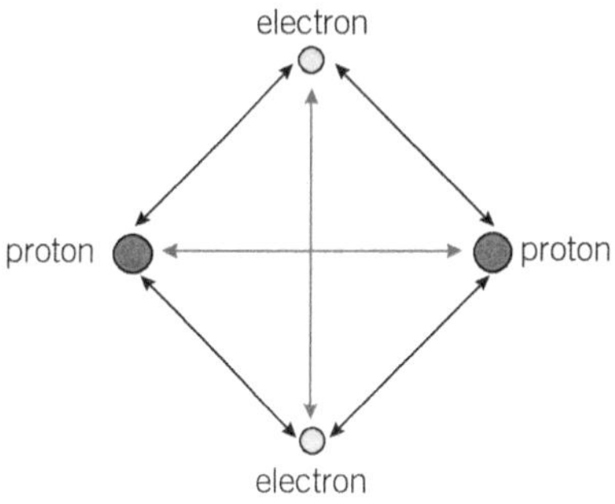

▲ **Figure 2** *The electrostatic forces within a hydrogen molecule*

Double covalent bonds

In a double bond, four electrons are shared. The two atoms in an oxygen molecule share two pairs of electrons so that the oxygen atoms have a double bond between them (Figure 3). You can represent the two pairs of shared electrons in a covalent bond by a double line, O=O.

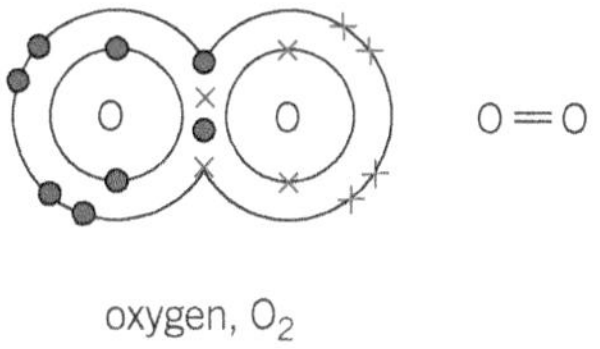

▲ **Figure 3** *An oxygen molecule has a double bond which shares two 2p electrons from each atom*

When you are drawing covalent bonding diagrams you may leave out the inner main shells because these are not involved at all. Other examples of molecules with covalent bonds are shown in Table 1.

All the examples in Table 1 are neutral molecules. The atoms within the molecules are strongly bonded together with covalent bonds within the molecule. However, the molecules are *not* strongly attracted to each other.

▼ **Table 1** *Examples of covalent molecules. Only the outer shells are shown.*

Formula	Name	Formula	Name
H_2 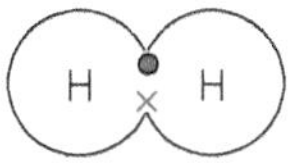	hydrogen Each hydrogen atom has a full outer main shell with just two electrons	NH_3 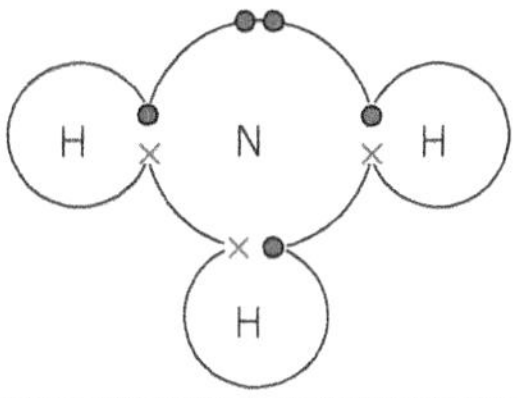	ammonia
HCl 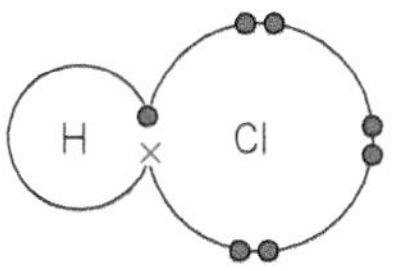	hydrogen chloride	C_2H_4 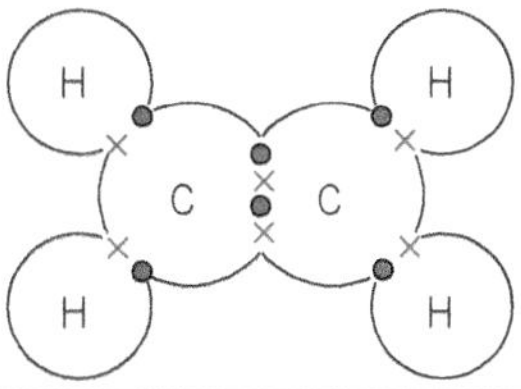	ethene There is a carbon–carbon double bond in this molecule
H_2O 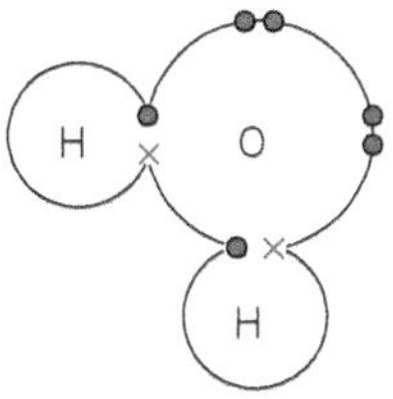	water	CO_2 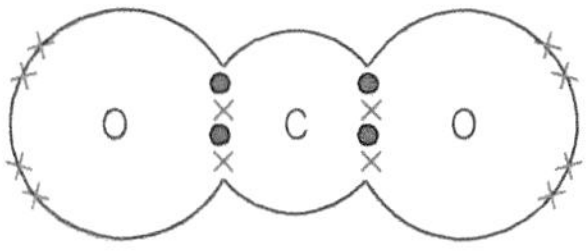	carbon dioxide There are two carbon–oxygen double bonds in this molecule

Properties of substances with molecular structures

Substances composed of molecules are gases, liquids, or solids with low melting temperatures. This is because the strong covalent bonds are only *between the atoms* within the molecules. There is only weak attraction between the molecules so the molecules do not need much energy to move apart from each other.

They are poor conductors of electricity because the molecules are neutral overall. This means that there are no charged particles to carry the current.

If they dissolve in water, and remain as molecules, the solutions do not conduct electricity. Again, this is because there are no charged particles.

Hint

Some covalent compounds react with water to form ions. In such cases the resulting solution *will* conduct electricity. Hydrogen chloride is an example of this:

$HCl(g) + aq \rightarrow H^+(aq) + Cl^-(aq)$

Hint

Triple bonds are also possible where three electrons from each atom are shared, e.g., N_2 is N≡N

Synoptic link

Co-ordinate (dative) bonding is very important in the chemistry of transition metal complexes. You will learn more about it in Chapter 23, The transition metals.

Co-ordinate bonding

A single covalent bond consists of a pair of electrons shared between two atoms. In most covalent bonds, each atom provides one of the electrons. But, in some bonds, one atom provides both the electrons. This is called **co-ordinate bonding**. It is also called **dative covalent bonding**.

In a co-ordinate or dative covalent bond:

- the atom that *accepts* the electron pair is an atom that does not have a filled outer shell of electrons – the atom is electron-deficient
- the atom that is *donating* the electrons has a pair of electrons that is not being used in a bond, called a **lone pair**.

Example: the ammonium ion

For example, ammonia, NH_3, has a lone pair of electrons. In the ammonium ion, NH_4^+, the nitrogen uses its lone pair of electrons to form a co-ordinate bond with an H^+ ion (a bare proton with no electrons at all and therefore electron-deficient).

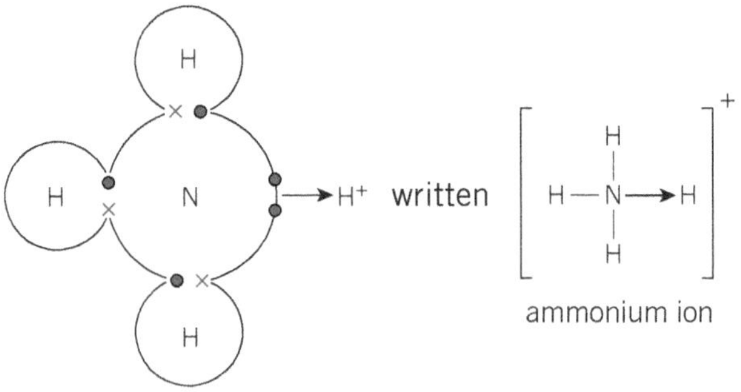

Co-ordinate covalent bonds are represented by an arrow. The arrow points towards the atom that is accepting the electron pair. However, this is only to show how the bond was made. The ammonium ion is completely symmetrical and all the bonds have exactly the same strength and length.

- Co-ordinate bonds have exactly the same strength and length as ordinary covalent bonds between the same pair of atoms.

The ammonium ion has *covalently* bonded atoms but is a charged particle.

Summary questions

1 State what a covalent bond is.

2 Identify which of the following have covalent bonding and explain your answer.
   a Na_2O
   b CF_4
   c $MgCl_2$
   d C_2H_4

3 Draw a dot-and-cross diagram for hydrogen sulfide, a compound of hydrogen and sulfur.

4 Draw a dot-and-cross diagram to show a water molecule forming a co-ordinate bond with an H^+ ion.

3.3 Metallic bonding

Metals are shiny elements made up of atoms that can easily lose up to three outer electrons, leaving positive metal ions. For example, sodium, Na, 2,8,1 ($1s^2\ 2s^2\ 2p^6\ 3s^1$) loses its one outer electron, aluminium, Al, 2,8,3 ($1s^2\ 2s^2\ 2p^6\ 3s^2\ 3p^1$) loses its three outer electrons.

Learning objectives:

→ Describe the nature of bonding in a metal.

→ Describe the properties of metals.

Specification reference: 3.1.3

Metallic bonding

The atoms in a metal element cannot transfer electrons (as happens in ionic bonding) unless there is a non-metal atom present to receive them. In a metal element, the outer main shells of the atoms merge. The outer electrons are no longer associated with any one particular atom. A simple picture of **metallic bonding** is that metals consist of a lattice of positive ions existing in a 'sea' of outer electrons. These electrons are **delocalised**. This means that they are not tied to a particular atom. Magnesium metal is shown in Figure 1. The positive ions tend to repel one another and this is balanced by the electrostatic attraction of these positive ions for the negatively charged 'sea' of delocalised electrons.

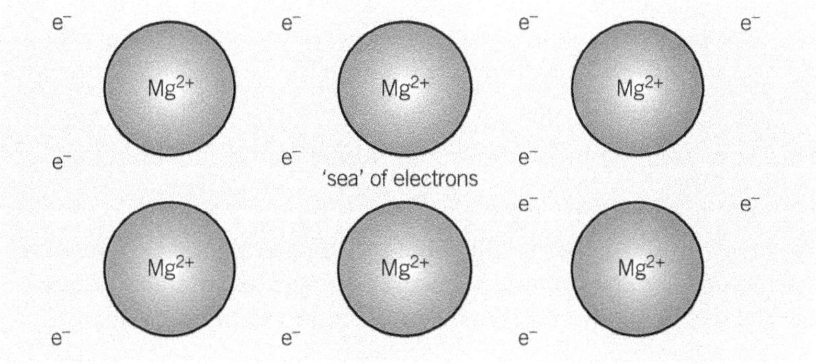

▲ **Figure 1** *The delocalised 'sea' of electrons in magnesium*

Hint

In Figure 1 the metal ions are shown spaced apart for clarity. In fact, metal atoms are more closely packed, and so metals tend to have high densities.

- The number of delocalised electrons depends on how many electrons have been lost by each metal atom.
- The metallic bonding spreads throughout so metals have giant structures.

Properties of metals

Metals are good conductors of electricity and heat

The delocalised electrons that can move throughout the structure explain why metals are such good conductors of electricity. An electron from the negative terminal of the supply joins the electron sea at one end of a metal wire while *at the same time* a different electron leaves the wire at the positive terminal, as shown in Figure 2.

Metals are also good conductors of heat – they have high thermal conductivities. The sea of electrons is partly responsible for this property, with energy also spread by increasingly vigorous vibrations of the closely packed ions.

Hint

The word 'delocalised' is often used to describe electron clouds that are spread over more than two atoms.

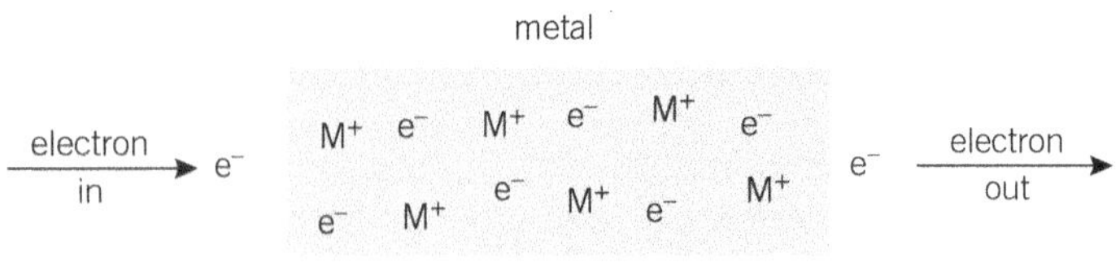

▲ **Figure 2** *The conduction of electricity by a metal*

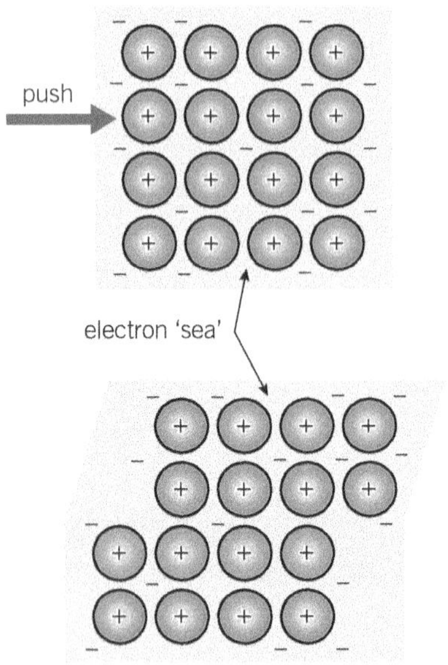

▲ **Figure 3** *The malleability and ductility of metals*

The strength of metals

In general, the strength of any metallic bond depends on the following:

- the charge on the ion – the greater the charge on the ion, the greater the number of delocalised electrons and the stronger the electrostatic attraction between the positive ions and the electrons.
- the size of ion – the smaller the ion, the closer the electrons are to the positive nucleus and the stronger the bond.

Metals tend to be strong. The delocalised electrons also explain this. These extend throughout the solid so there are no individual bonds to break.

Metals are malleable and ductile

Metals are malleable (they can be beaten into shape) and ductile (they can be pulled into thin wires). After a small distortion, each metal ion is still in exactly the same environment as before so the new shape is retained, see Figure 3.

Contrast this with the brittleness of ionic compounds in Topic 3.1.

Metals have high melting points

Metals generally have high melting and boiling points because they have giant structures. There is strong attraction between metal ions and the delocalised sea of electrons. This makes the atoms difficult to separate.

Summary questions

1 Give three differences in physical properties between metals and non-metals.

2 Write the electron arrangement of a calcium atom, Ca.

3 Which electrons will a calcium atom lose to gain a stable noble gas configuration?

4 State how many electrons each calcium atom will contribute to the delocalised sea of electrons that holds the metal atoms together.

5 Sodium forms +1 ions with a metallic radius of 0.191 nm. Magnesium forms +2 ions with a metallic radius of 0.160 nm. How would you expect the following properties of the two metals to compare? Explain your answers.

   a the melting point

   b the strength of the metals.

3.4 Bonding and physical properties

Learning objectives:

- → State the energy changes that occur when solids melt and liquids vaporise.
- → Explain the values of enthalpies of melting (fusion) and vaporisation.
- → Explain the physical properties of ionic solids, metals, macromolecular solids, and molecular solids in terms of their detailed structures and bonding.
- → List the three types of strong bonds.
- → List the three types of intermolecular forces.
- → Describe how melting temperatures and structure are related.
- → Describe how electrical conductivity is related to bonding.

Specification reference: 3.1.3

One of the key ideas of science is that matter, which is anything with mass, is made of tiny particles – it is *particulate*. These particles are in motion, which means they have kinetic energy. To understand the differences between the three states of matter – gas, liquid, and solid – you need to be able to explain the energy changes associated with changes between these physical states.

The three states of matter

Table 1 sets out the simple model used for the three states of matter.

▼ **Table 1** *The three states of matter*

	Solid	Liquid	Gas
arrangement of particles	regular	random	random
evidence	Crystal shapes have straight edges. Solids have definite shapes.	None direct but a liquid changes shape to fill the bottom of its container.	None direct but a gas will fill its container.
spacing	close	close	far apart
evidence	Solids are not easily compressed.	Liquids are not easily compressed.	Gases are easily compressed.
movement	vibrating about a point	rapid 'jostling'	rapid
evidence	Diffusion is very slow. Solids expand on heating.	Diffusion is slow. Liquids evaporate.	Diffusion is rapid. Gases exert pressure
models	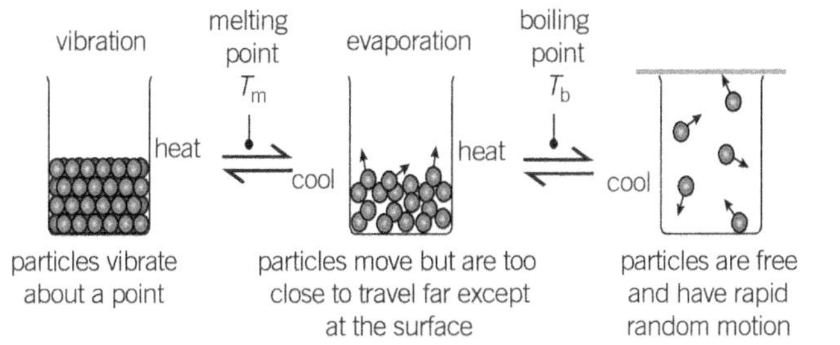		

Energy changes on heating

Heating a solid

When you first heat a solid and supply energy to the particles, it makes them vibrate more about a fixed position. This slightly increases the average distance between the particles and so the solid expands.

Turning a solid to liquid (melting – also called fusion)

In order to turn a solid – with its ordered, closely packed, vibrating particles – into a liquid – where the particles are moving randomly but still close together – you have to supply more energy. This energy is needed to weaken the forces that act between the particles, holding them together in the solid state. The energy needed is called the latent heat of melting, or more correctly the **enthalpy change of melting**. While a solid is melting, the temperature does not change because the

Hint

The enthalpy change of melting is sometimes called the enthalpy change of fusion.

Synoptic link

You will learn more about enthalpy in Topic 4.2, Enthalpy.

Hint

It is through an understanding of the relationship between bonding and physical properties that material scientists can engineer new materials with exciting properties. Examples include:

- carbon fibres
- materials based on carbon nanotubes
- materials that can self-repair.

Synoptic link

The forces between molecules, including Van der Waals forces are discussed in Topic 3.7, Forces acting between molecules.

heat energy provided is absorbed as the forces between particles are weakened.

Enthalpy is the heat energy change measured under constant pressure whilst *temperature* depends on the average kinetic energy of the particles and is therefore related to their speed – the greater the energy, the faster they go.

Heating a liquid

When you heat a liquid, you supply energy to the particles which makes them move more quickly – they have more kinetic energy. On average, the particles move a little further apart so liquids also expand on heating.

Turning a liquid to gas (boiling – also called vaporisation)

In order to turn a liquid into a gas, you need to supply enough energy to break all the intermolecular forces between the particles. A gas consists of particles that are far apart and moving independently. The energy needed is called the latent heat of vaporisation or more correctly the **enthalpy change of vaporisation**. As with melting, there is no temperature change during the process of boiling.

Heating a gas

As you heat a gas, the particles gain kinetic energy and move faster. They get much further apart and so gases expand a great deal on heating.

Crystals

Crystals are solids. The particles have a regular arrangement and are held together by forces of attraction. These could be strong bonds, such as covalent, ionic, or metallic, or weaker intermolecular forces, such as van der Waals, dipole–dipole, or hydrogen bonds. The strength of the forces of attraction between the particles in the crystal affects the physical properties of the crystals. For example, the stronger the force, the higher the melting temperature and the greater the enthalpy of fusion (the more difficult they are to melt). There are four basic crystal types – ionic, metallic, molecular, and macromolecular.

Ionic crystals

Ionic compounds have strong electrostatic attractions between oppositely charged ions. Sodium chloride, NaCl, is a typical ionic crystal, see Topic 3.1. Ionic compounds have high melting points. This is a result of the strong electrostatic attractions which extend throughout the structure. These require a lot of energy to break in order for the ions to move apart from each other. For example, the melting point of sodium chloride is 801 °C (1074 K).

Metallic crystals

Metals exist as a lattice of positive ions embedded in a delocalised sea of electrons, see Topic 3.3. Again the attraction of positive to negative extends throughout the crystal. The high melting temperature is a result of these strong metallic bonds. Magnesium is a typical example.

Molecular crystals

Molecular crystals consist of molecules held in a regular array by intermolecular forces. Covalent bonds *within* the molecules hold the atoms together but they do not act *between* the molecules.

Intermolecular forces are much weaker than covalent, ionic or metallic bonds, so molecular crystals have low melting temperatures and low enthalpies of melting.

Iodine (Figure 1) is an example of a molecular crystal. A strong covalent bond holds pairs of iodine atoms together to form I_2 molecules. Since iodine molecules have a large number of electrons, the van der Waals forces are strong enough to hold the molecules together as a solid. But van der Waals forces are much weaker than covalent bonds, giving iodine the following properties:

- crystals are soft and break easily
- low melting temperature (114 °C, 387 K) and sublimes readily to form gaseous iodine molecules.
- does not conduct electricity because there are no charged particles to carry charge.

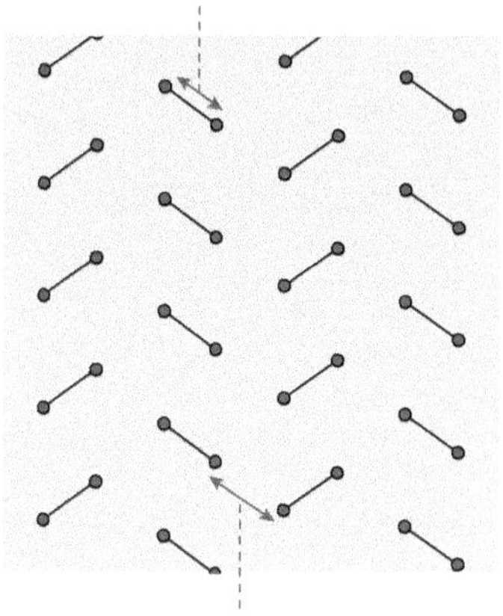

▲ **Figure 1** *The arrangement of an iodine crystal*

Macromolecular crystals

Covalent compounds are not always made up of small molecules. In some substances the covalent bonds extend throughout the compound and have the typical property of a giant structure held together with strong bonds – a high melting temperature. There are many examples of macromolecular crystals, including diamond and graphite.

Diamond and graphite

Diamond and graphite are both made of the element carbon only. They are called polymorphs or allotropes of carbon. They are very different materials because their atoms are differently bonded and arranged. They are examples of macromolecular structures.

Diamond

Diamond consists of pure carbon with covalent bonding between every carbon atom. The bonds spread throughout the structure, which is why it is a giant structure.

A carbon atom has four electrons in its outer shell. In diamond, each carbon atom forms four single covalent bonds with other carbon atoms, as shown in Figure 2. These four electron pairs repel each other, following the rules of the electron pair repulsion theory. See Topic 3.5. In three dimensions the bonds actually point to the corners of a tetrahedron (with bond angles of 109.5°).

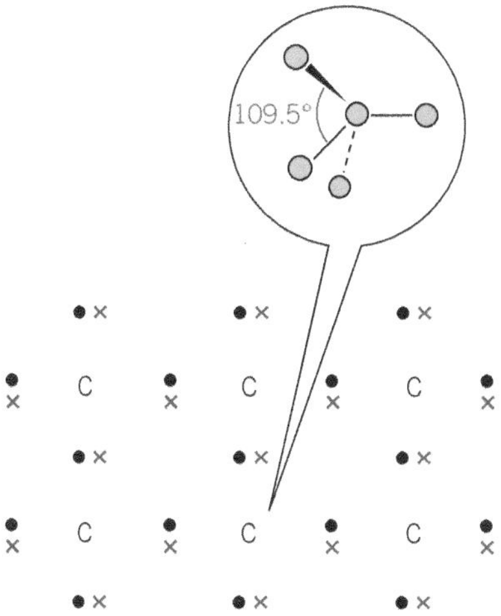

▲ **Figure 2** *A dot-and-cross diagram showing the bonding in diamond*

Each carbon atom is in an identical position in the structure, surrounded by four other carbon atoms. Figure 3 shows this three-dimensional arrangement.

The atoms form a giant three-dimensional lattice of strong covalent bonds, which is why diamond has the following properties:

- very hard material (one of the hardest known)
- very high melting temperature, over 3700 K
- does not conduct electricity because there are no free charged particles to carry charge.

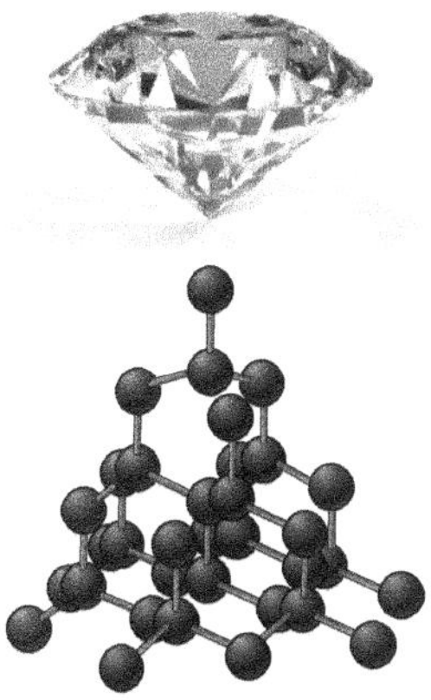

▲ **Figure 3** *A three-dimensional diagram of diamond*

Graphite

Graphite also consists of pure carbon, but the atoms are bonded and arranged differently from diamond. Graphite has two sorts of bonding – strong covalent and the weaker van der Waals forces.

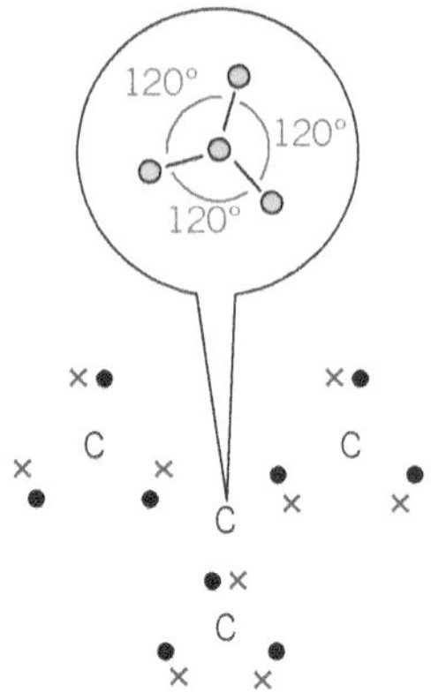

▲ **Figure 4** *A dot-and-cross diagram showing the three covalent bonds in graphite*

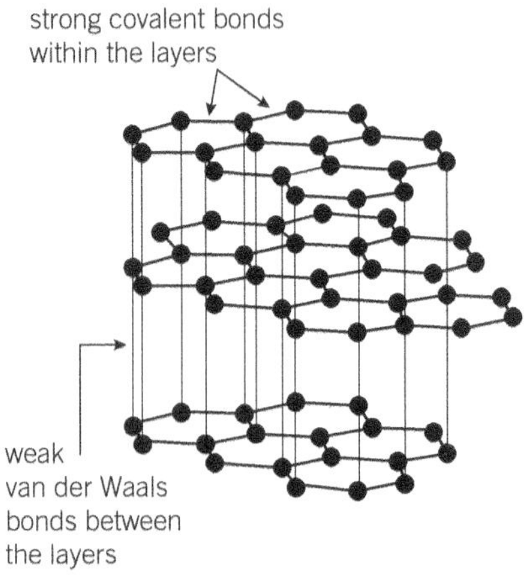

▲ **Figure 5** *Van der Waals forces between the layers of carbon atoms in graphite*

Hint

It is now believed that molecules such as oxygen can slide in between the layers of carbon.

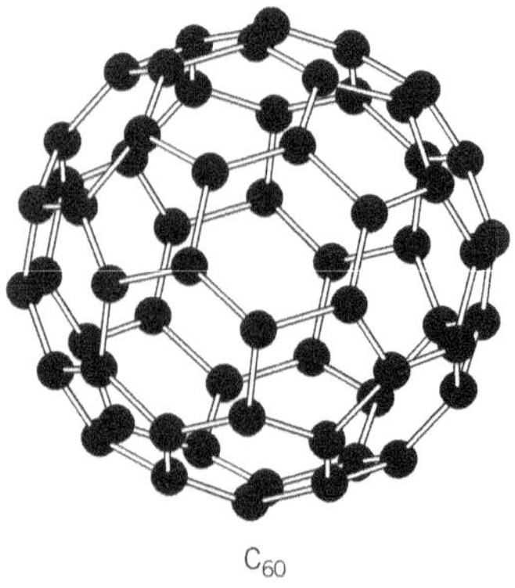

▲ **Figure 6** *Buckminsterfullerene – also called buckyballs*

In graphite, each carbon atom forms three single covalent bonds to other carbon atoms. As predicted by bonding electron pair repulsion theory, these form a flat trigonal arrangement, sometimes called **trigonal planar**, with a bond angle of 120° (Figure 4). This leaves each carbon atom with a 'spare' electron in a p-orbital that is not part of the three single covalent bonds.

This arrangement produces a two-dimensional layer of linked hexagons of carbon atoms, rather like a chicken-wire fence (Figure 5).

The p-orbitals with the 'spare' electron merge above and below the plane of the carbon atoms in each layer. These electrons can move anywhere within the layer. They are delocalised. This adds to the strength of the bonding and is rather like the **delocalised** sea of electrons in a metal, but in two dimensions only.

These delocalised electrons are what make graphite conduct electricity (very rare for a non-metal). They can travel freely through the material, though graphite will only conduct along the hexagonal planes, not at right angles to them.

There is no covalent bonding *between* the layers of carbon atoms. They are held together by the much weaker van der Waals forces, see Figure 5. This weak intermolecular force of attraction means that the layers can slide across one another making graphite soft and flaky. It is the 'lead' in pencils. The flakiness allows the graphite layers to transfer from the pencil to the paper.

- Graphite is a soft material.
- It has a very high melting temperature and in fact it breaks down before it melts. This is because of the strong network of covalent bonds, which make it a giant structure.
- It conducts electricity along the planes of the hexagons.

Giant footballs

More recently a number of other forms of pure carbon have been discovered. Chemists found the first one whilst they were looking for molecules in outer space. The structures of these new forms of carbon include closed cages of carbon atoms and also tubes called nanotubes. The most famous is buckminsterfullerene, C_{60}, in which atoms are arranged in a football-like shape (Figure 6). Harry Kroto and colleagues received the Nobel Prize for the discovery. Now, scientists are investigating many uses for these new materials.

Bonding – summary

There are three types of *strong* bonding that hold atoms together – ionic, covalent, and metallic. All three involve the outer electrons of the atoms concerned.

- In covalent bonding, the electrons are shared between atoms.
- In ionic bonding, electrons are transferred from metal atoms to non-metal atoms.
- In metallic bonding, electrons are spread between metal atoms to form a lattice of ions held together by delocalised electrons.

If you know what the compound is, you can usually tell the type of bonding from the types of atoms that it contains:

- metal atoms only – metallic bonding
- metal and non-metal – ionic bonding
- non-metal atoms only – covalent bonding.

The three types of bonding give rise to different properties.

Electrical conductivity

The property that best tells us what sort of bonding you have is electrical conductivity. Metals and alloys (an alloy is a mixture of metals) conduct electricity well, in both the solid and liquid states due to their metallic bonding. The current is carried by the delocalised electrons that hold the metal ions together, see Figure 7.

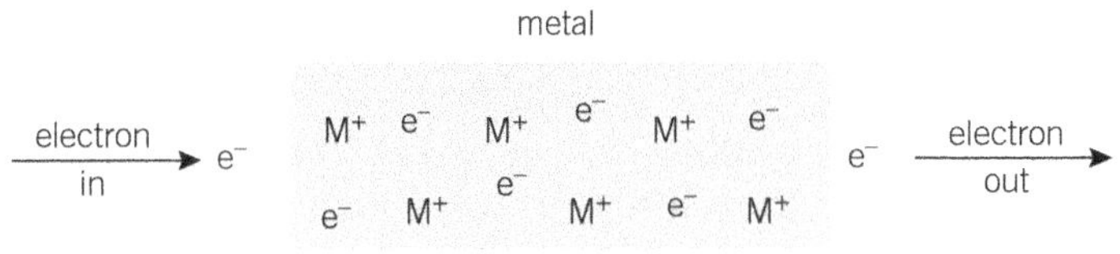

▲ **Figure 7** *The conduction of electricity by a metal*

Ionic compounds only conduct electricity in the liquid state (or when dissolved in water). They do *not* conduct when they are solid. The current is carried by the movement of ions towards the electrode of opposite charge. The ions are free to move when the ionic compound is liquid or dissolved in water. In the solid state they are fixed rigidly in position in the ionic lattice, Figure 8.

Generally, convalently bonded substances do not conduct electricity in either the solid or liquid state. This is because there are no charged particles to carry the current. Covalent compounds are often insoluble in water but some react to form ions, for example, ethanoic acid (present in vinegar). The solutions can then conduct electricity.

You can therefore decide what type of bonding a substance has by looking at how it conducts electricity. This is summarised in Table 2.

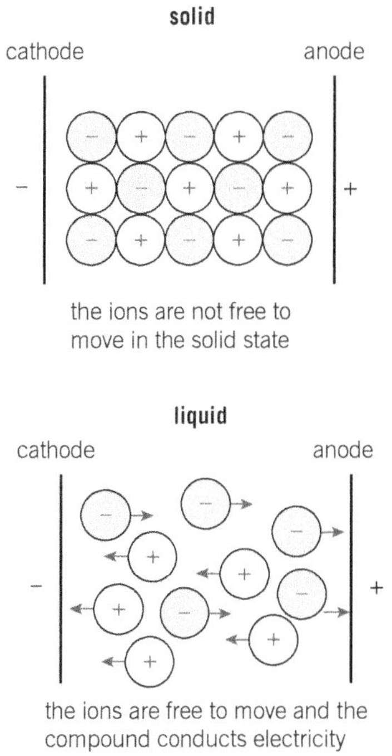

▲ **Figure 8** *Ionic liquids conduct electricity, ionic solids do not*

▼ **Table 2** *The pattern of electrical conductivity tells us about the type of bonding*

Type of bonding	Electrical conductivity		
	solid	liquid	aqueous solution
metallic	✓	✓	does not dissolve but may react
ionic	✗	✓	✓
covalent	✗	✗	✗ (but may react)

Structure – summary

Structure describes the arrangement in which atoms, ions, or molecules are held together in space. There are four main types – simple molecular, macromolecular (giant covalent), giant ionic, and metallic.

Hint

Note there are some covalently bonded substances that *do* conduct electricity, for example, graphite.

- A **simple molecular** structure is composed of small molecules – small groups of atoms strongly held together by covalent bonding. The forces of attraction *between* molecules are much weaker (often over 50 times weaker than a covalent bond) and are called intermolecular forces. Examples of molecules include Cl_2, H_2O, H_2SO_4, and NH_3.
- A **macromolecular** structure is one in which large numbers of atoms are linked in a regular three-dimensional arrangement by covalent bonds. Examples include diamond and silicon dioxide (silica), the main constituent of sand.
- A **giant ionic** structure consists of a lattice of positive ions each surrounded in a regular arrangement by negative ions and vice versa.
- A **metallic** structure consists of a regular lattice of positively charged metal ions held together by a cloud of delocalised electrons.

Macromolecular, giant ionic, and metallic structures are often called **giant structures** because they have regular three-dimensional arrangements of atoms in contrast to simple molecular structures.

Melting and boiling points

The property that best tells us if a structure is giant or simple molecular is the melting (or boiling) point.

- Simple molecular compounds have low melting (and boiling) points.
- Giant structures generally have high melting (and boiling) points.

Hint

Generally any substance with a high melting point also has a high boiling point. However, there are some substances, such as iodine, that sublime – they turn directly from solid to vapour.

If a compound has a low melting (and boiling) point, it has a simple molecular structure. All molecular compounds are covalently bonded. So all compounds with low melting (and boiling) points must have covalent bonding.

However, a compound with covalent bonding may have either a giant structure or a simple molecular structure and therefore may have either a high or low melting (and boiling) point.

Intermolecular forces

When you melt and boil simple molecular compounds, you are breaking the intermolecular forces *between* the molecules, not the covalent bonds *within* them. So the strength of the intermolecular forces determines the melting (and boiling) points.

There are three types of intermolecular force. In order of increasing strength, these are:

- van der Waals, which act between all atoms
- dipole–dipole forces, which act between molecules with permanent dipoles: $X^{\delta+}–Y^{\delta-}$
- hydrogen bonds, which act between the molecules formed when highly electronegative atoms (oxygen, nitrogen, and fluorine) and hydrogen atoms are covalently bonded.

Table 3 is a summary of the different properties of substances with covalent, ionic, and metallic bonding.

▼ **Table 3** *Summary of properties of substances with covalent, ionic, and metallic bonding*

	Structure	Bond	Melting point, T_m	Electrical conductivity		
				Solid	Liquid	Aqueous solution
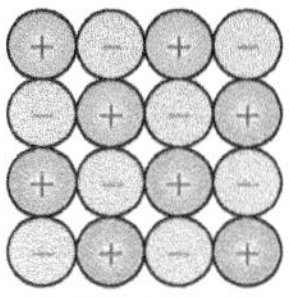	giant	ionic	high	no	yes	yes
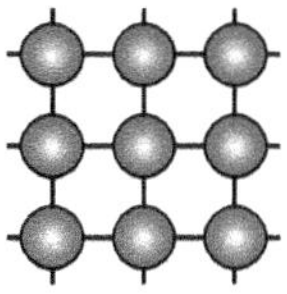	giant (macromolecular)	covalent	high	no (except graphite and graphene)	no	no
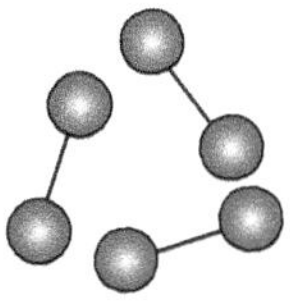	simple molecular	covalent	low	no	no	no (but may react)
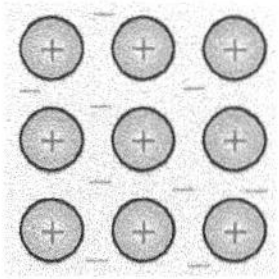	giant	metallic	high	yes	yes	– does not dissolve but may react

Summary questions

1 Describe what is the difference between a macromolecular crystal and a molecular crystal in terms of the following.

   a bonding **b** properties

2 Explain why graphite can be used as a lubricant.

3 Explain how graphite conducts electricity. How does it conduct differently from metals?

4 Explain why both diamond and graphite have high melting points.

5 The table below gives some information about four substances.

   a Identify which substances have giant structures.

   b Identify which substance is a gas at room temperature.

   c Identify which substance is a metal.

   d Identify which substances are covalently bonded.

   e Identify which substance has ionic bonding.

   f Identify which substance is a macromolecule.

Substance	Melting point / K (°C)	Boiling point / K (°C)	Electrical conductivity	
			solid	liquid
A	1356 (1083)	2840 (2567)	good	good
B	91 (–182)	109 (–164)	poor	poor
C	1996 (1723)	2503 (2230)	poor	good
D	1266 (993)	1968 (1695)	poor	poor

3.5 The shapes of molecules and ions

Learning objectives:

- → State the rules that govern the shapes of simple molecules.
- → Describe how the number of electron pairs around an atom affects the shape of the molecule.
- → Describe what happens to the shape of a molecule when a bonding pair of electrons is replaced by a non-bonding pair.

Specification reference: 3.1.3

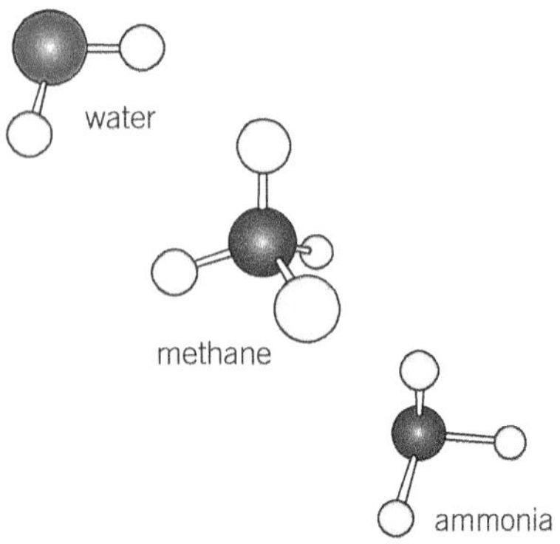

▲ **Figure 1** *The shapes of water, methane, and ammonia molecules*

Hint

It is acceptable to draw electron diagrams that show electrons in the outer shells only.

Hint

You can also think of electron pairs as clouds of negative charge.

Hint

Notice that in neither $BeCl_2$ nor BF_3 does the central atom have a full outer main electron shell.

Molecules are three-dimensional and they come in many different shapes (Figure 1).

Electron pair repulsion theory

You have seen that electrons in molecules exist in pairs in volumes of space called orbitals. You can predict the shape of a simple covalent molecule, for example, one consisting of a central atom surrounded by a number of other atoms, by using the ideas that:

- each pair of electrons around an atom will repel all other electron pairs
- the pairs of electrons will therefore take up positions as far apart as possible to minimise repulsion.

This is called the **electron pair repulsion theory**.

Electron pairs may be a shared pair or a lone pair.

The shape of a simple molecule depends on the number of pairs of electrons that surround the central atom. To work out the shape of any molecule you first need to draw a dot-and-cross diagram to find the number of pairs of electrons.

Two pairs of electrons

If there are two pairs of electrons around the atom, the molecule will be *linear*. The furthest away from each other the two pairs can get is *180°* apart. Beryllium chloride, which is a covalently bonded molecule in the gas phase, despite being a metal–non-metal compound, is an example of this.

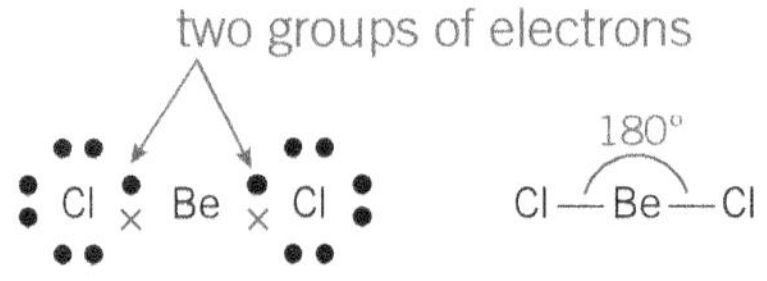

Three pairs of electrons

If there are three pairs of electrons around the central atom, they will be *120°* apart. The molecule is planar and is called *trigonal planar*. Boron trifluoride is an example of this.

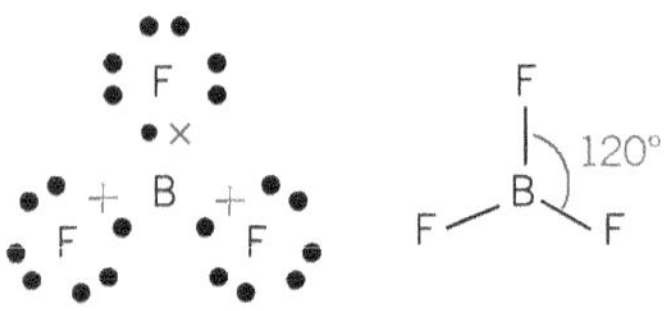

Four pairs of electrons

If there are four pairs of electrons, they are furthest apart when they are arranged so that they point to the four corners of a *tetrahedron*. This shape, with one atom positioned at the centre, is called *tetrahedral*, see Figure 2.

Methane, CH_4, is an example. The carbon atom is situated at the centre of the tetrahedron with the hydrogen atoms at the vertices. The angles here are *109.5°*. This is a three-dimensional, not planar, arrangement so the sum of the angles can be more than 360°.

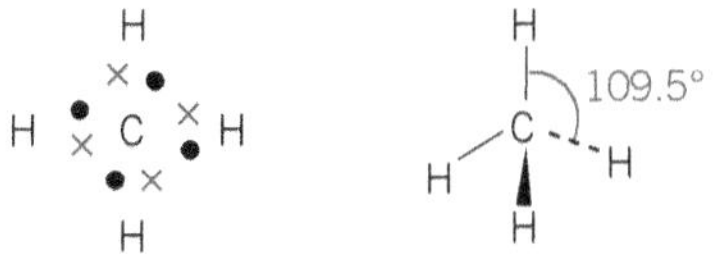

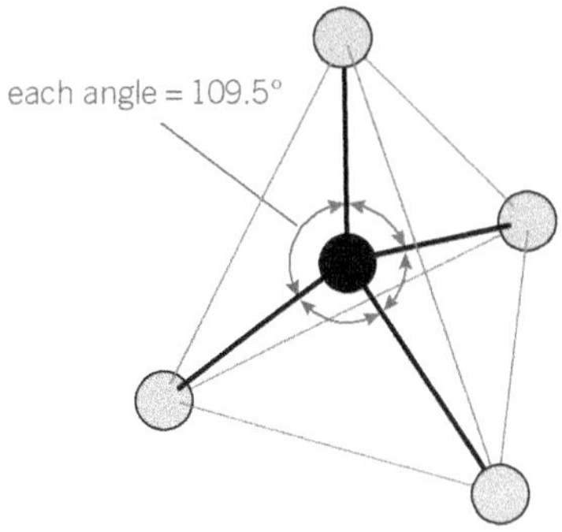

▲ **Figure 2** *A tetrahedron has four points and four faces*

The ammonium ion is also tetrahedral. It has four groups of electrons surrounding the nitrogen atom. The fact that the ion has an overall charge does not affect the shape.

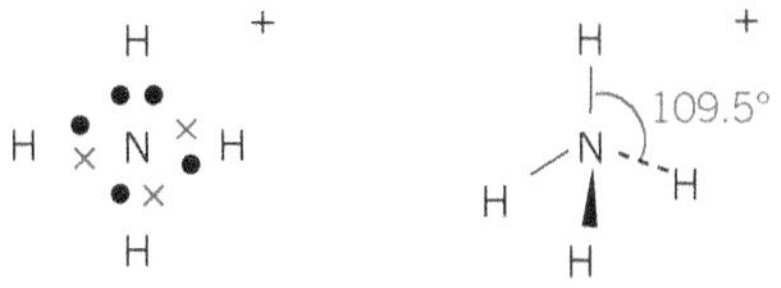

Hint

In three-dimensional representations of molecules, a wedge is used to represent a bond coming out of the paper and a dashed line represents one going into the paper, away from the reader, as shown in these figures.

Five pairs of electrons

If there are five pairs of electrons, the shape usually adopted is that of a *trigonal bipyramid*. Phosphorus pentachloride, PCl_5, is an example.

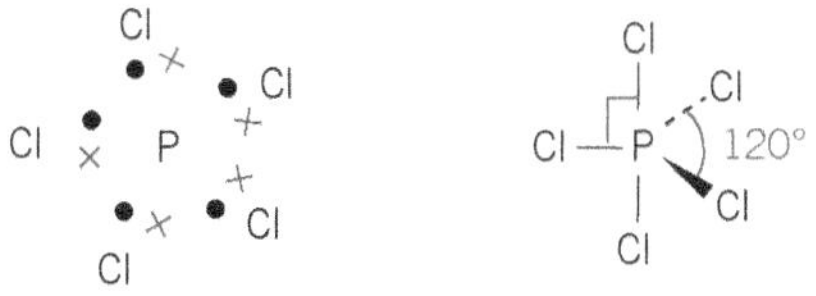

Six pairs of electrons

If there are six pairs of electrons, the shape adopted is *octahedral*, with bond angles of *90°*. The sulfur hexafluoride, SF_6, molecule is an example of this.

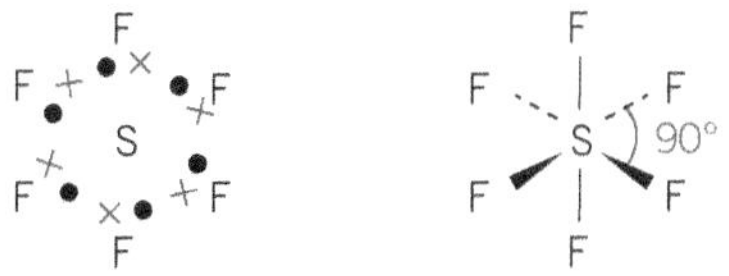

Hint

Take care. Octahedral sounds as if there should be eight electron groups, not six. Remember that an octahedron has eight *faces* but *six points*.

Molecules with lone pairs of electrons

Some molecules have unshared (lone) pairs of electrons. These are electrons that are not part of a covalent bond. The lone pairs affect the shape of the molecule. Always watch out for the lone pairs in your dot-and-cross diagram because otherwise you might overlook their effect. Ammonia and water are good examples of molecules where lone pairs affect the shape.

Study tip

Remember that electron pairs will get as far apart as possible.

> **Study tip**
>
> Draw structures showing bonds and lone electron pairs.

Ammonia, NH_3

Ammonia has four pairs of electrons and one of the groups is a lone pair.

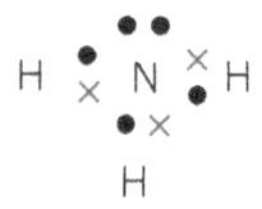

With its four pairs of electrons around the nitrogen atom, the ammonia molecule has a shape based on a tetrahedron. However, there are only three 'arms' so the shape is that of a *triangular pyramid*.

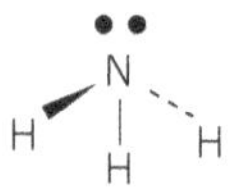

Another way of looking at this is that the *electron pairs* form a tetrahedron but the bonds form a triangular pyramid. (There is an atom at each vertex but, unlike the tetrahedral arrangement, no atom in the centre.)

Bonding pair–lone pair repulsion

The angles of a regular tetrahedron, see Figure 2, are all 109.5° but lone pairs affect these angles. In ammonia, for example, the *bonding* pairs of electrons are attracted towards the nitrogen nucleus and also the hydrogen nucleus. However, the *lone* pair is attracted only by the nitrogen nucleus and is therefore pulled closer to it than the shared pairs. So repulsion between a lone pair of electrons and a bonding pair of electrons is greater than that between two bonding pairs. This effect squeezes the hydrogen atoms together, reducing all the H–N–H angles. The approximate rule of thumb is 2° per lone pair, so the bond angles in ammonia are approximately 107°:

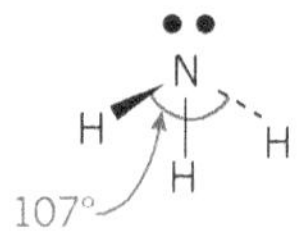

> **Study tip**
>
> It is difficult to fully appreciate the 3-D shapes of molecules without using models. Professional chemists routinely use powerful computer software (based on the principles described in this topic) to accurately model the shapes of complex molecules before they have even been synthesised. This can help to predict how they might act as drugs, for example.
>
> 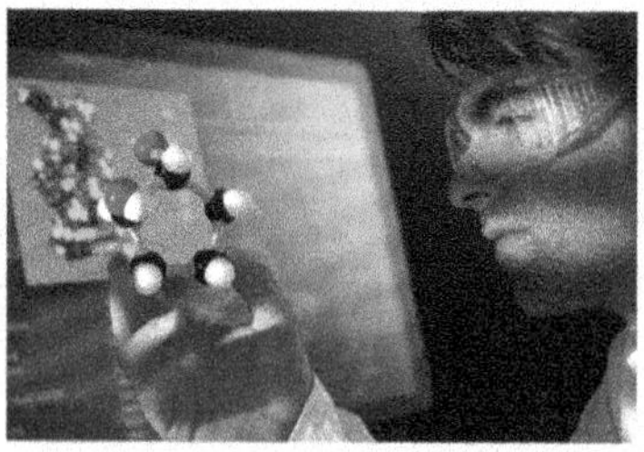
>
> You might consider buying a kit of molecular models (particularly if you intend to continue to study chemistry) or alternatively you can do a lot with matchsticks and a modelling material such as Plasticene©. Furthermore, there are websites and programmes where you can draw 3-D models of molecules and rotate them to view from different points in space.

Water, H_2O

Look at the dot-and-cross diagram for water.

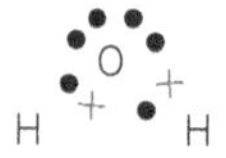

There are four pairs of electrons around the oxygen atom so, as with ammonia, the shape is based on a tetrahedron. However, two of the 'arms' of the tetrahedron are lone pairs that are not part of a bond. This results in a *V-shaped* or angular molecule. As in ammonia the electron pairs form a tetrahedron but the bonds form a V-shape. With two lone pairs, the H–O–H angle is reduced to 104.5°.

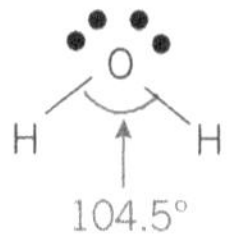

Chlorine tetrafluoride ion, ClF_4^-

The dot-and-cross diagram for this ion is as shown:

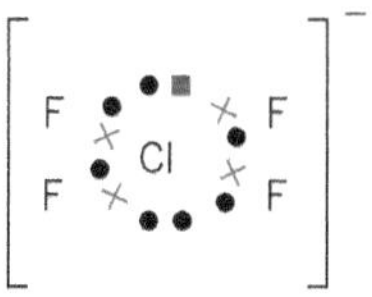

There are four bonding pairs of electrons and two lone pairs. One of the lone pairs contains an electron that has been donated to it, so the charge on the ion is negative (−1). This electron is shown as a square in the dot-and-cross diagram. This means that there are six pairs of electrons around the chlorine atom – four bonds and two lone pairs. The shape is therefore based on an octahedron in which two arms are not part of a bond.

As lone pairs repel the most, they adopt a position furthest apart. This leaves a flat square-shaped ion described as *square planar*. The lone pairs are above and below the plane, as shown here.

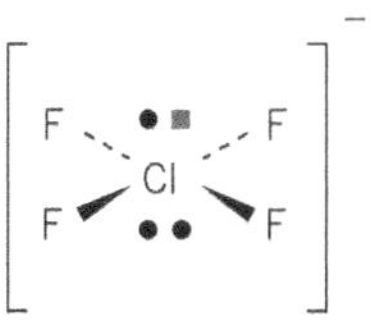

A summary of the repulsion between electron pairs

bonding pair–bonding pair	↓
lone pair–bonding pair	repulsion increases
lone pair–lone pair	↓

Summary questions

1. Draw a dot-and-cross diagram for NF_3 and predict its shape.
2. Explain why NF_3 has a different shape from BF_3.
3. Draw a dot-and-cross diagram for the molecule silane, SiH_4, and describe its shape.
4. State the H—Si—H angle in the silane molecule.
5. Predict the shape of the H_2S molecule *without* drawing a dot-and-cross diagram.

Hint

Some ions contain more than one atom. These are called compound ions. The atoms within the ion are covalently bonded but the ion has an overall charge. Ones you will come across include:

ammonium	NH_4^+
carbonate	CO_3^{2-}
sulfate	SO_4^{2-}
nitrate	NO_3^-
hydroxide	OH^-

3.6 Electronegativity – bond polarity in covalent bonds

Learning objectives:

- → State what is meant by the term electronegativity.
- → State what makes one atom more electronegative than another.
- → State what the symbols δ+ and δ– mean when placed above atoms in a covalent bond.

Specification reference: 3.1.3

Study tip

Learn the definition of electronegativity.

▼ **Table 1** *Some values for Pauling electronegativity*

H 2.1							He
Li 1.0	Be 1.5	B 2.0	C 2.5	N 3.0	O 3.5	F 4.0	Ne
Na 0.9	Mg 1.2	Al 1.5	Si 1.8	P 2.1	S 2.5	Cl 3.0	Ar
						Br 2.8	Kr

Hint

The Pauling scale is named after the US chemist Linus Pauling, who won the 1954 Nobel Prize in Chemistry for his work on chemical bonding.

Hint

Think of electronegative atoms as having more 'electron-pulling power'.

The forces that hold atoms together are all about the attraction of positive charges to negative charges. In ionic bonding there is complete transfer of electrons from one atom to another. But, even in covalent bonds, the electrons shared by the atoms will not be evenly spread if one of the atoms is better at attracting electrons than the other. This atom is more **electronegative** than the other.

Electronegativity

Flourine is better at attracting electrons than hydrogen. Fluorine is said to be more electronegative than hydrogen.

Electronegativity is the power of an atom to attract the electron density in a covalent bond towards itself.

When chemists consider the electrons as charge clouds, the term **electron density** is often used to describe the way the negative charge is distributed in a molecule.

The Pauling scale is used as a measure of electronegativity. It runs from 0 to 4. The greater the number, the more electronegative the atom, see Table 1. The noble gases have no number because they do not, in general, form covalent bonds.

Electronegativity depends on:

1. the nuclear charge
2. the distance between the nucleus and the outer shell electrons
3. the shielding of the nuclear charge by electrons in inner shells.

Note the following:

- The smaller the atom, the closer the nucleus is to the shared outer main shell electrons and the greater its electronegativity.
- The larger the nuclear charge (for a given shielding effect), the greater the electronegativity.

Trends in electronegativity

Going up a group in the Periodic Table, electronegativity increases (the atoms get smaller) and there is less shielding by electrons in inner shells.

Going across a period in the Periodic Table, the electronegativity increases. The nuclear charge increases, the number of inner main shells remain the same and the atoms become smaller.

So, the most electronegative atoms are found at the top right-hand corner of the Periodic Table (ignoring the noble gases which form few compounds). The most electronegative atoms are fluorine, oxygen, and nitrogen followed by chlorine.

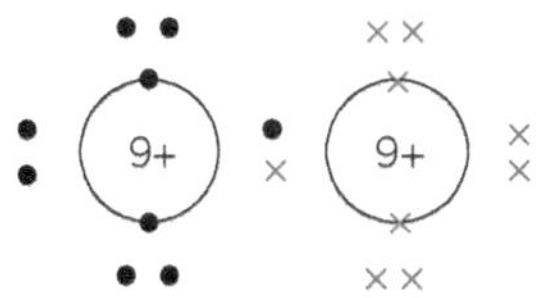

▲ **Figure 1** *Electron diagram of fluorine molecule*

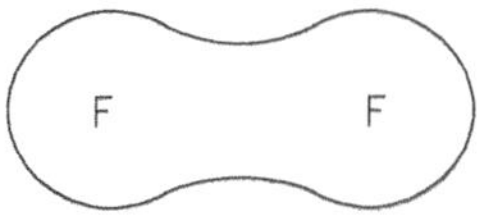

▲ **Figure 2** *Electron cloud around fluorine molecule*

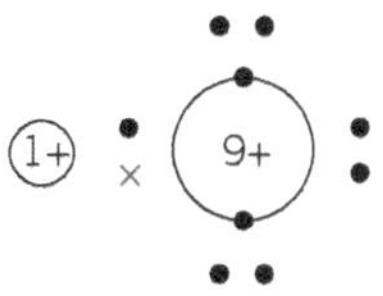

▲ **Figure 3** *Electron diagram of hydrogen fluoride molecule*

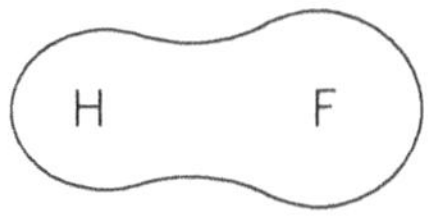

▲ **Figure 4** *Electron cloud around hydrogen fluoride molecule*

▼ **Table 2** *Trends in electronegativity*

Increasing electronegativity →

Li	Be	B	C	N	O	F
1.0	1.5	2.0	2.5	3.0	3.5	4.0
						Cl
						3.0
						Br
						2.8

↑ Increasing electronegativity

Polarity of covalent bonds

Polarity is about the unequal sharing of the electrons between atoms that are bonded together covalently. It is a property of the *bond*.

Covalent bonds between two atoms that are the same

When both atoms are the same, for example, in fluorine, F_2, the electrons in the bond *must* be shared equally between the atoms (Figure 1) – both atoms have exactly the same electronegativity and the bond is completely non-polar.

If you think of the electrons as being in a cloud of charge, then the cloud is uniformly spread between the two atoms, as shown in Figure 2.

Covalent bonds between two atoms that are different

In a covalent bond between two atoms of *different* electronegativity, the electrons in the bond will not be shared equally between the atoms. For example, the molecule hydrogen fluoride, HF, shown in Figure 3.

Hydrogen has an electronegativity of 2.1 and fluorine of 4.0. This means that the electrons in the covalent bond will be attracted more by the fluorine than the hydrogen. The electron cloud is distorted towards the fluorine, as shown in Figure 4.

The fluorine end of the molecule is therefore relatively negative and the hydrogen end relatively positive, that is, electron deficient. You show this by adding partial charges to the formula:

$$^{\delta+}H—F^{\delta-}$$

Covalent bonds like this are said to be **polar**. The greater the difference in electronegativity, the more polar is the covalent bond.

You could say that although the H—F bond is covalent, it has some ionic character. It is going some way towards the separation of the atoms into charged ions. It is also possible to have ionic bonds with some covalent character.

Hint

δ+ and δ− are pronounced 'delta plus' and 'delta minus'.

The + and – signs represent one 'electron's worth' of charge.

δ+ and δ− represent a small charge of less than one 'electron's worth'.

Summary questions

1. Explain why fluorine is more electronegative than chlorine.
2. Write δ+ and δ− signs to show the polarity of the bonds in a hydrogen chloride molecule.
3. Identify if these covalent bonds is/are non-polar, and explain your answer.
   - **a** H—H
   - **b** F—F
   - **c** H—F
4. **a** Arrange the following covalent bonds in order of increasing polarity:
   H—O, H—F, H—N
   b Explain your answer.

3.7 Forces acting between molecules

Learning objectives:

- → State the three types of intermolecular force.
- → Describe how dipole–dipole and van der Waals forces arise.
- → Describe how van der Waals forces affect boiling temperatures.
- → State what is needed for hydrogen bonding to occur.
- → Explain why NH_3, H_2O, and HF have higher boiling temperatures than might be expected.

Specification reference: 3.1.3

Hint

van der Waals is spelt with a small v, even at the beginning of a sentence. van der Waals forces are sometimes called **dispersion forces** or **London forces**.

Study tip

Do not confuse intermolecular forces with covalent bonds, which are at least 10 times stronger.

Atoms in molecules and in giant structures are held together by strong covalent, ionic, or metallic bonds. Molecules and separate atoms are attracted to one another by other, weaker forces called intermolecular forces. 'Inter' means *between*. If the intermolecular forces are strong enough, then molecules are held closely enough together to be liquids or even solids.

Intermolecular forces

There are three types of intermolecular forces:

- **van der Waals forces**
  act between *all* atoms and molecules. (weakest)
- **Dipole–dipole forces**
  act only between certain types of molecules.
- **Hydrogen bonding**
  acts only between certain types of molecules. (strongest)

weakest → strongest

Dipole–dipole forces

Dipole moments

Polarity is the property of a particular bond, see Topic 3.6, but molecules with polar bonds may have a dipole moment. This sums up the effect of the polarity of *all* the bonds in the molecule.

In molecules with more than one polar bond, the effects of each bond may cancel, leaving a molecule with no dipole moment. The effects may also add up and so reinforce each other. It depends on the shape of the molecule.

For example, carbon dioxide is a linear molecule and the dipoles cancel.

$$^{\delta-}O{=}C^{\delta+}{=}O^{\delta-}$$

Tetrachloromethane is tetrahedral and here too the dipoles cancel.

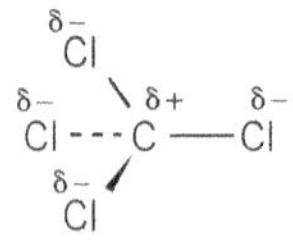

tetrachloromethane

But in dichloromethane the dipoles do not cancel because of the shape of the molecule.

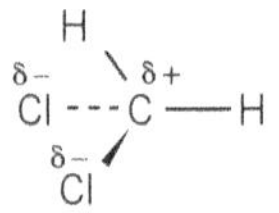

dichloromethane

Dipole–dipole forces act between molecules that have permanent dipoles. For example, in the hydrogen chloride molecule, chlorine is more electronegative than hydrogen. So the electrons are pulled towards the chlorine atom rather than the hydrogen atom. The molecule therefore has a dipole and is written $H^{\delta+}{-}Cl^{\delta-}$.

Two molecules which both have dipoles will attract one another, see Figure 1.

Whatever their starting positions, the molecules with dipoles will 'flip' to give an arrangement where the two molecules attract.

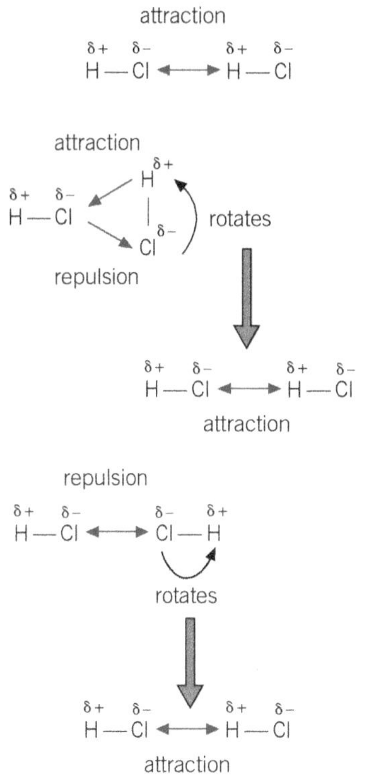

▲ **Figure 1** *Two polar molecules, such as hydrogen chloride, will always attract one another*

van der Waals forces

All atoms and molecules are made up of positive and negative charges even though they are neutral overall. These charges produce very weak electrostatic attractions between all atoms and molecules. These are called van der Waals forces.

How do van der Waals forces work?

Imagine a helium atom. It has two positive charges on its nucleus and two negatively charged electrons. The atom as a whole is neutral but at any moment in time the electrons could be anywhere, see Figure 2. This means the distribution of charge is changing *at every instant*.

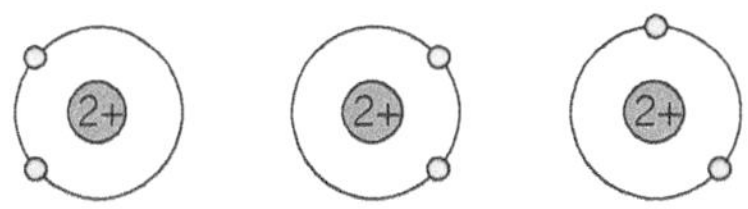

▲ **Figure 2** *These are just a few of the possible arrangements of the two electrons in helium. Remember, electrons are never in a fixed position*

Any of the arrangements in Figure 2 mean the atom has a dipole at that moment. An instant later, the dipole may be in a different direction. But, almost certainly the atom *will* have a dipole at any point in time, even though any particular dipole will be just for an instant – a temporary dipole. This dipole then affects the electron distribution in nearby atoms, so that they are attracted to the original helium atom for that instant. The original atom has induced dipoles in the nearby atoms, as shown in Figure 3 in which the electron distribution is shown as a cloud.

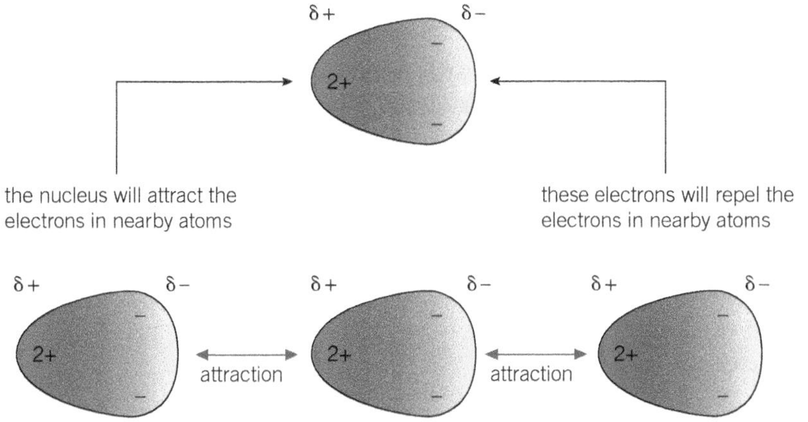

▲ **Figure 3** *Instantaneous dipoles induce dipoles in nearby atoms*

As the electron distribution of the original atom changes, it will induce new dipoles in the atoms around it, which will be attracted to the original one. These forces are sometimes called instantaneous dipole–induced dipole forces, but this is rather a mouthful. The more usual name is van der Waals forces after the Dutch scientist, Johannes van der Waals.

- van der Waals forces act between *all* atoms or molecules at all times.
- They are in addition to any other intermolecular forces.
- The dipole is caused by the changing position of the electron cloud, so the more electrons there are, the larger the instantaneous dipole will be.

Therefore the size of the van der Waals forces increases with the number of electrons present. This means that atoms or molecules with large atomic or molecular masses produce stronger van der Waals forces than atoms or molecules with small atomic or molecular masses.

This explains why:

- the boiling points of the noble gases increase as the atomic numbers of the noble gases increase
- the boiling points of hydrocarbons increase with increased chain length.

Hydrogen bonding

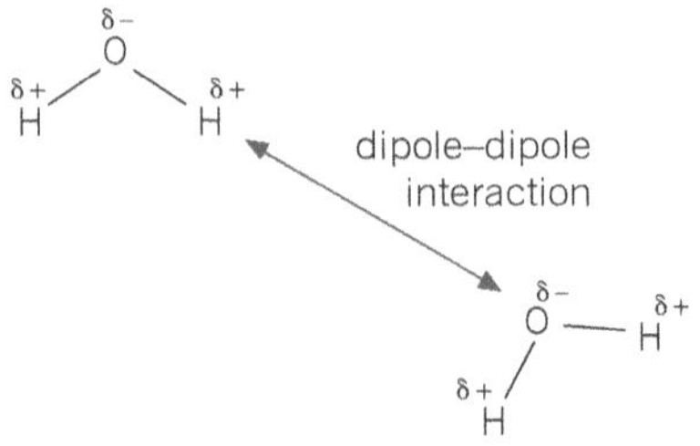

▲ **Figure 4** *Dipole attraction between water molecules*

Hydrogen bonding is a special type of intermolecular force with some characteristics of dipole–dipole attraction and some of a covalent bond. It consists of a hydrogen atom 'sandwiched' between two very electronegative atoms. There are conditions that have to be present for hydrogen bonding to occur. You need a very electronegative atom with a lone pair of electrons covalently bonded to a hydrogen atom. Water molecules fulfil these conditions. Oxygen is much more electronegative than hydrogen so water is polar, see Figure 4.

You would expect to find weak dipole–dipole attractions (as shown between hydrogen chloride in Figure 1) but in this case the intermolecular bonding is much stronger for two reasons:

1. The oxygen atoms in water have lone pairs of electrons.
2. In water the hydrogen atoms are highly electron deficient. This is because the oxygen is very electronegative and attracts the shared electrons in the bond towards it. The hydrogen atoms in water are positively charged and very small. These exposed protons have a very strong electric field because of their small size.

The lone pair of electrons on the oxygen atom of another water molecule is strongly attracted to the electron deficient hydrogen atom.

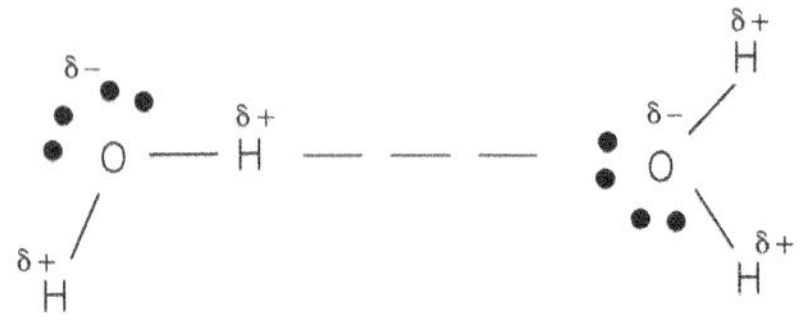

▲ **Figure 5** *Hydrogen bond between water molecules*

This strong intermolecular force is called a **hydrogen bond**. Hydrogen bonds are considerably stronger than dipole–dipole attractions, though much weaker than a covalent bond. They are usually represented by dashes – – –, as in Figure 5.

When do hydrogen bonds form?

Water is not the only example of hydrogen bonding. In order to form a hydrogen bond there must be the following:

- a hydrogen atom that is bonded to a very electronegative atom. This will produce a strong partial positive charge on the hydrogen atom.
- a very electronegative atom with a lone pair of electrons. These will be attracted to the partially charged hydrogen atom in another molecule and form the bond.

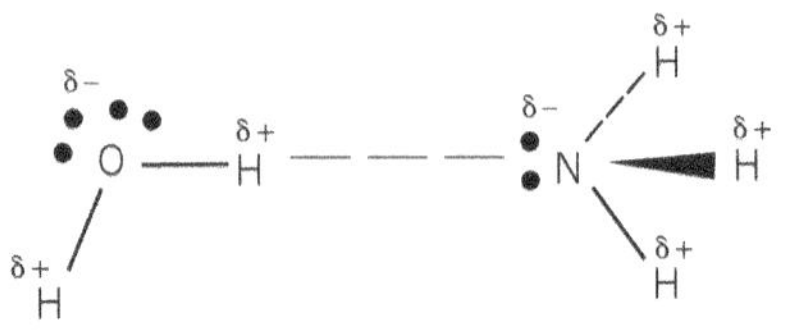

▲ **Figure 6** *Hydrogen bond between a water molecule and an ammonia molecule*

The only atoms that are electronegative enough to form hydrogen bonds are oxygen, O, nitrogen, N, and fluorine, F. For example, ammonia molecules, NH_3, form hydrogen bonds with water molecules, see Figure 6.

The nitrogen–hydrogen–nitrogen system is linear. This is because the pair of electrons in the O—H covalent bond repels those in the hydrogen bond between nitrogen and hydrogen. This linearity is always the case with hydrogen bonds.

The boiling points of the hydrides

The effect of hydrogen bonding between molecules can be seen if you look at the boiling points of hydrides of elements of Group 4, 5, 6, and 7 plotted against the period number, see Figure 7.

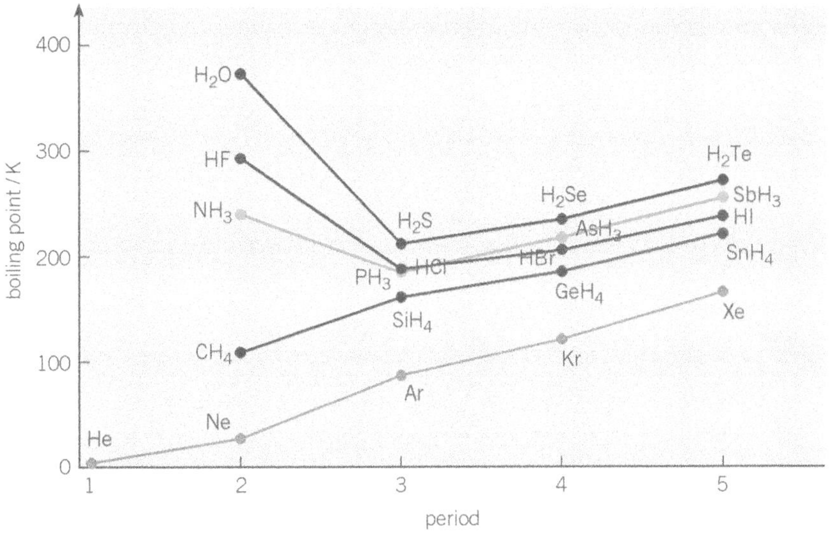

▲ **Figure 7** *Boiling points of the hydrides of Group 4, 5, 6, and 7 elements with the noble gases for comparison*

The noble gases show a gradual increase in boiling point because the only forces acting between the atoms are van der Waals forces, and these increase with the number of electrons present.

The boiling points of water, H_2O, hydrogen fluoride, HF, and ammonia, NH_3, are all higher than those of the hydrides of the next elements in their group, whereas you would expect them to be lower if only van der Waals forces were operating. This is because hydrogen bonding is present between the molecules in each of these compounds and these stronger intermolecular forces of attraction make the molecules more difficult to separate. Oxygen, nitrogen, and fluorine are the three elements that are electronegative enough to make hydrogen bonding possible.

The importance of hydrogen bonding

Although hydrogen bonds are only about 10% of the strength of covalent bonds, their effect can be significant – especially when there are a lot of them. The very fact that they are weaker than covalent bonds, and can break or make under conditions where covalent bonds are unaffected, is very significant.

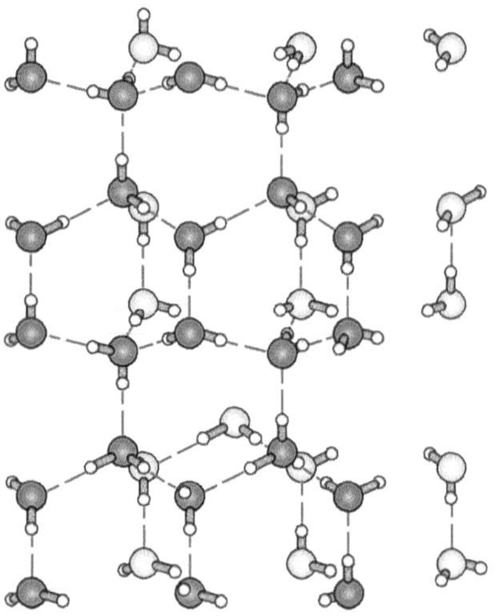

▲ **Figure 8** *The three-dimensional network of covalent bonds (grey) and hydrogen bonds (red) in ice. The blue lines are only construction lines*

The structure and density of ice

In water in its liquid state, the hydrogen bonds break and reform easily as the molecules are moving about. When water freezes, the water molecules are no longer free to move about and the hydrogen bonds hold the molecules in fixed positions. The resulting three-dimensional structure, shown in Figure 8, resembles the structure of diamond, see Topic 3.4.

In order to fit into this structure, the molecules are slightly less closely packed than in liquid water. This means that ice is less dense than water and forms on top of ponds rather than at the bottom. This insulates the ponds and enables fish to survive through the winter. This must have helped life to continue, in the relative warmth of the water under the ice, during the Ice Ages.

Living with hydrogen bonds

Proteins are a class of important biological molecules that fulfil a wide variety of functions in living things, including enzyme catalysts. The exact shape of a protein molecule is vital to its function. Proteins are long chain molecules with lots of C═O and N—H groups which can form hydrogen bonds. These hydrogen bonds hold the protein chains into fixed shapes. One common shape is the protein chain that forms a spiral (helix), as shown here.

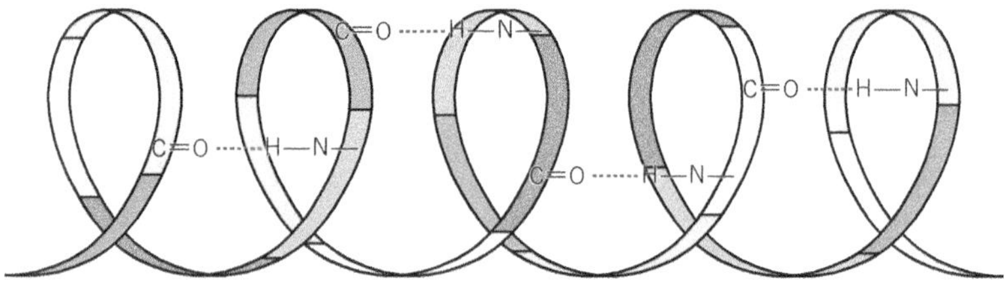

▲ **Figure 9**

Another example is the beta-pleated sheet. Here protein chains line up side by side, held in position by hydrogen bonds to form a two-dimensional sheet. The protein that forms silk has this structure.

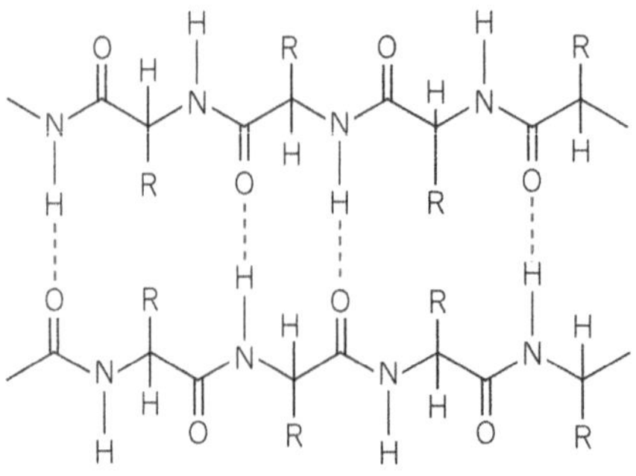

Synoptic link

The bonding in proteins and in DNA is discussed in more detail in Chapter 30, Amino acids and proteins.

The relative weakness of hydrogen bonds means that the shapes of proteins can easily be altered. Heating proteins much above body temperature starts to break hydrogen bonds and causes the protein to lose its shape and thus its function. This is why enzymes lose their effect as catalysts when heated – the protein is denatured. You can see this when frying an egg. The clear liquid protein albumen is transformed into an opaque white solid.

Ironing

When you iron clothes, the iron provides heat to break hydrogen bonds in the crumpled material and pressure to force the molecules into new positions so that the material is flat. When you remove the iron, the hydrogen bonds reform and hold the molecules in these new positions, keeping the fabric flat.

DNA

Another vital biological molecule is DNA (deoxyribonucleic acid) (Figure 10). It is the molecule that stores and copies genetic information that makes offspring resemble their parents. This molecule exists as a double-stranded helix. The two strands of the spiral are held together by hydrogen bonds. When cells divide or replicate, the hydrogen bonds break (but the covalently bonded main chains stay unchanged). The two separate helixes then act as templates for a new helix to form on each, so you end up with a copy of the original helix.

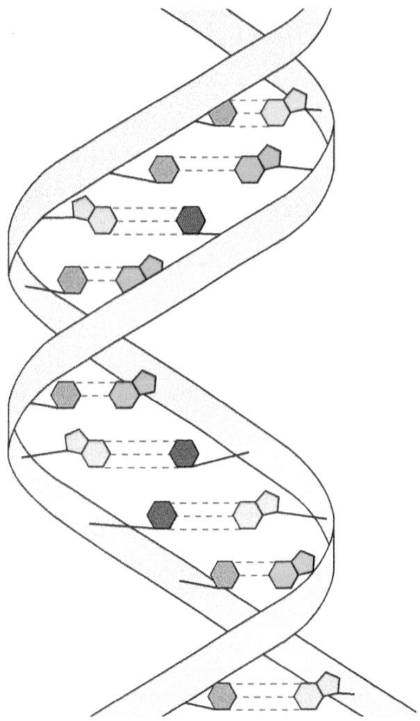

▲ **Figure 10** *The DNA double helix is held together by hydrogen bonds*

Summary questions

1 Place the following elements in order of the strength of the van der Waals forces between the atoms (weakest first): Ar, He, Kr, Ne. Explain your answer.

2 Identify which one of the following molecules *cannot* have dipole–dipole forces acting between them: H_2O, HCl, H_2

3 Explain why hexane is a liquid at room temperature whereas butane is a gas.

4 Explain why covalent molecules are gases, liquids, or solids with low-melting temperature.

5 Draw two hydrogen bromide molecules to show how they would be attracted together by dipole–dipole forces.

6 Identify in which of the following does hydrogen bonding *not* occur between molecules: H_2O, NH_3, HBr, HF

7 Explain why hydrogen bonds do not form between:

- **a** methane molecules, CH_4
- **b** tetrachloromethane molecules, CCl_4.

8 Draw a dot-and-cross diagram for a molecule of water.

- **a** State how many lone pairs it has.
- **b** State how many hydrogen atoms it has.
- **c** Explain why water molecules form on average two hydrogen bonds per molecule, whereas the ammonia molecule, NH_3, forms only one.

Practice questions

1 Phosphorus exists in several different forms, two of which are white phosphorus and red phosphorus. White phosphorus consists of P_4 molecules, and melts at 44 °C.

Red phosphorus is macromolecular, and has a melting point above 550 °C.

Explain what is meant by the term *macromolecular*. By considering the structure and bonding present in these two forms of phosphorus, explain why their melting points are so different.

(5 marks)
AQA, 2006

2 **(a)** Predict the shapes of the SF_6 molecule and the $AlCl_4^-$ ion. Draw diagrams of these species to show their three-dimensional shapes. Name the shapes and suggest values for the bond angles. Explain your reasoning.

(8 marks)

(b) Perfume is a mixture of fragrant compounds dissolved in a volatile solvent. When applied to the skin the solvent evaporates, causing the skin to cool for a short time. After a while, the fragrance may be detected some distance away. Explain these observations.

(4 marks)
AQA, 2003

3 Fritz Haber, a German chemist, first manufactured ammonia in 1909. Ammonia is very soluble in water.

(a) State the strongest type of intermolecular force between one molecule of ammonia and one molecule of water.

(1 mark)

(b) Draw a diagram to show how one molecule of ammonia is attracted to one molecule of water. Include all partial charges and all lone pairs of electrons in your diagram.

(3 marks)

(c) Phosphine, PH_3, has a structure similar to ammonia.
In terms of intermolecular forces, suggest the main reason why phosphine is almost insoluble in water.

(1 mark)
AQA, 2013

4 The following equation shows the reaction of a phosphine molecule, PH_3, with an H^+ ion.

$$PH_3 + H^+ \rightarrow PH_4^+$$

(a) Draw the shape of the PH_3 molecule. Include any lone pairs of electrons that influence the shape.

(1 mark)

(b) State the type of bond that is formed between the PH_3 molecule and the H^+ ion. Explain how this bond is formed.

(2 marks)

(c) Predict the bond angle in the PH_4^+ ion.

(1 mark)

(d) Although phosphine molecules contain hydrogen atoms, there is no hydrogen bonding between phosphine molecules.
Suggest an explanation for this.

(1 mark)
AQA, 2012

5 There are several types of crystal structure and bonding shown by elements and compounds.

(a) (i) Name the type of bonding in the element sodium.

(1 mark)

(ii) Use your knowledge of structure and bonding to draw a diagram that shows how the particles are arranged in a crystal of sodium.
You should identify the particles and show a minimum of six particles in a two-dimensional diagram.

(2 marks)
AQA, 2011

Answers to the Practice Questions and Section Questions are available at
www.oxfordsecondary.com/oxfordaqaexams-alevel-chemistry

4 Energetics

4.1 Endothermic and exothermic reactions

Most chemical reactions give out or take in energy as they proceed. The amount of energy involved when a chemical reaction takes place is important for many reasons. For example:

- you can measure the energy values of fuels
- you can calculate the energy requirements for industrial processes
- you can work out the theoretical amount of energy required to break bonds and the amount of energy released when bonds are made
- it helps to predict whether or not a reaction will take place.

The energy involved may be in different forms – light, electrical, or most usually heat.

Thermochemistry

Thermochemistry is the study of heat changes during chemical reactions.

- When a chemical reaction takes place, chemical bonds break and new ones are formed.
- Energy must be *put in* to break bonds and energy is *given out* when bonds are formed, so most chemical reactions involve an energy change.
- The overall change may result in energy being given out or taken in.
- At the end of the reaction, if energy has been given out, the reaction is **exothermic**.
- At the end of the reaction, if energy has been taken in, the reaction is **endothermic**.

Exothermic and endothermic reactions

Some reactions give out heat as they proceed. These are called *exothermic* reactions. Neutralising an acid with an alkali is an example of an exothermic reaction.

Some reactions take in heat from their surroundings to keep the reaction going. These are called *endothermic* reactions. The breakdown of limestone (calcium carbonate) to lime (calcium oxide) and carbon dioxide is an example of an endothermic reaction – it needs heat to proceed.

Another example of an endothermic reaction is heating copper sulfate. Blue copper sulfate crystals have the formula $CuSO_4 \cdot 5H_2O$. The water molecules are bonded to the copper sulfate. In order to break these bonds and make white, anhydrous copper sulfate, heat energy must be supplied (Figure 1). This reaction takes in heat so it is endothermic.

$$\underset{\text{blue copper sulfate}}{CuSO_4.5H_2O} \rightarrow \underset{\text{white anhydrous copper sulfate}}{CuSO_4} + \underset{\text{water}}{5H_2O}$$

Learning objectives:

→ Define the terms endothermic and exothermic.

Specification reference: 3.1.4

Hint

The unit of energy is the joule, J. One joule represents quite a small amount of heat energy. For example, in order to boil water for a cup of tea you would need about 80 000 J which is 80 kJ.

▲ **Figure 1** *Heating copper sulfate*

When you add water to anhydrous copper sulfate, the reaction gives out heat.

$CuSO_4$	+	$5H_2O$	→	$CuSO_4.5H_2O$
white anhydrous copper sulfate		water		blue copper sulfate

In this direction the reaction is exothermic.

It is *always* the case that a reaction that is endothermic in one direction is exothermic in the reverse direction.

Quantities

The amount of heat given out or taken in during a chemical reaction depends on the quantity of reactants. This energy is usually measured in kilojoules per mole, $kJ\,mol^{-1}$. To avoid any confusion about quantities you need to give an equation. For example, in the combustion of methane, CH_4, one mole of methane reacts with two moles of oxygen:

$$CH_4(g) + 2O_2(g) \rightarrow CO_2(g) + 2H_2O(l)$$

890 kJ are given out when one mole of methane burns in two moles of oxygen.

Hint

The expression mol^{-1} is a shorthand for 'per mole' and could also be written /mol. So kJ/mol has the same meaning as $kJ\,mol^{-1}$. Also note that the state symbols such as (g), meaning the gaseous state, are used. These are also important here.

Useful heat energy changes

When fuels are burnt there is a large heat output. These are very exothermic reactions.

For example, coal is mostly carbon. Carbon gives out 393.5 kJ when one mole, 12 g, is burnt completely so that the most highly oxidised product is formed. This is carbon dioxide and not carbon monoxide. Carbon dioxide is the only product.

$$C(s) + O_2(g) \rightarrow CO_2(g)$$

As you saw above, natural gas, methane, gives out 890 kJ when one mole is burnt completely to carbon dioxide and water.

Physiotherapists often treat sports injuries with cold packs. These produce 'coldness' by an endothermic reaction such as:

$$NH_4NO_3(s) + (aq) \rightarrow NH_4NO_3(aq)$$

This absorbs $26\,kJ\,mol^{-1}$ of heat energy.

The energy values of fuels

One important practical application of the study of thermochemistry is that it enables us to compare the efficiency of different fuels. Most of the fuels used today for transport (petrol for cars, diesel for cars and lorries, kerosene for aviation fuel, etc.) are derived from crude oil. This is a resource that will eventually run out so chemists are actively studying alternatives. Possible replacements include ethanol and methanol, both of which can be made from plant material, and hydrogen, which can be made by the electrolysis of water.

Theoretical chemists refer to the energy given out when a fuel burns completely as its heat (or enthalpy) of combustion. They measure this energy in kilojoules per mole ($kJ\,mol^{-1}$) because this compares the same number of ***molecules*** of each fuel. For use as fuels, the energy given out per ***gram*** of fuel burned, or the *energy density* of a fuel, is more important.

Some approximate values are given in the Table 1.

▼ **Table 1** *Enthalpy of combustion for various fuels*

Fuel	Enthalpy of combustion / $kJ\,mol^{-1}$	Mass of 1 mole / g	Energy density / $kJ\,g^{-1}$
petrol (pure octane)	−5500	114	48.2
ethanol	−1370	46	29.8
methanol	−730	32	22.8
hydrogen	−242	2	121.0

Notice that petrol stores significantly more energy per gram than either ethanol or methanol. This is a factor that will be significant for vehicles fuelled by either of these alcohols.

At first sight, hydrogen's energy density seems amazing. However, there is a catch. The other three fuels are liquids, whereas hydrogen is a gas. Although hydrogen stores lots of energy per gram, a gram of gaseous hydrogen takes up a lot of space because of the low density of gases. How to store hydrogen efficiently is a challenge for designers.

1 Write a balanced symbol equation for the combustion of methanol, CH_3OH.
2 How do the product(s) of combustion vary between hydrogen and the other fuels?
3 What environmental significance does this have?

Summary questions

1 Natural gas, methane, CH_4, gives out 890 kJ when one mole is burnt completely.

$CH_4(g) + 2O_2(g) \rightarrow CO_2(g) + 2H_2O(l)$

Calculate how much heat would be given out when 8 g of methane is burnt completely.

2 The following reaction does not take place under normal conditions.

$CO_2(g) + 2H_2O(l) \rightarrow CH_4(g) + 2O_2(g)$

If it did, would you expect it to be exothermic or endothermic?

3 Explain your answer to question **2**.

4 Approximately how much methane would have to be burnt to provide enough heat to boil a cup of tea? Choose from **a**, **b**, or **c**.

a 16 g **b** 1.6 g **c** 160 g

4.2 Enthalpy

Learning objectives:

→ Define what an enthalpy change is.

→ Describe what an enthalpy level diagram is.

Specification reference: 3.1.4

> **Hint**
>
> Chemists often use flasks open to the atmosphere to measure heat changes. The reaction is then carried out at atmospheric pressure. This varies slightly from day to day. Because these slight daily variations are small, this is only a small source of systematic error.

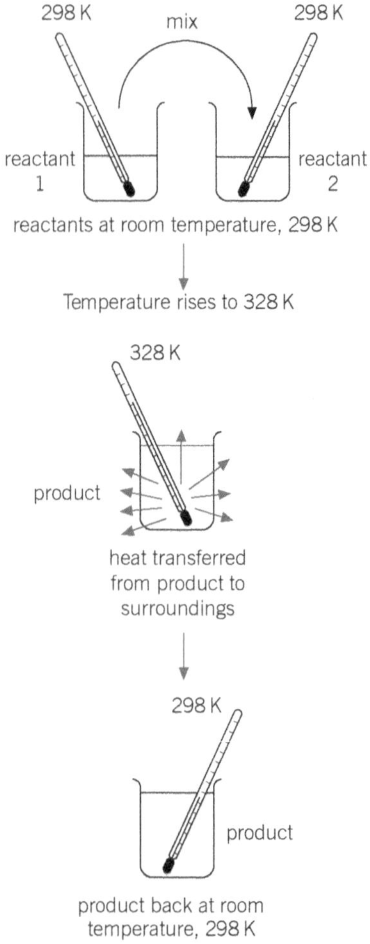

▲ **Figure 1** *A reaction giving out heat at 298 K*

The amount of heat given out or taken in by a reaction varies with the conditions – temperature, pressure, concentration of solutions, and so on. This means that you must state the conditions under which measurements are made. For example, you normally measure heat changes at constant atmospheric pressure.

Enthalpy change, ΔH

When you measure a heat change at constant pressure, it is called an **enthalpy change**.

Enthalpy has the symbol H so enthalpy changes are given the symbol ΔH. The Greek letter Δ (delta) is used to indicate a *change* in any quantity.

There are standard conditions for measuring enthalpy changes:

- pressure of 100 kPa (approximately normal atmospheric pressure)
- temperature of 298 K (around normal room temperature, 25 °C).

(The standard state of an element is the state in which it exists at 298 K and 100 kPa.)

When an enthalpy change is measured under standard conditions, it is written as $\Delta H^{\ominus}_{298}$ although usually the 298 is left out. $\Delta H^{\ominus}$ is pronounced 'delta H standard'.

It may seem strange to talk about measuring heat changes at a constant temperature because heat changes normally *cause* temperature changes. The way to think about this is to imagine the reactants at 298 K, see Figure 1. Mix the reactants and heat is produced (this is an exothermic reaction). This heat is given out to the surroundings.

A reaction is not thought of as being over until the *products* have *cooled back to 298 K*. The heat given out to the surroundings while the reaction mixture cools is the enthalpy change for the reaction, $\Delta H^{\ominus}$.

- In an exothermic reaction the products end up with less heat energy than the starting materials because they have lost heat energy when they heated up their surroundings. This means that ΔH is negative. It is therefore given a negative sign.

Some endothermic reactions that take place in aqueous solution absorb heat from the water and cool it down, for example, dissolving ammonium nitrate in water. Again you don't think of the reaction as being over until the *products* have *warmed up to the temperature at which they started*.

In this case the solution has to take in heat from the surroundings to do this. Unless you remember this, it can seem strange that a reaction that is absorbing heat, initially gets cold.

- In an endothermic reaction the products end up with more energy than the starting materials, so ΔH is positive. It is therefore given a positive sign.

Pressure affects the amount of heat energy given out by reactions that involve gases. If a gas is given out, some energy is required to push away the atmosphere. The greater the atmospheric pressure, the more energy is used for this. This means that less energy remains to be given out as heat by the reaction. This is why it is important to have a standard of pressure for measuring energy changes.

The physical states of the reactants and products

The physical states (gas, liquid, or solid) of the reactants and products also affect the enthalpy change of a reaction. For example, heat must be put in to change liquid to gas and is given out when a gas is changed to a liquid. This means that you must always include state symbols in your equations.

For example, hydrogen burns in oxygen to form water but there are two possibilities:

1 forming liquid water

$$H_2(g) + \tfrac{1}{2}O_2(g) \rightarrow H_2O(l) \qquad \Delta H\ -285.8\ \text{kJ mol}^{-1}$$

2 forming steam

$$H_2(g) + \tfrac{1}{2}O_2(g) \rightarrow H_2O(g) \qquad \Delta H\ -241.8\ \text{kJ mol}^{-1}$$

The difference in ΔH represents the amount of heat needed to turn one mole of water into steam.

Hint

Don't be confused by the different terms. *Heat* is a form of energy, so a heat change can also be described as an energy change. An *enthalpy change* is still an energy change, but it is measured under stated conditions of temperature and pressure.

Hint

One way of making sure that both reactants are at the same temperature is simply to leave them in the same room for some time.

Enthalpy level diagrams

Enthalpy level diagrams, sometimes called energy level diagrams, are used to represent enthalpy changes. They show the relative enthalpy levels of the reactants (starting materials) and the products. The vertical axis represents enthalpy, and the horizontal axis, the extent of the reaction. You are usually only interested in the beginning of the reaction, 100% reactants, and the end of the reaction, 0% reactants (and 100% products), so the horizontal axis is usually left without units.

Figure 2 shows a general enthalpy diagram for an exothermic reaction (the products have less enthalpy than the reactants) and Figure 3 shows an endothermic reaction (the products have more enthalpy than the reactants).

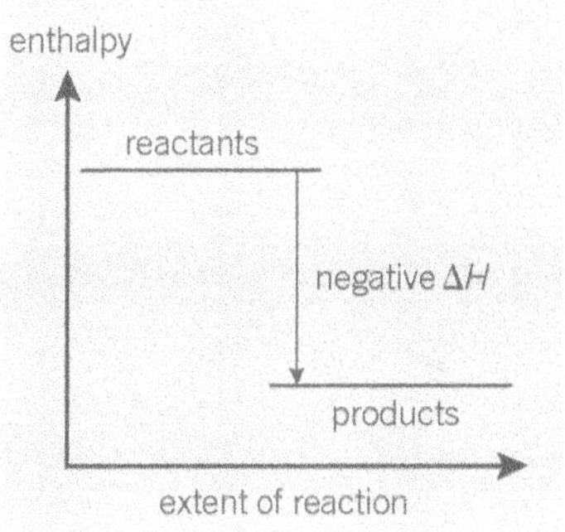

▲ **Figure 2** *Enthalpy diagram for an exothermic reaction*

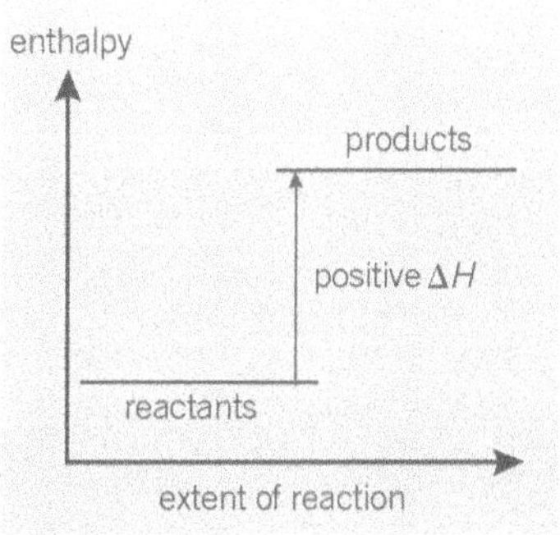

▲ **Figure 3** *Enthalpy diagram for an endothermic reaction*

Summary questions

1 Consider this reaction:

$$CH_4(g) + 2O_2(g) \rightarrow CO_2(g) + 2H_2O(l) \qquad \Delta H^{\ominus}_{298} = -890\ \text{kJ mol}^{-1}$$

a State what the symbol Δ means.

b State what the symbol H means.

c State what the 298 indicates.

d State what the minus sign indicates.

e Explain whether the reaction is exothermic or endothermic.

f Draw an enthalpy diagram to show the reaction.

4.3 Measuring enthalpy changes – calorimetry

Learning objectives:

- → Describe how enthalpy change is measured in a reaction.
- → Describe how you measure enthalpy changes more accurately.
- → Describe how you measure enthalpy changes in solution.

Specification reference: 3.1.4

Synoptic link

See Practical 2 on page 522.

Hint

The apparatus used to measure enthalpy changes is called a *calorimeter* and the process of measuring enthalpy changes is called *calorimetry*.

The general name for the enthalpy change for any reaction is the standard molar enthalpy change of reaction $\Delta H^{\ominus}$. It is measured in kilojoules per mole, $kJ\,mol^{-1}$ (molar means 'per mole'). You write a balanced symbol equation for the reaction and then find the heat change for the quantities in moles given by this equation.

For example, ΔH for $2NaOH + H_2SO_4 \rightarrow Na_2SO_4 + 2H_2O$ is the enthalpy change when two moles of NaOH react with one mole of H_2SO_4.

Standard enthalpies

Some commonly used enthalpy changes are given names, for example, the enthalpy change of formation $\Delta_f H^{\ominus}$ and the enthalpy change of combustion $\Delta_c H^{\ominus}$. Both of these quantities are useful when calculating enthalpy changes for reactions. In addition, $\Delta_c H^{\ominus}$s are relatively easy to measure for compounds that burn readily in oxygen. Their formal definitions are as follows:

The standard molar enthalpy of formation, $\Delta_f H^{\ominus}$, is the enthalpy change when one mole of substance is formed from its constituent elements under standard conditions, all reactants and products being in their standard states.

The standard molar enthalpy of combustion, $\Delta_c H^{\ominus}$, is the enthalpy change when one mole of substance is completely burnt in oxygen under standard conditions, all reactants and products being in their standard states.

Heat and temperature

Temperature is related to the *average* kinetic energy of the particles in a system. As the particles move faster, their average kinetic energy increases and the temperature goes up. But it doesn't matter how many particles there are, temperature is independent of the *number* present. Temperature is measured with a thermometer.

Heat is a measure of the *total* energy of all the particles present in a given amount of substance. It *does* depend on how much of the substance is present. The energy of every particle is included. So a bath of lukewarm water has much more heat than a red hot nail because there are so many more particles in it. Heat always flows from high to low temperature, so heat will flow from the nail into the bath water, even though the water has much more heat than the nail.

Measuring the enthalpy change of a reaction

The enthalpy change of a reaction is the heat given out or taken in as the reaction proceeds. There is no instrument that measures heat directly. To measure the enthalpy *change* you arrange for the heat to be transferred into a particular mass of a substance, often water. Then you need to know three things:

1. mass of the substance that is being heated up or cooled down
2. temperature change
3. specific heat capacity of the substance.

Hint

The words *heat* and *temperature* are often used to mean the same thing in daily conversation, but in science they are quite distinct and you must be clear about the difference.

The **specific heat capacity** c is the amount of heat needed to raise the temperature of 1 g of substance by 1 K. Its units are joules per gram per kelvin, or J g^{-1} K^{-1}. For example, the specific heat capacity of water is 4.18 J g^{-1} K^{-1}. This means that it takes 4.18 joules to raise the temperature of 1 gram of water by 1 kelvin. This is often rounded up to 4.2 J g^{-1} K^{-1}.

Then:

enthalpy change q = mass of substance m × specific heat capacity c × temperature change ΔT

The simple calorimeter or $q = mc\Delta T$

You can use the apparatus in Figure 1 to find the approximate enthalpy change when a fuel burns.

You burn the fuel to heat a known mass of water and then measure the temperature rise of the water. You assume that all the heat from the fuel goes into the water.

The apparatus used is called a **calorimeter** (from the Latin *calor* meaning heat).

Worked example: Working out the enthalpy change

The calorimeter in Figure 1 was used to measure the enthalpy change of combustion of methanol.

$$CH_3OH(l) + 1\frac{1}{2}O_2(g) \rightarrow CO_2(g) + 2H_2O(l)$$

0.32 g (0.01 mol) of methanol was burnt and the temperature of the 200.0 g of water rose by 4.0 K.

Heat change = $q = m \times c \times \Delta T$

$= 200.0 \times 4.2 \times 4.0 = 3360$ J

0.01 mol gives 3360 J

So 1 mol would give 336 000 J or 336 kJ

$\Delta_c H = -340$ kJ mol^{-1} (negative because heat is given out)

The simple calorimeter can be used to compare the $\Delta_c H$ values of a series of similar compounds because the errors will be similar for every experiment. However, you can improve the results by cutting down the heat loss, as shown in Figure 2, and reducing incomplete combustion by burning the fuel in oxygen rather than air.

The flame calorimeter

The flame calorimeter, shown in Figure 3 overleaf, is an improved version of the simple calorimeters used for measuring enthalpy changes of combustion. It incorporates the following features that are designed to reduce heat loss even further:

- the spiral chimney is made of copper
- the flame is enclosed
- the fuel burns in pure oxygen, rather than air.

Hint

The size of a kelvin is the same as the size of a degree Celsius. Only the starting point of the scale is different. A temperature *change* is numerically the same whether it is measured in Celsius or kelvin. To convert °C to K add 273.

Hint

Chemists normally report enthalpy changes in kJ mol^{-1}.

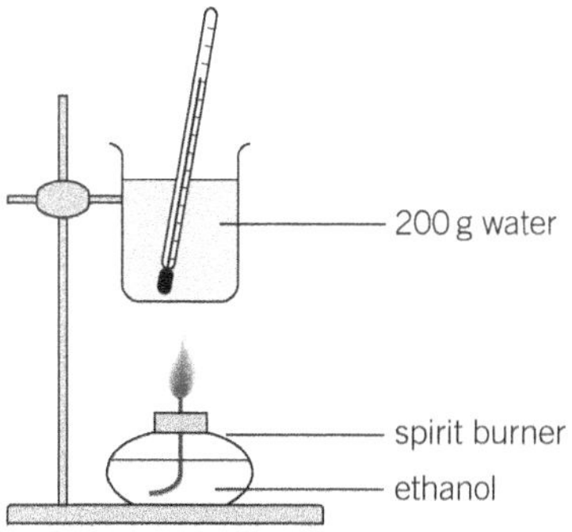

▲ **Figure 1** *A simple calorimeter*

Maths link

200.0 is 4 significant figures (s.f.), 0.32 is 2 significant figures, 4.0 is 2 significant figures. So you can only give the answer to 2 s.f. You round it up rather than down as 336 is nearer to 340 than 330. See Section 8, Mathematical skills, if you are not sure about significant figures.

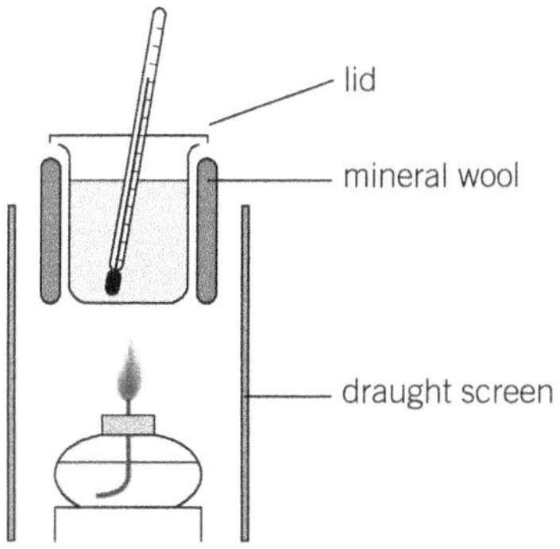

▲ **Figure 2** *An improved calorimeter*

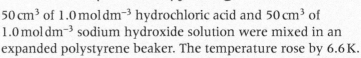

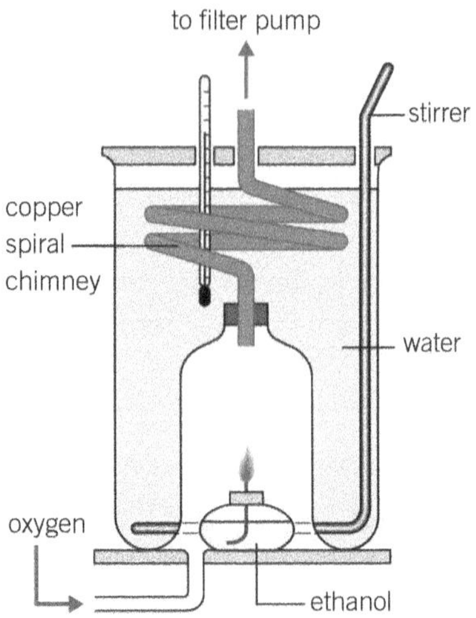

▲ **Figure 3** *A flame calorimeter*

Measuring enthalpy changes of reactions in solution

It is relatively easy to measure heat changes for reactions that take place in solution. The heat is generated in the solutions themselves and only has to be kept in the calorimeter. Expanded polystyrene beakers are often used for the calorimeters. These are good insulators (this reduces heat loss through their sides) and they have a low heat capacity so they absorb very little heat. The specific heat capacity of dilute solutions is usually taken to be the same as that of water, $4.2\,J\,g^{-1}\,K^{-1}$ (or more precisely $4.18\,J\,g^{-1}\,K^{-1}$).

Neutralisation reactions

Neutralisation reactions in solution are exothermic – they give out heat. When an acid is neutralised by an alkali the equation is:

$$\text{acid} + \text{alkali} \rightarrow \text{salt} + \text{water}$$

To find an enthalpy change for a reaction, you use the quantities in moles given by the balanced equation. For example, to find the molar enthalpy change of reaction for the neutralisation of hydrochloric acid by sodium hydroxide, the heat given out by the quantities in the equation needs to be found:

$HCl(aq)$	+	$NaOH(aq)$	→	$NaCl(aq)$	+	$H_2O(l)$
hydrochloric acid		sodium hydroxide		sodium chloride		water
1 mol		1 mol		1 mol		1 mol

Hint

Remember to use the *total* volume of the mixture, $100\,cm^3$. A common mistake is to use $50\,cm^3$.

Worked example: Enthalpy change for a reaction

$50\,cm^3$ of $1.0\,mol\,dm^{-3}$ hydrochloric acid and $50\,cm^3$ of $1.0\,mol\,dm^{-3}$ sodium hydroxide solution were mixed in an expanded polystyrene beaker. The temperature rose by 6.6 K.

The total volume of the mixture is $100\,cm^3$. This has a mass of approximately 100 g because the density of water and of dilute aqueous solutions is approximately $1\,g\,cm^{-3}$.

$$\text{enthalpy change } q = \text{mass of water } m \times \text{specific heat capacity of solution } c \times \text{temperature change } \Delta T$$

$$q = m \times c \times \Delta T$$

$$= 100 \times 4.2 \times 6.6 = 2772\,J$$

$$\text{number of moles of acid (and also of alkali) } n = \frac{\text{concentration } c\ (\text{mol dm}^{-3}) \times \text{volume } V\ (\text{cm}^3)}{1000}$$

$$= 1.0 \times \frac{50}{1000} = 0.05\,mol$$

$$\text{so 1 mol would give } \frac{2772}{0.05}\,J = 55\,440\,J = 55.44\,kJ$$

$$\Delta H = -55.44\,kJ\,mol^{-1}$$

$$\Delta H = -55\,kJ\,mol^{-1} \text{ (to 2 s.f.)}$$

The sign of ΔH is negative because heat is given out.

Displacement reactions

A metal that is more reactive than another will displace the less reactive one from a compound. If the compound will dissolve in water, this reaction can be investigated using a polystyrene beaker as before.

For example, zinc will displace copper from a solution of copper sulfate. The reaction is exothermic.

$$\underset{1\,\text{mol}}{Zn(s)} + \underset{1\,\text{mol}}{CuSO_4(aq)} \rightarrow \underset{1\,\text{mol}}{ZnSO_4(aq)} + \underset{1\,\text{mol}}{Cu(s)}$$

From the equation one mole of zinc reacts with one mole of copper sulfate.

Worked example: Enthalpy change in a displacement reaction

0.50 g of zinc was added to 25.0 cm^3 of 0.20 mol dm^{-3} copper sulfate solution. The temperature rose by 10 K.

$$q = m \times c \times \Delta T$$
$$= 25 \times 4.2 \times 10 = 1050\,J$$

A_r zinc = 65.4, so 0.50 g of zinc is $\frac{0.50}{65.4}$ moles = 0.0076 moles

number of moles of copper sulfate in solution = $\frac{c \times V}{1000}$

where c is concentration in mol dm^{-3} and V is volume in cm^3

$$= 0.20 \times \frac{25.0}{1000} = 0.005\ \text{mol}$$

This means that the zinc was in excess; 0.005 mol of each reactant has taken part in the reaction, leaving some unreacted zinc behind.

Therefore, 1 mole of zinc would produce $\frac{1050}{0.005}$ J = 210 000 J.

So, ΔH for this reaction is $-210\ kJ\,mol^{-1}$ (to 2 s.f.).

The sign of ΔH is negative because heat is given out.

Study tip

Rearrange $q = m \times c \times \Delta T$ to find any of the quantities in terms of the others:

$$\Delta T = \frac{q}{mc}$$

$$m = \frac{q}{c}\Delta T$$

$$c = \frac{q}{m}\Delta T$$

Allowing for heat loss

Although expanded polystyrene cups are good insulators, some heat will still be lost from the sides and top leading to low values for enthalpy changes measured by this method. This can be allowed for by plotting a cooling curve. As an example, the measurement of the heat of neutralisation of hydrochloric acid and sodium hydroxide is repeated using a cooling curve.

Before the experiment, all the apparatus and both solutions are left to stand in the laboratory for some time. This ensures that they all reach the same temperature, that of the laboratory itself.

Then proceed as follows:

1 Place 50 cm^3 of 1.0 mol dm^{-3} hydrochloric acid in one polystyrene cup and 50 cm^3 of 1.0 mol dm^{-3} sodium hydroxide solution in another.
2 Using a thermometer that reads to 0.1 °C, take the temperature of each solution every 30 seconds for four minutes to confirm

▲ **Figure 4** *Polystyrene beakers make good calorimeters because they are good insulators and have low heat capacities*

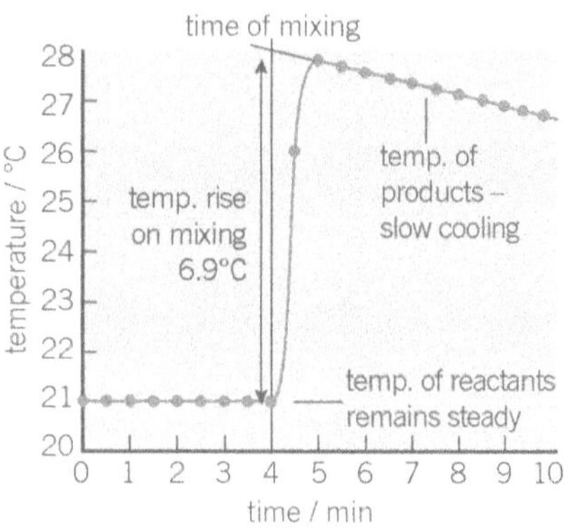

▲ **Figure 5** *Graph to show temperature as a neutralisation reaction proceeds*

that both solutions remain at the same temperature, that of the laboratory. A line of 'best fit' is drawn through these points. It is likely there will be very small variations around the line of best fit, indicating random errors.

3 Now pour one solution into the other and stir, continuing to record the temperature every 30 seconds for a further six minutes.

The results are shown on the graph in Figure 5. The experiment can also be done using an electronic temperature sensor and data logging software to plot the graph directly.

On mixing, the temperature rises rapidly as the reaction gives out heat, and then drops slowly and regularly as heat is lost from the polystyrene cup. To find the best estimate of the temperature immediately after mixing, you draw the best straight line through the graph points after mixing and extrapolate back to the time of mixing. This gives a temperature rise of 6.9 °C.

The calculation is as before.

$$q = m \times c \times \Delta T = 100 \times 4.2 \times 6.9 = 2898\,\text{J}$$

The number of moles of acid (and alkali) was 0.05 mol (as before).

So 1 mol would give $\frac{2898}{0.05}$ J = 57 960 J = 57.96 kJ

$$\Delta_{neut}H = -58\,\text{kJ}\,\text{mol}^{-1} \text{ (to 2 s.f.)}$$

The sign of ΔH is negative because heat is given out.

Summary questions

1 0.74 g (0.010 mol) of propanoic acid was burnt in the simple calorimeter like that described above for the combustion of methanol. The temperature rose by 8.0 K. Calculate the value this gives for the enthalpy change of combustion of propanoic acid.

2 50.0 cm^3 of 2.00 mol dm^{-3} sodium hydroxide and 50.0 cm^3 of 2.00 mol dm^{-3} hydrochloric acid were mixed in an expanded polystyrene beaker. The temperature rose by 11.0 K.

- **a** Calculate ΔH for the reaction.
- **b** Describe how this value will compare with the accepted value for this reaction.
- **c** Explain your answer to **b**.

3 Consider the expression $q = mc\Delta T$

- **a** State what the term q represents.
- **b** State what the term m represents.
- **c** State what the term c represents.
- **d** State what the term ΔT represents.

4.4 Hess's law

The enthalpy changes for some reactions cannot be measured directly. To find these you use an indirect approach. Chemists use enthalpy changes that they can measure to work out enthalpy changes that they cannot. In particular, it is often easy to measure enthalpies of combustion. To do this, chemists use Hess's law, first stated by Germain Hess, a Swiss-born Russian chemist, born in 1802.

Learning objectives:

- → Describe how to find enthalpy changes that cannot be measured directly.

Specification reference: 3.1.4

Hess's law

Hess's law states that the enthalpy change for a chemical reaction is the same, whatever route is taken from reactants to products.

This is a consequence of a more general scientific law, the Law of Conservation of Energy, which states that energy can never be created or destroyed. So, provided the starting and finishing points of a process are the same, the energy change must be the same. If not, energy would have been created or destroyed.

Using Hess's law

To see what Hess's law means, look at the following example where ethyne, C_2H_2, is converted to ethane, C_2H_6, by two different routes. How can we find the enthalpy of reaction?

Route 1: The reaction takes place directly – ethyne reacts with two moles of hydrogen to give ethane.

$$\underset{\text{ethyne}}{C_2H_2(g)} + 2H_2(g) \rightarrow \underset{\text{ethane}}{C_2H_6(g)} \qquad \Delta H_1 = ?$$

Route 2: The reaction takes place in two stages.

a Ethyne, C_2H_2, reacts with one mole of hydrogen to give ethene, C_2H_4.

$$\underset{\text{ethyne}}{C_2H_2(g)} + H_2(g) \rightarrow \underset{\text{ethene}}{C_2H_4(g)} \qquad \Delta H_2 = -176\,\text{kJ}\,\text{mol}^{-1}$$

b Ethene, C_2H_4, then reacts with a second mole of hydrogen to give ethane, C_2H_6.

$$\underset{\text{ethene}}{C_2H_4(g)} + H_2(g) \rightarrow \underset{\text{ethane}}{C_2H_6(g)} \qquad \Delta H_3 = -137\,\text{kJ}\,\text{mol}^{-1}$$

Hess's law tells us that the total energy change is the same whichever route you take – direct or via ethene (or, in fact, by any other route). You can show this on a diagram called a **thermochemical cycle**. The thermochemical cycle for converting ethyne to ethane is shown overleaf.

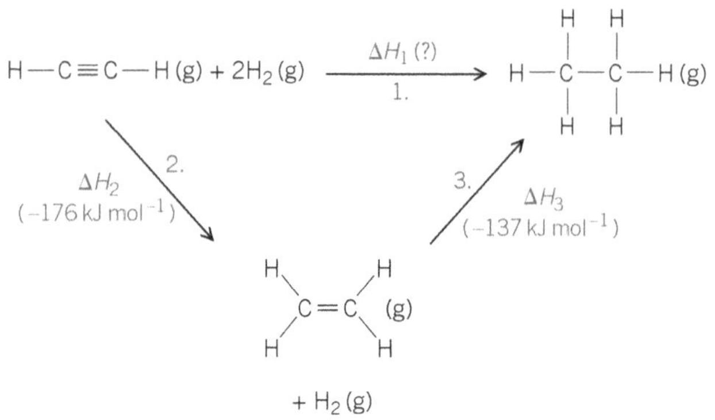

Hess's law means that: $\Delta H_1 = \Delta H_2 + \Delta H_3$

The actual figures are: $\Delta H_2 = -176\,\text{kJ}\,\text{mol}^{-1}$

$\Delta H_3 = -137\,\text{kJ}\,\text{mol}^{-1}$

So $\Delta H_1 = (-176) + (-137) = -313$ kJ mol^{-1}

This method of calculating ΔH_1 is fine if you know the enthalpy changes for the other two reactions. There are certain enthalpy changes that can be looked up for a large range of compounds. These include the enthalpy change of formation, $\Delta_f H^{\ominus}$, and enthalpy change of combustion, $\Delta_c H^{\ominus}$. In practice, many values of $\Delta_f H^{\ominus}$ are calculated from $\Delta_c H^{\ominus}$ via Hess's law cycles.

Using the enthalpy changes of formation $\Delta_f H^{\ominus}$

The enthalpy of formation, $\Delta_f H^{\ominus}$, is the enthalpy change when one mole of compound is formed from its constituent elements under standard conditions, all reactants and products being in their standard states.

Another theoretical way to convert ethyne to ethane could be via the elements carbon and hydrogen.

- Ethyne is first converted to its elements, carbon and hydrogen. This is the reverse of formation and the enthalpy change is the *negative* of the enthalpy of formation. This is a general rule. The reverse of a reaction has the negative of its ΔH value. It is in fact a consequence of Hess's law.
- Then the carbon and hydrogen react to form ethane. This is the enthalpy of formation for ethane.

Hess's law tells us that $\Delta H_1 = \Delta H_4 + \Delta H_5$

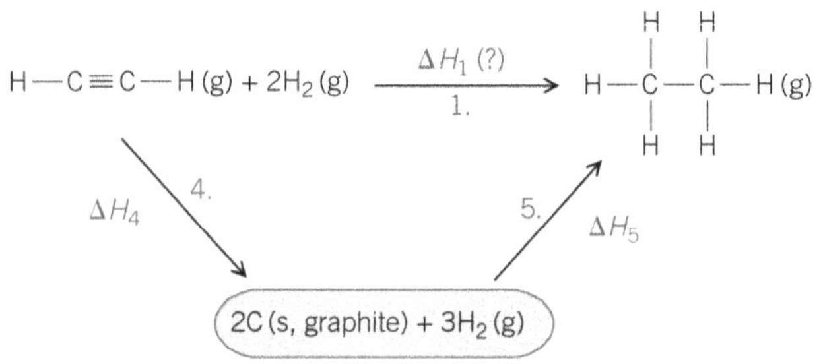

Hint

Graphite is the most stable form of carbon (another form is diamond). It has a special state symbol (s, graphite).

ΔH_5 is the enthalpy of formation, $\Delta_f H^\ominus$, of ethane whilst reaction 4 is the reverse of the formation of ethyne.

The values you need are: $\Delta_f H^\ominus(C_2H_2) = +228\,kJ\,mol^{-1}$

and $\Delta_f H^\ominus(C_2H_6) = -85\,kJ\,mol^{-1}$

So $\Delta H_4 = -228\,kJ\,mol^{-1}$

(remember to change the sign)

$\Delta H_5 = -85\ kJ\ mol^{-1}$

Therefore $\Delta H_1 = -228 + -85 = -313\,kJ\,mol^{-1}$

This was the result you got from the previous method, as you should expect from Hess's law.

Notice that in reaction 4 there are two moles of hydrogen 'spare' as only one of the three moles of hydrogen is involved. These two moles of hydrogen remain in their standard states and so no enthalpy change is invoved.

$C_2H_2(g) \rightarrow 2C(s, graphite) + H_2(g)$ is the reaction you are considering, but you have:

$$C_2H_2(g) + 2H_2(g) \rightarrow 2C(s, graphite) + 3H_2(g)$$

However, this makes no difference. The 'extra' hydrogen is *not* involved in the reaction and it does not affect ΔH.

Summary questions

1 Use the values of $\Delta_f H^\ominus$ in the table to calculate $\Delta H^\ominus$ for each of the reactions below using a thermochemical cycle.

a $CH_3COCH_3(l) + H_2(g) \rightarrow CH_3CH(OH)CH_3(l)$

b $C_2H_4(g) + Cl_2(g) \rightarrow C_2H_4Cl_2(l)$

c $C_2H_4(g) + HCl(g) \rightarrow C_2H_5Cl(l)$

d $Zn(s) + CuO(s) \rightarrow ZnO(s) + Cu(s)$

e $Pb(NO_3)_2(s) \rightarrow PbO(s) + 2NO_2(g) + \frac{1}{2}O_2(g)$

Compound	$\Delta_f H^\ominus$/ kJ mol^{-1}
$CH_3COCH_3(l)$	−248
$CH_3CH(OH)CH_3(l)$	−318
$C_2H_4(g)$	+52
$C_2H_4Cl_2(l)$	−165
$C_2H_5Cl(l)$	−137
$HCl(g)$	−92
$CuO(s)$	−157
$ZnO(s)$	−348
$Pb(NO_3)_2(s)$	−452
$PbO(s)$	−217
$NO_2(g)$	+33

4.5 Enthalpy changes of combustion

Learning objectives:

- Describe how the enthalpy change of combustion can be used to find the enthalpy change of a reaction.

Specification reference: 3.1.4

Study tip

Remember to multiply by the number of moles of reagents involved in each step.

The enthalpy change of combustion, $\Delta_c H^\ominus$, is the enthalpy change when one mole of substance is completely burnt in oxygen under standard conditions.

Thermochemical cycles using enthalpy changes of combustion

Look again at the thermochemical cycle used to find $\Delta H^\ominus$ for the reaction between ethyne and hydrogen to form ethane.

$$C_2H_2(g) + 2H_2(g) \rightarrow C_2H_6(g)$$

This time use enthalpy changes of combustion. In this case you can go via the combustion products of the three substances – carbon dioxide and water.

All three substances – ethyne, hydrogen, and ethane – burn readily. This means their enthalpy changes of combustion can be easily measured. The thermochemical cycle is:

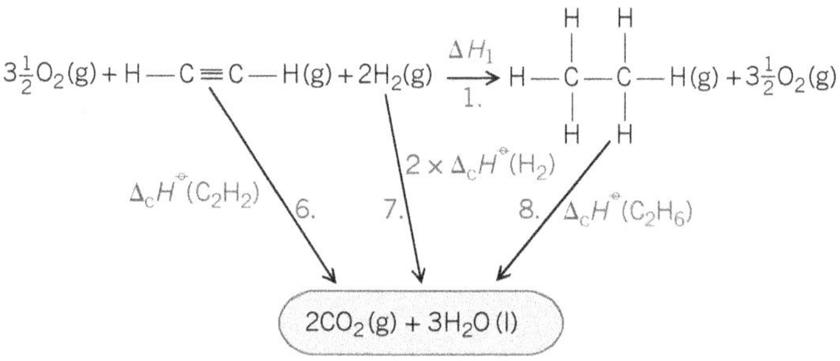

Putting in the values:

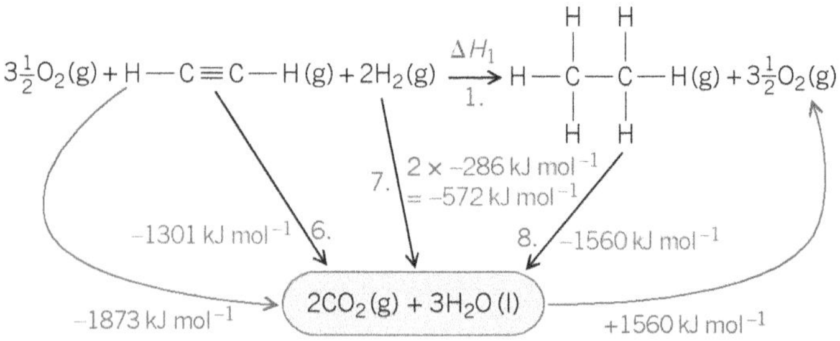

Hint

- Both reactions 6 and 7 have to occur to get from the starting materials to the combustion products. Do not forget the hydrogen.
- In this case there are two moles of hydrogen so you need *twice* the value of $\Delta_c H^\ominus$ which refers to one mole of hydrogen.

$\Delta_c H^\ominus(C_2H_2) = -1301\ kJ\,mol^{-1}$

$\Delta_c H^\ominus(H_2) = -286\ kJ\,mol^{-1}$

$\Delta_c H^\ominus(C_2H_6) = -1560\ kJ\,mol^{-1}$

To get the enthalpy change for reaction 1 you must go round the cycle in the direction of the red arrows. This means reversing reaction 8 so you must change its sign.

So $\Delta H_1 = -1873 + 1560\ kJ\,mol^{-1}$

$\Delta H_1 = -313\ kJ\,mol^{-1}$ once again, the same answer as before

Notice that in reaction 1 there are $3\frac{1}{2}$ moles of oxygen on either side of the equation. They take no part in reaction 1 and do not affect the value of ΔH.

Finding $\Delta_f H^\ominus$ from $\Delta_c H^\ominus$

Enthalpy changes of formation of compounds are often difficult or impossible to measure directly. This is because the reactants often do not react directly to form the compound that you are interested in.

For example, the following equation represents the formation of ethanol from its elements.

$$2C(s, \text{graphite}) + 3H_2(g) + \tfrac{1}{2}O_2(g) \rightarrow C_2H_5OH(l)$$

This does not take place. However, all the species concerned will readily burn in oxygen so their enthalpy changes of combustion can be measured. The thermochemical cycle you need is:

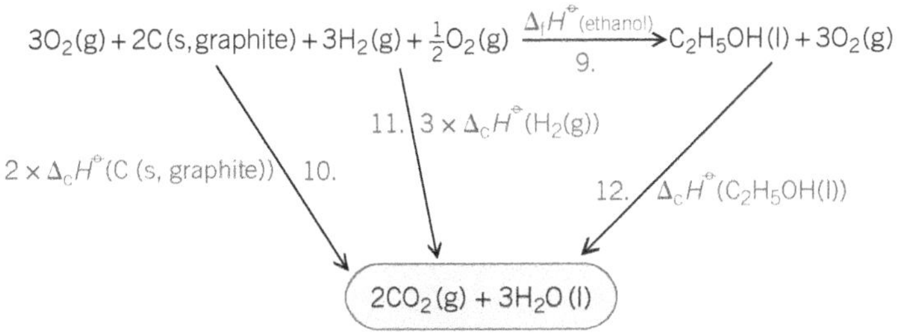

Putting in the values:

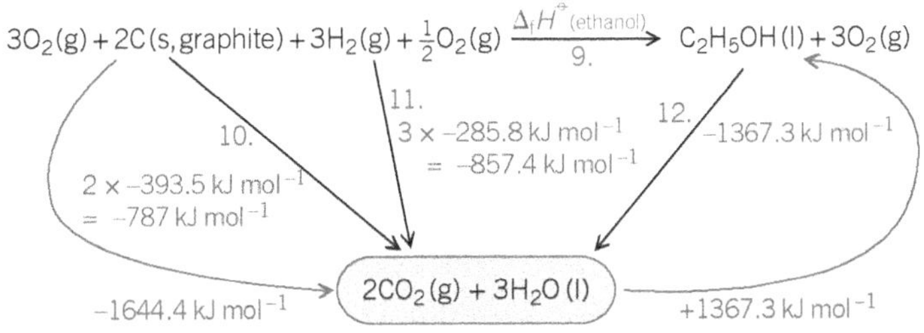

Note that in reaction 9 there are three moles of oxygen on either side of the equation that take no part in the reaction. This means that they do not affect the value of ΔH.

Note also that $\Delta_c H^\ominus$ (C(s, graphite)) is the same as $\Delta_f H^\ominus$ (CO$_2$(g)) and $\Delta_c H^\ominus$(H$_2$(g)) is the same as $\Delta_f H^\ominus$(H$_2$O(l)).

To get the enthalpy change for reaction 9, you must go round the cycle in the direction of the red arrows. This means reversing reaction 12 so you must change its sign.

So, $\Delta H_9 = -1644.4 + 1367.3 \, \text{kJ mol}^{-1} = -277.1 \, \text{kJ mol}^{-1}$

So, $\Delta_f H^\ominus$(C$_2$H$_5$OH(l)) $= -277.1 \, \text{kJ mol}^{-1}$

Hint

The values we need are:

$\Delta_c H^\ominus$(C(s, graphite)) $= -393.5 \, \text{kJ mol}^{-1}$

$\Delta_c H^\ominus$(H$_2$(g)) $= -285.8 \, \text{kJ mol}^{-1}$

$\Delta_c H^\ominus$(C$_2$H$_5$OH(l)) $= -1367.3 \, \text{kJ mol}^{-1}$

Summary questions

1 Calculate ΔH^\ominus for the reaction by thermochemical cycles:

H–C(H)(H)–C(=O)(H) (l) + H$_2$(g) ⟶ H–C(H)(H)–C(H)(H)–O–H(l)

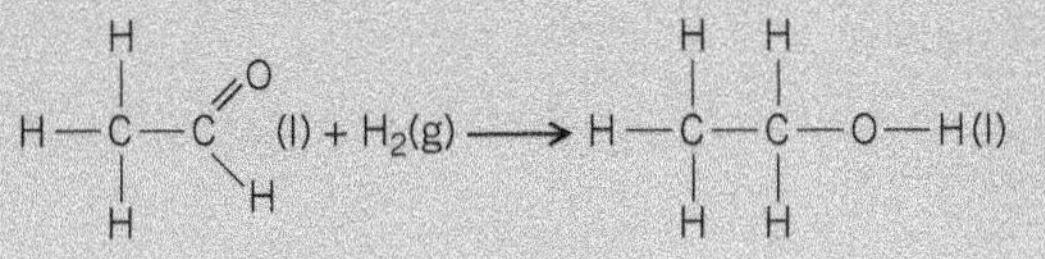

a via $\Delta_f H^\ominus$ values b via $\Delta_c H^\ominus$ values

Compound	ΔH_f^\ominus/ kJ mol^{-1}	ΔH_c^\ominus/ kJ mol^{-1}
CH$_3$CHO	−192	−1166
H$_2$	−	−286
CH$_3$CH$_2$OH	−277	−1367

4.6 Representing thermochemical cycles

Learning objectives:

- → Describe what an enthalpy diagram is.
- → State what is used as the zero for enthalpy changes.

Specification reference: 3.1.4

You can use **enthalpy diagrams** rather than thermochemical cycles to represent the enthalpy changes in chemical reactions. These show the energy (enthalpy) levels of the reactants and products of a chemical reaction on a vertical scale, so you can compare their energies. If a substance is of lower energy than another, you say it is energetically more stable.

The enthalpy of elements

So far you have considered enthalpy *changes*, not absolute values. When drawing enthalpy diagrams you need a zero to work from. You can then give absolute numbers to the enthalpies of different substances.

The enthalpies of all elements in their standard states (i.e., the states in which they exist at 298 K and 100 kPa) are taken as zero. (298 K and 100 kPa are approximately normal room conditions.)

This convention means that the standard state of hydrogen, for example, is H_2 and not H, because hydrogen exists as H_2 at room temperature and pressure.

Pure carbon can exist in a number of forms at room temperature including graphite, diamond, and buckminsterfullerene (buckyballs). These are called **allotropes**. Graphite is the most stable of these and is taken as the standard state of carbon. It is given the special state symbol (s, graphite), so C(s, graphite) represents graphite.

Thermochemical cycles and enthalpy diagrams

Here are two examples of reactions, with their enthalpy changes presented both as thermochemical cycles and as enthalpy diagrams.

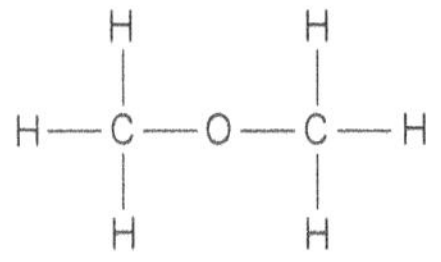

methoxymethane

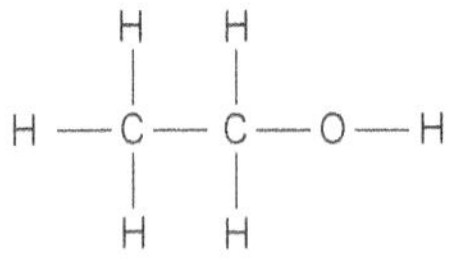

ethanol

▲ **Figure 1** *Isomers of C_2H_6O*

Example 1

What is $\Delta H^\ominus$ for the change from methoxymethane to ethanol? (The compounds are a pair of isomers – they have the same formula but different structures, see Figure 1.)

The standard molar enthalpy changes of formation of the two compounds are:

CH_3OCH_3 $\quad \Delta_f H^\ominus = -184\,kJ\,mol^{-1}$

C_2H_5OH $\quad \Delta_f H^\ominus = -277\,kJ\,mol^{-1}$

Using a thermochemical cycle

The following steps are shown in red on the thermochemical cycle.

1. Write an equation for the reaction.
2. Write down the elements in the two compounds with the correct quantities of each.
3. Put in the $\Delta_f H^\ominus$ values with arrows showing the direction – *from* elements *to* compounds.
4. Put in the arrows to go from starting materials to products via the elements (the red arrows).

5 Reverse the sign of $\Delta_f H^\ominus$ if the red arrow is in the opposite direction to the black arrow.
6 Go round the cycle in the direction of the red arrows and add up the $\Delta H^\ominus$ values as you go.

Hess's law states that this is the same as $\Delta H^\ominus$ for the direct reaction.

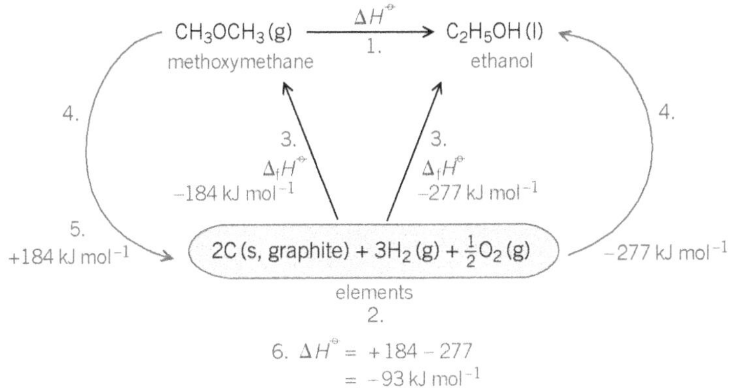

▲ **Figure 2** *Thermochemical cycle for the formation of ethanol from methoxymethane*

Using an enthalpy diagram

The following steps are shown in red on the enthalpy diagram.

1 Draw a line at level 0 to represent the elements.
2 Look up the values of $\Delta_f H^\ominus$ for each compound and enter these on the enthalpy diagrams, taking account of the signs – negative values are below 0, positive values are above.
3 Find the difference in levels between the two compounds. This represents the difference in their enthalpies.
4 $\Delta H^\ominus$ is the difference in levels *taking account of the direction of change.* Up is positive and down is negative. From methoxymethane to ethanol is *down* so the sign is negative. From ethanol to methoxymethane the sign of $\Delta H^\ominus$ would be positive.

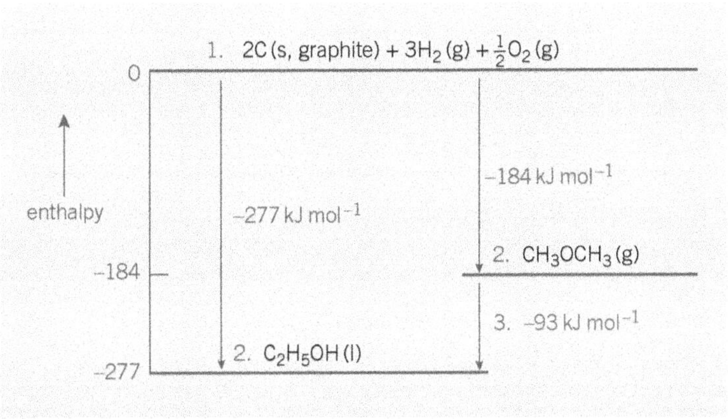

▲ **Figure 3** *The enthalpy diagram for the formation of ethanol from methoxymethane*

Notice how the enthalpy level diagram makes it much clearer than the thermochemical cycle that ethanol has less energy than methoxymethane. This means that it is the more energetically stable compound. The values of $\Delta H^\ominus$ for the reaction are the same whichever method you use.

Study tip

Remember that you do not need to draw these diagrams accurately to scale, but a rough scale is important to ensure that the relative levels are correct.

Hint

You can use a short cut to save drawing an enthalpy diagram or a thermochemical cycle. The enthalpy change of a reaction is the sum of the enthalpies of formation of all the products minus the sum of the enthalpies of formation of all the reactants. In this example

$\Delta H^{\ominus} = -314 - (-46 + (-92))$
$= -314 - (-138)$
$= -176\ \text{kJ mol}^{-1}$.

If you use this short cut, you must be *very* careful of the signs.

Example 2

To find $\Delta H^{\ominus}$ for the reaction $NH_3(g) + HCl(g) \rightarrow NH_4Cl(s)$

The standard molar enthalpy changes of formation of the compounds are:

NH_3	$\Delta_f H^{\ominus} = -46\ \text{kJ mol}^{-1}$
HCl	$\Delta_f H^{\ominus} = -92\ \text{kJ mol}^{-1}$
NH_4Cl	$\Delta_f H^{\ominus} = -314\ \text{kJ mol}^{-1}$

Using a thermochemical cycle

The thermochemical cycle for the formation of ammonium chloride is shown in Figure 4.

1 Write an equation for the reaction.
2 Write down the elements that make up the two compounds with the correct quantities of each.
3 Put in the $\Delta_f H^{\ominus}$ values with arrows showing the direction, that is, from elements to compounds.
4 Put in the arrows going from the starting materials to products via the elements (the red arrows).
5 Reverse the sign of $\Delta_f H^{\ominus}$ if the red arrow is in the opposite direction to the black arrow(s).
6 Go round the cycle in the direction of the red arrows and add up the values of $\Delta H^{\ominus}$ as you go.

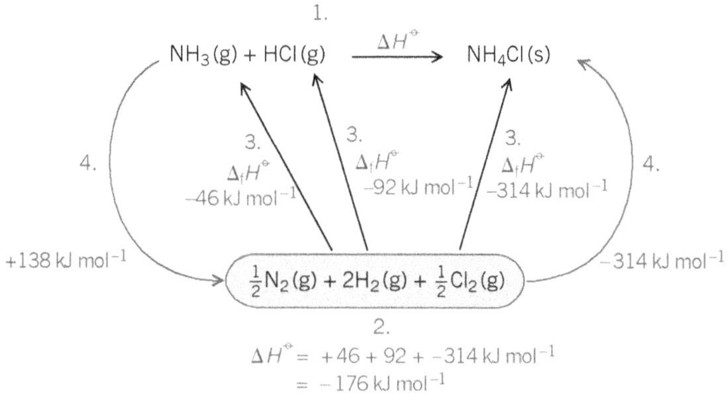

▲ **Figure 4** *Thermochemical cycle for the formation of ammonium chloride*

Using an enthalpy diagram

The following steps are shown in red on the enthalpy diagram.

1 Draw a line at level 0 to represent the elements.
2 Draw in NH_4Cl at the enthalpy 314 kJ mol^{-1} below this.
3 Draw a line representing ammonia 46 kJ mol^{-1} below the level of the elements. (There is still $\frac{1}{2}H_2$ and $\frac{1}{2}Cl_2$ left unused.)
4 Draw a line 92 kJ mol^{-1} below ammonia. This represents hydrogen chloride.
5 Find the difference in levels between the $(NH_3 + HCl)$ line and the NH_4Cl one. This represents $\Delta H^{\ominus}$ for the reaction. As the change from $(NH_3 + HCl)$ to NH_4Cl is down, $\Delta H^{\ominus}$ must be negative.

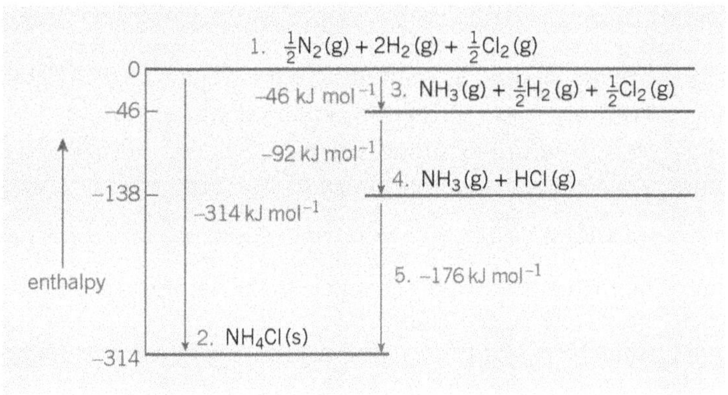

Notice how the enthalpy level diagram makes it much clearer than the thermochemical cycle that ammonium chloride is more energetically stable than the gaseous mixture of ammonia and hydrogen chloride. This is part of the reason why ammonia and hydrogen chloride react readily to form ammonium chloride. The values of $\Delta H^{\ominus}$ for the reaction are the same whichever method you use.

What would be the enthalpy change when solid ammonium chloride decomposes into the gases ammonia and hydrogen chloride?

+176 kJ mol^{-1}

Summary questions

1 Use the values of $\Delta_f H^{\ominus}$ in the table to calculate $\Delta H^{\ominus}$ for each of the reactions below using enthalpy diagrams.

a $CH_3COCH_3(l) + H_2(g) \rightarrow CH_3CH(OH)CH_3(l)$

b $C_2H_4(g) + Cl_2(g) \rightarrow C_2H_4Cl_2(l)$

c $C_2H_4(g) + HCl(g) \rightarrow C_2H_5Cl(l)$

d $Zn(s) + CuO(s) \rightarrow ZnO(s) + Cu(s)$

e $Pb(NO_3)_2(s) \rightarrow PbO(s) + 2NO_2(g) + \frac{1}{2}O_2(g)$

Compound	$\Delta_f H^{\ominus}$/ kJ mol^{-1}
$CH_3COCH_3(l)$	−248
$CH_3CH(OH)CH_3(l)$	−318
$C_2H_4(g)$	+52
$C_2H_4Cl_2(l)$	−165
$C_2H_5Cl(l)$	−137
$HCl(g)$	−92
$CuO(s)$	−157
$ZnO(s)$	−348
$Pb(NO_3)_2(s)$	−452
$PbO(s)$	−217
$NO_2(g)$	+33

4.7 Bond enthalpies

Learning objectives:

- → State what the definition of a bond enthalpy is.
- → Describe how mean bond enthalpies are worked out from given data.
- → Demonstrate how bond enthalpies are used in calculations.

Specification reference: 3.1.4

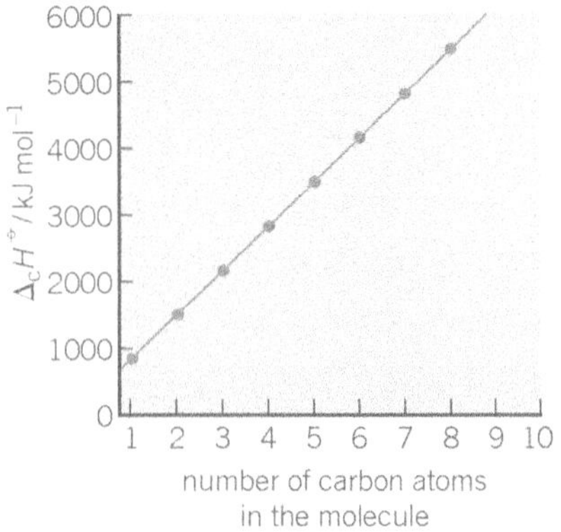

▲ **Figure 1** *$\Delta_c H^\ominus$ plotted against the number of carbon atoms in the alkane*

Hint

- If the bonds in the methane are broken one at a time, the energy required is not the same for each bond.
- The value of +416 kJ mol^{-1} is the C—H bond energy in methane. The value in other compounds will vary slightly. The average over many compounds is +413 kJ mol^{-1}.
- All mean bond energies are positive because we have to put energy in to break bonds – they are endothermic processes.

$\Delta_c H^\ominus$ is the enthalpy change of combustion. If you plot $\Delta_c H^\ominus$ against the number of carbon atoms in the molecule, for straight chain alkanes, you get a straight line graph, see Figure 1. For methane (with one carbon), $\Delta_c H^\ominus$ is the enthalpy change for:

$$CH_4(g) + 2O_2(g) \rightarrow CO_2(g) + 2H_2O(l)$$

The straight line means that $\Delta_c H^\ominus$ changes by the same amount for each extra carbon atom in the chain.

Each alkane differs from the previous one by one CH_2 group, that is, there is one extra C—C bond in the molecule and two extra C—H bonds. This suggests that you can assign a definite amount of energy to a particular bond. This is called the bond enthalpy.

Bond enthalpies

You have to put in energy to break a covalent bond – this is an endothermic change. **Bond dissociation enthalpy** is defined as the enthalpy change required to break a covalent bond with all species in the gaseous state. The same amount of energy is given out when the bond is formed – this is an exothermic change. However, the same bond, for example C—H, may have slightly different bond enthalpies in different molecules, but you usually use the average value. This value is called the **mean bond enthalpy** (often called the bond energy). The fact that you get out the same amount of energy when you make a bond, as you put in to break it, is an example of Hess's law.

As mean bond enthalpies are averages, calculations using them for specific compounds will only give approximate answers. However, they are useful, and quick and easy to use. Mean bond enthalpies have been calculated from Hess's law cycles. They can be looked up in data books and databases.

The H—H bond energy is the energy required to separate the two atoms in a hydrogen molecule in the gas phase into separate gaseous atoms.

$$H_2(g) \rightarrow 2H(g) \quad \Delta H^\ominus = +436\,\text{kJ mol}^{-1}$$

The C—H mean bond energy in methane is one quarter of the energy for the following process, in which four bonds are broken.

$$CH_4(g) \rightarrow C(g) + 4H(g) \quad \Delta H^\ominus = +1664\,\text{kJ mol}^{-1}$$

So the mean (or average) C—H bond energy in methane $= \dfrac{1664}{4}$

$= +416\,\text{kJ mol}^{-1}$

Using mean bond enthalpies to calculate enthalpy changes of reaction

You can use mean bond enthalpies to work out the enthalpy change of reactions, for example:

$$\underset{\text{ethane}}{C_2H_6(g)} + \underset{\text{chlorine}}{Cl_2(g)} \rightarrow \underset{\text{chloroethane}}{C_2H_5Cl(g)} + \underset{\text{hydrogen chloride}}{HCl(g)}$$

The mean bond enthalpies you will need for this example are given in Table 1.

▼ Table 1 *Mean bond enthalpies*

Bond	Bond enthalpy / $kJ\,mol^{-1}$
C—H	413
C—C	347
Cl—Cl	243
C—Cl	346
Cl—H	432
Br—Br	193
Br—H	366
C—Br	285

The steps are as follows:

1 First draw out the molecules and show all the bonds. (Formulae drawn showing all the bonds are called displayed formulae.)

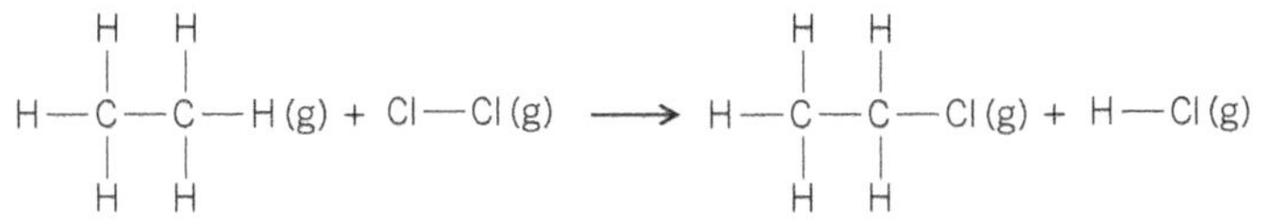

2 Now imagine that all the bonds in the *reactants* break leaving separate atoms. Look up the bond enthalpy for each bond and add them all up. This will give you the total energy that must be *put in* to break the bonds and form separate atoms.

You need to *break* these bonds:

6 × C—H	6 × 413 $kJ\,mol^{-1}$	= 2478 $kJ\,mol^{-1}$
1 × C—C	1 × 347 $kJ\,mol^{-1}$	= 347 $kJ\,mol^{-1}$
1 × Cl—Cl	1 × 243 $kJ\,mol^{-1}$	= 243 $kJ\,mol^{-1}$
		= 3068 $kJ\,mol^{-1}$

So 3068 $kJ\,mol^{-1}$ must be *put in* to convert ethane and chlorine to separate hydrogen, chlorine, and carbon atoms.

3 Next imagine the separate atoms join together to give the *products*. Add up the bond enthalpies of the bonds that must form. This will give you the total enthalpy *given out* by the bonds forming.

You need to *make* these bonds:

5 × C—H	5 × 413 $kJ\,mol^{-1}$	= 2065 $kJ\,mol^{-1}$
1 × C—C	1 × 347 $kJ\,mol^{-1}$	= 347 $kJ\,mol^{-1}$
1 × C—Cl	1 × 346 $kJ\,mol^{-1}$	= 346 $kJ\,mol^{-1}$
1 × Cl—H	1 × 432 $kJ\,mol^{-1}$	= 432 $kJ\,mol^{-1}$
		= 3190 $kJ\,mol^{-1}$

So 3190 $kJ\,mol^{-1}$ is *given out* when you convert the separate hydrogen, chlorine, and carbon atoms to chloroethane and hydrogen chloride.

The difference between the energy put in to break the bonds and the energy given out to form bonds is the approximate enthalpy change of the reaction.

The difference is 3190 – 3068 = 122 $kJ\,mol^{-1}$.

4 Finally work out the sign of the enthalpy change. If more energy was put in than was given out, the enthalpy change is positive (the reaction is endothermic). If more energy was given out than was put in the enthalpy change is negative (the reaction is exothermic).

In this case, more enthalpy is given out than put in, so the reaction is exothermic and $\Delta H = -122\,kJ\,mol^{-1}$

Note that in practice it would be impossible for the reaction to happen like this. However, Hess's law tells us that you will get the same answer whatever route you take, real or theoretical.

Synoptic link

Bond enthalpies give a measure of the strength of bonds, and can help to predict which bond in a molecule is most likely to break. However, this is not the only factor, the polarity of the bond is also important – see Topic 3.6, Electronegativity – bond polarity in covalent bonds, and Topic 13.2, Nucleophilic substitution in halogenoalkanes.

Summary questions

These questions are about the reaction:

$CH_3CH_3 + Br_2 \rightarrow CH_3CH_2Br + HBr$

1 Draw out the displayed structural formulae of all the products and reactants so that all the bonds are shown.

2 a Identify the bonds that have to be broken to convert the reactants into separate atoms.

b How much energy does this take?

3 a Identify the bonds that have to be made to convert separate atoms into the products.

b How much energy does this take?

4 Describe what the difference is between the energy put in to break bonds and the energy given out when the new bonds are formed.

5 a State what is $\Delta H^{\ominus}$ for the reaction (this requires a sign).

b Identify if the reaction in part **a** is endothermic or exothermic.

A short cut

You can often shorten mean bond enthalpy calculations:

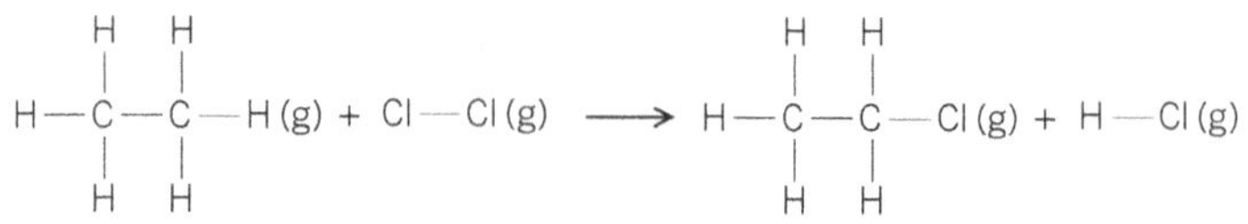

Only the bonds drawn in red make or break during the reaction so you only need to break: $1 \times$ C—H $= 413\,\text{kJ mol}^{-1}$

$1 \times$ Cl—Cl $= 243\,\text{kJ mol}^{-1}$

Total energy put in $=$ **656 kJ mol⁻¹**

You only need to make: $1 \times$ C—Cl $= 346\,\text{kJ mol}^{-1}$

$1 \times$ H—Cl $= 432\,\text{kJ mol}^{-1}$

Total energy given out $=$ **778 kJ mol⁻¹**

The difference is $778 - 656 = 122\,\text{kJ mol}^{-1}$

More energy is given out than taken in so

$\Delta H = -122\,\text{kJ mol}^{-1}$ (as before)

Comparing the result with that from a thermochemical cycle

This is only an approximate value. This is because the bond enthalpies are averages whereas in a compound any bond has a specific value for its enthalpy. You can find an accurate value for $\Delta H^{\ominus}$ by using a thermochemical cycle as shown here:

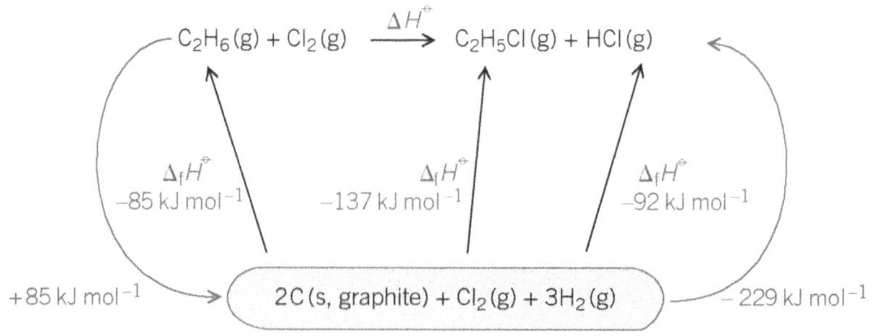

Remember $Cl_2(g)$ is an element so its $\Delta_f H^{\ominus}$ is zero.

$\Delta H^{\ominus} = 85 - 229\,\text{kJ mol}^{-1}$

$\Delta H^{\ominus} = -144\,\text{kJ mol}^{-1}$ (compared with $-122\,\text{kJ mol}^{-1}$ calculated from bond enthalpies)

This difference is typical of what might be expected using mean bond enthalpies. The answer obtained from the thermochemical cycle is the 'correct' one because all the $\Delta_f H^{\ominus}$ values have been obtained from the actual compounds involved.

Mean bond enthalpy calculations also allow us to calculate an approximate value for $\Delta_f H$ for a compound that has never been made.

1 A student used Hess's law to determine a value for the enthalpy change that occurs when anhydrous copper(II) sulfate is hydrated.
This enthalpy change was labelled ΔH_{exp} by the student in a scheme of reactions.

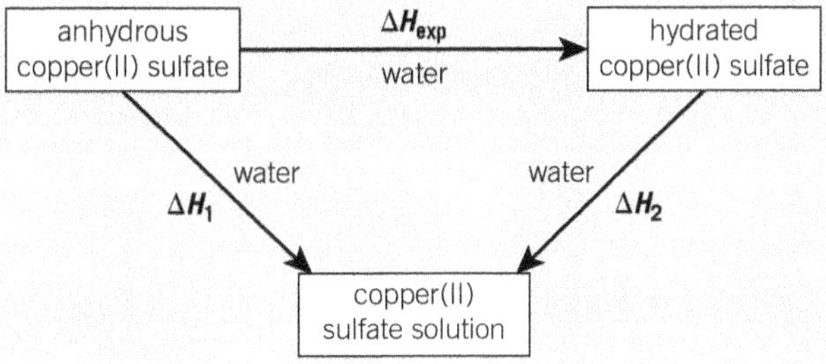

(a) State Hess's law. *(1 mark)*

(b) Write a mathematical expression to show how ΔH_{exp}, ΔH_1, and ΔH_2 are related to each other by Hess's law. *(1 mark)*

(c) Use the mathematical expression that you have written in part **(b)**, and the data book values for the two enthalpy changes ΔH_1 and ΔH_2 shown, to calculate a value for ΔH_{exp}

$$\Delta H_1 = -156 \text{ kJ mol}^{-1}$$

$$\Delta H_2 = +12 \text{ kJ mol}^{-1}$$ *(1 mark)*

(d) The student added 0.0210 mol of pure anhydrous copper(II) sulfate to 25.0 cm³ of deionised water in an open polystyrene cup. An exothermic reaction occurred and the temperature of the water increased by 14.0 °C.

 (i) Use these data to calculate the enthalpy change, in kJ mol⁻¹, for this reaction of copper(II) sulfate. This is the student value for ΔH_1

 In this experiment, you should assume that all of the heat released is used to raise the temperature of the 25.0 g of water. The specific heat capacity of water is 4.18 J K⁻¹ g⁻¹. *(3 marks)*

 (ii) Suggest **one** reason why the student value for ΔH_1 calculated in part **(d) (i)** is less accurate than the data book value given in part **(c)**. *(1 mark)*

(e) Suggest **one** reason why the value for ΔH_{exp} **cannot** be measured directly. *(1 mark)*

AQA, 2013

2 Hydrazine, N_2H_4, decomposes in an exothermic reaction. Hydrazine also reacts exothermically with hydrogen peroxide when used as a rocket fuel.

(a) Write an equation for the decomposition of hydrazine into ammonia and nitrogen only. *(1 mark)*

(b) State the meaning of the term *mean bond enthalpy*. *(2 marks)*

(c) Some mean bond enthalpies are given in the table.

Mean bond enthalpy/ kJ mol⁻¹	N—H	N—N	N≡N	O—H	O—O
	388	163	944	463	146

Use these data to calculate the enthalpy change for the gas-phase reaction between hydrazine and hydrogen peroxide. *(3 marks)*

$$
\begin{array}{c}
H \\ \diagdown \\ N{=}N \\ \diagup \quad \diagdown \\ H \qquad H
\end{array}
\quad + \ 2\ H{-}O{-}O{-}H \longrightarrow N{\equiv}N \ + \ 4\ H{-}O{-}H
$$

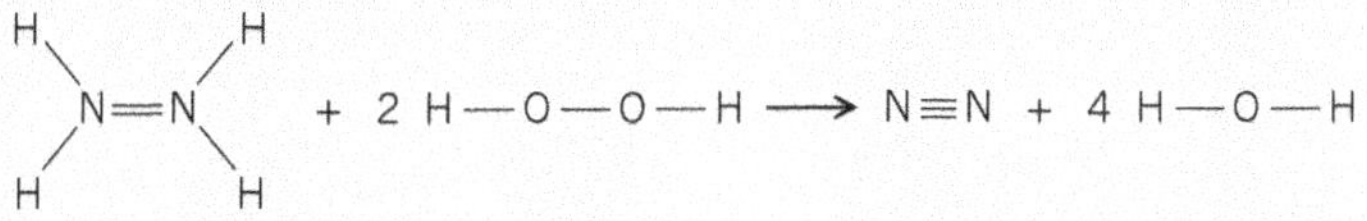

AQA, 2013

3 Hess's law is used to calculate the enthalpy change in reactions for which it is difficult
 to determine a value experimentally.
 (a) State the meaning of the term *enthalpy change*.

 (1 mark)

 (b) State Hess's law.

 (1 mark)

 (c) Consider the following table of data and the scheme of reactions.

Reaction	Enthalpy change/kJ mol^{-1}
$HCl(g) \rightarrow H^+(aq) + Cl^-(aq)$	−75
$H(g) + Cl(g) \rightarrow HCl(g)$	−432
$H(g) + Cl(g) \rightarrow H^+(g) + Cl^-(g)$	+963

$$H^+(g) \quad + \quad Cl^-(g) \quad \xrightarrow{\Delta_r H} \quad H^+(aq) \quad + \quad Cl^-(aq)$$
$$\uparrow \qquad\qquad\qquad\qquad\qquad\qquad \uparrow$$
$$H(g) \quad + \quad Cl(g) \quad \longrightarrow \quad HCl(g)$$

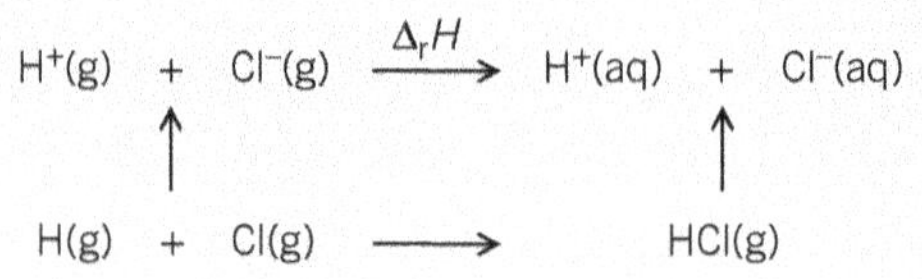

Use the data in the table, the scheme of reactions, and Hess's law to calculate a value for $\Delta_r H$.

(3 marks)

AQA, 2010

Answers to the Practice Questions and Section Questions are available at
www.oxfordsecondary.com/oxfordaqaexams-alevel-chemistry

5 Kinetics
5.1 Collision theory

Learning objectives:
→ Describe what must happen before a reaction will take place.
→ Explain why all collisions do not result in a reaction.

Specification reference: 3.1.5

Kinetics is the study of the factors that affect rates of chemical reactions – how quickly they take place. There is a large variation in reaction rates. 'Popping' a test tube full of hydrogen is over in a fraction of a second, whilst the complete rusting away of an iron nail could take several years. Reactions can be speeded up or slowed down by changing the conditions.

The **rate of a chemical reaction** is defined as the change in concentration of one of the reactants or products with unit time. It is usually measured in mol dm^{-3} s^{-1}

Collision theory

For a reaction to take place between two particles, they must collide with enough energy to break bonds. The collision must also take place between the parts of the molecule that are going to react together, so orientation also has a part to play. To get a lot of collisions you need a lot of particles in a small volume. For the particles to have enough energy to break bonds they need to be moving fast. So, for a fast reaction rate you need plenty of rapidly moving particles in a small volume.

Most collisions between molecules or other particles do not lead to reaction. They either do not have enough energy, or they are in the wrong orientation.

Factors that affect the rate of chemical reactions

The following factors will increase the rate of a reaction.

- **Increasing the temperature** This increases the speed of the molecules, which in turn increases both their energy and the number of collisions.
- **Increasing the concentration of a solution** If there are more particles present in a given volume then collisions are more likely and the reaction rate would be faster. However, as a reaction proceeds, the reactants are used up and their concentration falls. So, in most reactions the rate of reaction drops as the reaction goes on.
- **Increasing the pressure of a gas reaction** This has the same effect as increasing the concentration of a solution – there are more molecules or atoms in a given volume so collisions are more likely.
- **Increasing the surface area of solid reactants** The greater the *total* surface area of a solid, the more of its particles are available to collide with molecules in a gas or a liquid. This means that breaking a solid lump into smaller pieces increases the rate of its reaction because there are more sites for reaction.
- **Using a catalyst** A catalyst is a substance that can change the rate of a chemical reaction without being chemically changed itself.

Hint

A rough rule for many chemical reactions is that if the temperature goes up by 10 K (10 °C), the rate of reaction approximately doubles.

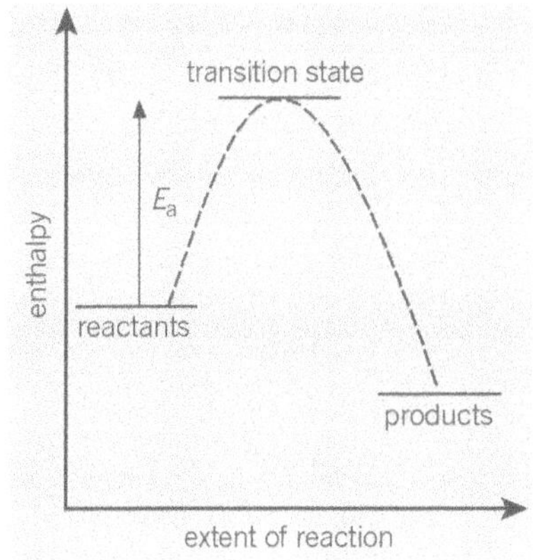

▲ **Figure 1** *An exothermic reaction with a large activation energy, E_a*

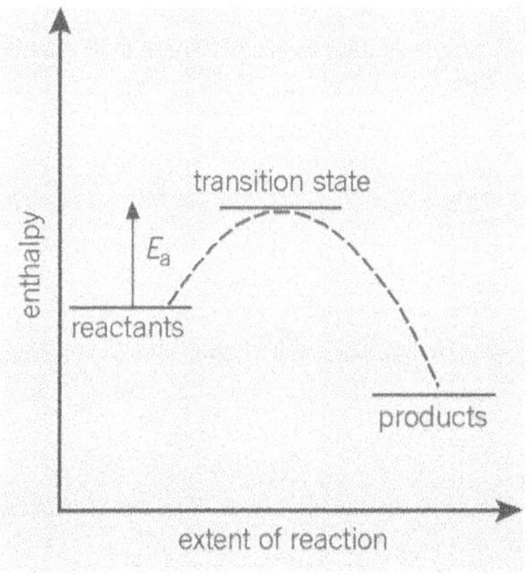

▲ **Figure 2** *An exothermic reaction with a small activation energy, E_a*

Activation energy

Only a very small proportion of collisions actually result in a reaction.

For a collision to result in a reaction, the molecules must have a certain minimum energy, enough to start breaking bonds.

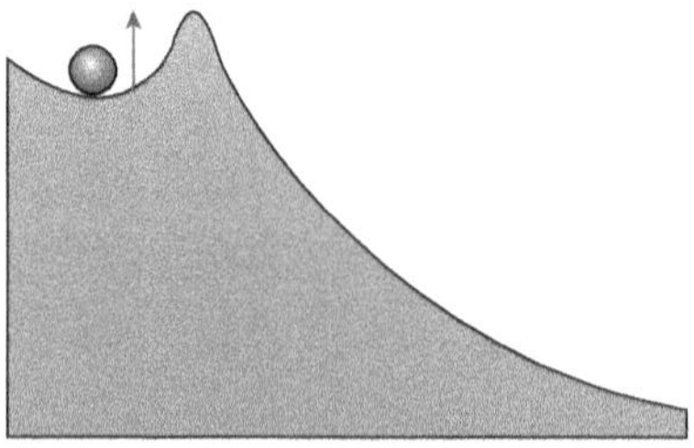

a with low activation energy

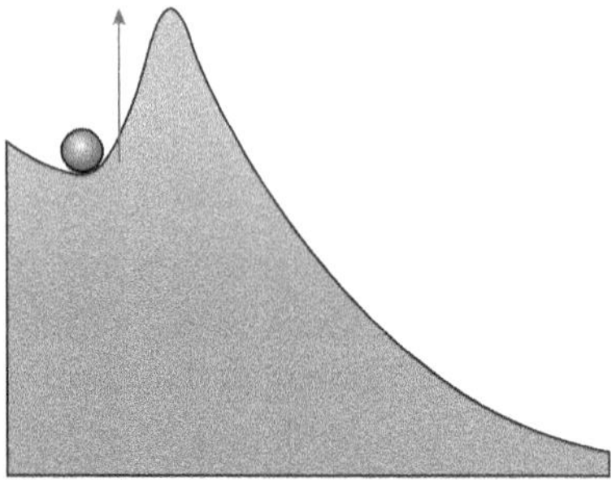

b with high activation energy

▲ **Figure 3** *Ball on a mountainside models*

Hint

'Species' is a term used by chemists to refer to an atom, molecule, or ion.

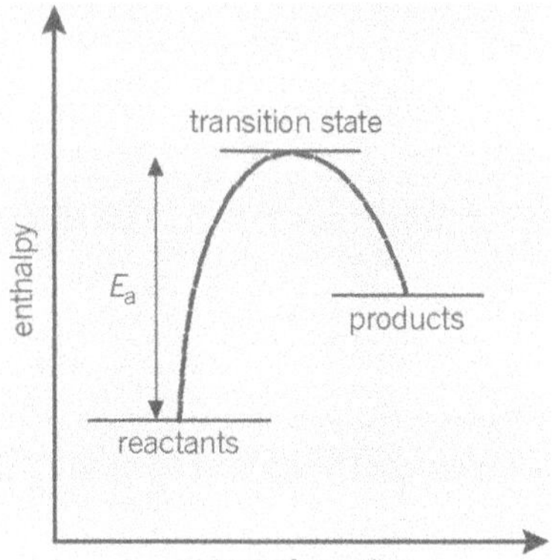

▲ **Figure 4** *An endothermic reaction with activation energy E_a*

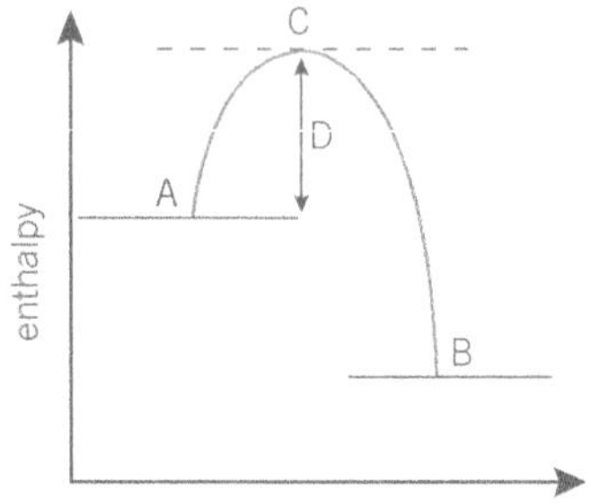

▲ **Figure 5** *A reaction profile*

The minimum energy needed to start a reaction is called the **activation energy** and has the abbreviation E_a.

You can include the idea of activation energy on an enthalpy diagram that shows the course of a reaction.

Exothermic reactions

Figure 1 shows the reaction profile for an exothermic reaction with a large activation energy. This reaction will take place extremely slowly at room temperature because very few collisions will have sufficient energy to bring about a reaction.

Figure 2 shows the reaction profile for an exothermic reaction with a small activation energy. This reaction will take place rapidly at room temperature because many collisions will have enough energy to bring about a reaction.

The situation is a little like a ball on a hill, see Figure 3. A small amount of energy is needed in Figure 3a, to set the ball rolling, whilst a large amount of energy is needed in Figure 3b.

The species that exists at the top of the curve of an enthalpy diagram is called a **transition state** or **activated complex**. Some bonds are in the process of being made and some bonds are in the process of being broken. Like the ball at the very top of the hill, it has extra energy and is unstable.

Endothermic reactions

Endothermic reactions are those in which the products have more energy than the reactants. An endothermic reaction, with activation energy E_a, is shown in Figure 4. The transition state has been labelled.

Notice that the activation energy is measured from the reactants to the top of the curve.

Summary questions

1 List five factors that affect the speed of a chemical reaction.

Use the reaction profile in Figure 5 to answer questions **2** and **3**:

2 a What is A?

b What is B?

c What is C?

d What is D?

3 a Identify whether the enthalpy profile represents an endothermic or an exothermic reaction.

b Explain your answer to part **a**.

5.2 The Maxwell–Boltzmann distribution

The particles in any gas (or solution) are all moving at different speeds – a few are moving slowly, a few very fast but most are somewhere in the middle. The energy of a particle depends on its speed so the particles also have a range of energies. If you plot a graph of energy against the fraction of particles that have that energy, you end up with the curve shown in Figure 1. This particular shape is called the **Maxwell–Boltzmann distribution** – it tells us about the distribution of energy amongst the particles.

- No particles have zero energy.
- Most particles have intermediate energies – around the peak of the curve.
- A few have very high energies (the right-hand side of the curve). In fact there is no upper limit.
- Note also that the average energy is not the same as the most probable energy.

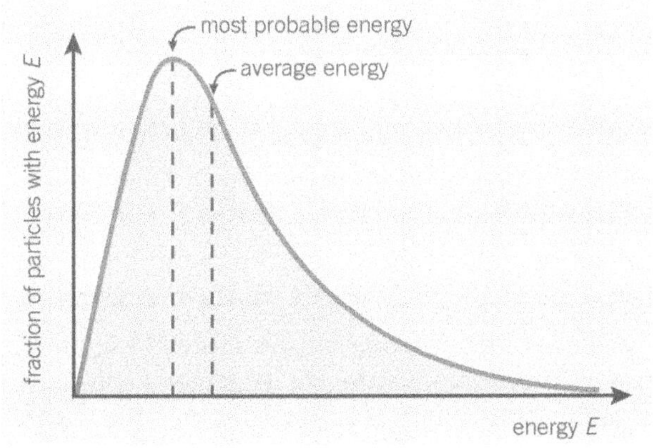

▲ **Figure 1** *The distribution of the energies of particles. The area under the graph represents the total number of particles*

Activation energy E_a

For a reaction to take place, a collision between particles must have enough energy to start breaking bonds, see Topic 5.1. This amount of energy is called the activation energy E_a. If you mark E_a on the Maxwell–Boltzmann distribution graph, Figure 2, then the area under the graph to the right of the activation energy line represents the number of particles with enough energy to react.

The need for the activation energy to be present before a reaction takes place explains why not all reactions that are exothermic occur spontaneously at room temperature.

For example, fuels are mostly safe at room temperature, as in a petrol station. But a small spark may provide enough energy to start the combustion reaction. The heat given out by the initial reaction is enough to supply the activation energy for further reactions. Similarly the chemicals in a match head are quite stable until the activation energy is provided by friction.

Learning objectives:

- → Define activation energy.
- → Explain how temperature affects the number of molecules with energy equal to or more than the activation energy.
- → Explain why a small increase in temperature has a large effect on the rate of a reaction.

Specification reference: 3.1.5

Hint

The rate of a chemical reaction is defined as the change in concentration of one of the reactants or products with unit time. It is usually measured in mol dm^{-3} per second or mol dm^{-3} s^{-1}.

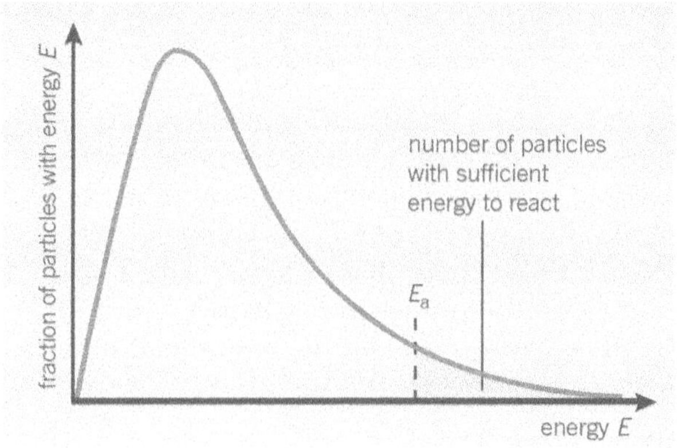

▲ **Figure 2** *Only particles with energy greater than E_a can react*

Even the high temperature of a single spark can set off a reaction. This is why if you smell gas, you must not even turn on a light. The electrical connection provided by the switch could produce enough energy to begin an explosion.

The effect of temperature on reaction rate

The shape of the Maxwell–Boltzmann graph changes with temperature, as shown in Figure 3.

At higher temperatures the peak of the curve is lower and moves to the right. The number of particles with very high energy increases. The total area under the curve is *the same* for each temperature because it represents the total number of particles.

The shaded areas to the right of the E_a line represent the number of molecules that have greater energy than E_a at each temperature.

The graphs show that at higher temperatures more of the molecules have energy greater than E_a so a higher percentage of collisions will result in reaction. This is why reaction rates increase with temperature. In fact, a small increase in temperature produces a large increase in the number of particles with energy greater than E_a.

Also, the total *number* of collisions in a given time increases a little as the particles move faster. However, this is not as important to the rate of reaction as the increase in the number of *effective* collisions (those with energy greater than E_a).

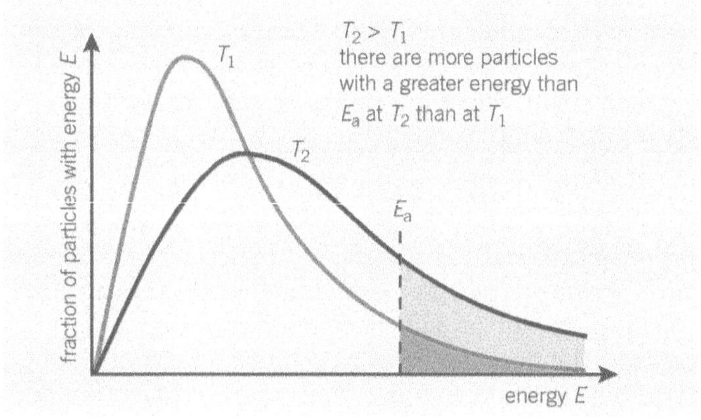

▲ **Figure 3** *The Maxwell–Boltzmann distribution of the energies of the same number of particles at two temperatures*

Summary questions

1 Use Figure 4 to answer the following questions:

  a What is the axis labelled A?

  b What is the axis labelled B?

  c What does area C represent?

  d If the temperature is increased, what happens to the peak of the curve?

  e If the temperature is increased, what happens to E_a?

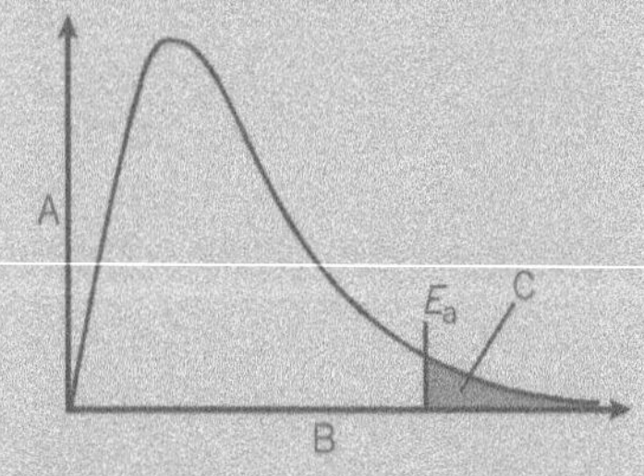

▲ **Figure 4** *The Maxwell–Boltzmann distribution of energies of particles at a particular temperature, with the activation energy, E_a marked*

5.3 Catalysts

Learning objectives:

- → State the definition of a catalyst.
- → Describe how a catalyst affects activation energy.
- → Describe how a catalyst affects enthalpy change.

Specification reference: 3.1.5

Catalysts are substances that affect the rate of chemical reactions without being chemically changed themselves at the end of the reaction. Catalysts are usually used to *speed up* reactions so they are important in industry. It is cheaper to speed up a reaction by using a catalyst than by using high temperatures and pressures. This is true, even if the catalyst is expensive, because it is not used up.

How catalysts work

Catalysts work because they provide a different pathway for the reaction, one with a lower activation energy. Therefore they reduce the activation energy of the reaction (the minimum amount of energy that is needed to start the reaction). You can see this on the enthalpy diagrams in Figure 1.

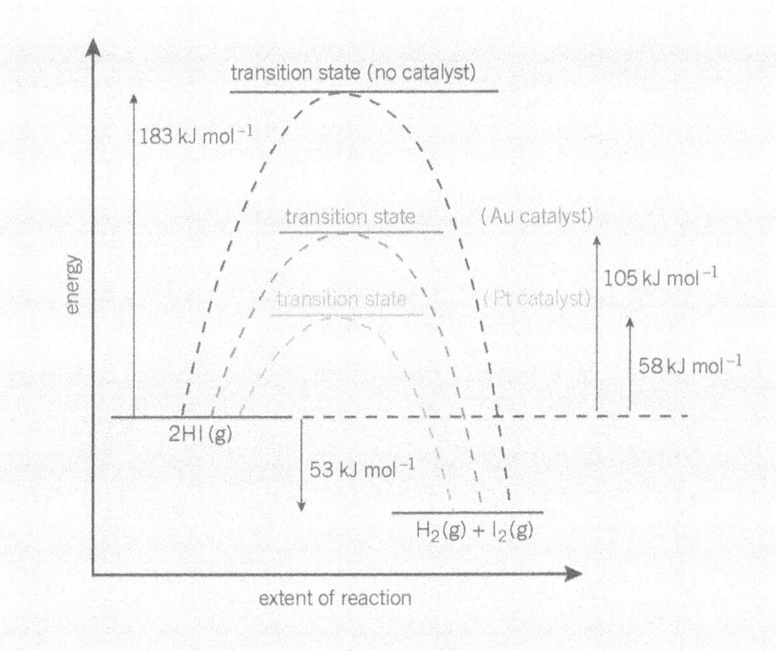

▲ **Figure 1** *The decomposition of hydrogen iodide with different catalysts*

For example, for the decomposition of hydrogen iodide:

$$2HI(g) \rightarrow H_2(g) + I_2(g)$$

E_a = 183 kJ mol^{-1} (without a catalyst)

E_a = 105 kJ mol^{-1} (with a gold catalyst)

E_a = 58 kJ mol^{-1} (with a platinum catalyst)

You can see what happens when you lower the activation energy if you look at the Maxwell–Boltzmann distribution curve in Figure 2. The area that is shaded pink represents the number of effective collisions that can happen without a catalyst. The area shaded blue, plus the area that is shaded pink, represents the number of effective collisions that can takes place with a catalyst.

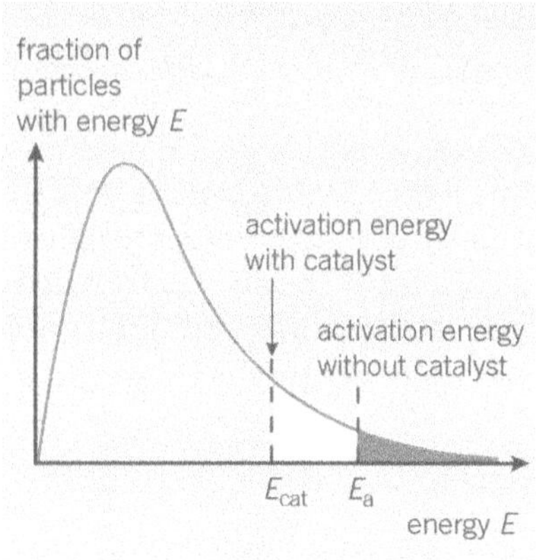

▲ **Figure 2** *With a catalyst the extra particles in the blue area react*

Catalysts do not affect the enthalpy change of the reactions, nor do they affect the position of equilibrium in a reversible reaction, see Topic 6.1.

▼ **Table 1** *Examples of catalysts*

Reaction	Catalyst	Use
$N_2(g) + 3H_2(g) \rightarrow 2NH_3(g)$ Haber process	iron	making fertilisers
$4NH_3 + 5O_2 \rightarrow 4NO + 6H_2O$ Ostwald process for making nitric acid	platinum and rhodium	making fertilisers and explosives
$H_2C{=}CH_2 + H_2 \rightarrow CH_3CH_3$ hardening of fats with hydrogen	nickel	making margarine
cracking hydrocarbon chains from crude oil	aluminium oxide and silicon dioxide zeolite	making petrol
catalytic converter reactions in car exhausts	platinum and rhodium	removing polluting gases
$H_2C{=}CH_2 + H_2O \rightarrow CH_3CH_2OH$ hydration of ethene to produce ethanol	H^+ absorbed on solid silica phosphoric acid, H_3PO_4	making ethanol – a fuel additive, solvent, and chemical feedstock
$CH_3CO_2H(l) + CH_3OH(l) \rightarrow CH_3CO_2CH_3(aq) + H_2O(l)$ esterification	H^+	making solvents

Different catalysts work in different ways – most were discovered by trial and error.

Catalytic converters

All new petrol-engine cars are now equipped with catalytic converters in their exhaust systems. These reduce the levels of a number of polluting gases.

The catalytic converter is a honeycomb, made of a ceramic material coated with platinum and rhodium metals – the catalysts. The honeycomb shape provides an enormous surface area, on which the reactions take place, so a little of these expensive metals goes a long way.

As they pass over the catalyst, the polluting gases react with each other to form less harmful products by the following reactions:

carbon monoxide + nitrogen oxides → nitrogen + carbon dioxide

hydrocarbons + nitrogen oxides → nitrogen + carbon dioxide + water

Synoptic link

You will learn more about catalytic converters in Topic 12.4, Combustion of alkanes.

The reactions take place on the surface of the catalyst in two steps:

1 The gases first form weak bonds with the metal atoms of the catalyst – this process is called **adsorption**. This holds the gas molecules in just the right position for them to react together. The gases then react on the surface.
2 The products then break away from the metal atoms – this process is called **desorption**. This frees up room on the catalyst surface for more gases to take their place and react.

The strength of the weak bonds holding the gases onto the metal surface is critical. They must be strong enough to hold the gases for long enough to react, but weak enough to release the products easily.

Zeolites

Zeolites are *minerals* that have a very open pore structure that ions or molecules can fit into. Zeolites confine molecules in small spaces, which causes changes in their structure and reactivity. More than 150 zeolite types have been synthesised and 48 naturally occurring zeolites are known. Synthetic zeolites are widely used as catalysts in the petrochemical industry.

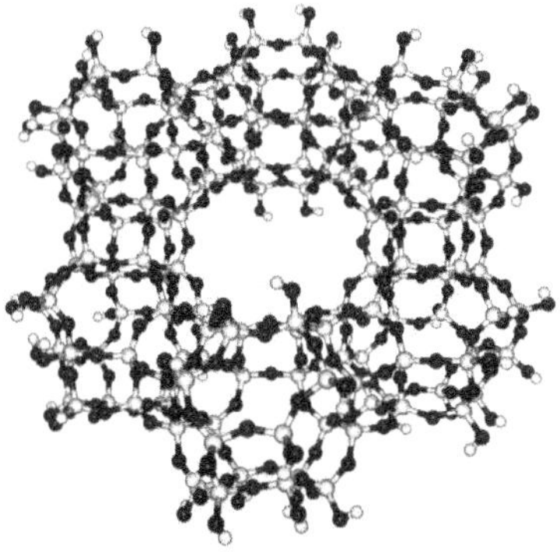

▲ **Figure 3** *Part of the structure of a synthetic zeolite*

Hardening fats

Unsaturated fats, used in margarines for example, are made more solid or hardened when hydrogen is added across some of the double bonds. This is done by bubbling hydrogen into the liquid fat which has a nickel catalyst mixed with it. The nickel is filtered off after the reaction. This allows the manufacturer to tailor the spreadability of the margarine.

▲ **Figure 4** *Margarine*

Catalysts and the ozone layer

Until recently, a group of apparently unreactive compounds called chlorofluorocarbons (CFCs) were used for a number of applications such as solvents, aerosol propellants, and in expanded polystyrene foams. They escaped high into the atmosphere where they remain because they are relatively unreactive. This is partly due to the strength of the carbon–halogen bonds.

CFCs do eventually decompose to produce separate chlorine atoms. These act as catalysts in reactions that bring about the destruction of ozone, O_3. Ozone is important because it forms a layer in the atmosphere of the Earth that acts as a shield. The layer prevents too much ultraviolet radiation from reaching the Earth's surface.

The overall reaction is shown below:

$$O_3(g) + O(g) \xrightarrow{\text{chlorine atom catalyst}} 2O_2(g)$$

Nitrogen monoxide acts as a catalyst in a similar way to chlorine atoms.

International agreements, such as the 1987 Montreal Protocol, have resulted in CFCs being phased out. Unfortunately there is still a reservoir of them remaining from before these agreements. Chemists have developed, and continue to work on, suitable substitutes for CFCs that do not result in damage to the upper atmosphere. These include hydrochlorofluorocarbons and hydrofluorocarbons. Former United Nations Secretary General, Kofi Annan, has referred to the Montreal Protocol as 'perhaps the single most successful international agreement to date'.

Summary questions

1 The following questions refer to Figure 5.

   a What are labels A, B, C, D, R, and P?

   b What do the distances from D to R and from C to R represent?

   c Is the reaction exothermic or endothermic?

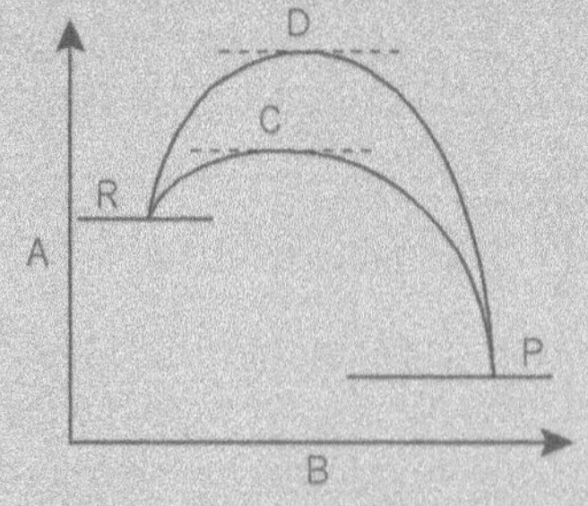

▲ **Figure 5** *A profile for a reaction with and without a catalyst*

Practice questions

1 The gas-phase reaction between hydrogen and chlorine is very slow at room temperature.

$$H_2(g) + Cl_2(g) \rightarrow 2HCl(g)$$

(a) Define the term *activation energy*.

(2 marks)

(b) Give **one** reason why the reaction between hydrogen and chlorine is very slow at room temperature.

(1 mark)

(c) Explain why an increase in pressure, at constant temperature, increases the rate of reaction between hydrogen and chlorine.

(2 marks)

(d) Explain why a small increase in temperature can lead to a large increase in the rate of reaction between hydrogen and chlorine.

(2 marks)

(e) Give the meaning of the term *catalyst*.

(1 mark)

(f) Suggest **one** reason why a solid catalyst for a gas-phase reaction is often in the form of a powder.

(1 mark)

AQA, 2006

2 The diagram below represents a Maxwell–Boltzmann distribution curve for the particles in a sample of a gas at a given temperature. The questions below refer to this sample of particles.

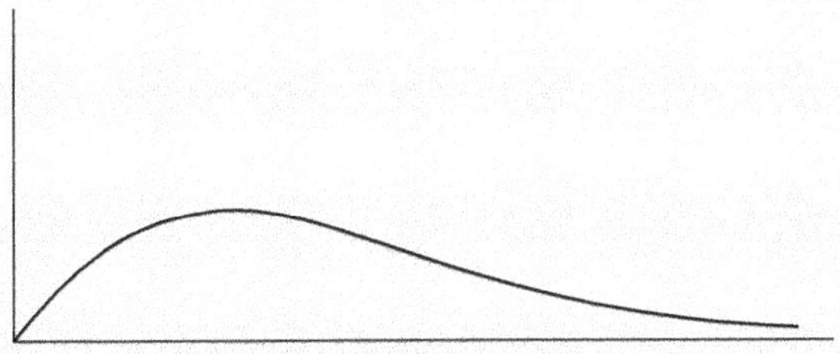

(a) Label the axes on a copy of the diagram.

(2 marks)

(b) On the diagram draw a curve to show the distribution for this sample at a **lower** temperature.

(2 marks)

(c) In order for two particles to react they must collide. Explain why most collisions do not result in a reaction.

(1 mark)

(d) State one way in which the collision frequency between particles in a gas can be increased without changing the temperature.

(1 mark)

(e) Suggest why a small increase in temperature can lead to a large increase in the reaction rate between colliding particles.

(2 marks)

(f) Explain in general terms how a catalyst works.

(2 marks)

AQA, 2004

3 The diagram shows the Maxwell–Boltzmann distribution of molecular energies in a gas at two different temperatures.

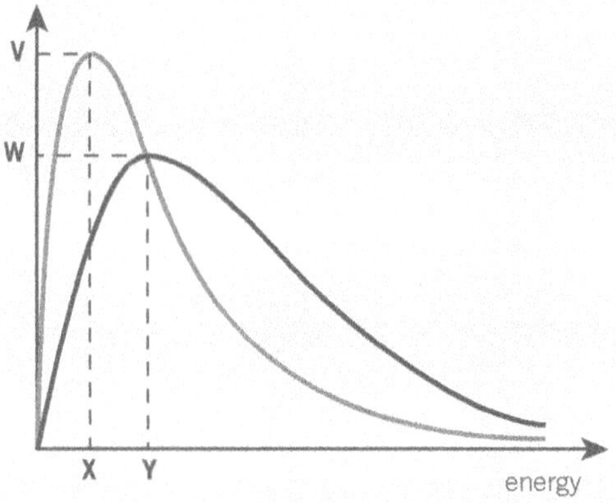

(a) One of the axes is labelled. Complete the diagram by labelling the other axis. *(1 mark)*

(b) State the effect, if any, of a solid catalyst on the shape of either of these distributions. *(1 mark)*

(c) State the letter, **V**, **W**, **X**, or **Y**, that represents the most probable energy of the molecules at the lower temperature. *(1 mark)*

(d) Explain what must happen for a reaction to occur between molecules of two different gases. *(2 marks)*

(e) Explain why a small increase in temperature has a large effect on the initial rate of a reaction. *(1 mark)*

AQA, 2012

4 The diagram shows the Maxwell–Boltzmann distribution for a sample of gas at a fixed temperature.

E_a is the activation energy for the decomposition of this gas.

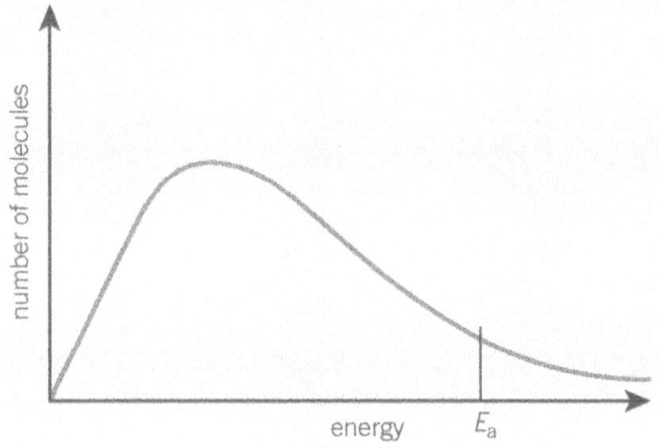

E_{mp} is the most probable value for the energy of the molecules.

(a) On the appropriate axis of this diagram, mark the value of E_{mp} for **this** distribution. On this diagram, sketch a new distribution for the same sample of gas at a **lower** temperature. *(3 marks)*

(b) With reference to the Maxwell–Boltzmann distribution, explain why a decrease in temperature decreases the rate of decomposition of this gas. *(2 marks)*

AQA, 2013

Answers to the Practice Questions and Section Questions are available at www.oxfordsecondary.com/oxfordaqaexams-alevel-chemistry

6 Equilibria

6.1 The idea of equilibrium

Learning objectives:

- → State the definition of a reversible reaction.
- → State what is meant by chemical equilibrium.
- → Explain why all reactions do not go to completion.
- → Explain what happens when equilibrium has been reached.

Specification reference: 3.1.6

Chemists usually think of a reaction as starting with the reactants and ending with the products.

$$\text{reactants} \rightarrow \text{products}$$

However, some reactions are reversible. For example, when you heat blue hydrated copper sulfate it becomes white anhydrous copper sulfate as the water of crystallisation is driven off. The white copper sulfate returns to blue if you add water.

$$\underset{\text{blue hydrated copper sulfate}}{CuSO_4.5H_2O} \rightleftharpoons \underset{\text{white anhydrous copper sulfate}}{CuSO_4 + 5H_2O}$$

However, something different would happen if you were to do this reaction in a closed container. As soon as the products are formed they react together and form the reactants again, so that instead of reactants *or* products you get a mixture of both. Eventually you get a mixture in which the proportions of all three components remain constant. This mixture is called an **equilibrium mixture**.

Setting up an equilibrium

You can understand how an equilibrium mixture is set up by thinking about what happens in a physical process, like the evaporation of water. This is easier to picture than a chemical change.

First imagine a puddle of water out in the open. Some of the water molecules at the surface will move fast enough to escape from the liquid and evaporate. Evaporation will continue until all the water is gone.

But think about putting some water into a *closed* container. At first the water will begin to evaporate as before. The volume of the liquid will get smaller and the number of vapour molecules in the gas phase will go up. But as more molecules enter the vapour, some gas-phase molecules will start to re-enter the liquid, see Figure 1.

After a time, the rate of evaporation and the rate of condensation will become equal. The level of the liquid water will then stay exactly the same and so will the number of molecules in the vapour and in the liquid. The evaporation and condensation are still going on but *at the same rate*. This situation is called a **dynamic equilibrium** and is one of the key ideas of this topic.

In fact, you could have started by filling the empty container with the same mass of water vapour as you originally had liquid water. The vapour would begin to condense and, in time, would reach exactly the same equilibrium position.

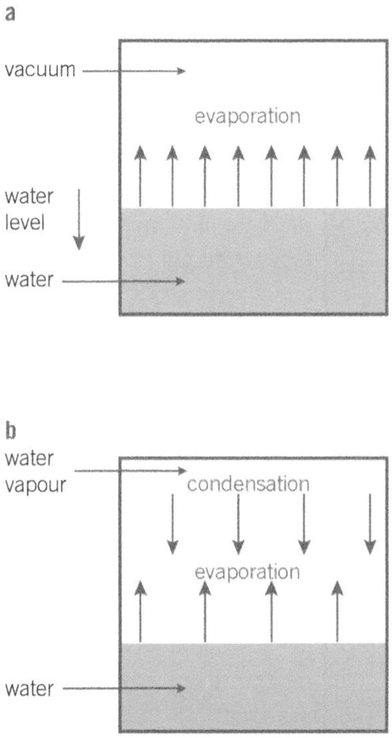

▲ **Figure 1 a** *Water will evaporate into an empty container. Eventually the rates of evaporation and condensation will be the same*
b *Equilibrium is set up*

Study tip

Remember that at equilibrium, both forward and backward reactions occur at the same rate so the concentrations of all the reactants and products remain constant.

The conditions for equilibrium

Although the system used here is very simple, you can pick out four conditions that apply to *all* equilibria:

- Equilibrium can only be reached in a **closed system** (one where the reactants and products can't escape). The system does not have to be sealed. For example, a beaker may be a closed system for a reaction that takes place in a solvent, as long as the reactants, products, and solvent do not evaporate.
- Equilibrium can be approached from *either direction* (in Figure 1, from liquid or from vapour) and the final equilibrium position will be the same (as long as conditions, such as temperature and pressure, stay the same).
- Equilibrium is a dynamic process. It is reached when the *rates* of two opposing processes, which are going on all the time (in Figure 1, evaporation and condensation), *are the same.*
- You know that equilibrium has been reached when the macroscopic properties of the system do not change with time. These are properties such as density, concentration, colour, and pressure – properties that do not depend on the total quantity of matter.

A reversible reaction that can reach equilibrium is denoted by the symbol $\rightleftharpoons$, for example:

$$\text{liquid water} \rightleftharpoons \text{water vapour}$$

$$H_2O(l) \rightleftharpoons H_2O(g)$$

Chemical equilibria

The same principles that you have found for a physical change also apply to chemical equilibria such as:

$$\underset{\text{reactants}}{A + B} \rightleftharpoons \underset{\text{products}}{C + D}$$

- Imagine starting with A and B only. At the start of the reaction the forward rate is fast, because A and B are plentiful. There is no reverse reaction because there is no C and D.
- Then as the concentrations of C and D build up, the reverse reaction speeds up. At the same time the concentrations of A and B decrease so the forward reaction slows down.
- A point is reached where exactly the same number of particles are changing from A + B to C + D as are changing from C + D to A + B. Equilibrium has been reached.

One important point to remember is that an equilibrium mixture can have *any* proportions of reactants and products. It is not necessarily half reactants and half products, though it could be. The proportions may be changed depending on the conditions of the reaction, such as temperature, pressure, and concentration. But at any given constant conditions the proportions of reactants and products do not change.

Summary questions

1 For each of the following statements about all equilibria, say whether it is true or false.
   - a Once equilibrium is reached the concentrations of the reactants and the products do not change.
   - b At equilibrium the forward and the backward reactions come to a halt.
   - c Equilibrium is only reached in a closed system.
   - d An equilibrium mixture always contains half reactants and half products.

2 What can be said about the rates of the forward and the backward reactions when equilibrium is reached?

6.2 Changing the conditions of an equilibrium reaction

Learning objectives:

- → State Le Châtelier's principle.
- → Explain how an equilibrium position is affected by concentration, temperature, pressure, or a catalyst.

Specification reference: 3.1.6

Some industrial processes, such as the production of ammonia or sulfuric acid, have reversible reactions as a key step. In closed systems these reactions would produce equilibrium mixtures containing both products and reactants. In principle, you would like to increase the proportion of products. For this reason it is important to understand how to control equilibrium reactions.

The equilibrium mixture

It is possible to change the proportion of reactants to products in an equilibrium mixture. In this way you are able to obtain a greater yield of the products. This is called changing the *position* of equilibrium.

- If the proportion of products in the equilibrium mixture is increased, we say that the equilibrium is moved to the right, or in the forward direction.
- If the proportion of reactants in the equilibrium mixture is increased, we say that the equilibrium is moved to the left, or in the backward direction.

You can often move the equilibrium position to the left or right by varying conditions such as temperature, the concentration of species involved, or the pressure (in the case of reactions involving gases).

Le Châtelier's principle

Le Châtelier's principle is useful because it gives us a rule. It tells us whether the equilibrium moves to the right or to the left when the conditions of an equilibrium mixture are changed.

It states:

If a system at equilibrium is disturbed, the equilibrium moves in the direction that tends to reduce the disturbance.

So in other words, if any factor is changed which affects the equilibrium mixture, the position of equilibrium will shift so as to oppose the change.

Le Châtelier's principle does not tell us *how far* the equilibrium moves so you cannot predict the *quantities* involved.

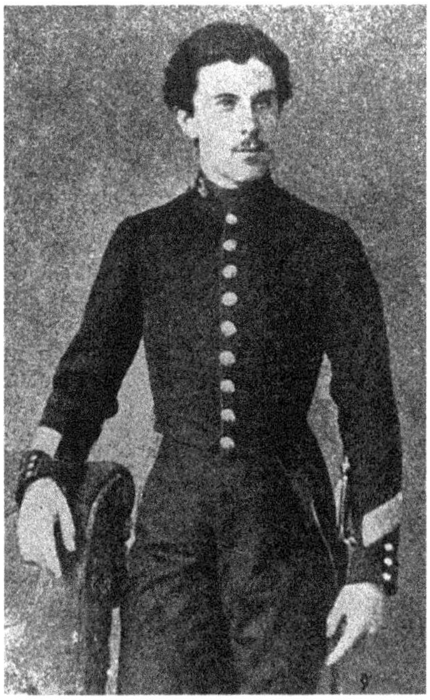

▲ **Figure 1** *Henri-Louis Le Châtelier was a French chemist who first put forward his 'Loi de stabilité d'équilibre chimique' in 1884*

Changing concentrations

If you *increase* the concentration of one of the reactants, Le Châtelier's principle says that the equilibrium will shift in the direction that tends to *reduce* the concentration of this reactant. Look at the reaction:

$$A(aq) + B(aq) \rightleftharpoons C(aq) + D(aq)$$

Suppose you add some extra A. This would increase the concentration of A. The only way that this system can reduce the concentration of A, is by some of A reacting with B (so forming more C and D). So, adding more A uses up more B, produces more C and D, and moves the equilibrium to the right. You end up with a greater proportion of

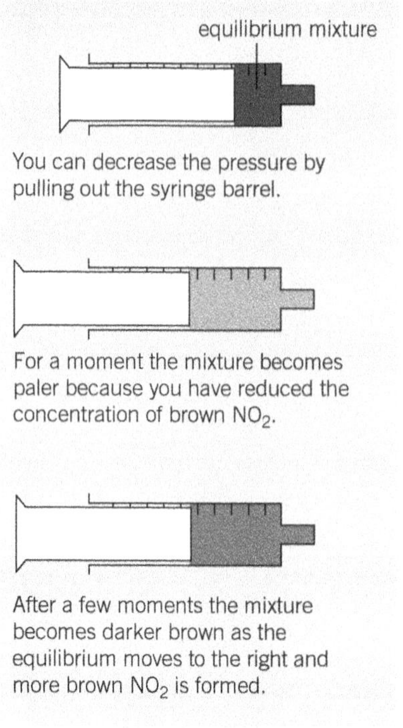

equilibrium mixture
You can decrease the pressure by pulling out the syringe barrel.
For a moment the mixture becomes paler because you have reduced the concentration of brown NO_2.
After a few moments the mixture becomes darker brown as the equilibrium moves to the right and more brown NO_2 is formed.

▲ **Figure 2** $N_2O_4(g) \rightleftharpoons 2NO_2(g)$ *The equilibrium moves to the right as you decrease the pressure*

Hint

Increasing the pressure or decreasing the volume of a mixture of gases increases the concentration of all the reactants and products by the same amount, not just one of them.

Hint

The *rate* at which equilibrium is reached *will* be speeded up by increasing the pressure, as there will be more collisions in a given time.

Study tips

- The terms 'move forwards' and 'move to the right' mean the same thing in this context.
- The terms 'move backwards' and 'move to the left' mean the same thing in this context.

products in the reaction mixture than before you added A. The same thing would happen if you added more B.

You could also remove C as it was formed. The equilibrium would move to the right to produce more C (and D) using up A and B. The same thing would happen if you removed D as soon as it was formed.

Changing the overall pressure

Pressure changes only affect reactions involving gases. Changing the overall pressure will only change the position of equilibrium of a gaseous reaction if there are a different number of molecules on either side of the equation.

An example of a such a reaction is:

$N_2O_4(g)$	$\rightleftharpoons$	$2NO_2(g)$
dinitrogen tetraoxide		nitrogen dioxide
1 mole		2 moles
colourless		brown

Increasing the pressure of a gas means that there are more molecules of it in a given volume – it is equivalent to increasing the concentration of a solution.

If you increase the pressure on this system, Le Châtelier's principle tells us that the position of equilibrium will move to decrease the pressure. This means that it will move to the left because fewer molecules exert less pressure. In the same way if you decrease the pressure, the equilibrium will move to the right – molecules of N_2O_4 will decompose to form molecules of NO_2, thereby increasing the pressure.

Dinitrogen tetraoxide is a colourless gas and nitrogen dioxide is brown. You can investigate this in the laboratory, by setting up the equilibrium mixture in a syringe. If you decrease the pressure, by pulling out the syringe barrel, you can watch as the equilibrium moves to the right because the colour of the mixture gets browner, see Figure 2.

Note that if there is the same number of moles of gases on both sides of the equation, then pressure has no effect on the equilibrium position. For example:

$$H_2(g) + I_2(g) \rightleftharpoons 2HI(g)$$

2 moles 2 moles

The equilibrium position will not change in this reaction when the pressure is changed so the proportions of the three gases will stay the same.

Changing temperature

Reversible reactions that are exothermic (give out heat) in one direction are endothermic (take in heat) in the other direction, see Topic 4.4. The size of the enthalpy is the same in both directions but the sign changes.

Example 1

Suppose you increase the temperature of an equilibrium mixture that is exothermic in the forward direction. An example is:

$$2SO_2(g) + O_2(g) \rightleftharpoons 2SO_3(g) \qquad \Delta H^{\ominus} = -197 \text{ kJ mol}^{-1}$$

The negative sign of $\Delta H^\ominus$ means that heat is given out when sulfur dioxide and oxygen react to form sulfur trioxide in the forward direction. This means that heat is absorbed as the reaction goes in the reverse direction, that is, to the left.

Le Châtelier's principle tells us that if you increase the temperature, the equilibrium moves in the direction that cools the system down. To do this it will move in the direction which absorbs heat (is endothermic), that is, to the left. The equilibrium mixture will then contain a greater proportion of sulfur dioxide and oxygen than before. In the same way, if we cool the mixture the equilibrium will move to the right and increase the proportion of sulfur trioxide.

Example 2

The effect of temperature on the dinitrogen tetraoxide/nitrogen dioxide equilibrium can also be investigated using the same apparatus you used to investigate the effect of pressure on this reaction. The reaction is endothermic as it proceeds from dinitrogen tetraoxide to nitrogen dioxide (the forward direction).

$$N_2O_4(g) \rightleftharpoons 2NO_2(g) \qquad \Delta H^\ominus = +58\text{ kJ mol}^{-1}$$

The gas mixture is contained in a syringe as before. The syringe is then immersed in warm water along with another syringe containing the same volume of air for comparison. The plunger of the syringe containing air will rise as the air expands. The plunger of the syringe containing the N_2O_4 / NO_2 mixture will also rise but by a greater amount. This indicates that more molecules of gas have been formed in this syringe. This is because the equilibrium has moved to the right; each molecule of N_2O_4 that disappears produces two molecules of NO_2. This is consistent with Le Châtelier's principle. When the mixture is warmed up, the equilibrium moves in the endothermic direction, that is, it absorbs heat which tends to cool the mixture down.

You should be able to predict the colour change that you would see during this experiment and also what would happen if the experiment were repeated in ice water.

Catalysts

Catalysts have no effect on the position of equilibrium so they do not alter composition of the equilibrium mixture. They work by producing an alternative route for the reaction, which has a lower activation energy of the reaction, see Topic 5.3. This affects the forward and back reactions equally.

Although catalysts have no effect on the position of equilibrium, that is, the yield of the reaction, they do allow equilibrium to be reached more quickly and are therefore important in industry.

Summary questions

1 In which of the following reactions will the position of equilibrium be affected by changing the pressure?

Explain your answers.

- **a** $2SO_2(g) + O_2(g) \rightleftharpoons 2SO_3(g)$
- **b** $CH_3CO_2H(aq) \rightleftharpoons CH_3CO_2^-(aq) + H^+(aq)$
- **c** $H_2(g) + CO_2(g) \rightleftharpoons H_2O(g) + CO(g)$

2 Consider the following equilibrium reaction.

$N_2(g) + 3H_2(g) \rightleftharpoons 2NH_3(g)$
$\Delta H^\ominus = -92\text{ kJ mol}^{-1}$

- **a** What would be the effect on the equilibrium position of heating the reaction? Choose from 'move to the right', 'move to the left', and 'no change'.
- **b** What would be the effect on the equilibrium position of adding an iron catalyst? Choose from 'move to the right', 'move to the left', and 'no change'.
- **c** What effect would an iron catalyst have on the reaction?
- **d** To get the maximum yield of ammonia in this reaction would a high or low pressure be best? Explain your answer.

6.3 The equilibrium constant K_c

Learning objectives:

- → Define the expression reversible reaction.
- → Define the term chemical equilibrium.
- → State the definition of an equilibrium constant and describe how it is determined.

Specification reference: 3.1.6

Synoptic link

There is more about titrations in Topic 2.5, Balanced equations and related calculations.

▲ **Figure 1** *Titrating the ethanoic acid to investigate the equilibrium position*

Hint

The concentration of a solution is the number of moles of solute dissolved in 1 dm^3 of solution. A square bracket around a formula is shorthand for 'concentration of that substance in mol dm^{-3}'.

Study tip

Practise calculating K_c from given data.

As you have seen, many reactions are reversible and do not go to completion, but instead end up as an equilibrium mixture of reactants and products. A reversible reaction that can reach equilibrium is indicated by the symbol $\rightleftharpoons$. In this topic you see how you can tackle equilibrium reactions mathematically. You will deal only with homogeneous systems – those where all the reactants and products are in the same phase, for example, all liquids.

The equilibrium constant K_c

Many reactions are reversible and will reach equilibrium with time. The reaction between ethanol, C_2H_5OH, and ethanoic acid, CH_3CO_2H, to produce ethyl ethanoate, $CH_3CO_2C_2H_5$, (an ester) and water is typical.

If ethanol and ethanoic acid are mixed in a flask (stoppered to prevent evaporation) and left for several days with a strong acid catalyst, an equilibrium mixture is obtained in which *all four* substances are present. You can write:

$$\underset{\text{ethanol}}{C_2H_5OH\,(l)} + \underset{\text{ethanoic acid}}{CH_3CO_2H\,(l)} \rightleftharpoons \underset{\text{ethyl ethanoate}}{CH_3CO_2C_2H_5(l)} + \underset{\text{water}}{H_2O(l)}$$

The mixture may be analysed by titrating the ethanoic acid with standard alkali (allowing for the amount of acid catalyst added). It is possible to do this without significantly disturbing the equilibrium mixture because the reversible reaction is much slower than the titration reaction.

The titration allows us to work out the number of moles of ethanoic acid in the equilibrium mixture. From this you can calculate the number of moles of the other components (and from this their concentrations if the total volume of the mixture is known).

If several experiments are done with different quantities of starting materials, it is always found that the ratio:

$$\frac{[CH_3CO_2C_2H_5(l)]_{eqm}\,[H_2O(l)]_{eqm}}{[CH_3CO_2H(l)]_{eqm}\,[C_2H_5OH(l)]_{eqm}}$$

has a constant value, provided the experiments are done at the same temperature. The subscript 'eqm' means that the concentrations have been measured when equilibrium has been reached.

For *any* reaction that reaches an equilibrium we can write the equation in the form:

$$aA + bB + cC \rightleftharpoons xX + yY + zZ$$

Then the expression $\dfrac{[X]_{eqm}{}^{x}\,[Y]_{eqm}{}^{y}\,[Z]_{eqm}{}^{z}}{[A]_{eqm}{}^{a}\,[B]_{eqm}{}^{b}\,[C]_{eqm}{}^{c}}$ is constant, provided the temperature is constant. We call this constant, K_c .This expression can be applied to *any* reversible reaction. K_c is called the **equilibrium constant** and is different for different reactions. It changes with temperature. The units of K_c vary, and you must work them out for each reaction by cancelling out the units of each term, for example:

$$2A + B \rightleftharpoons 2C \qquad K_c = \frac{[C]^2}{[A]^2[B]}$$

Units are: $$\frac{\cancel{(\text{mol dm}^{-3})^2}}{\cancel{(\text{mol dm}^{-3})^2}(\text{mol dm}^{-3})} = \frac{1}{\text{mol dm}^{-3}} = \text{mol}^{-1}\text{dm}^3$$

The value of K_c is found by experiment for any particular reaction at a given temperature.

Maths link

See Section 8, Mathematical skills, if you are not sure about cancelling units.

Study tip

It is acceptable to omit the 'eqm' subscripts unless they are specifically asked for.

To find the value of K_c for the reaction between ethanol and ethanoic acid

0.10 mol of ethanol is mixed with 0.10 mol of ethanoic acid and allowed to reach equilibrium. The total volume of the system is made up to 20.0 cm³ (0.020 dm³) with water. By titration, it is found that 0.033 mol ethanoic acid is present once equilibrium is reached.

From this you can work out the number of moles of the other components present at equilibrium:

At start

$C_2H_5OH\,(l)$ +	$CH_3CO_2H\,(l)$ $\rightleftharpoons$	$CH_3CO_2C_2H_5(l)$ +	$H_2O(l)$
0.10 mol	0.10 mol	0 mol	0 mol

You know that there are 0.033 mol of CH_3CO_2H at equilibrium. This means that:

- there must also be 0.033 mol of C_2H_5OH at equilibrium. (The equation tells you that they react 1:1 and you know we started with the same number of moles of each.)
- (0.10 – 0.033) = 0.067 mol of CH_3CO_2H has been used up. The equation tells you that when 1 mol of CH_3CO_2H is used up, 1 mol each of $CH_3CO_2C_2H_5$ and H_2O are produced. So, there must be 0.067 mol of each of these.

At equilibrium

$C_2H_5OH\,(l)$ +	$CH_3CO_2H\,(l)$ $\rightleftharpoons$	$CH_3CO_2C_2H_5(l)$ +	$H_2O(l)$
0.033 mol	0.033 mol	0.067 mol	0.067 mol

You need the *concentrations* of the components at equilibrium. As the volume of the system is 0.020 dm³ these are:

$C_2H_5OH\,(l)$ +	$CH_3CO_2H\,(l)$ $\rightleftharpoons$	$CH_3CO_2C_2H_5(l)$ +	$H_2O(l)$
$\frac{0.033}{0.020}$ mol dm⁻³	$\frac{0.033}{0.020}$ mol dm⁻³	$\frac{0.067}{0.020}$ mol dm⁻³	$\frac{0.067}{0.020}$ mol dm⁻³

Enter the concentrations into the equilibrium equation:

$$K_c = \frac{[CH_3CO_2C_2H_5(l)][H_2O(l)]}{[CH_3CO_2H(l)][C_2H_5OH(l)]}$$

$$K_c = \frac{[0.067/0.020\text{ mol dm}^{-3}][0.067/0.020\text{ mol dm}^{-3}]}{[0.033/0.020\text{ mol dm}^{-3}][0.033/0.020\text{ mol dm}^{-3}]} = 4.1$$

The units all cancel out, and the volumes (0.020 dm³) cancel out, so in this case you didn't need to know the volume of the system, so $K_c = 4.1$. In this case, K_c has no units.

Summary questions

1 Write down the expressions for the equilibrium constant for the following:

  a $A + B \rightleftharpoons C$

  b $2A + B \rightleftharpoons C$

  c $2A + 2B \rightleftharpoons 2C$

2 Work out the units for K_c for question **1a** to **c**.

3 For the reaction between ethanol and ethanoic acid, at a different temperature to the example above, the equilibrium mixture was found to contain 0.117 mol of ethanoic acid, 0.017 mol of ethanol, 0.083 mol ethyl ethanoate and 0.083 mol of water.

  a Calculate K_c

  b Why do you not need to know the volume of the system to calculate K_c in this example?

  c Is the equilibrium further to the right or to the left compared with the worked example above?

6.4 Calculations using equilibrium constant expressions

Learning objectives:

- Describe how K_c is used to work out the composition of an equilibrium mixture.

Specification reference: 3.1.6

A reaction that has reached equilibrium at a given temperature will be a mixture of reactants and products. You can use the equilibrium expression to calculate the composition of this mixture.

Worked example: Calculating the composition of a reaction mixture

The reaction of ethanol and ethanoic acid is:

$$\underset{\text{ethanol}}{C_2H_5OH\,(l)} + \underset{\text{ethanoic acid}}{CH_3CO_2H\,(l)} \rightleftharpoons \underset{\text{ethyl ethanoate}}{CH_3CO_2C_2H_5(l)} + \underset{\text{water}}{H_2O(l)}$$

You know that at equilibrium:

$$K_c = \frac{[CH_3CO_2C_2H_5(l)][H_2O(l)]}{[CH_3CO_2H(l)][C_2H_5OH(l)]}$$

Suppose that $K_c = 4.0$ at the temperature of our experiment and you want to know how much ethyl ethanoate you could produce by mixing one mol of ethanol and one mol of ethanoic acid. Set out the information as shown below:

Equation:	$C_2H_5OH\,(l)$ +	$CH_3CO_2H\,(l)$ $\rightleftharpoons$	$CH_3CO_2C_2H_5(l)$ +	$H_2O(l)$
	ethanol	ethanoic acid	ethyl ethanoate	water
At start:	1 mol	1 mol	1 mol	1 mol
At equilibrium:	$(1 - x)$ mol	$(1 - x)$ mol	x mol	x mol

You do not know how many moles of ethyl ethanoate will be produced, so you call this x. The equation tells us that x mol of water will also be produced and, in doing so, x mol of both ethanol and ethanoic acid will be used up. So the amount of each of these remaining at equilibrium is $(1 - x)$ mol.

These figures are in moles, but you need concentrations in mol dm^{-3} to substitute in the equilibrium law expression. Suppose the volume of the system at equilibrium was V dm^{-3}. Then:

$$[C_2H_5OH(l)]_{eqm} = \frac{(1-x)}{V}\ \text{mol dm}^{-3}$$

$$[CH_3CO_2H(l)]_{eqm} = \frac{(1-x)}{V}\ \text{mol dm}^{-3}$$

$$[CH_3CO_2C_2H_5(l)]_{eqm} = \frac{x}{V}\ \text{mol dm}^{-3}$$

$$[H_2O(l)]_{eqm} = \frac{x}{V}\ \text{mol dm}^{-3}$$

These figures may now be put into the expression for K_c:

$$K_c = \frac{x/\cancel{V} \times x/\cancel{V}}{(1-x)/\cancel{V} \times (1-x)/\cancel{V}}$$

The *V*s cancel, so *in this case* you do not need to know the actual volume of the system.

$$4.0 = \frac{x \times x}{(1-x) \times (1-x)}$$

$$4.0 = \frac{x^2}{(1-x)^2}$$

Taking the square root of both sides, you get:

$$2 = \frac{x}{(1-x)}$$

$$2(1-x) = x$$

$$2 - 2x = x$$

$$2 = 3x$$

$$x = \frac{2}{3}$$

So $\frac{2}{3}$ mol of ethyl ethanoate and $\frac{2}{3}$ mol of water are produced if the reaction reaches equilibrium, and the composition of the equilibrium mixture would be: ethanol $\frac{1}{3}$ mol, ethanoic acid $\frac{1}{3}$ mol, ethyl ethanoate $\frac{2}{3}$ mol, water $\frac{2}{3}$ mol.

Study tip

The volume of the reaction mixture will cancel for all systems with equal numbers of moles of products and reactants, so *V* is sometimes omitted. But, it is always better to include *V* and cancel it out later, so you will not forget it for systems where the *V*s do not cancel out.

You can also use K_c to find the amount of a reactant needed to give a required amount of product.

Worked example: Calculating the amount of a reactant needed

For the following reaction in ethanol solution, $K_c = 30.0\ \text{mol}^{-1}\ \text{dm}^3$:

$$\underset{\text{propanone}}{CH_3COCH_3} + \underset{\text{hydrogen cyanide}}{HCN} \rightleftharpoons \underset{\text{2-hydroxy-2-methylpropanenitrile}}{CH_3C(CN)(OH)CH_3}$$

$$K_c = \frac{[CH_3C(CN)(OH)CH_3]}{[CH_3COCH_3][HCN]} = 30.0\ \text{mol}^{-1}\ \text{dm}^3$$

Suppose you are carrying out this reaction in 2.00 dm³ of ethanol. How much hydrogen cyanide is required to produce 1.00 mol of product if you start with 4.00 mol of propanone?

Set out as before with the quantities at the start and at equilibrium.

At equilibrium, you want 1 mol of product. Let *x* be the number of moles of HCN required.

Equation:	CH_3COCH_3	+	HCN	$\rightleftharpoons$	$CH_3C(CN)(OH)CH_3$
At start:	4.00 mol		*x* mol		0 mol
At equilibrium:	(4.00 − 1.00) mol		(*x* − 1.00) mol		1.00 mol
	3.00 mol		(*x* − 1.00) mol		1.00 mol

Hint

Notice that this equilibrium constant *does* have units, which can be deduced by cancelling the units of the concentrations in the expression for K_c.

These are the numbers of moles, but we need the *concentrations* to put in the equilibrium law expression. The volume of the solution is 2.00 dm^3 and the units for concentration are mol dm^{-3} so you next divide each quantity by 2.00 dm^3.

So, at equilibrium $[CH_3COCH_3]_{eqm} = \dfrac{3.00}{2.00}$ mol dm^{-3}

$$[HCN]_{eqm} = \frac{(x - 1.00)}{2.00}\ \text{mol dm}^{-3}$$

$$[CH_3C(CN)(OH)CH_3]_{eqm} = \frac{1.00}{2.00}\ \text{mol dm}^{-3}$$

Putting the figures into the equilibrium expression:

$$30.0^3\ \text{mol}^{-1}\ \text{dm}^3 = \frac{1.00/2.00\ \cancel{\text{mol dm}^{-3}}}{3.00/2.00\ \cancel{\text{mol dm}^{-3}} \times (x - 1.00)/2.00\ \text{mol dm}^{-3}}$$

Cancelling through and rearranging we have:

$$30\left(\frac{3/2(x-1)}{2}\right) = \frac{1}{2}$$

$$45(x - 1) = 1$$

$$45x = 46$$

$$x = \frac{46}{45} = 1.02$$

So, to obtain 1 mol of product you must start with 1.02 mol hydrogen cyanide, if the volume of the system is 2.00 dm^3.

In this example the volume of the system *does* make a difference, because this reaction does not have the same number of moles of products and reactants.

Summary questions

1 Try reworking the problem above with the same conditions but:

a with a volume of 1.00 dm^3 of ethanol

b starting with 2.0 mol of propanone

c to produce 2.0 mol of product.

6.5 The effect of changing conditions on equilibria

As you have seen Le Châtelier's principle states that when a system at equilibrium is disturbed, the equilibrium position moves in the direction that will reduce the disturbance. You can use Le Châtelier's principle to predict the qualitative effect of changing temperature and concentration on the position of equilibrium.

In this topic you look at what underlies this by examining the effect of changing conditions on the equilibrium constant K_c.

Learning objectives:

- → Explain how Le Châtelier's principle can predict how changes in conditions affect the position of equilibrium.
- → Describe how the equilibrium constant is affected by changing the conditions of a reaction.

Specification reference: 3.1.6

The effect of changing temperature on the equilibrium constant

Changing the temperature changes the value of the equilibrium constant, K_c. Whether K_c increases or decreases depends on whether the reaction is exothermic or endothermic, see the summary in Table 1.

▼ Table 1 *The effect of changing temperature on equilibria*

Type of reaction	Temperature change	Effect on K_c	Effect on products	Effect on reactants	Direction of change of equilibrium
endothermic	decrease	decrease	decrease	increase	moves left
endothermic	increase	increase	increase	decrease	moves right
exothermic	increase	decrease	decrease	increase	moves left
exothermic	decrease	increase	increase	decrease	moves right

If the equilibrium constant K_c increases in value, the equilibrium moves to the right, that is, the forward direction (more product). If it decreases in value, the equilibrium moves to the left, that is, the backward direction (less product).

This is because the expression for K_c is always of the form $\frac{[\text{products}]}{[\text{reactants}]}$.

The general rule is that:

- For an exothermic reaction (ΔH is negative) increasing the temperature decreases the equilibrium constant.
- For an endothermic reaction (ΔH is positive) increasing the temperature increases the equilibrium constant.

So for an exothermic reaction, increasing the temperature will move the equilibrium to the left – for an endothermic reaction, increasing the temperature will move the equilibrium to the right.

Study tip

When the value for ΔH is given for a reversible reaction, it is taken to refer to the forward reaction, that is, left to right.

The effect of changing concentration on the position of equilibrium

First remember that the equilibrium constant does not change unless the temperature changes.

Study tip

Practise applying Le Châtelier's principle for all changes in conditions.

Look at the following example:

$$\underset{\text{ethanol}}{C_2H_5OH\,(l)} + \underset{\text{ethanoic acid}}{CH_3CO_2H\,(l)} \rightleftharpoons \underset{\text{ethyl ethanoate}}{CH_3CO_2C_2H_5(l)} + \underset{\text{water}}{H_2O(l)}$$

Le Châtelier's principle tells you that the equilibrium will react to any disturbance by moving in such a way as to reduce the disturbance.

Imagine you add more ethanol, thereby increasing its concentration. The only way this concentration can be reduced is by some of the ethanol reacting with ethanoic acid producing more ethyl ethanoate and water. Eventually a new equilibrium will be set up with relatively more of the products. The equilibrium has moved to the right (or in the forward direction).

Let us see how this works mathematically.

You know that:

$$K_c = \frac{[CH_3CO_2C_2H_5(l)][H_2O(l)]}{[CH_3CO_2H(l)][C_2H_5OH(l)]}$$

Remember that K_c remains constant, provided that temperature remains constant. Adding ethanol makes the bottom line of the **equilibrium law expression** larger. To restore the situation, some of the ethanol reacts with ethanoic acid reducing both the concentrations in the bottom line of the fraction. This produces more ethyl ethanoate and water, thus increasing the value in the top line of the fraction. The combined effect is to restore the fraction to the original value of K_c.

K_c and the position of equilibrium

The size of the equilibrium constant K_c can tell us about the composition of the equilibrium mixture. The equilibrium expression is always of the general form $\frac{[\text{products}]}{[\text{reactants}]}$. So:

- If K_c is much greater than 1, products predominate over reactants and the equilibrium position is over to the right.
- If K_c is much less than 1, reactants predominate and the equilibrium position is over to the left.

Reactions where the equilibrium constant is greater than 10^{10} are usually regarded as going to completion, whilst those with an equilibrium constant of less than 10^{-10} are regarded as not taking place at all.

Synoptic link

Look back at Topic 5.1, Collision theory, to revise activation energy for reactions.

Catalysts and the value of K_c

Catalysts have no effect whatsoever on the value of K_c and therefore the position of equilibrium. This is because they affect the rates of both forward and back reactions equally. They do this by reducing the activation energy for the reactions. They do however affect the *rate* at which equilibrium is attained – this is important in industrial processes.

Synoptic link

The equilibrium constant can also be calculated for gases using partial pressures. This equilibrium constant has the symbol K_p. The equilibrium constant K_p is covered in Chapter 19, Equilibrium constant K_p.

Gaseous equilibria

Reversible reactions may take place in the gas phase as well as in solution. These include many reactions of industrial importance such as the manufacture of ammonia by the Haber process and a key stage of the Contact process for making sulfuric acid. Gaseous equilibria also obey the equilibrium law, but usually their concentrations are expressed in a different way using partial pressures rather than concentrations.

Reversible reactions in industry

Many important industrial processes involve reversible reactions as a key step. One example is the Haber process for making ammonia from its elements:

$N_2(g) + 3H_2(g) \rightleftharpoons 2NH_3(g)$ $\Delta H = -92$ kJ mol^{-1}

To gain the maximum conversion ammonia, Le Châtelier's principle tells us that high pressure and low temperature are required. However, low temperature will slow down the reaction. High pressure requires expensive equipment to withstand it and also incurs energy costs in compressing the gases. So a set of compromise conditions is reached with medium pressure and temperature (by industrial standards) and a catalyst is used to speed up the reaction. Unconverted nitrogen and hydrogen are recycled.

Similar considerations apply to the Contact process for making sulfuric acid and the hydration of ethene to give ethanol.

Summary questions

1 A + B $\rightleftharpoons$ C + D represents an exothermic reaction and

$K_c = \frac{[C][D]}{[A][B]}$ In the above expression, what would happen to K_c:

a if the temperature were decreased

b if more A were added to the mixture

c if a catalyst were added?

2 The reaction of ethanol with ethanoic acid produces ethyl ethanoate and water.

$$C_2H_5OH(l) + CH_3COOH(l) \rightleftharpoons CH_3COOC_2H_5(l) + H_2O(l)$$

A student suggested that the yield of ethyl ethanoate, $CH_3COOC_2H_5$, could be increased by removing the water as it was formed.

Explain, using the idea of K_c, why this suggestion is sensible.

3 These questions are about reversible reactions. Give the correct word from **increases/decreases/does not change** to fill in the blank for each statement.

a In an endothermic reaction K_c ______ when the temperature is increased.

b In an endothermic reaction K_c ______ when the concentration of the reactants is decreased.

c In an exothermic reaction K_c ______ when the temperature is decreased.

d In an exothermic reaction K_c ______ when the concentration of the reactants is increased.

e If a suitable catalyst is added to the reaction K_c ______.

Practice questions

1 Methanol can be synthesised from carbon monoxide by the reversible reaction shown below.

$$CO(g) + 2H_2(g) \rightleftharpoons CH_3OH(g) \qquad \Delta H = -91\ kJ\ mol^{-1}$$

The process operates at a pressure of 5 MPa and a temperature of 700 K in the presence of a copper-containing catalyst. This reaction can reach dynamic equilibrium.

(a) By reference to rates and concentrations, explain the meaning of the term *dynamic equilibrium.*

(2 marks)

(b) Explain why a high yield of methanol is favoured by high pressure.

(2 marks)

(c) Suggest **two** reasons why the operation of this process at a pressure much higher than 5 MPa would be very expensive.

(2 marks)

(d) State the effect of an increase in temperature on the equilibrium yield of methanol and explain your answer.

(3 marks)

(e) If a catalyst were not used in this process, the operating temperature would have to be greater than 700 K. Suggest why an increased temperature would be required.

(1 mark)

AQA, 2003

2 At high temperatures, nitrogen is oxidised by oxygen to form nitrogen monoxide in a reversible reaction as shown in the equation below.

$$N_2(g) + O_2(g) \rightleftharpoons 2NO(g) \qquad \Delta H^{\ominus} = +180\ kJ\ mol^{-1}$$

(a) In terms of electrons, give the meaning of the term *oxidation*.

(1 mark)

(b) State and explain the effect of an increase in pressure, and the effect of an increase in temperature, on the yield of nitrogen monoxide in the above equilibrium.

(6 marks)

AQA, 2006

3 Hydrogen is produced on an industrial scale from methane as shown by the equation below.

$$CH_4(g) + H_2O(g) \rightleftharpoons CO(g) + 3H_2(g) \qquad \Delta H^{\ominus} = +205\ kJ\ mol^{-1}$$

(a) State Le Châtelier's principle.

(1 mark)

(b) The following changes are made to this reaction at equilibrium. In each case, predict what would happen to the yield of hydrogen from a given amount of methane. Use Le Châtelier's principle to explain your answer.

(i) The overall pressure is increased.

(ii) The concentration of steam in the reaction mixture is increased.

(6 marks)

(c) At equilibrium, a high yield of hydrogen is favoured by high temperature. In a typical industrial process, the operating temperature is usually less than 1200 K. Suggest two reasons why temperatures higher than this are not used.

(2 marks)

AQA, 2004

4 The equation for the formation of ammonia is shown below.

$$N_2(g) + 3H_2(g) \rightleftharpoons 2NH_3(g)$$

Experiment **A** was carried out starting with 1 mol of nitrogen and 3 mol of hydrogen at a constant temperature and a pressure of 20 MPa.

Curve **A** shows how the number of moles of ammonia present changed with time.

Curves **B**, **C**, and **D** refer to similar experiments, starting with 1 mol of nitrogen and 3 mol of hydrogen. In each experiment different conditions were used.

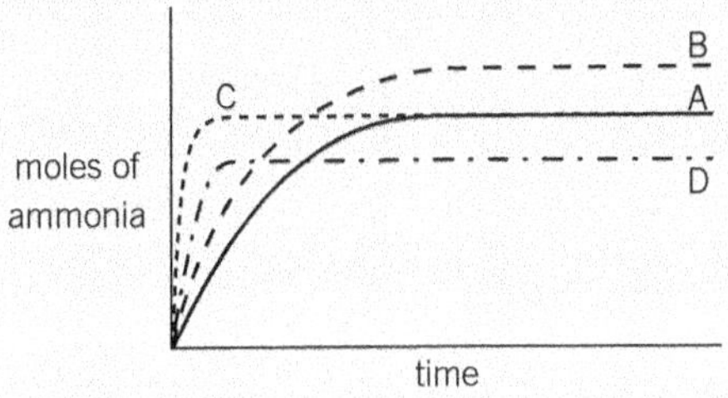

(a) On a copy of curve **A**, mark the point that represents the time at which equilibrium is first reached. Label this point **X**.

(1 mark)

(b) State Le Châtelier's principle.

(1 mark)

(c) Use Le Châtelier's principle to identify which one of the curves **B**, **C**, or **D** represents an experiment carried out at the same temperature as experiment **A** but at a higher pressure. Explain why this curve is different from curve **A**.

(4 marks)

(d) Identify which one of the curves **B**, **C**, or **D** represents an experiment in which the conditions are the same as in experiment **A** except that a catalyst is added to the reaction mixture. Explain your choice of curve.

(3 marks)

AQA, 2005

5 The reaction of methane with steam produces hydrogen for use in many industrial processes. Under certain conditions the following reaction occurs.

$$CH_4(g) + 2H_2O(g) \rightleftharpoons CO_2(g) + 4H_2(g) \qquad \Delta H = +165 \text{ kJ mol}^{-1}$$

(a) Initially, 1.0 mol of methane and 2.0 mol of steam were placed in a flask and heated with a catalyst until equilibrium was established. The equilibrium mixture contained 0.25 mol of carbon dioxide.

(i) Calculate the amounts, in moles, of methane, steam, and hydrogen in the equilibrium mixture.

(3 marks)

(ii) The volume of the flask was 5.0 dm^3. Calculate the concentration, in mol dm^{-3}, of methane in the equilibrium mixture.

(1 mark)

(b) The table below shows the equilibrium concentration of each gas in a different equilibrium mixture in the same flask and at temperature *T*.

gas	$CH_4(g)$	$H_2O(g)$	$CO_2(g)$	$H_2(g)$
concentration / mol dm^{-3}	0.10	0.48	0.15	0.25

(i) Write an expression for the equilibrium constant, K_c, for this reaction.

(1 mark)

(ii) Calculate a value for K_c at temperature *T* and give its units.

(3 marks)

(c) The mixture in part (**b**) was placed in a flask of volume greater than 5.0 dm^3 and allowed to reach equilibrium at temperature *T*.
State and explain the effect on the amount of hydrogen.

(3 marks)

(d) Explain why the amount of hydrogen decreases when the mixture in part (**b**) reaches equilibrium at a lower temperature.

(2 marks)

AQA, 2010

Answers to the Practice Questions and Section Questions are available at
www.oxfordsecondary.com/oxfordaqaexams-alevel-chemistry

7 Oxidation, reduction, and redox reactions

7.1 Oxidation and reduction

Learning objectives:

- → Define a redox reaction in terms of oxygen or hydrogen transfer.
- → Define a redox reaction in terms of electron transfer.
- → Define a half equation.

Specification reference: 3.1.7

Redox reactions

The word redox is short for reduction–oxidation. Historically, **oxidation** was used for reactions in which oxygen was added.

In this reaction copper has been oxidised to copper oxide. Oxygen is called an **oxidising agent**.

$$Cu(s) + \frac{1}{2}O_2(g) \rightarrow CuO(s)$$

Reduction described a reaction in which oxygen was removed.

In this reaction copper oxide has been reduced and hydrogen is the **reducing agent**.

$$CuO(s) + H_2(g) \rightarrow Cu(s) + H_2O(l)$$

As hydrogen was often used to remove oxygen, the addition of hydrogen was called reduction.

In this reaction chlorine has been reduced because hydrogen has been added to it.

$$Cl_2(g) + H_2(g) \rightarrow 2HCl(g)$$

The reverse, where hydrogen was removed, was called oxidation.

Gaining and losing electrons – redox reactions

By describing what happens to the *electrons* in the above reactions, you get a much more general picture. When something is oxidised it loses electrons, and when something is reduced it gains electrons. Since **redox reactions** always involve the movement of electrons they are also called **electron transfer reactions**. You can see the transfer of electrons by separating a redox reaction into two half equations that show the gain and loss of electrons.

Worked example 1: Half equations

Look again at the reaction between copper and oxygen to form copper oxide:

$$Cu + \frac{1}{2}O_2 \rightarrow CuO$$

Copper oxide is an ionic compound so you can write the balanced symbol equation using $(Cu^{2+} + O^{2-})$ (instead of CuO) to show the ions present in copper oxide:

$$Cu + \frac{1}{2}O_2 \rightarrow (Cu^{2+} + O^{2-})$$

Next look at the copper. It has lost two electrons so it has been oxidised.

$$Cu - 2e^- \rightarrow Cu^{2+} \qquad \text{or} \qquad Cu \rightarrow Cu^{2+} + 2e^-$$

This is a **half equation**. It is usual to write half equations with plus electrons rather than minus electrons, that is:

$$Cu \rightarrow Cu^{2+} + 2e^- \text{ rather than}$$

$$Cu - 2e^- \rightarrow Cu^{2+}$$

Next look at the oxygen. It has gained two electrons so it has been reduced:

$$\frac{1}{2}O_2(g) + 2e^- \rightarrow O^{2-}$$

If you add the two half equations together, you end up with the original equation. Notice that the numbers of electrons cancel out.

$$Cu \rightarrow Cu^{2+} + \cancel{2e^-}$$

$$\frac{1}{2}O_2(g) + \cancel{2e^-} \rightarrow O^{2-}$$

$$Cu(s) + \frac{1}{2}O_2(g) \rightarrow (Cu^{2+} + O^{2-})(s)$$

Worked example 2: Half equations

When copper oxide reacts with magnesium, copper and magnesium oxide are produced:

$$CuO(s) + Mg(s) \rightarrow MgO(s) + Cu(s)$$

Write the equation with copper oxide as $(Cu^{2+} + O^{2-})$ and magnesium oxide as $(Mg^{2+} + O^{2-})$ to show the ions present.

$$(Cu^{2+} + O^{2-}) + Mg \rightarrow Cu + (Mg^{2+} + O^{2-})$$

Look at the copper. It has gained two electrons so it has been reduced.

$$Cu^{2+} + 2e^- \rightarrow Cu$$

Look at the magnesium. It has lost electrons so it has been oxidised.

$$Mg \rightarrow Mg^{2+} + 2e^-$$

Notice that the O^{2-} ion takes no part in the reaction. It is called a **spectator ion**.

If you add these half equations you get:

$$Cu^{2+} + Mg \rightarrow Cu + Mg^{2+}$$

This is the ionic equation for the redox reaction.

The definition of oxidation and reduction now used is:

Oxidation Is Loss of electrons.
Reduction Is Gain of electrons.

By this definition, magnesium is oxidised by *anything* that removes electrons from it (not just oxygen) leaving a positive ion. For example, chlorine oxidises magnesium:

$$Mg(s) + Cl_2(g) \rightarrow (Mg^{2+} + 2Cl^-)(s)$$

Hint

The phrase OIL RIG makes the definition of oxidation and reduction easy to remember.

Oxidation
Is
Loss (of electrons)
Reduction
Is
Gain (of electrons)

Look at the magnesium. It has lost electrons and has therefore been oxidised.

$$Mg \rightarrow Mg^{2+} + 2e^-$$

Look at the chlorine. It has gained electrons and has therefore been reduced.

$$Cl_2 + 2e^- \rightarrow 2Cl^-$$

And adding the two half equations together, the electrons cancel out:

$$Mg(s) + Cl_2(g) \rightarrow (Mg^{2+} + 2Cl^-)(s)$$

You may find that adding arrows to the equation, which show the transfer of electrons, helps keep track of them, as shown in Figure 1.

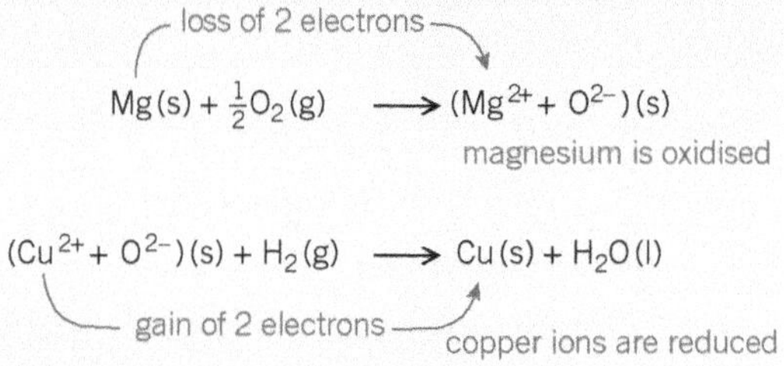

▲ **Figure 1** *Writing the electrons that are transferred helps to keep track of them*

In a chemical reaction, if one species is oxidised (loses electrons), another *must* be reduced (gains them).

Oxidising and reducing agents

It follows from the above that:

- reducing agents give away electrons – they are electron donors
- oxidising agents accept electrons.

Summary questions

1 The following questions are about the reaction:

$$Ca(s) + Br_2(g) \rightarrow (Ca^{2+} + 2Br^-)(s)$$

a Which element has gained electrons?

b Which element has lost electrons?

c Which element has been oxidised?

d Which element has been reduced?

e Write the half equations for these redox reactions.

f What is the oxidising agent?

g What is the reducing agent?

Oxidation states

Oxidation states are used to see what has been oxidised and what has been reduced in a redox reaction. Oxidation states are also called **oxidation numbers**.

The idea of oxidation states

Each element in a compound is given an oxidation state. In an ionic compound the oxidation state simply tells us how many electrons it has lost or gained, compared with the element in its uncombined state. In a molecule, the oxidation state tells us about the distribution of electrons between elements of different electronegativity. The more electronegative element is given the negative oxidation state.

- Every element in its uncombined state has an oxidation state of zero.
- A positive number shows that the element has lost electrons and has therefore been oxidised. For example, Mg^{2+} has an oxidation state of +2.
- A negative number shows that the element has gained electrons and has therefore been reduced. For example, Cl^- has an oxidation state of −1.
- The more positive the number, the more the element has been oxidised. The more negative the number, the more it has been reduced.
- The numbers always have a + or – sign unless they are zero.

Rules for finding oxidation states

The following rules will allow you to work out oxidation states:

1 Uncombined elements have oxidation state of 0.
2 Some elements always have the same oxidation state in all their compounds. Others usually have the same oxidation state. Table 1 gives the oxidation states of these elements.

▼ **Table 1** *The usual oxidation states of some elements*

Element	Oxidation state in compound	Example
hydrogen, H	+1 (except in metal hydrides, e.g., NaH, where it is −1)	HCl
Group 1	always +1	NaCl
Group 2	always +2	$CaCl_2$
aluminium, Al	always +3	$AlCl_3$
oxygen, O	−2 (except in peroxides where it is −1, and the compound OF_2, where it is +2)	Na_2O
fluorine, F	Always −1	NaF
chlorine, Cl	−1 (except in compounds with F and O, where it has positive values)	NaCl

3 The sum of all the oxidation states in a compound equals 0, since all compounds are electrically neutral.

Learning objectives:

→ Define an oxidation state.

→ Describe how oxidation states are worked out.

Specification reference: 3.1.7

4 The sum of the oxidation states of a complex ion, such as NH_4^+ or SO_4^{2-}, equals the charge on the ion.
5 In a compound the most electronegative element always has a negative oxidation state.

Working out oxidation states of elements in compounds

Start with the correct formula. Look for the elements whose oxidation states you know from the rules. Then deduce the oxidation states of any other element. Some examples are shown below.

Study tip

You should know the rules for finding oxidation states.

Phosphorus pentachloride, PCl_5

Chlorine has an oxidation state of –1, so the phosphorus must be +5, to make the sum of the oxidation states zero.

Ammonia, NH_3

Hydrogen has an oxidation state of +1, so the nitrogen must be –3, to make the sum of the oxidation states zero. Also, nitrogen is more electronegative than hydrogen, so hydrogen must have a positive oxidation state.

Nitric acid, HNO_3

Each oxygen has an oxidation state of –2, making –6 in total.

Hydrogen has an oxidation state of +1.

So the nitrogen must be +5, to make the sum of the oxidation states zero.

Notice that nitrogen may have different oxidation states in different compounds. Here nitrogen has a positive oxidation state because it is combined with a more electronegative element, oxygen.

Hydrogen sulfide, H_2S

Hydrogen has an oxidation state of +1, so the sulfur must be –2, to make the sum of the oxidation states zero.

Sulfate ion, SO_4^{2-}

Each oxygen has an oxidation state of –2, making –8 in total.

So the sulfur must be +6, to make the sum of the oxidation states equal to the charge on the ion.

Notice that sulfur may have different oxidation states in different compounds.

Black copper oxide, CuO

Oxygen has an oxidation state of –2, so the copper must be +2, to make the sum of the oxidation states zero.

Red copper oxide, Cu_2O

Oxygen has an oxidation state of –2, so each copper must be +1, to make the sum of the oxidation states zero.

Oxidation states are written in Roman numerals to distinguish between similar compounds in which the metal has a different oxidation state. So, black copper oxide is copper(II) oxide and red copper oxide is copper(I) oxide. These compounds are shown in Figure 1.

▲ **Figure 1** *The two oxides of copper – copper(II) oxide (left) and copper(I) oxide (right)*

Chlorine – an element with many oxidation states

The element chlorine, Cl_2, has an oxidation state of zero by definition. When combined with other elements, it can exhibit several different oxidation states from −1 to +7.

When combined with a metal, chlorine forms ionic compounds which contain the Cl^- ion whose oxidation state is −1. However, chlorine forms a number of compounds which also contain oxygen. Oxygen is more electronegative than chlorine and so the chlorine forms positive oxidation states. For example:

Formula	Oxidation state of chlorine	Name
Cl_2	0	Chlorine
NaCl	−1	Sodium chloride
NaClO	+1	Sodium hypochlorite (sodium chlorate(I))
$NaClO_2$	+3	Sodium chlorite (sodium chlorate(III))
ClO_2	+4	Chlorine dioxide
$NaClO_3$	+5	Sodium chlorate (sodium chlorate(V))
Cl_2O_6	+6	Dichlorine hexoxide
$NaClO_4$	+7	Sodium perchlorate (sodium chlorate(VII))

1 Suggest why fluorine does not form any compounds in which it has a positive oxidation state.

Flourine is the most electronegative element so it has a negative oxidation state in *all* its compounds.

Summary questions

1 Work out the oxidation states of each element in the following compounds:

a $PbCl_2$

b CCl_4

c $NaNO_3$

2 In the reaction: $CuO + Mg \rightarrow Cu + MgO$, what are the oxidation states of oxygen before and after the reaction?

3 In the reaction: $2Cu + O_2 \rightarrow 2CuO$, what are the oxidation states of oxygen before and after the reaction?

4 In the reaction: $FeCl_2 + \frac{1}{2}Cl_2 \rightarrow FeCl_3$, what are the oxidation states of iron before and after the reaction?

5 Give the oxidation state of the following:

a P in PO_4^{3-}

b N in NO_3^-

c N in NH_4^+

7.3 Redox equations

Learning objectives:

- Explain how half equations are used to balance an equation.
- Deduce half equations from a redox equation.

Specification reference: 3.1.7

Hint

Another way of working is to remember that when an element is reduced it gains electrons and its oxidation state is reduced.

For example, in: $M^{3+} \rightarrow M^{2+}$ the number of plusses has been reduced so M has been reduced.

It follows that for: $M^{2+} \rightarrow M^{3+}$ the number of plusses has been increased so M has been oxidised.

Using oxidation states in redox equations

You saw in Topic 7.1 that you can work out which element has been oxidised and which has been reduced in a redox reaction by considering electron transfer.

Remember that oxidation is loss of electrons (OIL) and reduction is gain of electrons (RIG).

You can also use oxidation states to help you to understand redox reactions.

When an element is reduced, it gains electrons and its oxidation state goes down. In the reaction below, iron is reduced because its oxidation state has gone down from +3 to +2, whilst iodide is oxidised:

$$\overset{+3}{Fe^{3+}} + \overset{-1}{I^-} \rightarrow \overset{+2}{Fe^{2+}} + \overset{0}{\tfrac{1}{2}I_2}$$

Even in complicated reactions, you can see which element has been oxidised and which has been reduced when you put in the oxidation states:

$$\overset{+5\ -2}{2IO_3^-} + \overset{+1\ +4\ -2}{5HSO_3^-} \rightarrow \overset{0}{I_2} + \overset{+6\ -2}{5SO_4^{2-}} + \overset{+1}{3H^+} + \overset{+1\ -2}{H_2O}$$

Iodine in IO_3^- is reduced (+5 to 0) and sulfur in HSO_3^- is oxidised (+4 to +6). The oxidation states of all the other atoms have not changed.

Balancing redox reactions

You can use the idea of oxidation states to help balance equations for redox reactions.

For an equation to be balanced:

- the numbers of atoms of each element on each side of the equation must be the same
- the total charge on each side of the equation must be the same.

Example 1: the thermite reaction

This is a strongly exothermic reaction in which aluminium reacts with iron(III) oxide to produce molten iron. It was used to weld railway lines.

The unbalanced equation is:

$$Fe_2O_3(s) + Al(s) \rightarrow Fe(l) + Al_2O_3(s)$$

Write the oxidation states above each element:

$$\overset{+3\ -2}{Fe_2O_3(s)} + \overset{0}{Al(s)} \rightarrow \overset{0}{Fe(l)} + \overset{+3\ -2}{Al_2O_3(s)}$$

If you look at the equation you can see that only the iron and aluminium have changed their oxidation state. The oxygen is unchanged.

Each iron atom has been reduced by gaining three electrons so you can write the half equation:

$$Fe^{3+} + 3e^- \rightarrow Fe$$

Each aluminium atom has been oxidised by losing three electrons:

$$Al \rightarrow Al^{3+} + 3e^-$$

In the reaction, the number of electrons gained must equal the number of electrons lost. This means that there must be the same number of aluminium atoms as iron atoms. (The oxygen is a spectator ion.) You started with two iron atoms, so you must also have two aluminium atoms. The balanced equation is therefore:

$$Fe_2O_3(s) + 2Al(s) \rightarrow 2Fe(l) + Al_2O_3(s)$$

▲ **Figure 1** *A demonstration of the thermite reaction*

Example 2: aqueous solutions

Sometimes in aqueous solutions, species take part in redox reactions but are neither oxidised nor reduced. You must balance them separately. These include water molecules, H^+ ions (in acid solution), and OH^- ions(in alkaline solution). Oxidation states only help us to balance the species that are oxidised or reduced.

Suppose you want to balance the following equation, where dark purple manganate(VII) ions react in acid solution with Fe^{2+} ions to produce pale pink Mn^{2+} ions and Fe^{3+} ions.

The unbalanced equation is:

$$MnO_4^- + Fe^{2+} + H^+ \rightarrow Mn^{2+} + Fe^{3+} + H_2O$$

1 Write the oxidation state above each element.

$$\overset{+7\ -2}{MnO_4^-} + \overset{+2}{Fe^{2+}} + \overset{+1}{H^+} \rightarrow \overset{+2}{Mn^{2+}} + \overset{+3}{Fe^{3+}} + \overset{+1\ -2}{H_2O}$$

2 Identify the species that has been oxidised and the species that has been reduced.

$\overset{+7}{MnO_4^-} \rightarrow \overset{+2}{Mn^{2+}}$ Manganese has been reduced from +7 to +2 therefore five electrons must be gained.

$MnO_4^- + 5e^- \rightarrow Mn^{2+}$ (this equation is not chemically balanced)

$\overset{+2}{Fe^{2+}} \rightarrow \overset{+3}{Fe^{3+}}$ Fe has been oxidised from +2 to +3 so one electron must be lost.

$$Fe^{2+} \rightarrow Fe^{3+} + e^-$$

In order to balance the number of electrons that are transferred, this step must be multiplied by 5:

$$5Fe^{2+} \rightarrow 5Fe^{3+} + 5e^-$$

So, you know that there are $5Fe^{2+}$ ions to every MnO_4^- ion.

3 Include this information in the unbalanced equation, to balance the redox process.

$$MnO_4^- + 5Fe^{2+} + H^+ \rightarrow Mn^{2+} + 5Fe^{3+} + H_2O$$

(this equation is still not chemically balanced)

4 Balance the remaining atoms, those that are neither oxidised nor reduced. In order to 'use up' the four oxygen atoms on the left-hand side, you need $4H_2O$ on the right-hand side, which will in turn require $8H^+$ on the left-hand side.

$$MnO_4^- + 5Fe^{2+} + 8H^+ \rightarrow Mn^{2+} + 5Fe^{3+} + 4H_2O$$

Notice that this equation is balanced for both atoms and charge.

Disproportionation

In some chemical reactions, atoms of the same element can be both oxidised and reduced. For example, hydrogen peroxide decomposes to oxygen and water.

$$\overset{-1}{2H_2O_2} \rightarrow \overset{-2}{2H_2O} + \overset{0}{O_2}$$

Check that you can work out the oxidation state of each oxygen (shown in red) using the rules in Topic 7.2.

Two of the oxygen atoms in the hydrogen peroxide have increased their oxidation state and two have reduced it.

1 Suggest why hairdressers, who use hydrogen peroxide as a bleach, store it in the fridge and in bottles with a small hole in the cap.

2 Here is another disproportionation reaction.
$Cu_2O \rightarrow Cu + CuO$
Work out the oxidation states of each atom using the rules in Topic 7.2. Which element disproportionates?

1 Slows down decomposition of H_2O_2. Gases produced from decomposition can escape through small hole.

2 Cu (+1 to 0 and +2)

Half equations from the balanced equation

Example 1

The reaction between copper and *cold dilute* nitric acid produces the gas nitrogen monoxide. The balanced symbol equation is shown:

$$3Cu + 8H^+ + 2NO_3^- \rightarrow 3Cu^{2+} + 2NO + 4H_2O$$

To work out the half equations, you first need to know which elements have been oxidised and which have been reduced.

1 Put in the numbers and look for a change in the oxidation states:

$$\overset{0}{3Cu} + \overset{+1}{8H^+} + 2\overset{+5}{N}\overset{-2}{O_3^-} \rightarrow \overset{+2}{3Cu^{2+}} + 2\overset{+2}{N}\overset{-2}{O} + \overset{+1}{H_2}\overset{-2}{O}$$

Copper has been oxidised and nitrogen has been reduced.

2 Now work out the half equations.

Each of the three copper atoms loses two electrons, a total of six electrons:

$$3Cu \rightarrow 3Cu^{2+} + 6e^-$$

The two nitrogen atoms NO_3^- have each gained three electrons so the half equation must be based on:

$$2NO_3^- + 6e^- \rightarrow 2NO$$

This half equation is not balanced for atoms or charge. There are six oxygen atoms on the left-hand side and only two on the right-hand side. The total charge on the left is –8 whereas the right-hand side has no charge. Look at the original equation. You need to include the eight H^+ ions on the left-hand side of our half equation (to use up the extra four oxygen atoms that are unaccounted for) and also the four H_2O on the right-hand side. This also accounts for the charge, so the complete half equation is:

$$2NO_3^- + 8H^+ + 6e^- \rightarrow 2NO + 4H_2O$$

This equation is balanced in both atoms and charge.

Example 2

The reaction between copper and *hot concentrated* nitric acid produces the gas nitrogen dioxide.

1 The balanced symbol equation is shown with the oxidation states included:

$$\overset{0}{Cu} + 4\overset{+1}{H^+} + 2\overset{+5\ -2}{NO_3^-} \rightarrow \overset{+2}{Cu^{2+}} + 2\overset{+1\ -2}{H_2O} + 2\overset{+4\ -2}{NO_2}$$

Copper has been oxidised and nitrogen has been reduced.

2 Now work out the half equations.
Copper has lost two electrons so the half equation is:

$$Cu \rightarrow Cu^{2+} + 2e^-$$

Nitrogen in NO_3^- has gained an electron so the half equation must be based on:

$$2NO_3^- + 2e^- \rightarrow 2NO_2$$

This is not balanced for charge or atoms. There are an extra two oxygens on the left-hand side and a total charge of −4 whereas the right-hand side is neutral. You need to add the four H^+ ions to the left-hand side to use up the extra oxygen. These will also balance the charge. You then need to add two H_2O to the right-hand side.

The half equation is:

$$2NO_3^- + 4H^+ + 2e^- \rightarrow 2H_2O + 2NO_2$$

Note that if you add the half equations together, the electrons cancel out and you get back to the original balanced equation.

Summary questions

1 The following questions are about the equation:

$Fe^{2+} + \frac{1}{2}Cl_2 \rightarrow Fe^{3+} + Cl^-$

a Write the oxidation states for each element.

b Which element has been oxidised? Explain your answer.

c Which element has been reduced? Explain your answer.

d Write the half equations for the reaction.

2 **a** Use oxidation states to balance the following equations:

i $Cl_2 + NaOH \rightarrow NaClO_3 + NaCl + H_2O$

ii $Sn + HNO_3 \rightarrow SnO_2 + NO_2 + H_2O$

b Write the half equations for **part a i** and **ii**.

Practice questions

1 **(a)** In terms of electron transfer, what does the reducing agent do in a redox reaction?

(1 mark)

(b) What is the oxidation state of an atom in an uncombined element?

(1 mark)

(c) Deduce the oxidation state of nitrogen in each of the following compounds.

(i) NCl_3
(ii) Mg_3N_2
(iii) NH_2OH

(3 marks)

(d) Lead(IV) oxide, PbO_2, reacts with concentrated hydrochloric acid to produce chlorine, lead(II) ions, Pb^{2+}, and water.

(i) Write a half equation for the formation of Pb^{2+} and water from PbO_2 in the presence of H^+ ions.
(ii) Write a half equation for the formation of chlorine from chloride ions.
(iii) Hence deduce an equation for the reaction which occurs when concentrated hydrochloric acid is added to lead(IV) oxide, PbO_2

(3 marks)

AQA, 2002

2 Chlorine and bromine are both oxidising agents.

(a) Define an *oxidising agent* in terms of electrons.

(1 mark)

(b) In aqueous solution, bromine oxidises sulfur dioxide, SO_2, to sulfate ions, SO_4^{2-}

(i) Deduce the oxidation state of sulfur in SO_2 and in SO_4^{2-}.
(ii) Deduce a half equation for the reduction of bromine in aqueous solution.
(iii) Deduce a half equation for the oxidation of SO_2 in aqueous solution forming SO_4^{2-} and H^+ ions.
(iv) Use these two half equations to construct an overall equation for the reaction between aqueous bromine and sulfur dioxide.

(5 marks)

AQA, 2004

3 **(a)** By referring to electrons, explain the meaning of the term *oxidising agent.*

(1 mark)

(b) For the element **X** in the ionic compound **MX**, explain the meaning of the term *oxidation state.*

(1 mark)

(c) Complete the table below by deducing the oxidation state of each of the stated elements in the given ion or compound.

	Oxidation state
carbon in CO_3^{2-}	
phosphorus in PCl_4^+	
nitrogen in Mg_3N_2	

(3 marks)

(d) In acidified aqueous solution, nitrate ions, NO_3^- react with copper metal forming nitrogen monoxide, NO, and copper(II) ions.

(i) Write a half equation for the oxidation of copper to copper(II) ions.
(ii) Write a half equation for the reduction, in an acidified solution, of nitrate ions to nitrogen monoxide.
(iii) Write an overall equation for this reaction.

(3 marks)

AQA, 2005

4 **(a)** Nitrogen monoxide, NO, is formed when silver metal reduces nitrate ions, NO_3^- in acid solution. Deduce the oxidation state of nitrogen in NO and in NO_3^-.

(b) Write a half equation for the reduction of NO_3^- ions in acid solution to form nitrogen monoxide and water.

(c) Write a half equation for the oxidation of silver metal to $Ag^+(aq)$ ions.
(d) Hence, deduce an overall equation for the reaction between silver metal and nitrate ions in acid solution.

(5 marks)
AQA, 2006

5 Iodine reacts with concentrated nitric acid to produce nitrogen dioxide, NO_2.
(a) (i) Give the oxidation state of iodine in each of the following.

I_2, HIO_3 *(2 marks)*

(ii) Complete the balancing of the following equation.

I2 + $10HNO_3 \rightarrow HIO_3 + NO_2 + H_2O$ *(1 mark)*

(b) In industry, iodine is produced from the $NaIO_3$ that remains after sodium nitrate has been crystallised from the mineral Chile saltpetre.
The final stage involves the reaction between $NaIO_3$ and NaI in acidic solution.
Half equations for the redox processes are given below.

$$IO_3^- + 5e^- + 6H^+ \rightarrow 3H_2O + \frac{1}{2}I_2$$

$$I^- \rightarrow \frac{1}{2}I_2 + e^-$$

Use these half equations to deduce an overall ionic equation for the production of iodine by this process. Identify the oxidising agent.

(2 marks)

(c) When concentrated sulfuric acid is added to potassium iodide, solid sulfur and a black solid are formed.
(i) Identify the black solid. *(1 mark)*
(ii) Deduce the half equation for the formation of sulfur from concentrated sulfuric acid.

(1 mark)

(d) When iodide ions react with concentrated sulfuric acid in a different redox reaction, the oxidation state of sulfur changes from +6 to –2. The reduction product of this reaction is a poisonous gas that has an unpleasant smell.
Identify this gas. *(1 mark)*

(e) A yellow precipitate is formed when silver nitrate solution, acidified with dilute nitric acid, is added to an aqueous solution containing iodide ions.
(i) Write the **simplest ionic** equation for the formation of the yellow precipitate.

(1 mark)

(ii) State what is observed when concentrated ammonia solution is added to this precipitate.

(1 mark)

(iii) State why the silver nitrate is acidified when testing for iodide ions.

(1 mark)

(f) Consider the following reaction in which iodide ions behave as reducing agents.

$$Cl_2(aq) + 2I^-(aq) \rightarrow I_2(aq) + 2Cl^-(aq)$$

(i) In terms of electrons, state the meaning of the term *reducing agent*.

(1 mark)

(ii) Write a half equation for the conversion of chlorine into chloride ions.

(1 mark)

(iii) Explain why iodide ions react differently from chloride ions.

(3 marks)
AQA, 2012

Answers to the Practice Questions and Section Questions are available at
www.oxfordsecondary.com/oxfordaqaexams-alevel-chemistry

1 Antimony is a solid element that is used in industry. The method used for the extraction of antimony depends on the grade of the ore.

(a) Antimony can be extracted by reacting scrap iron with low-grade ores that contain antimony sulfide, Sb_2S_3.

(i) Write an equation for the reaction of iron with antimony sulfide to form antimony and iron(II) sulfide.

(1 mark)

(ii) Write a half equation to show what happens to the iron atoms in this reaction.

(1 mark)

(b) In the first stage of the extraction of antimony from a high-grade ore, antimony sulfide is roasted in air to convert it into antimony(III) oxide (Sb_2O_3) and sulfur dioxide.

(i) Write an equation for this reaction.

(1 mark)

(ii) Identify **one** substance that is manufactured directly from the sulfur dioxide formed in this reaction.

(1 mark)

(c) In the second stage of the extraction of antimony from a high-grade ore, antimony(III) oxide is reacted with carbon monoxide at high temperature.

(i) Use the standard enthalpies of formation in **Table 1** and the equation given below **Table 1** to calculate a value for the standard enthalpy change for this reaction.

▼ **Table 1**

	$Sb_2O_3(s)$	$CO(g)$	$Sb(l)$	$CO_2(g)$
$\Delta_f H$ / kJ mol^{-1}	−705	−111	+20	−394

$$Sb_2O_3(s) + 3CO(g) \rightarrow 2Sb(l) + 3CO_2(g)$$

(3 marks)

(ii) Suggest why the value for the standard enthalpy of formation of liquid antimony, given in **Table 1**, is **not** zero.

(1 mark)

(iii) State the type of reaction that antimony(III) oxide has undergone in this reaction.

(1 mark)

(d) Deduce **one** reason why the method of extraction of antimony from a low-grade ore, described in part **(a)**, is a low-cost process. Do **not** include the cost of the ore.

(1 mark)

AQA, 2014

2 **(a)** Complete the following table.

	Relative mass	Relative charge
proton		
electron		

(2 mark)

(b) An atom has twice as many protons and twice as many neutrons as an atom of ^{19}F.

Deduce the symbol, including the mass number, of this atom. *(2 marks)*

(c) The Al^{3+} ion and the Na^+ ion have the same electron arrangement.

(i) Give the electron arrangement of these ions.

(ii) Explain why more energy is needed to remove an electron from the Al^{3+} ion than from the Na^+ ion. *(3 marks)*

AQA, 2007

3 Molecules of NH_3, H_2O, and HF contain covalent bonds. The bonds in these molecules are polar.

(a) (i) Explain why the H–F bond is polar.

(ii) State which of the molecules NH_3, H_2O, or HF contains the least polar bond.

(iii) Explain why the bond in your chosen molecule from part (**b**)(**ii**) is less polar than the bonds found in the other two molecules. *(4 marks)*

(iv) Explain why H_2O has a bond angle of 104.5°. *(2 marks)*

(b) The boiling points of NH_3, H_2O, and HF are all high for molecules of their size. This is due to the type of intermolecular force present in each case.

(i) Identify the type of intermolecular force responsible.

(ii) Draw a diagram to show how two molecules of ammonia are attracted to each other by this type of intermolecular force. Include partial charges and all lone pairs of electrons in your diagram. *(4 marks)*

(c) When an H^+ ion reacts with an NH_3 molecule, an NH_4^+ ion is formed.

(i) Give the name of the type of bond formed when an H^+ ion reacts with an NH_3 molecule.

(ii) Draw the shape, including any lone pairs of electrons, of an NH_3 molecule and of an NH_4^+ ion.

(iii) Name the shape produced by the arrangement of atoms in the NH_3 molecule.

(iv) Give the bond angle in the NH_4^+ ion. *(7 marks)*

AQA, 2007

Answers to the Practice Questions and Section Questions are available at
www.oxfordsecondary.com/oxfordaqaexams-alevel-chemistry

Section 1 summary

catalysts

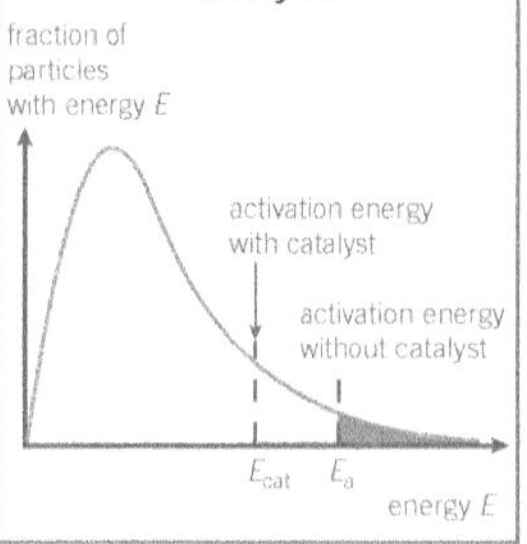

ionisation energy

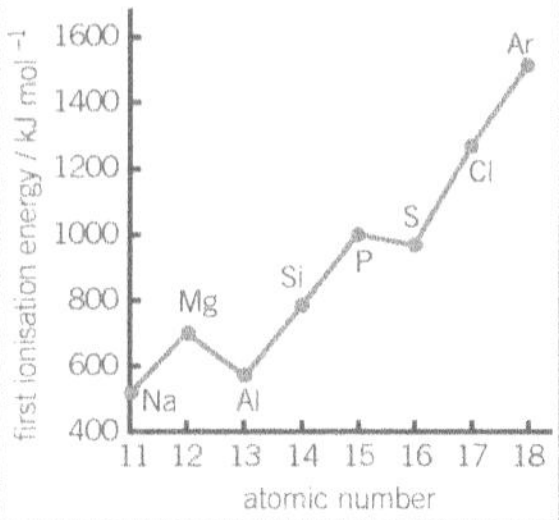

mass spectrometry in organic chemistry

trends in ionisation energy

mass spectrometry

equilibrium constant K_c
$aA + bB \rightleftharpoons cC + dD$
$K_c = \dfrac{(C)^c(D)^d}{(A)^a(B)^b}$

standard enthalpy of combustion $\Delta_c H^{\circ}$

equilibrium moves to reduce disturbance

equilibrium

Maxwell–Boltzmann distribution

atomic orbitals
- s-orbital: $2e^-$
- p-orbital: $6e^-$
- d-orbital: $10e^-$

isotopes
- same number of protons
- different number of neutrons

standard enthalpy of formation $\Delta_f H^{\circ}$

collision theory

enthalpy H

factors that affect rate
- temperature
- pressure
- concentration
- surface area

sub-atomic particles
- protons (+1)
- neutrons (neutral)
- electrons (–1)

measuring enthalpy change ΔH

Hess' law

Group 2 and the halogens

redox
Oxidation
Is
Loss
Reduction
Is
Gain

oxidation states

Section 1: Physical chemistry

electrons

fundamental particles

Group 2

bonding

periodicity and the halogens

ionic bonding
- metal and non-metal
- transfer of electrons to form ions

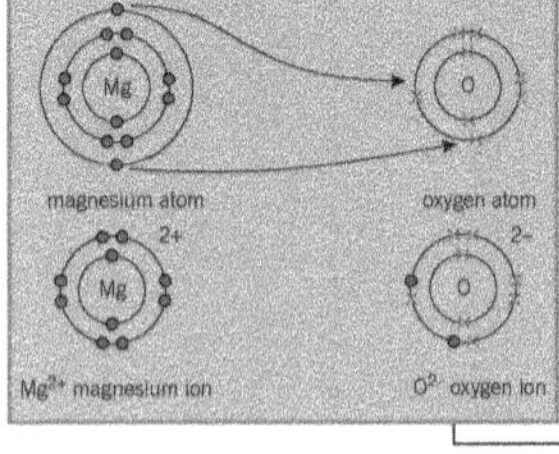

covalent bonding
- non-metal and non-metal
- shared pair of electrons

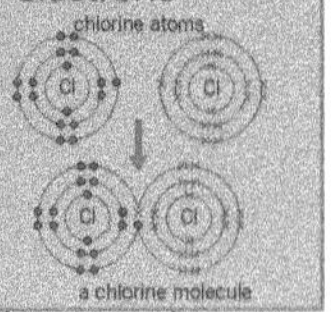

metallic bonding
- metals
- delocalised electrons around metal cations

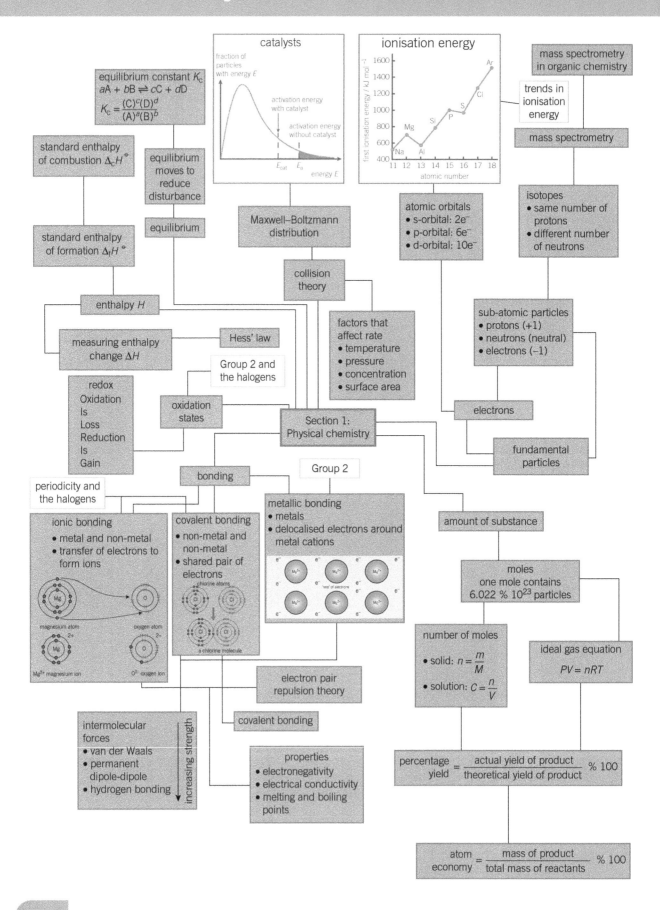

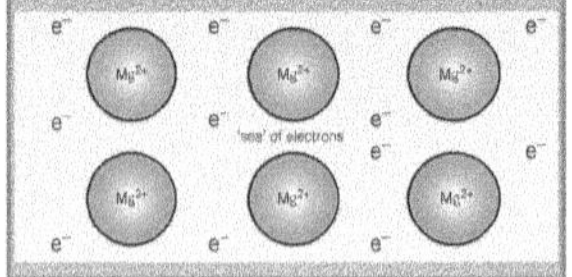

amount of substance

moles
one mole contains
6.022 % 10^{23} particles

number of moles
- solid: $n = \dfrac{m}{M}$
- solution: $C = \dfrac{n}{V}$

ideal gas equation
$PV = nRT$

electron pair repulsion theory

covalent bonding

intermolecular forces
- van der Waals
- permanent dipole-dipole
- hydrogen bonding

increasing strength

properties
- electronegativity
- electrical conductivity
- melting and boiling points

percentage yield $= \dfrac{\text{actual yield of product}}{\text{theoretical yield of product}}$ % 100

atom economy $= \dfrac{\text{mass of product}}{\text{total mass of reactants}}$ % 100

Practical skills

In this section you have met the following ideas:

- Finding the concentration of a solution by titration.
- Finding the yield of a reaction.
- Finding ΔH of reactions using calorimetry and Hess's law.
- Investigating the effect of temperature, concentration and a catalyst on the rate of reactions.
- Finding out K_c of a reaction.

Maths skills

In this section you have met the following maths skills:

- Using standard form in calculations.
- Carrying out calculations with the Avogradro constant.
- Carrying out calculations using Hess's law.
- Using appropriate significant figures.
- Calculating weighted means.
- Interpreting mass spectra.
- Working out the shape of molecules using ideas about electron pair repulsion.

Extension

Produce a timeline detailing how our understanding of atoms, atomic structure and chemical bonding has developed.

Suggested resources:

- Atkins, P. (2014), *Physical Chemistry: A very short Introduction.* Oxford University Press, UK. ISBN: 978-0-19-968909-5
- Dunmar, D., Sluckin, T., (2014), *Soap, Science and Flat-Screen TVs.* Oxford University Press, UK. ISBN: 978-0-19-870083-8
- Scerri, E. (2013), *The Tale of 7 Elements*. Oxford University Press, UK. ISBN: 978-0-19-539131-2.

Section 2

AS Inorganic chemistry 1

Chapters in this section

The Periodic Table of elements contains all the elements so far discovered. This unit examines how the properties of the elements are related to their electronic structures and how this determines their position in the Periodic Table.

Periodicity gives an overview of the Periodic Table and classifies blocks of elements in terms of s-, p-, d-, and f-orbitals. The chapter then concentrates on the properties of the elements in Period 3.

Group 2, the alkaline earth metals uses the ideas of electron arrangements to understand the bonding in compounds of these elements and the reactions and trends in reactivity in the group.

Group 7(17) the halogens deals with these reactive non-metal elements, explaining the trends in their reactivity in terms of electronic structure. It includes the reactions of the elements and their compounds using the ideas of redox reactions and oxidation states, and also the uses of chlorine and some of its compounds.

The concepts of the applications of science are found throughout the chapters, where they will provide you with an opportunity to apply your knowledge in a fresh context.

What you already know

The material in this section builds upon knowledge and understanding that you will have developed at GCSE, in particular the following:

- ☐ There are over 100 elements all made up of atoms.
- ☐ Each element has its own symbol and is part of the Periodic Table of elements.
- ☐ Groups in the Periodic Table are vertical and periods horizontal.
- ☐ Metals are on the left-hand side of the Periodic Table and non-metals are on the right.
- ☐ There are trends in the properties of elements both going down a group and across a period.

8 Periodicity
8.1 The Periodic Table

The Periodic Table is a list of all the elements in order of increasing atomic number. You can predict the properties of an element from its position in the table. You can use it to explain the similarities of certain elements and the trends in their properties, in terms of their electronic arrangements.

Learning objectives:

→ State the location of the s-, p-, and d-blocks of elements in the Periodic Table.

Specification reference: 3.2.1

The structure of the Periodic Table

The Periodic Table has been written in many forms including pyramids and spirals. The one shown below is one common layout. Some areas of the Periodic Table are given names. These are shown in Figure 1.

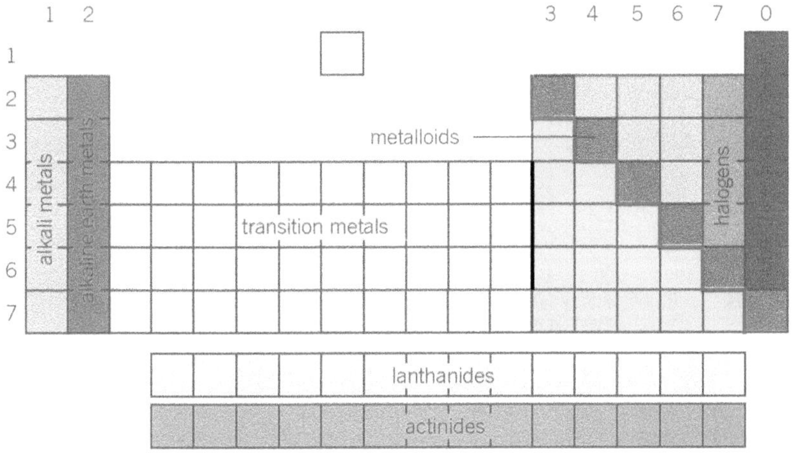

▲ **Figure 1** *Named areas of the Periodic Table*

Metals and non-metals

The red stepped line in Figure 1 (the 'staircase line') divides metals (on its left) from non-metals (on its right). Elements that touch this line, such as silicon, have a combination of metallic and non-metallic properties. They are called **metalloids** or semi-metals. Silicon, for example, is a non-metal but it looks quite shiny and conducts electricity, although not as well as a metal.

History of the Periodic Table

The development of the Periodic Table is one of the greatest achievements in chemistry. Credit for the final version goes firmly to a Russian, Dmitri Mendeleev, in 1869. He realised that there were undiscovered elements. He left spaces for some unknown elements, and arranged the known elements so that similar elements lined up in columns. Since then, new elements have been discovered that fit into the gaps he left. Mendeleev even accurately predicted the properties of the missing elements, confirming the success of his Periodic Table.

Many other scientists contributed to the Periodic Table. Research on the internet the parts played by Jons Jacob Berzelius, Robert Bunsen and Gustav Kirchoff, Alexandre Béguyer de Chancourtois, Marie Curie, Sir Humphry Davy, Julius Lothar Meyer, Henry Moseley, John Newlands, Sir William Ramsay, and Glenn T Seaborg.

A common form of the Periodic Table

A version of the Periodic Table is shown in Figure 2. The lanthanides and actinides are omitted and two alternative numbering schemes for groups are shown.

key: relative atomic mass / atomic symbol / name / atomic (proton) number

1 (1)	2 (2)	(3)	(4)	(5)	(6)	(7)	(8)	(9)	(10)	(11)	(12)	3 (13)	4 (14)	5 (15)	6 (16)	7 (17)	0 (18)
							1.0 H hydrogen 1										4.0 He helium 2
6.9 Li lithium 3	9.0 Be beryllium 4											10.8 B boron 5	12.0 C carbon 6	14.0 N nitrogen 7	16.0 O oxygen 8	19.0 F fluorine 9	20.2 Ne neon 10
23.0 Na sodium 11	24.3 Mg magnesium 12											27.0 Al aluminium 13	28.1 Si silicon 14	31.0 P phosphorus 15	32.1 S sulfur 16	35.5 Cl chlorine 17	39.9 Ar argon 18
39.1 K potassium 19	40.1 Ca calcium 20	45.0 Sc scandium 21	47.9 Ti titanium 22	50.9 V vanadium 23	52.0 Cr chromium 24	54.9 Mn manganese 25	55.8 Fe iron 26	58.9 Co cobalt 27	58.7 Ni nickel 28	63.5 Cu copper 29	65.4 Zn zinc 30	69.7 Ga gallium 31	72.6 Ge germanium 32	74.9 As arsenic 33	79.0 Se selenium 34	79.9 Br bromine 35	83.8 Kr krypton 36
85.5 Rb rubidium 37	87.6 Sr strontium 38	88.9 Y yttrium 39	91.2 Zr zirconium 40	92.9 Nb niobium 41	95.9 Mo molybdenum 42	[98] Tc technetium 43	101.1 Ru ruthenium 44	102.9 Rh rhodium 45	106.4 Pa palladium 46	107.9 Ag silver 47	112.4 Cd cadmium 48	114.8 In indium 49	118.7 Sn tin 50	121.8 Sb antimony 51	127.6 Te tellurium 52	126.9 I iodine 53	131.3 Xe xenon 54
132.9 Cs caesium 55	137.3 Ba barium 56	138.9 La* lanthanum 57	178.5 Hf hafnium 72	180.9 Ta tantalum 73	183.8 W tungsten 74	186.2 Re rhenium 75	190.2 Os osmium 76	192.2 Ir iridium 77	195.1 Pt platinum 78	197.0 Au gold 79	200.6 Hg mercury 80	204.4 Tl thallium 81	207.2 Pb lead 82	209.0 Bi bismuth 83	[209] Po polonium 84	[210] At astatine 85	[222] Rn radon 86
[223] Fr francium 87	[226] Ra radium 88	[227] Ac* actinium 89	[261] Rf rutherfordium 104	[262] Db dubnium 105	[266] Sg seaborgium 106	[264] Bh bohrium 107	[277] Hs hassium 108	[268] Mt meitnerium 109	[271] Ds darmstadtium 110	[272] Rg roentgenium 111	[285] Cn copernicum 112	[286] Uut ununtrium 113	[289] Fl flerovium 114	[293] Uup unenpentium 115	[293] Lv livermorium 116	[294] Uus ununseptium 117	[294] Uuo ununoctium 80

*the lanthanides (atomic numbers 58–71) and the actinides (atomic numbers 90–103) have been omitted

▲ **Figure 2** *The full form of the Periodic Table*

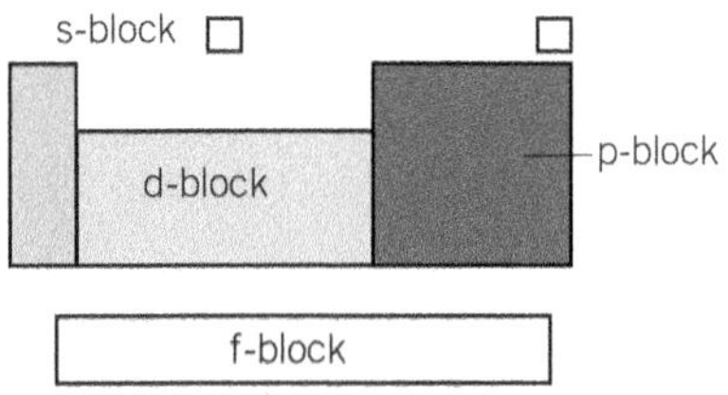

▲ **Figure 3** *The s-, p-, d-, and f-block areas of the Periodic Table*

The s-, p-, d-, and f-blocks of the Periodic Table

Figure 3 shows the elements described in terms of their electronic arrangement.

Areas of the table are labelled s-block, p-block, d-block, and f-block.

- All the elements that have their highest energy electrons in s-orbitals are in the s-block, for example, sodium, Na ($1s^2\ 2s^2\ 2p^6\ 3s^1$).
- All the elements that have their highest energy electrons in p-orbitals are called p-block, for example, carbon, C ($1s^2\ 2s^2\ 2p^2$).
- All the elements that have their highest energy electrons in d-orbitals are called d-block, for example, iron, Fe ($1s^2\ 2s^2\ 2p^6\ 3s^2\ 3p^6\ 4s^2\ 3d^6$) and so on.

Strictly speaking the transition metals and the d-block elements are not exactly the same. Scandium and zinc are not transition metals because they do not form any compounds in which they have partly filled d-orbitals, which is the characteristic of transition metals.

The origin of the terms s, p, d, and f is historical. When elements are heated they give out light energy at certain wavelengths, as excited electrons fall back from one energy level to a lower one. This causes

Synoptic link

You will learn more about the transition metals in Chapter 23, The transition metals.

lines to appear in the spectrum of light they give out. The letters s, p, d, and f stand for words that were used first to describe the lines – s for sharp, p for principal, d for diffuse, and f for fine.

Groups

A **group** is a vertical column of elements. The elements in the same group form a chemical 'family' – they have similar properties. Elements in the same group have the same number of electrons in the outer main shell. The groups were traditionally numbered I–VII in Roman numerals plus zero for the noble gases, missing out the transition elements. It is now common to number them in ordinary numbers as 1–7 and 0 (or 1–8) and sometimes as 1–18 including the transition metals.

Reactivity

In the s-block, elements (metals) get more reactive going down a group. To the right (non-metals), elements tend to get more reactive going up a group.

Transition elements are a block of rather unreactive metals. This is where most of the useful metals are found.

Lanthanides are metals which are not often encountered. They all tend to form +3 ions in their compounds and have broadly similar reactivity.

Actinides are radioactive metals. Only thorium and uranium occur naturally in the Earth's crust in anything more than trace quantities.

Periods

Horizontal rows of elements in the Periodic Table are called **periods**. The periods are numbered starting from Period 1, which contains only hydrogen and helium. Period 2 contains the elements lithium to neon, and so on. There are trends in physical properties and chemical behaviour as you go across a period, see Topic 8.2.

Placing hydrogen and helium

The positions of hydrogen and helium vary in different versions of the table. Helium is usually placed above the noble gases (Group 0) because of its properties. But, it is not a p-block element – its electronic arrangement is $1s^2$.

Hydrogen is sometimes placed above Group 1 but is often placed on its own. It usually forms singly charged +1 (H^+) ions like the Group 1 elements, but otherwise is not similar to them since they are all reactive metals and hydrogen is a gas. It is sometimes placed above the halogens because it can form H^- ions and also bond covalently.

Summary questions

1 From the elements, Br, Cl, Fe, K, Cs, and Sb, pick out:
- **a** two elements
  - **i** in the same period
  - **ii** in the same group
  - **iii** that are non-metals
- **b** one element
  - **i** that is in the d-block
  - **ii** that is in the s-block.

2 From the elements Tl, Ge, Xe, Sr, and W, pick out:
- **a** a noble gas
- **b** the element described by Group 4, Period 4
- **c** an s-block element
- **d** a p-block element
- **e** a d-block element.

8.2 Trends in the properties of elements of Period 3

Learning objectives:

→ Describe the trends in melting and boiling temperatures of the elements in Period 3.

→ Explain these trends in terms of bonding and structure.

Specification reference: 3.2.1

The Periodic Table reveals patterns in the properties of elements. For example, every time you go across a period you go from metals on the left to non-metals on the right. This is an example of **periodicity**. The word periodic means recurring regularly.

Periodicity and properties of elements in Period 3

Periodicity is explained by the electron arrangements of the elements.

- The elements in Groups 1, 2, and 3 (sodium, magnesium, and aluminium) are metals. They have giant structures. They lose their outer electrons to form ionic compounds.
- Silicon in Group 4 has four electrons in its outer shell with which it forms four covalent bonds. The element has some metallic properties and is classed as a semi-metal.
- The elements in Groups 5, 6, and 7 (phosphorus, sulfur, and chlorine) are non-metals. They either accept electrons to form ionic compounds, or share their outer electrons to form covalent compounds.
- Argon in Group 0 is a noble gas – it has a full outer shell and is unreactive.

Table 1 shows some trends across Period 3 (see Figure 1). Similar trends are found in other periods.

Study tip

- Remember that when a molecular substance melts, the covalent bonds remain intact but the van der Waals forces break.
- Learn the formulae P_4, S_8, Cl_2.

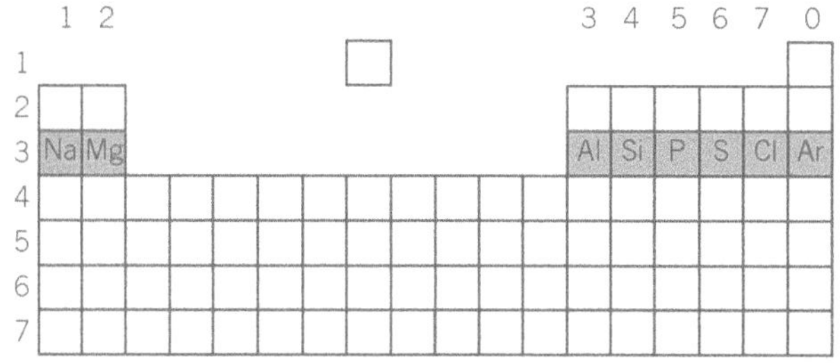

▲ **Figure 1** *The Periodic Table with Period 3 highlighted*

▼ **Table 1** *Some trends across Period 3*

Group	1	2	3	4	5	6	7	0
Element	sodium	magnesium	aluminium	silicon	phosphorus	sulfur	chlorine	argon
Electron arrangement	[Ne] $3s^1$	[Ne] $3s^2$	[Ne] $3s^2 3p^1$	[Ne] $3s^2 3p^2$	[Ne] $3s^2 3p^3$	[Ne] $3s^2 3p^4$	[Ne] $3s^2 3p^5$	[Ne] $3s^2 3p^6$
	s-block		p-block					
Structure of element	metals			semi-metal	non-metals			noble gas
	giant metallic			macromolecular (giant covalent)	molecular			atomic
					P_4	S_8	Cl_2	Ar
Melting point, T_m / K	371	922	933	1683	317 (white)	392 (monoclinic)	172	84
Boiling point, T_b / K	1156	1380	2740	2628	553 (white)	718	238	87

Trends in melting and boiling points

The trends in melting and boiling points in Period 3 are shown in Figure 2.

There is a clear break in the middle of the figure between elements with high melting points (on the left, with sodium, Na, in Group 1 as the exception) and those with low melting points (on the right). These trends are due to their structures.

- Giant structures (found on the left) tend to have high melting points and boiling points.
- Molecular or atomic structures (found on the right) tend to have low melting points and boiling points.

The melting points and boiling points of the metals increase from sodium to aluminium because of the strength of metallic bonding. As you go from left to right the charge on the ion increases so more electrons join the delocalised electron 'sea' that holds the giant metallic lattice together.

The melting points of the non-metals with molecular structures depend on the sizes of the van der Waals forces between the molecules. This in turn depends on the number of electrons in the molecule and how closely the molecules can pack together. As a result the melting points of these non-metals are ordered: $S_8 > P_4 > Cl_2$. Silicon with its giant structure has a much higher melting point. Boiling points follow a similar pattern.

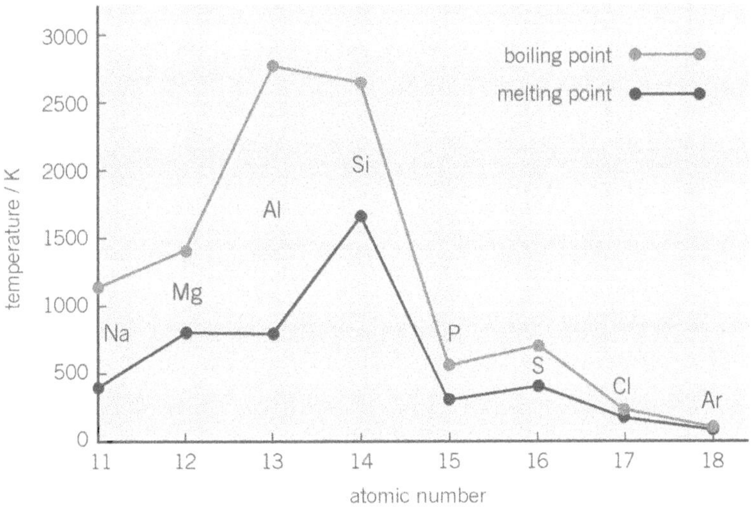

▲ **Figure 2** *Melting and boiling points of elements in Period 3*

Hint

The melting temperature of a substance is also the freezing temperature.

Synoptic link

To revise metallic bonding, look back at Topic 3.3, Metallic bonding.

Summary questions

1 Whereabouts in a period do you find the following? Choose from left, right, middle.

   a Elements that lose electrons when forming compounds

   b Elements that accept electrons when forming compounds

2 In what group do you find an element that exists as the following?

   a separate atoms

   b a macromolecule

3 A and B are both elements. Both conduct electricity – A well, B slightly. A melts at a low temperature, B at a much higher temperature. Suggest the identity of A and B and explain how their bonding and structure account for their properties.

8.3 More trends in the properties of the elements of Period 3

Learning objectives:

→ Describe the trends in atomic radius and first ionisation energy of the elements in Period 3.

→ Explain these trends.

Specification reference: 3.2.1

Some key properties of atoms, such as size and ionisation energy, are periodic, that is, there are similar trends as you go across each period in the Periodic Table.

Atomic radii

These tell us about the sizes of atoms. You cannot measure the radius of an isolated atom because there is no clear point at which the electron cloud density around it drops to zero. Instead half the distance between the centres of a pair of atoms is used, see Figure 1.

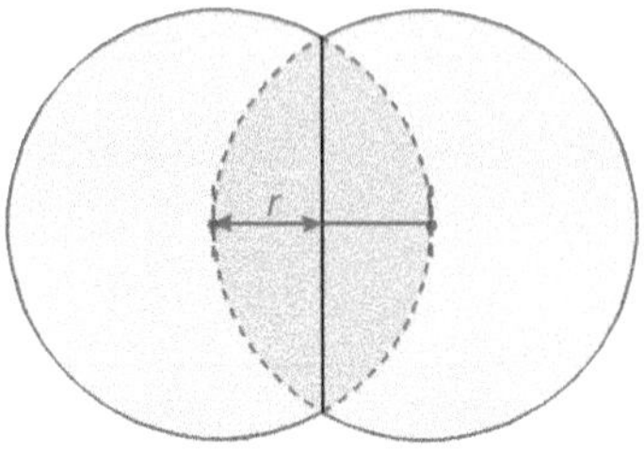

▲ **Figure 1** *Atomic radii are taken to be half the distance between the centres of a pair of atoms*

The atomic radius of an element can differ as it is a general term. It depends on the type of bond that it is forming – covalent, ionic, metallic, van der Waals, and so on. The covalent radius is most commonly used as a measure of the size of the atom. Figure 2 shows a plot of covalent radius against atomic number.

(Even metals can form covalent molecules such as Na_2 in the gas phase. Since noble gases do not bond covalently with one another, they do not have covalent radii and so they are often left out of comparisons of atomic sizes.)

Hint

1 nm is 1×10^{-9} m

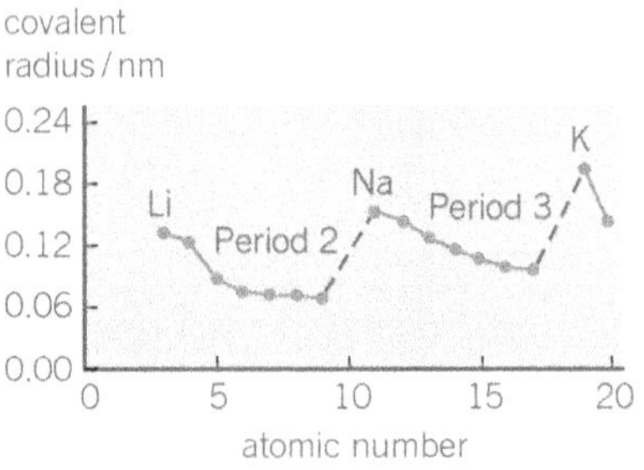

▲ **Figure 2** *The periodicity of covalent radii. The noble gases are not included because they do not form covalent bonds with one another*

The graph shows that:

- atomic radius is a periodic property because it decreases across each period and there is a jump when starting the next period
- atoms get larger down any group.

Why the radii of atoms decrease across a period

You can explain this trend by looking at the electronic structures of the elements in a period, for example, sodium to chlorine in Period 3, as shown in Figure 3.

As you move from sodium to chlorine you are adding protons to the nucleus and electrons to the outer main shell, which is the third shell. The charge on the nucleus increases from +11 to +17. This increased charge pulls the electrons in closer to the nucleus. There are no additional electron shells to provide more shielding. So the size of the atom *decreases* as you go across the period.

atom	Na	Mg	Al	Si	P	S	Cl
size of atom							
	2,8,1	2,8,2	2,8,3	2,8,4	2,8,5	2,8,6	2,8,7
atomic (covalent) radius / nm	0.156	0.136	0.125	0.117	0.110	0.104	0.099
nuclear charge	11+	12+	13+	14+	15+	16+	17+

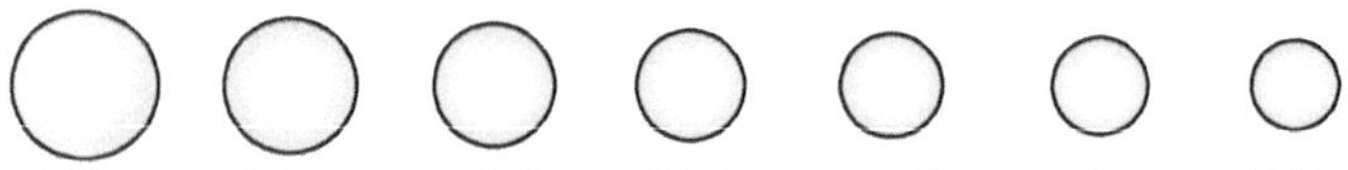

▲ **Figure 3** *The sizes and electronic structures of the elements sodium to chlorine*

Study tip

It is a common mistake to think that atoms increase in size as you cross a period. While the nuclei have more protons (and neutrons) the radius of the atom depends on the size of the electron shells.

Why the radii of atoms increase down a group

Going down a group in the Periodic Table, the atoms of each element have one extra complete main shell of electrons compared with the one before. So, for example, in Group 1 the outer electron in sodium is in main shell 3, whereas in potassium it is in main shell 4. So going down the group, the outer electron main shell is further from the nucleus and the atomic radii increase.

First ionisation energy

The first ionisation energy is the energy required to convert a mole of isolated gaseous atoms into a mole of singly positively charged gaseous ions, that is, to remove one electron from each atom.

$E(g) \rightarrow E^{+}(g) + e^{-}(g)$ where E stands for any element

The first ionisation energies also have periodic patterns. These are shown in Figure 4.

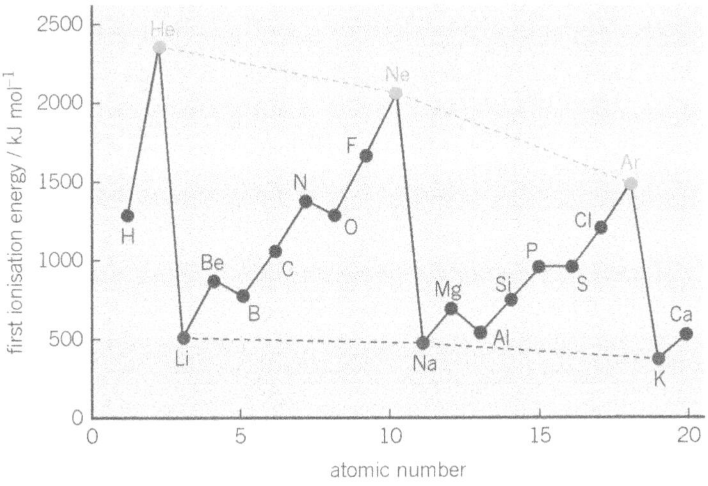

▲ **Figure 4** *The periodicity of first ionisation energies*

Synoptic link

You learnt about ionisation energy in Topic 1.6, Electron arrangements and ionistion energy.

Hint

Remember that electron levels are also referred to as electron shells.

The discovery of argon

When the Periodic Table was first put forward, none of the noble gases had been discovered. The first, argon, was discovered by Scottish chemist William Ramsay. He noticed that the density of nitrogen prepared by a chemical reaction was 1.2505 g dm^{-3} while nitrogen prepared by removing oxygen and carbon dioxide from air had a density of 1.2572 g dm^{-3}. He reasoned that there must be a denser impurity in the second sample, which he showed to be a previously unknown and very unreactive gas – argon. He later went on to discover the whole group of noble gases for which he won the Nobel (not noble!) Prize.

Chemists at the time had difficulty placing an unreactive gas of A_r, approximately 40 in the Periodic Table.

A suggestion was made that argon might be an allotrope of nitrogen, N_3, analogous to the O_3 allotrope of oxygen.

1. To how many significant figures were the two densities measured?
2. Using relative atomic masses from the Periodic Table, explain why argon is denser than nitrogen.
3. Suggest how oxygen could be removed from a sample of air.
4. What would be the M_r of N_3 (to the nearest whole number)?
5. Suggest why chemists were reluctant to regard argon as an element.

The first ionisation energy generally increases across a period (see Figure 4), alkali metals like sodium, Na, and lithium, Li, have the lowest values and the noble gases (helium, He, neon, Ne, and argon, Ar) have the highest values.

The first ionisation energy decreases going down any group. The trends for Group 1 and Group 0 are shown dotted in red and green, respectively on the graph.

You can explain these patterns by looking at electronic arrangements (Figure 5).

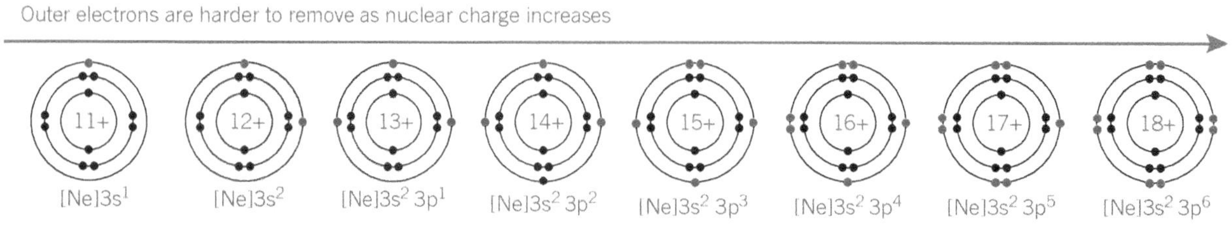

▲ **Figure 5** *The electronic structures of the elements sodium to argon*

Why the first ionisation energy increases across a period

As you go across a period from left to right, the number of protons in the nucleus increases but the electrons enter the same main shell, see Figure 5. The increased charge on the nucleus means that it gets increasingly difficult to remove an electron.

Hint

Filled inner electron shells are said to shield electrons in the outer shell from the nuclear charge.

Why the first ionisation energy decreases going down a group

The number of filled inner shells increases down the group. This results in an increase in shielding. Also, the electron to be removed is at an increasing distance from the nucleus and is therefore held less strongly. Thus the outer electrons get easier to remove going down a group because they are further away from the nucleus.

Why there is a drop in ionisation energy from one period to the next

Moving from neon in Period 0 (far right) with electron arrangement 2,8 to sodium, 2,8,1 (Period 1, far left) there is a sharp drop in the first ionisation energy. This is because at sodium a new main shell starts and so there is an increase in atomic radius, the outer electron is further from the nucleus, less strongly attracted and easier to remove.

Summary questions

1. What happens to the size of atoms as you go from left to right across a period? Choose from increase, decrease, no change.
2. What happens to the first ionisation energy as you go from left to right across a period? Choose from increase, decrease, no change.
3. What happens to the nuclear charge of the atoms as you go left to right across a period?
4. Why do the noble gases have the highest first ionisation energy of all the elements in their period?

8.4 A closer look at ionisation energies

This chapter revisits the trends in ionisation energies first dealt with in Topic 1.6, in the context of periodicity. The graph of first ionisation energy against atomic number across a period is not smooth. Figure 1 below shows the plot for Period 3.

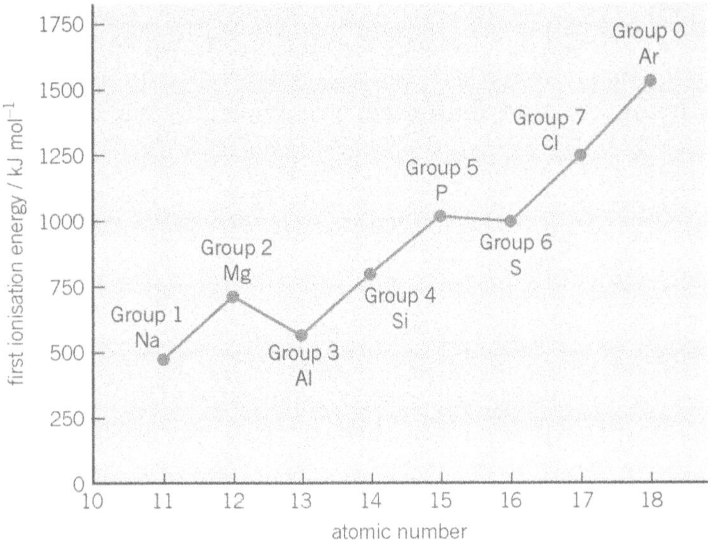

▲ **Figure 1** *Graph of first ionisation energy against atomic number for the elements of Period 3*

It shows that:

- the first ionisation energy actually drops between Group 2 and Group 3, so that aluminium has a lower ionisation energy than magnesium
- the ionisation energy drops again slightly between Group 5 (phosphorus) and Group 6 (sulfur).

Similar patterns occur in other periods. You can explain this if you look at the electron arrangements of these elements.

The drop in first ionisation energy between Groups 2 and 3

For the first ionisation energy:

- magnesium, $1s^2\ 2s^2\ 2p^6\ 3s^2$, loses a 3s electron
- aluminium, $1s^2\ 2s^2\ 2p^6\ 3s^2\ 3p^1$, loses the 3p electron.

The p-electron is already in a higher energy level than the s-electron, so it takes less energy to remove it, see Figure 2.

Learning objectives:

- ➔ Explain why the increase in ionisation energies across a period is not regular.
- ➔ Describe how successive ionisation energies explain electron arrangements.

Specification reference: 3.2.1

Hint

Ionisation energies are sometimes called **ionisation enthalpies.**

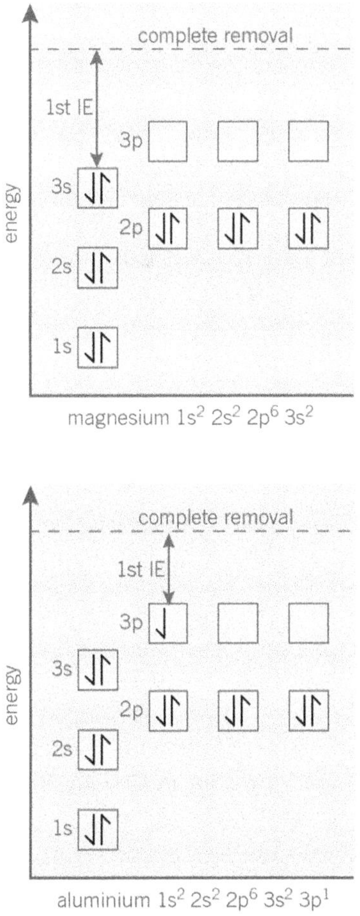

▲ **Figure 2** *The first ionisation energies of magnesium and aluminium (not to scale)*

The drop in first ionisation energy between Groups 5 and 6

An electron in a pair will be easier to remove that one in an orbital on its own because it is already being repelled by the other electron. As shown in Figure 3:

- phosphorus, $1s^2\ 2s^2\ 2p^6\ 3s^2\ 3p^3$, has no paired electrons in a p-orbital because each p-electron is in a different orbital
- sulfur, $1s^2\ 2s^2\ 2p^6\ 3s^2\ 3p^4$, has two of its p-electrons paired in a p-orbital so one of these will be easier to remove than an unpaired one due to the repulsion of the other electron in the same orbital.

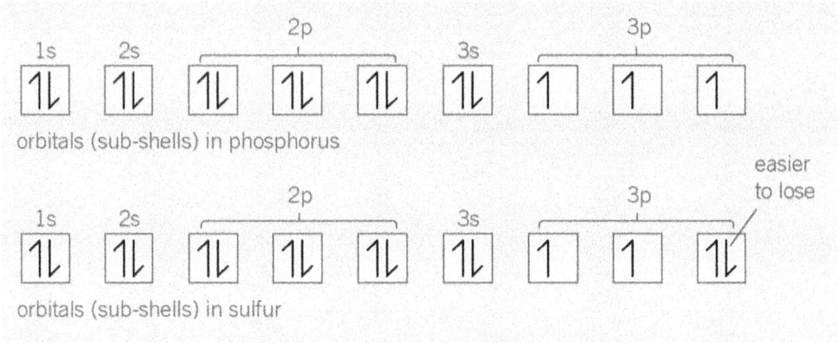

▲ **Figure 3** *Electron arrangements of phosphorus and sulfur*

Successive ionisation energies

If you remove the electrons from an atom one at a time, each one is harder to remove than the one before. Figure 4 is a graph of ionisation energy against number of electrons removed for sodium, electron arrangement 2,8,1.

You can see that there is a sharp increase in ionisation energy between the first and second electrons. This is followed by a gradual increase over the next eight electrons and then another jump before the final two electrons. Sodium, in Group 1 of the Periodic Table, has one electron in its outer main shell (the easiest one to remove), eight in the next main shell and two (very hard to remove) in the innermost main shell.

Figure 5 is a graph of successive ionisation energies against number of electrons removed for aluminium, electron arrangement 2,8,3.

It shows three electrons that are relatively easy to remove – those in the outer main shell – and then a similar pattern to that for sodium.

If you plotted a graph for chlorine, the first seven electrons would be relatively easier to remove than the next eight.

This means that the number of electrons that are relatively easy to remove tells us the group number in the Periodic Table. For example, the values of 906, 1763, 14 855, and 21 013 kJ mol^{-1} for the first five ionisation energies of an element, tell us that the element is in Group 2. This is because the big jump occurs after two electrons have been removed.

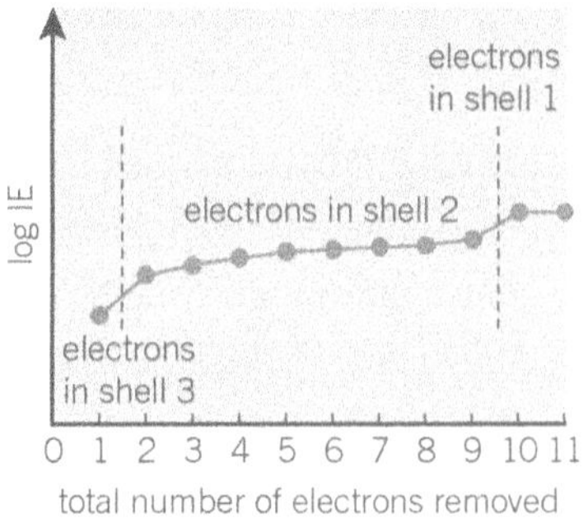

▲ **Figure 4** *Graph of successive ionisation energies against number of electrons removed for sodium. Note that the log of the ionisation energies is plotted in order to fit the large range of values onto the scale*

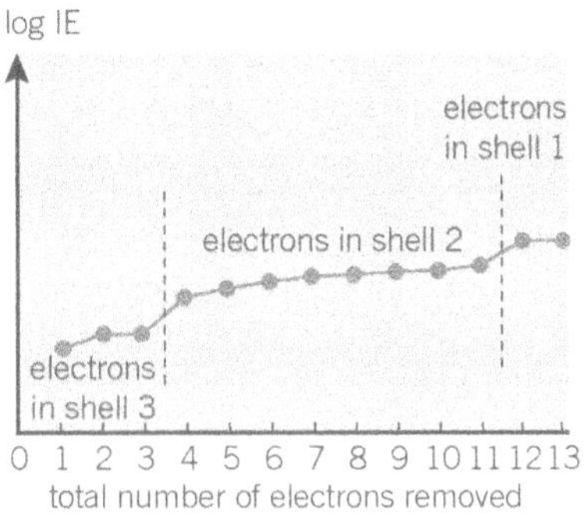

▲ **Figure 5** *Graph of successive ionisation energies against number of electrons removed for aluminium*

Summary questions

1 Write the electron arrangement in the form $1s^2$... for:
  a beryllium
  b boron.

2 If one electron is lost from for the following atoms, from what main shell does it come?
  a beryllium
  b boron

3 Why is the first ionisation energy of boron less than that of beryllium?

Practice questions

1 The following table gives the melting points of some elements in Period 3.

Element	Na	Al	Si	P	S
Melting point / K	371	933	1680	317	392

(a) State the type of structure shown by a crystal of silicon.
Explain why the melting point of silicon is very high.
(3 marks)

(b) State the type of structure shown by crystals of sulfur and phosphorus.
Explain why the melting point of sulfur is higher than the melting point of phosphorus.
(3 marks)

(c) Draw a diagram to show how the particles are arranged in aluminium and explain why aluminium is malleable.
(You should show a minimum of six aluminium particles arranged in two dimensions.)
(3 marks)

(d) Explain why the melting point of aluminium is higher than the melting point of sodium.
(3 marks)
AQA, 2011

2 Trends in physical properties occur across all periods in the Periodic Table.
This question is about trends in the Period 2 elements from lithium to nitrogen.

(a) Identify, from the Period 2 elements lithium to nitrogen, the element that has the largest atomic radius.
(1 mark)

(b) (i) State the general trend in first ionisation energies for the Period 2 elements lithium to nitrogen.
(1 mark)

(b) (ii) Identify the element that deviates from this general trend, from lithium to nitrogen, and explain your answer.
(3 marks)

(c) Identify the Period 2 element that has the following successive ionisation energies.

	First	Second	Third	Fourth	Fifth	Sixth
Ionisation energy / kJ mol^{-1}	1090	2350	4610	6220	37 800	47 000

(1 mark)
AQA, 2012

(d) Draw a cross on the diagram to show the melting point of nitrogen.
(1 mark)

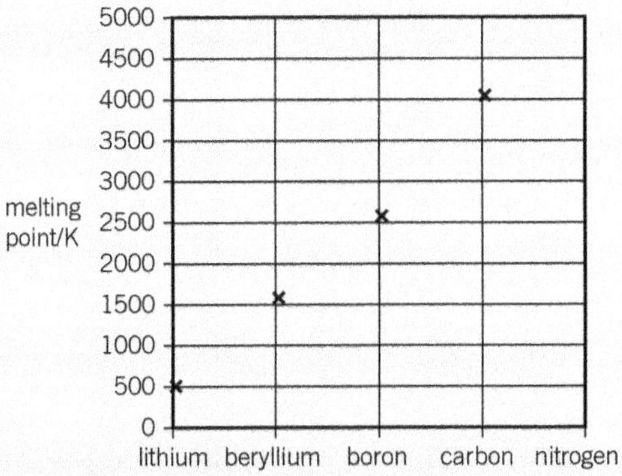

(e) Explain, in terms of structure and bonding, why the melting point of carbon is high.
(3 marks)

3 **(a)** Use your knowledge of electron configuration and ionisation energies to answer this question.
The following diagram shows the **second** ionisation energies of some Period 3 elements.

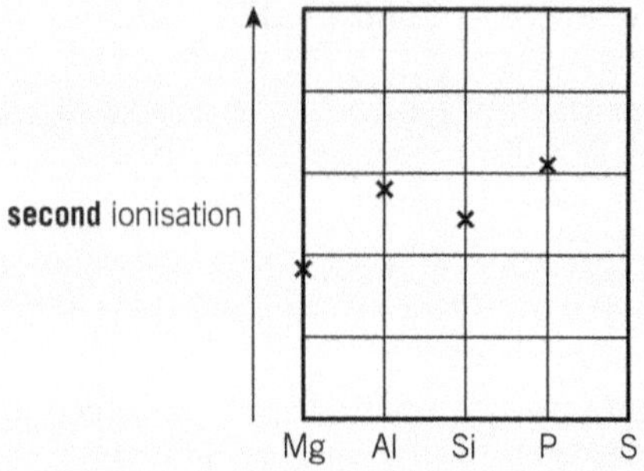

(i) Draw an **'X'** on the diagram to show the **second** ionisation energy of sulfur.

(1 mark)

(ii) Write the full electron configuration of the Al^{2+} ion.

(1 mark)

(iii) Write an equation to show the process that occurs when the **second** ionisation energy of aluminium is measured.

(1 mark)

(iv) Give **one** reason why the **second** ionisation energy of silicon is lower than the **second** ionisation energy of aluminium.

(1 mark)

(b) Predict the element in Period 3 that has the highest **second** ionisation energy.
Give a reason for your answer.

(2 marks)

(c) The following table gives the successive ionisation energies of an element in Period 3.

	First	Second	Third	Fourth	Fifth	Sixth
Ionisation energy / kJ mol^{-1}	786	1580	3230	4360	16 100	19 800

Identify this element.

(1 mark)

(d) Explain why the ionisation energy of every element is endothermic.

(1 mark)

AQA, 2013

4 The elements in Period 2 show periodic trends.

(a) Identify the Period 2 element, from carbon to fluorine, that has the largest atomic radius. Explain your answer.

(3 marks)

(b) State the general trend in first ionisation energies from carbon to neon.
Deduce the element that deviates from this trend and explain why this element deviates from the trend.

(4 marks)

(c) Write an equation, including state symbols, for the reaction that occurs when the first ionisation energy of carbon is measured.

(1 mark)

(d) Explain why the second ionisation energy of carbon is higher than the first ionisation energy of carbon.

(1 mark)

(e) Deduce the element in Period 2, from lithium to neon, that has the highest second ionisation energy.

(1 mark)

AQA, 2013

Answers to the Practice Questions and Section Questions are available at
www.oxfordsecondary.com/oxfordaqaexams-alevel-chemistry

9 Group 2, the alkaline earth metals

9.1 The physical and chemical properties of Group 2

The elements in Group 2 are sometimes called the **alkaline earth metals**. This is because their oxides and hydroxides are alkaline. Like Group 1, they are s-block elements. They are similar in many ways to Group 1 but they are less reactive. Beryllium is not typical of the group and is not considered here.

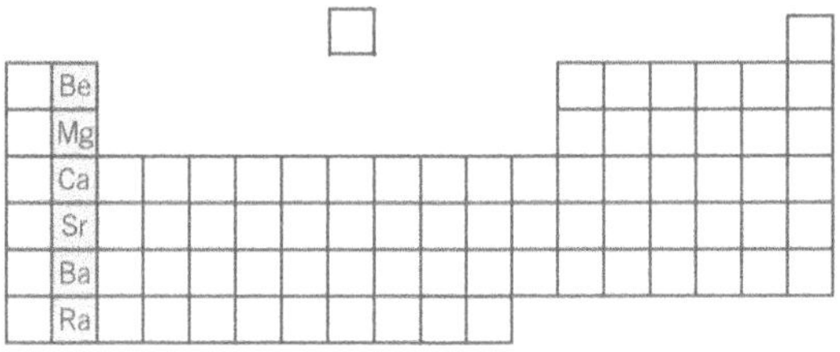

Learning objectives

- → Explain the changes in the atomic radius of the Group 2 elements from Mg to Ba.
- → Explain the changes in the first ionisation energy of the Group 2 elements from Mg to Ba.
- → Explain the changes in the melting point of the Group 2 elements from Mg to Ba.
- → State the trend in reactivity of the group.
- → State the trend in solubilities of a) the hydroxides b) the sulfates.

Specification reference: 3.2.2

The physical properties of the Group 2 elements, magnesium to barium

A summary of some of the physical properties of the elements from magnesium to barium is given in Table 1. Trends in properties are shown by the arrows, which show the direction of increase.

Electron arrangement

The elements all have two electrons in an outer s-orbital. This s-orbital becomes further away from the nucleus going down the group.

The sizes of the atoms

The atoms get bigger going down the group. The atomic (metallic) radii increase because each element has an extra filled main shell of electrons compared with the one above it.

Synoptic link

See Practical 3 on page 524.

Melting points

Group 2 elements are metals with high melting points, typical of a giant metallic structure. Going down the group, the electrons in the 'sea' of delocalised electrons are further away from the positive nuclei. As a result, the strength of the metallic bonds decreases going down the group. For this reason the melting points of Group 2 elements decrease slightly going down the group, starting with calcium.

Magnesium, with the lowest melting point, does not fit this trend but there is no simple explanation for this anomaly.

Synoptic link

Look back at Topic 3.3, Metallic bonding for more on the giant metallic structure.

Ionisation energies

In *all* their reactions, atoms of elements in Group 2 lose their two outer electrons to form ions with two positive charges.

$$M(g) \rightarrow M(g)^{2+} + 2e^-$$

So, an amount of energy equal to the sum of the first and the second ionisation energies is needed for complete ionisation.

$$M(g) \rightarrow M^+(g) + e^- \quad \text{plus} \quad M^+(g) \rightarrow M^{2+}(g) + e^-$$

Both the first ionisation energy and the second ionisation energy decrease going down the group – it takes less energy to remove the electrons as they become further and further away from the positive nucleus. The nucleus is more effectively shielded by more inner shells of electrons.

Hint

Remember the shorthand for writing electron arrangements using the previous inert gas. So [Ne]3s^2 is shorthand for 1s^2,2s^2,2p^6,3s^2.

▼ **Table 1** *The physical properties of Group 2, magnesium to barium*

	Atomic number Z	Electron arrangement	Metallic radius / nm	First + second IEs / kJ mol^{-1}	Melting point T_m / K	Boiling point T_b / K	Density ρ / g cm^{-3}
magnesium, Mg	12	[Ne]3s^2	0.160	738 + 1451 = 2189 ▲	922	1380	1.74
calcium, Ca	20	[Ar]4s^2	0.197	590 + 1145 = 1735	1112	1757	1.54
strontium, Sr	38	[Kr]5s^2	0.215	550 + 1064 = 1614	1042	1657	2.60
barium, Ba	56	[Xe]6s^2	0.224 ▼	503 + 965 = 1468	998	1913	3.51 ▼

In all their reactions, the metals get more reactive going down the group.

Lime kilns

Disused lime kilns can be found in many areas where there is limestone rock. Limestone is mainly calcium carbonate, $CaCO_3$, and it was heated in kilns fired by wood or coal to produce lime (calcium oxide, CaO) which was used to make building mortar, to treat acidic soils, and in making glass.

In the kiln, heat decomposes the limestone:

$$CaCO_3(s) \rightarrow CaO(s) + CO_2(g)$$

A typical kiln contained around 25 tonnes of limestone.

▲ **Figure 1** *A disused lime kiln*

1 Calculate how many tonnes of lime this would produce.

2 Give two reasons why lime kilns were significant emitters of carbon dioxide.

3 The limestone was broken into fist-sized lumps before firing. Suggest why.

In practice limestone is unlikely to be found 100% pure. One contaminant is silicon dioxide (sand), which is unaffected by heat. Imagine limestone that contains 15% sand, so 100 tonnes of limestone would contain 85 tonnes of calcium carbonate.

4 Rework the calculation above to calculate how much lime would actually be produced.

Both lime (calcium oxide, CaO) and slaked lime (calcium hydroxide, $Ca(OH)_2$) may be used to neutralise acids in soil.

The equations for their reactions with hydrochloric acid (for simplicity) are given below.

$$CaO(s) + 2HCl(aq) \rightarrow CaCl_2(aq) + H_2O(l)$$
$$Ca(OH)_2(s) + 2HCl(aq) \rightarrow CaCl_2(aq) + 2H_2O(l)$$

5 How many *moles* of **a** lime and **b** quicklime are needed to neutralise 2 mol HCl?

6 How many *grams* of **a** lime and **b** slaked lime are needed to neutralise 2 mol HCl?

7 What implications does this have for the farmer or gardener?

8 Suggest other factors to be considered when deciding which compound to use.

The chemical reactions of the Group 2 elements, magnesium to barium

Oxidation is loss of electrons so in all their reactions the Group 2 metals are oxidised. The metals go from oxidation state 0 to oxidation state +2. These are redox reactions.

Reaction with water

With water you see a trend in reactivity – the metals get more reactive going down the group. These are also redox reactions.
The basic reaction is as follows, where M is any Group 2 metal:

$$\overset{0}{M}(s) + 2\overset{+1}{H_2}O(l) \rightarrow \overset{+2}{M}(OH)_2(aq) + \overset{0}{H_2}(g)$$

Magnesium hydroxide is milk of magnesia and is used in indigestion remedies to neutralise excess stomach acid which causes heartburn, indigestion, and wind.

Magnesium reacts very slowly with cold water, but rapidly with steam to form an alkaline oxide and hydrogen.

$$Mg(s) + H_2O(g) \rightarrow MgO(s) + H_2(g)$$

Calcium reacts in the same way but more vigorously, even with cold water. Strontium and barium react more vigorously still. Calcium hydroxide is sometimes called **slaked lime** and is used to treat acidic soil. Most plants have an optimum level of acidity or alkalinity in which they thrive. For example, grass prefers a pH of around 6, so if the soil has a pH much below this it will not grow as well as it could. Crops such as wheat, corn, oats, and barley prefer soil that is nearly neutral.

The solubilities of the hydroxides and sulfates

Hydroxides

There is a clear trend in the solubilities of the hydroxides – going down the group they become more soluble. The hydroxides are all white solids.

Hint

Remember *lower* pH means *more* acidic.

▲ **Figure 2** *Two applications of Group 2 hydroxides.*

The extraction of titanium

Titanium is a metal with very useful properties – it is strong, low density, and has a high melting point. It is used in the aerospace industry and also for making replacement hip joints. It is a relatively common metal in the Earth's crust, but it is not easy to extract.

Most metals are found in the Earth as oxides, and the metal is extracted by reacting the oxide with carbon:

Metal oxide + carbon → metal + carbon dioxide

This method cannot be used for titanium as the metal reacts with carbon to form titanium carbide, TiC, which makes the metal brittle. So the titanium oxide is first reacted with chlorine and carbon (coke) to form titanium chloride, $TiCl_4$, and carbon monoxide.

The titanium chloride is then reduced to titanium by reaction with magnesium:

$$TiCl_4(l) + 2Mg(s) \rightarrow 2MgCl_2(s) + Ti(s)$$

1. Write a balanced symbol equation for the reaction of iron oxide, Fe_2O_3, with carbon.
2. What is unusual about titanium chloride as a metal compound and what does this suggest about the bonding in it?
3. Write a balanced symbol equation for the reaction of titanium oxide with chlorine and carbon.
4. Work out the oxidation state of each element in this equation before and after reaction. What has been oxidised and what has been reduced?

Hint

The trends in solubilities of the hydroxide and sulfates can be used as the basis of a test for Ca^{2+}, Sr^{2+}, and Ba^{2+} ions in compounds.

Hint

The symbol aq is used to represent an unspecified amount of water.

Hint

Magnesium sulfate is soluble in water and is used in bath salts. It is also used as a laxative, and by gardeners to treat magnesium-deficient soil.

Magnesium hydroxide, $Mg(OH)_2$ (milk of magnesia), is almost insoluble. It is sold as a suspension in water, rather than a solution.

- Calcium hydroxide, $Ca(OH)_2$, is sparingly soluble and a solution is used as lime water.
- Strontium hydroxide, $Sr(OH)_2$, is more soluble.
- Barium hydroxide, $Ba(OH)_2$, dissolves to produce a strongly alkaline solution:

$$Ba(OH)_2(s) + aq \rightarrow Ba^{2+}(aq) + 2OH^-(aq)$$

Sulfates

The solubility trend in the sulfates is exactly the opposite – they become less soluble going down the group. So, barium sulfate is virtually insoluble. This means that it can be taken by mouth as a barium meal to outline the gut in medical X-rays. (The heavy barium atom is very good at absorbing X-rays.) This test is safe, despite the fact that barium compounds are highly toxic, because barium sulfate is so insoluble.

The insolubility of barium sulfate is also used in a simple test for sulfate ions in solution. The solution is first acidified with nitric or hydrochloric acid. Then barium chloride solution is added to the solution under test and if a sulfate is present a white precipitate of barium sulfate is formed.

$$Ba^{2+}(aq) + SO_4^{2-}(aq) \rightarrow BaSO_4(s)$$

The addition of acid removes carbonate ions as carbon dioxide. (Barium carbonate is also a white insoluble solid, which would be indistinguishable from barium sulfate).

Flue gas desulfurisation

Coal and other fossil fuels contain sulfur. When this is burned in power stations the acidic gas sulfur dioxide is formed. This contributes to acid rain. Increasingly this is removed from the flue gases by passing them through a suspension of calcium oxide or calcium carbonate which reacts with sulfur dioxide to form calcium sulfite:

$CaO(s) + SO_2(g) \rightarrow CaSO_3(s)$ **or** $CaCO_3(s) + SO_2(g) \rightarrow CaSO_3 + CO_2(g)$

The calcium sulfite is then oxidised to calcium sulfate (gypsum) which can be sold for making builders' plaster among other things.

$$CaSO_3(s) + \frac{1}{2}O_2(g) \rightarrow CaSO_4(s)$$

Summary questions

1. **a** What is the oxidation number of all Group 2 elements in their compounds?
   b Explain your answer.
2. Why does it become easier to form +2 ions going down Group 2?
3. Explain why this is a redox reaction.
   $Ca + Cl_2 \rightarrow CaCl_2$
4. Write the equation for the reaction of calcium with water. Include the oxidation state of each element.
5. How would you expect the reaction of strontium with water to compare with those of the following? Explain your answers.
   a calcium **b** barium
6. Radium is below strontium in Group 2. How would you predict the solubilities of the following compounds would compare with the other members of the group? Explain your answers.
   a radium hydroxide **b** radium sulfate

Practice questions

1 State and explain the trend in melting point of the Group 2 elements Ca to Ba.

(3 marks)
AQA, 2006

2 State the trends in solubility of the hydroxides and of the sulfates of the Group 2 elements Mg to Ba.

Describe a chemical test you could perform to distinguish between separate aqueous solutions of sodium sulfate and sodium nitrate. State the observation you would make with each solution. Write an equation for any reaction which occurs.

(6 marks)
AQA, 2006

3 **(a)** For the elements Mg to Ba, state how the solubilities of the hydroxides and the solubilities of the sulfates change down Group 2.

(b) Describe a test to show the presence of sulfate ions in an aqueous solution. Give the results of this test when performed on separate aqueous solutions of magnesium chloride and magnesium sulfate. Write equations for any reactions occurring.

(c) State the trend in the reactivity of the Group 2 elements Mg to Ba with water. Write an ionic equation with state symbols to show the reaction of barium with an excess of water.

(9 marks)
AQA, 2005

4 Group 2 metals and their compounds are used commercially in a variety of processes and applications.

(a) State a use of magnesium hydroxide in medicine.

(1 mark)

(b) Calcium carbonate is an insoluble solid that can be used in a reaction to lower the acidity of the water in a lake.

Explain why the rate of this reaction decreases when the temperature of the water in the lake falls.

(3 marks)

(c) Strontium metal is used in the manufacture of alloys.

(i) Explain why strontium has a higher melting point than barium.

(2 marks)

(ii) Write an equation for the reaction of strontium with water.

(1 mark)

(d) Magnesium can be used in the extraction of titanium.

(i) Write an equation for the reaction of magnesium with titanium(IV) chloride.

(1 mark)

(ii) The excess of magnesium used in this extraction can be removed by reacting it with dilute sulfuric acid to form magnesium sulfate.

Use your knowledge of Group 2 sulfates to explain why the magnesium sulfate formed is easy to separate from the titanium.

(1 mark)
AQA, 2010

5 Group 2 elements and their compounds have a wide range of uses.

(a) For parts **(a)(i)** to **(a)(iii)**, choose the correct answer to complete each sentence.

(i) From $Mg(OH)_2$ to $Ba(OH)_2$, the solubility in water

decreases	increases	stays the same

(1 mark)

(ii) From Mg to Ba, the first ionisation energy

decreases	increases	stays the same

(1 mark)

(iii) From Mg to Ba, the atomic radius

decreases	increases	stays the same

(1 mark)

(b) Explain why calcium has a higher melting point than strontium.

(2 marks)

(c) Acidified barium chloride solution is used as a reagent to test for sulfate ions.

(i) State why sulfuric acid should **not** be used to acidify the barium chloride.

(1 mark)

(ii) Write the **simplest ionic** equation with state symbols for the reaction that occurs when acidified barium chloride solution is added to a solution containing sulfate ions.

(1 mark)

AQA, 2012

6 **(a)** There are many uses for compounds of barium.

(i) Write an equation for the reaction of barium with water.

(1 mark)

(ii) State the trend in reactivity with water of the Group 2 metals from Mg to Ba.

(1 mark)

(b) Give the formula of the **least** soluble hydroxide of the Group 2 metals from Mg to Ba.

(1 mark)

(c) State how barium sulfate is used in medicine.
Explain why this use is possible, given that solutions containing barium ions are poisonous.

(2 marks)

AQA, 2012

10 Group 7(17), the halogens

10.1 The halogens

Group 7, on the right-hand side of the Periodic Table, is made up of non-metals. As elements they exist as diatomic molecules, F_2, Cl_2, Br_2, and I_2, called the halogens. (Astatine is rare and radioactive.)

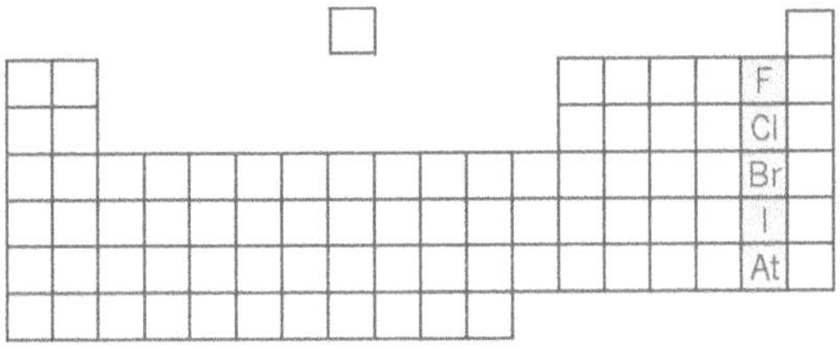

Learning objectives:

- → Explain how and why the atomic radius changes in Group 7 of the Periodic Table.
- → Explain how and why electronegativity changes in Group 7 of the Periodic Table.

Specification reference: 3.2.3

Physical properties

The gaseous halogens vary in appearance, as shown in Figure 1. At room temperature, fluorine is a pale yellow gas, chlorine a greenish gas, bromine a red-brown liquid, and iodine a black solid – they get darker and denser going down the group.

They all have a characteristic 'swimming-bath' smell.

A number of the properties of fluorine are untypical. Many of these untypical properties stem from the fact that the F—F bond is unexpectedly weak, compared with the trend for the rest of the halogens, see Table 1. The small size of the fluorine atom leads to repulsion between non-bonding electrons because they are so close together:

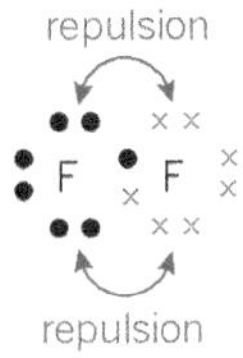

Hint

The word halogen means 'salt former'. The halogens readily react with many metals to form fluoride, chloride, bromide, and iodide salts.

▲ **Figure 1** *Fluorine chlorine, bromine and iodine in their gaseous states*

The physical properties of fluorine, chlorine, bromine, and iodine are shown in Table 2.

There are some clear trends shown by the red arrows. These can be explained as follows.

▼ **Table 1** *Bond energies for fluorine, chlorine, bromine, and iodine*

Bond	Bond energy / kJ mol^{-1}
F–F	158
Cl–Cl	243
Br–Br	193
I–I	151

Size of atoms

The atoms get bigger going down the group because each element has one extra filled main shell of electrons compared with the one above it, see Figure 2.

▼ **Table 2** *The physical properties of Group 7, fluorine to iodine*

Halogen	Atomic number, Z	Electron arrangement	Electronegativity	Atomic (covalent) radius / nm	Melting point T_m / K	Boiling point T_b / K
fluorine	9	[He] $2s^2 2p^5$	4.0 ↑	0.071	53	85
chlorine	17	[Ne] $3s^2 3p^5$	3.0	0.099	172	238
bromine	35	[Ar] $3d^{10} 4s^2 4p^5$	2.8	0.114	266	332
iodine	53	[Kr] $4d^{10} 5s^2 5p^5$	2.5	0.133 ↓	387 ↓	457 ↓

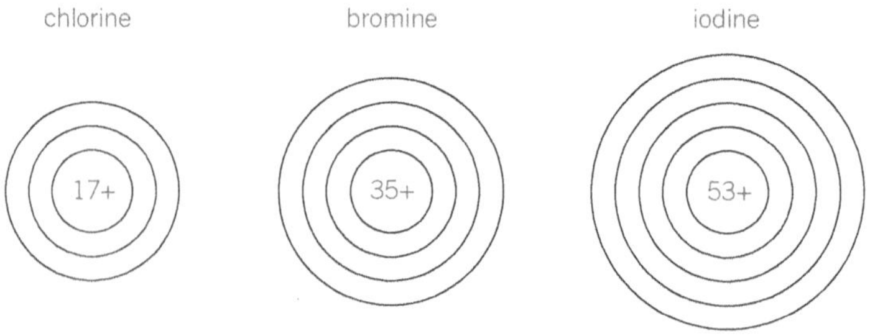

▲ **Figure 2** *The outer shell gets further from the nucleus going down the group*

Study tip

Remember that melting and boiling points involve weakening and breaking van der Waals forces only. The covalent bonds in the halogen molecules stay intact.

Study tip

Remember to write a halogen element as a diatomic molecule, for example, as F_2 not F.

Synoptic link

Look back at Topic 3.2, Covalent bonding.

Electronegativity

Electronegativity is a measure of the ability of an atom to attract electrons, or electron density, towards itself within a covalent bond.

Electronegativity depends on the attraction between the nucleus and bonding electrons in the outer shell. This, in turn, depends on a balance between the number of protons in the nucleus (nuclear charge) and the distance between the nucleus and the bonding electrons, plus the shielding effect of inner shells of electrons.

For example, consider the hydrogen halides, HX. The shared electrons in the H—X bond get further away from the nucleus as the atoms get larger going down the group. This makes the shared electrons further from the halogen nucleus and increases the shielding by more inner shells of electrons. These factors are more important than the increasing nuclear charge, so the electronegativity decreases going down the group.

Melting and boiling points

These increase as going down the group. This is because larger atoms have more electrons and this makes the van der Waals forces *between* the molecules stronger. The lower the boiling point, the more volatile the element. So chlorine, which is a gas at room temperature, is more volatile than iodine, which is a solid.

Hint

Melting and boiling involves breaking the forces *between* the hydrogen molecules, *not* the covalent bonds between the atoms.

Summary questions

1 Predict the properties of astatine compared with the other halogens in terms of:

- **a** physical state at room temperature, including colour
- **b** size of atom
- **c** electronegativity.

2 Explain your answers to question **1**.

3 **a** Use the data in Table 2 to make a rough estimate of the boiling point of astatine.

- **b** Why would you expect the boiling point of astatine to be the largest?

10.2 The chemical reactions of the halogens

Trends in oxidising ability

Halogens usually react by gaining electrons to become negative ions, with a charge of −1. These reactions are redox reactions – halogens are oxidising agents and are themselves reduced. For example:

$$Cl_2 + 2e^- \xrightarrow{\text{gain of electrons}} 2Cl^-$$

The oxidising ability of the halogens increases going up the group.

Fluorine is one of the most powerful oxidising agents known.

fluorine   chlorine   bromine   iodine

← increasing oxidising power ——

Learning objectives:

- → State the trend in oxidising ability of the halogens.
- → Describe the experimental evidence that confirms this trend.

Specification reference: 3.2.3

Hint

Remember OIL RIG: Oxidation Is Loss of electrons. Reduction Is Gain of electrons.

Displacement reactions

Halogens will react with metal halides in solution in such a way that the halide in the compound will be displaced by a more reactive halogen but not by a less reactive one. This is called a **displacement reaction**.

For example, chlorine will displace bromide ions, but iodine will not.

$$\overset{0}{Cl_2}(aq) + 2Na\overset{-1}{Br}(aq) \rightarrow \overset{0}{Br_2}(aq) + 2Na\overset{-1}{Cl}(aq)$$

The ionic equation for this reaction is:

$$Cl_2(aq) + \cancel{2Na^+(aq)} + 2Br^-(aq) \rightarrow Br_2(aq) + \cancel{2Na^+(aq)} + 2Cl^-(aq)$$

The sodium ions are spectator ions.

The virtually colourless starting materials react to produce the yellow-brown colour of bromine.

In this redox reaction the chlorine is acting as an oxidising agent, by removing electrons from Br^- and so oxidising $2Br^-$ to Br_2 (the oxidation number of the bromine increases from −1 to 0). In general, a halogen will always displace the ion of a halogen below it in the Periodic Table, see Table 1.

Hint

The colour of the halogen molecules in aqueous solution are as follows:

chlorine – *very* pale green (virtually colourless)

bromine – yellow brown

iodine – brown solution (which may form a black precipitate)

▼ **Table 1** *The oxidation of a halide by a halogen*

	F^-	Cl^-	Br^-	I^-
F_2	–	yes	yes	yes
Cl_2	no	–	yes	yes
Br_2	no	no	–	yes
I_2	no	no	no	–

You cannot investigate fluorine in an aqueous solution because it reacts with water.

▲ **Figure 1** *Iodine can be extracted from seaweed.*

The extraction of bromine from sea water

The oxidation of a halide by a halogen is the basis of a method for extracting bromine from sea water. Sea water contains small amounts of bromide ions which can be oxidised by chlorine to produce bromine:

$$Cl_2(aq) + 2Br^-(aq) \rightarrow Br_2(aq) + 2Cl^-(aq)$$

Extraction of iodine from kelp

Iodine was discovered in 1811. It was extracted from kelp, which is obtained by burning seaweed. Some iodine is still produced in this way. Salts such as sodium chloride, potassium chloride, and potassium sulfate are removed from the kelp by washing with water. The residue is then heated with manganese dioxide and concentrated sulfuric acid and iodine is liberated.

$$2I^- + MnO_2 + 4H^+ \rightarrow Mn^{2+} + 2H_2O + I_2$$

1 Is the reaction an oxidation or a reduction of the iodide ion? Explain your answer.
2 Find out why our table salt often has potassium iodide added to it.

Summary question

1 a Which of the following mixtures would react?
  i $Br_2(aq) + 2NaCl(aq)$
  ii $Cl_2(aq) + 2NaI(aq)$
 b Explain your answers.
 c Complete the equation for the mixture that reacts.

10.3 Reactions of halide ions

Halide ions as reducing agents

Halide ions can act as reducing agents. In these reactions the halide ions lose (give away) electrons and become halogen molecules. There is a definite trend in their reducing ability. This is linked to the size of the ions. The larger the ion, the more easily it loses an electron. This is because the electron is lost from the outer shell which is further from the nucleus as the ion gets larger so the attraction to the outer electron is less, see figure 1.

	increasing reducing power →			
	F^-	Cl^-	Br^-	I^-
Ionic radius	0.133 nm	0.180 nm	0.195 nm	0.215 nm

This trend can be seen in the reactions of solid sodium halides with concentrated sulfuric acid.

The reactions of sodium halides with concentrated sulfuric acid

Solid sodium halides react with concentrated sulfuric acid. The products are different and reflect the reducing powers of the halide ions shown above.

Sodium chloride (solid)

In this reaction, drops of concentrated sulfuric acid are added to solid sodium chloride. Steamy fumes of hydrogen chloride are seen. The solid product is sodium hydrogensulfate.

The reaction is:

$$NaCl(s) + H_2SO_4(l) \rightarrow NaHSO_4(s) + HCl(g)$$

This is *not* a redox reaction because no oxidation state has changed. The chloride ion is too weak a reducing agent to reduce the sulfur (oxidation state = +6) in sulfuric acid. It is an acid–base reaction.

$$\overset{+1}{Na}\,\overset{-1}{Cl}(s) + \overset{+1}{H_2}\,\overset{+6}{S}\,\overset{-2}{O_4}(l) \rightarrow \overset{+1}{Na}\,\overset{+1}{H}\,\overset{+6}{S}\,\overset{-2}{O_4}(s) + \overset{+1}{H}\,\overset{-1}{Cl}(g)$$

This reaction can be used to prepare hydrogen chloride gas which, because of this reaction, was once called salt gas.

A similar reaction occurs with sodium fluoride to produce hydrogen fluoride, an extremely dangerous gas that will etch glass. The fluoride ion is an even weaker reducing agent than the chloride ion.

Sodium bromide (solid)

In this case you will see steamy fumes of hydrogen bromide *and* brown fumes of bromine. Colourless sulfur dioxide is also formed.

Two reactions occur.

First sodium hydrogensulfate and hydrogen bromide are produced (in a similar acid–base reaction to sodium chloride).

$$NaBr(s) + H_2SO_4(l) \rightarrow NaHSO_4(s) + HBr(g)$$

Learning objectives:

- → State the trend in reducing ability of halide ions.
- → Explain how this trend is linked to ionic radius.
- → Describe how halide ions are identified using silver nitrate.

Specification reference: 3.2.3

Synoptic link

See Practical 3 on page 524.

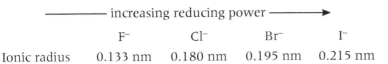

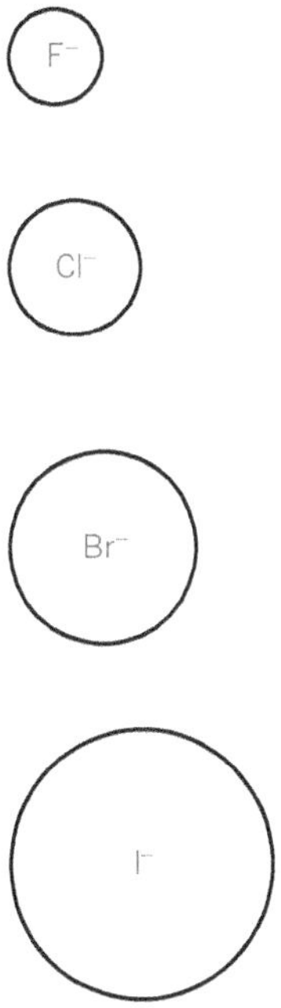

▲ **Figure 1** *the relative sizes of the halide ions*

Study tip

Remember that the reactions take place between *solid* halide salts and *concentrated* sulfuric acid.

However, bromide ions are strong enough reducing agents to reduce the sulfuric acid to sulfur dioxide. So, in a second step, the oxidation state of the sulfur is reduced by two from +6 to +4 and that of the bromine increases by one from −1 to 0.

$$\overset{-1}{2H^+ + 2Br^-} + \overset{+6}{H_2SO_4(l)} \rightarrow \overset{+4}{SO_2(g)} + 2H_2O(l) + \overset{0}{Br_2(l)}$$

This is a redox reaction. The reactions are exothermic and some of the bromine vaporises.

Sodium iodide (solid)

In this case you see steamy fumes of hydrogen iodide, the black solid of iodine, and the bad egg smell of hydrogen sulfide gas is present. Yellow solid sulfur may also be seen. Colourless sulfur dioxide is also evolved.

Several reactions occur. Hydrogen iodide is produced in an acid–base reaction as before.

$$NaI(s) + H_2SO_4(l) \rightarrow NaHSO_4(s) + HI(g)$$

Iodide ions are better reducing agents than bromide ions in the second step so they reduce the sulfur in sulfuric acid even further (from +6 to zero and −2) so that sulfur dioxide, sulfur, and hydrogen sulfide gas are produced. For example:

$$\overset{-1}{8H^+ + 8I^-} + \overset{+6}{H_2SO_4(l)} \rightarrow \overset{-2}{H_2S(g)} + 4H_2O(l) + \overset{0}{4I_2(s)}$$

During the reduction from +6 to −2, the sulfur passes through oxidation state 0 and some yellow, solid sulfur may be seen.

Further redox equations

As the sulfur in the sulfuric acid is reduced by the iodide ions from oxidation state +6 to −2, it passes through oxidation states +4 (sulfur dioxide, SO_2) and 0 (uncombined sulfur, S).

Use the oxidation state technique to help you to write equations for these two processes.

$$\overset{+6}{H_2SO_4} + \overset{-1}{6I^-} + 6H^+ \rightarrow \overset{0}{S} + \overset{0}{3I_2} + 4H_2O$$

$$\overset{+6}{H_2SO_4} + \overset{-1}{2I^-} + 2H^+ \rightarrow \overset{+4}{SO_2} + \overset{0}{I_2} + 2H_2O$$

Balancing the redox reactions

We can use half equations to help balance redox reactions, see Topic 7.3. In the redox reactions, the oxidising agent is the S(+6) in the sulfuric acid whose oxidation state is reduced.

Reaction with chloride ions, Cl^-

No redox reaction occurs as Cl^- ions are too difficult to be oxidised.

Reaction with bromide ions, Br^-

Here S(+6) drops to S(+4)

Down 2

$$2H^+ + 2e^- + \overset{+6}{H_2SO_4} \rightarrow \overset{+4}{SO_2} + 2H_2O$$

Up 1

$$\overset{-1}{Br} \rightarrow \overset{0}{\tfrac{1}{2} Br_2} + 2e^-$$

So the second equation must be multiplied by 2 so that the electrons cancel.

$$2Br^- \rightarrow Br_2 + 2e^-$$

Now add the two half equations to give

$$2H^+ + \cancel{2e^-} + 2Br^- + H_2SO_4 \rightarrow SO_2 + 2H_2O + Br_2 + \cancel{2e^-}$$

Now try this technique with iodide ions, I^-, where the sulfur goes from S(+6) to S(−2) (down 8).

Identifying metal halides with silver ions

All metal halides (except fluorides) react with silver ions in aqueous solution, for example, in silver nitrate, to form a precipitate of the insoluble silver halide. For example:

$$Cl^-(aq) + Ag^+(aq) \rightarrow AgCl(s)$$

Silver fluoride does not form a precipitate because it is soluble in water.

1 Dilute nitric acid HNO_3 or $(H^+(aq) + NO_3^-(aq))$ is first added to the halide solution to get rid of any soluble carbonate, $CO_3^{2-}(aq)$, or hydroxide, $OH^-(aq)$ impurities:

$$CO_3^{2-}(aq) + 2H^+(aq) + 2NO_3^-(aq) \rightarrow CO_2(g) + H_2O(l) + 2NO_3^-(aq)$$

$$OH^-(aq) + H^+(aq) + NO_3^-(aq) \rightarrow H_2O(l) + NO_3^-(aq)$$

These would interfere with the test by forming precipitates of insoluble silver carbonate:

$$2Ag^+(aq) + CO_3^{2-}(aq) \rightarrow Ag_2CO_3(s)$$

or insoluble silver hydroxide:

$$Ag^+(aq) + OH^-(aq) \rightarrow AgOH(s)$$

2 Then a few drops of silver nitrate solution are added and the halide precipitate forms.

The reaction can be used as a test for halides because you can tell from the colour of the precipitate which halide has formed, see Table 1. The colours of silver bromide and silver iodide are similar but if you add a few drops of concentrated ammonia solution, silver bromide dissolves but silver iodide does not.

▼ **Table 1** *Tests for halides*

Halide	silver fluoride	silver chloride	silver bromide	silver iodide
Colour	no precipitate	white ppt	cream ppt	pale yellow ppt
Further test		dissolves in dilute ammonia	dissolves in concentrated ammonia	insoluble in concentrated ammonia

Study tip

You should learn the colours of the silver halides, especially AgBr and AgI.

Hint

Sulfuric acid or hydrochloric acid cannot be used. Sulfuric acid would give a precipitate of silver sulfate and hydrochloric acid a precipitate of silver chloride. Either of these would invalidate the test.

Hint

Silver hydroxide in fact is converted into silver oxide – a brown precipitate:

$$2AgOH \rightarrow Ag_2O + H_2O$$

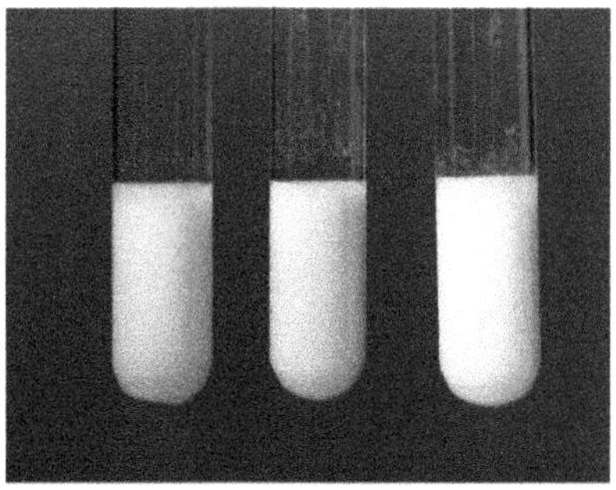

▲ **Figure 1** *The colours of the silver halides: (from left to right) AgCl, AgBr, AgI*

Study tip

Tests for anions

As well as the tests for halide ions, you should recall the following tests for other negative ions in solution that you will have met at GCSE.

Ion	Test	Result
Carbonate, CO_3^{2-}	Add a little dilute hydrochloric acid	Bubbles of gas (carbon dioxide) that turns limewater milky. NB hydrogencarbonates will also give a positive result
Sulfate, SO_4^{2-}	Add a little dilute hydrochloric acid them a few drops of barium chloride solution	A white precipitate of barium sulfate
Hydroxide, OH^-	pH indicator solution or paper Warm with an ammonium salt	Colour change indicates an alkali (colour change depends on indicator used) Ammonia gas evolved (an alkaline gas with a 'wet nappy' smell)

The test for OH^- ions can be used in reverse to test for ammonium ions.

Summary questions

1 The reaction between concentrated sulfuric acid and solid sodium fluoride is not usually carried out in the laboratory.

- **a** How does the reducing power of the fluoride ion compare with the other halide ions?
- **b** Explain why you would predict this.
- **c** Write a balanced symbol equation for the reaction between concentrated sulfuric acid and sodium fluoride.
- **d** Is this a redox reaction? Explain your answer.

2 A few drops of silver nitrate were added to an acidified solution, to show the presence of sodium bromide.

- **a** What would you see?
- **b** Write the equation for the reaction.
- **c** What would happen if you now added a few drops of concentrated ammonia solution?
- **d** Why is an acid added to sodium bromide solution initially?
- **e** Neither hydrochloric nor sulfuric acid may be used to acidify the solution. Explain why this is so.
- **f** Why can't this test be used to find out if fluoride ions are present?

10.4 Uses of chlorine

Chlorine is a poisonous gas and was notoriously used as such in the First World War. However, it is soluble in water and in this form and at low concentration has become an essential part of our life in the treatment of water both for drinking and in swimming pools.

Reaction with water

Chlorine reacts with water in a reversible reaction to form chloric(I) acid, HClO, and hydrochloric acid, HCl:

$$\overset{0}{Cl_2}(g) + H_2O(l) \rightleftharpoons H\overset{+1}{Cl}O(aq) + H\overset{-1}{Cl}(aq)$$

In this reaction, the oxidation number of one of the chlorine atoms increases from 0 to +1 and that of the other decreases from 0 to −1. This type of redox reaction, where the oxidation state of some atoms of the same element increase and others decrease, is called **disproportionation**.

This reaction takes place when chlorine is used to purify water for drinking and in swimming baths, to prevent life-threatening diseases. Chloric(I) acid is an oxidising agent and kills bacteria by oxidation. It is also a bleach.

The other halogens react similarly, but much more slowly going down the group.

In sunlight, a different reaction occurs:

$$\underset{\text{pale green}}{2Cl_2(g)} + 2H_2O(l) \rightarrow \underset{\text{colourless}}{4HCl(aq)} + O_2(g)$$

Chlorine is rapidly lost from pool water in sunlight so that shallow pools need frequent addition of chlorine.

An alternative to the direct chlorination of swimming pools is to add solid sodium (or calcium) chlorate(I). This dissolves in water to form chloric(I) acid, HClO(aq,) in a reversible reaction:

$$NaClO(s) + H_2O \rightleftharpoons Na^+(aq) + OH^-(aq) + HClO(aq)$$

In alkaline solution, this equilibrium moves to the left and the HClO is removed as ClO^- ions. To prevent this happening, swimming pools need to be kept slightly acidic. However, this is carefully monitored and the water never gets acidic enough to corrode metal components and affect swimmers.

Reaction with alkali

Chlorine reacts with cold, dilute sodium hydroxide to form sodium chlorate(I), NaClO. This is an oxidising agent and the active ingredient in household bleach. This is also a disproportionation reaction – see the oxidation numbers above the relevant species.

$$\overset{0}{Cl_2}(g) + 2NaOH(aq) \rightarrow Na\overset{+1}{Cl}O(aq) + Na\overset{-1}{Cl}(aq) + H_2O(l)$$

The other halogens except fluorine behave similarly.

Learning objectives:

→ Describe how chlorine reacts with water.

→ Describe how chlorine reacts with alkali.

Specification reference: 3.2.3

Hint

The use of chlorine to kill germs in drinking water is a major benefit to public health. The addition of flouride-containing chemicals to help eliminate dental cavities (bad teeth) has been more controversial.

Summary questions

1 Write the equations for bromine reacting with:
  a water
  b alkali.

2 Why is chlorine added to the domestic water supply?

3 a What products are obtained when an aqueous solution of chlorine is left in the sunlight?
  b Write the equation for the reaction, giving the oxidation states of every atom before and after reaction.
  c What has been oxidised?
  d What has been reduced?
  e What is the oxidising agent?
  f What is the reducing agent?

Practice questions

1 **(a)** State the trend in electronegativity of the elements down Group 7. Explain this trend.
(3 marks)

(b) (i) State the trend in reducing ability of the halide ions down Group 7.
(ii) Give an example of a reagent which could be used to show that the reducing ability of bromide ions is different from that of chloride ions.
(2 marks)

(c) The addition of silver nitrate solution followed by dilute aqueous ammonia can be used as a test to distinguish between chloride and bromide ions. For each ion, state what you would observe if an aqueous solution containing the ion was tested in this way.
(4 marks)

(d) Write an equation for the reaction between chlorine and cold, dilute aqueous sodium hydroxide. Give two uses of the resulting solution.
(3 marks)
AQA, 2006

2 **(a)** Explain, by referring to electrons, the meaning of the terms *reduction* and *reducing agent*.
(2 marks)

(b) Iodide ions can reduce sulfuric acid to three different products.
(i) Name the **three** reduction products and give the oxidation state of sulfur in each of these products.
(ii) Describe how observations of the reaction between solid potassium iodide and concentrated sulfuric acid can be used to indicate the presence of any two of these reduction products.
(iii) Write half equations to show how two of these products are formed by reduction of sulfuric acid.
(10 marks)

(c) Write an ionic equation for the reaction that occurs when chlorine is added to cold water. State whether or not the water is oxidised and explain your answer.
(3 marks)
AQA, 2006

3 **(a)** State the trend in the boiling points of the halogens from fluorine to iodine and explain this trend.
(4 marks)

(b) Each of the following reactions may be used to identify bromide ions. For each reaction, state what you would observe and, where indicated, write an appropriate equation.
(i) The reaction of aqueous bromide ions with chlorine gas
(ii) The reaction of aqueous bromide ions with aqueous silver nitrate followed by the addition of concentrated aqueous ammonia
(iii) The reaction of solid potassium bromide with concentrated sulfuric acid
(7 marks)

(c) Write an equation for the redox reaction that occurs when potassium bromide reacts with concentrated sulfuric acid.
(2 marks)
AQA, 2005

4 **(a)** State and explain the trend in electronegativity down Group 7 from fluorine to iodine.
(3 marks)

(b) State what you would observe when chlorine gas is bubbled into an aqueous solution of potassium iodide. Write an equation for the reaction that occurs.
(2 marks)

(c) Identify **two** sulfur-containing reduction products formed when concentrated sulfuric acid oxidises iodide ions. For each reduction product, write a half equation to illustrate its formation from sulfuric acid.
(4 marks)

(d) Write an equation for the reaction between chlorine gas and dilute aqueous sodium hydroxide. Name the **two** chlorine-containing products of this reaction and give the oxidation state of chlorine in each of these products.
(5 marks)
AQA, 2005

5 A student investigated the chemistry of the halogens and the halide ions.

(a) In the first two tests, the student made the following observations.

Test	Observation
1 Add chlorine water to aqueous potassium iodide solution.	The colourless solution turned a brown colour.
2 Add silver nitrate solution to aqueous potassium chloride solution.	The colourless solution produced a white precipitate.

(i) Identify the species responsible for the brown colour in Test **1**.
Write the **simplest ionic** equation for the reaction that has taken place in Test **1**.
State the type of reaction that has taken place in Test **1**.

(3 marks)

(ii) Name the species responsible for the white precipitate in Test **2**.
Write the **simplest ionic** equation for the reaction that has taken place in Test **2**.
State what would be observed when an excess of dilute ammonia solution is added to the white precipitate obtained in Test **2**.

(3 marks)

(b) In two further tests, the student made the following observations.

Test	Observation
3 Add concentrated sulfuric acid to solid potassium chloride.	The white solid produced misty white fumes which turned blue litmus paper to red.
4 Add concentrated sulfuric acid to solid potassium iodide.	The white solid turned black. A gas was released that smelled of rotten eggs. A yellow solid was formed.

(i) Write the **simplest ionic** equation for the reaction that has taken place in Test **3**.
Identify the species responsible for the misty white fumes produced in Test **3**.

(2 marks)

(ii) The student had read in a textbook that the equation for one of the reactions in Test **4** is as follows.

$$\mathbf{8H^+ + 8I^- + H_2SO_4 \rightarrow 4I_2 + H_2S + 4H_2O}$$

Write the **two** half equations for this reaction.
State the role of the sulfuric acid and identify the yellow solid that is also observed in Test **4**.

(4 marks)

(iii) The student knew that bromine can be used for killing microorganisms in swimming pool water.
The following equilibrium is established when bromine is added to cold water.

$$\mathbf{Br_2(l) + H_2O(l) \rightleftharpoons HBrO(aq) + H^+(aq) + Br^-(aq)}$$

Use Le Châtelier's principle to explain why this equilibrium moves to the right when sodium hydroxide solution is added to a solution containing dissolved bromine.
Deduce why bromine can be used for killing microorganisms in swimming pool water, even though bromine is toxic.

(3 marks)
AQA, 2012

Answers to the Practice Questions and Section Questions are available at
www.oxfordsecondary.com/oxfordaqaexams-alevel-chemistry

1 For each of the following reactions, select from the list below, the **formula** of a sodium halide that would react as described.

NaF NaCl NaBr NaI

Each **formula** may be selected once, more than once or not at all.

(a) This sodium halide is a white solid that reacts with concentrated sulfuric acid to give a brown gas.

(1 mark)

(b) When a solution of this sodium halide is mixed with silver nitrate solution, no precipitate is formed.

(1 mark)

(c) When this solid sodium halide reacts with concentrated sulfuric acid, the reaction mixture remains white and steamy fumes are given off.

(1 mark)

(d) A colourless aqueous solution of this sodium halide reacts with orange bromine water to give a dark brown solution.

(1 mark)

AQA, 2010

2 There are many uses for Group 2 metals and their compounds.

(a) State a medical use of barium sulfate.

State why this use of barium sulfate is safe, given that solutions containing barium ions are poisonous.

(2 marks)

(b) Magnesium hydroxide is used in antacid preparations to neutralise excess stomach acid.

Write an equation for the reaction of magnesium hydroxide with hydrochloric acid.

(1 mark)

(c) Solutions of barium hydroxide are used in the titration of weak acids.

State why magnesium hydroxide solution could **not** be used for this purpose.

(1 mark)

(d) Magnesium metal is used to make titanium from titanium(IV) chloride.

Write an equation for this reaction of magnesium with titanium(IV) chloride.

(1 mark)

(e) Magnesium burns with a bright white light and is used in flares and fireworks.

Use your knowledge of the reactions of Group 2 metals with water to explain why water should **not** be used to put out a fire in which magnesium metal is burning.

(2 marks)

AQA, 2014

3 **(a)** The diagram below shows the melting points of some of the elements in Period 3.

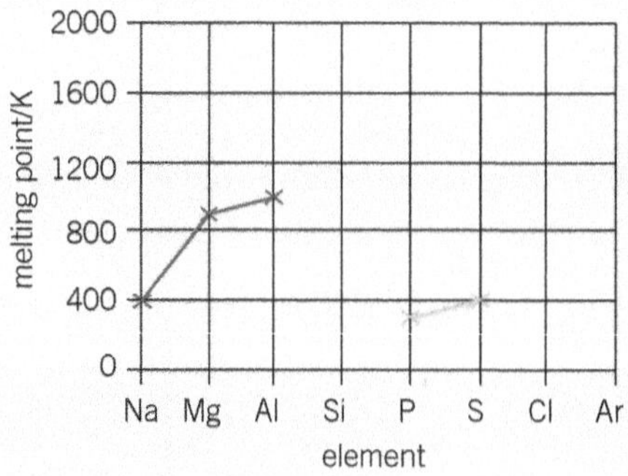

(i) On a copy of the diagram, use crosses to mark the approximate positions of the melting points for the elements silicon, chlorine, and argon. Complete the diagram by joining the crosses.

(ii) By referring to its structure and bonding, explain your choice of position for the melting point of silicon.

(iii) Explain why the melting point of sulfur, S_8, is higher than that of phosphorus, P_4.

(8 marks)

(b) State and explain the trend in melting point of the Group 2 elements Ca–Ba.

(3 marks)

AQA, 2006

reactions with water

- $M(s) + H_2O(l) \longrightarrow M(OH)_2(aq) + H_2(g)$
- reactivity increases down Group 2

solubility

- Group 2 sulfates: solubility decreases down Group 2
- Group 2 hydroxides: solubility increases down Group 2

properties

- reactivity increases down group
- giant metallic structure
- high melting point
- melting point decreases down group
- ionisation energy decreases down group

metallic bonding

Group 2

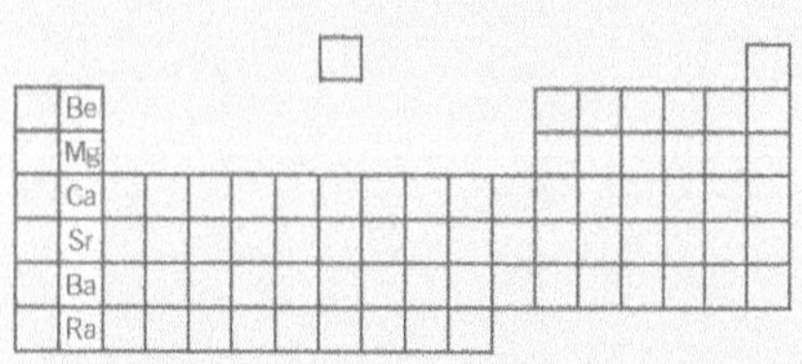

Group 7

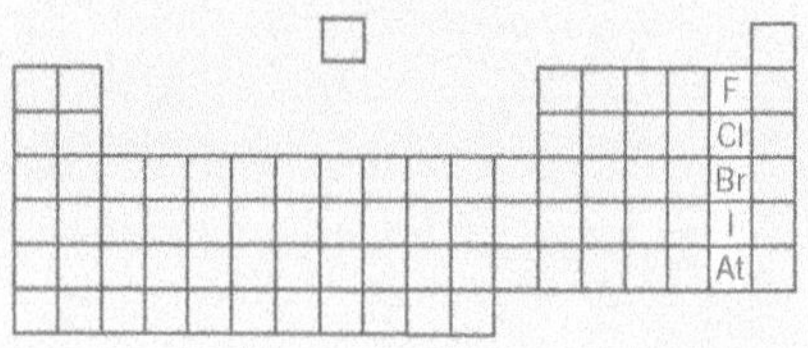

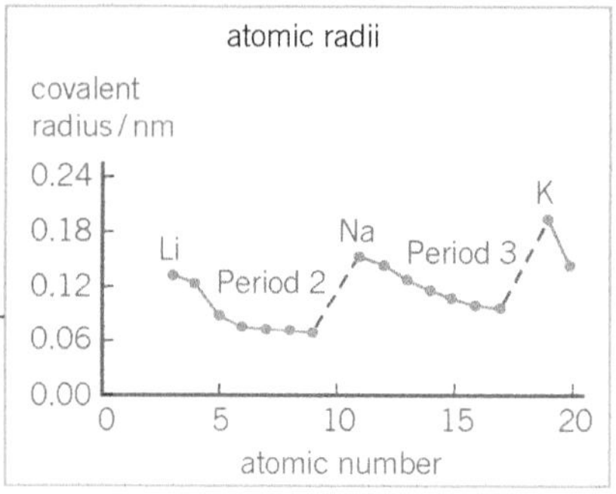

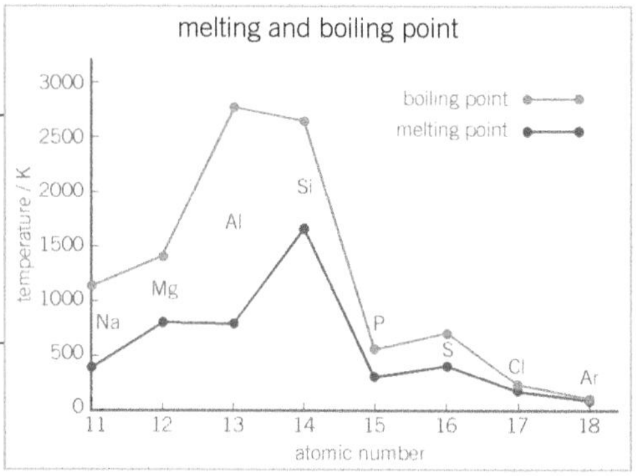

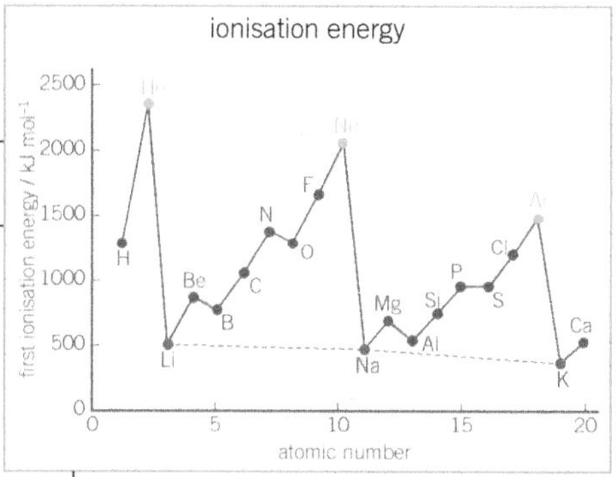

arrangement of electrons

trends in Period 3

Section 2 Inorganic chemistry

Periodic Table

covalent bonding in X_2

properties

- electronegativity decreases down group
- reactivity decreases down group
- melting and boiling point increase down group

reactions

- displacement reactions

	F^-	Cl^-	Br^-	I^-
F_2	–	yes	yes	yes
Cl_2	no	–	yes	yes
Br_2	no	no	–	yes
I_2	no	no	no	–

- halide ions and sulfuric acid

$NaX(s) + H_2SO_4(aq) \longrightarrow NaHSO_4(s) + HX(g)$

blocks

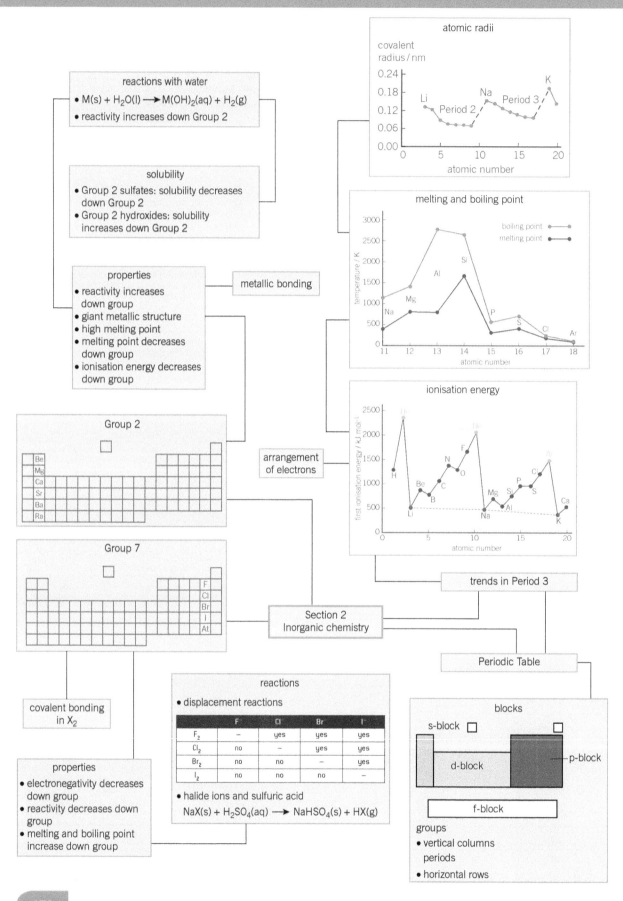

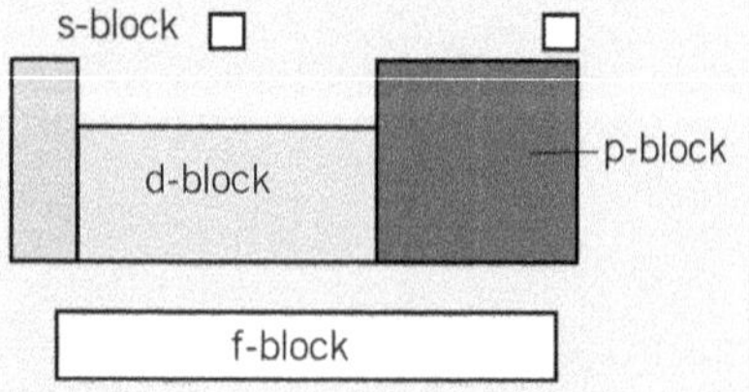

groups
- vertical columns

periods
- horizontal rows

Practical skills

In this section you have met the following ideas:

- Testing reactions of Group 2 metals with water.
- Testing solubility of Group 2 hydroxides and sulfates. and thus identifying Group 2 metal ions in solutions.
- Testing for sulfate ions using acidified barium chloride.
- Investigating the reactions of halogens and their halide ions.
- Testing for halide ions using acidified silver nitrate and ammonia, to identify the halide present.

Maths skills

In this section you have met the following maths skills:

- Identify trends and patterns in data.
- Balancing symbol equations.

Extension

Although Dimitri Mendeleev is credited with establishing the Periodic Table, it was really developed over time by a number of different scientists. One of the key achievements of Mendeleev's work was that he was able to predict the properties of elements that were yet to be discovered and was even able to predict the properties of those missing elements accurately.

1. Suggest how Mendeleev could have correctly predicted the properties of gallium, almost five years before it had been discovered.
2. Investigate the research on X-ray spectra by Henry Moseley and its impact on Mendeleev's Periodic Table.
3. Research the reaction between chlorine and sodium hydroxide. Using your knowledge and understanding of redox reactions explain why this reaction is particularly interesting.

Section 3

AS Organic chemistry 1

Chapters in this section

Carbon atoms have the ability to bond in chains which may be straight, branched, or in rings, forming millions of compounds. Organic chemistry is the study of compounds based on carbon chains. Hydrogen is almost always present.

Introduction to organic chemistry looks at the nature of carbon compounds and explains the different types of formulae that can be used to represent a compound, and also the IUPAC naming system, used to describe organic compounds. It looks at the different sorts of isomers that are possible in some organic compounds. (Isomers have the same formula but a different arrangement of atoms.)

Alkanes is about crude oil and its fractional distillation. It also looks at the different ways that large alkane molecules can be cracked into smaller, more useful molecules. It deals with the combustion of carbon compounds.

Halogenoalkanes looks at how these compounds are formed, how they react and their role in the problem of depletion of the ozone layer.

Alkenes describes the reactions of these compounds which have one or more double bonds.

Alcohols shows the importance of ethanol and describes the primary, secondary, and tertiary structures of alcohols and their reactions.

Organic analysis revisits the mass spectrometer and describes its use in determining the relative molecular masses of compounds and also their molecular formulae. Infrared spectroscopy is introduced as a vital tool for identifying the functional groups in organic compounds. Some test-tube reactions that may be used to help identify organic compounds are described.

The concepts of the applications of science are found throughout the chapters, where they will provide you with an opportunity to apply your knowledge in a fresh context.

What you already know

The material in this section builds upon knowledge and understanding that you will have developed at GCSE, in particular the following:

- ☐ Carbon atoms have four outer electrons and can form four single bonds.
- ☐ Organic compounds are based on chains of carbon atoms.
- ☐ Double covalent bonds can be formed by sharing four electrons between a pair of atoms.

11 Introduction to organic chemistry
11.1 Carbon compounds

Learning objectives:

→ State what is meant by the terms empirical formula, molecular formula, skeletal formula, and structural formula.

Specification reference: 3.3.1

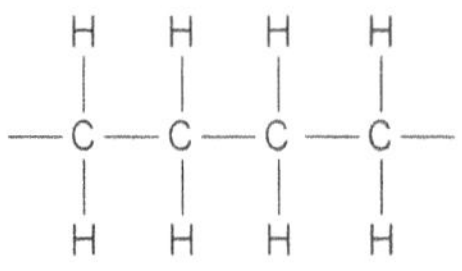

▲ **Figure 1** *Part of a hydrocarbon chain*

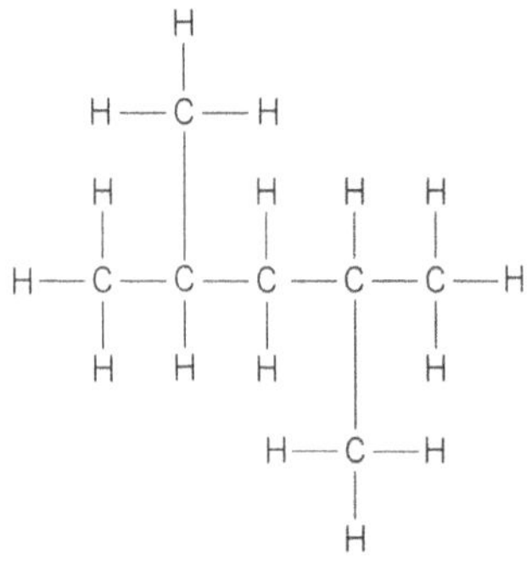

▲ **Figure 2** *A branched hydrocarbon chain*

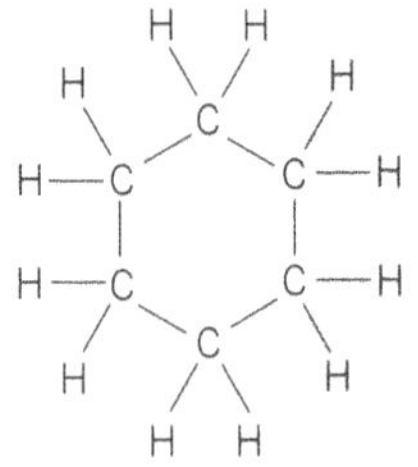

▲ **Figure 3** *A hydrocarbon ring*

Organic chemistry is the chemistry of carbon compounds. Life on our planet is based on carbon, and 'organic' means to do with living beings. Nowadays, many carbon-based materials, like plastics and drugs, are made synthetically and there are large industries based on synthetic materials. There are far more compounds of carbon known than those of all the other elements put together, well over 10 million.

What is special about carbon?

Carbon can form rings and very long chains, which may be branched. This is because:

- a carbon atom has four electrons in its outer shell, so it forms four covalent bonds
- carbon–carbon bonds are relatively strong (347 kJ mol^{-1}) and non-polar.

The carbon–hydrogen bond is also strong (413 kJ mol^{-1}) and relatively non-polar. Hydrocarbon chains form the skeleton of most organic compounds, see Figures 1, 2, and 3.

From inorganic to organic

Organic compounds were originally thought to be produced by living things only. This was disproved by Friedrich Wöhler in 1828. He made urea (an organic compound found in urine) from ammonium cyanate (an inorganic compound).

$NH_4^+(NCO)^- \rightarrow (NH_2)_2CO$

ammonium cyanate → urea

He reported to a fellow chemist

"I cannot, so to say, hold my chemical water and must tell you that I can make urea without thereby needing to have kidneys, or anyhow, an animal, be it human or dog".

This reaction is an isomerism reaction.

1 What is meant by an isomerism reaction?
2 What is the atom economy of this reaction? Explain your answer.

Bonding in carbon compounds

In *all* stable carbon compounds, carbon forms four covalent bonds and has eight electrons in its outer shell. It can do this by forming bonds in different ways.

- By forming four single bonds as in methane:

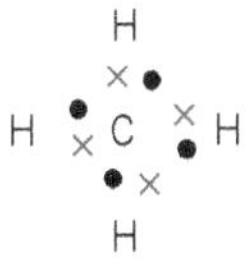

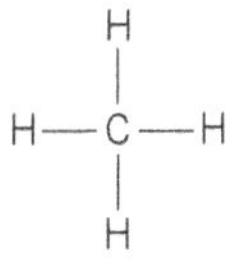

- By forming two single bonds and one double bond as in ethene:

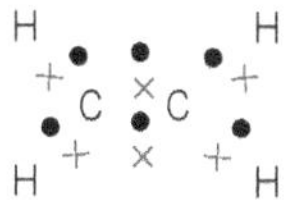

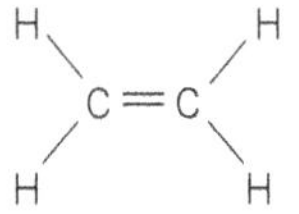

- By forming one single bond and one triple bond as in ethyne:

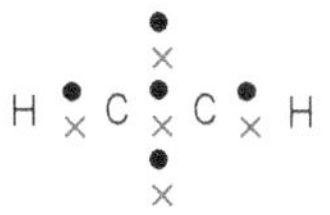

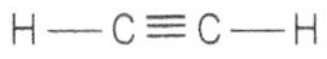

Formulae of carbon compounds

The empirical formula

Worked example: Empirical formula

The empirical formula is a formula that shows the simplest ratio of the atoms of each element present in a compound. For example, ethane:

3.00 g of ethane contains 2.40 g of carbon, $A_r = 12.0$, and 0.60 g of hydrogen, $A_r = 1.0$. What is its empirical formula?

number of moles of carbon $= \frac{2.40}{12.0} = 0.20$ mol of carbon

number of moles of hydrogen $= \frac{0.60}{1} = 0.60$ mol of hydrogen

Dividing through by the smaller number (0.20) gives C : H as 1 : 3

So the empirical formula of ethane is CH_3.

Hint

The mass of carbon plus the mass of hydrogen adds up to the mass of ethane so no other element is present.

The molecular formula

The molecular formula is the formula that shows the actual number of atoms of each element in the molecule. It is found from:

- the empirical formula
- the relative molecular mass of the empirical formula
- the relative molecular mass of the molecule.

Synoptic link

Look back at Topic 2.4, Empirical and molecular formulae, if you are not sure about this calculation.

Worked example: Molecular formula

The empirical formula of ethane is CH_3 and this group of atoms has a relative molecular mass of 15.0.

The relative molecular mass of ethane is 30.0, which is 2 × 15.0. So, there must be two units of the empirical formula in every molecule of ethane.

The molecular formula is therefore $(CH_3)_2$ or C_2H_6.

Other formulae

Other, different types of formulae are used in organic chemistry because, compared with inorganic compounds, organic molecules are more varied. The type of formula required depends on the information that you are dealing with. You may want to know about the way the atoms are arranged within the molecule, as well as just the number of each atom present. There are different ways of doing this.

The displayed formula

This shows every atom and every bond in the molecule:

— is a single bond

= is a double bond

≡ is a triple bond

For ethene, C_2H_4, the displayed formula is:

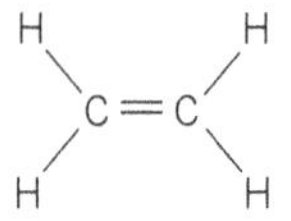

For ethanol, C_2H_6O, the displayed formula is:

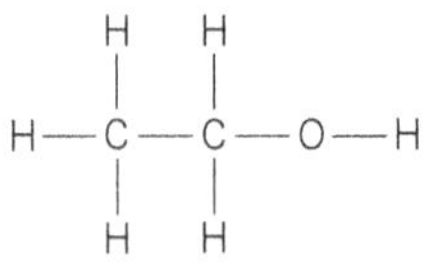

The structural formula

This shows the unique arrangement of atoms in a molecule in a simplified form, without showing all the bonds.

Each carbon is written separately, with the atoms or groups that are attached to it.

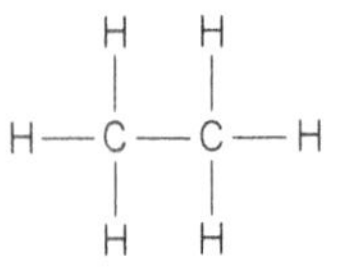

is written CH_3CH_3

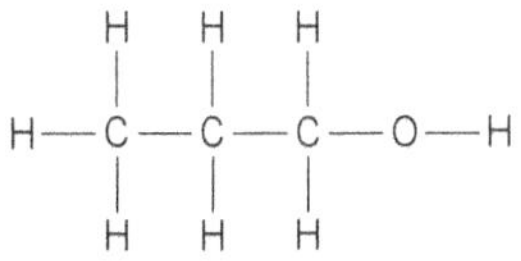 is written $CH_3CH_2CH_2OH$

Branches in the carbon chains are shown in brackets:

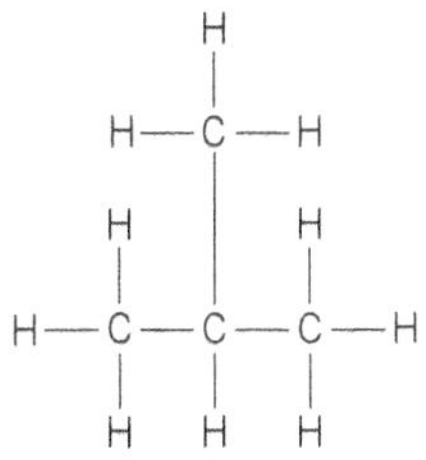 is written $CH_3CH(CH_3)CH_3$

Skeletal formulae

With more complex molecules, displayed structural formulae become time-consuming to draw. In skeletal notation the carbon atoms are not drawn at all. Straight lines represent carbon–carbon bonds and carbon atoms are assumed to be where the bonds meet. Neither hydrogen atoms nor C—H bonds are drawn. Each carbon is assumed to form enough C—H bonds to make a total of four bonds (counting double bonds as two).

Synoptic link

You will learn more about reaction mechanisms and free radicals in Topic 13.2, Nucleophilic substitution in halogenoalkanes, and Topic 12.3, Industrial cracking.

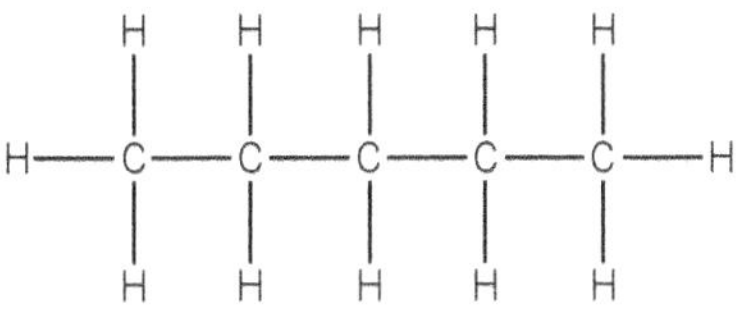 would be written 

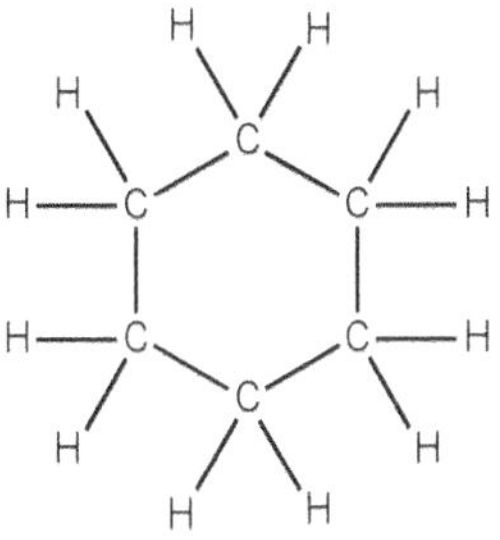 would be written 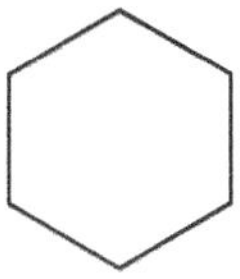

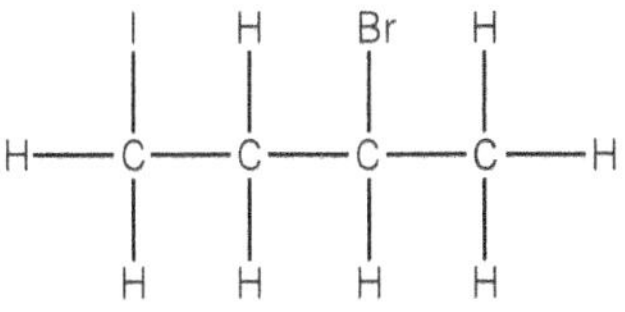 would be written 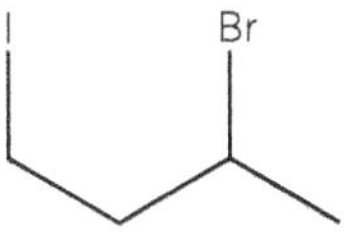

The choice of type of formula to use depends on the circumstances and the type of information you need to give. Notice that skeletal formulae give a rough idea of the bond angles. In an unbranched alkane chain these are 109.5°.

Three-dimensional structural formulae

These attempt to show the three-dimensional structure of the molecule. Bonds coming out of the paper are shown by wedges  and bonds going into the paper by dotted lines .

Some examples of different types of formulae are given in Table 4.

▼ **Table 4** *Different types of formulae*

Empirical formula	Molecular formula/name	Structural formula: Shorthand	Structural formula: Displayed	Structural formula: Skeletal	Structural formula: Three-dimensional
CH_2	C_6H_{12} hex-1-ene	$CH_2CHCH_2CH_2CH_2CH_3$	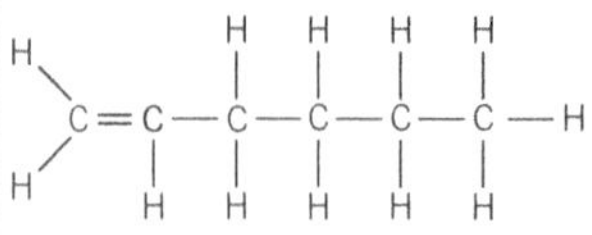	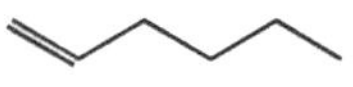	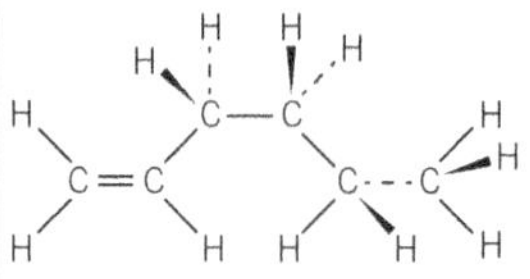
C_2H_6O	C_2H_6O ethanol	CH_3CH_2OH	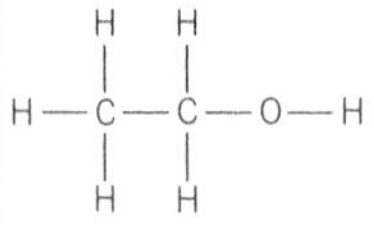	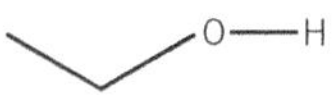	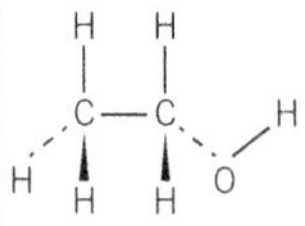
C_3H_7Cl	C_3H_7Cl 2-chloropropane	$CH_3CHClCH_3$	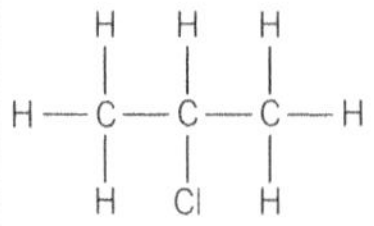	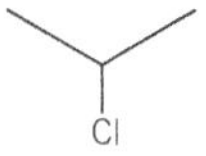	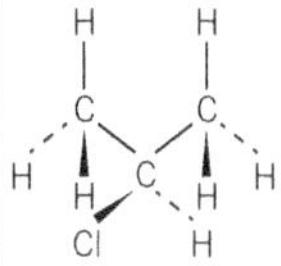
C_3H_6O	C_3H_6O propanone	CH_3COCH_3	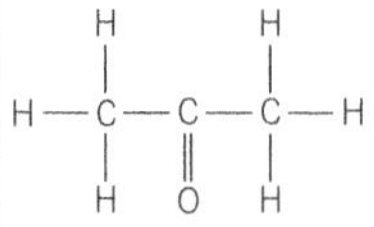	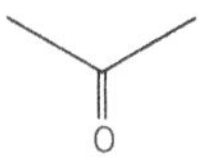	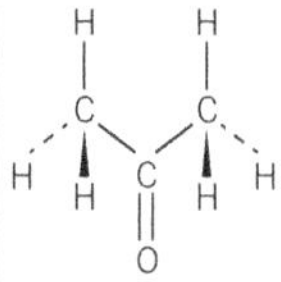

Summary questions

1 A compound comprising only carbon and hydrogen, in which 4.8 g of carbon combine with 1.0 g of hydrogen, has a relative molecular mass of 58.

- **a** How many moles of carbon are there in 4.8 g?
- **b** How many moles of hydrogen are there in 1.0 g?
- **c** What is the empirical formula of this compound?
- **d** What is the molecular formula of this compound?
- **e** Draw the structural formula and the skeletal formula of the compound that has a straight chain.
- **f** Draw the displayed formula and the skeletal formula of the compound that has a branched chain.

Reaction mechanisms

Curly arrows

You can often explain what happens in organic reactions by considering the movement of electrons. As electrons are negatively charged, they tend to move from areas of high electron density to more positively charged areas. For example, a lone pair of electrons will be attracted to the carbon atom at the positive end of a polar bond, written as $C^{\delta+}$. The movement of a pair of electrons is shown by a curly arrow that starts from a lone pair of electrons or from a covalent bond and moves towards a positively charged area of a molecule to form a new bond.

Free radicals

Sometimes a covalent bond (which consists of a pair of electrons shared between two atoms) may break in such a way that one electron goes to each atom that originally formed the bond. These fragments of the original molecule have an unpaired electron and are called **free radicals**. They are usually extremely reactive. The unpaired electron is represented by a dot, so $CH_3\bullet$ is a methyl radical.

11.2 Nomenclature – naming organic compounds

The system used for naming compounds was developed by the International Union of Pure and Applied Chemistry or **IUPAC**. This is an international organisation of chemists that draws up standards so that chemists throughout the world use the same conventions – rather like a universal language of chemistry. Systematic names tell us about the structures of the compounds rather than just the formula. Only the basic principles are covered here.

Roots

A systematic name has a *root* that tells us the longest unbranched hydrocarbon chain or ring, see Table 1.

The syllable after the root tells us whether there are any double bonds.

-ane means no double bonds. For example, ethane

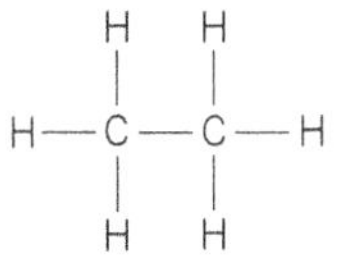

has two carbon atoms and no double bond.

-ene means there is a double bond. For example, ethene

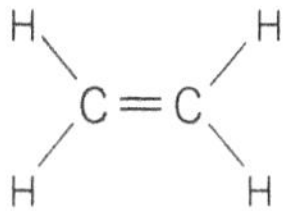

has two carbon atoms and one double bond.

Prefixes and suffixes

Prefixes and suffixes describe the changes that have been made to the root molecule.

• Prefixes are added to the beginning of the root.

For example, side chains are shown by a prefix, whose name tells us the number of carbons:

| methyl | CH_3- | ethyl | C_2H_5- |
| propyl | C_3H_7- | butyl | C_4H_9- |

For example:

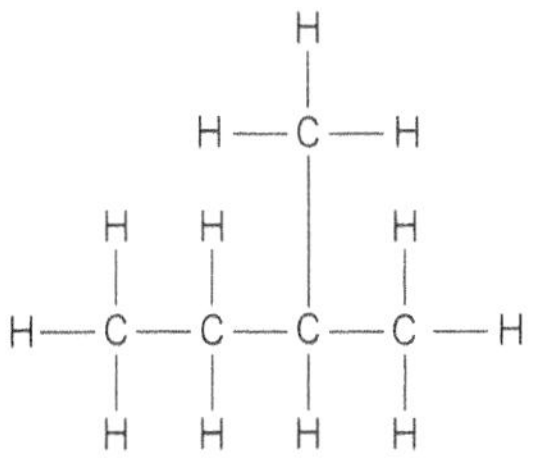

is called *methyl*butane. The longest unbranched chain is four carbons long, which gives us butane (as there are no double bonds) and there is a side chain of one carbon, a methyl group.

Learning objectives:
→ Explain the IUPAC rules for naming alkanes and alkenes.
→ State what is meant by a functional group.
→ State what is meant by a homologous series.

Specification reference: 3.3.1

▼ **Table 1** *The first six roots used in naming organic compounds*

Number of carbons	Root
1	meth
2	eth
3	prop
4	but
5	pent
6	hex

Study tip

It is important to learn the root names from C1 to C6.

Hydrocarbon ring molecules have the additional prefix 'cyclo'. So the compound below would be named cyclohexane:

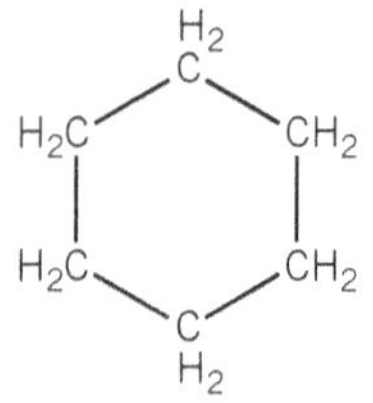

- Suffixes are added to the end of the root.

For example, alcohols, —OH, have the suffix -ol, as in methanol, CH_3OH.

Functional groups

Most organic compounds are made up of a hydrocarbon chain that has one or more reactive groups attached to it. These reactive groups are called **functional groups**. The functional group reacts in the same way, whatever the length of the hydrocarbon chain. So, for example, if you learn the reactions of one alkene, such as ethene, you can apply this knowledge to any alkene.

Functional groups are named by using a suffix or prefix as shown in Table 2.

▼ **Table 2** *The suffixes of some functional groups*

Family	Formula	Suffix	Example
alkanes	C_nH_{2n+2}	-ane	ethane
alkenes	R—CH═CH—R	-ene	propene
halogenoalkanes	R—X (X = F, Cl, Br, or I)	none	chloromethane CH_3Cl
alcohols	R—OH	-ol	ethanol C_2H_5OH
aldehydes	R—C(═O)—H	-al	ethanal CH_3CHO
ketones	R—C(═O)—R'	-one	propanone CH_3COCH_3
carboxylic acids	R—C(═O)—OH	-oic acid	ethanoic acid CH_3COOH

Note that the halogenoalkanes are named using a prefix (fluoro-, chloro-, bromo-, iodo-) rather than a suffix. R is often used to represent a hydrocarbon chain (of any length). Think of it as representing the rest of the molecule.

Study tip

At this stage, you will not need to learn how to name all the functional groups in the table but they are useful to illustrate the principles of naming.

Synoptic link

Aldehydes, ketones, and carboxylic acids are discussed in Chapter 26, Compounds containing the carbonyl group.

Examples

eth indicates that the molecule has a chain of two carbon atoms, *ane* that it is has no double or triple bonds, and *bromo* that one of the hydrogen atoms of ethane is replaced by a bromine atom.

bromoethane 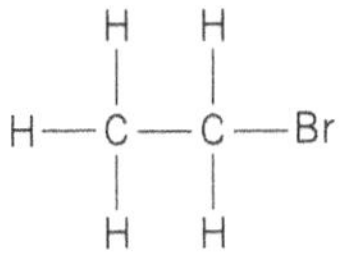

Prop indicates a chain of three carbon atoms and *ene* that there is one C=C (double bond).

propene 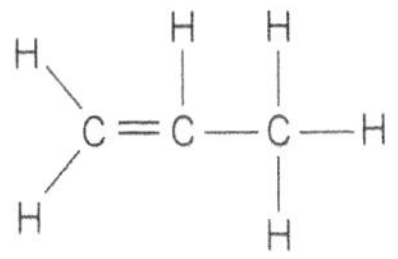

Meth indicates a single carbon, *an* that there are no double bonds and *ol* that there is an OH group (an alcohol).

methanol 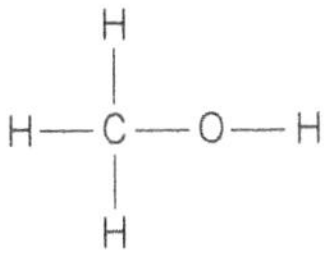

Chain and position isomers

With longer chains, you need to say where a side chain or a functional group is located on the main chain. For example, methylpentane could refer to:

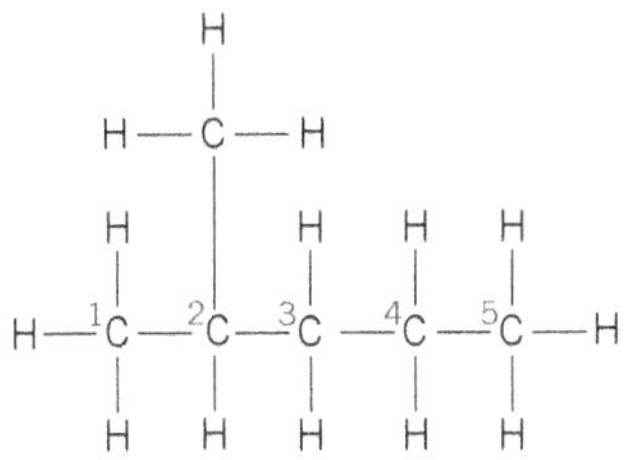

or

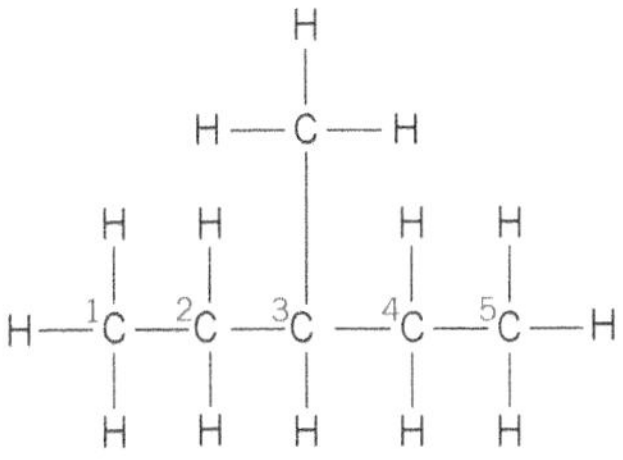

2–methylpentane 3–methylpentane

A number (sometimes called a locant) is used to tell us the position of any branching in a chain and the position of any functional group. The examples above are structural isomers. Structural isomers have the same molecular formula but different structural formulae, see Topic 11.3.

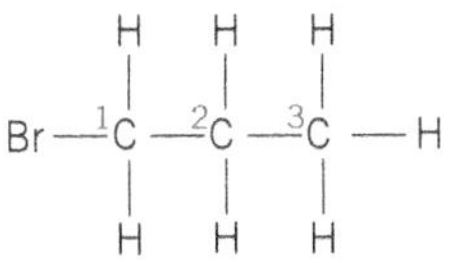

and

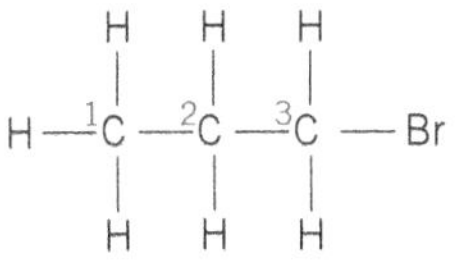

Both molecules are 1-bromopropane. The right-hand one is not 3-bromopropane because the smallest possible number is always used.

1-bromopropane may also be represented by either of the structural formulae below because all the hydrogens on carbon 1 are equivalent.

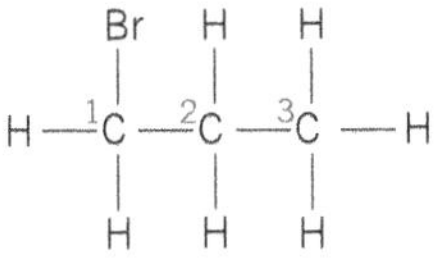

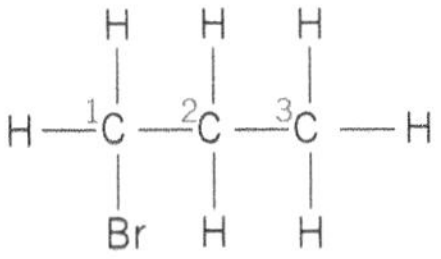

Hint

Take care. Don't get confused by the way the formula is drawn. These are the same molecule because the shape around each carbon atom is tetrahedral.

▲ **Figure 1** *Buckminster fullerene (top) and a geodesic dome (bottom)*

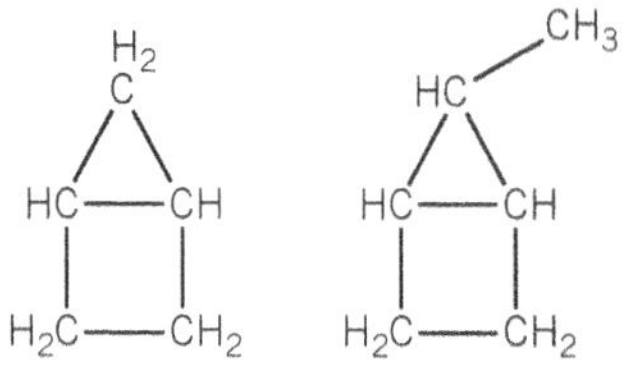

▲ **Figure 2** *Housane (left) and roof-methylhousane (right)*

What's in a name?

Before the advent of systematic naming, chemicals could be given any old name. For example, methanoic acid was called formic acid because it is produced by ants as a defence against predators. The Latin for ant is *formica*. However, the name tells us nothing about the chemical structure of formic acid (except that it is acidic), whereas *meth*anoic acid tells us that it has one carbon atom. Acetic acid's name comes from the Latin *acetum*, meaning vinegar, but the systematic name *eth*anoic acid tells us that it has two carbon atoms.

Sometimes, though, systematic names are just too long. Buckminsterfullerene was named from the similarity of its structure to the geodesic domes designed by the American architect Richard Buckminster Fuller (Figure 1).

Housane is named after the resemblance of its structural formula to a house – its systematic name is bicyclo[2.1.0]pentane.

If we replace one of the hydrogen atoms with a methyl group —CH_3, as shown, we get roof-methylhousane.

Draw the formulae of the two positional isomers of roof-methylhousane which might be called eave-methylhousane and floor-methylhousane.

Molecules with more than one functional group or side chain

A molecule may have more than one functional group. For example:

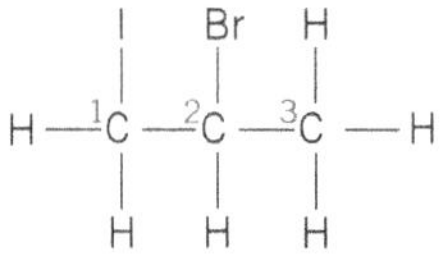

2-bromo-1-iodopropane

Even though iodine is on carbon 1 and bromine is on carbon 2, *bromo* is written before *iodo* because the substituting groups are put in *alphabetical order* rather than in the numerical order of the functional groups.

You can show that you have more than one of the same substituting group by adding prefixes as well as functional groups. di-, tri-, and tetra- mean two, three, and four, respectively.

So, 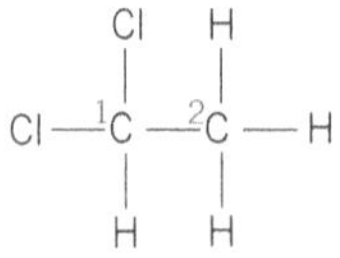 is called 1,1-dichloroethane

and 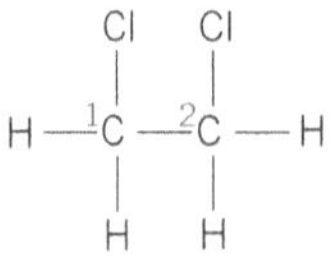 is called 1,2-dichloroethane.

Hint

In chemical names, strings of numbers are separated by commas. A hyphen is placed between words and numbers.

Homologous series

A **homologous series** is a family of organic compounds with the same functional group, but different carbon chain length.

- Members of a homologous series have a general formula. For example, the alkanes are C_nH_{2n+2} and alkenes, with one double bond, are C_nH_{2n}.
- Each member of the series differs from the next by CH_2.
- The length of the carbon chain has little effect on the *chemical* reactivity of the functional group.
- The length of the carbon chain affects physical properties, like melting point, boiling point, and solubility. Melting points and boiling points increase by a small amount as the number of carbon atoms in the chain increases. This is because the intermolecular forces increase. In general, small molecules are gases, larger ones liquids or solids.
- Chain branching generally reduces melting points because the molecules pack together less well.

▼ **Table 3** *Examples of systematic naming of organic compounds. Try covering up the name or structure to test yourself.*

Structural formula	Name
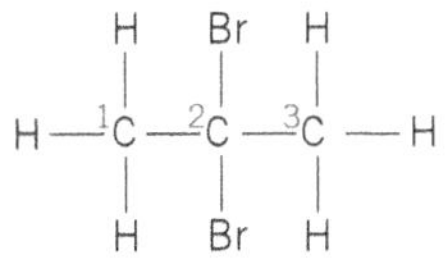	2,2-dibromopropane
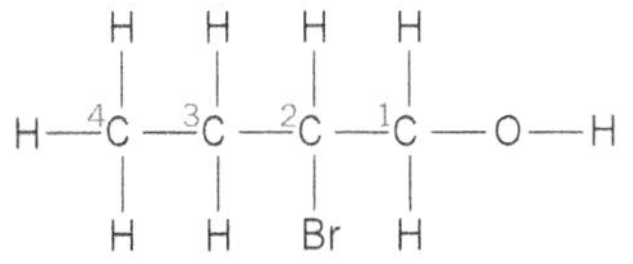	2-bromobutan-1-ol The suffix -ol defines the end of the chain you count from
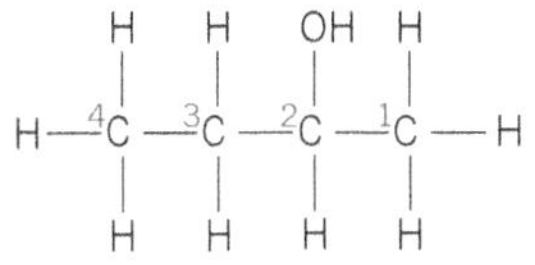	butan-2-ol
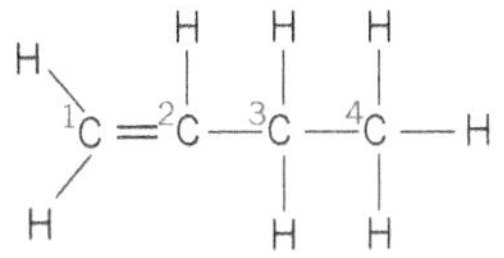	but-1-ene *Not* but-2-ene, but-3-ene, or but-4-ene as we use the smallest locant possible

Summary questions

1 What is the name of each of the following?

a $CH_3CH_2CH_2Cl$

b $CH_3CH_2CH_2CH_2CH_3$

c $CH_3CH_2CH{=}CHCH_3$

d $CH_3CH_2CH_2CH(CH_3)CH_3$

2 Draw the displayed formulae for:

a methylbutanone

b but-2-ene

c 2-chlorohexane

d but-1-ene.

11.3 Isomerism

Learning objectives:

- → State what is meant by structural isomers.
- → Describe the three ways in which structural isomerism can occur.

Specification reference: 3.3.1

Isomers

Isomers are molecules that have the same molecular formula but whose atoms are arranged differently. There are two basic types of isomerism in organic chemistry – structural isomerism and stereoisomerism.

Structural isomerism

Structural isomers are defined as having the same molecular formula but different structural formulae, see Topic 11.2. There are three sub-divisions. Structural isomers can have:

1. the same functional groups attached to the main chain at different points – this is called **positional isomerism**
2. functional groups that are different – this is called **functional group isomerism**
3. a different arrangement of the hydrocarbon chain (such as branching) – this is called **chain isomerism.**

Positional isomerism

The functional group is attached to the main chain at different points. For example, the molecular formula C_3H_7Cl could represent:

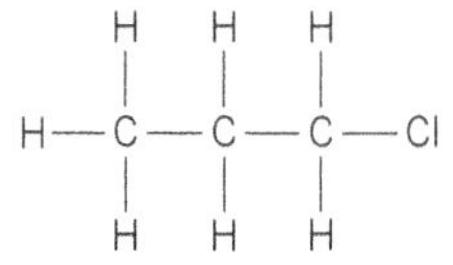
1-chloropropane

or

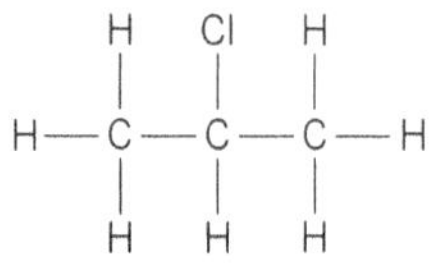
2-chloropropane

Functional group isomerism

There are different functional groups. For example, the molecular formula C_2H_6O could represent:

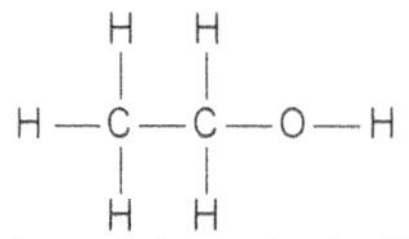
ethanol (an alcohol)

or

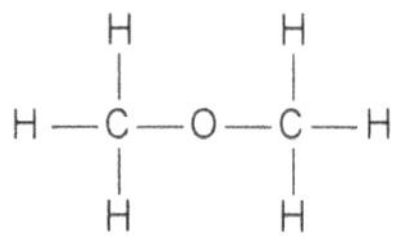
methoxymethane (an ether)

Chain isomerism

The hydrocarbon chain is arranged differently. For example, the molecular formula C_4H_9OH could represent:

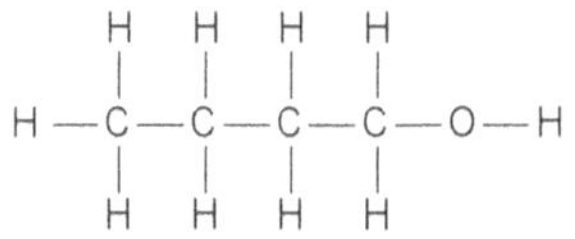
butan-1-ol

or

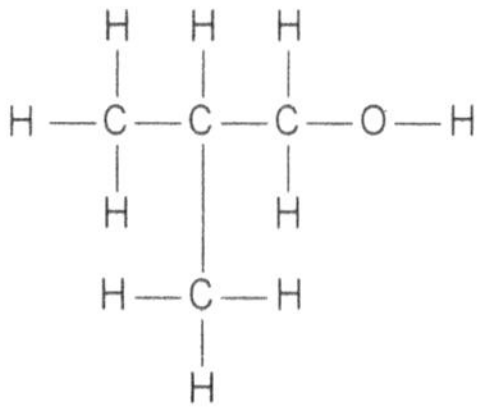
2-methylpropan-1-ol

These isomers are called chain-branching isomers.

The existence of isomers makes the task of identifying an unknown organic compound more difficult. This is because there may be a number of compounds with different structures that all have the same molecular formula. So, you have to use analytical methods that tell you about the structure.

Stereoisomerism

Stereoisomerism is where two (or more) compounds have the same structural formula. They differ in the *arrangement* of the bonds in space.

There are two types:

- *E-Z* isomerism and
- optical isomerism.

E-Z isomerism

E-Z isomerism tells us about the positions of substituents at either side of a carbon–carbon double bond. Two substituents may either be on the same side of the bond *Z* (*cis*) or on opposite sides *E* (*trans*).

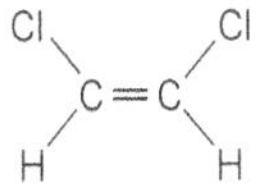

Z-1,2-dichloroethene

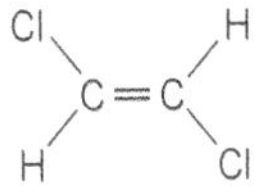

E-1,2-dichloroethene

Substituted groups joined by a single bond can rotate around the single bond, so there are no isomers (Figure 1) but there is no rotation around a double bond. So, *Z*- and *E*-isomers are separate compounds and are not easily converted from one to the other.

E-Z is from the German *Entgegen* (opposite – *trans*) and *Zusammen* (together – *cis*).

Synoptic link

Another type of isomerism, optical isomerism, will be covered in Topic 25.2, Optical isomerism.

Hint

You will also see the terms *cis* and *trans* used to describe the positions of groups attached to a double bond.

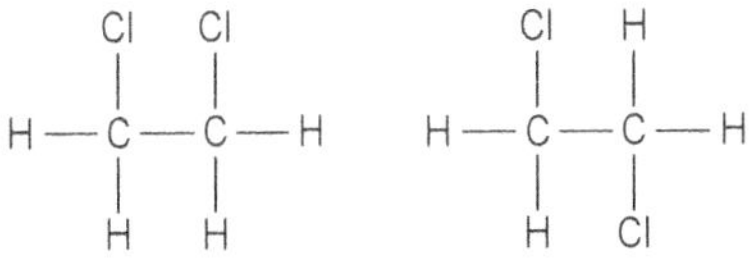

▲ **Figure 1** *Groups can rotate around a single bond. These are representations of the same molecule and are not isomers*

Synoptic link

E-Z isomerism is discussed further in Topic 14.2, Reactions of alkenes.

Summary questions

1 What type of structural isomerism is shown by the following pairs of molecules? Choose from: A = functional groups at different points, B = different functional groups, C = chain branching.

  a $CH_3CH_2OCH_3$ and $CH_3CH_2CH_2OH$

  b $CH_3CH_2CH_2OH$ and $CH_3CH(OH)CH_3$

  c $CH_3CH_2CH_2CH_2CH_3$ and $CH_3CH(CH_3)CH_2CH_3$

2 a Write the displayed and structural formulae for all the five isomers of hexane, C_6H_{14}.

  b Name these isomers.

3 Which of these molecules can show *E-Z* (*cis-trans*) isomerism?

  A $CH_2{=}CH_2$

  B $CH_3{-}CH_3$

  C $RCH{=}CH_2$

  D $RCH{=}CHR$

4 a Give the name of this:

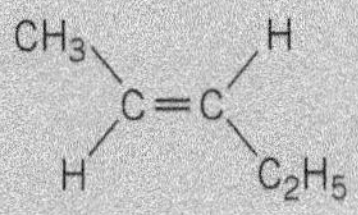

  b What is the name of its geometrical isomer?

Practice questions

1 The alkanes form a homologous series of hydrocarbons. The first four straight-chain alkanes are shown below.

methane	CH_4
ethane	CH_3CH_3
propane	$CH_3CH_2CH_3$
butane	$CH_3CH_2CH_2CH_3$

(a) (i) State what is meant by the term *hydrocarbon*.
(ii) Give the general formula for the alkanes.
(iii) Give the molecular formula for hexane, the sixth member of the series.

(3 marks)

(b) Each homologous series has its own general formula. State **two** other characteristics of a homologous series.

(2 marks)

(c) Branched-chain structural isomers are possible for alkanes which have more than three carbon atoms.
(i) State what is meant by the term *structural isomers*.
(ii) Name the **two** isomers of hexane shown below.

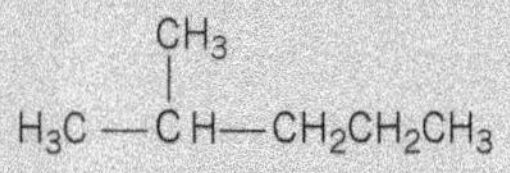

isomer 1

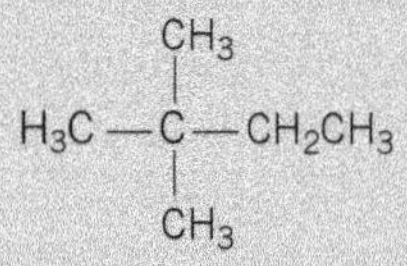

isomer 2

(iii) Give the structures of **two** other branched-chain isomers of hexane.

(6 marks)

(d) A hydrocarbon, **W**, contains 92.3% carbon by mass. The relative molecular mass of **W** is 78.0.
(i) Calculate the empirical formula of **W**.
(ii) Calculate the molecular formula of **W**.

(4 marks)
AQA, 2003

2 **(a)** Give the systematic chemical name of CCl_2F_2. *(1 mark)*
(b) Give the systematic chemical name of CCl_4. *(1 mark)*
(c) Give the systematic chemical name of $CHCl_2$ $CHCl_2$. *(1 mark)*
AQA, 2001

3 There are five structural isomers of the molecular formula C_5H_{10} which are alkenes. The displayed formulae of two of these isomers are given.

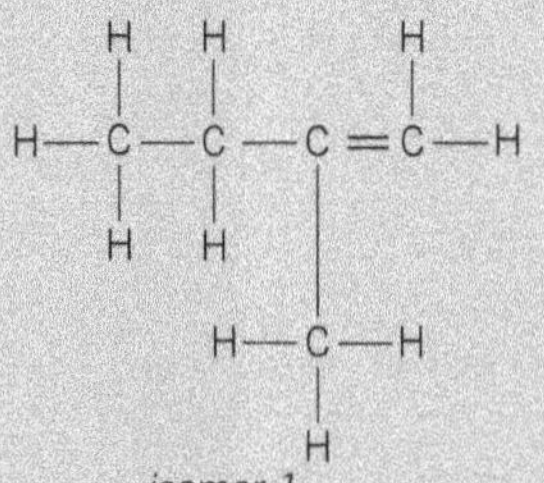

isomer 1

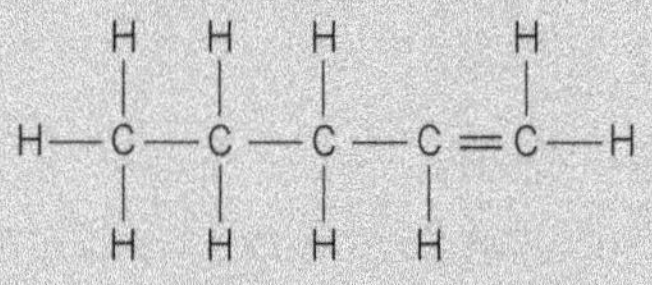

isomer 2

(a) Draw the displayed formulae of two of the remaining alkene structural isomers.

(2 marks)

(b) Consider the reaction scheme shown below and answer the question that follows.

$$\text{Isomer 1} \xrightarrow{\text{HBr}} \underset{\textbf{Y}}{CH_3CH_2CBr(CH_3)_2}$$

Give the name of compound **Y**.

(1 mark)
AQA, 2000

4 There are four structural isomers of molecular formula C_4H_9Br. The structural formulae of two of these isomers are given below.

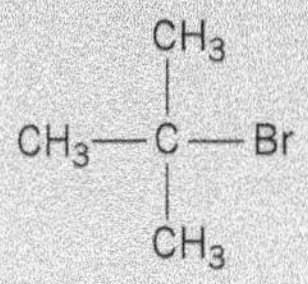

$$CH_3-\underset{\displaystyle \underset{CH_3}{|}}{\overset{\displaystyle \overset{CH_3}{|}}{C}}-Br \qquad\qquad CH_3CH_2CH_2CH_2Br$$

isomer 1 *isomer 2*

(i) Draw the structural formulae of the remaining two isomers. *(3 marks)*
(ii) Name isomer 1.

AQA, 2001

Answers to the Practice Questions and Section Questions are available at
www.oxfordsecondary.com/oxfordaqaexams-alevel-chemistry

12 Alkanes
12.1 Alkanes

Learning objectives:

- → State the definition of an alkane.
- → Explain how alkanes are named.
- → Describe their properties.

Specification reference: 3.3.1 and 3.3.2

Alkanes are **saturated hydrocarbons** – they contain only carbon–carbon and carbon–hydrogen *single* bonds. They are among the least reactive organic compounds. They are used as fuels and lubricants and as starting materials for a range of other compounds. This means that they are very important to industry. The main source of alkanes is crude oil.

The general formula

The general formula for all chain alkanes is C_nH_{2n+2}. Hydrocarbons may be unbranched chains, branched chains, or rings.

Unbranched chains

Unbranched chains are often called straight chains but the C—C—C angle is 109.5°. This means that the chains are not actually straight. In an unbranched alkane, each carbon atom has two hydrogen atoms except the end carbons which have one extra.

For example, pentane, C_5H_{12}:

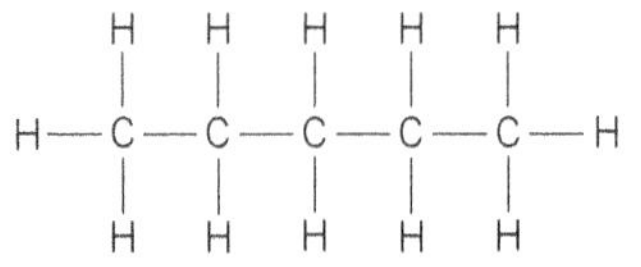

displayed

$CH_3CH_2CH_2CH_2CH_3$

structural

Synoptic link

You learnt about bond angles in alkanes in Topic 3.5, The shapes of molecules and ions.

Branched chains

For example, methylbutane, C_5H_{12}, which is an isomer of pentane:

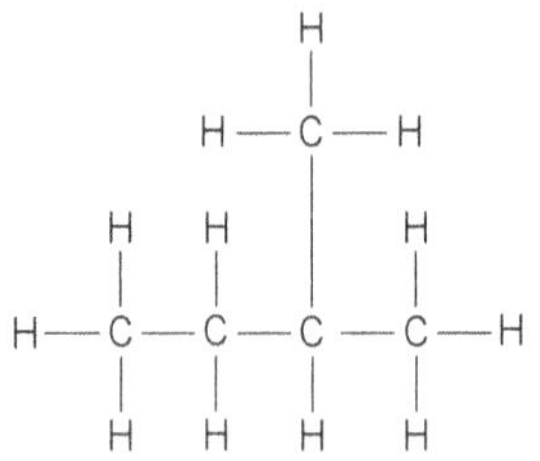

displayed

$CH_3CH_2CH(CH_3)CH_3$

structural

Ring alkanes

Ring alkanes have the general molecular formula C_nH_{2n} because the end hydrogens are not required.

How to name alkanes

Straight chains

Alkanes are named from the root, which tells us the number of carbon atoms, and the suffix -ane, denoting an alkane, see Table 1.

▼ **Table 1** *Names of the first six alkanes*

methane	CH_4
ethane	C_2H_6
propane	C_3H_8
butane	C_4H_{10}
pentane	C_5H_{12}
hexane	C_6H_{14}

Branched chains

When you are naming a hydrocarbon with a branched chain, you must first find the longest unbranched chain which can sometimes be a bit tricky, see the example below. This gives the root name. Then name the branches or *side chains* as prefixes: methyl-, ethyl-, propyl-, and so on. Finally, add numbers to say which carbon atoms the side chains are attached to.

Example

Both the hydrocarbons below are the same, though they seem different at first sight.

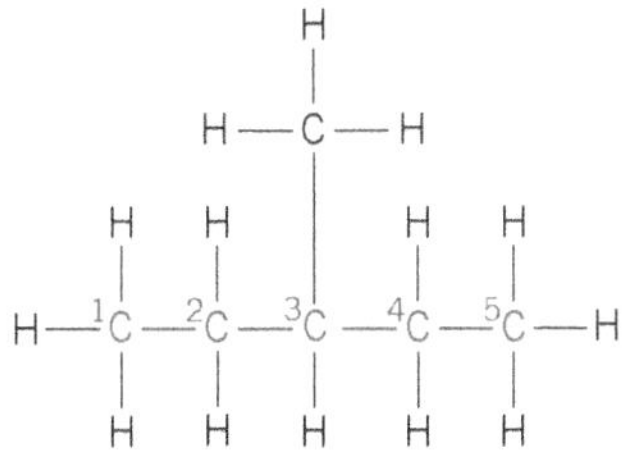

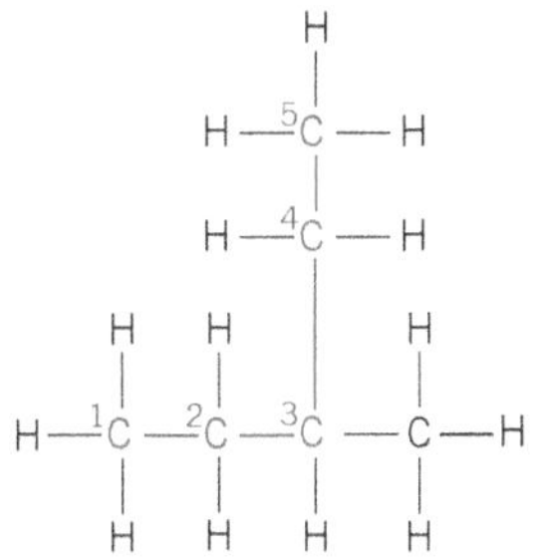

A skeletal formulae for 3-methylpentane is:

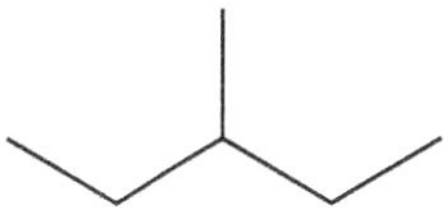

In both representations, the longest unbranched chain (in red) is five carbons, so the root is pentane. The only side chain has one carbon so it is methyl-. It is attached at carbon 3 so the full name is 3-methylpentane.

Structure

Isomerism

Methane, ethane, and propane have no isomers but after that, the number of possible isomers increases with the number of carbons in the alkane. For example, butane, with four carbons, has two isomers whilst pentane has three:

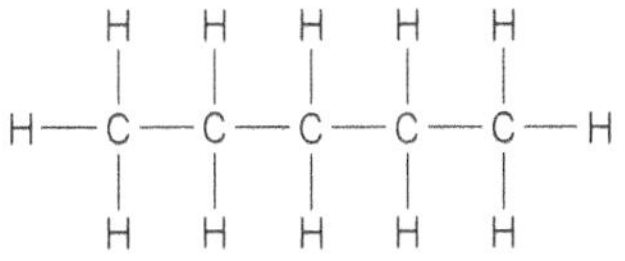

pentane

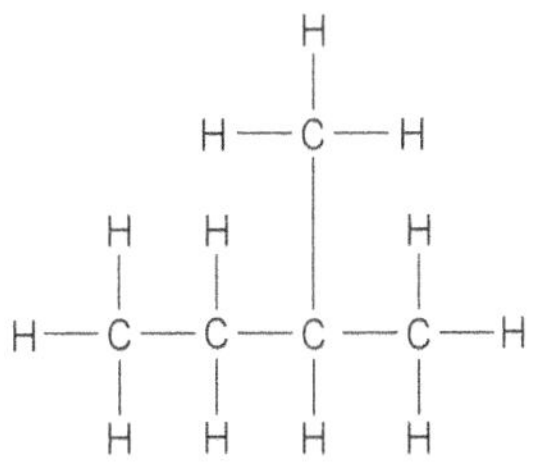

methylbutane

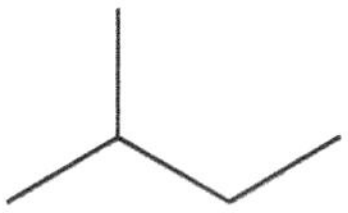

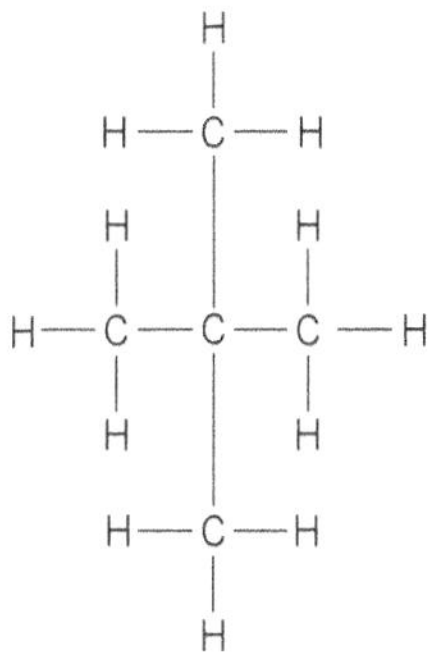

2,2-dimethylpropane

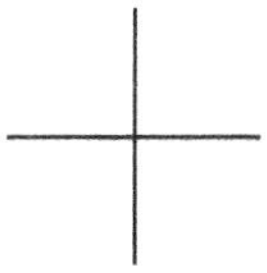

The number of isomers rises rapidly with chain length. Decane, $C_{10}H_{22}$, has 75 and $C_{30}H_{62}$ has over 4 billion.

Synoptic link

You learnt about electronegativity in Topic 3.6, Electronegativity – bond polarity in covalent bonds.

▲ **Figure 1** *Camping Gaz is a mixture of propane and butane. Polar expeditions use special gas mixtures with a higher proportion of propane, because butane is liquid below 272 K (−1 °C)*

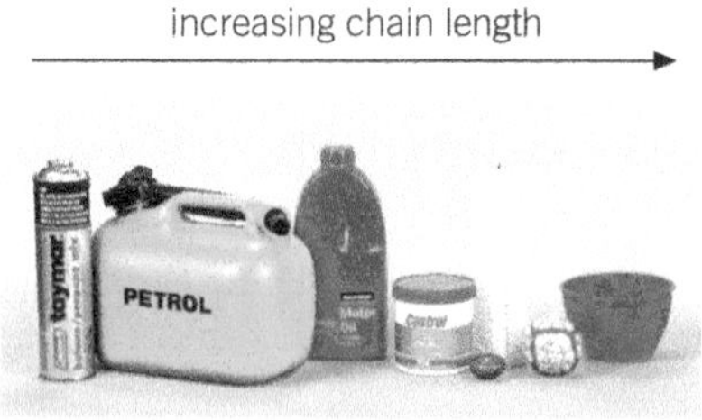

▲ **Figure 2** *The effect of increasing chain length on the physical properties of alkanes*

Physical properties

Polarity

Alkanes are almost non-polar because the electronegativities of carbon (2.5) and hydrogen (2.1) are so similar. As a result, the only intermolecular forces between their molecules are weak van der Waals forces, and the larger the molecule, the stronger the van der Waals forces.

Boiling points

This increasing intermolecular force is why the boiling points of alkanes increase as the chain length increases. The shorter chains are gases at room temperature. Pentane, with five carbons, is a liquid with a low boiling point of 309 K (36 °C). At a chain length of about 18 carbons, the alkanes become solids at room temperature. The solids have a waxy feel.

Alkanes with branched chains have lower melting points than straight chain alkanes with the same number of carbon atoms. This is because they cannot pack together as closely as unbranched chains and so the van der Waals forces are not so effective.

Solubility

Alkanes are insoluble in water. This is because water molecules are held together by hydrogen bonds which are much stronger than the van der Waal's forces that act between alkane molecules. However, alkanes do mix with other relatively non-polar liquids.

How alkanes react

Alkanes are relatively unreactive. They have strong carbon–carbon and carbon–hydrogen bonds. They do not react with acids, bases, oxidising agents, and reducing agents. However, they do burn and they will react with halogens under suitable conditions. They burn in a plentiful supply of oxygen to form carbon dioxide and water (or, in a restricted supply of oxygen, to form carbon monoxide or carbon).

Summary questions

1. Name the alkane $CH_3CH_2CH(CH_3)CH_3$ and draw its displayed formula.
2. Draw the displayed formula and structural formula of 2-methylhexane.
3. Name an isomer of 2-methylhexane that has a straight chain.
4. Which of the two isomers in question **3** will have the higher melting point? Explain your answer.

12.2 Fractional distillation of crude oil

Crude oil is at present the world's main source of organic chemicals. It is called a fossil fuel because it was formed millions of years ago by the breakdown of plant and animal remains at the high pressures and temperatures deep below the Earth's surface. Because it forms very slowly, it is effectively non-renewable.

Crude oil is a mixture mostly of alkanes, both unbranched and branched. Crude oils from different sources have different compositions. The composition of a typical North Sea oil is given in Table 1.

Learning objectives:

→ State the origin of crude oil.

→ Explain how crude oil is separated into useful fractions on an industrial scale.

Specification reference: 3.3.2

▼ **Table 1** *The composition of a typical North Sea crude oil*

	Gases	Petrol	Naphtha	Kerosene	Gas oil	Fuel oil and wax
Approximate boiling temperature / K	310	310–450	400–490	430–523	590–620	above 620
Chain length	1–5	5–10	8–12	11–16	16–24	25+
Percentage present	2	8	10	14	21	45

Crude oil contains small amounts of other compounds dissolved in it. These come from other elements in the original plants and animals the oil was formed from, for example, some contain sulfur. These produce sulfur dioxide, SO_2, when they are burnt. This is one of the causes of acid rain – sulfur dioxide reacts with oxygen high in the atmosphere to form sulfur trioxide. This reacts with water in the atmosphere to form sulfuric acid.

+ Combustion of sulfur

Write balanced equations for the three steps in which sulfur is converted into sulfuric acid.

$S + O_2 \rightarrow SO_2$

$2SO_2 + O_2 \rightarrow 2SO_3$

$SO_3 + H_2O \rightarrow H_2SO_4$

Fractional distillation of crude oil

To convert crude oil into useful products you have to separate the mixture. This is done by heating it and collecting the **fractions** that boil over different ranges of temperatures. Each fraction is a mixture of hydrocarbons of similar chain length and therefore similar properties, see Figure 1. The process is called **fractional distillation** and it is done in a **fractionating tower**.

- The crude oil is first heated in a furnace.
- A mixture of liquid and vapour passes into a tower that is cooler at the top than at the bottom.
- The vapours pass up the tower via a series of trays containing bubble caps until they arrive at a tray that is sufficiently cool (at a lower temperature than their boiling point). Then they condense to liquid.
- The mixture of liquids that condenses on each tray is piped off.
- The shorter chain hydrocarbons condense in the trays nearer to the top of the tower, where it is cooler, because they have lower boiling points.
- The thick residue that collects at the base of the tower is called tar or bitumen. It can be used for road surfacing but, as supply often exceeds demand, this fraction is often further processed to give more valuable products.

Hint

Crude oil is being produced now, but accumulation of a deposit of this oil is a very slow process.

Study tip

You do not need to memorise the various fractions and their boiling temperatures, but you should understand the principle of fractional distillation.

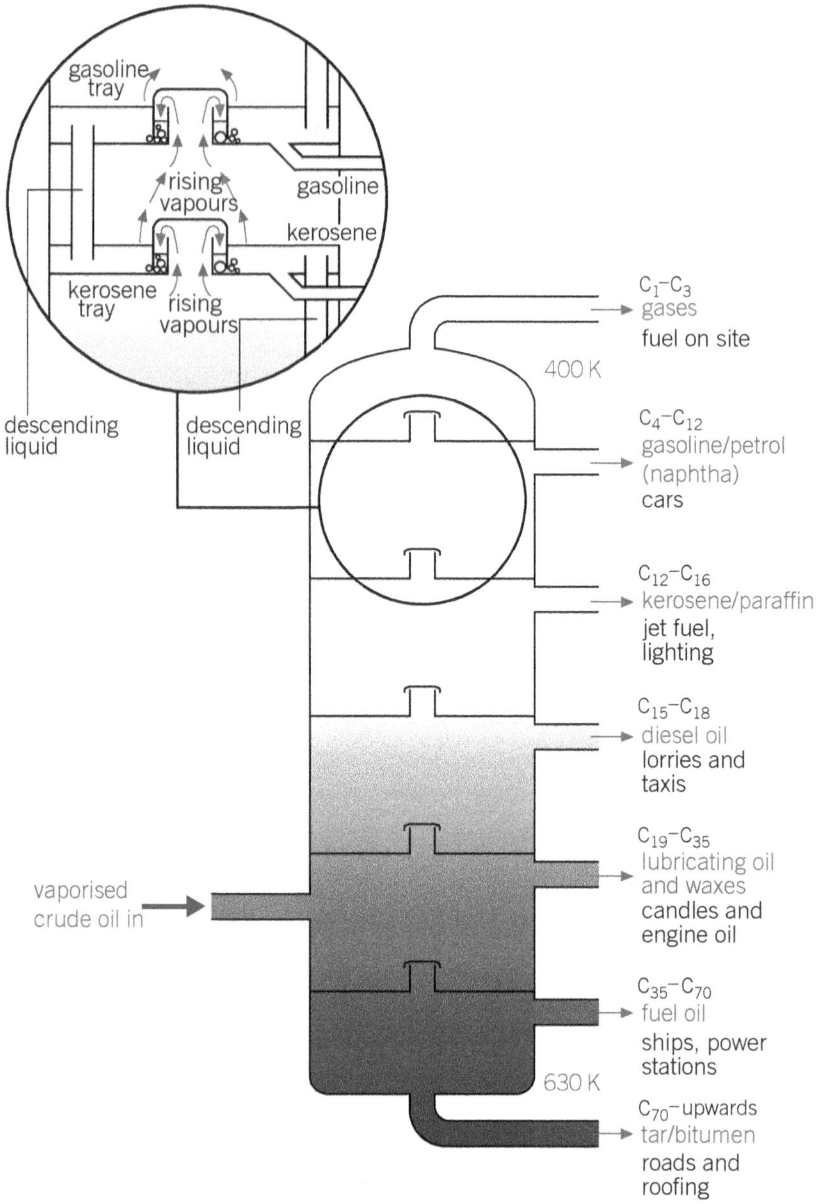

▲ **Figure 1** *The fractional distillation of crude oil. The chain length ranges are approximate*

Hint

Fractional distillation is a physical process so no covalent bonds within the molecules are broken. It is the van der Waals forces between the molecules that are broken during vaporisation and reform on condensing.

▲ **Figure 2** *Crude oil is separated into fractions by distillation in cylindrical towers, typically 8 m in diameter and 40 m high. Oil refineries vary but a typical one might process 3.5 million tonnes of crude oil per year*

Fracking

Almost half the UK's electricity is generated from natural gas (largely methane). Around half of this comes from the North Sea, but this percentage is decreasing as these wells become depleted and more and more gas is being imported via pipeline from Europe.

However, many areas of the UK have resources of natural gas – not caught under impervious rock layers as in the North Sea but trapped within shale rock rather like water in a sponge. This gas can be extracted by drilling into the shale and forcing pressurised water mixed with sand into the shale. This causes the rather soft shale rock to break up or fracture (giving the term 'fracking', short for hydraulic fracturing) releasing the trapped gas which flows to the surface. A number of chemicals are added to the water, such as hydrochloric acid, to help break up the shale and methanol to prevent corrosion in the system.

Many people are opposed to fracking for a variety of reasons:

- they do not like the infrastructure of wells and the associated traffic in their 'backyard'
- there is concern about the amount of water used
- they worry about the chemical additives polluting water supplies
- occasionally fracking appears to have caused small earthquakes
- burning natural gas produces carbon dioxide – a cause of global warming.

Set against these objections is the appeal of gas supplies for many years which are not subject to control by other countries.

Balance the equation for the combustion of methane.

$CH_4 + O_2 \rightarrow CO_2 + H_2O$

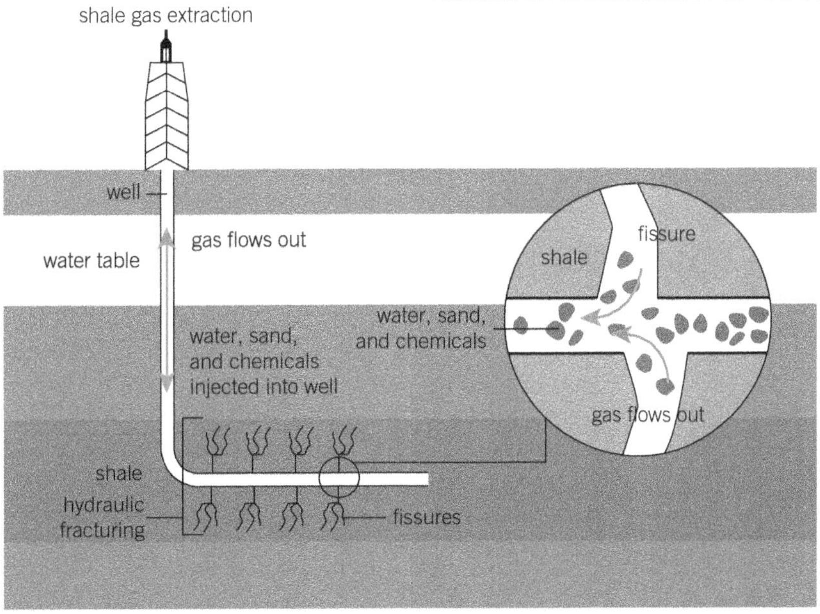

▲ **Figure 3** *The process of fracking*

$CH_4 + 2O_2 \rightarrow CO_2 + 2H_2O$

Summary questions

1 Draw the displayed formula and structural formula of hexane.

2 In which of the crude oil fractions named in Table 1 is hexane most likely to be found?

3 What is fractional distillation and how is it different from distillation?

4 Give the names of two gases produced in fractional distillation.

12.3 Industrial cracking

Learning objectives:

- → Describe what cracking is.
- → Describe what the conditions and products of thermal cracking are.
- → Describe the conditions and products of catalytic cracking.
- → Explain the economic reasons for cracking.

Specification reference: 3.3.2

The naphtha fraction from the fractional distillation of crude oil is in huge demand, for petrol and by the chemical industry. The longer chain fractions are not as useful and therefore of lower value economically. Most crude oil has more of the longer chain fractions than is wanted and not enough of the naphtha fraction.

The shorter chain products are economically more valuable than the longer chain material. To meet the demand for the shorter chain hydrocarbons, many of the longer chain fractions are broken into shorter lengths (cracked). This has two useful results:

- shorter, more useful chains are produced, especially petrol
- some of the products are alkenes, which are more reactive than alkanes.

Note that petrol is a mixture of mainly alkanes containing between four and twelve carbon atoms.

Study tip

It is useful to understand the commercial benefits of cracking.

Alkenes are used as chemical feedstock (which means they supply industries with the starting materials to make different products) and are converted into a huge range of other compounds, including polymers and a variety of products from paints to drugs. Perhaps the most important alkene is ethene, which is the starting material for poly(ethene) (also called polythene) and a wide range of other everyday materials.

Alkanes are very unreactive and harsh conditions are required to break them down. There are a number of different ways of carrying out cracking.

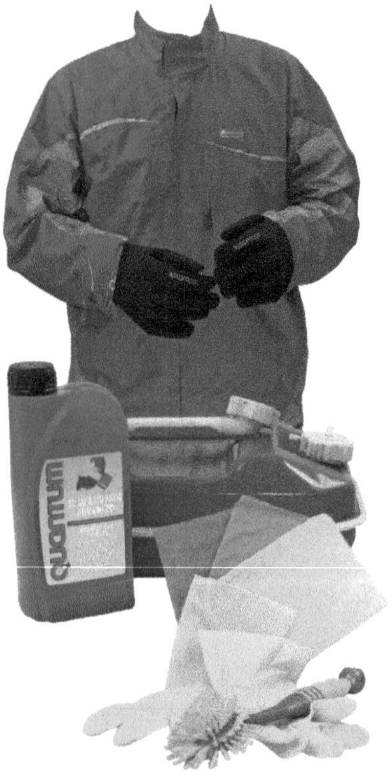

▲ **Figure 1** *A range of products obtained from crude oil*

Thermal cracking

This reaction involves heating alkanes to a high temperature, 700–1200 K, under high pressure, up to 7000 kPa. Carbon–carbon bonds break in such a way that one electron from the pair in the covalent bond goes to each carbon atom. So initially two shorter chains are produced, each ending in a carbon atom with an unpaired electron. These fragments are called **free radicals**. Free radicals are highly reactive intermediates and react in a number of ways to form a variety of shorter chain molecules.

As there are not enough hydrogen atoms to produce two alkanes, one of the new chains must have a C=C, and is therefore an alkene:

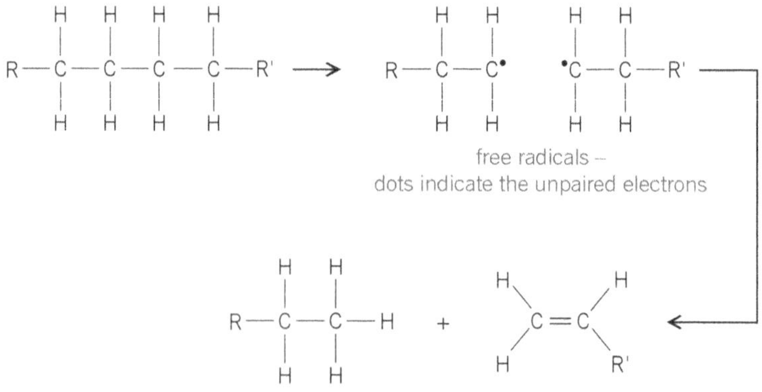

▲ **Figure 2** *Thermal cracking*

Any number of carbon–carbon bonds may break and the chain does not necessarily break in the middle. Hydrogen may also be produced. Thermal cracking tends to produce a high proportion of alkenes. To avoid too much decomposition (ultimately to carbon and hydrogen) the alkanes are kept in these conditions for a very short time, typically one second. The equation in Figure 2 shows cracking of a long chain alkane to give a shorter chain alkane and an alkene. The chain could break at any point.

Catalytic cracking

Catalytic cracking takes place at a lower temperature (approximately 720 K) and lower pressure (but more than atmospheric), using a zeolite catalyst, consisting of silicon dioxide and aluminium oxide (aluminosilicates). Zeolites have a honeycomb structure with an enormous surface area. They are also acidic. This form of cracking is used mainly to produce motor fuels. The products are mostly branched alkanes, cycloalkanes (rings), and aromatic compounds, see Figure 3.

The products obtained from cracking are separated by fractional distillation.

In the laboratory, catalytic cracking may be carried out in the apparatus shown in Figure 4, using lumps of aluminium oxide as a catalyst.

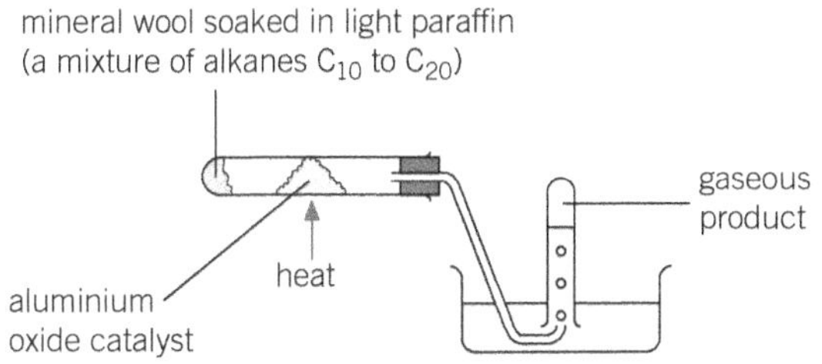

▲ **Figure 4** *Laboratory cracking of alkanes*

The products are mostly gases, showing that they have chain lengths of less than C_5 and the mixture decolourises bromine solution. This is a test for a carbon–carbon double bond showing that the product contains alkenes.

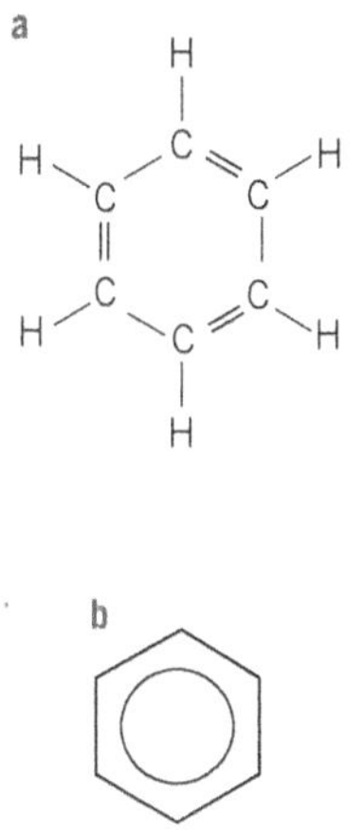

▲ **Figure 3** *Aromatic compounds are based on the benzene ring C_6H_6. Although it appears to have three double bonds as in* **a**, *the electrons are spread around the ring, making it more stable than expected. It is usually represented as in* **b**

Synoptic link

The chemistry of aromatic compounds is dealt with in Chapter 27, Aromatic chemistry.

Study tip

It is important to be able to predict the products and write equations for typical thermal and catalytic cracking reactions.

Summary questions

1 Complete the word equation for one possibility for the thermal cracking of decane.

decane → octane + ...

2 In the laboratory cracking of alkanes, how can you tell that the products have shorter chains than the starting materials?

3 Why would we not crack octane industrially?

4 How can the temperature required for cracking be reduced?

5 Give two economic reasons for cracking long chain alkanes.

12.4 Combustion of alkanes

Learning objectives:

- → Describe what a fuel is.
- → Explain why alkanes are good fuels.
- → Describe the environmental problems associated with the use of alkanes as fuels.
- → Describe how these problems may be tackled.

Specification reference: 3.3.2

Synoptic link

You learnt about enthalpy of combustion in Topic 4.5, Enthalpy changes of combustion.

Study tip

You should be able to balance combustion equations.

▲ **Figure 1** *Incomplete combustion is potentially dangerous because of carbon monoxide formation. Carbon monoxide detectors in kitchens can warn of dangerous levels of this gas.*

Study tip

Try to ensure you know the problems with using carbon-based fuels.

Alkanes are quite unreactive. They do not react with acids, bases, oxidising agents, or reducing agents. However, they do burn and they will react with halogens under suitable conditions.

Combustion

The shorter chain alkanes burn completely in a plentiful supply of oxygen to give carbon dioxide and water.

For example, methane:

$$CH_4(g) + 2O_2(g) \rightarrow CO_2(g) + 2H_2O(l) \qquad \Delta H = -890 \text{ kJ mol}^{-1}$$

Or ethane:

$$C_2H_6(g) + 3\frac{1}{2}O_2(g) \rightarrow 2CO_2(g) + 3H_2O(l) \qquad \Delta H = -15560 \text{ kJ mol}^{-1}$$

Combustion reactions give out heat and have large negative enthalpies of combustion. The more carbons present, the greater the heat output. For this reason they are important as fuels. **Fuels** are substances that release heat energy when they undergo combustion. They also store a large amount of energy for a small amount of weight. For example, octane produces approximately 48 kJ of energy per gram when burnt, which is about twice the energy output per gram of coal. Examples of alkane fuels include:

- methane (the main component of natural or North Sea gas)
- propane (camping gas)
- butane (Calor gas)
- petrol (a mixture of hydrocarbons of approximate chain length C_8)
- paraffin (a mixture of hydrocarbons of chain lengths C_{10} to C_{18}).

Incomplete combustion

In a limited supply of oxygen, the poisonous gas carbon monoxide, CO, is formed. For example, with propane:

$$C_3H_8(g) + 3\frac{1}{2}O_2(g) \rightarrow 3CO(g) + 4H_2O(l)$$

This is called **incomplete combustion**.

With even less oxygen, carbon (soot) is produced. For example, when a Bunsen burner is used with a closed air hole, the flame is yellow and a black sooty deposit appears on the apparatus. Incomplete combustion often happens with longer chain hydrocarbons, which need more oxygen to burn compared with shorter chains.

Polluting the atmosphere

All hydrocarbon-based fuels derived from crude oil may produce polluting products when they burn. They include the following:

- carbon monoxide, CO, a poisonous gas produced by incomplete combustion

- nitrogen oxides, NO, NO_2, and N_2O_4 (often abbreviated to NO_x) produced when there is enough energy for nitrogen and oxygen in the air to combine, for example:

$$N_2(g) + O_2(g) \rightarrow 2NO(g)$$

This happens in a petrol engine at the high temperatures present, when the sparks ignite the fuel. These oxides may react with water vapour and oxygen in the air to form nitric acid. They are therefore contributors to acid rain and photochemical smog

- sulfur dioxide is another contributor to acid rain. It is produced from sulfur-containing impurities present in crude oil. This oxide combines with water vapour and oxygen in the air to form sulfuric acid
- carbon particles, called **particulates**, which can exacerbate asthma and cause cancer
- unburnt hydrocarbons may also enter the atmosphere and these are significant greenhouse gases. They contribute to photochemical smog which can cause a variety of health problems (Figure 2)
- carbon dioxide, CO_2, is a greenhouse gas. It is always produced when hydrocarbons burn. Although carbon dioxide is necessary in the atmosphere, its level is rising and this is a cause of the increase in the Earth's temperature and consequent climate change
- water vapour which is also a greenhouse gas.

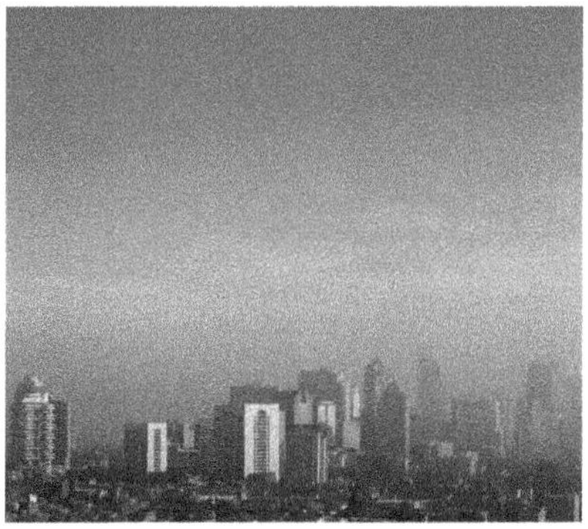

▲ **Figure 2** *Photochemical smog is the chemical reaction of sunlight, nitrogen oxides, NO_x, and volatile organic compounds in the atmosphere, which leaves airborne particles (called particulate matter) and ground-level ozone.*

Flue gas desulfurisation

Large numbers of power stations generate electricity by burning fossil fuels such as coal or natural gas. These fuels contain sulfur compounds and one of the products of their combustion is sulfur dioxide, SO_2, a gas that causes acid rain by combining with oxygen and water in the atmosphere to form sulfuric acid.

$$SO_2(g) + \frac{1}{2}O_2(g) + H_2O(l) \rightarrow H_2SO_4(l)$$

The gases given out by power stations are called **flue gases** so the process of removing the sulfur dioxide is called flue gas desulfurisation. In one method, a slurry of calcium oxide (lime) and water is sprayed into the flue gas which reacts with the calcium oxide and water to form calcium sulfite, which can be further oxidised to calcium sulfate, also called gypsum. The overall reaction is:

$$CaO(s) + 2H_2O(l) + SO_2(g) + \frac{1}{2}O_2 \rightarrow CaSO_4{\cdot}2H_2O(s)$$

Gypsum is a saleable product as it is used to make builders' plaster and plasterboard.

An alternative process uses calcium carbonate (limestone) rather than calcium oxide:

$$CaCO_3(s) + \frac{1}{2}O_2(g) + SO_2(g) \rightarrow CaSO_4(s) + CO_2(g)$$

Catalytic converters

The internal combustion engine produces most of the pollutants listed above, though sulfur is now removed from petrol so that sulfur dioxide has become less of a problem.

▲ **Figure 3** *A catalytic converter*

Visible radiation (sunlight) readily penetrates the atmosphere and warms the Earth.

Invisible infra-red radiation is emitted by the Earth and cools it down. But some of this infrared is trapped by greenhouse gases in the atmosphere which act as a blanket, keeping the heat in.

Both the Earth and a greenhouse trap heat in the same way.

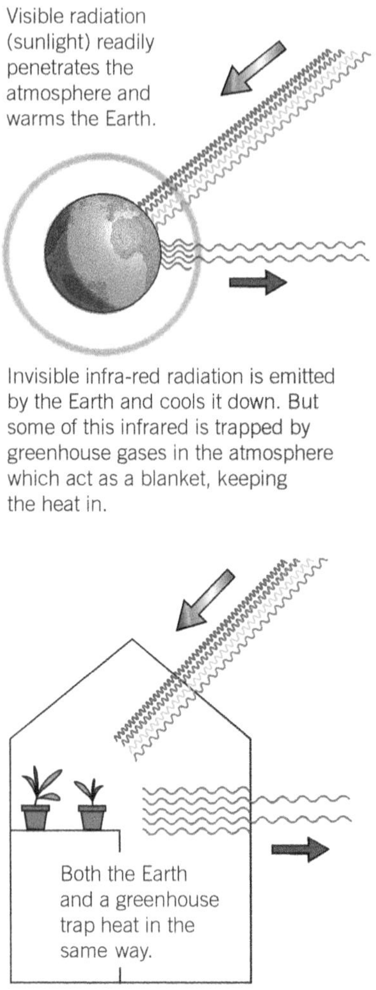

▲ **Figure 4** *The greenhouse effect*

Summary questions

1 Write word and balanced symbol equations for:

a the complete combustion of propane

b the incomplete combustion of butane to produce carbon monoxide and water.

2 a What are the problems with using carbon-based fuels?

b What steps are taken to reduce these problems?

3 Pick two alternative sources of power which do not use carbon-based fuels and discuss the pros and cons of each.

All new cars with petrol engines are now equipped with catalytic converters in their exhaust systems (Figure 3). These reduce the output of carbon monoxide, nitrogen oxides, and unburnt hydrocarbons in the exhaust gas mixture.

The catalytic converter is a honeycomb made of a ceramic material coated with platinum and rhodium metals. These are the catalysts. The honeycomb shape provides an enormous surface area, so a little of these expensive metals goes a long way. As the polluting gases pass over the catalyst, they react with each other to form less harmful products by the following reactions:

- $$2CO(g) + 2NO(g) \rightarrow N_2(g) + 2CO_2(g)$$
 carbon monoxide nitrogen oxide nitrogen carbon dioxide

- hydrocarbons + nitrogen oxide \rightarrow nitrogen + carbon dioxide + water

For example, $C_8H_{18} + 25NO \rightarrow 12\frac{1}{2}N_2 + 8CO_2 + 9H_2O$

The reactions take place on the surface of the catalyst, on the layer of platinum and rhodium metals.

Global warming and the greenhouse effect

Greenhouses become very warm inside. This is because the visible rays from the Sun pass through the glass. Rather than escaping, their energy is absorbed by everything inside the greenhouse and re-radiated as infrared energy, which is heat. Infrared energy has a longer wavelength and cannot pass back out through the glass.

Carbon dioxide behaves rather like glass. It traps infrared radiation so that the Earth's atmosphere heats up. This is important for life because without carbon dioxide and other greenhouse gases, the Earth would be too cold to sustain life. Other greenhouse gases are water vapour and methane. These are even more effective than carbon dioxide, but there has not been as much change in the level of these gases in the atmosphere in recent years. However, since the industrial revolution fossil fuels have been used to fuel industrial plants and the level of carbon dioxide has been rising. Gradually, the Earth's temperature has been rising too and the majority of scientists believe that the increasing level of carbon dioxide is the cause of global warming.

The concentration of water vapour, the most abundant greenhouse gas, in the atmosphere tends to stay roughly the same (except locally – by waterfalls, for example) because of the equilibrium that exists between water vapour and liquid water. However, if the temperature of the atmosphere rises, there will be more water vapour in the air and therefore more greenhouse warming. This may be offset by greater cloud formation and clouds reflect solar radiation. The role of water is therefore recognised as very important but as yet not fully understood.

Carbon neutral activities

Many people are concerned about activities, such as airline flights, that produce large amounts of carbon dioxide. A flight from London to Paris produces about 350 kg of carbon dioxide per passenger (from burning hydrocarbon fuels). Activities that produce no carbon dioxide emissions overall are referred to as **carbon neutral**.

12.5 The formation of halogenoalkanes

When you put a mixture of an alkane and a halogen into bright sunlight, or shine a photoflood lamp onto the mixture, the alkane and the halogen will react to form a halogenoalkane. The ultraviolet component of the light starts the reaction. Alkanes do not react with halogens in the dark at room temperature.

For example, if you put a mixture of hexane and a little liquid bromine into a test tube and leave it in the dark, it stays red-brown (the colour of bromine). However, if you shine ultraviolet light onto it, the mixture becomes colourless and misty fumes of hydrogen bromide appear.

A substitution reaction has taken place. One or more of the hydrogen atoms in the alkane has been replaced by a bromine atom and hydrogen bromide is given off as a gas. The main reaction is:

$$\underset{\text{hexane}}{C_6H_{14}(g)} + \underset{\text{bromine}}{Br_2(l)} \rightarrow \underset{\text{bromohexane}}{C_6H_{13}Br(l)} + \underset{\text{hydrogen bromide}}{HBr(g)}$$

Bromohexane is a halogenoalkane.

Chain reactions

The reaction above is called a free-radical substitution. It starts off a **chain reaction** which takes place in three stages – **initiation**, **propagation**, and **termination**.

The reaction between any alkane and a halogen goes by the same mechanism.

For example, methane and chlorine:

$$CH_4(g) + Cl_2(g) \rightarrow CH_3Cl(g) + HCl(g)$$

Initiation

- The first, or initiation, step of the reaction is breaking the Cl—Cl bond to form two chlorine atoms.
- The chlorine molecule absorbs the energy of a single quantum of ultraviolet (UV) light. The energy of one quantum of UV light is greater than the Cl—Cl bond energy, so the bond will break.
- Since both atoms are the same, the Cl—Cl bond breaks homolytically, that is, one electron going to each chlorine atom.
- This results in two separate chlorine atoms, written Cl•. They are called free radicals. The dot is used to show the unpaired electron.

$$Cl{—}Cl \xrightarrow{\text{UV light}} 2Cl\bullet$$

- Free radicals are highly reactive.
- The C—H bond in the alkane needs more energy to break than is available in a quantum of ultraviolet radiation, so this bond does not break.

Learning objectives:

→ Define what a free radical is.

→ Describe the reaction mechanism for the free-radical substitution of methane.

Specification reference: 3.3.2 and 3.3.1

Hint

In fact a mixture of many bromoalkanes is formed.

Hint

You can test for hydrogen bromide by wafting the fumes from a bottle of ammonia over the mouth of the test tube; white fumes of ammonium bromide are formed. This test will also give a positive result for other hydrogen halides.

▲ **Figure 1** *Sunscreen is important to prevent skin damage caused by UV in sunlight*

Propagation

This takes place in two stages:

1 The chlorine free radical takes a hydrogen atom from methane to form hydrogen chloride, a stable compound. This leaves a methyl free radical, $\bullet CH_3$.

$$Cl\bullet + CH_4 \rightarrow HCl + \bullet CH_3$$

2 The methyl free radical is also very reactive and reacts with a chlorine molecule. This produces another chlorine free radical and a molecule of chloromethane – a stable compound.

$$\bullet CH_3 + Cl_2 \rightarrow CH_3Cl + Cl\bullet$$

The effect of these two steps is to produce hydrogen chloride, chloromethane, and a new Cl• free radical. This is ready to react with more methane and repeat the two steps. This is the chain part of the chain reaction. These steps may take place thousands of times before the radicals are destroyed in the termination step.

Termination

Termination is the step in which the free radicals are removed. This can happen in any of the following three ways:

$Cl\bullet + Cl\bullet \rightarrow Cl_2$	Two chlorine free radicals react together to give chlorine.
$\bullet CH_3 + \bullet CH_3 \rightarrow C_2H_6$	Two methyl free radicals react together to give ethane.
$Cl\bullet + \bullet CH_3 \rightarrow CH_3Cl$	A chlorine free radical and a methyl free radical react together to give chloromethane.

Notice that in every case, two free radicals react to form a stable compound with no unpaired electrons.

Other products of the chain reaction

Other products are formed as well as the main ones, chloromethane and hydrogen chloride.

- Some ethane is produced at the termination stage, as shown above.
- Dichloromethane may be made at the propagation stage, if a chlorine radical reacts with some chloromethane that has already formed.

$$CH_3Cl + Cl\bullet \rightarrow \bullet CH_2Cl + HCl$$

followed by $\bullet CH_2Cl + Cl_2 \rightarrow CH_2Cl_2 + Cl\bullet$

- With longer-chain alkanes there will be many isomers formed because the Cl• can replace *any* of the hydrogen atoms.
- Chain reactions are not very useful because they produce such a mixture of products. They will also occur without light at high temperatures.

Why are chain reactions important?

It is believed that chloroflurorocarbons (CFCs) in the stratosphere are destroying the ozone layer.

Ozone is a molecule made from three oxygen atoms, O_3. It decomposes to oxygen. Too much ozone at ground level causes lung irritation and degradation of paints and plastics, but high in the atmosphere it has a vital role.

The ozone layer is important because it protects the Earth from the harmful exposure to too many ultraviolet (UV) rays. Without this protective layer, life on Earth would be very different. For example, plankton in the sea, which are at the very bottom of the food chain of the oceans, need protection from too much UV radiation. Also, too much UV radiation causes skin cancer in people by damaging DNA.

Chlorine free radicals are formed from CFCs because the C—Cl bond breaks homolytically in the presence of UV radiation to produce chlorine free radicals, Cl•. Ozone molecules are then attacked by these:

$$Cl\bullet + O_3 \rightarrow \underset{\text{free radical}}{ClO\bullet} + O_2$$

The resulting free radicals also attack ozone and regenerate Cl•:

$$ClO\bullet + O_3 \rightarrow 2O_2 + Cl\bullet$$

Adding the two equations, you can see that the chlorine free radical is not destroyed in this process. It acts as a catalyst in the breakdown of ozone to oxygen.

$$2O_3 \rightarrow 3O_2$$

Summary questions

1 What stage of a free-radical reaction of bromine with methane is represented by the following?

a $Br\bullet + Br\bullet \rightarrow Br_2$

b $CH_4 + Br\bullet \rightarrow CH_3\bullet + HBr$

c $\bullet CH_3 + Br_2 \rightarrow CH_3Br + Br\bullet$

d $Br_2 \rightarrow 2Br\bullet$

2 Look at the equations for the destruction of ozone in the last section of this topic.

a Which two are propagation steps?

b Suggest three possible termination steps.

1 Octane is the eighth member of the alkane homologous series.

(a) State **two** characteristics of a homologous series. *(2 marks)*

(b) Name a process used to separate octane from a mixture containing several different alkanes. *(1 mark)*

(c) The structure shown below is one of several structural isomers of octane.

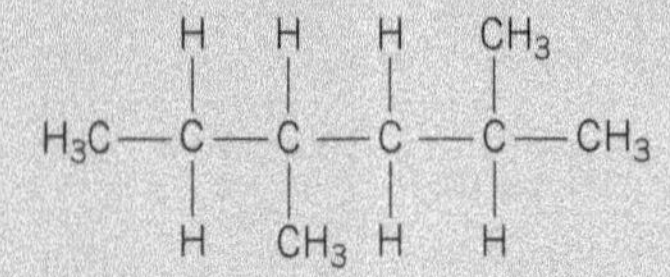

Give the meaning of the term *structural isomerism*. Name this isomer and state its empirical formula. *(4 marks)*

(d) Suggest why the branched chain isomer shown above has a lower boiling point than octane. *(2 marks)*

AQA, 2011

2 Cetane, $C_{16}H_{34}$, is a major component of diesel fuel.

(a) Write an equation to show the complete combustion of cetane. *(1 mark)*

(b) Cetane has a melting point of 18°C and a boiling point of 287°C. In polar regions vehicles that use diesel fuel may have ignition problems. Suggest **one** possible cause of this problem with the diesel fuel. *(1 mark)*

(c) The pollutant gases NO and NO_2 are sometimes present in the exhaust gases of vehicles that use petrol fuel.

(i) Write an equation to show how NO is formed and give a condition needed for its formation. *(2 marks)*

(ii) Write an equation to show how NO is removed from the exhaust gases in a catalytic converter. Identify a catalyst used in the converter. *(2 marks)*

(iii) Deduce an equation to show how NO_2 reacts with water and oxygen to form nitric acid, HNO_3. *(1 mark)*

(d) Cetane, $C_{16}H_{34}$, can be cracked to produce hexane, butene, and ethene.

(i) State **one** condition that is used in this cracking reaction. *(1 mark)*

(ii) Write an equation to show how one molecule of cetane can be cracked to form hexane, butene, and ethene. *(1 mark)*

(iii) State **one** type of useful solid material that could be formed from alkenes. *(1 mark)*

AQA, 2011

3 Hexane, C_6H_{14}, is a member of the homologous series of alkanes.

(a) (i) Name the raw material from which hexane is obtained. *(1 mark)*

(ii) Name the process used to obtain hexane from this raw material. *(1 mark)*

(b) C_6H_{14} has structural isomers.

(i) Deduce the number of structural isomers with molecular formula C_6H_{14} *(1 mark)*

(ii) State **one** type of structural isomerism shown by the isomers of C_6H_{14} *(1 mark)*

(c) One molecule of an alkane **X** can be cracked to form one molecule of hexane and two molecules of propene.

(i) Deduce the molecular formula of **X**. *(1 mark)*

(ii) State the type of cracking that produces a high percentage of alkenes. State the conditions needed for this type of cracking. *(2 marks)*

(iii) Explain the main economic reason why alkanes are cracked. *(1 mark)*

(d) Hexane can react with chlorine under certain conditions as shown in the following equation.

$$C_6H_{14} + Cl_2 \rightarrow C_6H_{13}Cl + HCl$$

(i) Both the products are hazardous. The organic product would be labelled 'flammable'. Suggest the most suitable hazard warning for the other product. *(1 mark)*

(ii) Calculate the percentage atom economy for the formation of $C_6H_{13}Cl$ (M_r = 120.5) in this reaction. *(1 mark)*

(e) A different chlorinated compound is shown below. Name this compound and state its empirical formula.

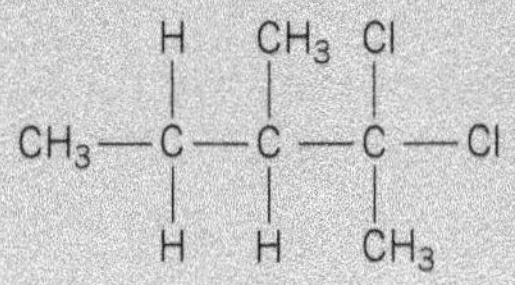

(2 marks)
AQA, 2012

4 Alkanes are used as fuels. A student burned some octane, C_8H_{18}, in air and found that the combustion was incomplete.

(a) (i) Write an equation for the incomplete combustion of octane to produce carbon monoxide as the only carbon-containing product. *(1 mark)*

(ii) Suggest **one** reason why the combustion was incomplete. *(1 mark)*

(b) Catalytic converters are used to remove the toxic gases NO and CO that are produced when alkane fuels are burned in petrol engines.

(i) Write an equation for a reaction between these two toxic gases that occurs in a catalytic converter when these gases are removed. *(1 mark)*

(ii) Identify a metal used as a catalyst in a catalytic converter.
Suggest **one** reason, other than cost, why the catalyst is coated on a ceramic honeycomb. *(2 marks)*

(c) If a sample of fuel for a power station is contaminated with an organic sulfur compound, a toxic gas is formed by complete combustion of this sulfur compound.

(i) State **one** environmental problem that can be caused by the release of this gas. *(1 mark)*

(ii) Identify **one** substance that could be used to remove this gas.
Suggest **one** reason, other than cost, why this substance is used. *(2 marks)*
AQA, 2012

5 Chlorine can be used to make chlorinated alkanes such as dichloromethane.

(a) Write an equation for each of the following steps in the mechanism for the reaction of chloromethane, CH_3Cl, with chlorine to form dichloromethane, CH_2Cl_2.
Initiation step
First propagation step
Second propagation step
The termination step that forms a compound with empirical formula CH_2Cl *(4 marks)*

(b) When chlorinated alkanes enter the upper atmosphere, carbon–chlorine bonds are broken. This process produces a reactive intermediate that catalyses the decomposition of ozone. The overall equation for this decomposition is $2O_3 \rightarrow 3O_2$

(i) Name the type of reactive intermediate that acts as a catalyst in this reaction. *(1 mark)*

(ii) Write **two** equations to show how this intermediate is involved as a catalyst in the decomposition of ozone. *(2 marks)*
AQA, 2013

13 Halogenoalkanes
13.1 Halogenoalkanes – introduction

Learning objectives:
- → Explain why halogenoalkanes are more reactive than alkanes.
- → Explain why carbon–halogen bonds are polar.
- → Explain the trends in bond enthalpy and bond polarity of the carbon–halogen bond.

Specification reference: 3.3.3

Hint

Halogenoalkanes are sometimes called **haloalkanes**.

▲ **Figure 1** *Applications of halogenoalkanes*

▼ **Table 1** *Electronegativities of carbon and the halogens*

Element	Electronegativity
carbon	2.5
fluorine	4.0
chlorine	3.5
bromine	2.8
iodine	2.6

Not many halogenoalkanes occur naturally but they are the basis of many synthetic compounds. Some examples of these are PVC (used to make drainpipes), Teflon (the non-stick coating on pans), and a number of anaesthetics and solvents. Halogenoalkanes have an alkane skeleton with one or more halogen (fluorine, chlorine, bromine, or iodine) atoms in place of hydrogen atoms.

The general formula

The general formula of a halogenoalkane with a single halogen atom is $C_nH_{2n+1}X$ where X is the halogen. This is often shortened to R—X.

How to name halogenoalkanes

- The prefixes fluoro-, chloro-, bromo-, and iodo- tell us which halogen is present.
- Numbers are used, if needed, to show on which carbon the halogen is bonded:

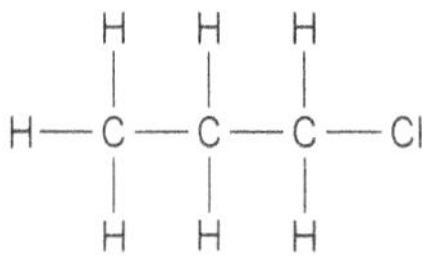

1-chloropropane

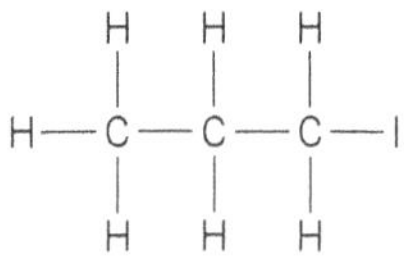

1-iodopropane

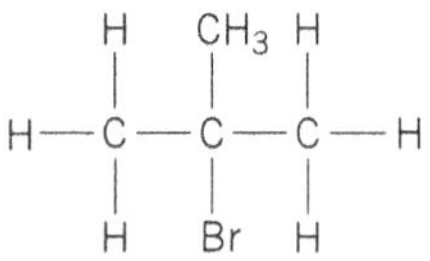

2-bromo-2-methylpropane

- The prefixes di-, tri-, tetra-, and so on, are used to show *how many* atoms of each halogen are present.
- When a compound contains different halogens they are listed in alphabetical order, *not* in order of the number of the carbon atom to which they are bonded. For example:

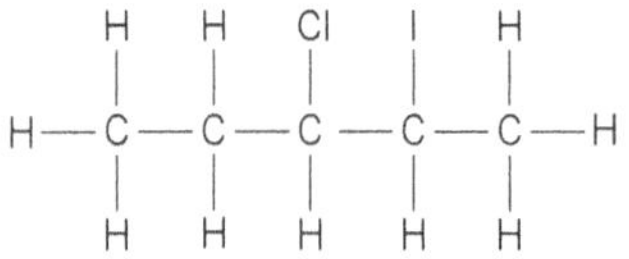

is 3-chloro-2-iodopentane not 2-iodo-3-chloropentane. (C is before I in the alphabet.)

Bond polarity

Halogenoalkanes have a C—X bond. This bond is polar, $C^{\delta+}$—$X^{\delta-}$, because halogens are more electronegative than carbon. The electronegativities of carbon and the halogens are shown in Table 1. Notice that as you go down the group, the bonds get less polar.

Physical properties of halogenoalkanes

Solubility

- The polar $C^{\delta+}$—$X^{\delta-}$ bonds are not polar enough to make the halogenoalkanes soluble in water.
- The main intermolecular forces of attraction are dipole–dipole attractions and van der Waal forces.

- Halogenoalkanes mix with hydrocarbons so they can be used as dry-cleaning fluids and to remove oily stains. (Oil is a mixture of hydrocarbons.)

Boiling point

The boiling point depends on the number of carbon atoms and halogen atoms.

- Boiling point *increases* with *increased* chain length.
- Boiling point *increases* going *down* the halogen group.

Both these effects are caused by increased van der Waals forces because the larger the molecules, the greater the number of electrons (and therefore the larger the van der Waals forces).

As in other homologous series, increased branching of the carbon chain will tend to lower the melting point.

Halogenoalkanes have higher boiling points than alkanes with similar chain lengths because they have higher relative molecular masses and they are more polar.

How the halogenoalkanes react – the reactivity of the C—X bond

When halogenoalkanes react it is almost always the C—X bond that breaks. There are two factors that determine how readily the C—X bond reacts. These are:

- the $C^{\delta+}$—$X^{\delta-}$ bond polarity
- the C—X bond enthalpy.

Bond polarity

The halogens are more electronegative than carbon so the bond polarity will be $C^{\delta+}$—$X^{\delta-}$. This means that the carbon bonded to the halogen has a partial positive charge – it is electron deficient. This means that it can be attacked by reagents that are electron rich or have electron-rich areas. These are called **nucleophiles**. A nucleophile is an electron pair donor, see Topic 13.2.

The polarity of the C—X bond would predict that the C—F bond would be the most reactive. It is the most polar, so the $C^{\delta+}$ has the most positive charge and is therefore most easily attacked by a nucleophile. This argument would make the C—I bond least reactive because it is the least polar.

Bond enthalpies

C—X bond enthalpies are listed in Table 2. The bonds get weaker going down the group. Fluorine is the smallest atom of the halogens and the shared electrons in the C—F bond are strongly attracted to the fluorine nucleus. This makes a strong bond. Going down the group, the shared electrons in the C—X bond get further and further away from the halogen nucleus, so the bond becomes weaker.

The bond enthalpies would predict that iodo-compounds, with the weakest bonds, are the most reactive, and fluoro-compounds, with the strongest bonds, are the least reactive.

Experiments confirm that reactivity increases going down the group. This means that bond enthalpy is a more important factor than bond polarity.

▼ **Table 2** *Carbon–halogen bond enthalpies*

Bond	Bond enthalpy / $kJ\,mol^{-1}$
C—F	467
[C—H	413]
C—Cl	346
C—Br	290
C—I	228

(arrow upwards: stronger)

Summary questions

1 These questions are about the following halogenoalkanes:

 i $CH_3CH_2CH_2CH_2I$
 ii $CH_3CHBrCH_3$
 iii $CH_2ClCH_2CH_2CH_3$
 iv $CH_3CH_2CHBrCH_3$

 a Draw the displayed formula for each halogenoalkane and mark the polarity of the C—X bond.
 b Name each halogenoalkane.
 c Predict which of them would have the highest boiling point and explain your answer.

2 Why do the halogenoalkanes get less reactive going up the halogen group?

13.2 Nucleophilic substitution in halogenoalkanes

Learning objectives:

→ State the definition of a nucleophile.

→ Describe nucleophilic substitution.

→ Explain why ^-OH, ^-CN, and NH_3 behave as nucleophiles.

→ Describe the mechanism of nucleophilic substitution.

Specification reference: 3.3.3

Most reactions of organic compounds take place via a series of steps. You can often predict these steps by thinking about how electrons are likely to move. This can help you understand why reactions take place as they do, and this can save a great deal of rote learning.

Nucleophiles

Nucleophiles are reagents that attack and form bonds with positively or partially positively charged carbon atoms.

- A **nucleophile** is either a negatively charged ion or has an atom with a $\delta-$ charge.
- A nucleophile has a lone (unshared) pair of electrons which it can use to form a covalent bond.
- The lone pair is situated on an electronegative atom.

So, in organic chemistry a nucleophile is a species that has a lone pair of electrons with which it can form a bond by donating its electrons to an electron-deficient carbon atom. Some common nucleophiles are:

- the hydroxide ion, $^-$:OH
- ammonia, :NH_3
- the cyanide ion, $^-$:CN.

They will each replace the halogen in a halogenoalkane. These reactions are called **nucleophilic substitutions** and they all follow essentially the same reaction mechanism.

A reaction mechanism describes a route from reactants to products via a series of theoretical steps. These may involve short-lived intermediates.

Nucleophilic substitution

The general equation for nucleophilic substitution, using :Nu^- to represent any negatively charged nucleophile and :X^- to represent a halogen atom, is:

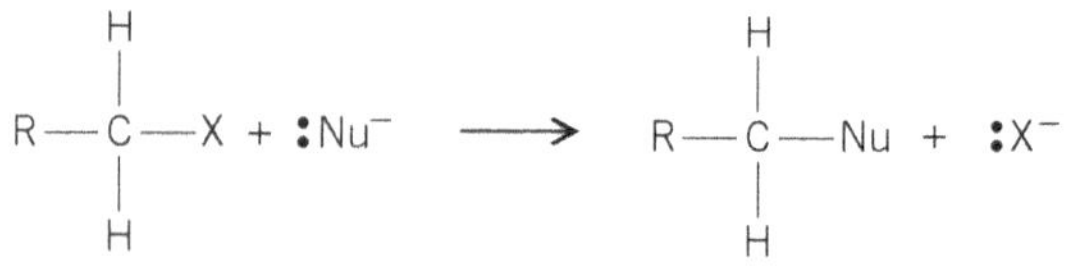

Reaction mechanisms and curly arrows

Curly arrows are used to show how electron pairs move in organic reactions. These are shown here in red for clarity. You can write the above reaction as:

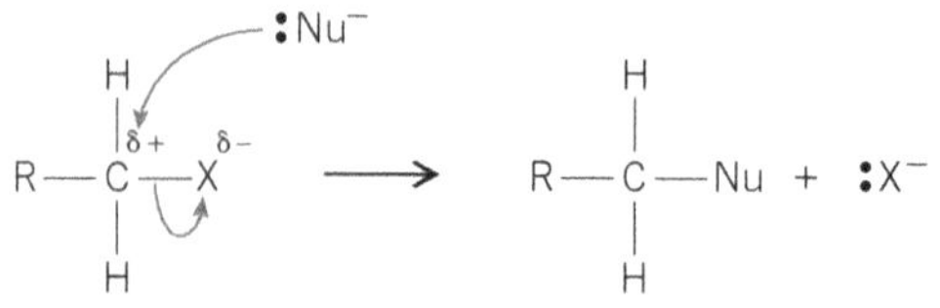

The lone pair of electrons of a nucleophile is attracted towards a partially positively charged carbon atom. A curly arrow starts at a lone pair of electrons and moves towards $C^{\delta+}$.

Study tip

It is important to remember that a curly arrow indicates the movement of an electron pair.

The lower curly arrow shows the electron pair in the C—X bond moving to the halogen atom, X, and making it a halide ion. The halide ion is called the **leaving group**.

The rate of substitution depends on the halogen. Fluoro-compounds are unreactive due to the strength of the C—F bond. Then, going down the group, the rate of reaction increases as the C—X bond strength decreases, see Topic 13.1.

Examples of nucleophilic substitution reactions

All these reactions are similar. Remember the basic pattern, shown above. Then work out the product with a particular nucleophile. This is easier than trying to remember the separate reactions.

Halogenoalkanes with aqueous sodium (or potassium) hydroxide

The nucleophile is the hydroxide ion, $^{-}$:OH.

This reaction occurs very slowly at room temperature. To speed up the reaction it is necessary to warm the mixture. Halogenoalkanes do not mix with water, so ethanol is used as a solvent in which the halogenoalkane and the aqueous sodium (or potassium) hydroxide both mix. This is called a **hydrolysis** reaction.

The overall reaction is:

$$R—X + OH^- \rightarrow ROH + X^-$$

so an alcohol, ROH, is formed.

For example:

$$\underset{\text{bromoethane}}{C_2H_5Br} + OH^- \rightarrow \underset{\text{ethanol}}{C_2H_5OH} + Br^-$$

This is the mechanism:

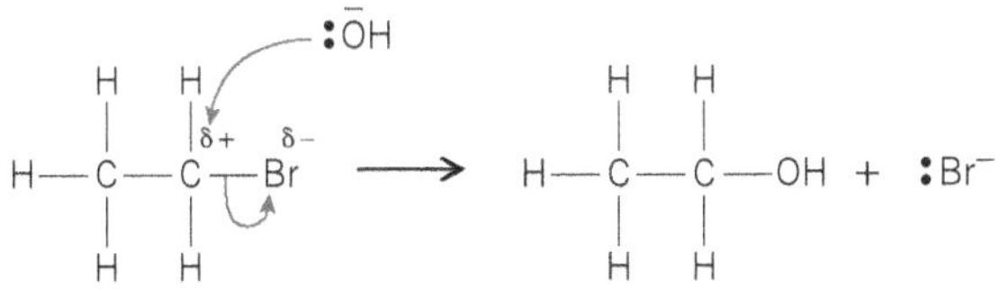

The rate of the reaction depends on the strength of the carbon–halogen bond C—F > C—Cl > C—Br > C—I (see Table 2 in Topic 13.1). Fluoroalkanes do not react at all whilst iodoalkanes react rapidly.

Halogenoalkanes with cyanide ions

When halogenoalkanes are warmed with an aqueous alcoholic solution of potassium cyanide, nitriles are formed. The nucleophile is the cyanide ion, $^{-}$:CN.

The reaction is:

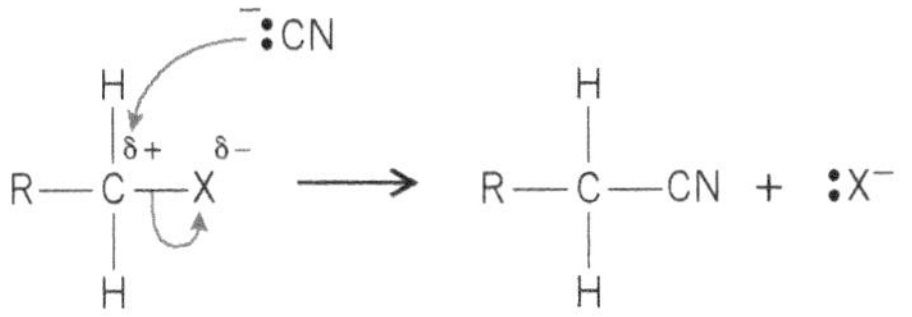

Hint

Nitriles have the functional group −C≡N. They are named using the suffix 'nitrile'. The carbon of the −CN group is counted as part of the root, so CH_3CH_2CN is propanenitrile, *not* ethanenitrile.

Hint

Primary amines have the functional group $-NH_2$, and are named with the suffix 'amine' attached to the appropriate side chain stem, rather than the usual root name, so $C_2H_5NH_2$ is ethylamine not ethanamine.

Study tip

Each step of a reaction mechanism must balance for atoms and charges. The sum of the steps equals the balanced equation.

Summary questions

1 This equation represents the hydrolysis of a halogenoalkane by sodium hydroxide solution:

$R{-}X + OH^- \rightarrow ROH + X^-$

a Why is the reaction carried out in ethanol?

b What is the nucleophile?

c Why is this a substitution?

d Which is the leaving group?

e Which would have the fastest reaction: R—F, R—Cl, R–Br, or R—I?

2 a Starting with bromoethane, what nucleophile will produce a product with three carbon atoms?

b Give the equation for this, using curly arrows to show the mechanism of the reaction.

c Name the product.

The product is called a **nitrile**. It has one extra carbon in the chain than the starting halogenoalkane. This is useful if you want to make a product that has one carbon more than the starting material.

Halogenoalkanes with ammonia

The nucleophile is ammonia, $:NH_3$.

The reaction of halogenoalkanes with an excess of concentrated solution of ammonia in ethanol is carried out under pressure. The reaction produces an amine, RNH_2.

$$R{-}X + 2NH_3 \rightarrow RNH_2 + NH_4X$$

This is the mechanism:

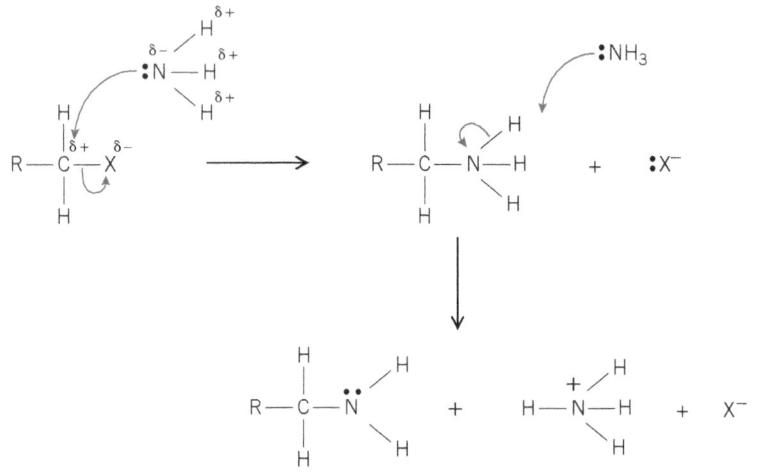

Ammonia is a nucleophile because it has a lone pair of electrons that it can donate (although it has no negative charge) and the nitrogen atom has a δ– charge. Because ammonia is a neutral nucleophile, a proton, H^+, must be lost to form the neutral product, called an **amine**. The H^+ ion reacts with a second ammonia molecule to form an NH_4^+ ion.

The uses of nucleophilic substitution

Nucleophilic substitution reactions are useful because they are a way of introducing new functional groups into organic compounds. Halogenoalkanes can be converted into alcohols, amines, and nitriles. These in turn can be converted to other functional groups.

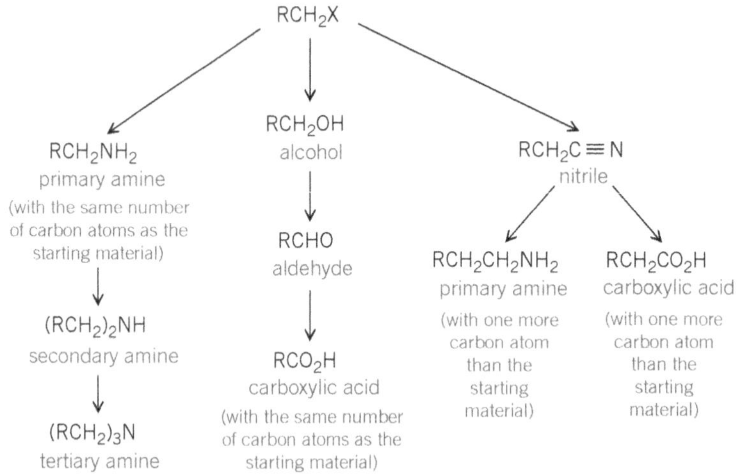

▲ **Figure 1** *Uses of nucleophilic reactions*

13.3 Elimination reactions in halogenoalkanes

Halogenoalkanes typically react by nucleophilic substitution. But, under different conditions they react by **elimination**. A hydrogen halide is eliminated from the molecule, leaving a double bond in its place so that an alkene is formed.

OH⁻ ion acting as a base

You saw in Topic 13.2 that the OH^- ion, from aqueous sodium or potassium hydroxide, is a nucleophile and its lone pair will attack a halogenoalkane at $C^{\delta+}$ to form an alcohol.

Under different conditions, the OH^- ion can act as a **base**, removing an H^+ ion from the halogenoalkane. In this case it is an elimination reaction rather than a substitution. In the example below, bromoethane reacts with potassium hydroxide to form ethene. A molecule of hydrogen bromide, HBr, is eliminated then the hydrogen bromide reacts with the potassium hydroxide. The reaction produces ethene, potassium bromide, and water.

$$
\begin{array}{c}
\underset{\substack{|\\H}}{\overset{\substack{Br\\|}}{H-C}}-\underset{\substack{|\\H}}{\overset{\substack{H\\|}}{C}}-H \; + \; KOH \longrightarrow \underset{\substack{\diagup\\H}}{\overset{\substack{H\\\diagdown}}{}}C=C\underset{\substack{\diagdown\\H}}{\overset{\substack{\diagup\\H}}{}} \; + \; KBr \; + \; H_2O
\end{array}
$$

The conditions of reaction

The sodium (or potassium) hydroxide is dissolved in ethanol and mixed with the halogenoalkane. *There is no water present.* The mixture is heated. The experiment can be carried out using the apparatus shown in Figure 1.

The product is ethene. Ethene burns and also decolourises bromine solution, showing that it has a C=C bond.

The mechanism of elimination

Hydrogen bromide is eliminated as follows. The curly arrows show the movement of electron pairs:

$$
\underset{\substack{|\\Br}}{\overset{\substack{:OH\\ \\H \quad H}}{H-C}}-\underset{\substack{|\\H}}{C}-H \longrightarrow H-\underset{\substack{|\\H}}{C}=C-H \; + \; H-O-H \; + \; :Br^-
$$

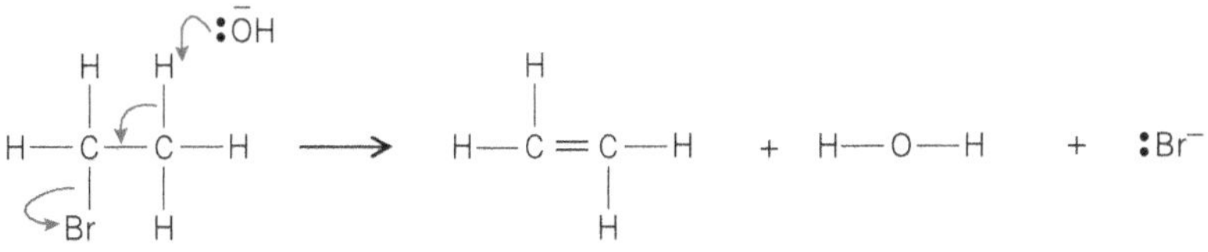

- The OH^- ion uses its lone pair to form a bond with one of the hydrogen atoms on the carbon next to the C—Br bond. These hydrogen atoms are very slightly δ+.
- The electron pair from the C—H bond now becomes part of a carbon–carbon double bond.
- The bromine takes the pair of electrons in the C—Br bond and leaves as a bromide ion (the leaving group).

This reaction is a useful way of making molecules with carbon–carbon double bonds.

Learning objectives:
- State the definition of an elimination reaction.
- Describe the mechanism for elimination reactions in halogenoalkanes.
- Describe the conditions that favour elimination rather than substitution.
- Show when and how isomeric alkenes are formed.

Specification reference: 3.3.3

mineral wool soaked in reaction mixture

heat

gas

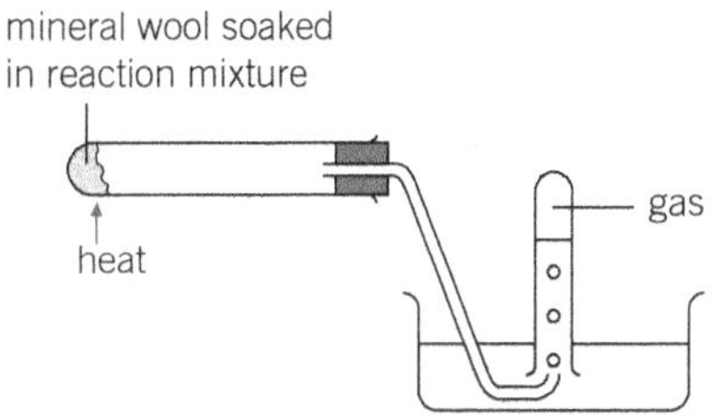

▲ **Figure 1** *Apparatus for elimination of hydrogen bromide from bromoethane*

Synoptic link

Decolourising bromine water is a test for an alkene. The bromine adds on across the double bond. See Topic 14.2, Reactions of alkenes.

$$
\underset{\substack{\diagup\\H}}{\overset{\substack{H\\\diagdown}}{}}C=C\underset{\substack{\diagdown\\H}}{\overset{\substack{\diagup\\H}}{}}
$$

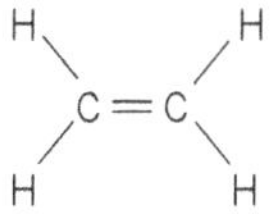

▲ **Figure 2** *A better representation of the shape of ethene*

Substitution or elimination?

Since the hydroxide ion will react with halogenoalkanes as a nucleophile *or* as a base, there is competition between substitution and elimination. In general a mixture of an alcohol and an alkene is produced. For example:

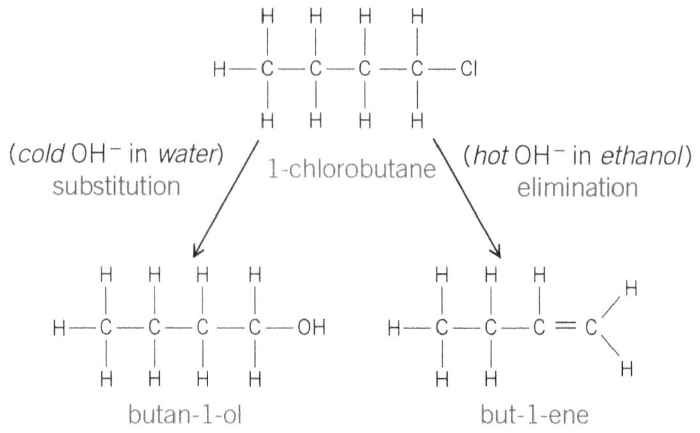

The reaction that predominates depends on two factors – the reaction conditions (aqueous or ethanolic solution) and the type of halogenoalkanes (primary, secondary, or tertiary).

The conditions of the reaction

- Hydroxide ions at room temperature, dissolved in water (aqueous), favour substitution.
- Hydroxide ions at high temperature, dissolved in ethanol, favour elimination.

The type of halogenoalkane

Primary halogenoalkanes tend to react by substitution and tertiary ones by elimination. Secondary will do both.

Hint

Halogenoalkanes are classified as primary, secondary, and tertiary according to whether the halogen atom is at the end of the hydrocarbon chain (primary), in the body of the chain (secondary), or at a branch in the chain (tertiary). The same system is used for alcohols, see Topic 15.1, Alcohols – introduction.

Isomeric products

In some cases a mixture of isomeric elimination products is possible.

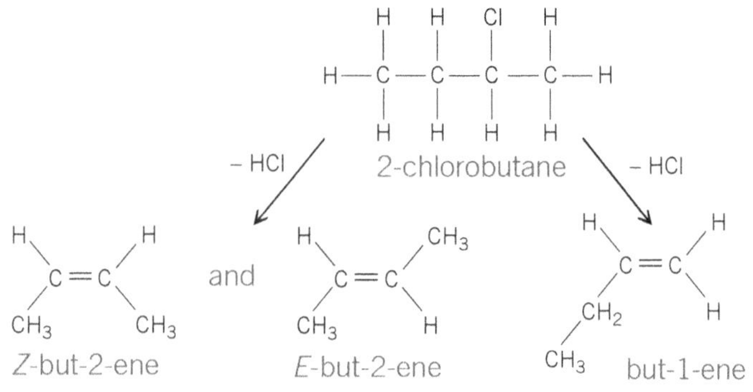

Synoptic link

The prefixes *Z* and *E* are explained in Topic 11.3, Isomerism.

Halogenoalkanes and the environment

Chlorofluorocarbons

Chlorofluorocarbons are halogenoalkanes containing both chlorine and fluorine atoms but no hydrogen, for example, trichlorofluoromethane, CCl_3F.

- They are also called CFCs.
- They are very unreactive under normal conditions.
- The short chain ones are gases and were used, for example, as aerosol propellants, refrigerants, and blowing agents for foams like expanded polystyrene.
- Longer chain ones are used as dry cleaning and de-greasing solvents.

CFC gases eventually end up in the atmosphere where they decompose to give chlorine atoms. Chlorine atoms decompose ozone, O_3, in the stratosphere, which has caused a hole in the Earth's ozone layer. Upper atmosphere research together with laboratory research showed how ozone is broken down. Politicians were influenced by scientists and, under international agreement, CFCs are being phased out and replaced by other, safer, compounds including hydrochlorofluorocarbons, HCFCs, such as CF_3CHCl_2. However, a vast reservoir of CFCs remains in the atmosphere and it will be many years before the ozone layer recovers.

CFCs

CFCs were introduced in the 1930s by an American engineer, Thomas Midgley, for use in refrigerators. He famously demonstrated their non-toxicity and non-flammability to a scientific conference by breathing in a lungful and exhaling it to extinguish a lighted candle. It was not until long after MIdgley's death that it was realised that CFCs were involved in the depletion of the ozone layer because they release chlorine atoms in the stratosphere.

Chemists have developed less harmful replacements for CFCs. Initially these were HCFCs, which contain hydrogen, carbon, fluorine, and chlorine. One example is $CHFCl_2$. These decompose more easily than CFCs due to their C—H bonds, and the chlorine atoms are released lower in the atmosphere where they do not contribute to the destruction of the ozone layer.

The so-called second generation replacements are called HFCs (hydrofluorocarbons) such as CHF_2CF_3. These contain no chlorine and therefore do not damage the ozone layer. They are not wholly free of environmental problems though, and chemists are working on third generation compounds. Some are considering reverting to refrigerants such as ammonia, which were used before the advent of CFCs.

1 Draw a 3-D representation of the formula of $CHFCl_2$ .What shape is this molecule? Does it have any isomers? Explain your answer.
2 What sort of formula is CHF_2CF_3? Draw its displayed formula. What is its molecular formula?

Summary questions

1 In elimination reactions of halogenoalkanes, the OH^- group is acting as which of the following?
   - **A** A base
   - **B** An acid
   - **C** A nucleophile
   - **D** An electrophile

2 Which of the following molecules is a CFC?
   - **A** CH_3CH_2Cl
   - **B** $CF_2{=}CF_2$
   - **C** CF_3CH_2Cl
   - **D** CCl_2F_2

3 **a** Name the two possible products when 2-bromopropane reacts with hydroxide ions.
   b How could you show that one of the products is an alkene?
   c Give the mechanism (using curly arrows) of the reaction that is an elimination.

Practice questions

1 Haloalkanes are used in the synthesis of other organic compounds.

(a) Hot concentrated ethanolic potassium hydroxide reacts with 2-bromo-3-methylbutane to form two alkenes that are structural isomers of each other. The major product is 2-methylbut-2-ene.

(i) Name and outline a mechanism for the conversion of 2-bromo-3-methylbutane into 2-methylbut-2-ene according to the equation.

$(CH_3)_2CHCHBrCH_3 + KOH \rightarrow (CH_3)_2C{=}CHCH_3 + KBr + H_2O$

(4 marks)

(ii) Draw the **displayed formula** for the other isomer that is formed.

(1 mark)

(iii) State the type of structural isomerism shown by these two alkenes.

(1 mark)

(b) A small amount of another organic compound, **X**, can be detected in the reaction mixture formed when hot concentrated ethanolic potassium hydroxide reacts with 2-bromo-3-methylbutane.

Compound **X** has the molecular formula $C_5H_{12}O$ and is a secondary alcohol.

(i) Draw the **displayed formula** for **X**.

(1 mark)

(ii) Suggest **one** change to the reaction conditions that would increase the yield of **X**.

(1 mark)

(iii) State the type of mechanism for the conversion of 2-bromo-3-methylbutane into **X**.

(1 mark)

AQA, 2013

2 (a) Consider the following reaction.

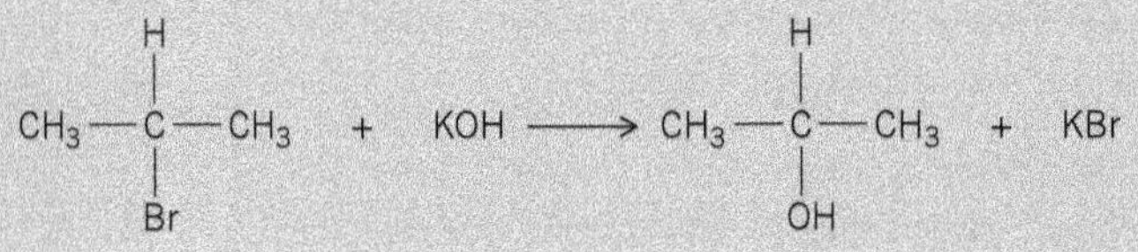

(i) Name and outline a mechanism for this reaction.

(3 marks)

(ii) Name the halogenoalkane in this reaction.

(1 mark)

(iii) Identify the characteristic of the halogenoalkane molecule that enables it to undergo this type of reaction.

(1 mark)

(iv) A student predicted that the yield of this reaction would be 90%. In an experiment 10.0 g of the halogenoalkane was used and 4.60 g of the organic product was obtained. Is the student correct? Justify your answer with a calculation using these data.

(1 mark)

(b) An alternative reaction can occur between this halogenoalkane and potassium hydroxide as shown by the following equation.

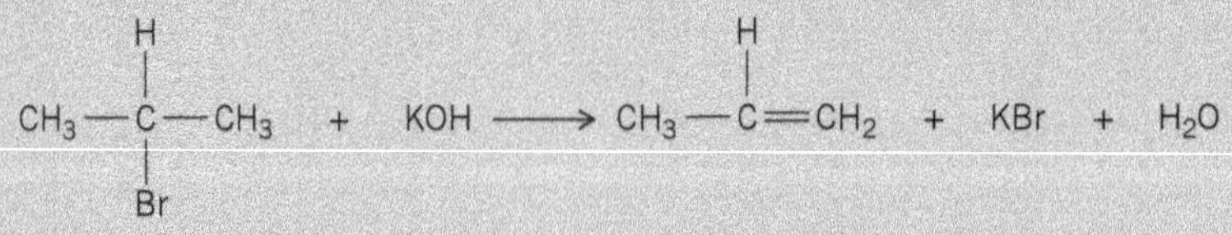

Name and outline a mechanism for this reaction.

(4 marks)

(c) Give **one** condition needed to favour the reaction shown in part (**b**) rather than that shown in part (**a**).

(1 mark)

AQA, 2010

3 **(a)** Write a balanced symbol equation for the reaction of $CH_3CH_2CH_2Br$ with aqueous hydroxide ions.
(2 marks)

(b) Name the starting material and the product.
(2 marks)

(c) Give the formula of the leaving group in this reaction.
(1 mark)

(d) Classify the reaction as substitution, elimination, or addition.
(1 mark)

(e) The hydroxide ion acts as a nucleophile in this reaction. State two features of the hydroxide ion that allow it to act as a nucleophile.
(2 marks)

(f) Draw the mechanism of the reaction using 'curly arrows' to show the movement of electron pairs.
(2 marks)

(g) How would you expect the rate of a similar reaction with $CH_3CH_2CH_2I$ to compare with that of $CH_3CH_2CH_2Br$? Explain your answer.
(2 marks)

(h) Water molecules can act as nucleophiles in a similar reaction. How do they compare with hydroxide ions as nucleophiles? Explain your answer.
(2 marks)

(i) What extra step has to occur in the reaction of a neutral nucleophile such as water compared with the reaction with a negatively charged ion such as the hydroxide ion?
(1 mark)

Answers to the Practice Questions and Section Questions are available at
www.oxfordsecondary.com/oxfordaqaexams-alevel-chemistry

14 Alkenes

14.1 Alkenes

Learning objectives:

- → Define an alkene.
- → Describe the isomerism that alkenes display.
- → Explain why they are reactive.

Specification reference: 3.3.1 and 3.3.4

Alkenes are **unsaturated** hydrocarbons. They are made of carbon and hydrogen only and have one or more carbon–carbon double bonds. This means that alkenes have fewer than the maximum possible number of hydrogen atoms. The double bond makes them more reactive than alkanes because of the high concentration of electrons (high electron density) between the two carbon atoms. Ethene, the simplest alkene, is the starting material for a large range of products, including polymers such as polythene, PVC, polystyrene, and terylene fabric, as well as products like antifreeze and paints. Alkenes are produced in large quantities when crude oil is thermally cracked.

The general formula

The homologous series of alkenes with one double bond has the general formula C_nH_{2n}.

How to name alkenes

There cannot be a C═C bond if there is only one carbon. So, the simplest alkene is ethene, $CH_2{=}CH_2$ followed by propene, $CH_3CH{=}CH_2$.

Structure

The shape of alkenes

Ethene is a planar (flat) molecule. This makes the angles between each bond roughly 120°.

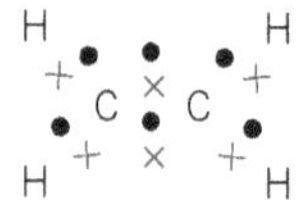

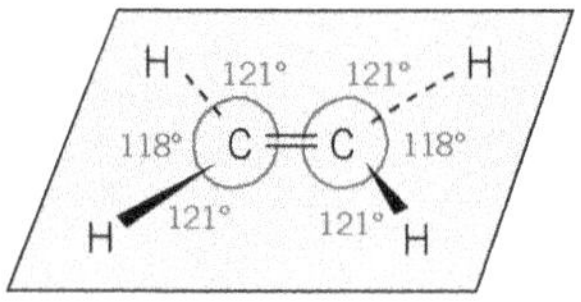

Hint

The H—C—H angle is slightly less than 120° because the group of four electrons in the C═C double bond repels more strongly than the groups of two in the C—H single bonds.

Unlike the C—C bonds in alkanes, there is no rotation about the double bond. This is because of the make-up of a double bond. Any molecules in which a hydrogen atom in ethene is replaced by another atom or group will have the same flat shape around the carbon–carbon double bond.

Why a double bond cannot rotate

As well as a normal C—C single bond, there is a p-orbital (which contains a single electron) on each carbon. These two orbitals overlap to form an orbital with a cloud of electron density above and below the single bond, see Figure 1 and Figure 2. This is called a **π-orbital** (pronounced pi) and its presence means the bond cannot rotate. This is sometimes called restricted rotation.

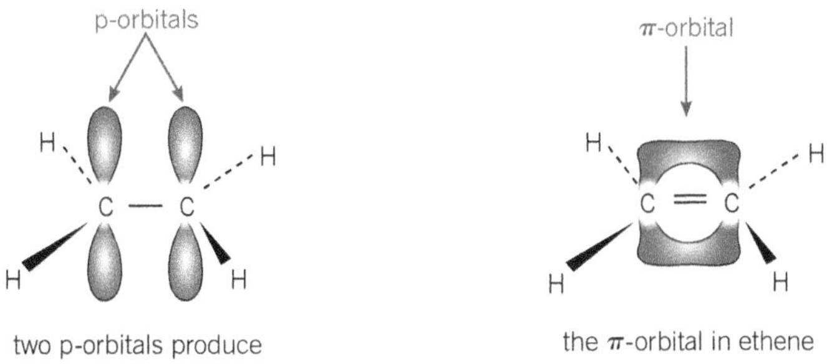

▲ **Figure 1** *The double bond in ethene*

Isomers

Alkenes with more than three carbons can form different types of isomers and they are named according to the IUPAC system, using the suffix -ene to indicate a double bond.

As well as chain isomers like those found in alkanes, alkenes can form two types of isomer that involve the double bond:

- position isomers
- geometrical isomers.

Position isomers

These are isomers with the double bond in different positions, that is, between a pair of adjacent carbon atoms in different positions in the carbon chain.

but-2-ene   but-1-ene

The longer the carbon chain, the more possibilities there will be and therefore the greater the number of isomers.

Geometrical isomers

Geometrical isomerism is a form of stereoisomerism. The two stereoisomers have the same structural formula but the bonds are arranged differently in space. It occurs only around C=C double bonds. For example, but-2-ene, above, can exist as shown below.

Z-but-2-ene   *E*-but-2-ene

The isomer in which both $-CH_3$ groups are on the same side of the double bond is called *Z*-but-2-ene and the one in which they are on opposite sides is called *E*-but-2-ene. This type of isomerism is often called *E-Z* isomerism.

Hint

The C—C single bond is formed by overlap of two p-orbitals along the line joining the two carbon atoms. The orbital formed is called a σ-orbital.

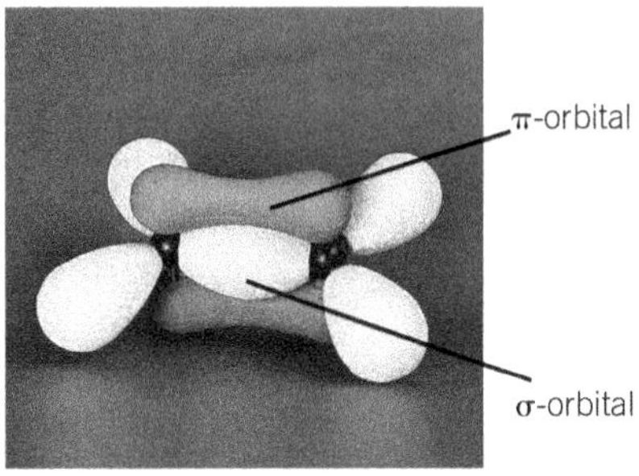

▲ **Figure 2** *Model of ethene showing orbitals*

Synoptic link

The IUPAC system of naming compounds in covered in Topic 11.2, Nomenclature – naming organic compounds.

Hint

The *E-Z* system of naming is often called the CIP system after the names of its inventors – Robert **C**ahn, Christopher **I**ngold, and Vladimir **P**relog.

Nomenclature – Cahn–Ingold–Prelog (CIP) notation

The number of known organic compounds is huge and increasing all the time. Finding information about these in databases, books, and journals would be almost impossible if chemists did not agree on how they should be named. This is why the Interntional Union of Pure and Applied Chemistry (IUPAC) produces rules for nomenclature. The *E-Z* notation is one example.

E-Z isomerism, until fairly recently, used to be known as *cis–trans* isomerism and the prefixes *cis-* and *trans-* were used instead of *Z-* and *E-*, respectively. So, for example, *Z*-but-2-ene was named *cis*-but-2-ene and *E*-but-2-ene was *trans*-but-2-ene. This notation is still often found in older books. However, a disadvantage of the older notation was that it did not work when there were more than two *different* substituents around a double bond. For example:

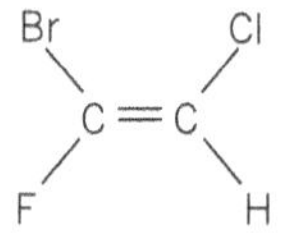
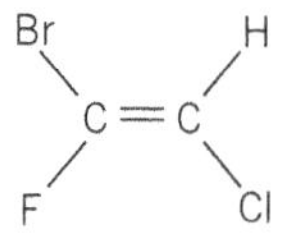

To give these two isomers different and unambiguous names the *E-Z* notation is used.

Simply, the *E-Z* notation is based on atomic numbers. Look at the atoms attached to each of the carbon atoms in the double bond. When the two atoms of higher atomic number (bromine and chlorine) are on the same side of the C═C, the isomer is described as *Z*, from the German word for together, *zusammen*.

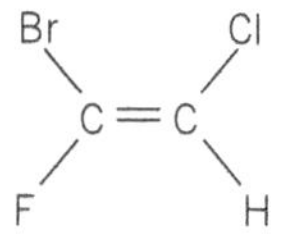

So this is *Z*-1-bromo-2-chloro-1-fluoroethene.

Study tip

Remember, when writing a systematic name, the groups are listed alphabetically.

The other isomer has the positions of the hydrogen and chlorine atoms reversed.

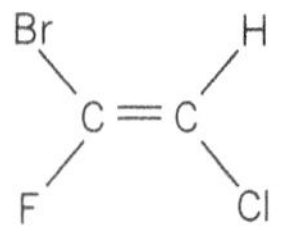

So this is *E*-1-bromo-2-chloro-1-fluoroethene. See how this fits the IUPAC naming system.

The simplest interpretation of this naming system is that if the two atoms with the greatest atomic number are on the *same* side of the double bond, the name has the prefix *Z*-. If not, it has the prefix *E*-, from the German word for opposite, *entgegen*. However, the *cis–trans* notation is still commonly used when there is no possibility of confusion.

Physical properties of alkenes

The double bond does not greatly affect properties such as boiling and melting points. van der Waals forces are the only intermolecular forces that act between the alkene molecules. This means that the physical properties of alkenes are very similar to those of the alkanes. The melting and boiling points increase with the number of carbon atoms present. Alkenes are not soluble in water.

How alkenes react

The double bond makes a big difference to the reactivity of alkenes compared with alkanes. The bond enthalpy for C—C is 347 kJ mol^{-1} and that for C═C is 612 kJ mol^{-1}, so you might predict that alkenes would be less reactive than alkanes. In fact alkenes are *more* reactive than alkanes.

The C═C forms an electron-rich area in the molecule, which can easily be attacked by positively charged reagents. These reagents are called **electrophiles** (electron liking). They are electron pair acceptors. An example of a good electrophile is the H^+ ion. As alkenes are unsaturated they can undergo addition reactions.

In conclusion, most of the reactions of alkenes are **electrophilic additions**.

Study tip

Learn the definition: an electrophile is an electron pair acceptor.

Bond energies

Remember that a C═C bond consists of a σ-bond and a π-bond.

1. Use the bond energies in the text to calculate the strength of the π part of the bond alone.
2. Explain why the π part of the bond is weaker than the σ part.

Summary questions

1. What is the name of $CH_3CH{=}CHCH_2CH_2CH_3$?
2. Draw the structural formula for hex-1-ene.
3. There are six isomeric pentenes. Draw their displayed formulae.
4. Which of these attacks the double bond in an alkene? Choose from **a**, **b**, or **c**.
   - **a** electrophiles
   - **b** nucleophiles
   - **c** alkanes
5. The double bond in an alkene can best be described by which of the following? Choose from **a**, **b**, **c**, or **d**.
   - **a** electron-rich
   - **b** electron-deficient
   - **c** positively charged
   - **d** acidic

14.2 Reactions of alkenes

Learning objectives:

- Describe electrophilic addition reactions.
- Outline the mechanism for these reactions.

Specification reference: 3.3.4

Combustion

Alkenes will burn in air:

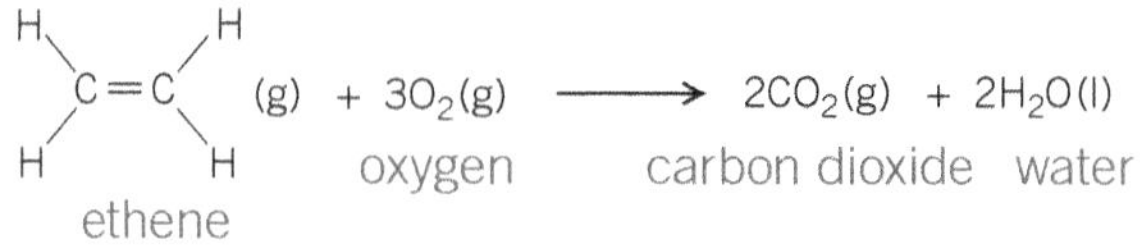

However, they are not used as fuels. This is because their reactivity makes them very useful for other purposes.

Electrophilic addition reactions

The reactions of alkenes are typically electrophilic additions. The four electrons in the carbon–carbon double bond make it a centre of high electron density. Electrophiles are attracted to it and can form a bond by using two of the four electrons in the carbon–carbon double bond (of the four electrons, the two that are in a π-bond, see Topic 14.1 Alkenes).

The mechanism is always essentially the same:

1. The electrophile is attracted to the double bond.
2. Electrophiles are positively charged and accept a pair of electrons from the double bond. The electrophile may be a positively charged ion or have a positively charged area.
3. A positive ion (a **carbocation**) is formed.
4. A negatively charged ion forms a bond with the carbocation.

See how the examples below fit this general mechanism.

Hint

Remember cations are positively charged.

Reaction with hydrogen halides

Hydrogen halides, HCl, HBr, and HI, add across the double bond to form a halogenoalkane. For example:

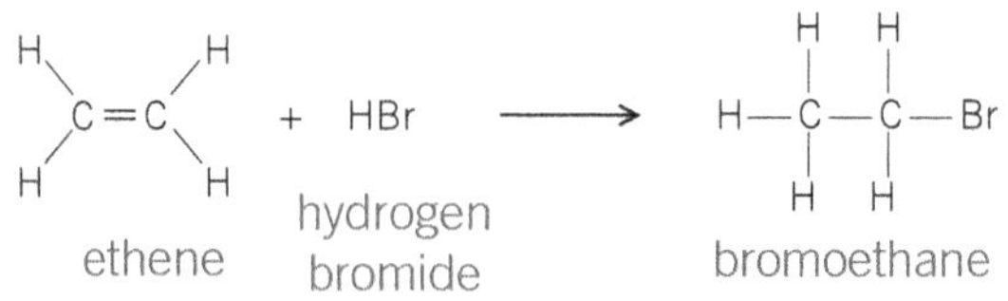

- Bromine is more electronegative than hydrogen, so the hydrogen bromide molecule is polar, $H^{\delta+}$—$Br^{\delta-}$.
- The electrophile is the $H^{\delta+}$ of the $H^{\delta+}$—$Br^{\delta-}$.
- The $H^{\delta+}$ of HBr is attracted to the C=C bond because of the double bond's high electron density.
- One of the pairs of electrons from the C=C forms a bond with the $H^{\delta+}$ to form a positive ion (called a carbocation), whilst at the same time the electrons in the $H^{\delta+}$—$Br^{\delta-}$ bond are drawn towards the $Br^{\delta-}$.

- The bond in hydrogen bromide breaks heterolytically. Both electrons from the shared pair in the bond go to the bromine atom because it is more electronegative than hydrogen leaving a Br^- ion.

  The Br^- ion attaches to the positively charged carbon of the carbocation forming a bond with one of its electron pairs.

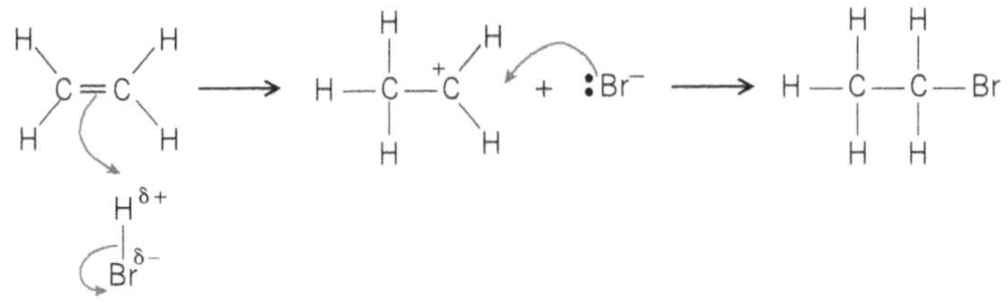

Hint

Heterolytic bond breaking means that when a covalent bond breaks, both the electrons go to one of the atoms involved in the bond and none to the other. This results in the formation of a negative ion and a positive ion.

Asymmetrical alkenes

When hydrogen bromide adds to ethene, bromoethane is the only possible product.

However, when the double bond is not exactly in the middle of the chain, there are two possible products – the bromine of the hydrogen bromide could bond to either of the carbon atoms of the double bond.

For example, propene could produce:

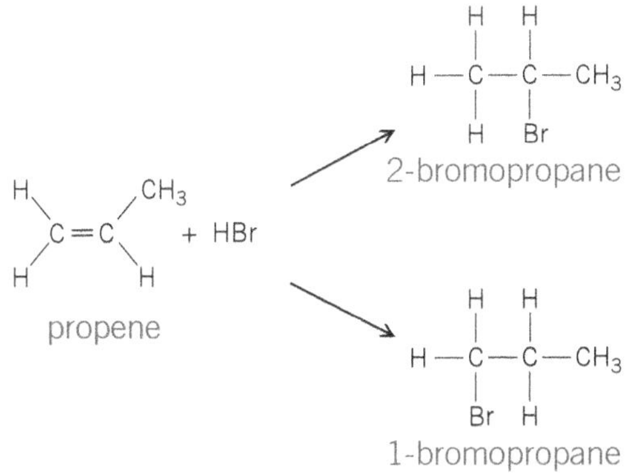

Hint

There is simple way to work out the product. When hydrogen halides add on to alkenes, the hydrogen adds on to the carbon atom which already has the most hydrogens. This is called Markovnikov's rule.

In fact the product is almost entirely 2-bromopropane.

To explain this, you need to know that alkyl groups, for example, $-CH_3$ or $-C_2H_5$, have a tendency to release electrons. This is known as a **positive inductive effect** and is sometimes represented by an arrow along their bonds to show the direction of the release.

This electron-releasing effect tends to stabilise the positive charge of the intermediate carbocation. The more alkyl groups there are attached to the positively charged carbon atom, the more stable the carbocation is. So, a positively charged carbon atom which has three alkyl groups (called a tertiary carbocation) is more stable than one with two alkyl groups (a secondary carbocation) which is more stable than one with just one (a primary carbocation), see Figure 1.

The product will tend to come from the more stable carbocation.

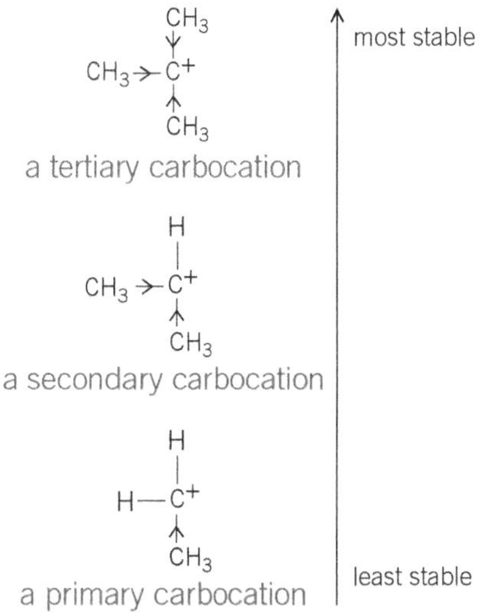

▲ **Figure 1** *Stability of primary, secondary, and tertiary carbocations*

So, the two possible carbocations when propene reacts with HBr are:

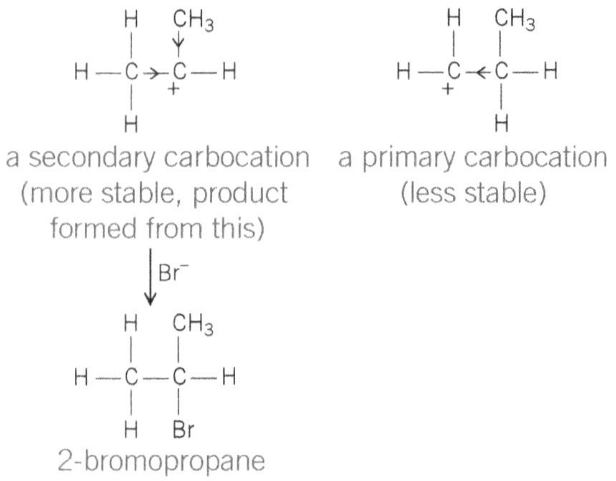

The secondary carbocation is more stable because it has two methyl groups releasing electrons towards the positive carbon. The majority of the product is formed from this.

Reaction of alkenes with halogens

Alkenes react rapidly with chlorine gas, or with solutions of bromine and iodine in an organic solvent, to give dihalogenoalkanes.

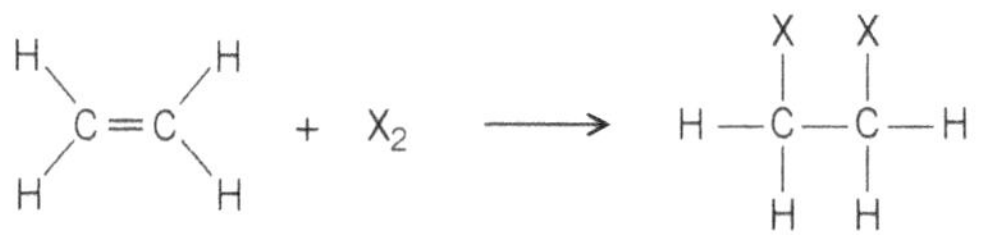

The halogen atoms *add* across the double bond.

In this case the halogen molecules act as electrophiles:

- At any instant, a bromine (or any other halogen) molecule is likely to have an instantaneous dipole, $Br^{\delta+}$—$Br^{\delta-}$. (An instant later, the dipole could be reversed $Br^{\delta-}$—$Br^{\delta+}$.) The δ+ end of this dipole is attracted to the electron-rich double bond in the alkene – the bromine molecule has become an electrophile.
- The electrons in the double bond are attracted to the $Br^{\delta+}$. They repel the electrons in the Br—Br bond and this strengthens the dipole of the bromine molecule.

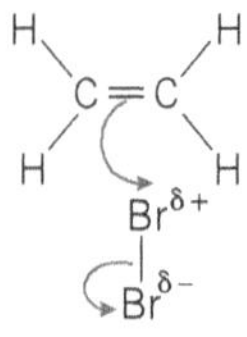

- Two of the electrons from the double bond form a bond with the $Br^{\delta+}$ and the other bromine atom becomes a Br^- ion. This leaves a carbocation, in which the carbon atom that is not bonded to the bromine has the positive charge.
- The Br^- ion now forms a bond with the carbocation.

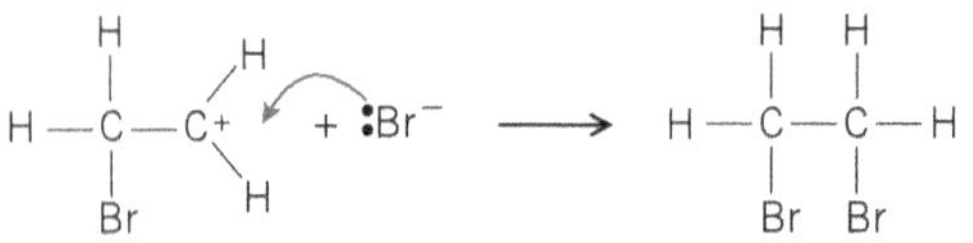

Hint

This dipole is also induced when a bromine molecule collides with the electron-rich double bond.

Hint

The carbocation will react with any nucleophile that is present. In aqueous solution, such as bromine water, water reacts with the carbocation, forming some CH_2BrCH_2OH, 2-bromoethanol.

So the addition takes place in two steps:

1 formation of the carbocation by electrophilic addition
2 rapid reaction with a negative ion.

The test for a double bond

This addition reaction is used to test for a carbon–carbon double bond. When a few drops of bromine solution, sometimes called bromine water (which is reddish-brown), are added to an alkene the solution is decolourised because the products are colourless.

Reaction with concentrated sulfuric acid

Concentrated sulfuric acid also adds across the double bond. The reaction occurs at room temperature and is exothermic.

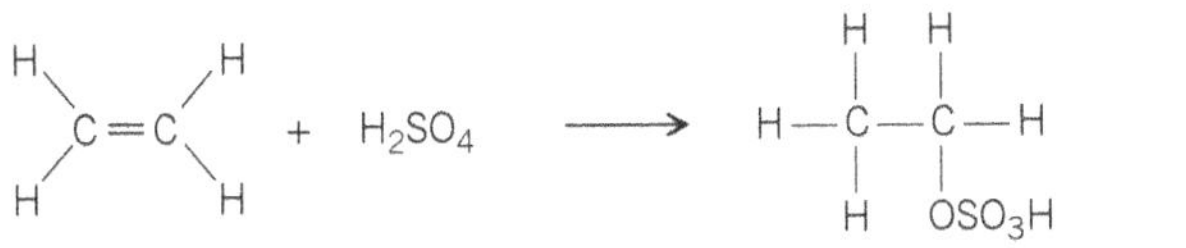

ethene ethyl hydrogensulfate

The electrophile is a partially positively charged hydrogen atom in the sulfuric acid molecule. This can be shown as $H^{\delta+}—O^{\delta-}—SO_3H$

The carbocation which forms then reacts rapidly with the negatively charged hydrogensulfate ion.

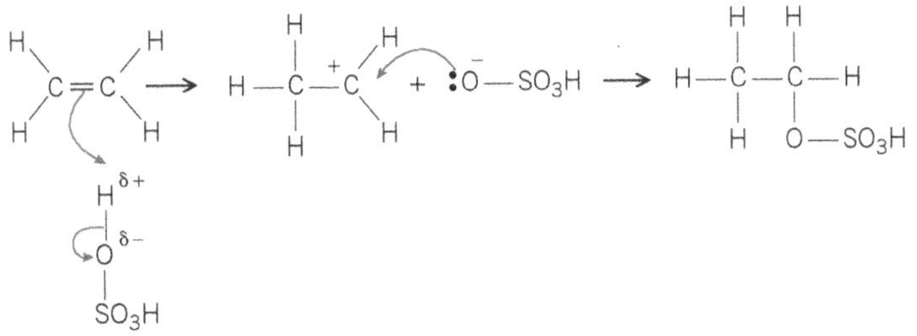

When water is added to the product an alcohol is formed and sulfuric acid reforms.

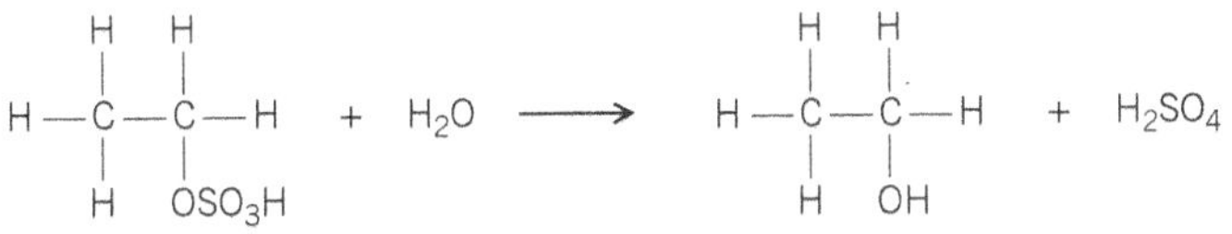

ethyl hydrogensulfate ethanol

The overall effect is to add water H—OH across the double bond, and the sulfuric acid is a catalyst for the process.

Asymmetrical alkenes

In an asymmetrical alkene, such as propene, the carbocation is exactly the same as that found in the reaction with hydrogen bromide. This means that you can predict the products by looking at the relative stability of the possible carbocations that could form (or by using Markovnikov's rule).

Reaction with water

Water also adds on across the double bond in alkenes. The reaction is used industrially to make alcohols and is carried out with steam, at a suitable temperature and pressure, using an acid catalyst such as phosphoric acid, H_3PO_4.

$$CH_2{=}CH_2(g) + H_2O(g) \rightarrow CH_3CH_2OH(g)$$

Hint

The structure of sulfuric acid is:

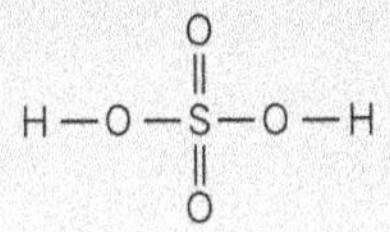

Summary questions

1 Write the equation for the complete combustion of propene.

2 Which of the following are typical reactions of alkenes?
 a Electrophilic additions
 b Electrophilic substitutions
 c Nucleophilic substitutions

3 a What are the two possible products of the reaction between propene and hydrogen bromide?
 b Which is the main product?
 c Explain why this product is more likely.

4 What is the product of the reaction between ethene and hydrogen chloride?

5 Which of the following is the test for a carbon–carbon double bond?
 a Forms a white precipitate with silver nitrate
 b Turns limewater milky
 c Decolourises bromine solution

Synoptic link

The manufacture of ethanol is discussed in Topic 15.2, The reactions of alcohols.

14.3 Addition polymers

Learning objectives:

→ Describe an addition polymer

→ Explain what sort of molecules react to form addition polymers

Specification reference: 3.3.4

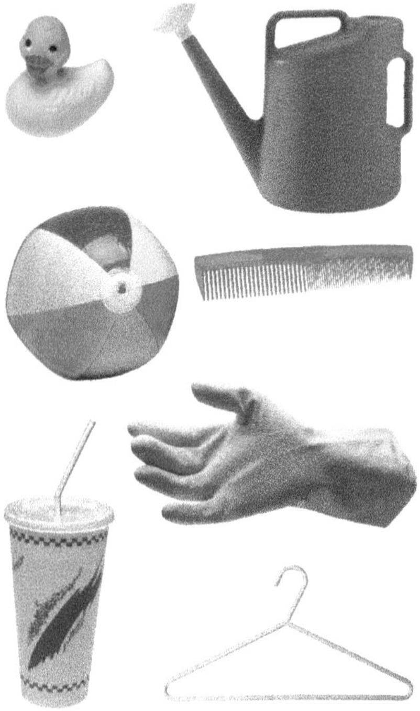

▲ **Figure 1** *Polymers around us*

Polymers are very large molecules that are built up from small molecules, called **monomers**. They occur naturally everywhere: starch, proteins, cellulose, and DNA are all polymers. The first completely synthetic polymer was Bakelite, which was patented in 1907. Since then, many synthetic polymers have been developed with a range of properties to suit them for very many applications, see Figure 1.

One way of classifying polymers is by the type of reaction by which they are made.

Addition polymers are made from a monomer or monomers with a carbon–carbon double bond (alkenes). Addition polymers are made from monomers based on ethene. The monomer has the general formula:

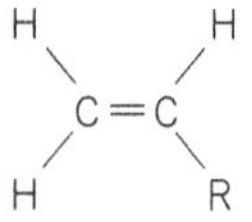

When the monomers polymerise, the double bond opens and the monomers bond together to form a backbone of carbon atoms as shown:

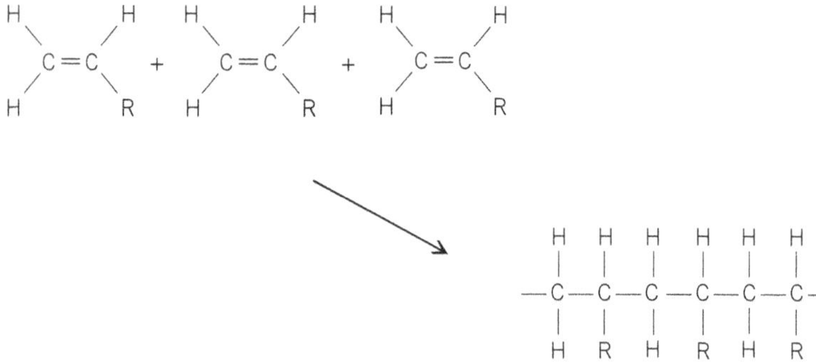

This may also be represented by equations such as:

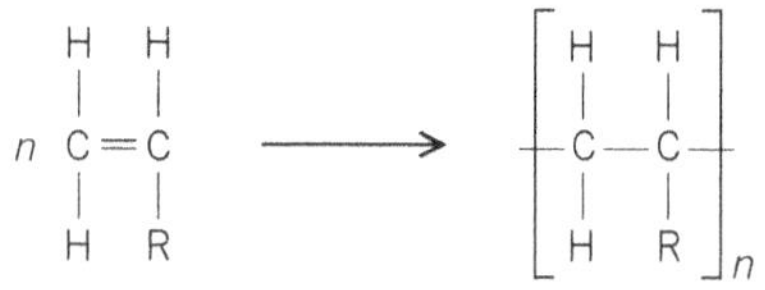

R may be an alkyl or an aryl group.

For example, ethene polymerises to form poly(ethene)

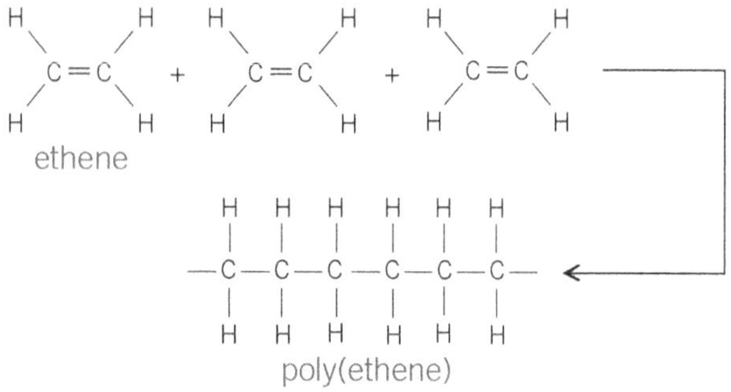

Study tip

Addition polymers are named systematically with the prefix 'poly' followed by the name of the monomer in brackets. They are often called poly(alkenes). Non-systematic names are often used for everyday and commercial purposes.

and phenylethene polymerises to poly(phenylethene):

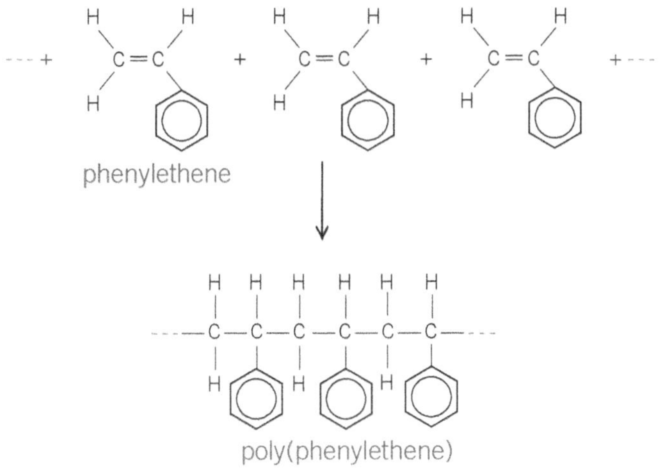

▲ **Figure 2** *Both the model and the packaging are made from polystyrene*

Phenylethene is sometimes called styrene, which is why poly(phenylethene) is usually called polystyrene.

Table 1 gives some examples of addition polymers based on different substituents.

▼ **Table 1** *Some addition polymers made from the monomer $H_2C{=}CHR$*

R	Monomer	Polymer	Name of polymer	Common or trade name	Typical uses
$-H$	$CH_2=CH_2$	$-[CH_2-CH_2]_n-$	poly(ethene)	polythene	carrier bags, washing up bowls
$-CH_3$	CH_3 $CH=CH_2$	CH_3 $-[CH-CH_2]_n-$	poly(propene)	polypropylene	yoghurt containers car bumpers
$-Cl$	Cl $CH=CH_2$	Cl $-[CH-CH_2]_n-$	poly (chloroethene)	PVC (polyvinyl chloride)	aprons, 'vinyl' records, drainpipes
$-C{\equiv}N$	CN $CH=CH_2$	CN $-[CH-CH_2]_n-$	poly (propenenitrile)	acrylic (Acrilan, Courtelle)	clothing fabrics
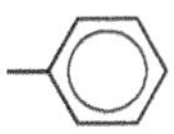	 $CH=CH_2$	 $-[CH-CH_2]_n-$	poly (phenylethene)	polystyrene	packing materials, electrical insulation

Identifying the addition polymer formed from the monomer

The best way to think about this is to remember that an addition polymer is formed from monomers with carbon–carbon double bonds.

There is usually only one monomer (though it is possible to have more), and the double bond opens to form a single bond, see Table 1. This will give the repeat unit for the polymer.

Hint

The repeating unit of an addition polymer is the smallest group of atoms that produce the polymer when repeated over and over. It normally corresponds to the monomer, shown in brackets in Table 1. In poly(ethene) the repeating unit is just $-CH_2-$ but it is more usual to quote the repeating unit $-CH_2-CH_2-$ based on the monomer.

▲ **Figure 3** *Bottles made from HD and LD polythene*

Study tip

It is important that you understand why polyalkenes are chemically inert.

Study tip

Poly(alkane) chain molecules attract each other by van der Waals forces. Although these are individually weak, there are thousands of them which add up to produce strong materials.

Identifying the monomer(s) used to make an addition polymer

An addition polymer must have a backbone of carbon atoms and the monomer must contain at least two carbons, so that there can be a carbon–carbon double bond. So, in the molecule below the monomer is shown in the red brackets:

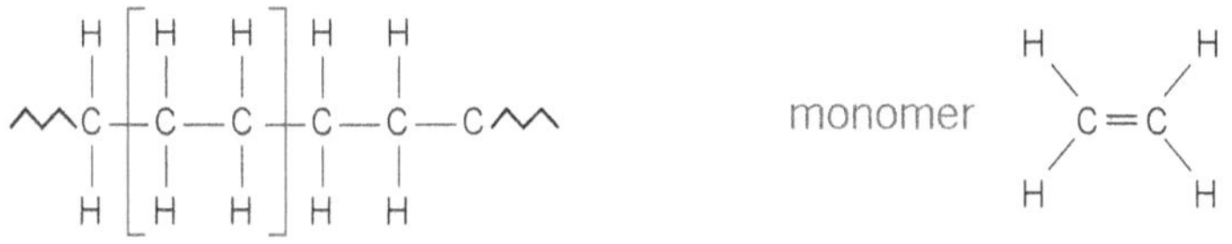

Where some of the carbon atoms have substituents, the monomer must have the substituent, as well as a double bond:

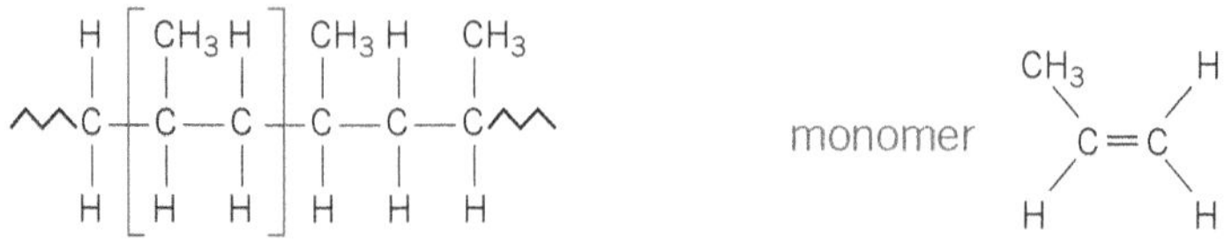

Modifying the plastics

The properties of polymers materials can be considerably modified by the use of additives such as plasticisers. These are small molecules that get between the polymer chains forcing them apart and allowing them to slide across each other. For example, PVC is rigid enough for use as drainpipes, but with the addition of a plasticiser it becomes flexible enough for making aprons.

Biodegradability

Polyalk*enes*, in spite of their name, have a backbone which is a long chain saturated alk*ane* molecule. Alkanes have strong non-polar C—C and C—H bonds. So, they are very unreactive molecules, which is a useful property in many ways. However, this does mean that they are not attacked by biological agents – like enzymes – and so they are not biodegradable. This is an increasing problem in today's world, where waste disposal is becoming more and more difficult.

High and low density polythene

Low density poly(ethene) (polythene) is made by polymerising ethene at high pressure and high temperature via a free-radical mechanism. This produces a polymer with a certain amount of chain branching. This is a consequence of the rather random nature of free-radical reactions. The branched chains do not pack together particularly well and the product is quite flexible, stretches well and has a fairly low density. These properties make it suitable for packaging (plastic bags), sheeting and insulation for electrical cables.

High density polythene is made at temperatures and pressures little greater than room conditions and uses a Ziegler–Natta catalyst, named after its developers. This results in a polymer with much less chain branching (around one branch for every 200 carbons on the main chain). The chains can pack together well. This makes the density of the plastic greater and its melting temperature higher. Typical uses are milk crates, buckets, and bottles for which low density polythene would be insufficiently rigid.

The solutions to pollution by plastics

To reduce the amount of plastic it can be reused or recycled.

Mechanical recycling

The simplest form of recycling is called mechanical recycling. The first step is to separate the different types of plastics. (Plastic containers are now collected in recycling facilities for this purpose.) The plastics are then washed and once they are sorted they may be ground up into small pellets. These can be melted and remoulded. For example, recycled soft drinks bottles made from PET (polyethylene terephthalate) are used to make fleece clothes.

Feedstock recycling

Here, the plastics are heated to a temperature that will break the polymer bonds and produce monomers. These can then be used to make new plastics.

There are problems with recycling. Poly(propene), for example, is a thermoplastic polymer. This means that it will soften when heated so it can be melted and re-used. However, this can only be done a limited number of times because at each heating some of the chains break and become shorter thus degrading the plastic's properties.

Properties of polymers

Since the first major commercial syntheses of plastics like polyethylene and polyvinyl chloride (PVC) in the 1930s, over time, their properties have been tailored to meet demand. The addition of plasticisers makes polymers more flexible because the polymer chains can slide over each other. Adding branches to the chains will also make them more flexible as the branches weaken the intermolecular forces between the chains. Cross linking will produce a more rigid plastic.

Summary questions

1 Which of the following monomers could form an addition polymer?

A
$$\begin{array}{c} H \quad\quad H \\ \diagdown \quad\quad \diagup \\ C = C \\ \diagup \quad\quad \diagdown \\ H \quad\quad H \end{array}$$

B
$$\begin{array}{c} F \quad\quad F \\ \diagdown \quad\quad \diagup \\ C = C \\ \diagup \quad\quad \diagdown \\ F \quad\quad F \end{array}$$

C
$$\begin{array}{c} NH_2 \\ | \\ CH_3 - C - COOH \\ | \\ H \end{array}$$

D
$$\begin{array}{c} H \quad\quad H \\ \diagdown \quad\quad \diagup \\ C = C \\ \diagup \quad\quad \diagdown \\ H \quad\quad CH_3 \end{array}$$

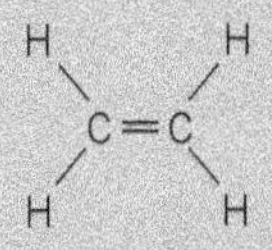

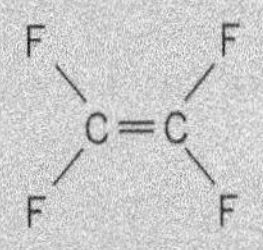

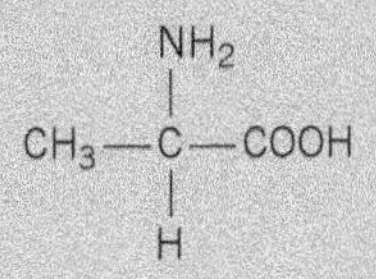

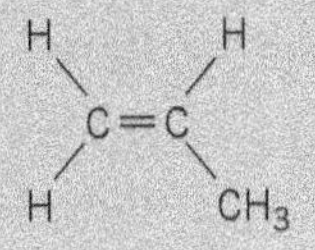

2 a Draw a section of the polymer formed from the monomer

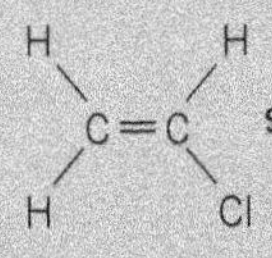

$$\begin{array}{c} H \quad\quad H \\ \diagdown \quad\quad \diagup \\ C = C \\ \diagup \quad\quad \diagdown \\ H \quad\quad Cl \end{array}$$ showing six carbon atoms.

b What is the common name of the monomer?

c What is the systematic name of the polymer?

3
$$\begin{array}{c} F \;\; F \;\; F \;\; F \;\; F \;\; F \\ | \;\; | \;\; | \;\; | \;\; | \;\; | \\ -C-C-C-C-C-C- \\ | \;\; | \;\; | \;\; | \;\; | \;\; | \\ F \;\; F \;\; F \;\; F \;\; F \;\; F \end{array}$$

This is a section of the polymer that non-stick pans are coated with. It's trade name Teflon.

What is the monomer?

4
$$\begin{array}{c} Cl \;\; H \;\; Cl \;\; H \;\; Cl \;\; H \\ | \;\; | \;\; | \;\; | \;\; | \;\; | \\ -C-C-C-C-C-C- \\ | \;\; | \;\; | \;\; | \;\; | \;\; | \\ H \;\; H \;\; H \;\; H \;\; H \;\; H \end{array}$$

This is a section of the polymer that drainpipes are made from, trade name polyvinylchloride (PVC)

What is the monomer?

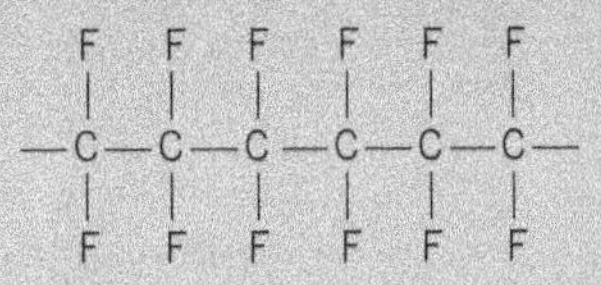

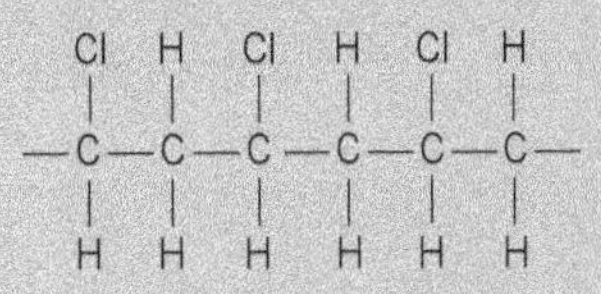

14.4 Epoxyethane

Learning objectives

- State the formula of epoxyethane.
- State how epoxyethane is manufactured.
- Explain why epoxyethane is highly reactive.
- Explain the hazards associated with epoxyethane.
- Write equations for the reactions of epoxyethane with water and with alcohols.
- Explain the economic importance of epoxyethane.

Specification reference 3.3.4

Epoxyethane is an important chemical produced industrially to the tune of 17 million tonnes each year worldwide. It is not produced for sale itself; rather it is an intermediate chemical from which many useful products can be made. Epoxyethane is made by partially oxidising ethane – a mixture of ethene and oxygen is passed over a catalyst of finely-divided silver:

$$CH_2{=}CH_2 + \frac{1}{2}O_2 \rightarrow CH_2OCH_2$$

A second reaction is possible

$$CH_2{=}CH_2 + 3O_2 \rightarrow 2CO_2 + 2H_2O$$

But with a suitable catalyst and control of conditions, 90% of the first reaction can be obtained.

The formula of epoxyethane is

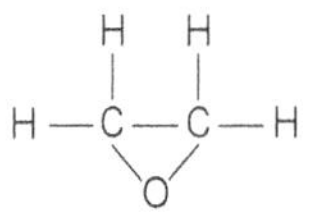

Displayed

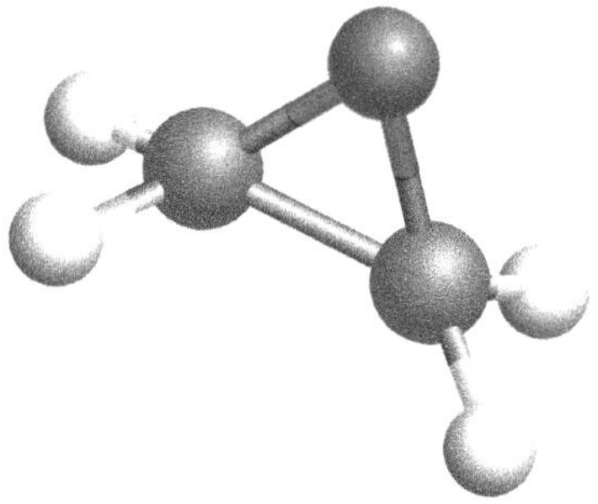

3D

The three-membered ring is important in the chemistry of epoxyethane. All the bond angles in the ring are approximately 60°, while the normal angle for C—C—O and C—O—C bonds is 109.5°. This means that the ring tends to spring open (we say the molecule has 'ring strain') and this makes epoxyethane highly reactive and potentially hazardous. Because of this, epoxyethane is not normally stored – it is converted to other useful products on the site where it is made.

Reactions of epoxyethane

Epoxyethane reacts with water to form ethane-1,2-diol which is used as antifreeze in cars and other vehicles.

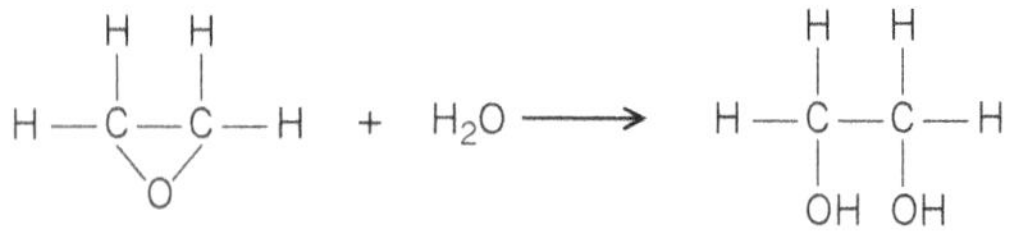

The mechanism of the reaction involves protonation of the oxygen followed by a nucleophilic attack of an OH^- ion.

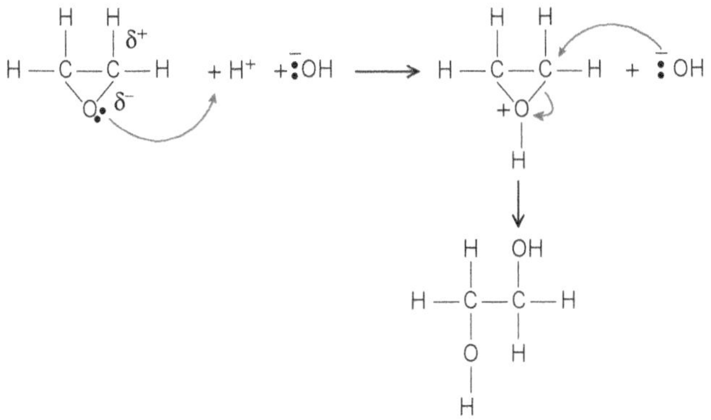

Further reaction can produce a range of polymers of different chain lengths which have a range of applications.

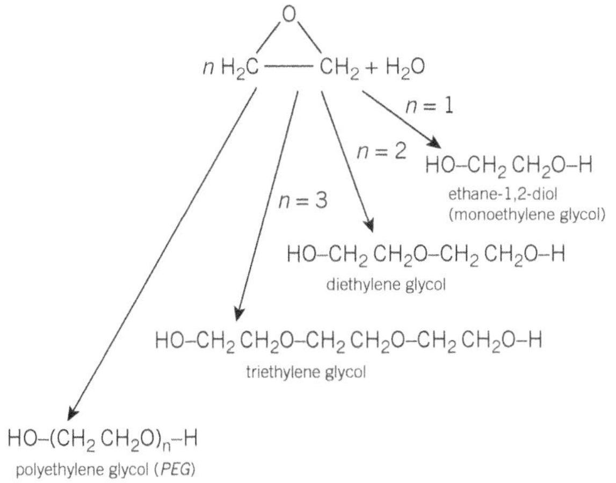

These include

n =1: production of polyesters such as PET, the plastic used for soft drinks bottles

n = 2: used to make polyurethane plastics

n = 3: used as a plasticiser to make plastics such as PVC more flexible

n = 4+: (polyethylene glycols (PEGs)) used in detergents, cosmetics, and paints

Epoxyethane will also react with alcohols, RO—H to form compounds such as RO—CH_2—CH_2—O—H, etc.

Similar reactions with ammonia produce a range of compounds called **ethanolamines**.

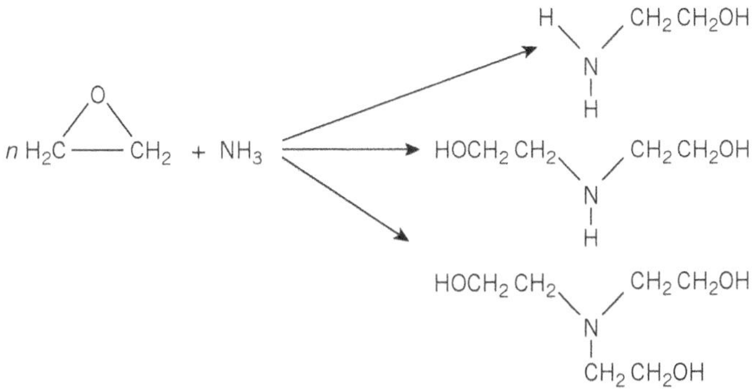

These, too have a range of uses in cosmetics, soaps, and detergents.

Summary questions

1 What is the atom economy of the reaction of epoxyethane plus water to give ethane-1, 2-diol? What type of reaction is this that produces this atom economy?

2 Epoxyethane can also react with alcohols in a similar way to its reaction with water. Draw the displayed formula of the product of the reaction of 1 molecule of epoxyethane with one molecule of methanol, CH_3OH.

3 What features of the ammonia and water molecules make them into nucleophiles?

4 Suggest one way of ensuring that the reaction of ethene with oxygen produces epoxyethane rather than carbon dioxide and water.

Practice questions

1 The table below gives the names and structures of three isomeric alkenes.

Name	Structure
but-1-ene	$CH_3CH_2CH{=}CH_2$
but-2-ene	$CH_3CH{=}CHCH_3$
methylpropene	$H_3C-\underset{\underset{CH_3}{\vert}}{C}{=}CH_2$

(a) Give the molecular formula and the empirical formula of but-2-ene.

(2 marks)

(b) Methylpropene reacts with hydrogen bromide to produce 2-bromo-2-methylpropane as the major product.
 (i) Name and outline the mechanism for this reaction.
 (ii) Draw the structure of another product of this reaction and explain why it is formed in smaller amounts.

(8 marks)

(c) Draw the structures and give the names of the two stereoisomers of but-2-ene.

(2 marks)

AQA, 2004

2 Copy and complete the mechanism below by drawing appropriate curly arrows.

$H\overset{..}{O}{:}^-$

$H_3C-\underset{\underset{H}{\vert}}{\overset{\overset{H}{\vert}}{C}}-\underset{\underset{H}{\vert}}{\overset{\overset{H}{\vert}}{C}}-\underset{\underset{Br}{\vert}}{\overset{\overset{H}{\vert}}{C}}-CH_3 \rightarrow CH_3CH_2CH{=}CHCH_3 + H_2O + Br^-$

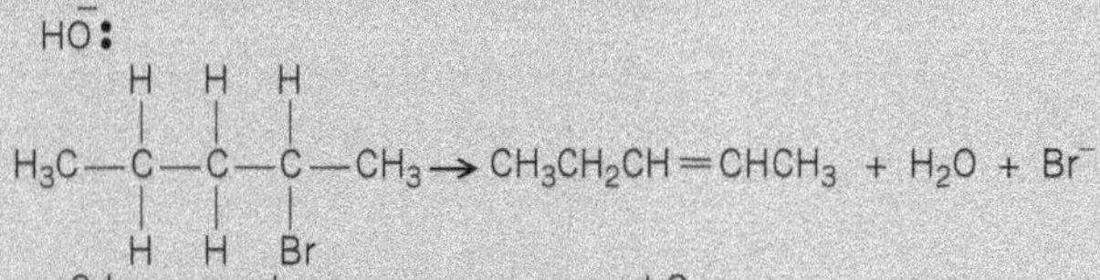

2-bromopentane pent-2-ene

(3 marks)

AQA, 2005

3 (a) Propene reacts with hydrogen bromide by an electrophilic addition mechanism forming 2-bromopropane as the major product.

The equation for this reaction is shown below.

$\underset{H}{\overset{H_3C}{>}}C{=}C\underset{H}{\overset{H}{<}} + HBr \longrightarrow H_3C-\underset{\underset{H}{\vert}}{\overset{\overset{Br}{\vert}}{C}}-\underset{\underset{H}{\vert}}{\overset{\overset{H}{\vert}}{C}}-H$

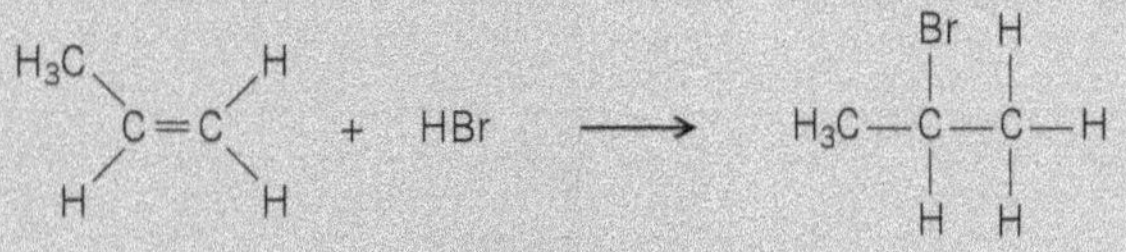

 (i) Outline the mechanism for this reaction, showing the structure of the intermediate carbocation formed.
 (ii) Give the structure of the alternative carbocation which could be formed in the reaction between propene and hydrogen bromide.

(5 marks)

AQA, 2003

4 The reaction scheme below shows the conversion of compound **A**, 2-methylbut-1-ene, into compound **B** and then into compound **C**.

$\underset{A}{CH_2{=}\underset{\underset{CH_3}{\vert}}{C}-CH_2CH_3} \xrightarrow[\underset{H_2SO_4}{concentrated}]{Step\ 1} \underset{B}{CH_3-\underset{\underset{OSO_2OH}{\vert}}{\overset{\overset{CH_3}{\vert}}{C}}-CH_2CH_3}$

\downarrow Step 2

$\underset{C}{CH_3-\underset{\underset{OH}{\vert}}{\overset{\overset{CH_3}{\vert}}{C}}-CH_2CH_3}$

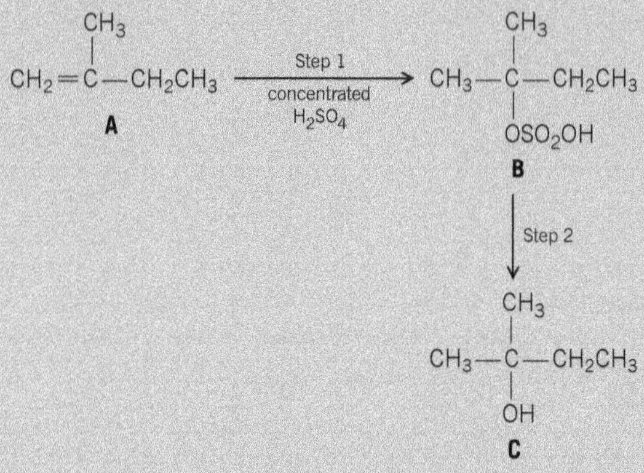

(a) The structure of **A** is shown below. Circle those carbon atoms which must lie in the same plane.

(1 mark)

$\underset{H}{\overset{H}{>}}C{=}C\underset{CH_2-CH_3}{\overset{CH_3}{<}}$

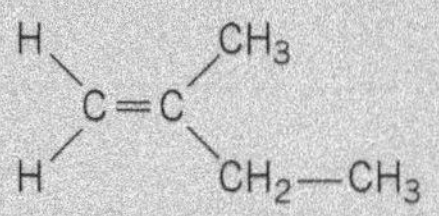

(b) Outline a mechanism for the reaction in Step 1.

(4 marks)

AQA, 2002

5 It is possible to convert but-1-ene into its structural isomer but-2-ene.
(a) State the type of structural isomerism shown by but-1-ene and but-2-ene.

(1 mark)

(b) The first stage in this conversion involves the reaction of hydrogen bromide with but-1-ene.

$CH_3CH_2CH{=}CH_2 + HBr \rightarrow CH_3CH_2CHBrCH_3$

Outline a mechanism for this reaction.

(4 marks)

(c) The second stage is to convert 2-bromobutane into but-2-ene.

$CH_3CH_2CHBrCH_3 + KOH \rightarrow CH_3CH{=}CHCH_3 + KBr + H_2O$

Outline a mechanism for this reaction.

(3 marks)

AQA, 2012

6 Alkenes are useful intermediates in the synthesis of organic compounds.

(a) (i) Complete the elimination mechanism by drawing appropriate curly arrows.

HO:

$$CH_2=\overset{\overset{\displaystyle H}{|}}{C}-\overset{\overset{\displaystyle H}{|}}{\underset{\underset{\displaystyle H}{|}}{C}}-\overset{\overset{\displaystyle H}{|}}{\underset{\underset{\displaystyle Br}{|}}{C}}-CH_2CH_3$$

3-bromohexane

↓

$$CH_3CH_2CH=CHCH_2CH_3 + H_2O + Br^-$$

hex-3-ene

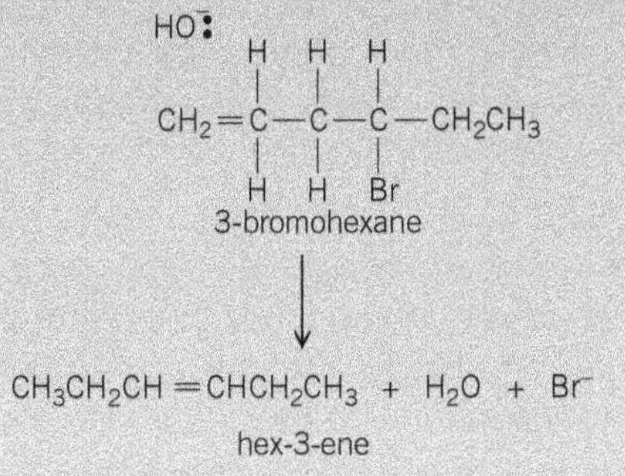

(3 marks)

(ii) Draw structures for the *E* and *Z* stereoisomers of hex-3-ene.

E isomer of hex-3-ene

Z isomer of hex-3-ene

(2 marks)

(iii) State the meaning of the term *stereoisomers*.

(2 marks)

(b) The equation for the first reaction in the conversion of hex-3-ene into hexan-3-ol is shown below.

$$CH_3CH_2CH{=}CHCH_2CH_3 + H_2SO_4 \rightarrow$$
$$CH_3CH_2CH_2CH(OSO_2OH)CH_2CH_3$$

Outline a mechanism for this reaction.

(4 marks)

AQA, 2012

7 Propene reacts with bromine by a mechanism known as electrophilic addition.

(a) Explain what is meant by the term *electrophile* and by the term *addition*.

(2 marks)

(b) Outline the mechanism for the electrophilic addition of bromine to propene. Give the name of the product formed.

(5 marks)

AQA, 2002

8 (a) (i) Name the alkene $CH_3CH_2CH{=}CH_2$

(ii) Explain why $CH_3CH_2CH{=}CH_2$ does **not** show stereoisomerism.

(iii) Draw an isomer of $CH_3CH_2CH{=}CH_2$ which does show *E–Z* isomerism.

(iv) Draw another isomer of $CH_3CH_2CH{=}CH_2$ which does **not** show *E–Z* isomerism.

(4 marks)

(b) (i) Name the type of mechanism for the reaction shown by alkenes with concentrated sulfuric acid.

(ii) Write a mechanism showing the formation of the major product in the reaction of concentrated sulfuric acid with $CH_3CH_2CH{=}CH_2$.

(iii) Explain why this compound is the major product.

(4 marks)

9 The alkene (*Z*)-3-methylpent-2-ene reacts with hydrogen bromide as shown below.

$$\underset{H}{\overset{CH_3}{>}}C{=}C\underset{CH_3}{\overset{CH_2CH_3}{<}}$$

HBr →

$$CH_3CH_2-\overset{\overset{\displaystyle Br}{|}}{\underset{\underset{\displaystyle CH_3}{|}}{C}}-CH_2CH_3$$

major product, **P**

HBr →

minor product, **Q**

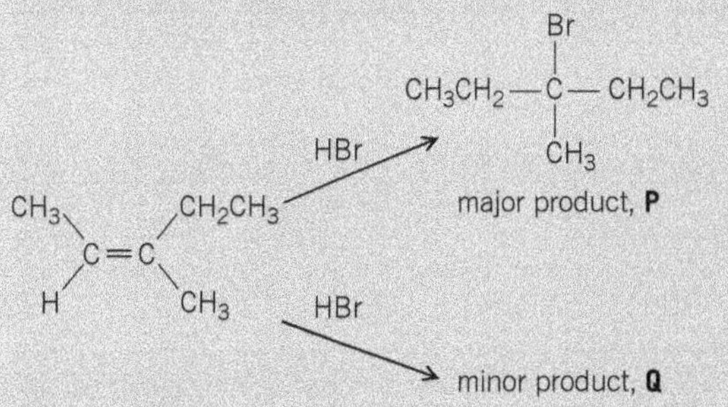

(a) (i) Name the major product **P**.

(1 mark)

(ii) Name the mechanism for these reactions.

(1 mark)

(iii) Draw the displayed formula for the minor product **Q** and state the type of structural isomerism shown by **P** and **Q**.

(2 marks)

(iv) Draw the structure of the (*E*)-stereoisomer of 3-methylpent-2-ene.

(1 mark)

AQA, 2010

10 An organic compound A is shown below.

$$\underset{CH_3CH_2}{\overset{CH_3CH_2CH_2}{>}}C{=}C\underset{H}{\overset{CH_2CH_2OH}{<}}$$

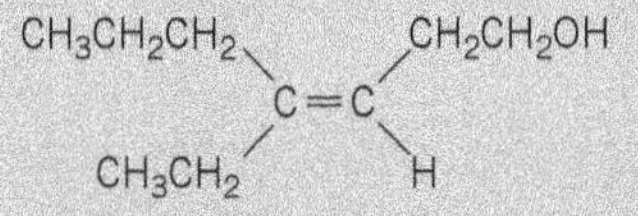

Explain how the Cahn–Ingold–Prelog (CIP) priority rules can be used to deduce the full IUPAC name of this compound.

(6 marks)

Answers to the Practice Questions and Section Questions are available at
www.oxfordsecondary.com/oxfordaqaexams-alevel-chemistry

15 Alcohols

15.1 Alcohols – introduction

Learning objectives:

- → State the general formula of an alcohol.
- → Describe how alcohols are classified.
- → Describe the physical properties of alcohols.

Specification reference: 3.3.5

Alcohols have many scientific, medicinal, and industrial uses. The most important alcohol is ethanol which is used as a solvent, as an intermediate in the manufacture of many other organic compounds and increasingly as an additive to motor fuels. It is present in many cosmetic products and is also the alcohol in alcoholic drinks.

The general formula

Alcohols have the functional group –OH attached to a hydrocarbon chain. They are relatively reactive. The alcohol most commonly encountered in everyday life is ethanol.

The general formula of an alcohol is $C_nH_{2n+1}OH$. This is often shortened to ROH.

How to name alcohols

The name of the functional group (the –OH group) is normally given by the suffix -ol. (The prefix hydroxy- is used if some other functional groups are present.)

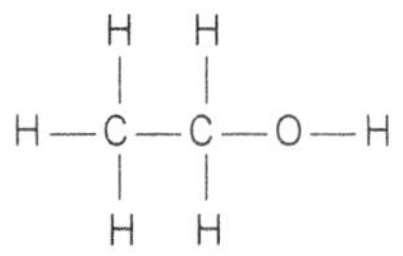

ethanol

With chains longer than ethanol, you need a number to show where the –OH group is.

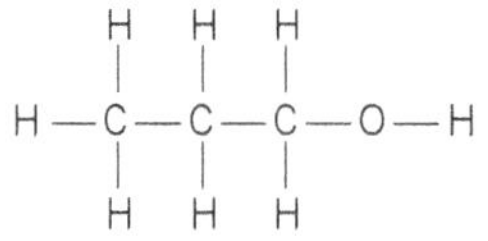

propan-1-ol

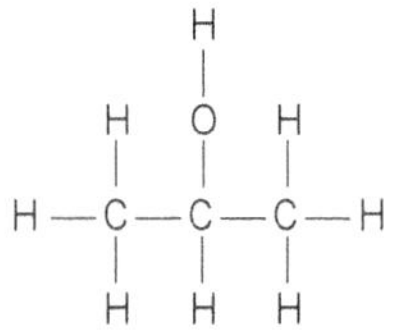

propan-2-ol

If there is more than one –OH group, di-, tri-, tetra-, and so on are used to say *how many* –OH groups there are and numbers to say *where* they are.

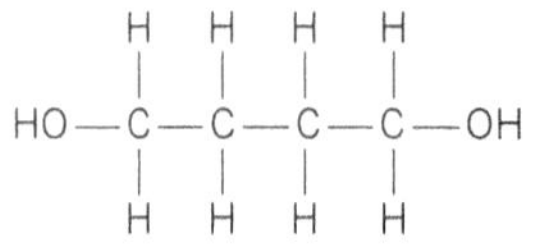

butane-1,4-diol

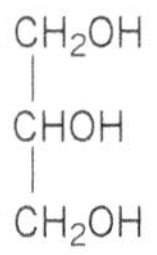

propane-1,2,3-triol

Propane-1,2,3-triol is also known as glycerol, which may be obtained from the fats and oils found in living organisms.

Antifreeeze

Ethane-1,2-diol is the main ingredient in most antifreezes. These are added to the water in the cooling systems of motor vehicles so that the resulting coolant mixture does not freeze at winter temperatures.

1 Draw the structural formula of ethane-1,2-diol and indicate where hydrogen bonding with water may take place.
2 Suggest why solutions of ethane-1,2-diol have lower freezing temperatures than pure water.

▲ **Figure 1** *Ethanol is used as a fuel in camping stoves*

The reactivity of alcohols

How an organic molecule reacts depends, among other things, on the strength and polarity of the bonds within it.

1 Draw a displayed formula for ethanol and mark on it the bond energy (mean bond enthalpy) of each bond (the amount of energy required to break it). Use a data book or database or the table of bond enthalpies in Topic 4.7 and the values C–O = 336 kJ mol^{-1} and O–H = 464 kJ mol^{-1}. There is a table of electronegativities in Topic 13.1, Halogenoalkanes – introduction, and the values for hydrogen and oxygen are 2.1 and 3.5 respectively.
2 Use your diagram and the electronegativity values to explain why:
   a the hydrocarbon skeleton remains intact in most reactions of ethanol
   b why the typical reactions of ethanol are nucleophilic substitutions in which the –OH group is lost.

Shape

In alcohols, the oxygen atom has two bonding pairs of electrons and two lone pairs. The C—O—H angle is about 105° because the 109.5° angle of a perfect tetrahedron is 'squeezed down' by the presence of the lone pairs. These two lone pairs will repel each other more than the pairs of electrons in a covalent bond.

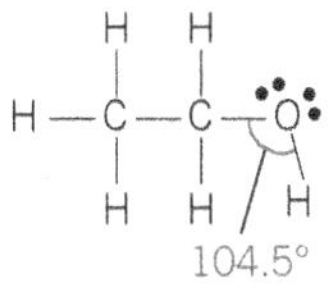

Synoptic link

You learnt about the repulsion of lone pairs of electrons compared to the repulsion of bonded pairs of electrons in Topic 3.5, The shapes of molecules and ions.

Classification of alcohols

Alcohols are classified as **primary** (1°), **secondary** (2°), or **tertiary** (3°) according to how many other groups (R) are bonded to the carbon that has the –OH group.

Hint

Halogenoalkanes may be classified as primary, secondary, or tertiary in the same way.

Primary alcohols

In a primary alcohol, the carbon with the –OH group has one R group (and therefore two hydrogen atoms).

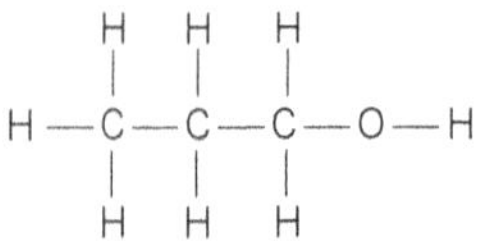

propan-1-ol is a primary alcohol

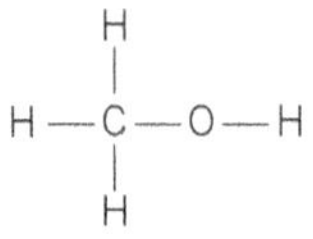

methanol, where the carbon has no R groups is counted as a primary alcohol

A primary alcohol has the –OH group at the end of a chain.

Secondary alcohols

In a secondary alcohol, the –OH group is attached to a carbon with two R groups (and therefore one hydrogen atom).

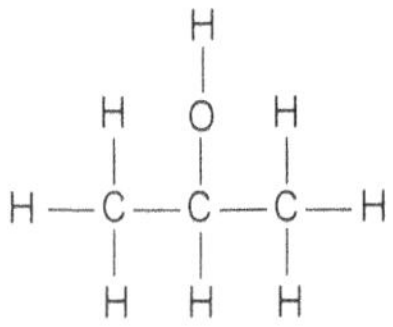

propan-2-ol is a secondary alcohol

A secondary alcohol has the –OH group in the body of the chain.

Tertiary alcohols

Tertiary alcohols have three R groups attached to the carbon that is bonded to the –OH (so this carbon has no hydrogen atoms).

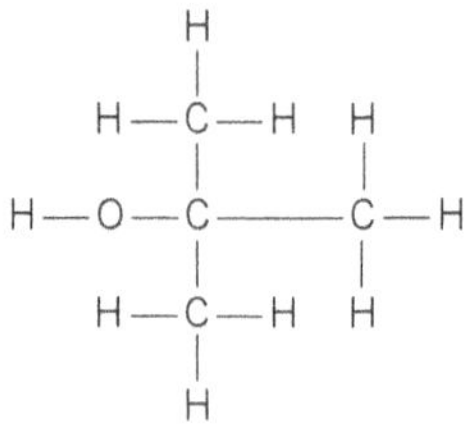

2-methylpropan-2-ol is a tertiary alcohol

A tertiary alcohol has the –OH group at a branch in the chain.

Physical properties

The –OH group in alcohols means that hydrogen bonding occurs between the molecules. This is the reason that alcohols have higher melting and boiling points than alkanes of similar relative molecular mass.

The –OH group of alcohols can hydrogen bond to water molecules, but the non-polar hydrocarbon chain cannot. This means that the alcohols with short hydrocarbon chains are soluble in water because the hydrogen bonding predominates. In longer-chain alcohols the non-polar hydrocarbon chain dominates and the alcohols become insoluble in water.

Summary questions

1. Draw the displayed formula and name the alcohol $C_2H_5CH(OH)CH_3$.
2. Sort these alcohols into primary, secondary, and tertiary:
   butan-2-ol
   2-methylpentan-2-ol
   methanol
3. Why is the C—O—H angle in alcohols less than 109.5°?

15.2 The reactions of alcohols

Learning objectives:

- → State the products when primary, secondary, and tertiary alcohols are oxidised.
- → Explain how the oxidation of a primary alcohol is controlled.
- → State what is meant by aldehydes and ketones.
- → Describe how a mild oxidising agent can be used to distinguish between an aldehyde and a ketone.
- → Describe what elimination reactions are.
- → Describe how alcohols are dehydrated to form alkenes.

Specification reference: 3.3.5

Combustion

Alcohols burn completely to carbon dioxide and water if there is enough oxygen available. (Otherwise there is incomplete combustion and carbon monoxide or even carbon is produced.) This is the equation for the complete combustion of ethanol:

$$C_2H_5OH(l) + 3O_2(g) \rightarrow 2CO_2(g) + 3H_2O(l)$$

Ethanol is often used as a fuel, for example, in picnic stoves that burn methylated spirits. Methylated spirits is ethanol with a small percentage of poisonous methanol added to make it unfit to drink. A purple dye is also added to show that it should not be drunk.

Elimination reactions

Elimination reactions are ones in which a small molecule leaves the parent molecule. In the case of alcohols, this molecule is always water. The water is made from the –OH group and a hydrogen atom from the carbon next to the –OH group. So, the elimination reactions of alcohols are always dehydrations.

Dehydration

Alcohols can be dehydrated with excess hot concentrated sulfuric acid or by passing their vapours over heated aluminium oxide. An alkene is formed. For example, propan-1-ol is dehydrated to propene:

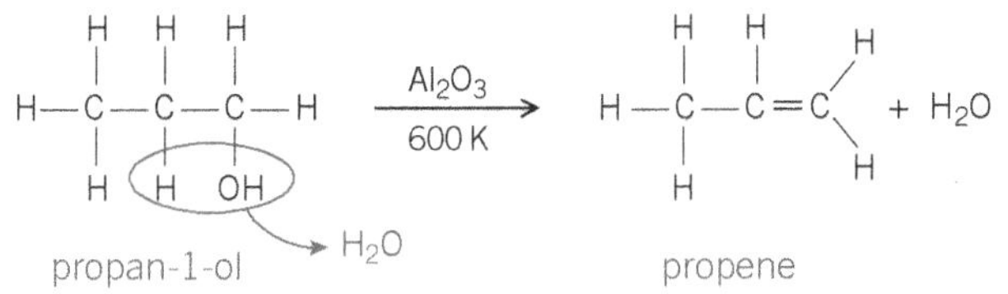

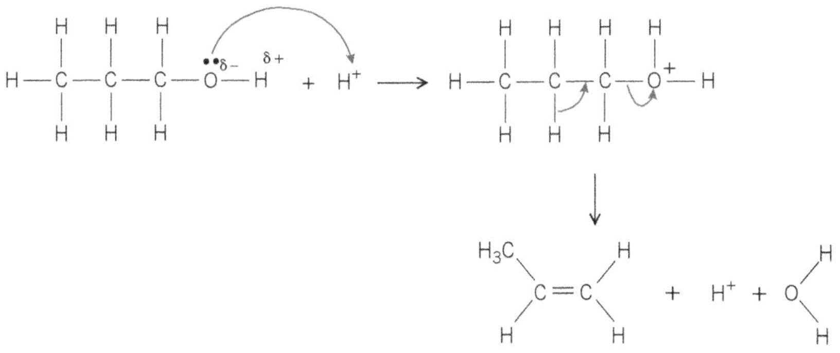

▲ **Figure 1** *The mechanism for the dehydration of propan-1-ol*

The apparatus used in the laboratory is shown in Figure 2.

Phosphoric(V) acid is an alternative dehydrating agent.

Hint

Alkenes can be polymerised to form plastics. Since ethanol can be produced by fermentation, this process is a route to produce poly(ethene) without the use of crude oil as a starting material.

Hint

In order to be dehydrated an alcohol must have a hydrogen atom on a carbon next to the —OH group.

Synoptic link

See Practicals 4 and 5 on page 524

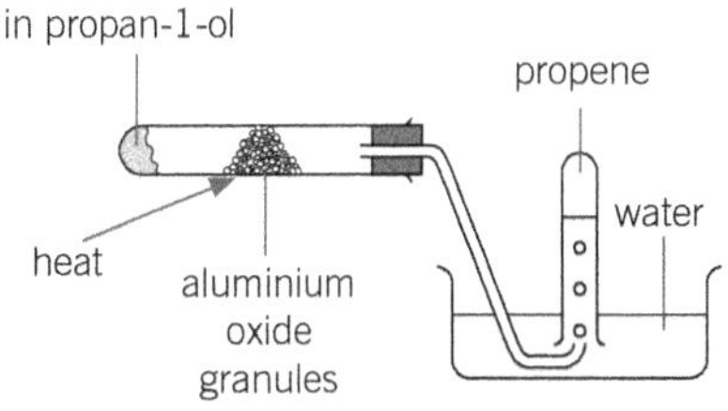

▲ **Figure 2** *Dehydration of an alcohol*

Isomeric alkenes

Dehydration of longer chain or branched alcohols may produce a mixture of alkenes, including ones with *Z* and *E* isomers, see Topic 11.3.

For example, with butan-2-ol there are three possible products:
but-1-ene, *Z*-but-2-ene, and *E*-but-2-ene.

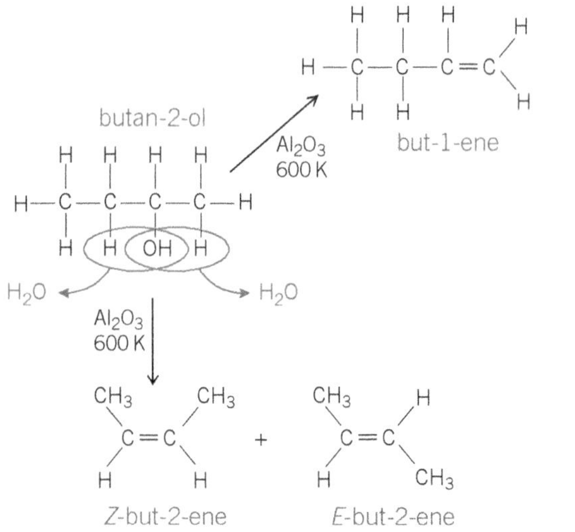

Name the two isomeric alkenes that are formed using the *cis/trans* notation.

Z isomer: *cis*-but-2-ene and E isomer: *trans*-but-2-ene respectively

Oxidation

Combustion is usually complete oxidation. Alcohols can also be oxidised gently and in stages. Primary alcohols are oxidised to **aldehydes**, RCHO. Aldehydes can be further oxidised to carboxylic acids, RCOOH. For example:

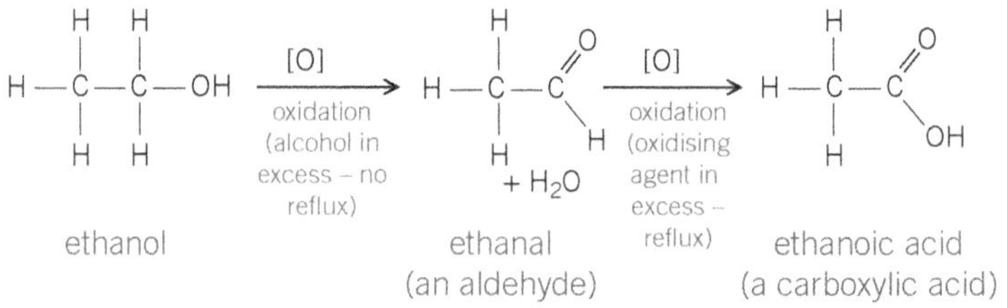

Secondary alcohols are oxidised to **ketones**, R_2CO. Ketones are not oxidised further.

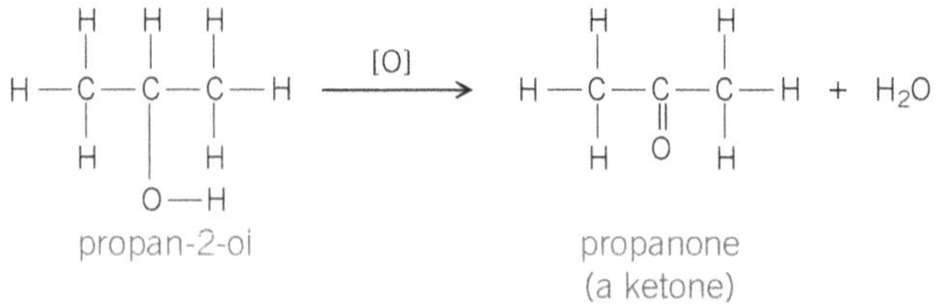

Tertiary alcohols are not easily oxidised. This is because oxidation would need a C—C bond to break, rather than a C—H bond (which is what happens when an aldehyde is oxidised). Ketones are not oxidised further for the same reason.

Many aldehydes and ketones have pleasant smells.

The experimental details

A solution of potassium dichromate, acidified with dilute sulfuric acid, is often used to oxidise alcohols to aldehydes and ketones. It is the oxidising agent. In the reaction, the orange dichromate(VI) ions are reduced to green chromium(III) ions.

To oxidise ethanol (1° alcohol) to ethanal – an aldehyde

Dilute acid and less potassium dichromate(VI) than is needed for complete oxidation to carboxylic acid are used. The mixture is heated gently with the addition of anti-bumping granules so that it boils gently, in apparatus like that shown in Figure 3, but with the receiver cooled in ice to reduce evaporation of the product. Ethanal (boiling temperature 294 K, 21 °C) vaporises as soon as it is formed and distils off. This stops it from being oxidised further to ethanoic acid. Unreacted ethanol remains in the flask. Anti-bumping granules are used to ensure gentle boiling.

The notation [O] is often used to represent oxygen from the oxidising agent. The reaction is given by the equation:

$$\underset{\text{ethanol}}{CH_3CH_2OH(l)} + [O] \longrightarrow \underset{\text{ethanal}}{CH_3CHO(g)} + H_2O(l)$$

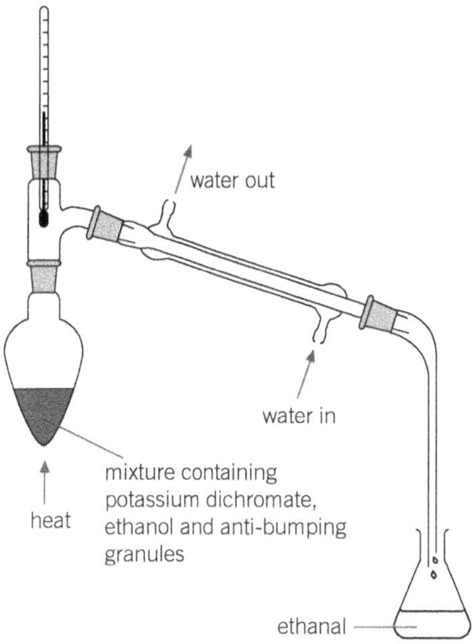

▲ **Figure 3** *Apparatus to oxidise ethanol to produce ethanal. Ethanal is then distilled to prevent further oxidation*

Hint

Notice that even if you use the notation [O] for oxidation, the equation must still balance.

To oxidise ethanol (1° alcohol) to ethanoic acid – a carboxylic acid

Concentrated sulfuric acid and more than enough potassium dichromate(VI) is used for complete reaction (the dichromate(VI) is in excess). The mixture is refluxed in the apparatus shown in Figure 4. **Reflux** means that vapour condenses and drips back into the reaction flask.

Whilst the reaction mixture is refluxing, any ethanol or ethanal vapour will condense and drip back into the flask until, eventually, it is all oxidised to the acid. After refluxing for around 20 minutes, you can distil off the ethanoic acid (boiling temperature 391 K, 118 °C), along with any water, by rearranging the apparatus to that shown in Figure 3.

Using [O] to represent oxygen from the oxidising agent, the equation is:

$$\underset{\text{ethanol}}{CH_3CH_2OH(l)} + 2[O] \rightarrow \underset{\text{ethanoic acid}}{CH_3COOH(g)} + H_2O(l)$$

Notice that twice as much oxidising agent is used in this reaction compared with the oxidation to ethanal.

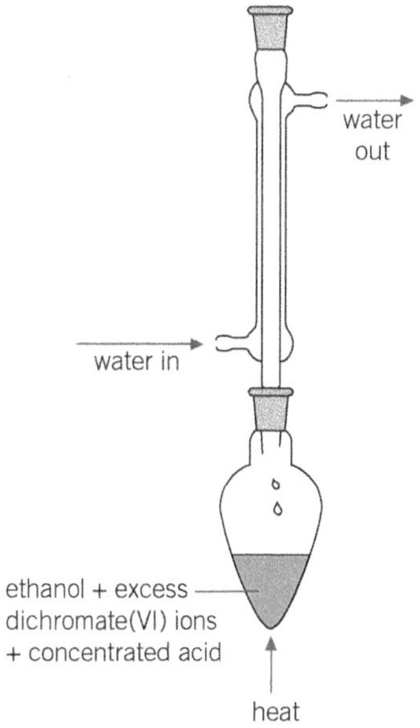

▲ **Figure 4** *Reflux apparatus for oxidation of ethanol to ethanoic acid*

Hint

Acidified dichromate is a mixture of potassium dichromate solution, acidified with dilute sulfuric acid.

Oxidising a secondary alcohol to a ketone

Secondary alcohols are oxidised to ketones by acidified dichromate. You do not have to worry about further oxidation of the ketone.

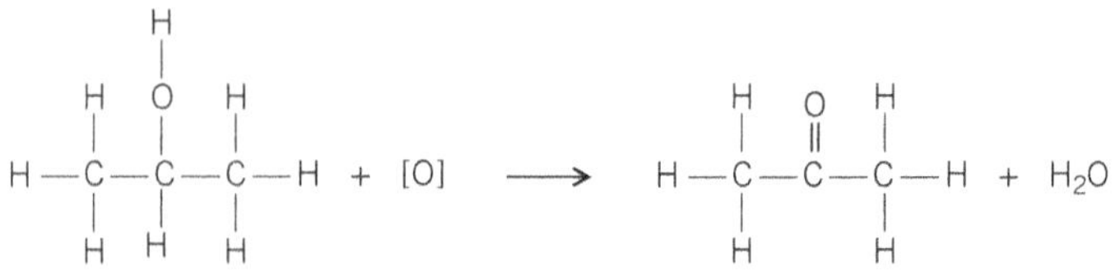

$$H-\underset{\underset{H}{|}}{\overset{\overset{H}{|}}{C}}-\underset{\underset{H}{|}}{\overset{\overset{O-H}{|}}{C}}-\underset{\underset{H}{|}}{\overset{\overset{H}{|}}{C}}-H \;+\; [O] \;\longrightarrow\; H-\underset{\underset{H}{|}}{\overset{\overset{H}{|}}{C}}-\overset{\overset{O}{\parallel}}{C}-\underset{\underset{H}{|}}{\overset{\overset{H}{|}}{C}}-H \;+\; H_2O$$

Aldehydes and ketones

Aldehydes and ketones both have the C=O group. This is called the **carbonyl** group.

In aldehydes it is at the end of the hydrocarbon chain:

$$\underset{H}{\overset{R}{>}}C=O \qquad \text{or RCHO}$$

In ketones it is in the body of the hydrocarbon chain:

$$\underset{R'}{\overset{R}{>}}C=O \qquad \text{or RCOR'}$$

Aldehydes are usually named using the suffix –al and ketones with the suffix –one.

So CH_3CHO is ethan*al* (two carbons) and CH_3COCH_3 is propan*one* (three carbons).

$$\underset{H}{\overset{H_3C}{>}}C=O \qquad\qquad \underset{H_3C}{\overset{H_3C}{>}}C=O$$

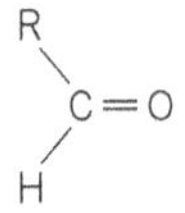

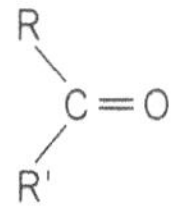

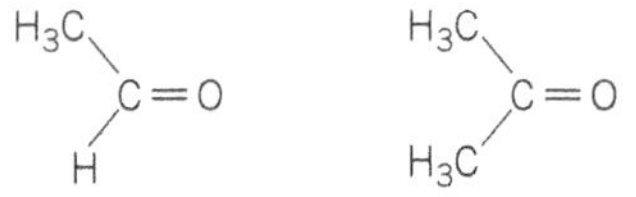

ethanal propanone

Tests for aldehydes and ketones

Aldehydes and ketones have similar physical properties, but there are two tests that can tell them apart. Both these tests involve gentle oxidation.

- Aldehydes are oxidised to carboxylic acids: RCHO + [O] → RCOOH (This is the second stage of the oxidation of a primary alcohol.)
- Ketones are not changed by gentle oxidation.

The Tollens' (silver mirror) test

Tollens' reagent is a gentle oxidising agent. It is a solution of silver nitrate in aqueous ammonia. It oxidises aldehydes but has no effect on ketones. It contains colourless silver(I) complex ions, containing Ag^+, which are reduced to metallic silver, Ag, as the aldehyde is oxidised.

On warming an aldehyde with Tollens' reagent, a deposit of metallic silver is formed on the inside of the test tube – the silver mirror, see Figure 5. This reaction was once used commercially for making mirrors.

▲ **Figure 5** *The silver mirror test*

The Fehling's test

The Fehling's reagent and is a gentle oxidising agent. It contains blue copper(II) complex ions which will oxidise aldehydes but not ketones. During the oxidation, the blue solution gradually changes to a brick-red precipitate of copper(I) oxide:

$$Cu^{2+} + e^- \rightarrow Cu^+.$$

On warming an aldehyde with blue Fehling's solution a brick-red precipitate gradually forms.

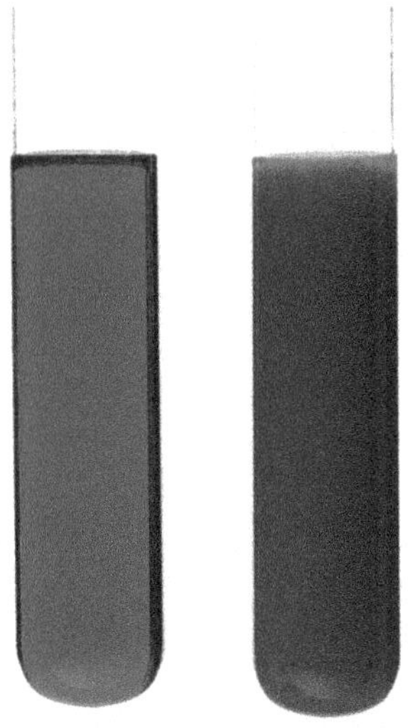

▲ **Figure 6** *The Fehling's test*

Summary questions

1. State what happens in each case when the following alcohols are oxidised as much as possible, by acidified potassium dichromate.
   - **a** a primary alcohol
   - **b** a secondary alcohol
2. Why is a tertiary alcohol not oxidised by the method outlined in question **1**?
3. What is the difference between distilling and refluxing?
4. Suggest how you would distinguish between a primary alcohol and a secondary alcohol, using Tollens' reagent or Fehling's solution.
5. Write the equation for the elimination of water from ethanol and name the product.
6. What are the possible products of dehydrating pentan-2-ol?

Practice questions

1 Ethanol can be oxidised by acidified potassium dichromate(VI) to ethanoic acid in a two-step process.

$$ethanol \rightarrow ethanal \rightarrow ethanoic\ acid$$

In order to ensure that the oxidation to ethanoic acid is complete, the reaction is carried out under reflux.

(a) Describe what happens when a reaction mixture is refluxed and why it is necessary, in this case, for complete oxidation to ethanoic acid.

(3 marks)

(b) Write a half equation for the overall oxidation of ethanol into ethanoic acid.

(1 mark)

Oxford International AQA Examinations specimen paper 2, 2016

2 Consider the following pairs of structural isomers.

Molecular formula	Structure	Structure
$C_4H_{10}O$	Isomer **A** $$H_3C - \overset{\overset{\displaystyle CH_3}{\vert}}{\underset{\underset{\displaystyle OH}{\vert}}{C}} - CH_3$$	Isomer **B** $CH_3CH_2CH_2CH_2OH$
	Isomer **C** $$CH_3CH_2 - \overset{}{\underset{\underset{\displaystyle H}{\vert}}{C}} = O$$	Isomer **D** $$H_3C - \overset{}{\underset{\underset{\displaystyle O}{\Vert}}{C}} - CH_3$$
C_6H_{12}	Isomer **E** $$\begin{array}{ccc} & CH_2 & \\ H_2C & & CH_2 \\ H_2C & & CH_2 \\ & CH_2 & \end{array}$$	Isomer **F** $CH_3CH_2CH=CHCH_2CH_3$

(a) (i) Explain what is meant by the term *structural isomers*.
 (ii) Complete the table to show the molecular formula of isomers **C** and **D**.
 (iii) Give the empirical formula of isomers **E** and **F**.

(4 marks)

(b) A simple chemical test can be used to distinguish between separate samples of isomer **A** and isomer **B**. Suggest a suitable test reagent and state what you would observe in each case.

(3 marks)

(c) A simple chemical test can be used to distinguish between separate samples of isomer **C** and isomer **D**. Suggest a suitable test reagent and state what you would observe in each case.

(3 marks)

(d) A simple chemical test can be used to distinguish between separate samples of isomer **E** and isomer **F**. Suggest a suitable test reagent and state what you would observe in each case.

(3 marks)
AQA, 2006

3 (a) Pentanal, $CH_3CH_2CH_2CH_2CHO$, can be oxidised to a carboxylic acid.
 (i) Write an equation for this reaction. Use [O] to represent the oxidising agent.
 (ii) Name the carboxylic acid formed in this reaction.

(4 marks)

(b) Pentanal can be formed by the oxidation of an alcohol.
 (i) Identify this alcohol.
 (ii) State the class to which this alcohol belongs.

(2 marks)
AQA, 2006

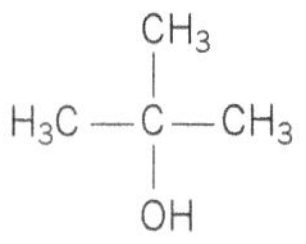

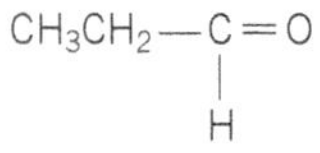

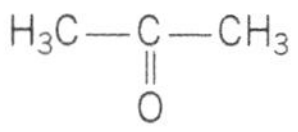

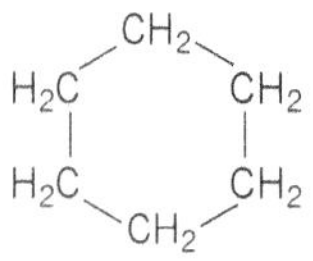

4 Some alcohols can be oxidised to form aldehydes, which can then be oxidised further to form carboxylic acids.
Some alcohols can be oxidised to form ketones, which resist further oxidation. Other alcohols are resistant to oxidation.

(a) Draw the structures of the **two** straight-chain isomeric alcohols with molecular formula, $C_4H_{10}O$.

(2 marks)

(b) Draw the structures of the oxidation products obtained when the two alcohols from part (**a**) are oxidised separately by acidified potassium dichromate(VI). Write equations for any reactions which occur, using [O] to represent the oxidising agent.

(6 marks)

(c) Draw the structure and give the name of the alcohol with molecular formula $C_4H_{10}O$ which is resistant to oxidation by acidified potassium dichromate(VI).

(2 marks)
AQA, 2005

5 Consider the following reaction schemes involving two alcohols, **A** and **B**, which are position isomers of each other.

$$\underset{\textbf{A}}{CH_3CH_2CH_2CH_2OH} \rightarrow \underset{\text{butanal}}{CH_3CH_2CH_2CHO} \rightarrow \underset{\text{butanoic acid}}{CH_3CH_2CH_2COOH}$$

$$\underset{\textbf{B}}{CH_3CH_2CH(OH)CH_3} \rightarrow \underset{\textbf{C}}{CH_3CH_2COCH_3}$$

(a) State what is meant by the term *position isomers*.

(2 marks)

(b) Name compound **A** and name the class of compounds to which **C** belongs.

(2 marks)

(c) Each of the reactions shown in the schemes above is of the same type and uses the same combination of reagents.
(i) State the type of reaction.
(ii) Identify a suitable combination of reagents.
(iii) State how you would ensure that compound **A** is converted into butanoic acid rather than into butanal.
(iv) Draw the structure of an isomer of compound **A** which does not react with this combination of reagents.
(v) Draw the structure of the carboxylic acid formed by the reaction of methanol with this combination of reagents.

(6 marks)

(d) (i) State a reagent which could be used to distinguish between butanal and compound **C**.
(ii) Draw the structure of another aldehyde which is an isomer of butanal.

(2 marks)
AQA, 2005

6 Epoxyethane is formed when ethane is oxidised by air in the presence of a catalyst. The reaction is exothermic, $\Delta H = -210$ kJ mol^{-1}

(a) Write an equation for the reaction. Name the catalyst used. Identify the hazards associated with the process

(4 marks)

(b) Draw the structure of epoxyethane and explain why the compound is very reactive.

(2 marks)

16 Organic analysis
16.1 Test-tube reactions

Learning objectives:

→ Describe how organic groups can be identified.

Specification reference: 3.3.6

When you are identifying an organic compound you need to know the functional groups present.

Chemical reactions

Some tests are very straightforward.

- Is the compound acidic (suggests carboxylic acid)?
- Is the compound solid (suggests long carbon chain or ionic bonding), liquid (suggests medium length carbon chain or polar or hydrogen bonding), or gas (suggests short carbon chain, little or no polarity)?
- Does the compound dissolve in water (suggests polar groups) or not (suggests no polar groups)?

Some specific chemical tests are listed in Table 1.

▼ **Table 1** *Chemical tests for functional groups*

Functional group	Test	Result
alkene —C═C—	shake with bromine water	orange colour disappears
halogenoalkane R—X	1. add NaOH(aq) and warm 2. acidify with HNO_3 3. add $AgNO_3$(aq)	precipitate of AgX
alcohol R—OH	add acidified $K_2Cr_2O_7$	orange colour turns green with primary or secondary alcohols (also with aldehydes)
aldehydes R—CHO	warm with Fehling's solution or warm with Tollens' solution	blue colour turns to red precipitate silver mirror forms
carboxylic acids R—COOH	add $NaHCO_3$(aq)	CO_2 given off

Synoptic link

Which halogen is present can be determined by the solubility of the precipitate of AgX in ammonia, see Topic 10.3, Reactions of halide ions.

Summary questions

1 How could you tell if R—X was a chloroalkane, a bromoalkane, or an iodoalkane?

2 In the test for a halogenoalkane:

  a Explain why it is necessary to acidify with dilute acid before adding silver nitrate.

  b Why would acidifying with hydrochloric acid not be suitable?

3 A compound decolourises bromine solution and fizzes when sodium hydrogencarbonate solution is added:

  a What two functional groups does it have?

  b Its relative molecular mass is 72. What is its structural formula?

  c Give equations for the two reactions.

16.2 Mass spectrometry

You saw in Topic 1.3 how mass spectrometry is used to measure the relative *atomic* masses of atoms. It is also the main method for finding the relative *molecular* mass of organic compounds. The compound enters the mass spectrometer in solution. It is ionised and the positive ions are accelerated through the instrument as a beam of ionised molecules. These then fly through the instrument towards a detector.

Their times of flight are measured. These depend on the mass to charge ratio *m/z* of the ion.

The output is then presented as a graph of relative abundance (vertical axis) against mass/charge ratio (horizontal axis). However, since the charge on the ions is normally +1, the horizontal axis is effectively relative mass. This graph is called a mass spectrum.

Learning objectives:

- → State what is meant by the term molecular ion.
- → Describe what the mass of a molecular ion shows.
- → Explain what a high resolution mass spectrum can tell us.

Specification reference: 3.3.6

Fragmentation

There are many techniques for mass spectrometry. In some of these the ions of the sample break up or fragment as they pass through the instrument.

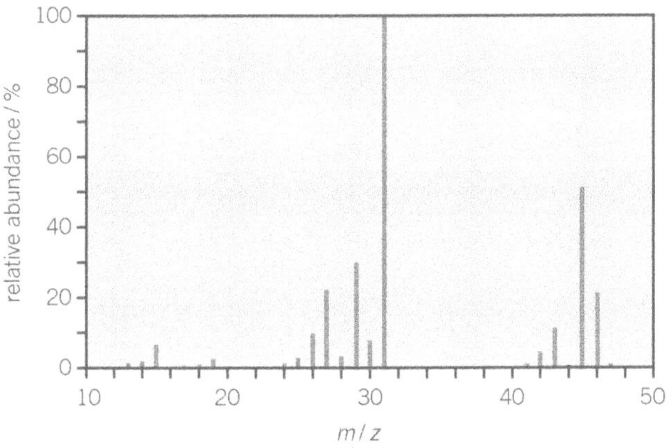

▲ **Figure 1** *The mass spectrum of ethanol*

A mass spectrum of ethanol is shown in Figure 1. Notice that it contains many lines and not just one as we might expect. When ethanol is ionised it forms the ion $C_2H_5OH^+$ ($CH_3CH_2OH^+$). This is called the **molecular ion**. Many of these ions will then break up because some of their bonds break as they are ionised, so there are other ions of smaller molecular mass. This process is called **fragmentation**. Each of these fragment ions produces a line in the mass spectrum. These can provide information that will help to deduce the structure of the compound. They also act as a 'fingerprint' to help identify it.

Bearing in mind the fact that the ethanol molecule, CH_3CH_2OH is breaking up, suggest formulae for the fragments represented by the peaks at *m/z* 46, 45, 31, and 29.

46: $CH_3CH_2OH^+$, 45: $CH_3CH_2O^+$ or $CH_2CH_2OH^+$ or CH_3CHOH^+, 31: CH_2OH^+, 29: $CH_3CH_2^+$

Mass spectrometry and sport

One of the many applications of mass spectrometry is testing athletes for the presence of drugs in urine samples. It is also used in forensic work.

Hint

- In any mass spectrum of an organic compound there will be a tiny peak one mass unit to the right of the molecular ion. This is caused by ions containing the ^{13}C isotope.

GCMS

Gas Chromatography Mass Spectrometry (GCMS) is one of the most powerful analytical techniques used currently. It is used in forensic work and also to detect drugs used by athletes and doping of racehorses. It is a combination of two techniques.

Gas chromatography is a technique for separating mixtures which uses a stream of gas to carry a mixture of vapours through a tube packed with a powdered solid. The different components of the mixture emerge from the tube (called a column) at different times. As the components emerge from the column, their amounts are measured and they are fed straight into a mass spectrometer which produces the mass spectrum of each and allows them to be identified. So the amount and identity of each component in a complex mixture can be found.

Hint

Remember that the mass spectrometer detects isotopes separately.

High resolution mass spectrometry

Mass spectra often show masses to the nearest whole number only. However, many mass spectrometers can measure masses to three or even four decimal places. This method allows us to work out the molecular formula of the parent ion. It makes use of the fact that isotopes of atoms do not have exactly whole number atomic masses (except for carbon-12 which is exactly twelve by definition), for example, $^{16}O = 15.99491$ and $^{1}H = 1.007829$.

A parent ion of mass 200, to the nearest whole number, could have the molecular formulae of: $C_{10}H_{16}O_4$, or $C_{11}H_4O_4$, or $C_{11}H_{20}O_3$

Adding up the accurate atomic masses gives the following molecular masses:

$$C_{10}H_{16}O_4 = 200.1049$$

$$C_{11}H_4O_4 = 200.0110$$

$$C_{11}H_{20}O_3 = 200.1413$$

These can easily be distinguished by high resolution mass spectrometry. A computer database can be used to identify the molecular formula from the accurate relative molecular mass.

Water sampling

Water boards sample the water from the rivers in their areas to monitor pollutants. The pollutants are separated by chromatography and fed into a mass spectrometer. Each pollutant can be identified from its spectrum; a computer matches its spectrum with known compounds in a library of spectra.

Summary questions

1 a How are ions formed from molecules in a mass spectrometer?

b What sign of charge do the ions have as a result of this?

2 A compound was found to have a molecular ion with a mass to charge ratio of 136.125. Which of the following molecular formulae could it have? $C_9H_{12}O$ or $C_{10}H_{16}$

You will need to work out the accurate M_r of each of these molecules.

16.3 Infrared spectroscopy

Infrared (IR) spectroscopy is often used by organic chemists to help them identify compounds.

Learning objectives:

- → Describe how the absorption of infrared radiation can be used to indicate the presence of certain functional groups in an organic molecule.
- → State what the fingerprint region is and where is it found.
- → Describe how infrared spectroscopy can be used to confirm the identity of a compound.
- → Explain how an infrared spectrum can be used to show the presence of impurities.

Specification reference: 3.3.6

How infrared spectroscopy works

A pair of atoms joined by a chemical bond is always vibrating. The system behaves rather like two balls (the atoms) joined by a spring (the bond). Stronger bonds vibrate faster (at higher frequency) and heavier atoms make the bond vibrate more slowly (at lower frequency). Every bond has its own unique natural frequency that is in the infrared region of the electromagnetic spectrum.

When you shine a beam of infrared radiation (heat energy) through a sample, the bonds in the sample can absorb energy from the radiation and vibrate more. However, any particular bond can only absorb radiation that has the same frequency as the natural frequency of the bond. Therefore, the radiation that emerges from the sample will be missing the frequencies that correspond to the bonds in the sample, see Figure 1.

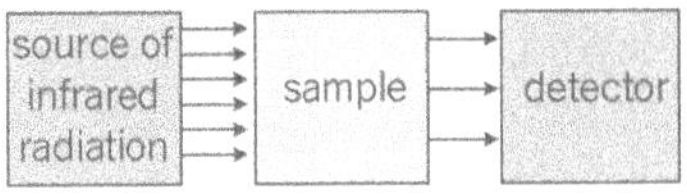

▲ **Figure 1** *Schematic diagram of an infrared spectrometer*

The infrared spectrometer

This is what happens in an infrared spectrometer:

1. A beam of infrared radiation containing a spread of frequencies is passed through a sample.
2. The radiation that emerges is missing the frequencies that correspond to the types of bonds found in the sample.
3. The instrument plots a graph of the intensity of the radiation emerging from the sample, called the transmittance, against the frequency of radiation.
4. The frequency is expressed as a wavenumber, measured in cm^{-1}.

Hint

Wavenumber is proportional to frequency.

The infrared spectrum

A typical graph, called an infrared spectrum is shown in Figure 2. The dips in the graph (confusingly, they are usually called peaks) represent particular bonds. Figure 3 and Table 1 show the wavenumbers of some bonds commonly found in organic chemistry.

These can help us to identify the functional groups present in a compound. For example:

- the O—H bond produces a broad peak between 3230 and 3550 cm^{-1} and this is found in alcohols, ROH; and a very broad O—H peak between 2500 and 3000 cm^{-1} in carboxylic acids, RCOOH.
- the C=O bond produces a peak between 1680 and 1750 cm^{-1}. This bond is found in aldehydes, RCHO, ketones, R_2CO, and carboxylic acids, RCOOH.

▼ **Table 1** *Characteristic infrared absorptions in organic molecules*

Bond	Location	Wavenumber / cm^{-1}
C—O	alcohols, esters	1000–1300
C=O	aldehydes, ketones, carboxylic acids, esters	1680–1750
O—H	hydrogen bonded in carboxylic acids	2500–3000 (broad)
N—H	primary amines	3100–3500
O—H	hydrogen bonded in alcohols, phenols	3230–3550

Study tip

You should be able to interpret infrared spectra using infrared absorption data.

Hint

Notice that the wavenumber decreases from left to right on an infrared spectrum.

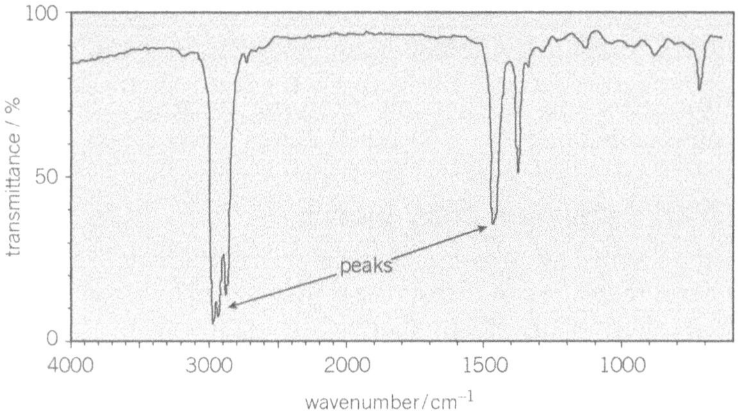

▲ **Figure 2** *A typical infrared spectrum. Note that wavenumber gets smaller going from left to right*

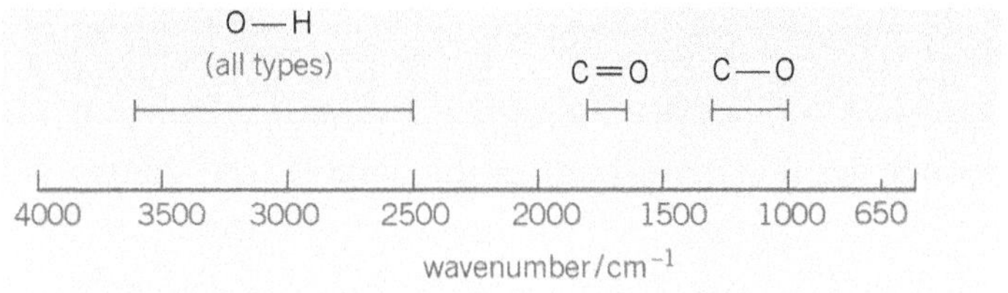

▲ **Figure 3** *The ranges of wavenumbers at which some bonds absorb infrared radiation*

Data about the frequencies that correspond to different bonds can be found on the data sheet at the back of the book.

Figures 4, 5, and 6 show the infrared spectra of ethanal, ethanol, and ethanoic acid with the key peaks marked.

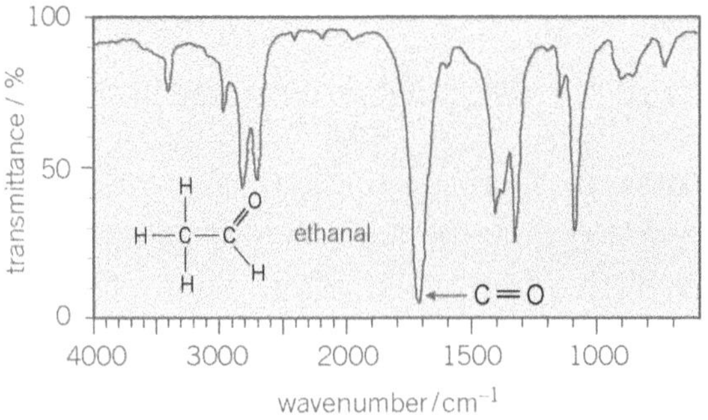

▲ **Figure 4** *Infrared spectrum of ethanal*

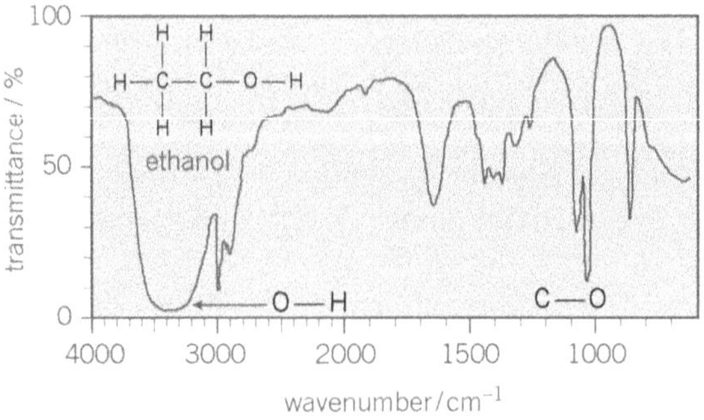

▲ **Figure 5** *Infrared spectrum of ethanol*

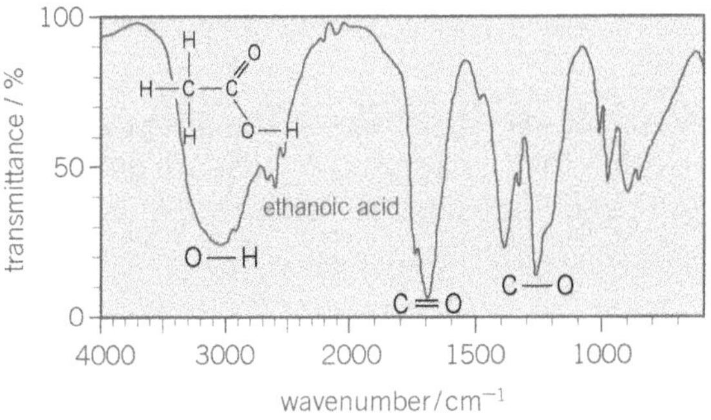

▲ **Figure 6** *Infrared spectrum of ethanoic acid*

Greenhouse gases

The greenhouse effect, which contributes to global warming, is caused by gases in the atmosphere that absorb the infrared radiation given off from the surface of the Earth and would otherwise be lost into space. The table gives some data about some of these gases. The infrared radiation is absorbed by bonds in these gases in the same way as in an infrared spectrometer. Carbon dioxide has two C═O bonds which absorb in the infrared region of the spectrum.

Gas	Relative greenhouse effect per molecule	Concentration in the atmosphere / parts per million (ppm)
carbon dioxide, CO_2	1	350
methane, CH_4	30	1.7
nitrous oxide (dinitrogen monoxide, NO)	160	0.31
ozone, O_3	2000	0.06
trichlorofluoromethane (a CFC)	21 000	0.000 26
dichlorodifluoromethane (a CFC)	25 000	0.000 24

Water vapour is a powerful greenhouse gas, absorbing IR via its O—H bonds. It is not included in the table because its concentration in the atmosphere is very variable.

1 Write the displayed formulae of trichlorofluoromethane and of dichlorodifluoromethane (showing all the atoms and the bonds).
2 What bonds are present in these compounds? Suggest why their relative effects are so similar.
3 Suggest why the concentration of water vapour is so variable.
4 One way of comparing the overall greenhouse contribution of a gas would be to multiply its concentration by its relative effect. Use this method to compare the contribution of carbon dioxide and methane.

The fingerprint region

The area of an infrared spectrum below about 1500 cm^{-1} usually has many peaks caused by complex vibrations of the whole molecule. This shape is unique for any particular substance. It can be used to identify the chemical, just as people can be identified by their fingerprints. It is therefore called the **fingerprint region**.

Chemists can use a computer to match the fingerprint region of a sample with those on a database of compounds. An exact match confirms the identification of the sample.

Figures 7 and 8 show the IR spectra of two very similar compounds, propan-1-ol and propan-2-ol.

They are as expected, very similar overall. However, superimposing the spectra, Figure 9, shows that their fingerprint regions are quite distinct. This is shown more clearly in Figure 10, where the fingerprint region has been enlarged.

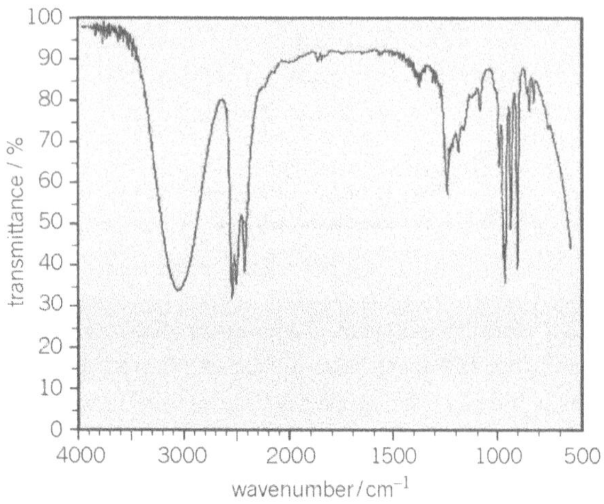

▲ **Figure 7** *Infrared spectrum of propan-1-ol*

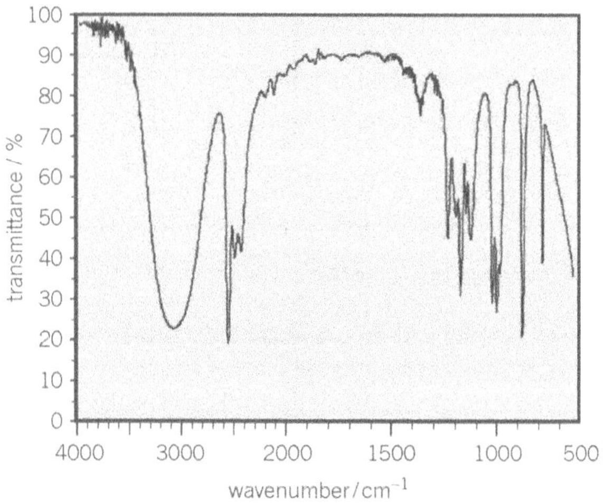

▲ **Figure 8** *Infrared spectrum of propan-2-ol*

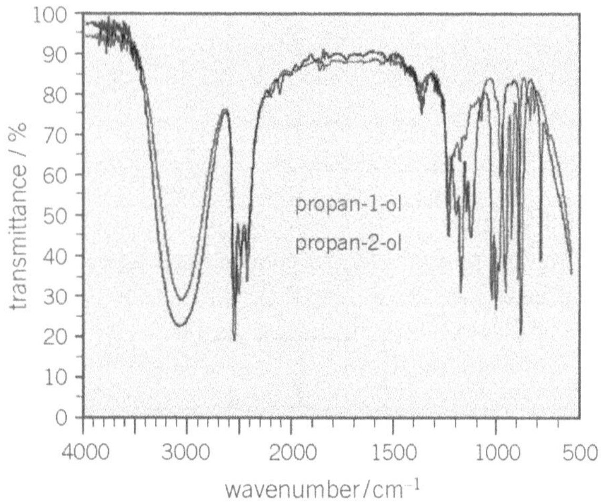

▲ **Figure 9** *Infrared spectra of propan-1-ol superimposed on propan-2-ol*

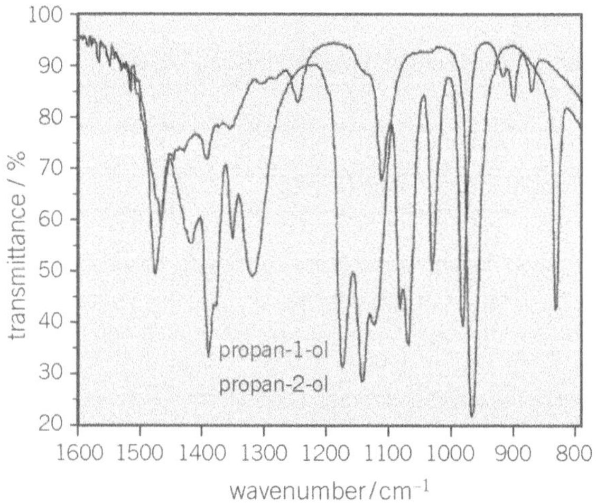

▲ **Figure 10** *The fingerprint region of the infrared spectra of propan-1-ol superimposed on propan-2-ol enlarged*

Identifying impurities

Infrared spectra can also be used to show up the presence of impurities. These may be revealed by peaks that should not be there in the pure compound. Figures 11 and 12 show the spectrum of a sample of pure caffeine and that of caffeine extracted from tea. The broad peak at around 3000 cm^{-1} in the impure sample (Figure 12) is an O—H stretch caused by water in the sample that has not been completely dried. Notice that there are no O—H bonds in caffeine (Figure 13).

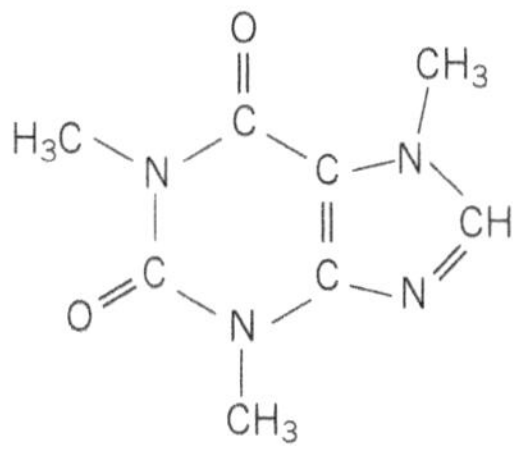

▲ **Figure 13** *The structural formula of caffeine. It has no O—H bonds*

In practice, analytical chemists will often use a combination of spectroscopic techniques to identify unknown compounds.

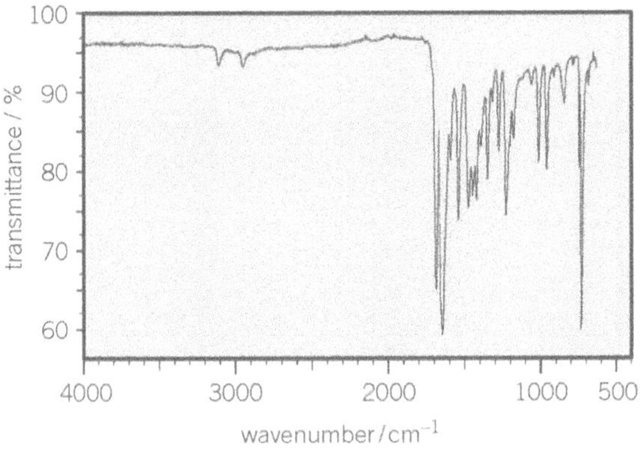

▲ **Figure 11** *The infrared spectrum of pure caffeine*

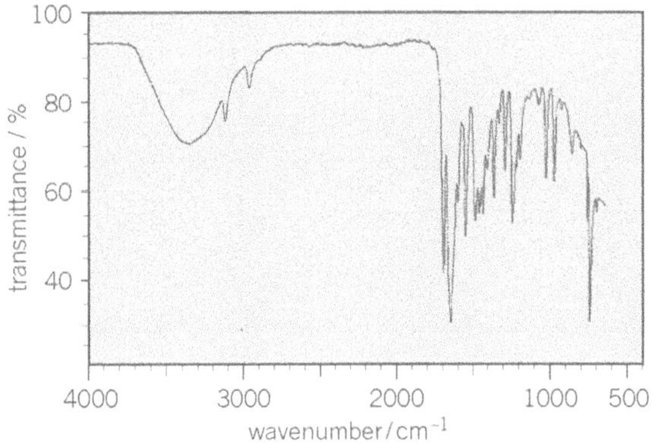

▲ **Figure 12** *The infrared spectrum of impure caffeine*

Summary questions

1 An organic compound has a peak in the IR spectrum at about 1725 cm^{-1}. Which of the following compounds could it be?

a 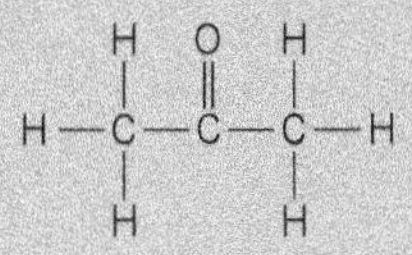 **b** 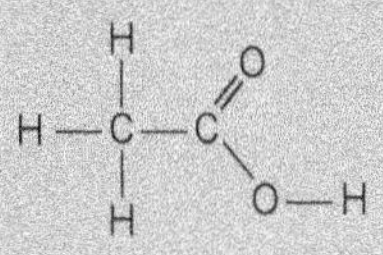 **c** 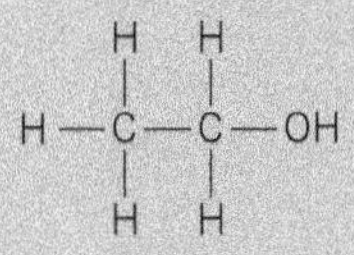

2 Explain your answer to question **1**.

3 An organic compound has a peak in the IR spectrum at about 3300 cm^{-1}. Which of the compounds in question **1** could it be?

4 Explain your answer to question **3**.

5 An organic compound has a peak in the IR spectrum at 1725 and 3300 cm^{-1}. Which of the compounds in question **1** could it be?

6 Explain your answer to question **5**.

1 (a) The infrared spectrum of compound A, $C_3H_6O_2$, is shown below.

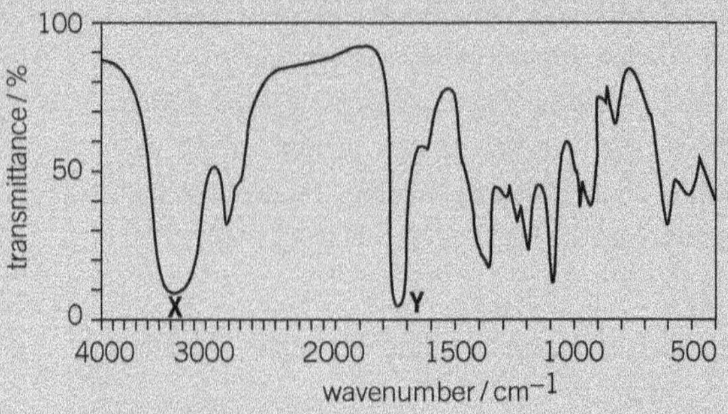

Identify the functional groups which cause the absorptions labelled **X** and **Y**.
Using this information draw the structures of the three possible structural isomers
for A.

(5 marks)

AQA, 2006

2 The table below shows the structures of three isomers with the molecular formula $C_5H_{10}O$.

Isomer 1 $\begin{array}{c}H_3C \qquad\quad H \\ \diagdown \quad / \\ C{=}C \\ / \qquad \diagdown \\ H \qquad CH(OH)CH_3\end{array}$	(*E*)-pent-3-en-2-ol
Isomer 2 $\begin{array}{c}CH_3CH_2CH_2CH_2 \\ \diagdown \\ C{=}O \\ / \\ H\end{array}$	pentanal
Isomer 3 $\begin{array}{c}CH_3CH_2CH_2 \\ \diagdown \\ C{=}O \\ / \\ H_3C\end{array}$	

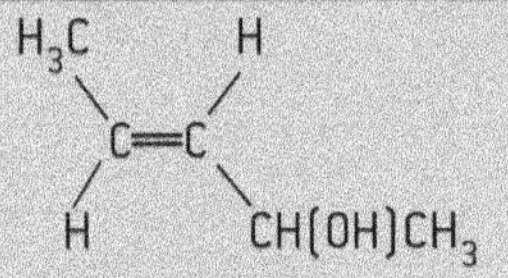

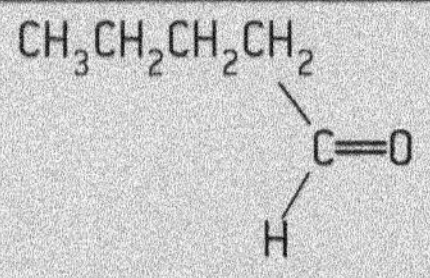

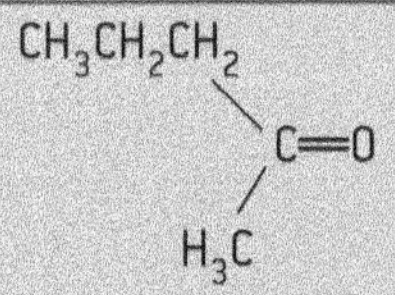

(a) Complete the table by naming Isomer **3**.

(1 mark)

(b) State the type of structural isomerism shown by these three isomers.

(1 mark)

(c) The compound (*Z*)-pent-3-en-2-ol is a stereoisomer of (*E*)-pent-3-en-2-ol.
 (i) Draw the structure of (*Z*)-pent-3-en-2-ol.

(1 mark)

 (ii) Identify the feature of the double bond in (*E*)-pent-3-en-2-ol and that in
 (*Z*)-pent-3-en-2-ol that causes these two compounds to be stereoisomers.

(1 mark)

(d) A chemical test can be used to distinguish between separate samples of Isomer **2** and
 Isomer **3**.
 Identify a suitable reagent for the test.
 State what you would observe with Isomer **2** and with Isomer **3**.

(3 marks)

(e) The following is the infrared spectrum of one of the isomers **1**, **2**, or **3**.

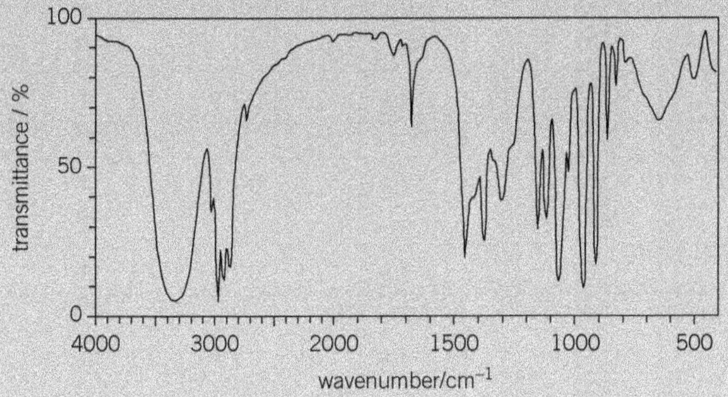

(i) Deduce which of the isomers (**1**, **2**, or **3**) would give this infrared spectrum.

(1 mark)

(ii) Identify two features of the infrared spectrum that support your deduction. In each case, identify the functional group responsible.

(2 marks)

AQA, 2011

Section 3 practice questions

1 Trifluoromethane, CHF_3, can be used to make the refrigerant chlorotrifluoromethane, $CClF_3$.

(a) Chlorotrifluoromethane is formed when trifluoromethane reacts with chlorine.

$CHF_3 + Cl_2 \rightarrow CClF_3 + HCl$

The reaction is a free-radical substitution reaction similar to the reaction of methane with chlorine.

(i) Write an equation for each of the following steps in the mechanism for the reaction of CHF_3 with Cl_2 *(4 marks)*

Initiation step

First propagation step

Second propagation step

Termination step to form hexafluoroethane

(ii) Give **one** essential condition for this reaction. *(1 mark)*

AQA, 2014

(b) A small amount of $CClF_3$ with a mass of 2.09×10^{-4} kg escaped from a refrigerator into a room with a volume of $200\,m^3$. Calculate the number of $CClF_3$ molecules in a volume of $500\,cm^3$. Assume that the $CClF_3$ molecules are evenly distributed throughout the air in the room. Give your answer to the appropriate number of significant figures.

The Avogadro constant = $6.02 \times 10^{23}\,mol^{-1}$. *(3 marks)*

2 Some oil-fired heaters use paraffin as a fuel.

One of the compounds in paraffin is the straight-chain alkane, dodecane, $C_{12}H_{26}$.

(a) Give the name of the substance from which paraffin is obtained.

State the name of the process used to obtain paraffin from this substance. *(2 marks)*

(b) The combustion of dodecane produces several products.

Write an equation for the **incomplete** combustion of dodecane to produce gaseous products only. *(1 mark)*

(c) Oxides of nitrogen are also produced during the combustion of paraffin in air.

(i) Explain how these oxides of nitrogen are formed. *(2 marks)*

(ii) Write an equation to show how nitrogen monoxide in the air is converted into nitrogen dioxide. *(1 mark)*

(iii) Nitric acid, HNO_3, contributes to acidity in rainwater.

Deduce an equation to show how nitrogen dioxide reacts with oxygen and water to form nitric acid. *(1 mark)*

(d) Dodecane, $C_{12}H_{26}$, can be cracked to form other compounds.

(i) Give the general formula for the homologous series that contains dodecane. *(1 mark)*

(ii) Write an equation for the cracking of one molecule of dodecane into equal amounts of two different molecules each containing the same number of carbon atoms.

State the empirical formula of the straight-chain alkane that is formed. Name the catalyst used in this reaction. *(3 marks)*

(iii) Explain why the melting point of dodecane is higher than the melting point of the straight-chain alkane produced by cracking dodecane. *(2 marks)*

(e) Give the IUPAC name for the following compound and state the type of structural isomerism shown by this compound and dodecane. *(2 marks)*

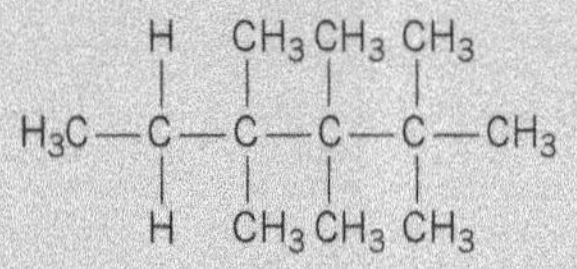

(f) Dodecane can be converted into halododecanes.
Deduce the formula of a substance that could be reacted with dodecane to produce 1-chlorododecane and hydrogen chloride only. *(1 mark)*
AQA, 2014

3 The following table shows the boiling points of some straight-chain alkanes.

	CH_4	C_2H_6	C_3H_8	C_4H_{10}	C_5H_{12}
Boiling point / °C	−162	− 88	− 42	−1	36

(a) State a process used to separate an alkane from a mixture of these alkanes. *(1 mark)*
(b) Both C_3H_8 and C_4H_{10} can be liquefied and used as fuels for camping stoves.
Suggest, with a reason, which of these two fuels is liquefied more easily. *(1 mark)*
(c) Write an equation for the complete combustion of C_4H_{10} *(1 mark)*
(d) Explain why the complete combustion of C_4H_{10} may contribute to environmental problems. *(1 mark)*
(e) Balance the following equation that shows how butane is used to make the compound called maleic anhydride.

$$\_\_\_\_ CH_3CH_2CH_2CH_3 + \_\_\_\_ O_2 \rightarrow \_\_\_\_ C_2H_2(CO)_2O + \_\_\_\_ H_2O$$

(1 mark)

(f) Ethanethiol, C_2H_5SH, a compound with an unpleasant smell, is added to gas to enable leaks from gas pipes to be more easily detected.
(i) Write an equation for the combustion of ethanethiol to form carbon dioxide, water, and sulfur dioxide. *(1 mark)*
(ii) Identify a compound that is used to react with the sulfur dioxide in the products of combustion before they enter the atmosphere.
Give **one** reason why this compound reacts with sulfur dioxide. *(2 marks)*
(iii) Ethanethiol and ethanol molecules have similar shapes.
Explain why ethanol has the higher boiling point. *(2 marks)*

(g) The following compound **X** is an isomer of one of the alkanes in the table above.

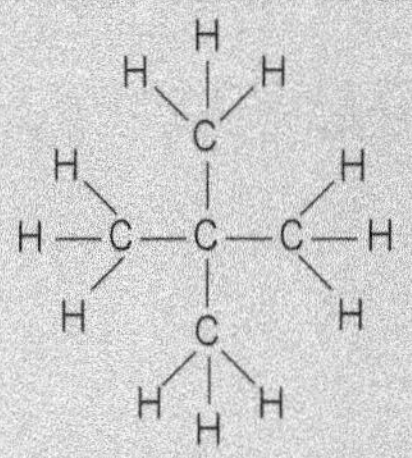

(i) Give the IUPAC name of **X**. *(1 mark)*
(ii) **X** has a boiling point of 9.5 °C.
Explain why the boiling point of **X** is lower than that of its straight-chain isomer. *(2 marks)*
(iii) The following compound **Y** is produced when **X** reacts with chlorine.

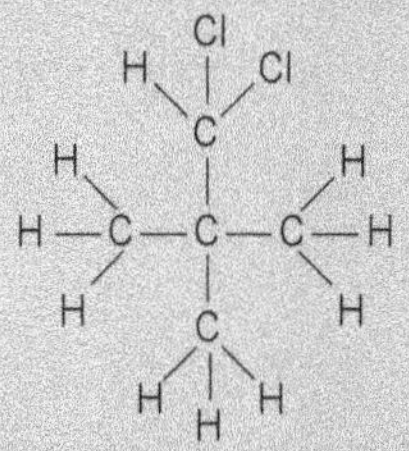

Deduce how many **other** position isomers of **Y** can be formed. Write the number of **other** position isomers. *(1 mark)*

(h) Cracking of one molecule of an alkane **Z** produces one molecule of ethane, one molecule of propene, and two molecules of ethene.
(i) Deduce the molecular formula of **Z**. *(1 mark)*
(ii) State the type of cracking that produces a high proportion of ethene and propene.
Give the **two** conditions for this cracking process. *(2 marks)*
AQA, 2013

Section 3 summary

polymerisation

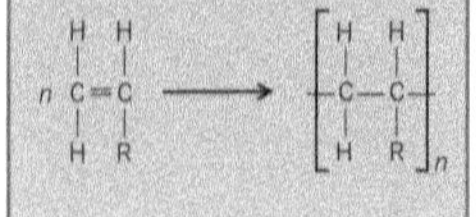

oxidation

- 1° alcohols oxidised to aldehydes then carboxylic acids
- 2° alcohols oxidised to ketones
- 3° alcohols not oxidised

formation

- chain reactions

initiation
↓
propagation
↓
termination

fractional distillation

catalytic cracking

elimination

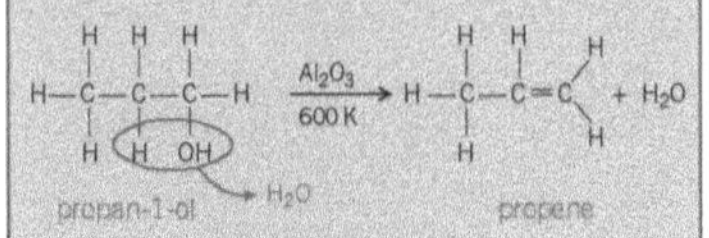

elimination reaction

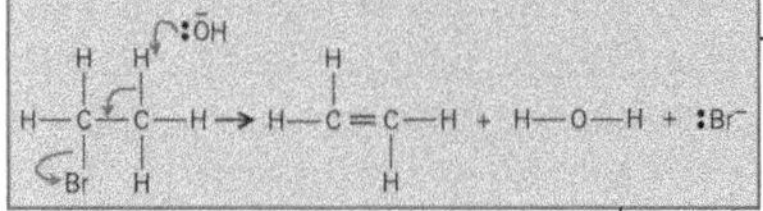

alcohols

combustion

- complete combustion

$CH_4 + 2O_2 \longrightarrow CO_2 + 2H_2O$

- incomplete combustion

$C_2H_6 + 2\frac{1}{2}O_2 \longrightarrow 2CO + 3H_2O$

nucleophilic substitution

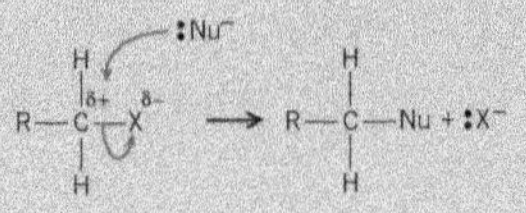

electrophilic addition

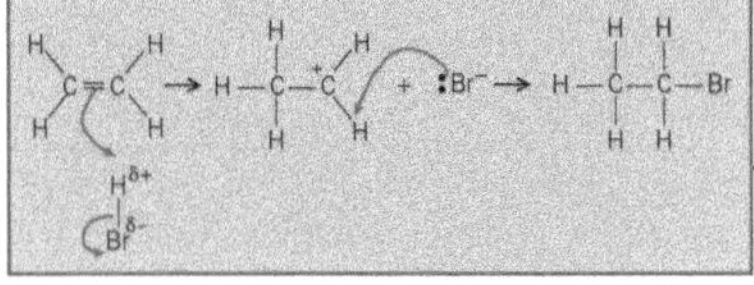

halogenoalkanes

alkenes

alkanes

Section 3 Organic chemistry

formula and nomenclature

analysis

mass number

mass spectrometry

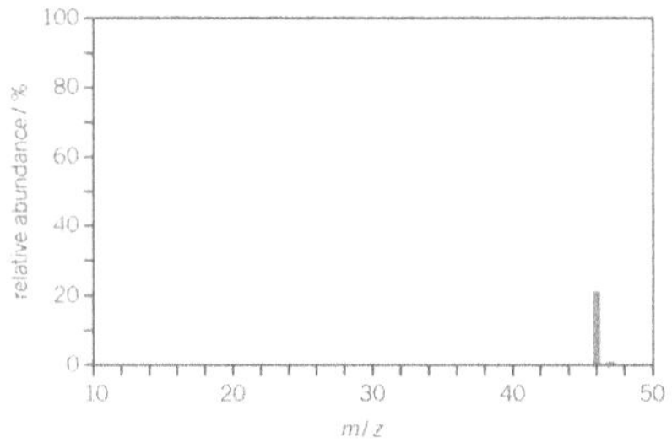

formula

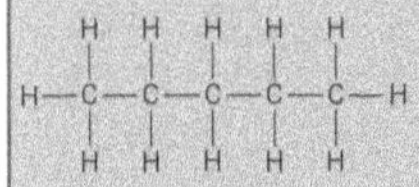

displayed

$CH_3CH_2CH_2CH_2CH_3$

structural

skeletal

infrared spectroscopy

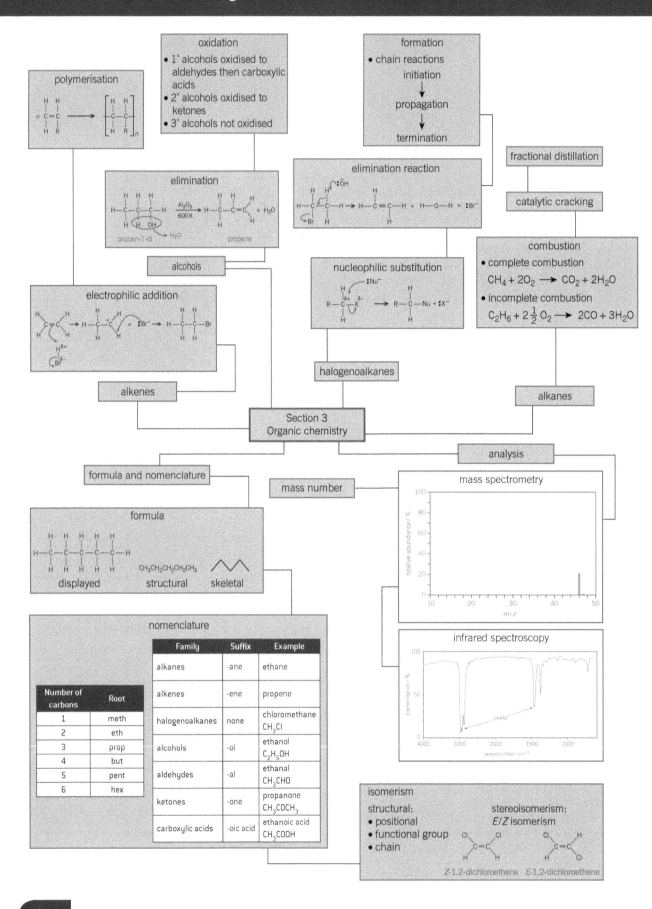

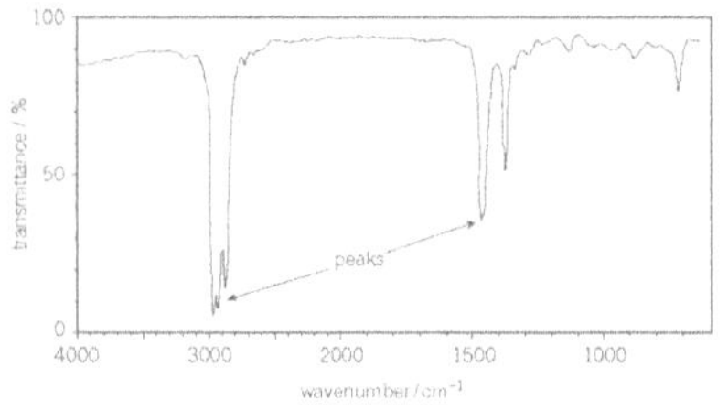

nomenclature

Number of carbons	Root
1	meth
2	eth
3	prop
4	but
5	pent
6	hex

Family	Suffix	Example
alkanes	-ane	ethane
alkenes	-ene	propene
halogenoalkanes	none	chloromethane CH_3Cl
alcohols	-ol	ethanol C_2H_5OH
aldehydes	-al	ethanal CH_3CHO
ketones	-one	propanone CH_3COCH_3
carboxylic acids	-oic acid	ethanoic acid CH_3COOH

isomerism

structural:
- positional
- functional group
- chain

stereoisomerism:
E/*Z* isomerism

Z-1,2-dichloroethene *E*-1,2-dichloroethene

Practical skills

In this section you have met the following ideas:

- Performing fractional distillation
- Carrying out hydrolysis of halogenoalkanes to find their relative rates of reaction
- Testing organic compounds for unsaturation using bromine water
- Making a polymer, such as poly(phenylethene) from its monomer
- Purifying ethanol by distillation
- Preparing aldehydes and carboxylic acids by the oxidation of a primary alcohol
- Carry out tests to identify alcohols, aldehydes, alkenes, and carboxylic acids

Maths skills

In this section you have met the following maths skills:

- Balancing symbol equations
- Representing 2-D and 3-D forms using two-dimensional diagrams
- Identifying and drawing different isomers of a substance by its formula

Extension

1 The production of ethene from ethanol by dehydration is considered more sustainable than by cracking naphtha. Write:

  a a summary of its use for industrial applications

  b current and future challenges. These could be, for example, related to sustainability or new applications.

2 Research how advances in computer technology and robotics have changed the way in which research of organic compounds is conducted, particularly the development of new medicines.

Section 4

A2 Physical chemistry 2

Chapters in this section

Thermodynamics builds on the ideas introduced in AS Chemistry and looks at how you can calculate enthalpy changes that are hard, or impossible, to measure directly. It applies this idea to the enthalpy changes involved in forming ionic compounds. It introduces the idea of **entropy** – a measure of disorder – that drives chemical reactions and the idea of free energy, a way of predicting whether a reaction will take place at a particular temperature.

Kinetics is about rates of reactions. The rate equation is an expression that links the rate of a reaction to the concentrations of different species in the reaction mixture. The idea of a reaction mechanism as a series of simple steps is introduced along with the concept of the rate-determining step.

Equilibria shows how to apply the equilibrium law and Le Châtelier's principle to reversible reactions that take place in the gas phase. The idea of partial pressure is introduced as well as the gaseous equilibrium constant K_p, which is expressed in terms of partial pressures.

In **Electrode potentials and electrochemical cells**, the idea of half cells which can be joined to generate an electrical potential difference is introduced. This leads on to a method of predicting the course of redox reactions and also to a description of how a number of types of batteries work.

Acids, bases, and buffers extends the definition of acids and bases and gives an expression to find the pH of a solution. The idea of strong and weak acids and bases is introduced and applied quantitatively. Titrations between strong and weak acids and bases are discussed and the operation of buffer solutions, which resist changes of pH, is explained.

What you already know:

The material in this unit builds on knowledge and understanding that you have built up at AS level. In particular the following:

- ▢ It is possible to measure energy (enthalpy) changes in chemical reactions.
- ▢ The rates of chemical reactions are governed by collisions between particles that occur with sufficient energy.
- ▢ Reversible reactions may reach equilibrium.
- ▢ Le Châtelier's principle can be used to make predictions about the position of an equilibrium.
- ▢ The equilibrium law expression for K_c allows you to make calculations about the position of equilibrium for reactions in solution.
- ▢ Redox reactions involve the transfer of electrons and can be kept track of using the idea of oxidation state (oxidation numbers).

17 Thermodynamics
17.1 Enthalpy change

Learning objective:

→ List the enthalpy changes that are relevant to the formation of ionic compounds.

Specification reference: 3.1.8

Synoptic link

You will need to know the energetics, states of matter, ionic bonding, and change of state studied in Chapter 3, Bonding, and Chapter 4, Energetics.

Hint

You may also refer to the enthalpy of atomisation of a *compound*.

Hint

Ionisation enthalpies are *always* positive because energy has to be put in to pull an electron away from the attraction of the positively charged nucleus of the atom.

Study tip

The second ionisation energy of sodium is *not* the energy change for

$Na(g) \rightarrow Na^{2+}(g) + 2e^-$

It is the energy change for

$Na^+(g) \rightarrow Na^{2+}(g) + e^-$

Hint

Hess's law states that the enthalpy change for a chemical reaction is always the same, whatever route is taken from reactants to products.

Hess's law

You have already seen how to use Hess's law to construct enthalpy cycles and enthalpy diagrams. In this chapter you will return to Hess's law and use it to investigate the enthalpy changes when an ionic compound is formed.

Definition of terms

When you measure a heat change at constant pressure, you call it an **enthalpy change**.

Standard conditions chosen are 100 kPa and a stated temperature, usually 298 K.

The standard molar enthalpy of formation $\Delta_f H^\ominus$ is the enthalpy change when one mole of a compound is formed from its constituent elements under standard conditions, all reactants and products in their standard states.

For example: $H_2(g) + \frac{1}{2}O_2(g) \rightarrow H_2O(l)$ $\Delta_f H^\ominus = -286\ kJ\ mol^{-1}$

The standard enthalpy of formation of an element is, by definition, zero.

The standard molar enthalpy change of combustion $\Delta_c H^\ominus$ is the enthalpy change when one mole of substance is completely burnt in oxygen.

For example: $CH_4(g) + 2O_2(g)$ $CO_2(g) + 2H_2O(l)$

$\Delta_c H^\ominus = -890\ kJ\ mol^{-1}$

The standard enthalpy of atomisation $\Delta_{at} H^\ominus$ is the enthalpy change which accompanies the formation of one mole of gaseous atoms from the element in its standard state under standard conditions.

For example: $Mg(s) \rightarrow Mg(g)$ $\Delta_{at} H^\ominus = +147.7\ kJ\ mol^{-1}$

$\frac{1}{2}Br_2(l) \rightarrow Br(g)$ $\Delta_{at} H^\ominus = +111.9\ kJ\ mol^{-1}$

$\frac{1}{2}Cl_2(g) \rightarrow Cl(g)$ $\Delta_{at} H^\ominus = +121.7\ kJ\ mol^{-1}$

This is given per mole of chlorine or bromine *atoms* and not per mole of chlorine or bromine *molecules*.

First ionisation energy (first IE) $\Delta_i H^\ominus$ is the standard enthalpy change when one mole of gaseous atoms is converted into a mole of gaseous ions each with a single positive charge.

For example: $Mg(g) \rightarrow Mg^+(g) + e^-$ $\Delta_i H^\ominus = +738\ kJ\ mol^{-1}$

or first IE = +738 kJ mol^{-1}

The **second ionisation energy** (second IE) refers to the loss of a mole of electrons from a mole of singly positively charged ions.

For example: $Mg^+(g) \rightarrow Mg^{2+}(g) + e^-$ $\Delta_i H^\ominus = +1451\ kJ\ mol^{-1}$

or second IE = + 1451 kJ mol^{-1}

The first electron affinity $\Delta_{ea} H^\ominus$ is the standard enthalpy change when a mole of gaseous atoms is converted to a mole of gaseous ions, each with a single negative charge.

For example: $O(g) + e^- \rightarrow O^-(g)$ $\Delta_{ea}H^\ominus = -141.1\,kJ\,mol^{-1}$

or first EA = $-141.1\,kJ\,mol^{-1}$

This refers to single atoms, not to oxygen molecules O_2.

The second electron affinity $\Delta_{ea}H^\ominus$ is the enthalpy change when a mole of electrons is added to a mole of gaseous ions each with a single negative charge to form ions each with two negative charges.

For example: $O^-(g) + e^- \rightarrow O^{2-}(g)$ $\Delta_{ea}H^\ominus = +798\,kJ\,mol^{-1}$

or second electron affinity = $+798\,kJ\,mol^{-1}$

Lattice enthalpy of formation $\Delta_L H^\ominus$ is the standard enthalpy change when one mole of solid ionic compound is formed from its gaseous ions.

For example: $Na^+(g) + Cl^-(g) \rightarrow NaCl(s)$ $\Delta_L H^\ominus = -788\,kJ\,mol^{-1}$

When a lattice forms, new bonds are formed, resulting in energy being given out, so $\Delta H^\ominus$ is always negative for this process.

The opposite process, when one mole of ionic compound separates into its gaseous ions, is called the **enthalpy of lattice dissociation**.

Lattice enthalpy of dissociation is the standard enthalpy change when one mole of solid ionic compound dissociated into its gaseous ions.

Lattice enthalpies cannot be measured directly – they need to be calculated (Topic 17.2). The enthalpy of lattice dissociation has the same numerical value as the lattice enthalpy, but $\Delta H^\ominus$ is always positive for this process.

Enthalpy of hydration $\Delta_{hyd}H^\ominus$ is the standard enthalpy change when water molecules surround one mole of gaseous ions.

For example: $Na^+(g) + aq \rightarrow Na^+(aq)$ $\Delta_{hyd}H^\ominus = -406\,kJ\,mol^{-1}$

or $Cl^-(g) + aq \rightarrow Cl^-(aq)$ $\Delta_{hyd}H^\ominus = -363\,kJ\,mol^{-1}$

Enthalpy of solution $\Delta_{sol}H^\ominus$ is the standard enthalpy change when one mole of solute dissolves completely in sufficient solvent to form a solution in which the molecules or ions are far enough apart not to interact with each other.

For example: $NaCl(s) + aq \rightarrow Na^+(aq) + Cl^-(aq)$ $\Delta_{sol}H^\ominus = +19\,kJ\,mol^{-1}$

Mean bond enthalpy is the enthalpy change when one mole of gaseous molecules each breaks a covalent bond to form two free radicals, averaged over a range of compounds.

For example: $CH_4(g) \rightarrow C(g) + 4H(g)$ $\Delta_{diss}H^\ominus = +1664\,kJ\,mol^{-1}$

So the mean (or average) C—H bond energy in methane is:

$$\frac{1664}{4} = +416\,kJ\,mol^{-1}$$

It is important to have an equation to refer to for enthalpy changes.

Ionic bonding

In a simple model of ionic bonding, electrons are transferred from metal atoms to non-metal atoms. Positively charged metal ions and negatively charged non-metal ions are formed that all have stable outer shells of electrons. These ions arrange themselves into a lattice so that ions of opposite charge are next to one another (Figure 1).

Synoptic link

Successive ionisation energies can help us to understand the arrangement of electrons in atoms – see Topic 1.6, Electron arrangements and ionisation energy.

Hint

First electron affinities are *always* negative as energy is given out when an electron is attracted to the positively charged nucleus of an atom. However, second electron affinities are *always* positive as energy must be put in to overcome the repulsion between an electron and a negatively charged ion.

Hint

Standard enthalpy changes are written in full as, for example, $\Delta_f H^\ominus_{298}$, but the 298 is often omitted – the symbol $\ominus$ being taken to imply that the figure refers to a temperature of 298 K (25 °C), which is around room temperature.

Synoptic link

Ionic bonding is discussed in Topic 3.1, The nature of ionic bonding.

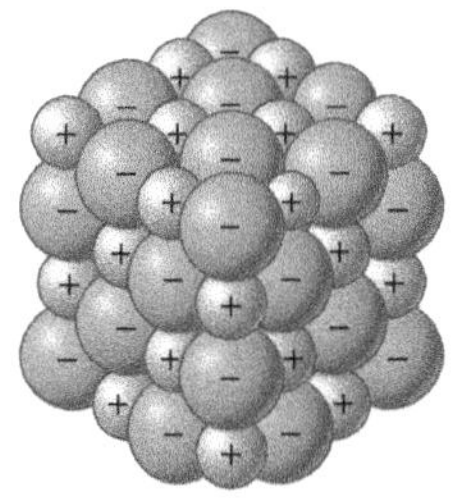

▲ **Figure 1** *Part of an ionic lattice*

Enthalpy changes on forming ionic compounds

If a cleaned piece of solid sodium is placed in a gas jar containing chlorine gas, an exothermic reaction takes place, forming solid sodium chloride:

$$Na(s) + \frac{1}{2}Cl_2(g) \rightarrow NaCl(s) \qquad \Delta_f H^\ominus = -411\,kJ\,mol^{-1}$$

You can think of it as taking place in several steps:

- The reaction involves solid sodium, not gaseous, and chlorine *molecules*, not separate atoms, so you must start with the enthalpy changes for atomisation:

$$Na(s) \rightarrow Na(g) \qquad \Delta_{at} H^\ominus = +108\,kJ\,mol^{-1}$$
$$\frac{1}{2}Cl_2(g) \rightarrow Cl(g) \qquad \Delta_{at} H^\ominus = +122\,kJ\,mol^{-1}$$

Energy has to be *put in* to pull apart the atoms ($\Delta_{at}H^\ominus$ is positive in both cases).

- The gaseous sodium atoms must each give up an electron to form gaseous Na^+ ions:

$$Na(g) \rightarrow Na^+(g) + e^-$$

The enthalpy change for this process is the enthalpy change of first ionisation (ionisation energy, first IE) of sodium and is +496 kJ mol^{-1}.

- The chlorine atoms must gain an electron to form gaseous Cl^- ions:

$$Cl(g) + e^- \rightarrow Cl^-(g)$$

The enthalpy change for this process of electron *gain* is the first electron affinity. The first electron affinity for the chlorine atom is −349 kJ mol^{-1} (i.e., energy is given out when this process occurs).

There is a further energy change. At room temperature sodium chloride exists as a solid lattice of alternating positive and negative ions, and not as separate gaseous ions. If positively charged ions come together with negatively charged ions, they form a solid lattice and energy is given out due to the attraction between the oppositely charged ions. This is called the lattice formation enthalpy $\Delta_L H^\ominus$ and it refers to the process:

$$Na^+(g) + Cl^-(g) \rightarrow NaCl(s) \qquad \Delta_L H^\ominus = -788\,kJ\,mol^{-1}$$

The following five processes lead to the formation of NaCl(s) from its elements.

- Atomisation of Na:
  $$Na(s) \rightarrow Na(g) \qquad \Delta_{at} H^\ominus = +108\,kJ\,mol^{-1}$$
- Atomisation of chlorine:
  $$\frac{1}{2}Cl(g) \rightarrow Cl(g) \qquad \Delta_{at} H^\ominus = +122\,kJ\,mol^{-1}$$
- Ionisation (e^- loss) of Na:
  $$Na(g) \rightarrow Na^+(g) + e^- \qquad \text{first IE} = +496\,kJ\,mol^{-1}$$
- Electron affinity of Cl:
  $$Cl(g) + e^- \rightarrow Cl^-(g) \qquad \text{first electron affinity} = -349\,kJ\,mol^{-1}$$
- Formation of lattice:
  $$Na^+(g) + Cl^-(g) \rightarrow NaCl(s) \qquad \Delta_L H^\ominus = -788\,kJ\,mol^{-1}$$

Hess's law tells us that the total energy (or enthalpy) change for a chemical reaction is the same *whatever route is taken*, provided that the initial and final conditions are the same. It does not matter whether the reaction actually takes place *via* these steps or not.

So the sum of the first five energy changes (taking the signs into account) is equal to the enthalpy change of formation of sodium chloride. You can calculate any one of the quantities, provided all the others are known. You do this by using a thermochemical cycle, called a **Born–Haber cycle**.

Hint

Do not confuse ionisation energy that refers to electron *loss* with electron affinity, which refers to electron *gain*.

Summary questions

1. a What is the value of ΔH for this process?
   $NaCl(s) \rightarrow Na^+(g) + Cl^-(g)$
   b Explain your answer.
   c What is the term that describes this process?
2. Explain why:
   a Loss of an electron from a sodium atom (ionisation) is an endothermic process.
   b Gain of an electron by a chlorine atom is an exothermic process.
3. a Write the equation to represent:
   i the first ionisation energy of aluminium
   ii the second ionisation energy of aluminium.
   b In terms of the enthalpy changes for the two processes in **a**, what is the enthalpy change when a $Al^{2+}(g)$ ion is formed from a Al(g) atom?

17.2 Born–Haber cycles

A **Born–Haber cycle** is a thermochemical cycle that includes all the enthalpy changes involved in the formation of an ionic compound. Born–Haber cycles are constructed by starting with the elements in their standard states. All elements in their standard states have zero enthalpy by definition.

The Born–Haber cycle for sodium chloride

There are six steps in the Born–Haber cycle for the formation of sodium chloride. Here you will use the cycle to calculate the lattice formation enthalpy ($\Delta_L H$ or LE). The other five steps are shown in Figure 1. (Remember that if you know any five, you can calculate the other). Figure 1 shows you how each step is added to the one before, starting from the elements in their standard state. Positive (endothermic changes) are shown upwards, and negative exothermic changes downwards.

$Na(s) \rightarrow Na(g)$ $\qquad \Delta_{at}H^{\ominus}$ Na = +108 kJ mol^{-1}

$\frac{1}{2}Cl_2(g) \rightarrow Cl(g)$ $\qquad \Delta_{at}H^{\ominus}$ Cl = +122 kJ mol^{-1}

$Na(g) \rightarrow Na^+(g) + e^-$ $\qquad$ first IE = +496 kJ mol^{-1}

$Cl(g) + e^- \rightarrow Cl^-(g)$ $\qquad$ first electron affinity = −349 kJ mol^{-1}

$Na(s) + \frac{1}{2}Cl_2(g) \rightarrow NaCl(s)$ $\qquad \Delta_f H^{\ominus}$ = −411 kJ mol^{-1}

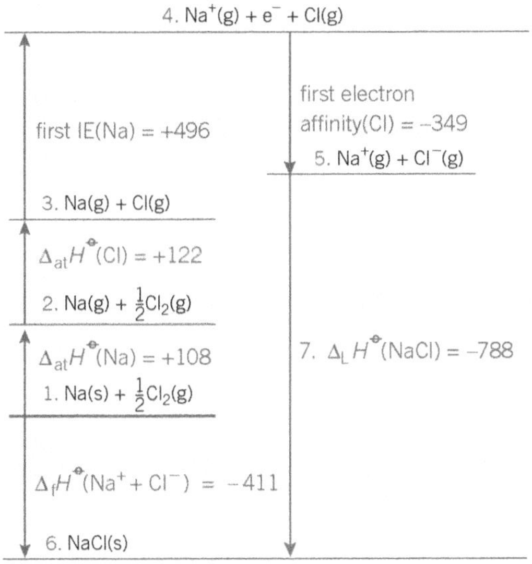

▲ **Figure 1** *Stages in the construction of the Born–Haber cycle for sodium chloride, NaCl, to find the lattice enthalpy. All enthalpies are in kJ mol^{-1}*

Using a Born–Haber cycle you can see why the formation of an ionic compound from its elements is an exothermic process. This is mainly due to the large amount of energy given out when the lattice forms.

1 Start with elements in their standard states. This is the energy zero of the diagram.

Learning objectives:

- → Illustrate how a Born–Haber cycle is constructed for a simple ionic compound.
- → Describe how Born–Haber cycles can be used to predict enthalpy changes of formation of theoretical compounds.

Specification reference: 3.1.8

Study tip

Remember, the standard enthalpy of atomisation is the enthalpy change which accompanies the *formation* of one mole of gaseous *atoms*.

To form one mole of Cl(g) you need $\frac{1}{2}Cl_2(g)$.

Study tip

Most errors in Born–Haber cycle calculations result either from lack of knowledge of the enthalpy change definitions, or lack of care with signs.

2 Add in the atomisation of sodium. This is positive, so it is drawn 'uphill'.
3 Add in the atomisation of chlorine. This too is positive, so draw 'uphill'.
4 Add in the ionisation of sodium, also positive and so drawn 'uphill'.
5 Add in the electron affinity of chlorine. This is a negative energy change and so it is drawn 'downhill'.
6 Add in the enthalpy of formation of sodium chloride, also negative and drawn 'downhill'.
7 The final unknown quantity is the lattice formation enthalpy of sodium chloride. The size of this is 788 kJ mol^{-1} from the diagram. Lattice formation enthalpy is the change from separate ions to solid lattice and you must therefore go 'downhill', so $LE(Na^+ + Cl^-)(s) = -788\,kJ\,mol^{-1}$.

When drawing Born–Haber cycles:

- make up a rough scale, for example, one line of lined paper to 100 kJ mol^{-1}
- plan out roughly first to avoid going off the top or bottom of the paper. (The zero line representing elements in their standard state will need to be in the middle of the paper.)
- remember to put in the sign of each enthalpy change and an arrow to show its direction. Positive enthalpy changes go up, negative enthalpy changes go down.

Worked example: The Born–Haber cycle for magnesium chloride

Figure 2 shows the complete Born–Haber cycle for the formation of magnesium chloride, $MgCl_2$, from its elements, together with notes on how it is constructed.

Since two chlorine atoms are involved all the quantities related to chlorine are doubled, that is, $\Delta_{at}H^{\ominus}(Cl)$ and the first electron affinity are both *multiplied* by two.

Also notice that the ionisation of magnesium, $Mg \rightarrow Mg^{2+}$, is the first ionisation enthalpy *plus* the second ionisation enthalpy. The second ionisation enthalpy is larger because it is more difficult to lose an electron from a positively charged ion than from a neutral atom.

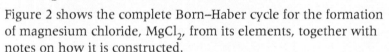

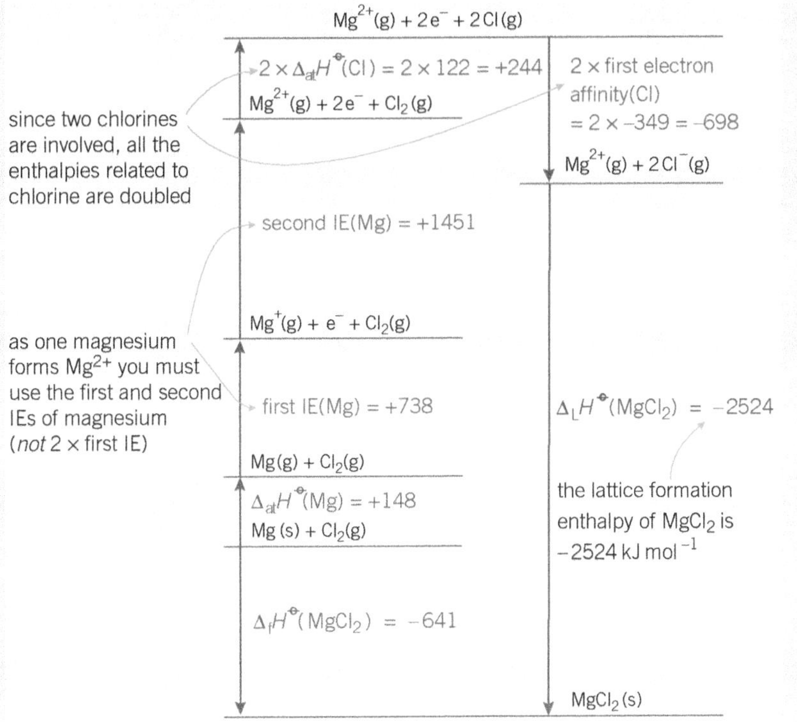

▲ **Figure 2** *The Born–Haber cycle for magnesium chloride, $MgCl_2$. All enthalpies are in kJ mol^{-1}*

Trends in lattice enthalpies

The lattice formation enthalpies of some simple ionic compounds of formula M^+X^- are given in Table 1.

▼ **Table 1** *Some values of lattice formation enthalpies in kJ mol^{-1} for compounds M^+X^-*

		Larger negative ions (anions) →			
		F^-	Cl^-	Br^-	I^-
Larger positive ions (cations) ↓	Li^+	−1031	−848	−803	−759
	Na^+	−918	−788	−742	−705
	K^+	−817	−711	−679	−651
	Rb^+	−783	−685	−656	−628
	Cs^+	−747	−661	−635	−613

Larger ions lead to smaller lattice enthalpies. This is because the opposite charges do not approach each other as closely when the ions are larger.

Table 2 shows lattice enthalpies for some compounds $M^{2+}X^{2-}$. You can see the same trend related to size of ions as before in Table 1.

Comparing Table 1 with Table 2 shows that for ions of approximately similar size (i.e., formed from elements in the same group of the Periodic Table, such as Na^+ and Mg^{2+} or F^- and O^{2-}) the lattice enthalpy increases with the size of the charge. This is because ions with double the charge give out roughly twice as much energy when they come together.

Hint

For chlorine, you multiply the first electron affinity and $\Delta_{at}H^{\ominus}$ by two.

For magnesium, you add together the first and second ionisation energy.

This is because there are two chlorine atoms gaining *one* electron (so two first electron affinities), but *one* magnesium atom losing *two* electrons (so the first *and* second ionisation energy).

▼ **Table 2** *Some values of lattice formation enthalpies for compounds $M^{2+}X^{2-}$*

		Larger anions →	
		O^{2-}	S^{2-}
Larger cations ↓	Be^{2+}	−4443	−3832
	Mg^{2+}	−3791	−3299
	Ca^{2+}	−3401	−3013
	Sr^{2+}	−3223	−2848
	Ba^{2+}	−3054	−2725

Predicting enthalpies of formation of theoretical compounds – worked example

Born–Haber cycles can be used to investigate the enthalpy of formation of theoretical compounds to see if they might be expected to exist. The cycles in Figure 3 are for CaF, CaF_2, and CaF_3. They use lattice enthalpies that have been calculated using sensible assumptions about the crystal structures of these compounds and the sizes of the Ca^+ and Ca^{3+} ions.

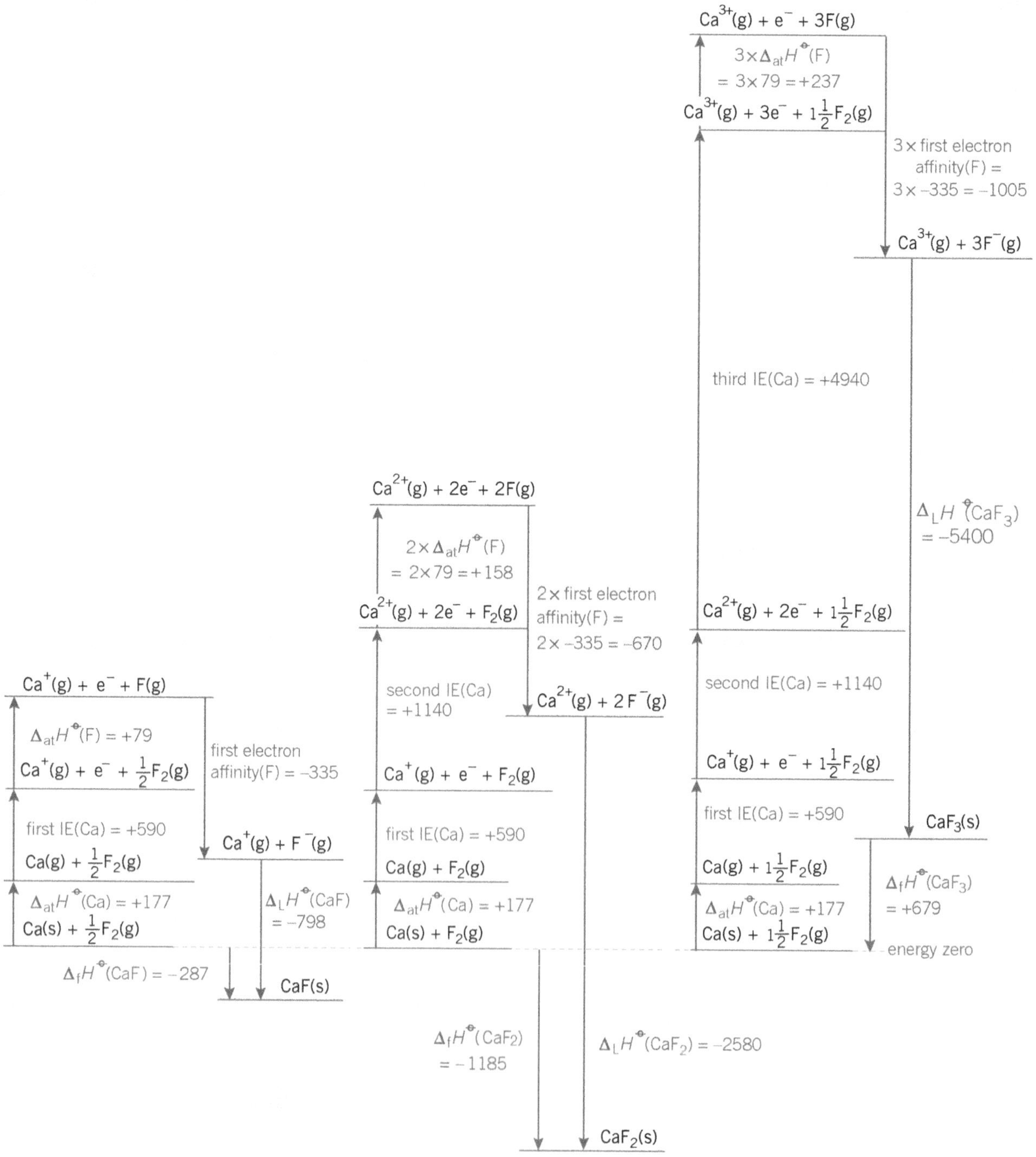

▲ **Figure 3** *Born–Haber cycles for CaF, CaF_2, and CaF_3. All enthalpies in kJ mol^{-1}*

Look at the enthalpies of formation. A large amount of energy would have to be put in to form CaF_3. The formation of CaF would give out energy but not as much as CaF_2. This explains why only CaF_2 has been prepared as a stable compound. CaF has indeed been made but readily turns into CaF_2 and Ca. It is unstable with respect to CaF_2 and Ca.

You can see from the relative enthalpy levels of CaF and CaF_2 on Figure 3 that CaF_2 is $1185 - 287 = 898\,kJ\,mol^{-1}$ below CaF.

You can draw a thermochemical cycle to calculate ΔH for the reaction:

$$\Delta H^{\ominus} = +574 - 1185 = -611\ kJ\ mol^{-1}$$

$2CaF(s) \longrightarrow CaF_2(s) + Ca(s)$

$+574\ kJ\ mol^{-1}$

$2 \times \Delta_f H^{\ominus}(CaF)$
$= 2 \times -287$
$= -574\ kJ\ mol^{-1}$

$\Delta_f H^{\ominus}(CaF_2) =$
$-1185\ kJ\ mol^{-1}$

$-1185\ kJ\ mol^{-1}$

$2\ Ca(s) + F_2(g)$

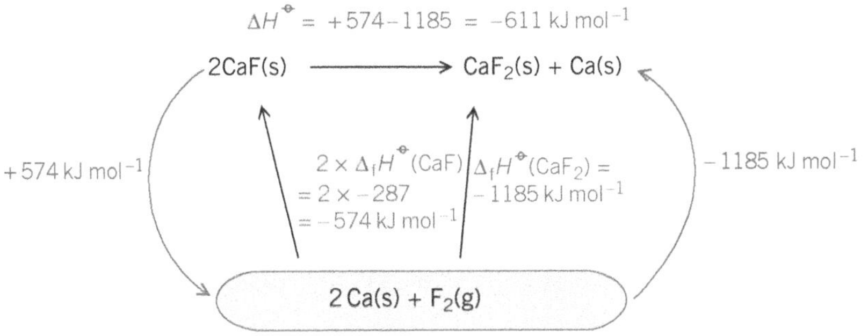

+ The first noble gas compound

The noble gases are often called the inert gases and, until 1962, they seemed to be just that – inert. There were no known compounds of them at all. This was explained on the basis of their stable electron arrangements. It had been predicted that there might be compounds of krypton and xenon with fluorine, but no one took much notice. However, in 1962 British chemist Neil Bartlett created a chemical sensation when he announced that he had prepared the first noble gas compound, xenon hexafluoroplatinate(V). Although the name may seem exotic, Bartlett predicted that the compound had a good chance of existing by using a very simple piece of chemical theory.

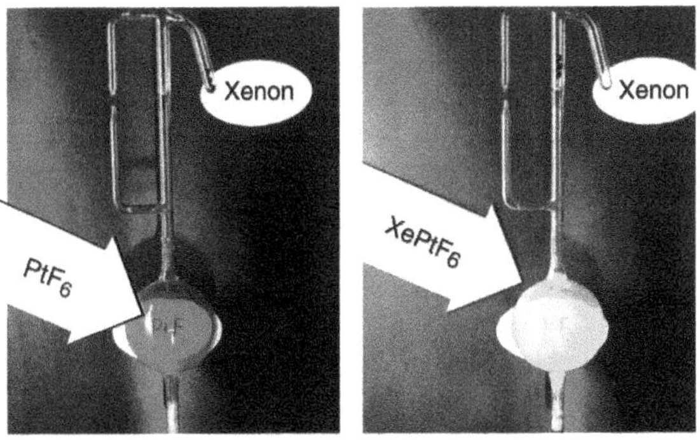

▲ **Figure 4** *Formation of XePtF$_6$*

He had previously found that the powerful oxidising agent platinum(VI) fluoride gas, PtF_6, would oxidise oxygen molecules to form the compound dioxygenyl hexfluoroplatinate(V), $O_2^+PtF_6^-$, in which the oxidising agent has removed an electron from an oxygen molecule.

He then realised that the first ionisation energy of xenon (the energy required to remove an electron from an atom of it) was a little less positive than that of the oxygen molecule, so that if platinum(VI) fluoride could remove an electron from oxygen, it should also be able to remove one from xenon. The values are:

$Xe(g) \rightarrow Xe^+(g) + e^-$ $\Delta H = +1170\ kJ\ mol^{-1}$ (first IE of xenon)

$O_2(g) \rightarrow O_2^+(g) + e^-$ $\Delta H = +1183\ kJ\ mol^{-1}$ (first IE of an oxygen molecule)

This is not the same as the first ionisation energy of an oxygen *atom*.

The experiment itself was surprisingly simple, as soon as the two gases came into contact, the compound was formed immediately – no heat or catalyst was required. In Bartlett's own words 'When I broke the seal between the red PtF_6 gas and the colourless xenon gas, there was an immediate interaction, causing an orange-yellow solid to precipitate. At once I tried to find someone with whom to share the exciting finding, but it appeared that everyone had left for dinner!'

This was one of those moments when all the textbooks had to be re-written.

The reaction can be represented:

$$Xe(g) + PtF_6(g) \rightarrow Xe^+PtF_6^-(s)$$

More recently it has been realised that the formula of the product may be a little more complex than this.

There are now over 100 noble gas compounds known, although most are highly unstable.

1 Write an equation to represent the first ionisation energy of an oxygen *atom*.
2 If you assume that noble gas compounds are formed with positive noble gas ions, suggest why compounds of xenon and krypton were predicted rather than ones of helium or neon.
3 Why might you *not* expect platinum(VI) fluoride to be a gas?
4 Explain the oxidation states of the elements in $Xe^+PtF_6^-$.

Summary question

1 a Draw a Born–Haber cycle to find the lattice formation enthalpy for sodium fluoride, NaF. The values for the relevant enthalpy terms are given below.

b What is the lattice formation enthalpy of NaF, given these values?

$Na(s) \rightarrow Na(g)$ $\Delta_{at}H^\ominus = +108\ kJ\ mol^{-1}$

$\frac{1}{2}F_2(g) \rightarrow F(g)$ $\Delta_{at}H^\ominus = +79\ kJ\ mol^{-1}$

$Na(g) \rightarrow Na^+(g) + e^-$ first IE $= +496\ kJ\ mol^{-1}$

$F(g) + e^- \rightarrow F^-(g)$ first electron affinity $= -328\ kJ\ mol^{-1}$

$Na(s) + \frac{1}{2}F_2(g) \rightarrow NaF(s)$ $\Delta_f H^\ominus = -574\ kJ\ mol^{-1}$

17.3 More enthalpy changes

Learning objectives:

→ Describe how to find the enthalpy change of solution.

→ Describe the evidence that theoretical calculations for lattice enthalpies provide about bonding.

Specification reference: 3.1.8

Finding the enthalpy of solution

Ionic solids can only dissolve well in polar solvents. In order to dissolve an ionic compound the lattice must be broken up. This requires an input of energy – the lattice enthalpy. The separate ions are then solvated by the solvent molecules, usually water. These cluster round the ions so that the positive ions are surrounded by the negative ends of the dipole of the water molecules and the negative ions are surrounded by the positive ends of the dipoles of the water molecules. This is called **hydration** when the solvent is water (Figure 1).

The enthalpy change of hydration shows the same trends as lattice enthalpy – it is more negative for more highly charged ions and less negative for bigger ions.

You can think of dissolving an ionic compound in water as the sum of three processes:

1 Breaking the ionic lattice to give separate gaseous ions – the lattice dissociation enthalpy has to be put in.

2 Hydrating the positive ions (cations) – the enthalpy of hydration is given out.

3 Hydrating the negative ions (anions) – the enthalpy of hydration is given out.

For ionic compounds the enthalpy change of hydration has rather a small value and may be positive or negative. For example, the enthalpy of hydration $\Delta_{hyd}H^{\ominus}$ for sodium chloride is given by the equation:

$$NaCl(s) + aq \rightarrow Na^+(aq) + Cl^-(aq)$$

It may be calculated via a thermochemical cycle as shown below.

These are the steps that are needed:

4 $NaCl(s) \rightarrow Na^+(g) + Cl^-(g)$ $\quad \Delta_L H^{\ominus} = +788\,kJ\,mol^{-1}$

This is the enthalpy change for lattice dissociation.

5 $Na^+(g) + aq + Cl^-(g) \rightarrow Na^+(aq) + Cl^-(g)$ $\quad \Delta_{hyd}H^{\ominus} = -406\,kJ\,mol^{-1}$

This is the enthalpy change for the hydration of the sodium ion.

6 $Na^+(aq) + Cl^-(g) + aq \rightarrow Na^+(aq) + Cl^-(aq)$ $\quad \Delta_{hyd}H^{\ominus} = -363\,kJ\,mol^{-1}$

This is the enthalpy change for the hydration of the chloride ion.

7 So $\Delta_{hyd}H^{\ominus}(NaCl) = \Delta_L H^{\ominus}(NaCl) + \Delta_{hyd}H^{\ominus}(Na^+) + \Delta_{hyd}H^{\ominus}(Cl^-)$

$+788 \qquad -406 \qquad -363 = +19\,kJ\,mol^{-1}$

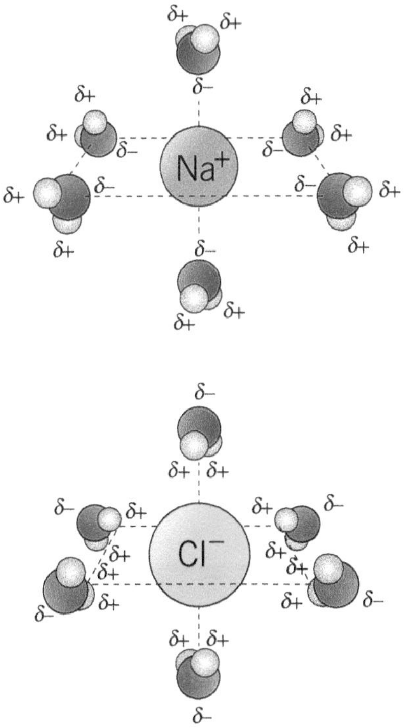

▲ **Figure 1** *The hydration of sodium and chloride ions by water molecules*

The process of dissolving can be represented on an enthalpy diagram (Figure 2) or calculated directly as above. Either method is equally acceptable.

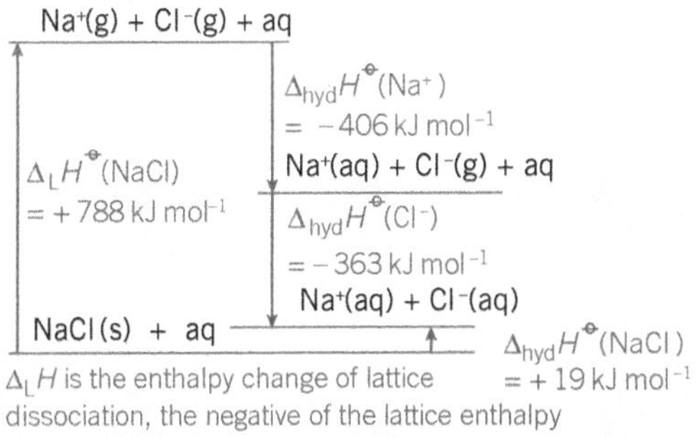

▲ **Figure 2** *Thermochemical cycle for the enthalpy of hydration of sodium chloride*

Lattice enthalpies and bonding

It is possible to work out a theoretical value for the lattice formation enthalpy of an ionic compound if you know the charge on the ions, their distance apart, and the geometry of its structure.

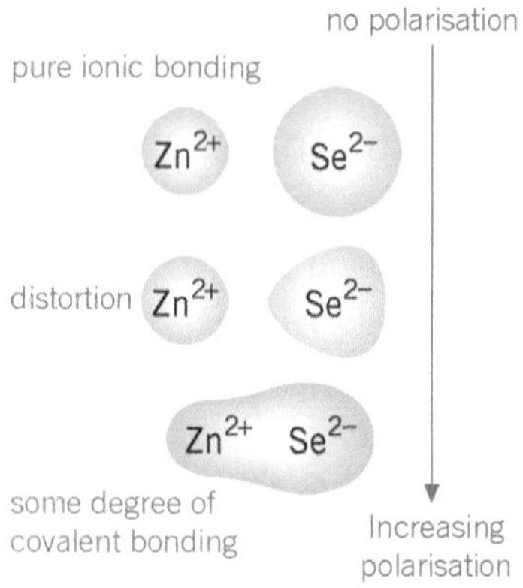

▲ **Figure 3** *Polarisation in zinc selenide*

For many ionic compounds, the lattice formation enthalpy determined from experimental values via a Born–Haber cycle agrees very well with that calculated theoretically, and this confirms that you have the correct model of ionic bonding in that compound. However, there are some compounds where there is a large discrepancy between the two values for lattice formation enthalpy because the bond in question has some covalent character.

For example, zinc selenide, ZnSe, has an experimental lattice formation enthalpy of −3611 kJ mol^{-1}. The theoretical value, based on the model of complete ionisation (Zn^{2+} + Se^{2-}) is −3305 kJ mol^{-1}, some 10% lower. The greater experimental value implies some extra bonding is present.

This can be explained as follows. The ion Zn^{2+} is relatively small and has a high positive charge, whilst Se^{2-} is relatively large and has a high negative charge. The small Zn^{2+} can approach closely to the electron clouds of the Se^{2-} and distort them by attracting them towards it (Figure 3). The Se^{2-} is fairly easy to distort, because its large size means the electrons are far from the nucleus and its double charge means there is plenty of negative charge to distort. This distortion means there are more electrons than expected concentrated *between* the Zn and Se nuclei, and represents a degree of electron sharing or covalency which accounts for the lattice enthalpy discrepancy. The Se^{2-} ion is said to be **polarised**.

The factors which increase polarisation are:

- positive ion (cation) – small size, high charge
- negative ion (anion) – large size, high charge.

Table 1 shows some values of experimentally determined lattice enthalpies, compared with those calculated assuming pure ionic bonding. The biggest discrepancy (the most extra covalent-type bonding) is cadmium iodide. The cadmium ion is small and doubly charged, whilst the iodide ion is large and easily polarised.

▼ **Table 1** *Some values of experimental and calculated lattice enthalpies*

Compound	Experimental value of LE / kJ mol^{-1}	Calculated value of LE assuming ionic bonding / kJ mol^{-1}
LiF	−1031	−1021
NaCl	−780	−777
KBr	−679	−667
CaF_2	−2611	−2586
CdI_2	−2435	−1986
AgCl	−890	−769

So all ionic and covalent bonds can be seen as part of a continuum from purely ionic to purely covalent. For example, caesium fluoride, Cs^+F^-, which has a large singly charged positive ion and a small singly charged negative ion, is hardly polarised at all and is almost completely ionic, whereas a bond between two identical atoms *must* be 100% covalent.

Study tip

It is important that you know why lattice enthalpies obtained by theoretical calculation and from a Born–Haber cycle can differ.

Summary questions

1. Draw a diagram to calculate the enthalpy of hydration of potassium bromide using the experimental value of lattice enthalpy in Table 1.
   $\Delta_{hyd}H^\ominus(K^+) = -322$ kJ mol^{-1}
   $\Delta_{hyd}H^\ominus(Br^-) = -335$ kJ mol^{-1}
   Why is the value relatively small?
2. Explain why compounds of beryllium and aluminium are nearly all significantly covalent. Would you expect the calculated value of lattice enthalpy to be greater or smaller than the experimental value? Explain your answer.

17.4 Why do chemical reactions take place?

Chemists use the terms **feasible** or **spontaneous** to describe reactions which could take place of their own accord. The terms take no account of the rate of the reaction, which could be so slow as to be unmeasurable at room temperature.

You may have noticed that many of the reactions that occur of their own accord are exothermic (ΔH is negative). For example, if you add magnesium to copper sulfate solution, the reaction to form copper and magnesium sulfate takes place and the solution gets hot. Negative ΔH is a factor in whether a reaction is spontaneous, but it does not explain why a number of endothermic reactions are spontaneous.

For example, both the following reactions, which occur spontaneously, are endothermic (ΔH is positive):

$$\underset{\text{citric acid}}{C_6H_8O_7(aq)} + \underset{\text{sodium hydrogencarbonate}}{3NaHCO_3(aq)} \rightarrow \underset{\text{sodium citrate}}{Na_3C_6H_5O_7(aq)} + \underset{\text{water}}{3H_2O(l)} + \underset{\text{carbon dioxide}}{3CO_2(g)}$$

$$\underset{\text{ammonium nitrate}}{NH_4NO_3(s)} + aq \rightarrow \underset{\text{aqueous ammonium nitrate}}{NH_4NO_3(aq)}$$

Entropy or randomness

Many processes which take place spontaneously involve mixing or spreading out, for example, liquids evaporating, solids dissolving to form solutions, or gases mixing.

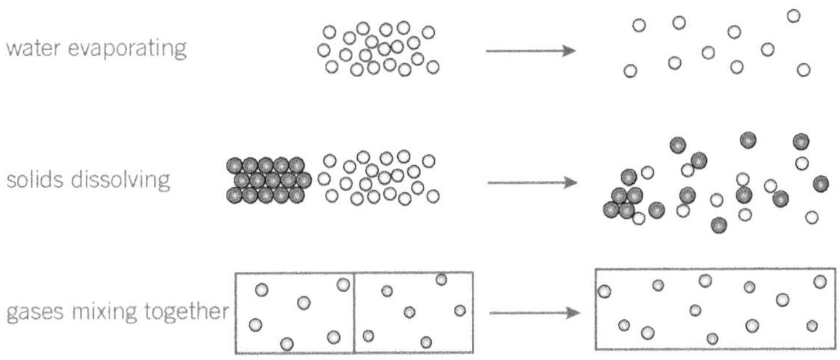

▲ **Figure 1** *Spontaneous processes*

This is the clue to the second factor which drives chemical processes – a tendency towards randomising or disordering, that is, towards chaos. Gases are more random than liquids, and liquids are more random than solids, because of the arrangement of their particles.

So endothermic reactions may be spontaneous if they involve spreading out, randomising, or disordering. This is true of the two reactions above – the arrangement of the particles in the products is more random than in the reactants.

The randomness of a system, expressed mathematically, is called the **entropy** of the system and is given the symbol S. A reaction like the two above, in which the products are more disordered than the reactants, will have positive values for the entropy change ΔS.

Learning objectives:

- → Explain why endothermic reactions occur.
- → Explain how a temperature change affects feasibility.

Specification reference: 3.1.8

Hint

The reaction between citric acid and sodium hydrogencarbonate takes place on your tongue when you eat sherbet – you can feel your tongue getting cold.

Study tip

In chemistry the words 'spontaneous' and 'feasible' mean exactly the same thing – that a reaction has a tendency to happen.

Entropies have been determined for a vast range of substances and can be looked up in databases. They are usually quoted for standard conditions: 298 K and 100 kPa pressure. Table 1 gives some examples.

Study tip

Take care – the units of entropy are $J K^{-1} mol^{-1}$, *not* $kJ K^{-1} mol^{-1}$.

▼ **Table 1** *Some values of entropy*

Substance	State at standard conditions	Entropy S / $J K^{-1} mol^{-1}$
carbon (diamond)	solid	2.4
carbon (graphite)	solid	5.7
copper	solid	33.0
iron	solid	27.0
ammonium chloride	solid	95.0
calcium carbonate	solid	93.0
calcium oxide	solid	40.0
iron(III) oxide	solid	88.0
water (ice)	solid	48.0
water (liquid)	liquid	70.0
mercury	liquid	76.0
water (steam)	gas	189.0
hydrogen chloride	gas	187.0
ammonia	gas	192.0
carbon dioxide	gas	214.0

Study tip

Unlike enthalpies, the entropies of elements in their standard states are *not* zero. All entropies are positive numbers, but it is possible to have a negative entropy *change* for a reaction.

In general, gases have larger values than liquids, which have larger values than solids.

Table 1 shows that the entropy increases when water turns to steam. Entropies increase with temperature, largely because at higher temperatures particles spread out and randomness increases.

Calculating entropy changes

The entropy change for a reaction can be calculated by adding all the entropies of the products and subtracting the sum of the entropies of the reactants. For example:

$$CaCO_3(s) \rightarrow CaO(s) + CO_2(g)$$

Using the values from Table 1:

$$\text{entropy of products} = 40 + 214 = 254\,J K^{-1} mol^{-1}$$

$$\text{entropy of reactant} = 93\,J K^{-1} mol^{-1}$$

$$\Delta S = 254 - 93 = +161\,J K^{-1} mol^{-1}$$

This is a large positive value – a gas is formed from a solid.

The Gibbs free energy change ΔG

You have seen above that a combination of *two* factors govern the feasibility of a chemical reaction:

- the enthalpy change
- the entropy change.

These two factors are combined in a quantity called the **Gibbs free energy *G***. If the change in G, ΔG, for a reaction is negative, then this reaction is feasible. If ΔG is positive, the reaction is not feasible.

ΔG combines the enthalpy change ΔH and entropy change ΔS factors as follows:

$$\Delta G = \Delta H - T\Delta S$$

ΔG depends on temperature, because of the term $T\Delta S$. This means that some reactions may be feasible at one temperature and not at another. So an endothermic reaction can become feasible when temperature is increased if there is a large enough positive entropy change. (A *positive* value for ΔS will make ΔG more *negative* because of the negative sign in the $T\Delta S$ term.)

Here are some examples of how this works.

Take the reaction:

$$CaCO_3(s) \rightarrow CaO(s) + CO_2(g) \qquad \Delta H = +178\,kJ\,mol^{-1}$$

You have seen above that $\Delta S = +161\ J\,K^{-1}\,mol^{-1} = 0.161\,kJ\,K^{-1}\,mol^{-1}$

So at room temperature (298 K):

$$\Delta G = \Delta H - T\Delta S$$

$$\Delta G = 178 - (298 \times 0.161) = +130\,kJ\,mol^{-1}$$

This positive value means that the reaction is not feasible at room temperature. The reverse reaction will have $\Delta G = -130\,kJ\,mol^{-1}$ and will be feasible:

$$CaO(s) + CO_2(g) \rightarrow CaCO_3(s)$$

This is the reaction that occurs in desiccators to absorb carbon dioxide.

However, if you do the calculation for a temperature of 1500 K, you get a different result:

At 1500 K:

$\Delta G = \Delta H - T\Delta S$

$\Delta G = 178 - (1500 \times 0.161)$

$\Delta G = 178 - 242$

$\Delta G = -64\,kJ\,mol^{-1}$

ΔG is negative and the reaction is feasible at this temperature. This is the reaction that occurs in a lime kiln to make lime (calcium oxide) from limestone (calcium carbonate).

What happens when ΔG = 0?

There is a temperature at which $\Delta G = 0$ for this reaction. This is the point at which the reaction is just feasible. You can calculate this temperature for the reaction above:

$\Delta G = \Delta H - T\Delta S$

$0 = \Delta H - T\Delta S$

$\Delta H = T\Delta S$ where $\Delta H = +178\,kJ\,mol^{-1}$ and $\Delta S = 0.161\,kJ\,K^{-1}$

So $T = \dfrac{178}{0.161}$

$= 1105.6\,K$

The link between ΔG and K, the equilibrium constant

Strictly speaking, ΔG is related to K, the equilibrium constant of the reaction.

$$\Delta G = -RT\ln K$$

The negative sign in this equation means that if ΔG is negative, then $RT\ln K$ is positive, so the equilibrium constant is greater than one. This means the reaction is feasible. For example, if ΔG is $-85\ kJ\,mol^{-1}$, K is 10^{15}.

Hint

The symbol ln stands for natural log or log to the base e.

Study tip

Remember to convert the entropy units by dividing by 1000 because enthalpy is measured in $kJ\,mol^{-1}$ and entropy in $J\,K^{-1}\,mol^{-1}$.

In fact, the reaction does not suddenly flip from feasible to non-feasible. In a closed system an equilibrium exists around this temperature in which both products and reactants are present.

Calculating an entropy change

You can use the temperature at which $\Delta G = 0$ to calculate an entropy change. For example, a solid at its melting point is equally likely to exist as a solid or a liquid – an equilibrium exists between solid and liquid. So ΔG for the melting process must be zero and:

$$0 = \Delta H - T\Delta S$$

For example, the melting point for water is 273 K and the enthalpy change for melting is 6.0 kJ mol^{-1}. Putting these values into the equation:

$$0 = 6.0 - 273 \times \Delta S$$

$$\Delta S = \frac{6.0}{273} = 0.022\,\text{kJ}\,\text{K}^{-1}\,\text{mol}^{-1} = +\,22\,\text{J}\,\text{K}^{-1}\,\text{mol}^{-1}$$

This is the entropy change that occurs when ice changes to water. It is positive, which you would expect as the molecules in water are more disordered than those in ice.

Hint

At the melting point of a substance:
$\Delta H = T\Delta S$

Determining an entropy change

You can measure the abstract quantity of an entropy change using kitchen equipment.

A chemistry teacher set out to find the entropy change for the vaporisation of water at home using a household kettle and a top pan balance.

At its boiling point, water is equally likely to exist as liquid or vapour (water or steam), so for vaporisation, $\Delta G = 0$.

Inserting the ΔG value into $\Delta G = \Delta H - T\Delta S$ gives:

$0 = \Delta H - T\Delta S$

So $\Delta H = T\Delta S$

Rearrange to $\Delta S = \frac{\Delta H}{T}$

The boiling point of water (at atmospheric pressure) is 100 °C (373 K), $T = 373$ K so all they needed to measure was ΔH.

The kettle had a power rating of 2.4 kW, which means it supplies 2.4 kJ of energy per second.

They brought some water to the boil in an ordinary kitchen kettle, and weighed the kettle and its contents on a top pan balance that read to the nearest gram. They switched on the kettle again and allowed it to boil for 100 seconds holding down the automatic switch. They then reweighed the kettle to find how much water had boiled away. They found that 100 g of water had boiled away, that is, turned from water to steam (vaporised).

Calculate $\Delta_{vap}S$ by the following steps:

1. Calculate how many kilojoules of heat were supplied to the water in 100 s.
2. Calculate the value of M_r for water and hence find how many moles of water were vaporised.

3 Calculate $\Delta_{vap}H$ for the process in $kJ\ mol^{-1}$ and convert this into $J\ mol^{-1}$.

4 Use $\Delta_{vap}S = \frac{\Delta_{vap}H}{T_b}$ to calculate the entropy change of vaporisation.

5 To how many significant figures can you quote your answer?

6 What systematic error (experimental design error) is there in this experiment? How could you reduce it?

7 If the top pan balance weighs to the nearest gram, what is the percentage error in the weighing?

8 The value of $\Delta_{vap}S$ for water is higher than for most liquids. Suggest why. Hint – you are measuring the increase in *disorder* between the liquid and vapour phases. Think about what causes order in the liquid state of water.

Extracting metals

A good way of extracting metals from their oxide ores is to heat them with carbon, which removes the oxygen as carbon dioxide and leaves the metal. This has the advantage that carbon (in the form of coke) is cheap. The gaseous carbon dioxide simply diffuses away, so there is no problem separating it from the metal (although it does contribute to global warming as it is a greenhouse gas). You can use ΔG to investigate under what conditions the reaction might be feasible for different metals.

One of the most important metals is iron and its ore is largely iron(III) oxide, Fe_2O_3. You can calculate $\Delta G^\ominus$ from an energy level diagram.

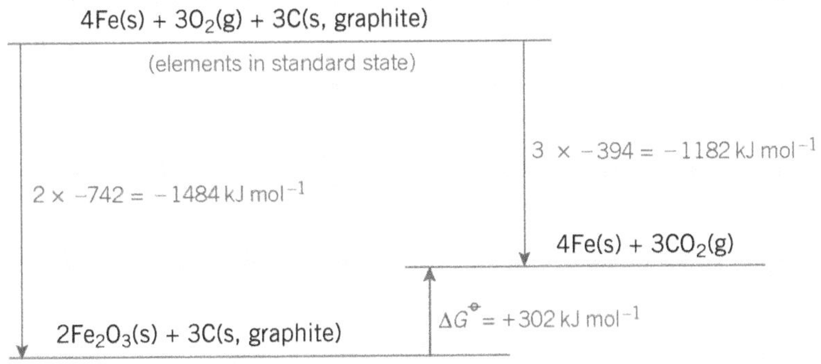

▲ **Figure 2** *Free energy diagram for the reduction of iron(III) oxide by graphite*

$2Fe_2O_3(s) + 3C(s, graphite) \rightarrow 4Fe(s) + 3CO_2(g)\ \Delta G^\ominus = +302\ kJ\ mol^{-1}$

So this reaction is not feasible under standard conditions (298 K).

Will the reaction take place at a higher temperature?

You can work out the temperature at which the reaction just becomes feasible (this is when $\Delta G = 0$) using:

$$\Delta G = \Delta H - T\Delta S$$

Synoptic link

Look back at Chapter 4, Energetics to revise thermochemical cycles.

Calculating ΔH

ΔH for the reaction can be calculated from the following thermochemical cycle.

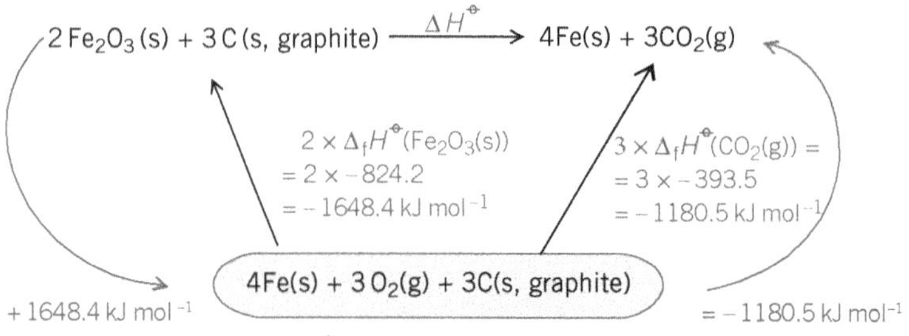

$\Delta H^{\circ} = +1648.4 - 1180.5\ \text{kJ mol}^{-1}$
$\Delta H^{\circ} = 467.9\ \text{kJ mol}^{-1}$

Calculating ΔS

You can calculate the entropy change of the reaction by finding the difference between the sum of the entropies of all the products and the sum of the entropies of all the reactants:

$$2Fe_2O_3(s) + 3C(s, graphite) \rightarrow 4Fe(s) + 3CO_2(g)$$

$(2 \times 87.4) + (3 \times 5.7)$ $\quad$ $(4 \times 27.3) + (3 \times 213.6)$

191.9 $\quad$ 750.0

$$\Delta S = +558.1\ \text{J K}^{-1}\ \text{mol}^{-1}$$

Study tip

Remember, to convert entropy in $\text{J K}^{-1}\ \text{mol}^{-2}$ to $\text{kJ K}^{-1}\ \text{mol}^{-1}$, divide by 1000.

This value is large and positive, as you would expect from a reaction in which two solids produce a gas.

Putting these values into $\Delta G = \Delta H - T\Delta S$:

$$0 = +467.9 - T \times \frac{558.1}{1000}$$

$$T = 467.9 \times \frac{1000}{558.1}$$

$$= 838.4\ \text{K}$$

The reaction is not feasible below this temperature.

Hint

In fact, in the blast furnace a higher temperature is used, above the melting point of iron (1808 K), so that the iron is formed as a liquid. Also, the carbon is not pure graphite, but coke.

Calculate ΔG at 2000 K using the values above.

−648.3 kJ mol−1

Kinetic factors

Neither enthalpy changes nor entropy changes tell us *anything* about how quickly or slowly a reaction is likely to go. So, you might predict that a certain reaction should occur spontaneously because of enthalpy and entropy changes, but the reaction might take place so slowly that for practical purposes it does not occur at all. In other words, there is a large activation energy barrier for the reaction.

Synoptic link

Look back at Topic 5.1, Collision theory, to revise activation energy.

Carbon gives an interesting example:

$$C(s, graphite) + O_2(g) \rightarrow CO_2(g) \qquad \Delta H^{\ominus} = -393.5\ \text{kJ mol}^{-1}$$

The reaction is exothermic and you can calculate the actual value of ΔS and so find ΔG.

Calculating ΔS

ΔS for the reaction is the sum of the entropies of the product minus the sum of the entropies of the reactants.

$$\underset{5.7}{C(s, graphite)} + \underset{205.0}{O_2(g)} \rightarrow \underset{213.6}{CO_2(g)} \qquad \Delta H^\ominus = -394\,kJ\,mol^{-1}$$

So $\Delta S = 213.6 - (5.7 + 205.0)$

$\Delta S = +2.9\ JK^{-1}\ mol^{-1}$, positive as predicted.

Calculating ΔG

$$\Delta G = \Delta H - T\Delta S$$

So under standard condition (approximately room temperature and pressure):

$$\Delta G = -394 - \left(298 \times \frac{2.9}{1000}\right)$$

Remember to divide the entropy value by 1000 to convert from $JK^{-1}\ mol^{-1}$ to $kJK^{-1}\ mol^{-1}$.

$$\Delta G = -394 - 0.86$$

$\Delta G = -394.86\,kJ\,mol^{-1}$, negative so the reaction is feasible.

However, experience with graphite (the 'lead' in pencils) tells you that the reaction does not take place at room temperature – although it will take place at higher temperatures. At room temperature, the reaction is so slow that in practice it doesn't take place at all.

Since the branch of chemistry dealing with enthalpy and entropy changes is called **thermodynamics**, and that dealing with rates is called **kinetics**, graphite is said to be thermodynamically unstable but kinetically stable.

Study tip

Entropies of elements in their standard states are *not* zero.

▼ **Table 2** *Entropy values for some substances*

Substance	$S^\ominus/JK^{-1}\ mol^{-1}$
$Mg(s)$	32.7
$MgO(s)$	26.9
$MgCO_3(s)$	65.7
$Zn(s)$	41.6
$ZnO(s)$	43.6
$Pb(NO_3)_2(s)$	213.0
$PbO(s)$	68.7
$NO_2(g)$	240.0
$O_2(g)$	205.0
$CO_2(g)$	213.6
$H_2O(l)$	69.7
$H_2O(g)$	188.7

Summary questions

1 a Without doing a calculation, predict whether the entropy change for the following reactions will be significantly positive, significantly negative, or approximately zero and explain your reasoning.

i $Mg(s) + ZnO(s) \rightarrow MgO(s) + Zn(s)$

ii $2Pb(NO_3)_2(s) \rightarrow 2PbO(s) + 4NO_2(g) + O_2(g)$

iii $MgO(s) + CO_2(g) \rightarrow MgCO_3(s)$

iv $H_2O(l) \rightarrow H_2O(g)$

b Calculate $\Delta S^\ominus$ for each reaction using data in Table 2. Comment on your answers.

2 For the reaction:

$MgO(s) \rightarrow Mg(s) + \frac{1}{2}O_2(g)$

$\Delta H^\ominus = +602\ kJ\ mol^{-1}$

$\Delta S^\ominus = +109\ J\ K^{-1}\ mol^{-1}$

a Using the equation $\Delta G = \Delta H - T\Delta S$, calculate ΔG at:

i 1000 K

ii 6000 K

iii At which temperature is the reaction feasible?

b Calculate the temperature when $\Delta G = 0$.

3 Calculate the entropy change for:

$NH_3(g) + HCl(g) \rightarrow NH_4Cl(s)$

ammonia, hydrogen chloride, ammonium chloride

The entropy values are:
$S^\ominus$ NH_3 $192\ J\ K^{-1}\ mol^{-1}$
$S^\ominus$ HCl $187\ J\ K^{-1}\ mol^{-1}$
$S^\ominus$ NH_4Cl $95\ J\ K^{-1}\ mol^{-1}$

1 **(a)** Figure 1 shows how the entropy of a molecular substance X varies with temperature.

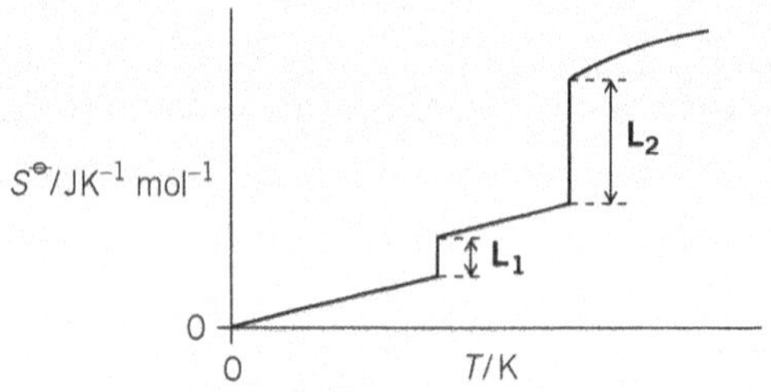

Figure 1

(i) Explain, in terms of molecules, why the entropy is zero when the temperature is zero kelvin.
(2 marks)

(ii) Explain, in terms of molecules, why the first part of the graph in Figure 1 is a line that slopes up from the origin.
(2 marks)

(iii) On Figure 1, mark on the appropriate axis the boiling point T_b of substance X.
(1 mark)

(iv) In terms of the behaviour of molecules, explain why L_2 is longer than L_1 in Figure 1.
(2 marks)

(b) Figure 2 shows how the free-energy change for a particular gas-phase reaction varies with temperature.

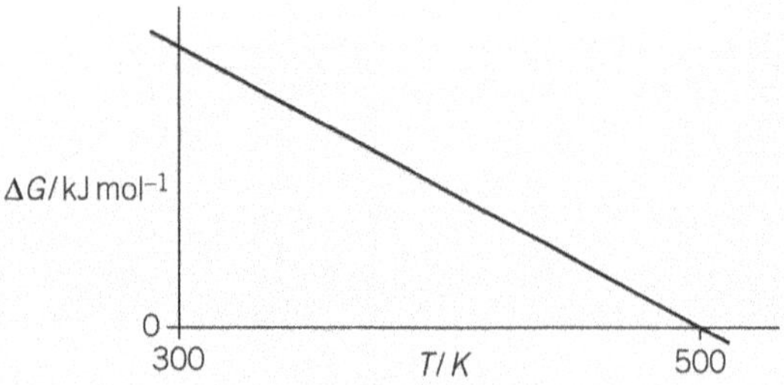

Figure 2

(i) Explain, with the aid of a thermodynamic equation, why this line obeys the mathematical equation for a straight line, $y = mx + c$.
(2 marks)

(ii) Explain why the magnitude of ΔG decreases as T increases in this reaction.
(1 mark)

(iii) State what you can deduce about the feasibility of this reaction at temperatures lower than 500 K.
(1 mark)

(c) The following reaction becomes feasible at temperatures above 5440 K.

$$H_2O(g) \rightarrow H_2(g) + \frac{1}{2}O_2(g)$$

The entropies of the species involved are shown in the following table.

	$H_2O(g)$	$H_2(g)$	$O_2(g)$
S / J K^{-1} mol^{-1}	189	131	205

(i) Calculate the entropy change ΔS for this reaction.
(1 mark)

(ii) Calculate a value, with units, for the enthalpy change for this reaction at 5440 K.
(3 marks)

AQA, 2013

2 The following equation shows the formation of ammonia.

$$\frac{1}{2}N_2(g) + \frac{3}{2}H_2(g) \rightarrow NH_3(g)$$

The graph shows how the free-energy change for this reaction varies with temperature above 240 K.

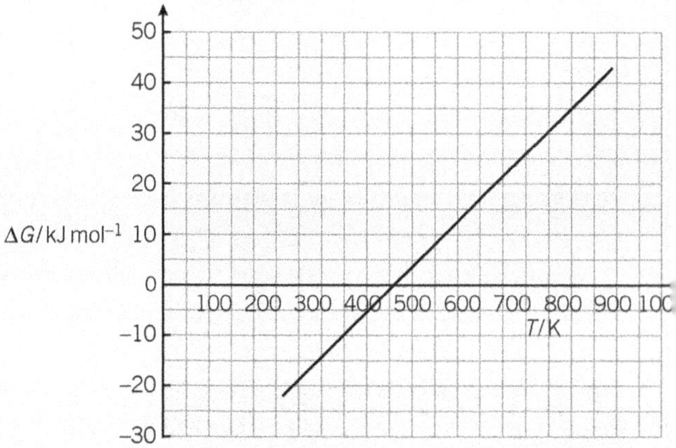

Figure 3

(a) Write an equation to show the relationship between ΔG, ΔH, and ΔS.
(1 mark)

(b) Use the graph to calculate a value for the slope (gradient) of the line. Give the units of this slope and the symbol for the thermodynamic quantity that this slope represents.
(3 marks)

(c) Explain the significance, for this reaction, of temperatures below the temperature value where the line crosses the temperature axis.
(2 marks)

(d) The line is not drawn below a temperature of 240 K because its slope (gradient) changes at this point.
Suggest what happens to the ammonia at 240 K that causes the slope of the line to change.
(1 mark)

AQA, 2012

3 This question is about magnesium oxide. Use data from the table below, where appropriate, to answer the following questions.

	$\Delta H^{\ominus}$/kJ mol^{-1}
First electron affinity of oxygen (formation of $O^-(g)$ from $O(g)$)	−142
Second electron affinity of oxygen (formation of $O^{2-}(g)$ from $O^-(g)$)	+844
Atomisation enthaply of oxygen	+248

(a) Define the term *enthalpy of lattice dissociation*.

(3 marks)

(b) In terms of the forces acting on particles, suggest one reason why the first electron affinity of oxygen is an exothermic process.

(1 mark)

(c) Complete the Born–Haber cycle for magnesium oxide by drawing the missing energy levels, symbols, and arrows.
The standard enthalpy change values are given in kJ mol^{-1}.

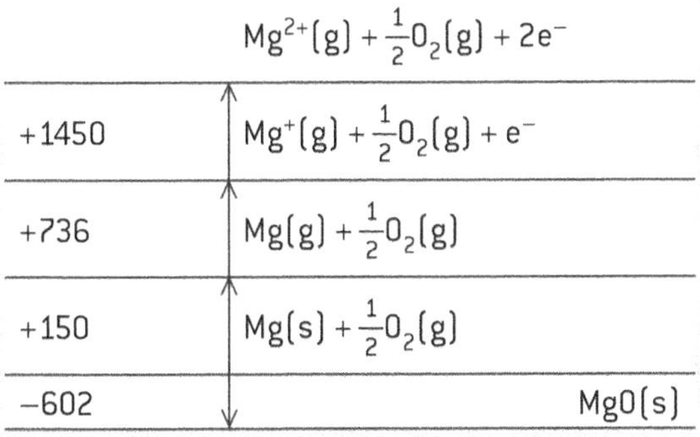

(4 marks)

(d) Use your Born–Haber cycle from part **c** to calculate a value for the enthalpy of lattice dissociation for magnesium oxide.

(2 marks)

(e) The standard free-energy change for the formation of magnesium oxide from magnesium and oxygen, $\Delta_f G^{\ominus} = -570$ kJ mol^{-1}. Suggest **one** reason why a sample of magnesium appears to be stable in air at room temperature, despite this negative value for $\Delta_f G^{\ominus}$.

(1 mark)

(f) Use the value of $\Delta_f G^{\ominus}$ given in part **e** and the value of $\Delta_f H^{\ominus}$ from part **c** to calculate a value for the entropy change $\Delta S^{\ominus}$ when one mole of magnesium oxide is formed from magnesium and oxygen at 298 K. Give the units of $\Delta S^{\ominus}$.

(3 marks)

(g) In terms of the reactants and products and their physical states, account for the sign of the entropy change that you calculated in part **f**.

(2 marks)

AQA, 2012

Answers to the Practice Questions and Section Questions are available at www.oxfordsecondary.com/oxfordaqaexams-alevel-chemistry

18 Kinetics

18.1 The rate of chemical reactions

Learning objective:

→ Define the rate of a reaction.

Specification reference: 3.1.9

Hint

Square brackets round a chemical symbol, [ ], are used to indicate its concentration in mol dm^{-3}.

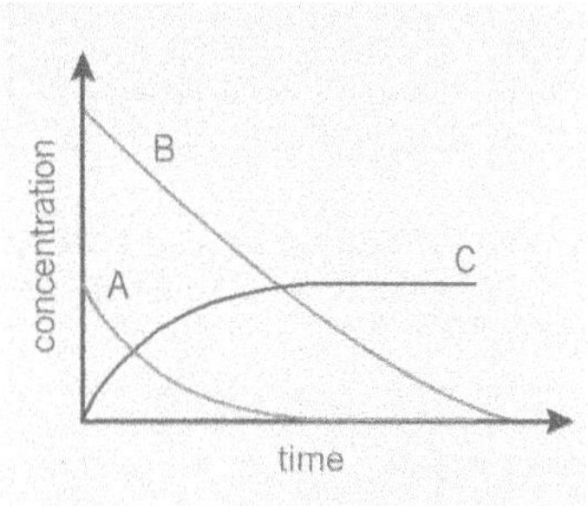

▲ **Figure 1** *Changes of concentration with time for A, B, and C*

Synoptic link

You will need to know the kinetics studied in Chapter 5, Kinetics.

Synoptic link

See Practical 6 on page 525.

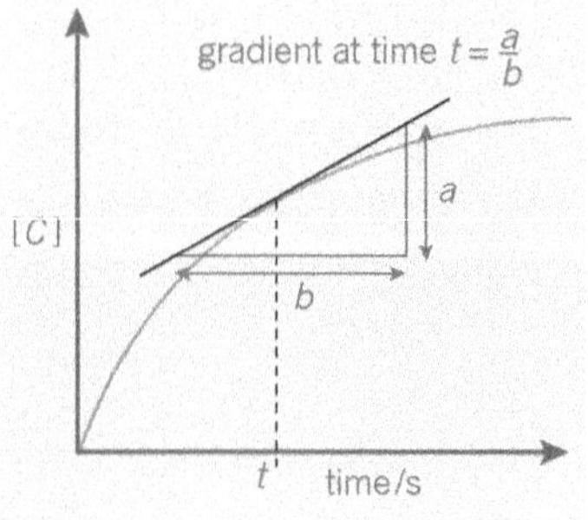

▲ **Figure 2** *The rate of change of [C] at time t, is the gradient of the concentration–time graph at t*

The main factors that affect the rate of chemical reactions are temperature, concentration, pressure, surface area, and catalysts. In this topic you will look at the measurement of reaction rates.

What is a reaction rate?

As a reaction, A + 2B → C, takes place, the concentrations of the reactants A and B decrease with time and the concentration of product C increases with time. You could measure the concentration of A, B, or C with time and plot the results (Figure 1).

The **rate of the reaction** is defined as the change in concentration (of any of the reactants or products) with unit time, but notice how different the graphs are for A, B, and C. As [C] (the product) increases, [A] and [B] (the reactants) decrease. However, as the equation tells us, for every A that reacts there are two of B, so [B] decreases twice as fast as [A]. For this reason it is important to state whether you are following A, B, or C. Usually it is assumed that a rate is measured by following the concentration of a product, because the concentration of the product increases with time.

The rate of reaction at any instant

You are often interested in the rate at a *particular* instant in time rather than over a period of time. To find the rate of change of [C] at a particular instant, draw a tangent to the curve at that time and then find its gradient (slope), as in Figure 2.

$$\text{rate} = \frac{a}{b} = \frac{\text{change in concentration}}{\text{time}}$$

Worked example: Measuring a reaction rate

To measure a reaction rate, you need a method of measuring the concentration of one of the reactants or products over a period of time (keeping the temperature constant, because rate varies with temperature). The method chosen will depend on the substance whose concentration is being measured and also on the speed of the reaction.

Reaction rates are measured in mol dm^{-3} s^{-1}.

For example, in the reaction between bromine and methanoic acid, the solution starts off brown (from the presence of bromine) and ends up colourless:

$$Br_2(aq) + HCO_2H(aq) \rightarrow 2Br^-(aq) + 2H^+(aq) + CO_2(g)$$

So, a colorimeter can be used to measure the decreasing concentration of bromine. The reaction is slow enough to enable the colorimeter to be read every half a minute and the measurements recorded. A computer or data logger could also be used to measure the readings, and this may be essential for faster reactions.

Table 1 shows some typical results.

In order to find the reaction rate at different times, the results are plotted on a graph and then you can measure the gradients of the tangents at the times required. For example, when $t = 0$, 300 s, and 600 s (Figure 3).

At $t = 0$ s, rate of reaction $= \frac{0.010}{240} = 0.000\,041\,6\,\text{mol dm}^{-3}\,\text{s}^{-1}$

At $t = 300$ s, rate of reaction $= \frac{0.0076}{540} = 0.000\,014\,\text{mol dm}^{-3}\,\text{s}^{-1}$

At $t = 600$ s, rate of reaction $= \frac{0.0046}{840} = 0.000\,005\,5\,\text{mol dm}^{-3}\,\text{s}^{-1}$

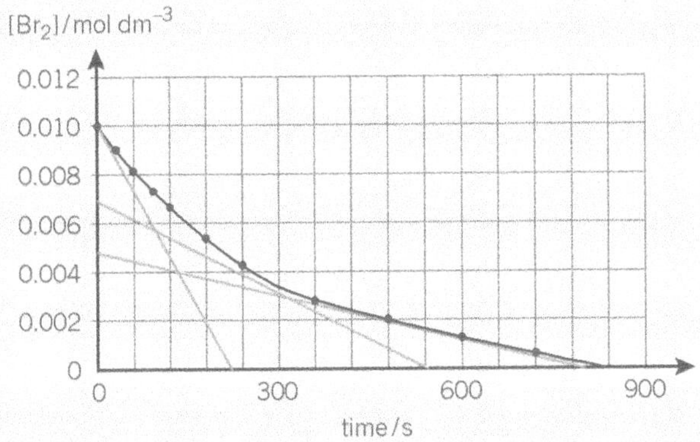

▲ **Figure 3** *Finding the rate of reaction at t = 0, t = 300, and t = 600 s*

▼ **Table 1** *$[Br_2]$ measured over time*

Time / s	$[Br_2]$ / mol dm^{-3}
0	0.0100
30	0.0090
60	0.0081
90	0.0073
120	0.0066
180	0.0053
240	0.0044
360	0.0028
480	0.0020
600	0.0013
720	0.0007

Study tip

The rate at $t = 0$ is also called the initial rate – the rate at the start of the reaction.

Hint

The reaction can also be monitored by collecting the carbon dioxide gas.

Fast reactions

Measuring the rate of a chemical reaction requires the experimenter to measure the concentration of one of the reactants or products several times over the course of the reaction. This is fine for reactions that take a few hours or a few minutes, for example, the reaction mixture can be sampled every so often and a titration can be carried out to find the concentration of one of the components. However, some reactions can be over in a few seconds or less.

The British chemists George Porter and Ronald Norrish received the 1967 Nobel Prize for chemistry for devising a technique to follow reactions that are over in a microsecond (10^{-6} s). They shared the prize with Manfred Eigen. Their method is called flash photolysis and involves starting a reaction by firing a powerful pulse of light (the photolysis flash) into a reaction mixture. This breaks chemical bonds and produces highly reactive free radicals which react rapidly with each other and with other molecules. Shortly after the first flash, further flashes of light (the probe flashes) are shone through the reaction vessel at carefully timed intervals down to as little as a microsecond (the timing is done electronically). The probe flashes are used to record the amount of light absorbed by one of the species involved in the reaction and thereby measure its concentration.

Synoptic link

Free radicals are species with unpaired electrons. See Topic 12.3, Industrial cracking, and Topic 12.5, The formation of halogenoalkanes.

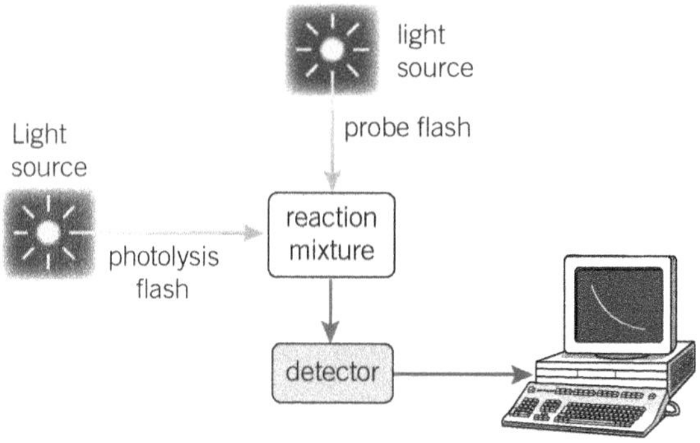

▲ **Figure 5** *Flash photolysis*

In the 1950s, Norrish and Porter used their new technique to measure the reaction of the chlorine monoxide free radical produced in the flash:

$$ClO_2 \rightarrow ClO\bullet + O\bullet$$

Pairs of these radicals reacted to give chlorine and oxygen:

$$2ClO\bullet \rightarrow Cl_2 + O_2$$

At the time, this reaction (over in about $\frac{1}{1000}$ s) was thought to be of academic interest only. However, 30 years on, it was realised that it was involved in the breakdown of ozone in the atmosphere catalysed by chlorine resulting from CFC molecules in aerosol propellants and other items. So Norrish and Porter's work has come to have immense practical importance in enabling understanding this environmental problem.

In recent years, the use of lasers for the flashes has allowed chemists to measure even faster reactions, down to picoseconds and less (a picosecond is 10^{-12} s).

Suggest what might happen to the O• radical produced in the original reaction.

Summary questions

Answer the following questions about the reaction rate graph in Figure 4.

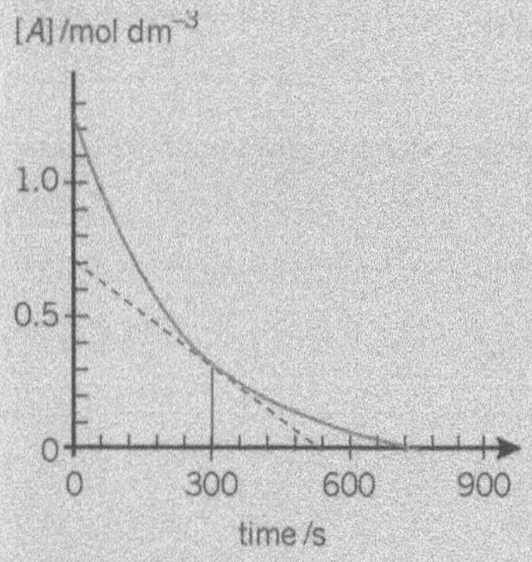

▲ **Figure 4**

1 Is the concentration being plotted that of a reactant or a product? Explain your answer.

2 The tangent to the curve at the time 300 seconds is drawn on the graph. Find the gradient of the tangent. Remember to include units.

3 What does this gradient represent?

4 Without drawing tangents, what can be said about the gradients of the tangents at time 0 seconds and time 600 seconds?

5 Explain your answer to question **4**.

18.2 The rate expression and order of reaction

The rate of a chemical reaction depends on the concentrations of some or all of the species in the reaction vessel – reactants and catalysts. But these do not necessarily all make the same contribution to how fast the reaction goes. The **rate expression** tells us about the contributions of the species that do affect the reaction rate.

For example, in the reaction $X + Y \rightarrow Z$, the concentration of X, [X], may have more effect than the concentration of Y, [Y]. Or, it may be that [X] has no effect on the rate and only [Y] matters. The detail of how each species contributes to the rate of the reaction can only be found out by experiment. A species that does not appear in the chemical equation may also affect the rate, for example, a catalyst.

Learning objectives:

- → Define the expressions order of reaction and overall order of reaction.
- → Define the expression rate equation.
- → State what a rate equation is.
- → Define the term rate constant of a rate equation.

Specification reference: 3.1.9

The rate expression

The rate expression is the result of experimental investigation. It is an equation that describes how the rate of the reaction at a particular temperature depends on the concentration of species involved in the reaction. It is quite possible that one (or more) of the species that appear in the chemical equation will not appear in the rate expression. This means that they do not affect the rate. For example, the reaction:

$$X + Y \rightarrow Z$$

This reaction *might* have the rate expression:

$$\text{rate} \propto [X][Y]$$

The symbol $\propto$ means proportional to.

This would mean that both [X] and [Y] have an equal effect on the rate. Doubling either [X] or [Y] would double the rate of the reaction. Doubling the concentration of both would quadruple the rate.

But it might be that the rate expression for the reaction is:

$$\text{rate} \propto [X][Y]^2$$

This would mean that doubling [X] would double the rate of the reaction, but doubling [Y] would quadruple the rate.

A species that is not in the chemical equation may appear in the rate equation.

Study tip

'Species' in chemistry is a general term that includes molecules, ions, and atoms that might be involved in a chemical reaction.

Hint

When $X \propto Y$, if X is doubled then Y also doubles.

The rate constant *k*

By introducing a constant into the expression you can get rid of the proportionality sign. For example, suppose the rate expression were:

$$\text{rate} \propto [X][Y]^2$$

This can be written:

$$\text{rate} = k[X][Y]^2$$

k is called the **rate constant** for the reaction. *k* is different for every reaction and varies with temperature, so the temperature at which it was measured needs to be stated. If the concentrations of all the species in the rate equation are 1 mol dm^{-3}, then the rate of reaction is equal to the value of *k*.

The order of a reaction

Suppose the rate expression for a reaction is:

$$\text{rate} = k[X][Y]^2$$

This means that [Y], which is raised to the power of 2, has double the effect on the rate than that of [X]. The **order of reaction**, with respect to one of the species, is the *power* to which the concentration of that species is raised in the rate expression. It tells us how the rate depends on the concentration of that species.

So, for rate = $k[X][Y]^2$ the order with respect to X is *one* ([X] and $[X]^1$ are the same thing), and the order with respect to Y is *two*.

The overall order of the reaction is the sum of the orders of all the species, which appear in the rate expression. In this case the overall order is *three*. So this reaction is said to be *first order* with respect to X, *second order* with respect to Y, and *third order* overall.

So if the rate expression for a reaction is rate = $k[A]^m[B]^n$, where m and n are the orders of the reaction with respect to A and B, the overall order of the reaction is $m + n$.

Study tip

Remember that the rate expression is entirely derived from experimental evidence and that it *cannot* be predicted from the chemical equation for the reaction. It is therefore quite unlike the equilibrium law expression that you met in Topic 6.3, The equilibrium constant K_c, (although it looks similar to it at first sight).

The chemical equation and the rate expression

The rate expression tells us about the species that affect the rate. Species that appear in the chemical equation do not necessarily appear in the rate equation. Also, the coefficient of a species in the chemical equation – the number in front of it – has no relevance to the rate expression. But catalysts, which do not appear in the chemical equation, *may* appear in the rate expression.

For example, in the reaction:

$$\underset{\text{propanone}}{CH_3COCH_3(aq)} + \underset{\text{iodine}}{I_2(aq)} \xrightarrow{H^+\ \text{catalyst}} \underset{\text{iodopropanone}}{CH_2ICOCH_3(aq)} + \underset{\text{hydrogen iodide}}{HI(aq)}$$

The rate expression has been found *by experiment* to be:

$$\text{rate} = k[CH_3COCH_3(aq)][H^+(aq)]$$

So the reaction is first order with respect to propanone, first order with respect to H^+ ions, and second order overall. The rate does not depend on $[I_2(aq)]$, so you can say the reaction is *zero* order with respect to iodine. The H^+ ions act as a catalyst in this reaction.

Synoptic link

You will need to know the influence of carbon–halogen bond enthalpy on the rate of hydrolysis of halogenoalkanes and the mechanism of substitution and elimination reactions covered in Topic 13.1, Halogenalkanes – introduction, and Topic 13.2, Nucleophilic substitution in halogenoalkanes.

Units of the rate constant

The units of the rate constant vary depending on the overall order of reaction.

For a **zero** order reaction:

$$\text{rate} = k$$

The units of rate are mol dm^{-3} s^{-1}.

For a **first** order reaction where:

$$\text{rate} = k[A]$$

$$k = \frac{\text{rate}}{[A]}$$

Maths link

See Section 8, Mathematical skills, if you are not sure about cancelling units.

The units of rate are $mol\,dm^{-3}\,s^{-1}$ and the units of [A] are $mol\,dm^{-3}$, so the units of k are s^{-1} obtained by cancelling:

$$k = \frac{\cancel{mol\,dm^{-3}}\,s^{-1}}{\cancel{mol\,dm^{-3}}}$$

Therefore, the units of k for a first order rate constant are s^{-1}.

For a **second** order reaction where:

$$\text{rate} = k[B]\,[C]$$

$$k = \frac{\text{rate}}{[B]\,[C]}$$

The units of rate are $mol\,dm^{-3}\,s^{-1}$ and the units of both [B] and [C] are $mol\,dm^{-3}$, so the units of k are s^{-1} obtained by cancelling:

$$k = \frac{\cancel{mol\,dm^{-3}}\,s^{-1}}{\cancel{mol\,dm^{-3}}\,mol\,dm^{-3}}$$

Therefore the units of k for a second order rate constant are $mol^{-1}\,dm^{3}\,s^{-1}$.

For a **third** order reaction:

$$\text{rate} = k[D]\,[E]^2$$

$$k = \frac{\text{rate}}{[D]\,[E]^2}$$

The unit of rate is $mol\,dm^{-3}\,s^{-1}$, the unit of [D] is $mol\,dm^{-3}$, and the unit of $[E]^2$ is $(mol\,dm^{-3})^2$.

$$k = \frac{\cancel{mol\,dm^{-3}}\,s^{-1}}{\cancel{mol\,dm^{-3}}\,(mol\,dm^{-3})^2}$$

Therefore, the units of k for a third order rate constant are $mol^{-2}\,dm^{6}\,s^{-1}$.

Hint

It is better to work out the units rather than try to remember them.

Study tip

It is important you understand the terms 'rate of reaction', 'order of reaction', and 'rate constant'.

Summary questions

1 Write down the rate expression for a reaction that is first order with respect to [A], first order with respect to [B], and second order with respect to [C].

2 Consider the reaction:

$$BrO_3^-(aq) + 5Br^-(aq) + 6H^+(aq) \rightarrow 3Br_2(aq) + 3H_2O(l)$$

bromate ions, bromide ions, hydrogen ions → bromine, water

The rate expression is:

$$\text{rate} = k[BrO_3^-(aq)][Br^-(aq)][H^+(aq)]^2$$

a What is the order with respect to:
i $BrO_3^-(aq)$ ii $Br^-(aq)$ iii $H^+(aq)$?

b What would happen to the rate if you doubled the concentration of:
i $BrO_3^-(aq)$ ii $Br^-(aq)$ iii $H^+(aq)$?

c What are the coefficients of the following in the chemical equation above?
i $BrO_3^-(aq)$ ii $Br^-(aq)$ iii $H^+(aq)$
iv $Br_2(aq)$ v $H_2O(l)$

d Work out the units for the rate constant.

3 In the reaction $L + M \rightarrow N$ the rate expression is found to be:

$$\text{rate} = k[L]^2[H^+]$$

a What is k?

b What is the order of the reaction with respect to:
i L ii M iii N iv H^+?

c What is the overall order of the reaction?

d The rate is measured in $mol\,dm^{-3}\,s^{-1}$. What are the units of k?

e Suggest the function of H^+ in the reaction.

4 In the reaction $G + 2H \rightarrow I + J$, which is the correct rate expression?

A rate $= k[G][H]^2$

B rate $= \dfrac{k[G][H]}{[I][J]}$

C rate $= k[G][H]$

D It is impossible to tell without experimental data.

18.3 The Arrhenius equation

Learning objectives:

- Perform calculations using the Arrhenius equation.
- Rearrange the Arrhenius equation into the form $\ln k = -\frac{E_a}{RT} + \ln A$
- Use experimental data to determine the activation energy of a reaction.

Specification reference: 3.1.9

Synoptic link

Look back at Topic 5.2, where the Maxwell Boltzmann curve is discussed.

The rates of chemical reactions increase greatly for relatively small rises in temperature. As the temperature rises, the number of collisions between reactant molecules increases. However, the increase in the number of collisions is not great enough to account for the increase in the rate of the reaction.

Increasing the temperature of a reaction also increases the number of collisions that have energy greater than the activation energy. The shape of a Maxwell–Boltzmann distribution curve shows how the distribution of energies of the reactants is different at different temperatures.

The fraction of molecules with energy greater than the activation energy is given by $e^{-\frac{E_a}{RT}}$ where E_a is the activation energy in J, R is the gas constant ($8.3\,\mathrm{J\,K^{-1}\,mol^{-1}}$), and T the temperature in K. Also e is a mathematical constant numerically equal to 2.718.

The activation energy can be linked to the rate constant by the Arrhenius equation:

$$k = A\,e^{-\frac{E_a}{RT}}$$

k is the rate constant, which is proportional to reaction rate.

A is the pre-exponential factor, which is related to the number of collisions between reactant molecules.

$e^{-\frac{E_a}{RT}}$ is the fraction of collisions with enough energy to react.

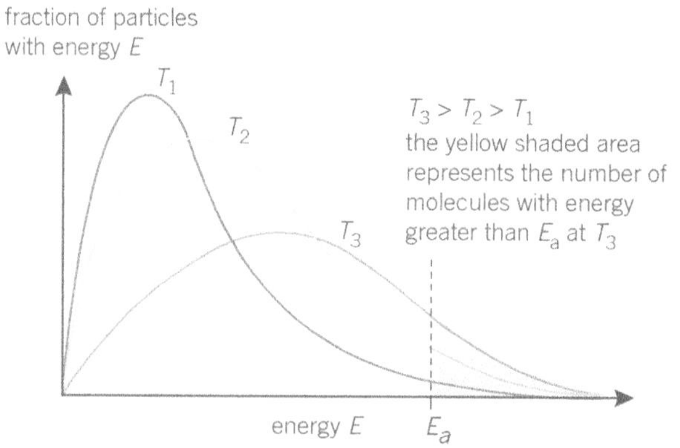

▲ **Figure 1** *Maxwell–Boltzmann distribution curve showing the distribution of molecular energies at three different temperatures*

Logarithmic form of the Arrhenius equation

The Arrhenius equation is easier to use when you take logs of both sides to the base *e*, called natural logs.

$$\ln k = -\frac{E_a}{RT} + \ln A$$

or

$$\ln k = -\frac{E_a}{R} \times \frac{1}{T} + \ln A$$

This means that a graph of ln*k* against $\frac{1}{T}$ will be a straight line of gradient $-\frac{E_a}{RT}$ (Figure 2). You can use this to find a value for E_a experimentally by measuring the rate of a reaction at different temperatures.

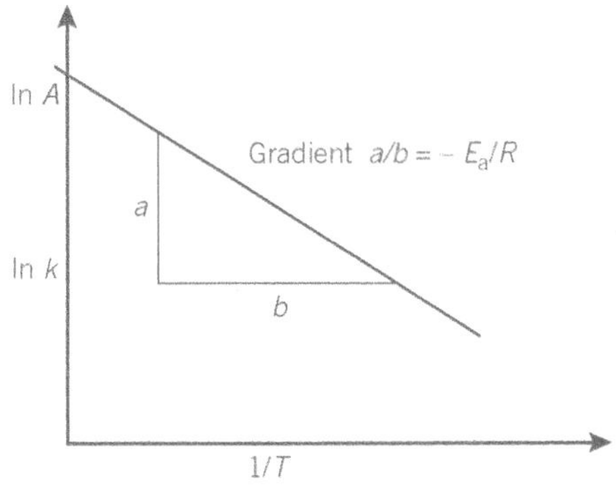

▲ **Figure 2** *An Arrhenius plot*

Using the Arrhenius equation

Look at the reaction of the decomposition of hydrogen iodide, HI:

$2HI(g) \rightarrow I_2(g) + H_2(g)$

The experimentally determined values of the rate constant *k* at different temperatures are shown in Table 1, and you can use these to calculate ln*k* and $\frac{1}{T}$.

Study tip

Natural logs are written 'ln' to distinguish them from logs to the base 10, which are written $\log_{10}$ or just log.

Hint

This is probably the most useful form of the Arrhenius equation.

Hint

Check with your calculator that you get the values for ln*k* and $\frac{1}{T}$ shown in Table 1 on the next page.

Remember that ln*k* has no units, it is just a number. The calculated value of E_a will be in J mol^{-1} and must be converted to kJ mol^{-1}.

Hint

In principle we could take logs to any base number. e represents the number 2.718. It is used as the base for taking natural logs, represented by the function ln. In the same way that the logs to the base 10 (log, or $\log_{10}$) of a number represent the power that 10 is raised to give the number, natural logs are the power to which e is raised to give the number. Natural logs have certain mathematical advantages. Make sure that you can use your calculator to do natural logs and antilogs.

▼ **Table 1** *Experimentally determined values for* ln*k and* $\frac{1}{T}$ *for the decomposition of hydrogen iodide*

$\frac{T}{\text{K}}$	$\frac{1}{T}/10^{-3}\,\text{K}^{-1}$	$k/\text{dm}^3\,\text{mol}^{-1}\,\text{s}^{-1}$	ln*k*
633	1.579	1.78×10^{-5}	−10.936
666	1.501	1.07×10^{-4}	−9.142
697	1.434	5.01×10^{-4}	−7.599
715	1.398	1.05×10^{-3}	−6.858
781	1.280	1.51×10^{-2}	−4.193

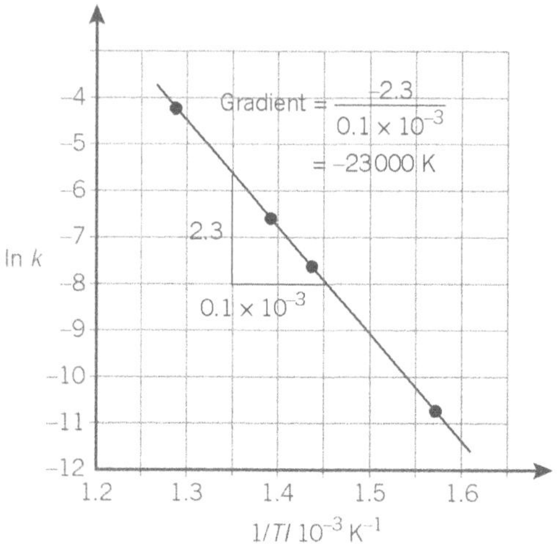

▲ **Figure 3** *Graph of* ln*k against* $\frac{1}{T}$

Using this data, you can plot a graph of ln*k* against $\frac{1}{T}$ and measure the gradient (Figure 3). From this you can calculate E_a.

$$\text{gradient} = -\frac{E_a}{R}$$

$$E_a = \text{gradient} \times R$$

$$E_a = 23\,000\,\text{K} \times 8.3\,\text{J}\,\text{K}^{-1}\,\text{mol}^{-1}$$

$$E_a = 190\,000\,\text{J}\,\text{mol}^{-1}$$

$$\mathbf{E_a = 190\,kJ\,mol^{-1}}$$

The activation energy for the decomposition of hydrogen iodide is $190\,\text{kJ}\,\text{mol}^{-1}$.

This value is realistic as it represents the energy required to break a covalent bond. Typical values for activation energy are between 40 and $400\,\text{kJ}\,\text{mol}^{-1}$.

The effect of increasing the temperature

Imagine a reaction with activation energy 50 kJ mol^{-1} (50 000 J mol^{-1}), a fairly typical value.

At a temperature of 300 K (that of a warm laboratory), the fraction of molecules with energy greater than 50 kJ mol^{-1} is:

$$= e^{-\frac{50\,000}{8.3 \times 300}}$$
$$= e^{-20.08}$$
$$= 1.9 \times 10^{-9}$$
$$= 19 \times 10^{-10} = \frac{19}{10^{10}}$$

This means that only 19 molecules out of every 10^{10} molecules have enough energy to react.

At a temperature of 310 K (10 K or 10 °C) higher, the fraction of molecules with energy greater than 50 kJ mol^{-1} $= e^{-\frac{50\,000}{8.3 \times 310}}$

$$= e^{-19.43}$$
$$= 3.6 \times 10^{-9}$$
$$= 36 \times 10^{-10} = \frac{36}{10^{10}}$$

Now 36 molecules out of every 10^{10} molecules have enough energy to react. The 10 K rise in temperature has almost doubled the number of molecules that have energy greater than the activation energy, and therefore the rate will almost double. This is a general rule of thumb often used by chemists. However, it is only applicable to reactions where E_a is around 50 kJ mol^{-1} and around room temperature.

Summary questions

1 Calculate the proportion of molecules with energy greater than an activation energy of 100 kJ mol^{-1} at:

a 300 K

b 310 K

2 The activation energy for the reaction $2HI(g) \rightarrow I_2(g) + H_2(g)$ without a catalyst is 190 kJ mol^{-1}. The reaction is catalysed by a number of metals. Using a metal the following data was obtained:

Temperature / K	Rate constant k / mol dm^{-3} s^{-1}
625	2.27
667	7.56
727	24.87
767	91.18
833	334.59

Plot a graph of ln$k$ against $\frac{1}{T}$ and find the activation energy with this catalyst.

18.4 Determining the rate equation

Learning objectives:

- → Describe how the order of a reaction with respect to a reagent is found experimentally.
- → Describe how to find the value of the rate constant.
- → Describe how a change in temperature affects the value of the rate constant.

Specification reference: 3.1.9

The rate expression tells you how the rate of a reaction depends on the concentration of the species involved. It only includes the species that affect the rate of the reaction.

- If the rate is not affected by the concentration of a species, the reaction is *zero order* with respect to that species. The species is not included in the rate expression.
- If the rate is directly proportional to the concentration of the species, the reaction is *first order* with respect to that species.
- If the rate is proportional to the square of the concentration of the species, the reaction is *second order* with respect to that species, and so on.

Finding the order of a reaction by using rate–concentration graphs

One method of finding the order of a reaction with respect to a particular species, A, is by plotting a graph of rate against concentration.

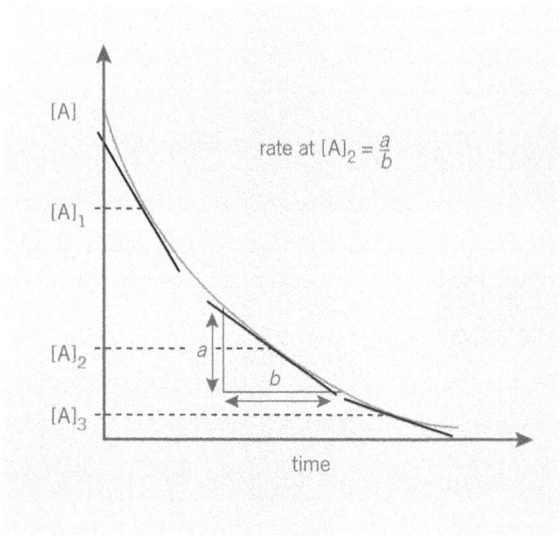

▲ **Figure 1** *Finding the rates of reaction for different values of [A]*

Plot the original graph of [A] against time, and draw tangents at different values of [A]. The gradients of these tangents are the reaction rates (the changes in concentration over time) at different concentrations (Figure 1). The values for these rates can then be used to construct a second graph of rate against concentration (Figure 2).

- If the graph is a horizontal straight line (Figure 2a), this means that the rate is unaffected by [A] so the order is zero.
- If the graph is a sloping straight line through the origin (Figure 2b), then rate $\propto [A]^1$ so the order is 1.
- If the graph is not a straight line (Figure 2c), the order cannot be found directly – it could be two. Try plotting rate against $[A]^2$. If this is a straight line, then the order is two.

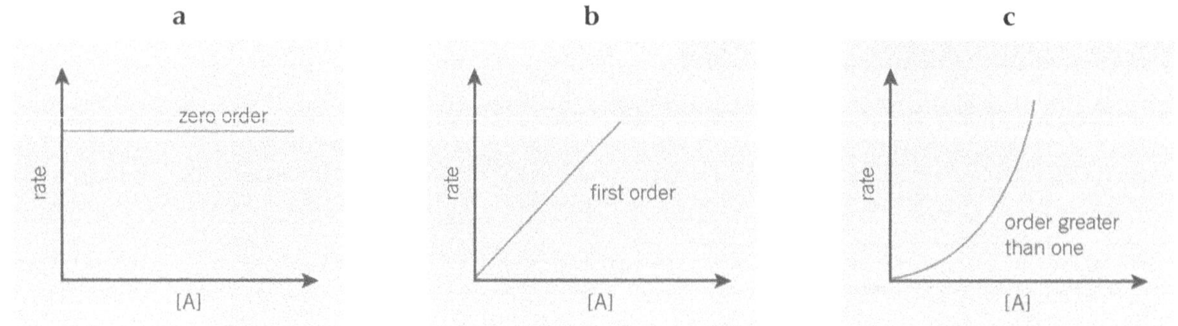

▲ **Figure 2** *Graphs of rate against concentration*

The initial rate method

With the initial rate method, a series of experiments is carried out at constant temperature. Each experiment starts with a different combination of initial concentrations of reactants, catalyst, and so on. The experiments are planned so that, between any pair of experiments, the concentration of only one species varies – the rest stay the same. Then, for each experiment, the concentration of one reactant is followed and a concentration–time graph plotted (Figure 3). The tangent to the graph at time = 0 is drawn. The gradient of this tangent is the **initial rate**. By measuring the *initial* rate, the concentrations of all substances in the reaction mixture are known *exactly* at this time.

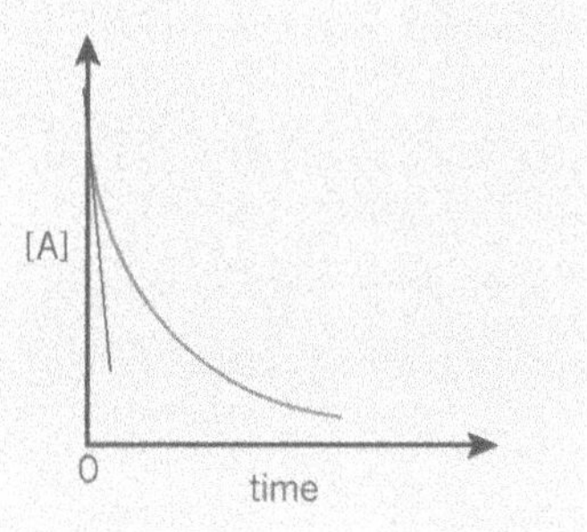

▲ **Figure 3** *Finding the initial rate of a reaction. The initial rate is the gradient at time = 0*

Comparing the initial concentration and the initial rates for pairs of experiments allows the order with respect to each reactant to be found. For example, for the reaction:

$$\underset{\text{nitrogen monoxide}}{2NO(g)} + \underset{\text{oxygen}}{O_2(g)} \rightarrow \underset{\text{nitrogen dioxide}}{2NO_2(g)}$$

The initial rates are shown in Table 1.

▼ **Table 1** *Results obtained for the reaction $2NO(g) + O_2(g) \rightarrow 2NO_2(g)$*

Experiment number	Initial [NO] / $mol\,dm^{-3}$	Initial $[O_2]$ / $mol\,dm^{-3}$	Initial rate / $mol\,dm^{-3}\,s^{-1}$
1	1.0×10^{-3}	1.0×10^{-3}	7.0×10^{-4}
2	2.0×10^{-3}	1.0×10^{-3}	28.0×10^{-4}
3	3.0×10^{-3}	1.0×10^{-3}	63.0×10^{-4}
4	2.0×10^{-3}	2.0×10^{-3}	56.0×10^{-4}
5	3.0×10^{-3}	3.0×10^{-3}	189.0×10^{-4}

Comparing Experiment 1 with Experiment 2, [NO] is doubled whilst $[O_2]$ stays the same. The rate quadruples (from $7.0 \times 10^{-4}\,mol\,dm^{-3}\,s^{-1}$ to $28.0 \times 10^{-4}\,mol\,dm^{-3}\,s^{-1}$) which suggests rate $\propto [NO]^2$. This is confirmed by considering Experiments 1 and 3 where [NO] is trebled whilst $[O_2]$ stays the same. Here the rate is increased ninefold, as would be expected if rate $\propto [NO]^2$. So the order with respect to nitrogen monoxide is two.

Now compare Experiment 2 with Experiment 4. Here [NO] is constant but $[O_2]$ doubles. The rate doubles (from $28.0 \times 10^{-4}\,mol\,dm^{-3}\,s^{-1}$ to $56.0 \times 10^{-4}\,mol\,dm^{-3}\,s^{-1}$) so it looks as if rate $\propto [O_2]$. This is confirmed by considering Experiments 3 and 5. Again [NO] is constant, but $[O_2]$ triples. The rate triples too, confirming that the order with respect to oxygen is one.

So rate $\propto [NO]^2$ and rate $\propto [O_2]^1$

That is, rate $\propto [NO]^2[O_2]^1$

Provided that no other species affect the reaction rate, the overall order is three and the rate expression is:

$$\text{rate} = k[NO]^2[O_2]^1$$

Hint

It is easier to apply the technique to problems than to read about it. However, you can always work out the answer mathematically.

You know that in the example, rate is $= k[NO]^x[O_2]^y$, where x is the order with respect to NO and y is the order with respect to O_2.

So, $\dfrac{\text{rate of Experiment 2}}{\text{rate of Experiment 1}} = \dfrac{k[NO]_2^{\,x}[O_2]_2^{\,y}}{k[NO]_1^{\,x}[O_2]_1^{\,y}}$

Putting in the numbers from the table:

$$\frac{28.0 \times 10^{-4}}{7.0 \times 10^{-4}} = \frac{k[2.0 \times 10^{-3}]^x[1.0 \times 10^{-3}]^y}{k[1.0 \times 10^{-3}]^x[1.0 \times 10^{-3}]^y}$$

$$\frac{28.0 \times 10^{-4}}{7.0 \times 10^{-4}} = \frac{\cancel{k}[2.0 \times 10^{-3}]^x\cancel{[1.0 \times 10^{-3}]^y}}{\cancel{k}[1.0 \times 10^{-3}]^x\cancel{[1.0 \times 10^{-3}]^y}}$$

$$4 = 2^x$$
$$x = 2$$

So the order with respect to NO is 2.

Hint

You have already come across the blue-black colour with iodine as a test for starch.

The iodine clock reaction

The iodine clock reaction is used to measure the rate of the reaction between hydrogen peroxide and potassium iodide in acidic conditions to form iodine (Reaction 1).

$H_2O_2(aq) + 2I^-(aq) + 2H^+(aq) \rightarrow I_2(aq) + 2H_2O(l)$ **Reaction 1**

This reaction can be timed by adding a known number of moles of sodium thiosulfate to the reaction mixture along with a little starch. In effect, you are measuring the initial rate of the reaction.

As soon as the iodine is produced by the reaction above, it reacts immediately with the thiosulfate ions in a 1 : 2 ratio by Reaction 2.

$I_2(aq) + 2S_2O_3^{2-}(aq) \rightarrow S_4O_6^{2-}(aq) + 2I^-(aq)$ **Reaction 2**

Reaction 2 acts solely as a timing device for Reaction 1. Once half the number of moles of iodine has been produced as the number of moles of thiosulfate added, the free iodine produced by Reaction 1 reacts immediately with the starch to give a dark blue/black colour. The appearance of this colour is sudden, and the time it takes to appear after the reactants have been mixed can be timed accurately. The shorter the time t for the iodine to appear, the faster the rate of Reaction 1. So the value $\frac{1}{t}$ is proportional to the reaction rate.

To carry out the reaction, a solution of hydrogen peroxide is added to one containing potassium iodide, sodium thiosulfate, and starch.

▲ **Figure 4** *Timing the iodine clock reaction*

Some results for an iodine clock reaction are shown below.

Concentration of hydrogen peroxide / mol dm^{-3}	Time for blue colour to appear / s	$\frac{1}{t}$ / s^{-1}
1.0	20	0.05
0.5	40	0.025
0.25	80	0.0125

1. This is a redox reaction. What species has been oxidised?
2. What is the oxidising agent?
3. What happens to the rate of reaction when the concentration of hydrogen peroxide is doubled?
4. What is the order of the reaction with respect to hydrogen peroxide?
5. What can you say about the order of the reaction with respect to iodide ions *from these results*?

6 Suggest how the concentration of hydrogen peroxide can easily be varied.
7 State three factors that must be kept constant in order to find the order of the reaction with respect to hydrogen peroxide.

Finding the rate constant *k*

To find k in the reaction of NO and O_2 substitute any set of values of rate, [NO], and $[O_2]$ in the equation.

Taking the values for Experiment 2:

$$28.0 \times 10^{-4} = k(2 \times 10^{-3})^2 \times 1 \times 10^{-3}$$

$$28.0 \times 10^{-4} = k \times 4 \times 10^{-9}$$

$$k = \frac{28.0}{4} \times 10^5$$

$$k = 7.0 \times 10^5$$

But you need to work out the units for k, as these are different for reactions of different overall order. Putting in the units gives:

$$28.0 \times 10^{-4}\,\text{mol dm}^{-3}\,\text{s}^{-1} = k\,(2 \times 10^{-3})^2\,(\text{mol dm}^{-3})\,(\text{mol dm}^{-3}) \times 1 \times 10^{-3}\,\text{mol dm}^{-3}$$

Units can be cancelled in the same way as numbers, so cancelling the units gives:

$$28.0 \times 10^{\cancel{-4}}\,\cancel{\text{mol dm}}^{-3}\,\text{s}^{-1} =$$

$$k\,(4 \times 10^{-6})\,(\cancel{\text{mol dm}^{-3}})\,(\text{mol dm}^{-3}) \times 1 \times 10^{-3}\,\text{mol dm}^{-3}$$

$$28.0 \times 10^{-4} = k \times 4 \times 10^{-9}\,\text{mol}^2\,\text{dm}^{-6}\,\text{s}^1$$

$$k = \frac{28.0}{4} \times 10^5\,\text{mol}^{-2}\,\text{dm}^6\,\text{s}^{-1}$$

$$k = 7.0 \times 10^5\,\text{dm}^6\,\text{mol}^{-2}\,\text{s}^{-1}$$

Since the units of k vary for reactions of different orders, it is important to put the units for rate and the concentrations in and then cancel them to make sure you have the correct units for k.

The effect of temperature on *k*

Small changes in temperature produce large changes in reaction rates. A rough rule is that for every 10 K rise in temperature, the rate of a reaction doubles. Suppose the rate expression for a reaction is rate = k[A][B]. You know that [A] and [B] do not change with temperature, so the rate constant k must increase with temperature.

In fact, the rate constant k allows you to compare the speeds of different reactions at a given temperature. It is an inherent property of a particular reaction. It is the rate of the reaction at a particular temperature when the concentrations of all the species in the rate expression are 1 mol dm^{-3}. The larger the value of k, the faster the reaction. Look at Table 2. You can see that the value of k increases with temperature. This is true for all reactions.

Hint

You should get the same value of k using the figures for any of the experiments.

Study tip

Practise working out the units for a rate constant.

▼ **Table 2** *The values of the rate constant, k, at different temperatures for the reaction* $2HI(g) \rightarrow I_2(g) + H_2(g)$

Temperature / K	k / mol^{-1} dm^3 s^{-1}
633	0.0178×10^{-3}
666	0.107×10^{-3}
697	0.501×10^{-3}
715	1.05×10^{-3}
781	15.1×10^{-3}

Study tip

Remember, increasing temperature always increases the rate of reaction and the value of the rate constant k.

Synoptic link

Look back at Topic 5.1, Collision theory, to revise activation energy.

Summary question

1 For the reaction A + B → C, the following data were obtained:

Initial [A] / $mol\,dm^{-3}$	Initial [B] / $mol\,dm^{-3}$	Initial rate / $mol\,dm^{-3}\,s^{-1}$
1	1	3
1	2	12
2	2	24

a What is the order of reaction with respect to:
   i A ii B?

b What is the overall order?

c What would be the initial rate if the initial [A] were 1 $mol\,dm^{-3}$ and [B] were 3 $mol\,dm^{-3}$?

d What do these results suggest is the rate expression for this reaction?

e Can we be certain that this is the full rate expression? Explain your answer.

18.5 The rate-determining step

Most reactions take place in more than one step. The separate steps that lead from reactants to products are together called the **reaction mechanism**. For example, the reaction below involves 12 ions:

$$BrO_3^-(aq) + 6H^+(aq) + 5Br^-(aq) \rightarrow 3Br_2(aq) + 3H_2O(l)$$

This reaction *must* take place in several steps – it is very unlikely indeed that the 12 ions of the reactants will all collide at the same time. The steps in-between will involve very short-lived intermediates. These intermediate species, which give information about the mechanism of the reaction, are usually difficult or impossible to isolate and therefore identify. So, other ways of working out the mechanism of the reaction are used.

Learning objectives:

→ Define the expression rate-determining step of a reaction.

→ Describe the connection between the rate equation for a reaction and the reaction mechanism.

Specification reference: 3.1.9

The rate-determining step

In a multi-step reaction, the steps nearly always follow *after* each other, so that the product of one step is the starting material for the next. Therefore the rate of the slowest step will govern the rate of the whole process. The slowest step may form a 'bottleneck', called the **rate-determining step** or **rate-limiting step**. Suppose you had everything you needed to make a cup of coffee, starting with cold water. The rate of getting your drink will be governed by the slowest step – waiting for the kettle to boil – no matter how quickly you get the cup out of the cupboard, the coffee out of the jar, or add the milk.

In a chemical reaction, any step that occurs *after* the rate-determining step will not affect the rate, provided that it is fast compared with the rate-determining step. So species that are involved in the mechanism after the rate-determining step do not appear in the rate expression. For example, the reaction:

$$A + B + C \rightarrow Y + Z$$

This reaction might occur in the following steps:

1 $A + B \xrightarrow{\text{fast}} D$ (first intermediate)

2 $D \xrightarrow{\text{slow}} E$ (second intermediate)

3 $E + C \xrightarrow{\text{fast}} Y + Z$

Step 2 is the slowest step and so determines the rate. Then, as soon as some E is produced, it rapidly reacts with C to produce Y and Z.

But the rate of Step 1 might affect the overall rate – the concentration of D depends on this. So, any species involved in or *before* the rate-determining step could affect the overall rate and therefore appear in the rate expression.

So, for the reaction $A + B + C \rightarrow Y + Z$, the rate equation may be:

rate $\propto$ [A][B][D]

The reaction between iodine and propanone demonstrates this.

The overall reaction is:

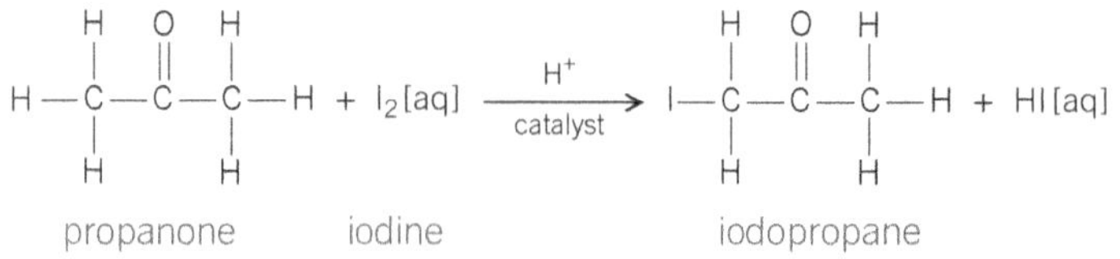

The rate expression is found to be rate = $k[CH_3COCH_3][H^+]$
The mechanism is:

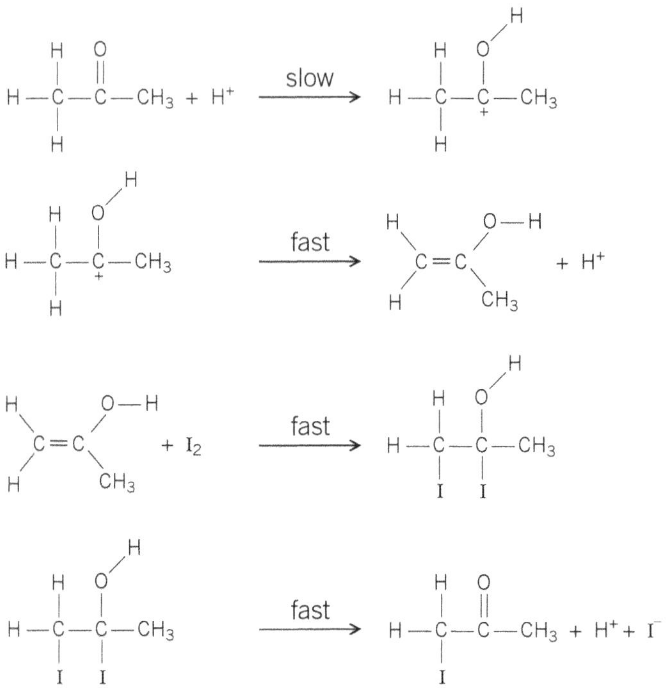

The rate-determining step is the first one, which explains why $[I_2]$ does *not* appear in the rate expression.

Using the order of a reaction to find the rate-determining step

Here is a simple example. The three structural isomers with formula C_4H_9Br all react with alkali. The overall reaction is represented by the following equation:

$$C_4H_9Br + OH^- \rightarrow C_4H_9OH + Br^-$$

Two mechanisms are possible.

a A two-step mechanism:

Step 1: $C_4H_9Br \xrightarrow{\text{slow}} C_4H_9^+ + Br^-$

Step 2: $C_4H_9^+ + OH^- \xrightarrow{\text{fast}} C_4H_9OH$

The slow step involves breaking the C—Br bond whilst the second (fast) step is a reaction between oppositely charged ions.

b A one-step mechanism:

$$C_4H_9Br + OH^- \xrightarrow{\text{slow}} C_4H_9OH + Br^-$$

The C—Br bond breaks at the same time as the C—OH bond is forming.

The three isomers of formula C_4H_9Br are:

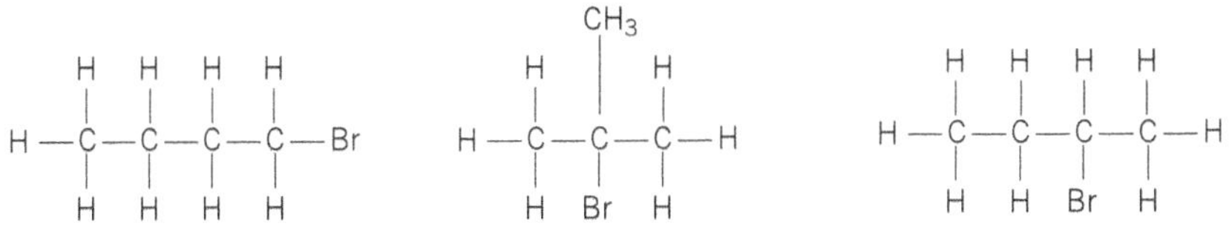

1-bromobutane 2-bromo-2-methylpropane 2-bromobutane

Experiments show that 1-bromobutane reacts by a second order mechanism:

$$\text{rate} = k[C_4H_9Br][OH^-]$$

The rate depends on the concentration of *both* the bromobutane *and* the OH^- ions, suggesting mechanism **b**, a one-step reaction.

Experiments show that 2-bromo-2-methylpropane reacts by a first order mechanism:

$$\text{rate} = k[C_4H_9Br]$$

This suggests mechanism **a** in which a slow step, breaking the C—Br bond, is followed by a rapid step in which two oppositely charged ions react together. So, the breaking of the C—Br bond is the rate-determining step.

The compound 2-bromobutane reacts by a mixture of both mechanisms and has a more complex rate expression.

Study tip

The species in the rate equation are the reactants involved in reactions occurring before the rate-determining step.

Summary question

1 The following reaction schemes show possible mechanisms for the overall reaction:

$A + E \xrightarrow{\text{catalyst}} G$

Scheme 1	Scheme 2	Scheme 3
(i) $A + B \xrightarrow{\text{slow}} C$	(i) $A + B \xrightarrow{\text{fast}} C$	(i) $A + B \xrightarrow{\text{fast}} C$
(ii) $C \xrightarrow{\text{fast}} D + E$	(ii) $C \xrightarrow{\text{fast}} D + E$	(ii) $C \xrightarrow{\text{slow}} D + E$
(iii) $D + E \xrightarrow{\text{fast}} F$	(iii) $D + E \xrightarrow{\text{slow}} F$	(iii) $D + E \xrightarrow{\text{fast}} F$
(iv) $F \xrightarrow{\text{fast}} G$	(iv) $F \xrightarrow{\text{slow}} G$	(iv) $F \xrightarrow{\text{fast}} G$

a In Scheme 2, which species is the catalyst?

b Which species *cannot* appear in the rate expression for Scheme 1?

c Which is the rate-determining step in Scheme 3?

Practice questions

1 This question involves the use of kinetic data to calculate the order of a reaction and also a value for a rate constant.

(a) The data in this table were obtained in a series of experiments on the rate of the reaction between compounds **E** and **F** at a constant temperature.

Experiment	Initial concentration of E / mol dm^{-3}	Initial concentration of F / mol dm^{-3}	Initial rate of reaction / mol dm^{-3} s^{-1}
1	0.15	0.24	0.42×10^{-3}
2	0.45	0.24	3.78×10^{-3}
3	0.90	0.12	7.56×10^{-3}

(i) Deduce the order of reaction with respect to **E**. *(1 mark)*

(ii) Deduce the order of reaction with respect to **F**. *(1 mark)*

(b) The data in the following table were obtained in two experiments on the rate of the reaction between compounds **G** and **H** at a constant temperature.

Experiment	Initial concentration of G / mol dm^{-3}	Initial concentration of H / mol dm^{-3}	Initial rate of reaction / mol dm^{-3} s^{-1}
4	3.8×10^{-2}	2.6×10^{-2}	8.6×10^{-4}
5	6.3×10^{-2}	7.5×10^{-2}	To be calculated

The rate equation for this reaction is

$$\text{rate} = k[\mathbf{G}]^2[\mathbf{H}]$$

(i) Use the data from Experiment **4** to calculate a value for the rate constant *k* at this temperature. Deduce the units of *k*.

(3 marks)

(ii) Calculate a value for the initial rate of reaction in Experiment **5**.

(1 mark)

AQA, 2013

2 Gases **P** and **Q** react as shown in the following equation.

$$2P(g) + 2Q(g) \rightarrow R(g) + S(g)$$

The initial rate of the reaction was measured in a series of experiments at a constant temperature. The following rate equation was determined.

$$\text{rate} = k[\mathbf{P}]^2[\mathbf{Q}]$$

(a) Complete the table of data for the reaction between **P** and **Q**.

Experiment	Initial [G] / mol dm^{-3}	Initial [Q] / mol dm^{-3}	Initial rate / mol dm^{-3} s^{-1}
1	2.5×10^{-2}	1.8×10^{-2}	5.0×10^{-5}
2	7.5×10^{-2}	1.8×10^{-2}	
3	5.0×10^{-2}		5.0×10^{-5}
4		5.4×10^{-2}	4.5×10^{-4}

(3 marks)

(b) Use the data from Experiment **1** to calculate a value for the rate constant *k* at this temperature. Deduce the units of *k*.

(3 marks)

AQA, 2012

3 Propanone and iodine react in acidic conditions according to the following equation.

$$CH_3COCH_3 + I_2 \longrightarrow ICH_2COCH_3 + HI$$

A student studied the kinetics of this reaction using hydrochloric acid and a solution containing propanone and iodine. From the results the following rate equation was deduced.

$$\text{rate} = k[CH_3COCH_3][H^+]$$

(a) Give the overall order for this reaction.

(1 mark)

(b) When the initial concentrations of the reactants were as shown in the table below, the initial rate of reaction was found to be 1.24×10^{-4} mol dm^{-3} s^{-1}.

	Initial concentration / mol dm^{-3}
CH_3COCH_3	4.40
I_2	5.00×10^{-3}
H^+	0.820

Use these data to calculate a value for the rate constant, *k* for the reaction and give its units.

(3 marks)

(c) Deduce how the initial rate of reaction changes when the concentration of iodine is doubled but the concentrations of propanone and of hydrochloric acid are unchanged.

(1 mark)

(d) The following mechanism for the overall reaction has been proposed.

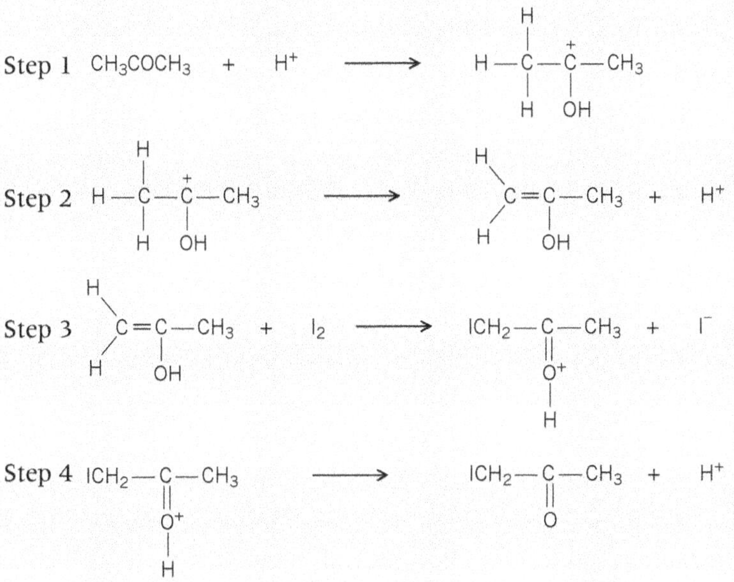

Use the rate equation to suggest which of the four steps could be the rate-determining step. Explain your answer.

(2 marks)

(e) Use your understanding of reaction mechanisms to predict a mechanism for Step 2 by adding one or more curly arrows as necessary to the structure of the carbocation below.

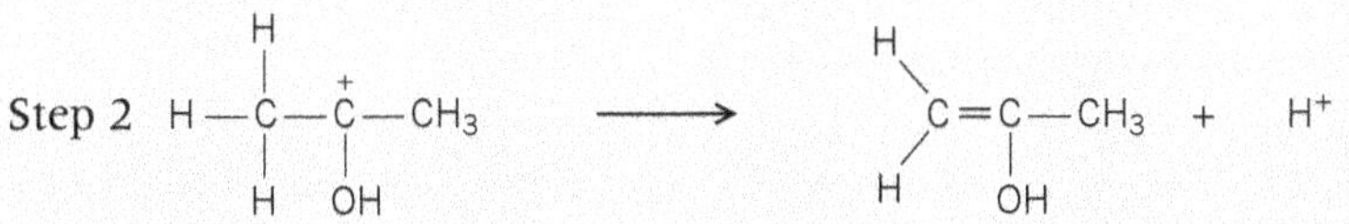

(1 mark)

AQA, 2010

Answers to the Practice Questions and Section Questions are available at www.oxfordsecondary.com/oxfordaqaexams-alevel-chemistry

19 Equilibrium constant K_p

19.1 Equilibrium constant K_p for homogeneous systems

Learning objectives:

- State what is meant by partial pressure.
- Apply the equilibrium law to gaseous equilibria.
- Predict the effect of changing pressure and temperature on a gaseous equilibrium.

Specification reference 3.1.10

Synoptic link

Revise the material on equilibrium in Chapter 6, Equilibria, in particular the equilibrium law.

Hint

The SI unit of pressure is the pascal (Pa). 1 pascal is a pressure of 1 newton per square metre ($N\,m^{-2}$). This is roughly the weight of an apple spread over the area of an opened newspaper.

Gaseous equilibria

You have seen how you can apply the equilibrium law to reversible reactions that occur in solution and how to derive an expression for an equilibrium constant K_c for such reactions. Many reversible reactions take place in the gas phase. These include many important industrial reactions such as the synthesis of ammonia and a key stage of the contact process for making sulfuric acid. Gaseous equilibria also obey the equilibrium law but it is usual to express their concentrations in a different way using the idea of partial pressure.

Partial pressure

In a mixture of gases, each gas contributes to the total pressure. This contribution is called its partial pressure p and is the pressure that the gas would exert if it occupied the container on its own. The sum of the partial pressures of all the gases in a mixture is the total pressure. For example, air is a mixture of approximately 20% oxygen molecules and 80% nitrogen molecules and has a pressure (at sea level) of approximately 100 kPa (kilopascals).

So the approximate partial pressure of oxygen in the air is 20 kPa and that of nitrogen is 80 kPa.

Mathematically, the partial pressure of a gas in a mixture is given by its mole fraction multiplied by the total pressure and is given the symbol p.

partial pressure p of A = mole fraction of A × total pressure

$$\text{The mole fraction of a gas A} = \frac{\text{number of moles of gas A in the mixture}}{\text{total number of moles of gas in the mixture}}$$

Applying the equilibrium law to gaseous equilibria

An equilibrium constant can be found in the same way as for a reaction in solution but it is given the symbol K_p rather than K_c.

For a reaction $\quad aA(g) + bB(g) \rightleftharpoons yY(g) + zZ(g)$

$$K_p = \frac{p^y Y(g)_{eqm}\, p^z Z(g)_{eqm}}{p^a A(g)_{eqm}\, p^b B\,(g)_{eqm}}$$

Note how this corresponds to the equilibrium law expressed in terms of concentration that you have seen earlier.

Worked example: Equilibrium law in gaseous equilibria

1 For the equilibrium

$H_2(g) + I_2(g) \rightleftharpoons 2HI(g)$

$$K_p = \frac{p^2HI(g)_{eqm}}{pH_2(g)_{eqm}\, pI_2(g)_{eqm}}$$

This particular K_p has no units as they cancel.

2 For the equilibrium

$3H_2(g) + N_2(g) \rightleftharpoons 2NH_3(g)$ (the key step in the Haber process)

$$K_p = \frac{p^2NH_3(g)_{eqm}}{p^3H_2(g)_{eqm}\, pN_2(g)_{eqm}}$$

Here K_p would have units of Pa^{-2}

Worked example: Calculating partial pressure

This example shows how you can use the expression for K_p to calculate the composition of an equilibrium mixture

K_p is 0.020 for the reaction

$2HI(g) \rightleftharpoons H_2(g) + I_2(g)$

If the reaction started with pure HI, and the initial pressure of HI was 100 kPa, what would be the partial pressure of hydrogen when equilibrium is reached?

Set out the problem in the same way as when using K_c.

	$2HI(g)$	$\rightleftharpoons$	$H_2(g)$	+	$I_2(g)$
Start:	100 kPa		0 kPa		0 kPa
At eqm:	$(100 - 2x)$ kPa		x kPa		x kPa

The chemical equation tells us:

- that there will be the same number of moles of H_2 and I_2 at equilibrium, therefore $pH_{2eqm} = pI_{2eqm} = x$
- that for each mole of hydrogen (and of iodine) that is produced, *two* moles of hydrogen iodide are used up so that if $pH_{2eqm} = x$, $pHI_{eqm} = (100 - 2x)$

$$K_p = \frac{pH_2(g)_{eqm}\, pI_2(g)_{eqm}}{p^2HI(g)_{eqm}}$$

Putting in the figures gives:

$$0.02 = \frac{x^2}{(100 - 2x)^2}$$

Taking the square root of each side gives:

$$0.141 = \frac{x}{100 - 2x}$$

$$0.141 \times (100 - 2x) = x$$

$$14.1 - 0.282x = x$$

$$14.1 = 1.282x$$

$$x = \frac{14.1}{1.282}$$

$$x = 10.99$$

$pH_2(g) = 11$ kPa (to 2 s.f.)

The chemical equation tells us that $pI_2(g)$ must be the same as $pH_2(g)$ and that pHI(g) must be $100 - (2 \times 11) = 78$ kPa.

The effect of changing temperature and pressure on a gaseous equilibrium

Le Châtelier's principle applies to gaseous equilibria in the same way as to equilibria in solution. The only difference is that the partial pressure of reactants and products replace concentration.

So, for a reaction that is exothermic going left to right, increasing the temperature forces the equilibrium to the left, so that the reaction absorbs heat. In other words, increasing the temperature decreases K_p. So for the Haber process reaction

$$3H_2(g) + N_2(g) \rightleftharpoons 2NH_3(g) \quad \Delta H = -92\ \text{kJ mol}^{-1}$$

So increasing the temperature *decreases* the yield of ammonia at equilibrium.

Increasing the pressure forces the equilibrium in such a way as to reduce the total pressure, that is, to the side with fewer molecules. In other words, K_p increases. So increasing the total pressure *increases* the yield of ammonia.

Changing the total pressure only affects the equilibrium position when there is a change in the total number of molecules on either side of the reaction. So for the equilibrium

$$2HI(g) \rightleftharpoons H_2(g) + I_2(g)$$

Pressure will have no effect on the equilibrium position.

Increasing the pressure on a gas phase reaction will increase the rate at which equilibrium is reached as there will be more collisions between molecules. Increasing temperature will also increase the rate at which equilibrium is attained as will the use of a catalyst.

Summary questions

1 Using Le Châtelier's principle, predict the effect of increasing: (i) the pressure and (ii) the temperature on the following reactions:

a $2SO_2(g) + O_2(g) \rightleftharpoons 2SO_3(g)$

$\Delta H = -197\ \text{kJ mol}^{-1}$

b $N_2O_4(g) \rightleftharpoons 2NO_2(g) \quad \Delta H = +58\ \text{kJ mol}^{-1}$

c $H_2(g) + CO_2(g) \rightleftharpoons H_2O(g) + CO(g)$

$\Delta H = +40\ \text{kJ mol}^{-1}$

2 $A(g) + B(g) \rightleftharpoons C(g) + D(g)$ represents an exothermic reaction and $K_p = pC(g)\,pD(g) / pA(g)\,pB(g)$.

In the above expression, what would happen to K_p:

a if the temperature were decreased

b if more A were added to the mixture

c if a catalyst were added?

3 State how the position of each of the following gaseous equilibria will be affected by:

i Increasing temperature

ii Decreasing total pressure

iii Using a catalyst

Give the units of K_p in each case:

a $2SO_2(g) + O_2(g) \rightleftharpoons 2SO_3(g)$

$\Delta H = -192\ \text{kJ mol}^{-1}$

b $H_2(g) + CO_2(g) \rightleftharpoons H_2O(g) + CO(g)$

$\Delta H = +40\ \text{kJ mol}^{-1}$

c $N_2O_4(g) \rightleftharpoons 2NO_2(g) \quad \Delta H = +57\ \text{kJ mol}^{-1}$

Practice questions

1 Consider the equilibrium system below.
$2SO_2(g) + O_2(g) \rightleftharpoons 2SO_3(g)$
The partial pressures for the gases in the equilibrium, mixture are
$pSO_2 = 0.080$ atm
$pO_2 = 0.90$ atm
$pSO_3 = 5.0$ atm
Calculate K_p for this system. Give your answer to an appropriate number of significant figures. Include the unit.

(3 marks)

2 Calculate the value of K_p for the system shown below.
$2NO_2(g) \rightleftharpoons N_2O_4(g)$
At 65 °C the partial pressures of the gases at equilibrium are
$pNO_2 = 0.80$ atm
$pN_2O_4 = 0.25$ atm
Give your answer to three significant figures and include the unit.

(3 marks)

3 A chemist analysed the equilibrium system below
$3H_2(g) + N_2(g) \rightleftharpoons 2NH_3(g)$
and found that there was 26.0 moles of NH_3, 13.0 moles of H_2, and 65.0 moles of N_2 present in the equilibrium mixture.
The total pressure of the system was 12.0 atm.

(i) Calculate the mole fraction of each gas at equilibrium.

(1 mark)

(ii) Calculate the partial pressure of each gas at equilibrium.

(1 mark)

(iii) Calculate K_p for this system. Give your answer to three decimal places and include any units.

(1 mark)

4 A chemist investigated the equilibrium system below.
$H_2(g) + I_2(g) \rightleftharpoons 2HI(g)$
At 450 °C and a pressure of 3.00 atm.
At equilibrium there was 0.30 mol of H_2, 0.40 mol of I_2, and 1.40 mol of HI.
Calculate the K_p. Give your answer to two significant figures and include any units.

(5 marks)

5 Phosphorus pentachloride, PCl_5 decomposes on heating to form phosphorus trichloride PCl_3 and chlorine, Cl_2 according to the equation below.
$PCl_5 \rightleftharpoons PCl_3 + Cl_2$
At a temperature of 350 °C and a pressure of 12.0 atm the amount of gas present at equilibrium was 0.40 mol of PCl_5, 0.75 mol of PCl_3, and 0.90 mol of Cl_2.
Calculate the value of K_p. Give your answer to two significant figures and include any units.

(5 marks)

6 Calculate the value of K_p for the system shown below.
$3H_2(g) + N_2(g) \rightleftharpoons 2NH_3(g)$
At 80 °C the partial pressures of the gases at equilibrium are:
$pH_2 = 0.80$ atm
$pN_2 = 0.25$ atm
$pNH_3 = 0.35$ atm
Give your answer to two significant figures and include any units.

(3 marks)

Answers to the Practice Questions and Section Questions are available at
www.oxfordsecondary.com/oxfordaqaexams-alevel-chemistry

20 Electrode potentials and electrochemical cells

20.1 Electrode potentials and the electrochemical series

Learning objectives:

- Illustrate how half equations are written for the reactions at an electrode.
- Explain the term standard electrode potential.
- Describe how standard electrode potentials are measured.
- Describe the conventional representation of a cell.

Specification reference: 3.1.11

Synoptic link

You will need to know redox equations studied in Chapter 7, Oxidation, reduction, and redox reactions.

Synoptic link

See Practical 7 on page 525.

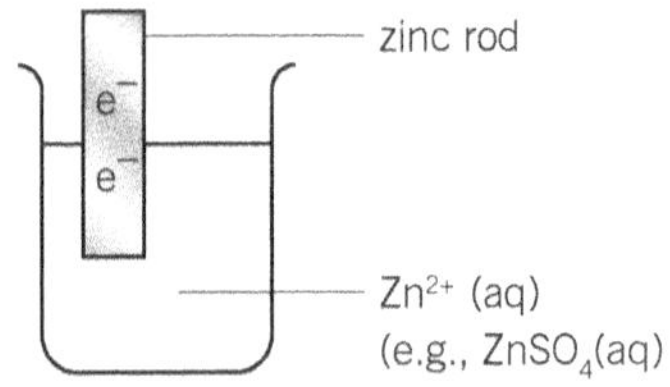

▲ **Figure 2** *A zinc electrode*

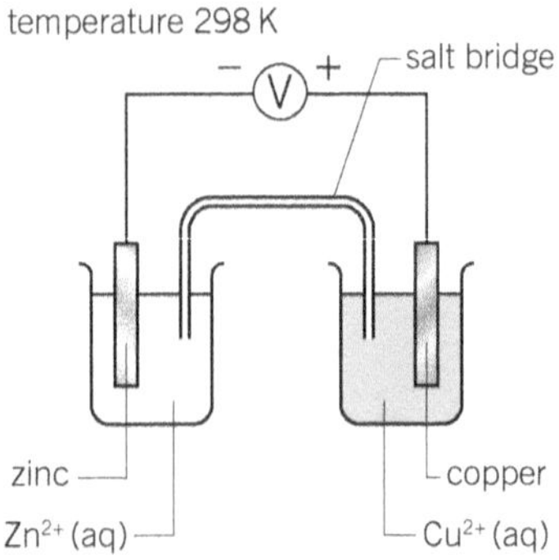

▲ **Figure 3** *Two electrodes connected together with a voltmeter to measure the potential difference*

If you place two different metals in a salt solution and connect them together (Figure 1) an electric current flows so that electrons pass from the more reactive metal to the less reactive. This is the basis of batteries that power everything from MP3 players to milk floats.

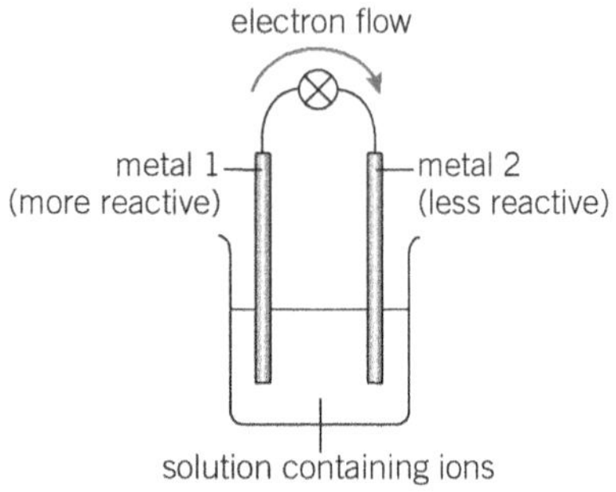

▲ **Figure 1** *Flow of electric current*

This topic and the following two look at how electricity is produced by electrochemical cells and how this can be used to explain and predict redox reactions (which are all about electron transfer).

Half cells

When a rod of metal is dipped into a solution of its own ions, an equilibrium is set up.

For example, dipping zinc into zinc sulfate solution sets up the following equilibrium:

$$Zn(s) \rightleftharpoons Zn^{2+}(aq) + 2e^-$$

This arrangement is called an electrode, or a half cell, as two half cells can be joined together to make an electrical cell (Figure 2).

If you could measure this potential, it would tell us how readily electrons are released by the metal, that is, how good a reducing agent the metal is. (Remember that reducing agents release electrons.)

However, electrical potential cannot be measured directly, only potential *difference* (often called voltage). What you *can* do is to connect together two different electrodes and measure the potential difference between them with a voltmeter (Figure 3) for copper and zinc electrodes.

The electrical circuit is completed by a **salt bridge**, the simplest form of which is a piece of filter paper soaked in a solution of a salt (usually saturated potassium nitrate). A salt bridge is used rather than a piece of wire, to avoid further metal/ion potentials in the circuit.

If you connect the two electrodes to the voltmeter (Figures 3 and 4) you get a potential difference (voltage) of 1.10 V (if the solutions are 1.00 mol dm^{-3} and the temperature 298 K). The voltmeter shows that the zinc electrode is the more negative.

The fact that the zinc electrode is negative tells you that zinc loses its electrons more readily than does copper – zinc is a better reducing agent. If the voltmeter were removed and electrons allowed to flow, they would do so from zinc to copper. The following changes would take place:

1 Zinc would dissolve to form $Zn^{2+}(aq)$, increasing the concentration of $Zn^{2+}(aq)$.
2 The electrons would flow through the wire to the copper rod where they would combine with $Cu^{2+}(aq)$ ions (from the copper sulfate solution) so depositing fresh copper on the rod and decreasing the concentration of $Cu^{2+}(aq)$.

The following two **half reactions** would take place:

$$Zn(s) \rightarrow Zn^{2+}(aq) + 2e^-$$

and $$Cu^{2+}(aq) + 2e^- \rightarrow Cu(s)$$

adding: $$Zn(s) + Cu^{2+}(aq) + \cancel{2e^-} \rightarrow Zn^{2+}(aq) + Cu(s) + \cancel{2e^-}$$

When the two half reactions are added together, the electrons cancel out and you get the overall reaction:

$$Zn(s) + Cu^{2+}(aq) \rightarrow Zn^{2+}(aq) + Cu(s)$$

This is the reaction you get on putting zinc directly into a solution of copper ions. It is a redox reaction with zinc being oxidised and copper ions reduced. If the two half cells are connected they generate electricity. This forms an electrical cell called the Daniell cell (Figure 5).

The hydrogen electrode

To compare the tendency of different metals to release electrons, a standard electrode is needed to which any other half cell can be connected for comparison. The half cell chosen is called the standard hydrogen electrode (Figure 6).

Hydrogen gas is bubbled into a solution of $H^+(aq)$ ions. Since hydrogen doesn't conduct, electrical contact is made via a piece of unreactive platinum metal (coated with finely divided platinum to increase the surface area and allow any reaction to proceed rapidly). The electrode is used under standard conditions of $[H^+(aq)] = 1.00\ mol\ dm^{-3}$, pressure 100 kPa, and temperature 298 K.

Hint

A perfect voltmeter does not allow any current to flow – it merely measures the electrical 'push' or pressure (the potential difference) which tends to make current flow.

Study tip

The salt chosen for the salt bridge must not react with either of the solutions in the half cells.

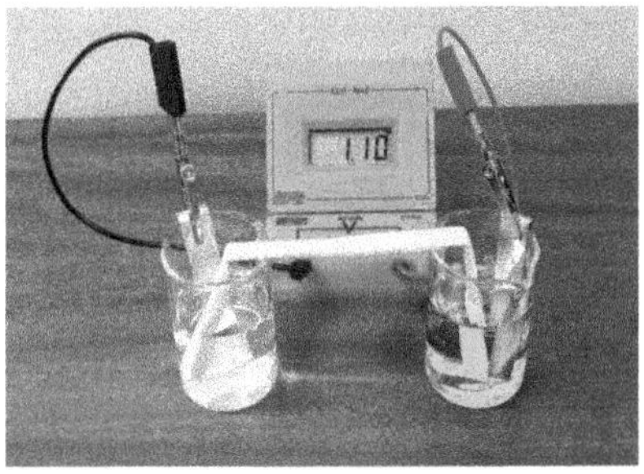

▲ **Figure 4** *Measuring the potential difference of zinc and copper half cells using a salt bridge of salt-soaked filter paper.*

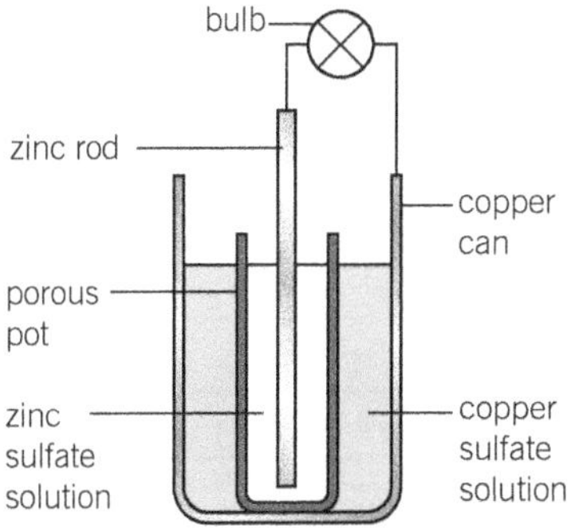

▲ **Figure 5** *A Daniell cell lighting a bulb. The porous pot acts like a salt bridge*

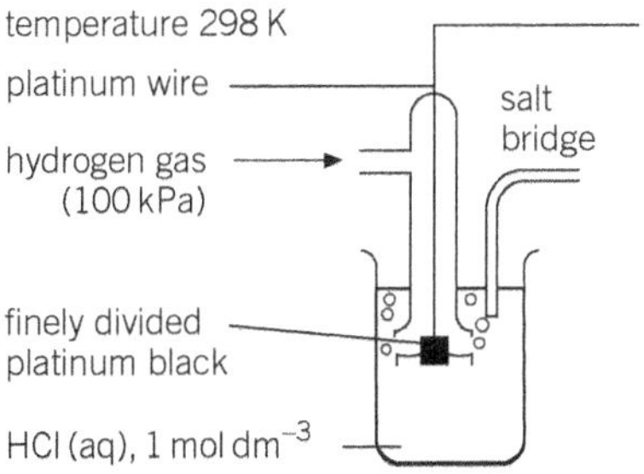

▲ **Figure 6** *The standard hydrogen electrode*

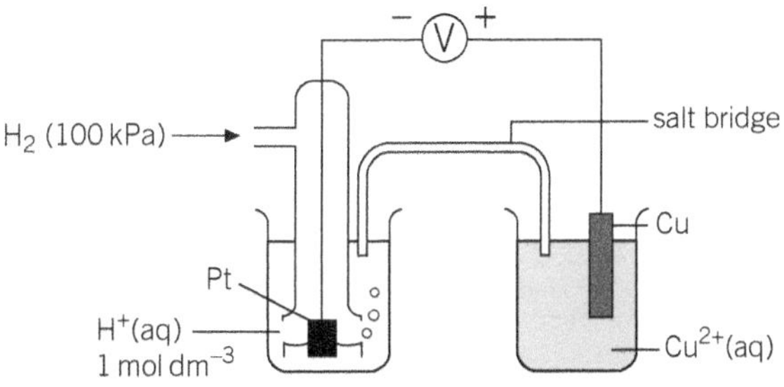

▲ **Figure 7** *Measuring $E^\ominus$ for a copper electrode*

Study tip

Remember the use of square brackets to represent concentration in mol dm^{-3}.

The potential of the standard hydrogen electrode is defined as zero, so if it is connected to another electrode (Figure 7), the measured voltage, called the electromotive force E (emf), is the electrode potential of that cell. If the second cell is at standard conditions ([metal ions] = 1.00 mol dm^{-3}, temperature = 298 K), then the emf is given the symbol $E^{\ominus}$. Electrodes with negative values of $E^{\ominus}$ are better at releasing electrons (better reducing agents) than hydrogen.

Changing the conditions, such as the concentration of ions or temperature, of an electrode will change its electrical potential.

The electrochemical series

A list of some $E^{\ominus}$ values for metal/metal ion standard electrodes is given in Table 1.

Study tip

Oxidation occurs at the negative electrode. Reduction occurs at the positive electrode.

In Table 1, the equilibria are written as reduction reactions (with the electrons on the left of the arrow). These are called electrode potentials, sometimes known as reduction potentials. (Remember from OIL RIG – Reduction Is Gain.)

Arranged in this order with the most negative values at the top, this list is called the electrochemical series. The number of electrons involved in the reaction has no effect on the value of $E^{\ominus}$.

The voltage obtained by connecting two standard electrodes together is found by the difference between the two $E^{\ominus}$ values. So connecting an Al^{3+}(aq)/Al(s) standard electrode to a Cu^{2+}(aq)/Cu(s) standard electrode would give a voltage of 2.00 V (Figure 8).

▼ **Table 1** *Some $E^{\ominus}$ values. Good reducing agents have negative values for $E^{\ominus}$*

Half reaction	$E^{\ominus}$/V
$Li^+(aq) + e^- \rightarrow Li(s)$	−3.03
$Ca^{2+}(aq) + 2e^- \rightarrow Ca(s)$	−2.87
$Al^{3+}(aq) + 3e^- \rightarrow Al(s)$	−1.66
$Zn^{2+}(aq) + 2e^- \rightarrow Zn(s)$	−0.76
$Pb^{2+}(aq) + 2e^- \rightarrow Pb(s)$	−0.13
$2H^+(aq) + 2e^- \rightarrow H_2(g)$	0.00
$Cu^{2+}(aq) + 2e^- \rightarrow Cu(s)$	+0.34
$Ag^+(aq) + e^- \rightarrow Ag(s)$	+0.80

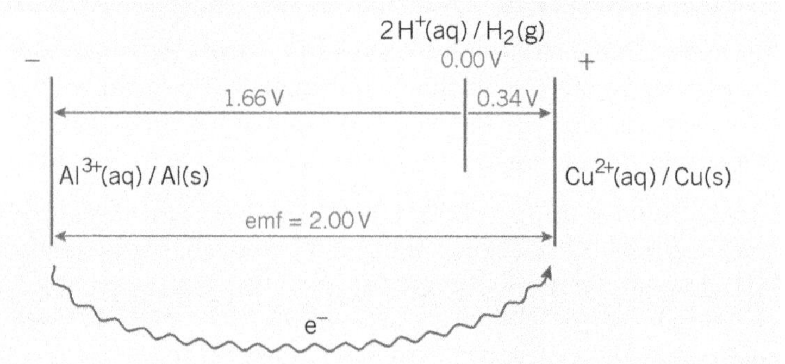

▲ **Figure 8** *Calculating the value of the voltage when two electrodes are connected*

Hint

What is $E^{\ominus}$ for this half reaction?
$2Li^+(aq) + 2e^- \rightarrow 2Li(s)$
−3.03 V, as the number of electrons makes no difference to electrical potential.

If you connect an Al^{3+}(aq)/Al(s) standard electrode to a Cu^{2+}(aq)/Cu(s) standard electrode, the emf will be 2.00 V and the Al^{3+}(aq)/Al(s) electrode will be negative.

If you connect an Al^{3+}(aq)/Al(s) standard electrode to a Pb^{2+}(aq)/Pb(s) standard electrode the emf will be 1.53 V and the Al^{3+}(aq)/Al(s) electrode will be the negative electrode of the cell (Figure 9).

It is worth sketching diagrams like the ones shown in Figure 8 and Figure 9. It will prevent you getting confused with signs. Remember that more negative values are drawn to the left on the diagrams.

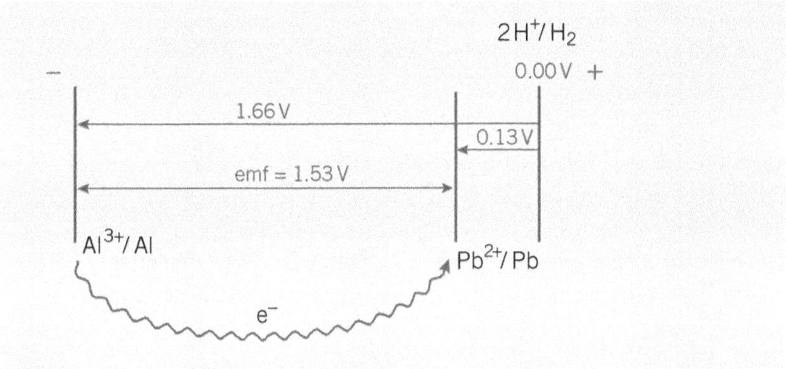

▲ **Figure 9** *Calculating the value of the emf for an Al^{3+}/Al electrode connected to a Pb^{2+}/Pb electrode*

For example, if you connect an Al^{3+}(aq)/Al(s) standard electrode to a Zn^{2+}(aq)/Zn(s) standard electrode (Figure 10) the voltmeter will read 0.90 V and the Al^{3+}(aq)/Al(s) electrode will be negative.

Representing cells

There is shorthand for writing down the cell formed by connecting two electrodes. The conventions are those recommended by IUPAC (International Union of Pure and Applied Chemistry). The usual apparatus diagram is shown in Figure 10 and the cell diagram is written using the following conventions:

- A vertical solid line indicates a **phase boundary**, for example, between a solid and a solution.
- A double vertical line shows a salt bridge.
- The species with the highest oxidation state is written next to the salt bridge.
- When giving the value of the emf $E^\ominus$ state the polarity (i.e., whether it is positive or negative) of the right-hand electrode, as the cell representation is written. In the case of the aluminium and copper cells in Figure 8 the copper half cell is more positive (it is connected to the positive terminal of the voltmeter) and, if allowed to flow, electrons would go from aluminium to copper.

Al(s)|Al^{3+}(aq)||Cu^{2+}(aq)|Cu(s) $\qquad E^\ominus_{cell} = +2.00\,V$

You could also have written the cell:

Cu(s)|Cu^{2+}(aq)||Al^{3+}(aq)|Al(s) $\qquad E^\ominus_{cell} = -2.00\,V$

This still tells us that electrons flow from aluminium to copper as the polarity of the right-hand electrode is always given.

So emf = $E^\ominus(R) - E^\ominus(L)$

where $E^\ominus$(R) represents the emf of the right-hand electrode and $E^\ominus$(L) that of the left-hand electrode.

The cell representation for a silver electrode connected to a Pb^{2+}(aq)/Pb(s) half cell would be:

Pb(s)|Pb^{2+}(aq)||Ag^+(aq)|Ag(s) $\qquad E^\ominus_{cell} = +0.93\,V$

Study tip

The phase indicates a state of matter such as solid, liquid, or gas. It also refers to aqueous solutions.

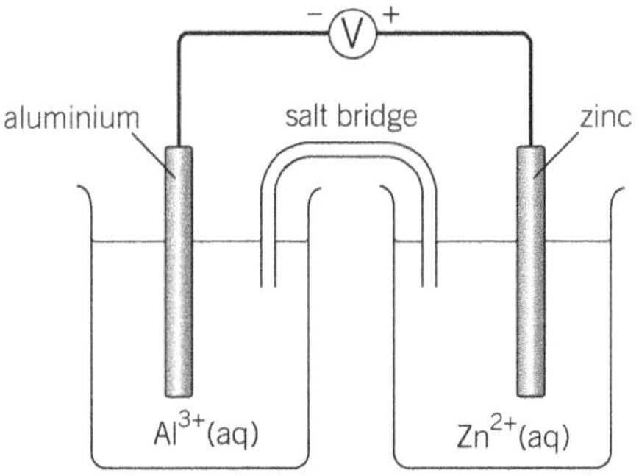

▲ **Figure 10** *Connecting a pair of electrodes*

Summary questions

1 **a** Represent the following on a conventional cell diagram:

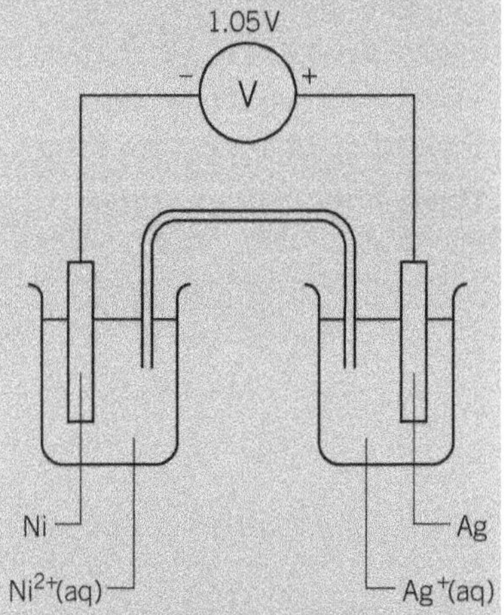

b If the voltmeter was replaced by a wire, in which direction would the electrons flow? Write equations for the reactions occurring in each beaker and write an equation for the overall cell reaction.

2 Calculate $E^\ominus_{cell}$ for:

a Zn(s)|Zn^{2+}(aq) || Pb^{2+}(aq)|Pb(s)

b Pb(s)|Pb^{2+}(aq) || Zn^{2+}(aq) Zn(s)

20.2 Predicting the direction of redox reactions

Learning objectives:

→ Describe how standard electrode potentials can be used to predict the direction of a redox reaction.

Specification reference: 3.1.11

It is possible to use standard electrode potentials to decide on the feasibility of a redox (i.e., electron transfer) reaction. When you connect a pair of electrodes, the electrons will flow from the more negative to the more positive and not in the opposite direction. So the signs of the electrodes tell us the direction of a redox reaction.

Think of the following electrodes (Figure 1):

$Zn^{2+}(aq) + 2e^- \rightleftharpoons Zn(s)$ written in short as $Zn^{2+}(aq)/Zn(s)$ $E^\ominus = -0.76\,V$ and

$Cu^{2+}(aq) + 2e^- \rightleftharpoons Cu(s)$ written in short as $Cu^{2+}(aq)/Cu(s)$ $E^\ominus = +0.34\,V$

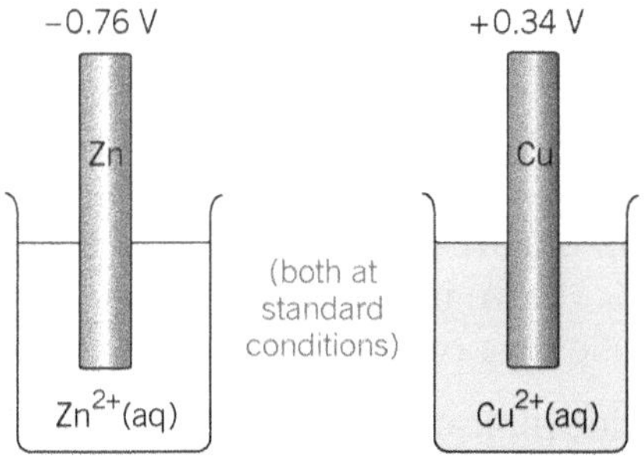

▲ **Figure 1** *The two electrodes. Their potentials are measured with respect to a standard hydrogen electrode*

Study tip

Electrons flow spontaneously from the negative electrode to the positive electrode.

Figure 2 shows these two electrodes connected together. Electrons will tend to flow from zinc (the more negative) to copper (the more positive).

So you know which way the two half reactions must go. You can use a diagram to represent the cell and if you then include $E^\ominus$ values for the two electrodes, you can find the emf for the cell. Figure 3 is the diagram for the zinc/copper cell. You can see how it is related to the apparatus.

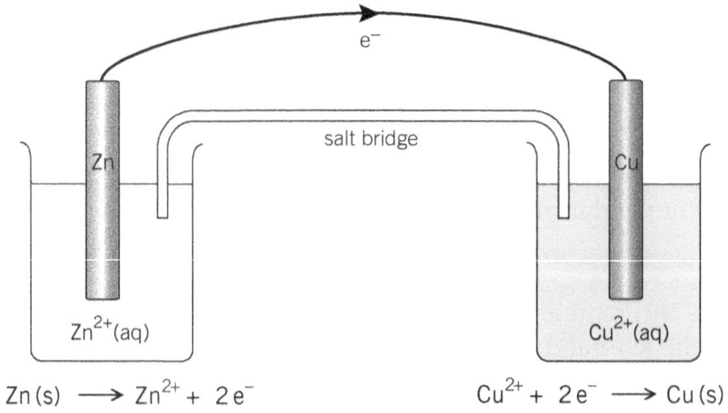

▲ **Figure 2** *Connecting $Zn^{2+}(aq)/Zn(s)$ and $Cu^{2+}(aq)/Cu(s)$ electrodes*

So the two equations are:

$$Zn(s) \rightarrow Zn^{2+}(aq) + 2e^-$$

$$Cu^{2+}(aq) + 2e^- \rightarrow Cu(s)$$

The overall effect is:

$$Zn(s) \rightarrow Zn^{2+}(aq) + 2e^-$$

$$Cu^{2+}(aq) + 2e^- \rightarrow Cu(s)$$

$$Cu^{2+}(aq) + Zn(s) + 2e^- \rightarrow Cu(s) + Zn^{2+}(aq) + 2e^-$$

So this reaction is feasible and is the reaction that actually happens, either by connecting the two electrodes or more directly by adding Zn to $Cu^{2+}(aq)$ ions in a test tube. The reverse reaction is *not* feasible and does not occur.

$$Cu(s) + Zn^{2+}(aq) \nrightarrow Zn(s) + Cu^{2+}(aq)$$

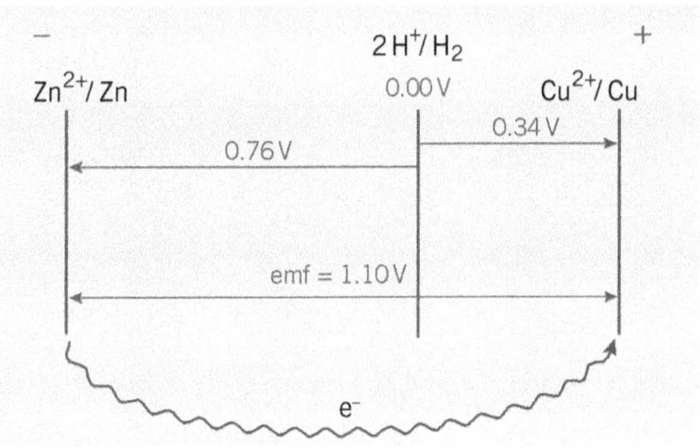

▲ **Figure 3** *Predicting the direction of electron flow when a Zn^{2+}/Zn electrode is connected to a Cu^{2+}/Cu electrode*

This reaction would require electrons to flow from positive to negative.

You can go through this process whenever you want to predict the outcome of a redox reaction.

With redox systems that only involve metal ions but no metal (e.g., Fe^{3+} / Fe^{2+}), a beaker containing all the relevant ions and a platinum electrode to make electrical contact is used in order to measure $E^\ominus$ by connecting to a hydrogen electrode (Figure 4).

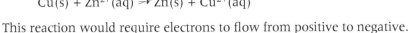

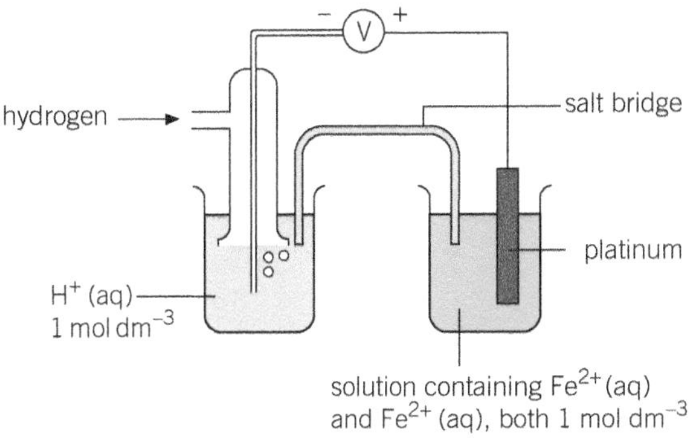

▲ **Figure 4**

The cell diagram would be written as follows:

$$Pt \mid H_2(g) \mid 2H^+(aq) \parallel Fe^{3+}(aq), Fe^{2+}(aq) \mid Pt$$

Remember the rule that the most oxidised species (in this case H^+ and Fe^{3+}) go next to the salt bridge.

Further examples of predicting the direction of redox reactions

You can extend the electrochemical series to systems other than simple metal / metal ion ones (Table 1).

▼ **Table 1** *$E^\ominus$ values for more reduction half equations*

Reduction half equation	$E^\ominus$/V
$Li^+(aq) + e^- \rightarrow Li(s)$	−3.03
$Ca^{2+}(aq) + 2e^- \rightarrow Ca(s)$	−2.87
$Al^{3+}(aq) + 3e^- \rightarrow Al(s)$	−1.66
$Zn^{2+}(aq) + 2e^- \rightarrow Zn(s)$	−0.76
$Cr^{3+}(aq) + e^- \rightarrow Cr^{2+}(aq)$	−0.41
$Pb^{2+}(aq) + 2e^- \rightarrow Pb(s)$	−0.13
$2H^+(aq) + 2e^- \rightarrow H_2(g)$	0.00
$Cu^{2+}(aq) + e^- \rightarrow Cu^+(aq)$	+0.15
$Cu^{2+}(aq) + 2e^- \rightarrow Cu(s)$	+0.34
$I_2(aq) + 2e^- \rightarrow 2I^-(aq)$	+0.54
$Fe^{3+}(aq) + e^- \rightarrow Fe^{2+}(aq)$	+0.77
$Ag^+(aq) + e^- \rightarrow Ag(s)$	+0.79
$Br_2(aq) + 2e^- \rightarrow 2Br^-(aq)$	+1.07
$Cl_2(aq) + 2e^- \rightarrow 2Cl^-(aq)$	+1.36
$MnO_4^- + 8H^+(aq) + 5e^- \rightarrow Mn^{2+}(aq) + 4H_2O(l)$	+1.51
$Ce^{4+}(aq) + e^- \rightarrow Ce^{3+}(aq)$	+1.70

Worked example: emf for an iron–chlorine electrochemical cell

Will the following reaction occur or not?

$$Fe^{3+}(aq) + Cl^-(aq) \rightarrow Fe^{2+}(aq) + \frac{1}{2}Cl_2(aq)$$

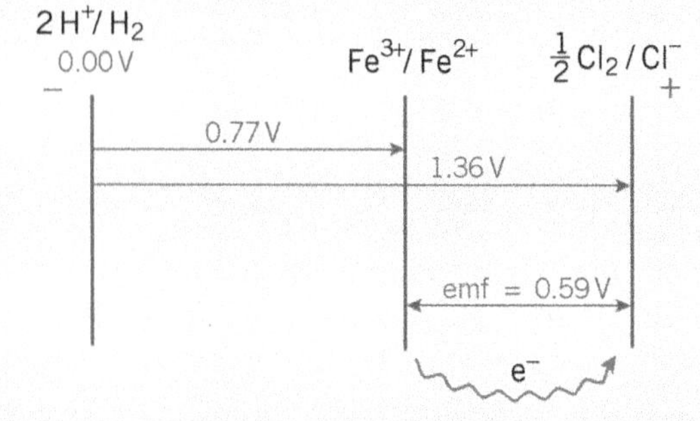

▲ **Figure 5** *Working out the emf for* $Fe^{3+}(aq) + Cl^-(aq) \rightarrow Fe^{2+}(aq) + \frac{1}{2}Cl_2(aq)$

Hint

$E^\ominus$ values tell you whether a reaction is feasible or not. They don't give any information about the speed of the reaction. A feasible reaction may be so slow that in practice it does not take place at all at room temperature.

Figure 5 shows that the emf is 0.59 V with iron the more negative. So, electrons will flow from the Fe^{3+}/Fe^{2+} standard electrode to the $\frac{1}{2}Cl_2/Cl^-$ standard electrode.

$$Fe^{2+} \rightarrow Fe^{3+} + e^-$$

$$\frac{1}{2}Cl_2 + e^- \rightarrow Cl^-$$

$$Fe^{2+} + \frac{1}{2}Cl_2 + \cancel{e^-} \rightarrow Fe^{3+} + Cl^- + \cancel{e^-}$$

This is the reaction that will occur. The following reaction is not feasible.

$$Fe^{3+}(aq) + Cl^-(aq) \not\rightarrow Fe^{2+}(aq) + \frac{1}{2}Cl_2(aq)$$

So, chlorine will oxidise iron(II) ions to iron(III) ions.

Worked example: emf for an iron–iodine electrochemical cell

Will the following reaction occur or not?

$$Fe^{3+}(aq) + I^-(aq) \rightarrow Fe^{2+}(aq) + \frac{1}{2}I_2(aq)$$

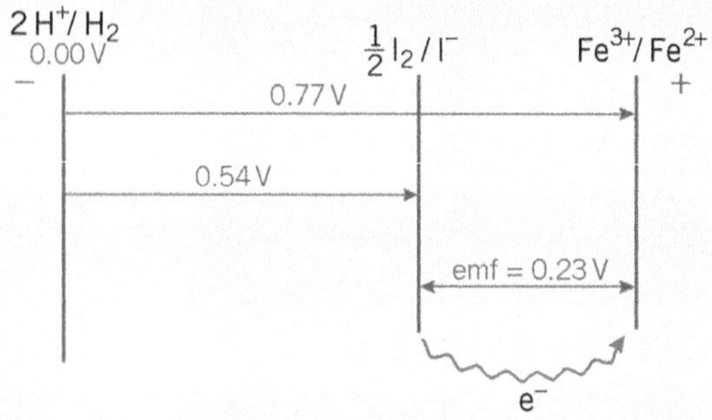

▲ **Figure 6** *Working out the emf for* $Fe^{3+}(aq) + I^-(aq) \rightarrow Fe^{2+}(aq) + \frac{1}{2}I_2(aq)$

➔

Figure 6 shows that the emf is 0.23 V with iodine the more negative.

So, electrons will flow from the $\frac{1}{2}I_2/I^-$ electrode to the Fe^{3+}/Fe^{2+} electrode.

$$I^- \rightarrow \frac{1}{2}I_2 + e^-$$
$$Fe^{3+} + e^- \rightarrow Fe^{2+}$$

$$I^- + Fe^{3+} + \cancel{e^-} \rightarrow \frac{1}{2}I_2 + Fe^{2+} + \cancel{e^-}$$

This is the reaction that will occur. The following reaction is not feasible.

$$Fe^{2+}(aq) + \frac{1}{2}I_2(aq) \not\rightarrow Fe^{3+}(aq) + I^-(aq)$$

So, iron(III) ions will oxidise iodide ions to iodine.

Hint

$\mathcal{E}(R) - \mathcal{E}(L) = \text{emf}$

If the emf is positive then the reaction is feasible.

Summary questions

1 What will be the value of the emf for an $Al^{3+}(aq)/Al(s)$ standard electrode connected to a $Zn^{2+}(aq)/Zn(s)$ standard electrode? Draw a diagram like Figure 3 to illustrate your answer.

2 Use the values of $E^\ominus$ in Table 1 to calculate the emf for the following:

a $Ce^{4+}(aq) + Fe^{2+}(aq) \rightarrow Ce^{3+}(aq) + Fe^{3+}(aq)$

b $I_2(aq) + 2Br^-(aq) \rightarrow Br_2(aq) + 2I^-(aq)$

c $MnO_4^-(aq) + 8H^+(aq) + 5I^-(aq) \rightarrow Mn^{2+}(aq) + 4H_2O(l) + 2\frac{1}{2}I_2(aq)$

d $2H^+(aq) + Pb(s) \rightarrow Pb^{2+}(aq) + H_2(g)$

3 Which of the halogens could possibly oxidise Ag(s) to $Ag^+(aq)$ ions?

4 a Is the reaction $Br_2(aq) + 2Cl^-(aq) \rightarrow Cl_2(aq) + 2Br^-(aq)$ feasible?

b Is the reaction $Fe^{3+}(aq) + Br^-(aq) \rightarrow Fe^{2+}(aq) + \frac{1}{2}Br_2(aq)$ feasible?

Learning objectives

- Describe the differences between non-rechargeable, rechargeable, and fuel cells.
- Describe the electrode reactions in a hydrogen–oxygen fuel cell.
- Describe the benefits and risks to society associated with each type of cell.

Specification reference: 3.1.11

▲ **Figure 1** *Tablet computers use rechargeable batteries*

▲ **Figure 2** *In the Victorian era, Morse code telegraphic communications used Daniell cells*

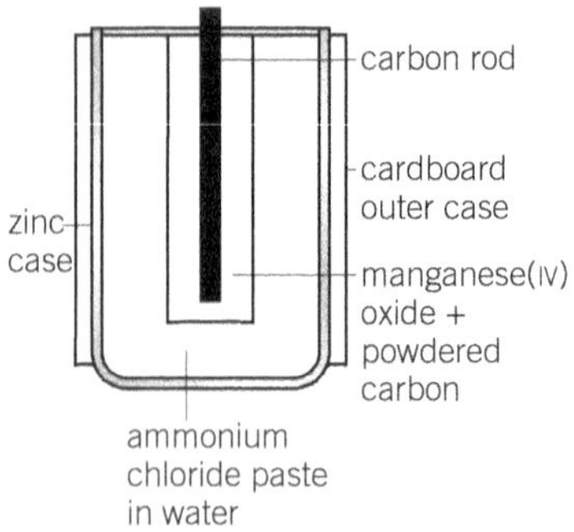

▲ **Figure 3** *A Leclanché cell*

Modern life would not be the same without batteries – both rechargeable and single-use for tablets, phones, MP3 players, and so on. Nowadays there is a huge variety of types and brands advertised with slogans like 'long life' and 'high power'. Batteries are based on the principles of electrochemical cells. Strictly, a battery refers to a number of cells connected together, but in everyday speech the word has come to mean almost any portable source of stored electricity. You will need to be able to apply the principles of electrochemical cells, but you will not be expected to learn the details of the construction of the cells described below.

Non-rechargeable cells

Zinc/copper cells

The Daniell cell, see Topic 20.1, provides an emf of 1.1 V. It was developed by the British chemist John Daniell in the 1830s, and was used to provide the electricity for old-fashioned telegraphs which sent messages by Morse code (Figure 2). However, it was not practical for portable devices, because of the liquids that it contained. It works on the general principle of electrons being transferred from a more reactive metal to a less reactive one. The voltage can be worked out from the difference between the electrode potentials in the electrochemical series.

Zinc/carbon cells

The electrodes can be made from materials other than metals. For example, in the Leclanché cell (which is named after the Frenchman George Leclanché), the positive electrode is carbon, which acts like the inert platinum electrode in the hydrogen electrode (Figure 3). The Leclanché cell is the basis of most ordinary disposable batteries. The electrolyte is a paste rather than a liquid.

The commercial form of this type of cell consists of a zinc canister filled with a paste of ammonium chloride, NH_4Cl, and water – the electrolyte. In the centre is a carbon rod. It is surrounded by a mixture of manganese(IV) oxide and powdered carbon. The half equations are:

$$Zn(s) \rightleftharpoons Zn^{2+}(aq) + 2e^- \qquad E \approx -0.8\,V$$

$$2NH_4^+(aq) + 2e^- \rightleftharpoons 2NH_3(g) + H_2(g) \qquad E \approx +0.7\,V$$

These are not $E^\ominus$ values as the conditions are far from standard.

The reactions that take place are:

- at the zinc:

$$Zn(s) \rightarrow Zn^{2+}(aq) + 2e^-$$

- at the carbon rod:

$$2NH_4^+(aq) + 2e^- \rightarrow 2NH_3(g) + H_2(g)$$

So the overall reaction as the cell discharges is:

$$2NH_4^+(aq) + Zn(s) \rightarrow 2NH_3(g) + H_2(g) + Zn^{2+}(aq)$$

emf ≈ 1.5 V with the zinc as the negative terminal

The hydrogen gas is oxidised to water by the manganese(IV) oxide (preventing a build up of pressure), whilst the ammonia dissolves in the water of the paste.

As the cell discharges, the zinc is used up and the walls of the zinc canister become thin and prone to leakage. The ammonium chloride electrolyte is acidic and can be corrosive. That is why you should remove spent batteries from equipment. This cell is ideal for doorbells, for example, which need a small current intermittently.

A variant of this cell is the zinc chloride cell. It is similar to the Leclanché but uses zinc chloride as the electrolyte. Such cells are better at supplying high currents than the Leclanché and are marketed as extra life batteries for radios, torches, and shavers.

Long life alkaline batteries are also based on the same system, but with an electrolyte of potassium hydroxide. Powdered zinc is used, whose greater surface area allows the battery to supply high currents. The cell is enclosed in a steel container to prevent leakage. These cells are suitable for equipment taking continuous high currents such as personal stereos. In this situation they can last up to 16 times as long as ordinary zinc/carbon batteries, but they are more expensive.

Many other electrode systems are in use, especially for miniature batteries such as those used in watches, hearing aids, cameras, and electronic equipment. These include zinc/air, mercury(II) oxide/zinc, silver oxide/zinc, and lithium/manganese(IV) oxide. Which is used for which application depends on the precise requirements of voltage, current, size, and cost.

Rechargeable batteries

These can be recharged by reversing the cell reactions. This is done by applying an external voltage greater than the voltage of the cell to drive the electrons in the opposite direction.

Lead–acid batteries

Lead–acid batteries are rechargeable batteries used to operate the starter motors of cars. They consist of six 2 V cells connected in series to give 12 V. Each cell consists of two plates dipped into a solution of sulfuric acid. The positive plate is made of lead coated with lead(IV) oxide, PbO_2, and the negative plate is made of lead (Figure 4).

On discharging, the following reactions occur as the battery drives electrons from the lead plate to the lead(IV) oxide coated one.

At the lead plate:

$$Pb(s) + SO_4^{2-}(aq) \rightarrow PbSO_4(s) + 2e^-$$

At the lead-dioxide-coated plate:

$$PbO_2(s) + 4H^+(aq) + SO_4^{2-}(aq) + 2e^- \rightarrow PbSO_4(s) + 2H_2O(l)$$

The overall reaction as the cell discharges is:

$$PbO_2(s) + 4H^+(aq) + 2SO_4^{2-}(aq) + Pb(s) \rightarrow 2PbSO_4(s) + 2H_2O(l)$$

$$\text{emf} \approx 2\,V$$

These reactions are reversed as the battery is charged up and electrons flow in the reverse direction, driven by the car's generator.

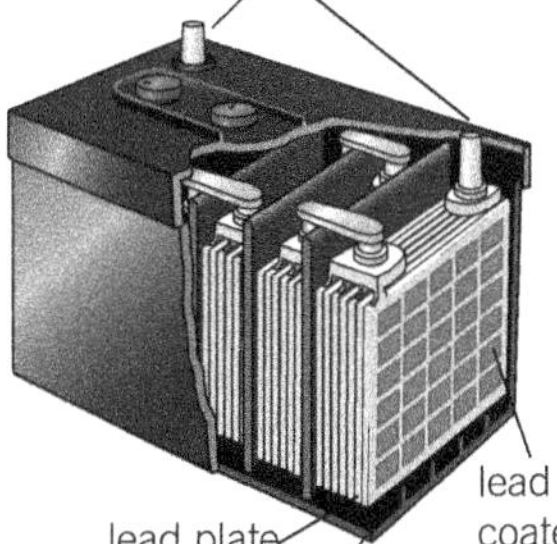

▲ **Figure 4** *A lead–acid car battery*

Hint

Technically, a battery consists of two or more simple cells connected together. So a 1.5 V zinc–carbon battery is really a cell, whilst a car battery is a true battery.

Study tip

You can use given data to determine the emf of any cell.

▲ **Figure 5** *Some of the enormous variety of batteries available today*

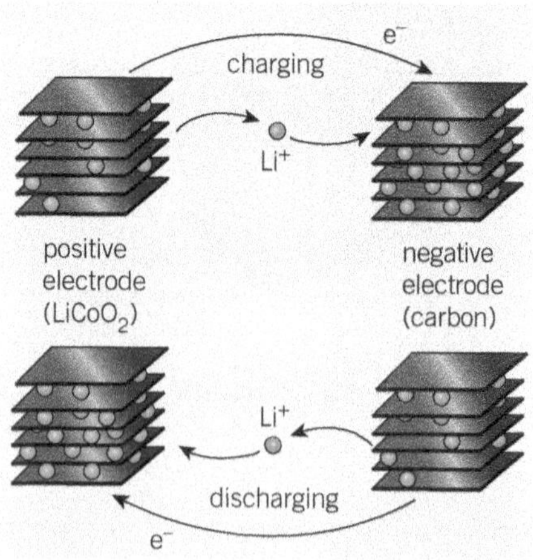

▲ **Figure 6** *The lithium ion cell*

Portable batteries

There are now rechargeable batteries that come in all shapes and sizes.

Nickel/cadmium

These are now available in standard sizes to replace traditional zinc–carbon batteries. Although more expensive to buy, they can be recharged up to 500 times, reducing the effective cost significantly. These cells are called nickel/cadmium and have an alkaline electrolyte. The two half equations are:

$$Cd(OH)_2(s) + 2e^- \rightleftharpoons Cd(s) + 2OH^-(aq)$$

$$NiO(OH)(s) + H_2O(l) + e^- \rightleftharpoons Ni(OH)_2(s) + OH^-(aq)$$

Overall:

$$2NiO(OH)(s) + Cd(s) + 2H_2O(l) \rightleftharpoons 2Ni(OH)_2(s) + Cd(OH)_2(s)$$

emf $\approx +1.2\,V$

The reaction goes from left to right on discharge (electrons flowing from Cd to Ni) and right to left on charging.

Lithium ion

The rechargeable lithium ion cell is used in laptops, tablets, smartphones, and other mobile gadgets. It is light because lithium is the least dense metal. The electrolyte is a solid polymer rather than a liquid or paste so it cannot leak, and its charge can be topped up at any time without the memory effect of some other rechargeable batteries, which can only be recharged efficiently when they have been fully discharged. The cell can even be bent or folded without leaking.

The positive electrode is made of lithium cobalt oxide, $LiCoO_2$, and the negative electrode is carbon. These are arranged in layers with a sandwich of solid electrolyte in between. On charging, electrons are forced through the external circuit from positive to negative electrode and at the same time lithium ions move through the electrolyte towards the positive electrode to maintain the balance of charge.

The reactions that occur on charging are:

negative electrode: $Li^+ + e^- \rightarrow Li$ $\quad E^\ominus = -3\,V$

positive electrode: $Li^+ + (CoO_2)^- \rightarrow Li^+ + CoO_2 + e^-$ $\quad E^\ominus = -1\,V$

On discharging, the processes are reversed so electrons flow from negative to positive. A single cell gives a voltage of between 3.5 V and 4.0 V, compared with around 1.5 V for most other cells.

Alkaline hydrogen–oxygen fuel cell

The reactions that occur in an alkaline hydrogen–oxygen fuel cell are:

$$2H_2(g) + 4OH^-(aq) \rightarrow 4H_2O(l) + 4e^- \quad E^\ominus = -0.83\,V$$

$$O_2(g) + 2H_2O(l) + 4e^- \rightarrow 4OH^-(aq) \quad E^\ominus = +0.40\,V$$

Overall:

$$2H_2(g) + O_2(g) \rightarrow 2H_2O(l) \quad E = 1.23\,V$$

The cell has two electrodes of a porous platinum-based material. They are separated by a semi-permeable membrane and the electrolyte is sodium hydroxide solution. Hydrogen enters at the negative electrode and the following half reaction takes place:

$$2H_2(g) + 4OH^-(aq) \rightarrow 4H_2O(l) + 4e^- \qquad E^\ominus = -0.83\,V$$

This releases electrons, which flow through the circuit to the other electrode where oxygen enters and the following reaction takes place:

$$O_2(g) + 2H_2O(l) + 4e^- \rightarrow 4OH^-(aq) \qquad E^\ominus = +0.40V$$

This accepts electrons from the other electrode and releases OH^- ions which travel through the semi-permeable membrane to that electrode.

The overall effect is for hydrogen and oxygen to react together to produce water and generate an emf of 1.23 V.

$$O_2(g) + 4H_2(g) \rightarrow 2H_2O(l) \qquad \text{emf} = 1.23\,V$$

This is the same reaction as burning hydrogen and oxygen but it takes place at a low temperature so there is no production of nitrogen oxides, which form if hydrogen is burnt directly.

This type of fuel cell is used to generate electricity on spacecraft because the only by-product is pure water, which can be used as drinking water by the astronauts. In terrestrial use it is important because unlike many other sources of electrical energy, it produces no carbon dioxide.

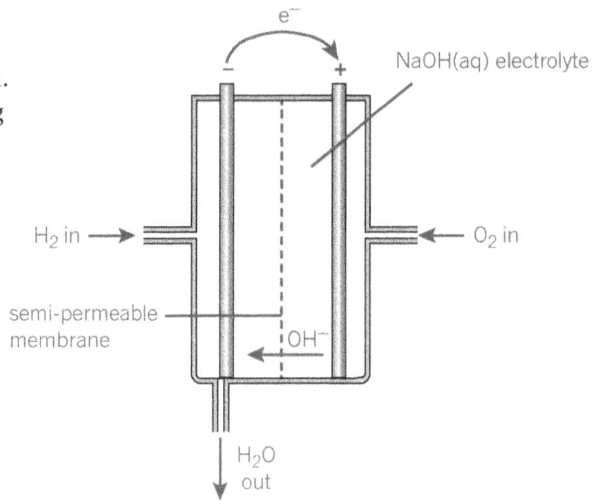

▲ **Figure 7**

Study tip

The hydrogen/oxygen fuel cell does not need to be electrically recharged.

The hydrogen economy

At first sight, this type of fuel cell appears to be very 'green' because the only product is water. However you have to consider the source of the hydrogen. At present, most hydrogen is made from crude oil – a non-renewable resource. It could be made by electrolysis of water but most electricity is made by burning fossil fuels, which emit carbon dioxide. Also, hydrogen-powered vehicles will need an infrastructure of hydrogen filling stations to be built, which also raises the issues of storing and transporting a highly flammable gas. For example, in the 1930s two airships which used hydrogen as their lifting gas – the Hindenberg and the R101 – exploded with serious loss of life. A major problem is storing hydrogen because it is a gas. For example, burning 1 g of hydrogen gives out around three times as much energy as burning 1 g of petrol but the hydrogen takes up around 8000 times as much space.

▲ **Figure 8** *A fuel-cell powered vehicle*

Hydrogen storage

One way of storing hydrogen safely is by absorbing it into solid compounds called **metal hydrides.** The hydrogen is absorbed under pressure and released by gentle heating.

Summary question

1 List the advantages and disadvantages of each cell, described with regard to cost, practicality, safety, and the environment.

Practice questions

1 Hydrogen–oxygen fuel cells can operate in acidic or in alkaline conditions but commercial cells use porous platinum electrodes in contact with concentrated aqueous potassium hydroxide. The table below shows some standard electrode potentials measured in acidic and in alkaline conditions.

Half-equation	$E^{\ominus}$/V
$O_2(g) + 4H^+(aq) + 4e^- \rightarrow 2H_2O(l)$	+1.23
$O_2(g) + 2H_2O(l) + 4e^- \rightarrow 4OH^-(aq)$	+0.40
$2H^+(aq) + 2e^- \rightarrow H_2(g)$	0.00
$2H_2O(l) + 2e^- \rightarrow 2OH^-(aq) + H_2(g)$	−0.83

(a) State why the electrode potential for the standard hydrogen electrode is equal to 0.00 V. *(1 mark)*

(b) Use data from the table to calculate the emf of a hydrogen–oxygen fuel cell operating in alkaline conditions. *(1 mark)*

(c) Write the conventional representation for an alkaline hydrogen–oxygen fuel cell. *(2 marks)*

(d) Use the appropriate half equations to construct an overall equation for the reaction that occurs when an alkaline hydrogen–oxygen fuel cell operates. Show your working. *(2 marks)*

(e) Give **one** reason, other than cost, why the platinum electrodes are made by coating a porous ceramic material with platinum rather than by using platinum rods. *(1 mark)*

(f) Suggest why the emf of a hydrogen–oxygen fuel cell, operating in acidic conditions, is exactly the same as that of an alkaline fuel cell. *(1 mark)*

(g) Other than its lack of pollution, state briefly the main advantage of a fuel cell over a rechargeable cell such as the nickel–cadmium cell when used to provide power for an electric motor that propels a vehicle. *(1 mark)*

(h) Hydrogen–oxygen fuel cells are sometimes regarded as a source of energy that is carbon neutral. Give **one** reason why this may **not** be true. *(1 mark)*

AQA, 2010

2 Where appropriate, use the standard electrode potential data in the table below to answer the questions which follow.

Standard electrode potential	$E^{\ominus}$/V
$Zn^{2+}(aq) + 2e^- \rightarrow Zn(s)$	−0.76
$V^{3+}(aq) + e^- \rightarrow V^{2+}(aq)$	−0.26
$SO_4^{2-}(aq) + 2H^+(aq) + 2e^- \rightarrow SO_3^{2-}(aq) + H_2O(l)$	+0.17
$VO^{2+}(aq) + 2H^+(aq) + e^- \rightarrow V^{3+}(aq) + H_2O(l)$	+0.34
$Fe^{3+}(aq) + e^- \rightarrow Fe^{2+}(aq)$	+0.77
$VO_2^+(aq) + 2H^+(aq) + e^- \rightarrow VO^{2+}(aq) + H_2O(l)$	+1.00
$Cl_2(aq) + 2e^- \rightarrow 2Cl^-(aq)$	+1.36

(a) From the table above select the species which is the most powerful reducing agent. *(1 mark)*

(b) From the table above select

(i) a species which, in acidic solution, will reduce $VO_2^+(aq)$ to $VO^{2+}(aq)$ but will **not** reduce $VO^{2+}(aq)$ to $V^{3+}(aq)$,

(ii) a species which, in acidic solution, will oxidise $VO^{2+}(aq)$ to $VO_2^+(aq)$. *(2 marks)*

(c) The cell represented below was set up under standard conditions.
$Pt|Fe^{2+}(aq),\ Fe^{3+}(aq)||Tl^{3+}(aq),\ Tl^{+}(aq)|Pt$ Cell emf = + 0.48 V

(i) Deduce the standard electrode potential for the following half reaction.
$Tl^{3+}(aq) + 2e^- \rightarrow Tl^{+}(aq)$

(ii) Write an equation for the spontaneous cell reaction. *(3 marks)*

AQA, 2005

3 The table shows some standard electrode potential data.

	$E^\ominus$/V
$ZnO(s) + H_2O(l) + 2e^- \rightarrow Zn(s) + 2OH^-(aq)$	−1.25
$Fe^{2+}(aq) + 2e^- \rightarrow Fe(s)$	−0.44
$O_2(g) + 2H_2O(l) + 4e^- \rightarrow 4OH^-(aq)$	+0.40
$2HOCl(aq) + 2H^+(aq) + 2e^- \rightarrow Cl_2(g) + 2H_2O(l)$	+1.64

(a) Give the conventional representation of the cell that is used to measure the standard electrode potential of iron as shown in the table. *(2 marks)*

(b) With reference to electrons, give the meaning of the term reducing agent. *(1 mark)*

(c) Identify the weakest reducing agent from the species in the table. Explain how you deduced your answer. *(2 marks)*

(d) When HOCl acts as an oxidising agent, one of the atoms in the molecule is reduced.

(i) Place a tick next to the atom that is reduced.

Atom that is reduced	Tick (√)
H	
O	
Cl	

(ii) Explain your answer to part **(d)**(i) in terms of the change in the oxidation state of this atom. *(1 mark)*

(e) Using the information given in the table, deduce an equation for the redox reaction that would occur when hydroxide ions are added to HOCl. *(2 marks)*

(f) The half equations from the table that involve zinc and oxygen are simplified versions of those that occur in hearing aid cells.
A simplified diagram of a hearing aid cell is shown below.

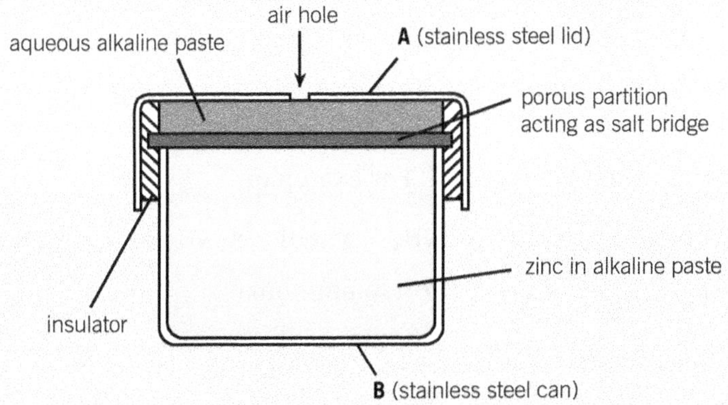

Use data from the table to calculate the emf of this cell. *(1 mark)*

AQA, 2014

Answers to the Practice Questions and Section Questions are available at
www.oxfordsecondary.com/oxfordaqaexams-alevel-chemistry

21 Acids, bases, and buffers
21.1 Defining an acid

Learning objectives:

- → State the Brønsted–Lowry definitions of an acid and a base.
- → Describe what happens in Brønsted–Lowry acid–base reactions.
- → State the expression for the ionic product of water K_w.

Specification reference: 3.1.12

Study tip

Water-soluble bases are called alkalis and produce OH^- ions in aqueous solution.

The Brønsted–Lowry description of acidity (developed in 1923 by Thomas Lowry and Johannes Brønsted independently) is the most generally useful current theory of acids and bases.

An acid is a substance that can donate a proton (H^+ ion) and a base is a substance that can accept a proton.

Proton transfer

Hydrogen chloride gas and ammonia gas react together to form ammonium chloride – a white ionic solid:

$$\underset{\text{hydrogen chloride}}{HCl(g)} + \underset{\text{ammonia}}{NH_3(g)} \rightarrow \underset{\text{ammonium chloride}}{NH_4Cl(s)}$$

Here, hydrogen chloride is acting as an acid by donating a proton to ammonia. Ammonia is acting as a base by accepting a proton. Acids and bases can only react in pairs – one acid and one base.

So, you could think of the reaction in these terms:

$$\underset{\text{acid}}{HCl(g)} + \underset{\text{base}}{NH_3(g)} \rightarrow [NH_4^+Cl^-](s)$$

Another example is a mixture of concentrated sulfuric acid, H_2SO_4, and concentrated nitric acid, HNO_3. They behave as an acid–base pair:

$$H_2SO_4 + HNO_3 \rightarrow H_2NO_3^+ + HSO_4^-$$

Sulfuric acid donates a proton to nitric acid, so is acting as the acid, whilst in this example nitric acid is acting as a base. In fact, whether a species is acting as an acid or a base depends on the reactants. Water is a good example of this.

Water as an acid and a base

Hydrogen chloride can donate a proton to water, so that water acts as a base:

$$HCl + H_2O \rightarrow H_3O^+ + Cl^-$$

H_3O^+ is called the **oxonium ion**, but the names hydronium ion and hydroxonium ion are also used.

Water may also act as an acid. For example:

$$H_2O + NH_3 \rightarrow OH^- + NH_4^+$$

Here water is donating a proton to ammonia.

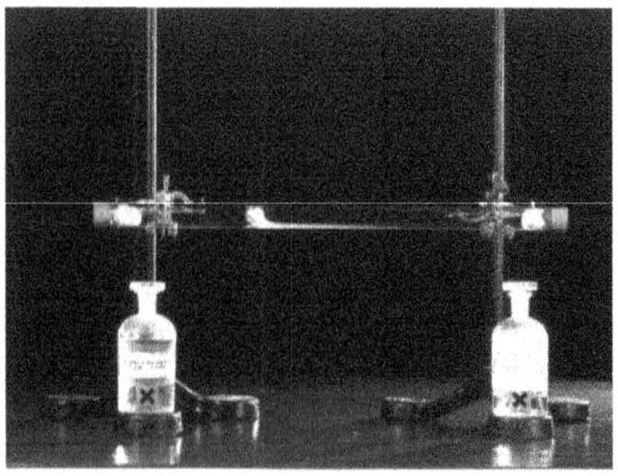

▲ **Figure 1** *The white ring of ammonium chloride is formed when hydrogen chloride (left) and ammonia (right) react*

The proton in aqueous solution

It is important to realise that the H^+ ion is just a proton. The hydrogen atom has only one electron and if this is lost all that remains is a proton (the hydrogen nucleus). This is about 10^{-15} m in diameter, compared to 10^{-10} m or more for any other chemical entity. This extremely small size and consequent intense electric field cause it to have unusual properties compared with other positive ions. It is never found isolated. In aqueous solutions it is always bonded to at least one water molecule to form the ion H_3O^+. For simplicity, protons are represented in an aqueous solution by $H^+(aq)$ rather than $H_3O^+(aq)$.

Since the H^+ ion has no electrons of its own, it can only form a bond with another species that has a lone pair of electrons.

The ionisation of water

Water is slightly ionised:

$$H_2O(l) \rightleftharpoons H^+(aq) + OH^-(aq)$$

This may be written:

$$H_2O(l) + H_2O(l) \rightleftharpoons H_3O^+(aq) + OH^-(aq)$$

This emphasises that this is an acid–base reaction in which one water molecule donates a proton to another.

This equilibrium is established in water and all aqueous solutions:

$$H_2O(l) \rightleftharpoons H^+(aq) + OH^-(aq)$$

You can write an equilibrium expression:

$$K_c = \frac{[H^+(aq)][OH^-(aq)]}{[H_2O(l)]}$$

The concentration of water $[H_2O(l)]$ is constant and is incorporated into a modified equilibrium constant K_w, where $K_w = K_c \times [H_2O(l)]$.

So, $$K_w = [H^+(aq)]\,[OH^-(aq)]$$

K_w is called the **ionic product** of water and at 298 K it is equal to $1.0 \times 10^{-14}\ mol^2\ dm^{-6}$. Each H_2O that dissociates (splits up) gives rise to one H^+ and one OH^- so, in pure water, at 298 K:

$$[OH^-(aq)] = [H^+(aq)]$$

So, $$1.0 \times 10^{-14} = [H^+(aq)]^2$$

$$[H^+(aq)] = 1.0 \times 10^{-7}\ mol\,dm^{-3} = [OH^-(aq)] \text{ at } 298\,K\ (25\,°C)$$

Summary questions

1 Identify which reactant is an acid and which a base in the following:

a $HNO_3 + OH^- \rightarrow NO_3^- + H_2O$

b $CH_3COOH + H_2O \rightarrow CH_3COO^- + H_3O^+$

2 At 298 K in an acidic solution, $[H^+]$ is $1 \times 10^{-4}\ mol\,dm^{-3}$. What is $[OH^-(aq)]$?

3 What species are formed when the following bases accept a proton?

a OH^- b NH_3 c H_2O d Cl^-

The concentration of water

What is the concentration of water? This question often catches out even experienced chemists. 1 dm^3 of water weighs 1000 g. The M_r of water is 18.0

1 How many moles of water is 1000 g
2 What does this make the concentration of water in $mol\,dm^{-3}$?

1 55.5 moles
2 55.5 $mol\,dm^{-3}$

Synoptic link

Look back at Topic 6.3, The Equilibrium constant K_c, to revise equilibrium constants.

Study tip

The important point to remember at this stage is that the product of $[H^+(aq)]_{eqm}$ and $[OH^-(aq)]_{eqm}$ is constant at any given temperature so that if the concentration of one of these ions increases, the other must decrease proportionately.

21.2 The pH scale

Learning objectives:

- → Define pH.
- → Explain why a logarithmic scale is used.
- → Describe how pH is measured.
- → Describe how the pH of a solution is used to find the concentration of H^+(aq) and OH^-(aq) ions.
- → Describe how the pH of a solution is calculated from the concentration of H^+(aq) ions.

Specification reference: 3.1.12

The acidity of a solution depends on the concentration of H^+(aq) and is measured on the pH scale.

$$\mathrm{pH} = -\log_{10}[\mathrm{H^+(aq)}]$$

Remember that square brackets, [ ], mean the concentration in mol dm^{-3}.

How the pH scale was invented

Did you know that the pH scale was first introduced by a brewer? In 1909, the Danish biochemist Søren Sørenson was working for the Carlsberg company studying the brewing of beer. Brewing requires careful control of acidity to produce conditions in which yeast (which aids the fermentation process) will grow but unwanted bacteria will not. The concentrations of acid with which Sørenson was working were very small, such as one ten-thousandth of a mole per litre, and so he looked for a way to avoid using numbers such as 0.0001 (1×10^{-4}). Taking the $\log_{10}$ of this number gave −4, and for further convenience he took the negative of it giving 4. So the pH scale was born.

This expression is more complicated than simply stating the concentration of H^+(aq). However, using the logarithm of the concentration does away with awkward numbers like 10^{-13}, etc., which occur because the concentration of H^+(aq) in most aqueous solutions is so small. The minus sign makes almost all pH values positive (because the logs of numbers less than 1 are negative).

On the pH scale:

- The *smaller* the pH, the *greater* the concentration of H^+(aq).
- A difference of *one* pH number means a *tenfold* difference in $[H^+]$ so that, for example, pH 2 has ten times the H^+ concentration of pH 3.

Remember that, at 298 K, $K_w = [\mathrm{H^+(aq)}][\mathrm{OH^-(aq)}] = 1.00 \times 10^{-14}$ mol^2 dm^{-6}. This means that in neutral aqueous solutions:

$$[\mathrm{H^+(aq)}] = [\mathrm{OH^-(aq)}] = 1.0 \times 10^{-7}\ \mathrm{mol\,dm^{-3}}$$

$$\mathrm{pH} = -\log_{10}[\mathrm{H^+(aq)}] = -\log_{10}[1.0 \times 10^{-7}] = 7.00$$

so the pH is 7.00.

Study tip

Always give pH values to two decimal places.

Maths link

You must be able to use your calculator to look up logs to the base 10 ($\log_{10}$) and antilogs. See Section 8, Mathematical skills, if you are not sure about these.

Mixing bathroom cleaners

▲ **Figure 1** *Household bleach products*

Bathroom cleaners come in essentially two types – bleach-based for removing coloured stains, and acid-based used, for example, for removing limescale in the toilet bowl.

1 Limescale is made up of calcium carbonate, $CaCO_3$. Write the equation for the reaction of hydrochloric acid with calcium carbonate.

Most bathroom cleaners have a warning on the label not to mix them with other types of cleaner. Why is this?

The active ingredient in household bleach is chloric(I) acid (HClO), whilst acid-based cleaners contain hydrochloric acid (HCl). These react together to form chlorine gas.

$$HClO(aq) + HCl(aq) \rightleftharpoons Cl_2(g) + H_2O(aq)$$

Imagine you have put a large amount of bleach in the toilet bowl (so that chloric(I) acid is in excess) and then you add a squirt (say 50 cm^3) of acid-based cleaner of concentration 1 mol dm^{-3}. Assume the equilibrium is forced completely to the right.

2 How many moles of HCl have you added?
3 How many moles of chlorine would be produced?
4 What volume of chlorine is this?
5 Why would there be less chlorine gas in the bathroom than you have calculated in **3**?

This is a significant amount of chlorine and, considering that it was used as a poisonous gas in the First World War, something to be avoided.

pH and temperature

The equilibrium reaction below is endothermic in the forward direction.

$$H_2O(l) \rightleftharpoons H^+(aq) + OH^-(aq) \qquad \Delta H = +57.3\ kJ\ mol^{-1}$$

Therefore the value of K_w increases with temperature and the pH of water is different at different temperatures (Table 1). So, for example, at 373 K (boiling point), the pH of water is about 6.

This does not mean that water is acidic (water is always neutral because it always has an equal number of H^+ ions and OH^- ions) but merely that the neutral value for the pH is 6 at this temperature, rather than pH 7 at room temperature. So in boiling water $[H^+]$ (and $[OH^-]$) are both about 1×10^{-6} mol dm^{-3}.

Table 1 shows that the pH of sea water at the poles and at the Equator will be different.

1 Use Table 1 to calculate the concentration of H^+ at 313 K.
2 What is the concentration of OH^- at the same pH?

2 OH^- concentration will be the same 1.7×10^{-7} mol dm^{-3}
1 1.7×10^{-7} mol dm^{-3}

Synoptic link

Look back at Topic 6.5, The effect of changing conditions on equilbria, to revise equilibrium constants and how they're affected by conditions, and Topic 6.2, Changing the conditions of an equilibrium reaction, to revise K_w.

▼ **Table 1** *The effect of temperature on the pH of water*

T / K	K_w / mol^2 dm^{-6}	pH (neutral)
273	0.114×10^{-14}	7.47
283	0.293×10^{-14}	7.27
293	0.681×10^{-14}	7.08
298	1.008×10^{-14}	7.00
303	1.471×10^{-14}	6.92
313	2.916×10^{-14}	6.77
323	5.476×10^{-14}	6.63
373	51.300×10^{-14}	6.14

Measuring pH

pH can be measured using an indicator paper or a solution, such as universal indicator. This is made from a mixture of dyes that change colour at different $[H^+(aq)]$. This is fine for measurements to the nearest whole number, but for more precision a pH meter is used. A pH meter has an electrode which dips into a solution and produces a voltage related to $[H^+(aq)]$. The pH readings can then be read directly on the meter or fed into a computer or data logger, for continuous monitoring of a chemical process or medical procedure, for example.

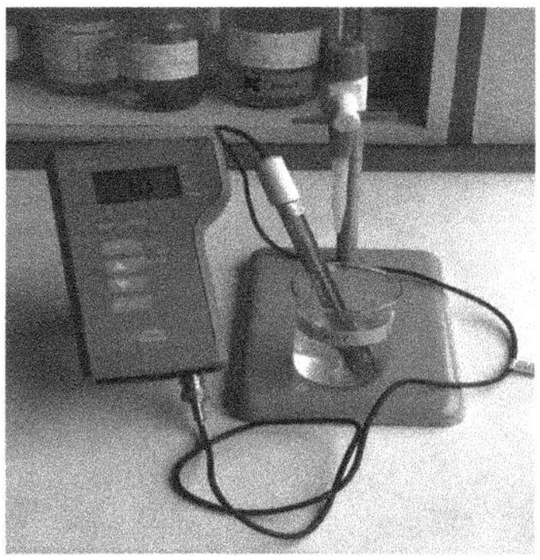

◀ **Figure 2** *Using a pH meter*

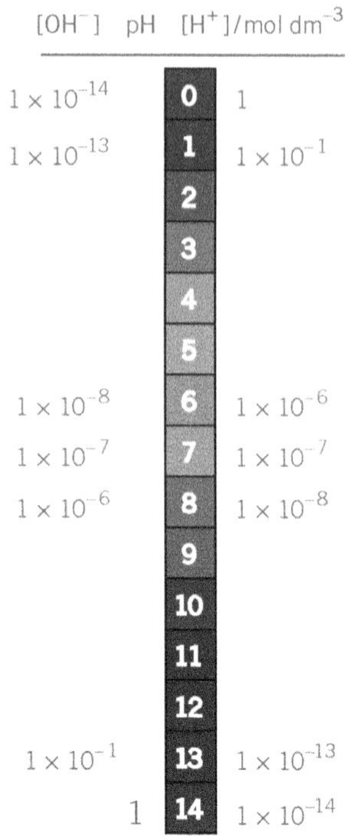

▲ **Figure 3** *The pH scale*

Study tip

A solution is neutral when $[H^+] = [OH^-]$.

Maths link

See Section 8, Mathematical skills, if you are not confident about handling numbers in standard form.

Study tip

Look out for how the $\log_{10}$ button is represented on your calculator and how to find antilogs.

pH measures alkalinity as well

pH measures alkalinity as well as acidity, because as $[H^+(aq)]$ goes up, $[OH^-(aq)]$ goes down. At 298 K, if a solution contains more $H^+(aq)$ than $OH^-(aq)$, its pH will be less than 7 and it is called acidic. If a solution contains more $OH^-(aq)$ than $H^+(aq)$, its pH will be greater than 7 and it is called alkaline (Figure 3).

Working with the pH scale

Finding $[H^+(aq)]$ from pH

You can work out the concentration of hydrogen ions $[H^+]$ in an aqueous solution if you know the pH. It is the antilogarithm of the pH value.

For example, an acid has a pH of 3.00:

$$pH = -\log_{10}[H^+(aq)]$$

$$3.00 = -\log_{10}[H^+(aq)]$$

$$-3.00 = \log_{10}[H^+(aq)]$$

Take the antilog of both sides:

$$[H^+(aq)] = 1.0 \times 10^{-3}\ mol\,dm^{-3}$$

Finding $[OH^-(aq)]$ from pH

With bases, you need two steps. Suppose the pH of a solution is 10.00:

$$pH = -\log_{10}[H^+(aq)]$$

$$10.00 = -\log_{10}[H^+(aq)]$$

$$-10.00 = \log_{10}[H^+(aq)]$$

Take the antilog of both sides:

$$[H^+(aq)] = 1.0 \times 10^{-10}$$

You know $[H^+(aq)]\,[OH^-(aq)] = 1.0 \times 10^{-14}\ mol^2\,dm^{-6}$

Substituting your value for $[H^+(aq)] = 1.0 \times 10^{-10}$ into the equation:

$$[1.0 \times 10^{-10}]\,[OH^-(aq)] = 1.0 \times 10^{-14}\ mol^2\,dm^{-6}$$

$$[OH^-(aq)] = \frac{1.0 \times 10^{-14}}{1.0 \times 10^{-10}} = 1.0 \times 10^{-4}\ mol\,dm^{-3}$$

The pH of strong acid solutions – worked example

HCl dissociates completely in dilute aqueous solution to $H^+(aq)$ ions and $Cl^-(aq)$ ions, that is, the reaction goes to completion:

$$HCl(aq) \rightarrow H^+(aq) + Cl^-(aq)$$

Acids that dissociate completely like this are called **strong acids**.

So in $1.00\ mol\,dm^{-3}$ HCl:

$$[H^+(aq)] = 1.00\ mol\,dm^{-3}$$

$$\log[H^+(aq)] = \log 1.00 = 0.00$$

$$-\log[H^+(aq)] = 0.00$$

So the pH of 1 mol dm^{-3} HCl = 0.00

In a 0.16 mol dm^{-3} solution of HCl:

$$[H^+(aq)] = 0.16\,mol\,dm^{-3}$$

$$\log\,[H^+(aq)] = \log 0.160 = -0.796$$

$$-\log[H^+(aq)] = 0.796$$

So the pH of 0.16 mol dm^{-3} HCl = 0.80 to 2 d.p.

Worked example: The pH of alkaline solutions

In alkaline solutions, it takes two steps to calculate $[H^+(aq)]$.

To find the $[H^+(aq)]$ of an alkaline solution at 298 K:

1 Calculate $[OH^-(aq)]$.
2 Then use $[H^+(aq)]\,[OH^-(aq)] = 1.00 \times 10^{-14}\,mol^2\,dm^{-6}$ to calculate $[H^+(aq)]$.

The pH can then be calculated.

For example, to find the pH of 1.00 mol dm^{-3} sodium hydroxide solution:

Sodium hydroxide is fully dissociated in aqueous solution – it is called a strong alkali.

$$NaOH(aq) \rightarrow Na^+(aq) + OH^-(aq)$$

$$[OH^-(aq)] = 1.00\,mol\,dm^{-3}$$

but $$[OH^-(aq)]\,[H^+(aq)] = 1.00 \times 10^{-14}\,mol\,dm^{-3}$$

$$[H^+(aq)] = 1.00 \times 10^{-14}\,mol\,dm^{-3}$$

and $$[H^+(aq)] = -\log\,(1.00 \times 10^{-14})$$

$$pH = 14.00$$

In a 0.200 mol dm^{-3} sodium hydroxide solution:

$$[OH^-(aq)] = 2.00 \times 10^{-1}\,mol\,dm^{-3}$$

$$[OH^-(aq)]\,[H^+(aq)] = 1.00 \times 10^{-14}\,mol\,dm^{-3}$$

$$[H^+(aq)] \times 2.00 \times 10^{-1} = 1.00 \times 10^{-14}\,mol\,dm^{-3}$$

$$[H^+(aq)] = 5.00 \times 10^{-14}\,mol\,dm^{-3}$$

$$\log\,[H^+(aq)] = -13.30$$

$$pH = 13.30$$

Study tip

In a very concentrated solution even strong acids are not fully ionised.

Study tip

The pH of solutions of strong acids which have a concentration greater than 1 mol dm^{-3} is negative. The pH of solutions of strong bases which have concentration greater than 1 mol dm^{-3} is larger than 14.00.

Study tip

When writing out calculations on pH it is acceptable to omit the 'aq' in expressions such as $[H^+(aq)]$, so we could write simply $pH = -\log_{10}[H^+]$.

Summary questions

1 What is the pH of a solution in which $[H^+]$ is 1.00×10^{-2} mol dm^{-3}?

2 What is $[H^+]$ in a solution of pH = 6.00?

3 At 298 K what is $[OH^-]$ in a solution of pH = 9.00?

4 Calculate the pH of a 0.020 mol dm^{-3} solution of HCl.

5 Calculate the pH of 0.200 mol dm^{-3} sodium hydroxide.

Learning objectives:

- → Define the terms weak acid and weak base.
- → Describe how the pH of a weak acid is calculated.

Specification reference: 3.1.12

Hint

Although in the gas phase, hydrogen chloride, HCl, is a covalent molecule, a solution of it in water is wholly ionic (i.e., $H^+(aq) + Cl^-(aq)$). You can assume that there are no molecules remaining, so hydrochloric acid is a strong acid.

Study tip

The *strength* of an acid and its *concentration* are completely independent, so use the two different words carefully.

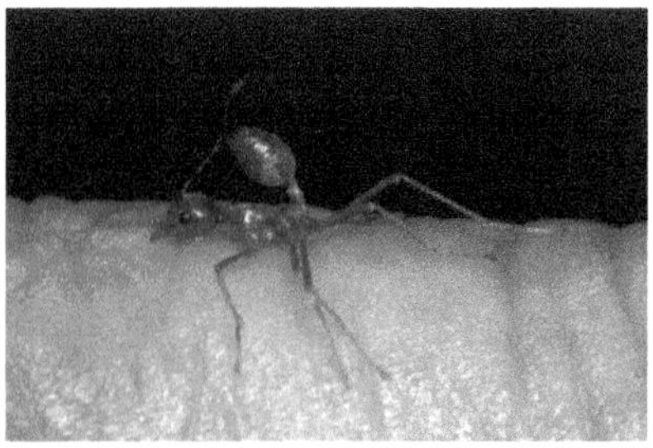

▲ **Figure 1** *Formic acid (methanoic acid) is quite concentrated when used as a weapon by the stinging ant and, although it is a weak acid, being sprayed with it can be a painful experience*

In Topic 21.2 you looked at the pH of acids such as hydrochloric acid which dissociate completely into ions when dissolved in water. Acids that completely dissociate into ions in aqueous solutions are called **strong acids**. The word strong refers *only* to the extent of dissociation and not in any way to the concentration. So it is perfectly possible to have a very dilute solution of a strong acid.

The same arguments apply to bases. Strong bases are completely dissociated into ions in aqueous solutions. For example, sodium hydroxide is a strong base:

$$NaOH(aq) \rightarrow Na^+(aq) + OH^-(aq)$$

Weak acids and bases

Many acids and bases are only slightly ionised (not fully dissociated) when dissolved in water. Ethanoic acid (the acid in vinegar, also known as acetic acid) is a typical example. In a 1 mol dm^{-3} solution of ethanoic acid, only about four in every thousand ethanoic acid molecules are dissociated into ions (so the degree of dissociation is $\frac{4}{1000}$) – the rest remain dissolved as wholly covalently bonded molecules. In fact an equilibrium is set up:

	$CH_3COOH(aq)$	$\rightleftharpoons$ $H^+(aq)$	+ $CH_3COO^-(aq)$
	ethanoic acid	hydrogen ions	ethanoate ions
Before dissociation:	1000	0	0
At equilibrium:	996	4	4

Acids like this are called **weak acids**. Weak refers *only* to the degree of dissociation. In a 5 mol dm^{-3} solution, ethanoic acid is still a weak acid, while in a 10^{-4} mol dm^{-3} solution, hydrochloric acid is still a strong acid.

When ammonia dissolves in water, it forms an alkaline solution. The equilibrium lies well to the left and ammonia is weakly basic:

$$NH_3(aq) + H_2O(l) \rightleftharpoons NH_4^+(aq) + OH^-(aq)$$

The dissociation of weak acids

Weak acids

Imagine a weak acid HA which dissociates:

$$HA(aq) \rightleftharpoons H^+(aq) + A^-(aq)$$

The equilibrium constant is given by:

$$K_c = \frac{[H^+(aq)]_{eqm}[A^-(aq)]_{eqm}}{[HA(aq)]_{eqm}}$$

For a weak acid, this is usually given the symbol K_a and called the **acid dissociation constant**.

$$K_a = \frac{[H^+(aq)]_{eqm}[A^-(aq)]_{eqm}}{[HA(aq)]_{eqm}}$$

The larger the value of K_a, the further the equilibrium is to the right, the more the acid is dissociated, and the stronger it is. Acid dissociation constants for some acids are given in Table 1.

K_a has units and it is important to state these. They are found by multiplying and cancelling the units in the expression for K_a.

$$K_a = \frac{\cancel{\text{mol dm}^{-3}} \times \text{mol dm}^{-3}}{\cancel{\text{mol dm}^{-3}}} = \text{mol dm}^{-3}$$

▼ **Table 1** *Values of K_a for some weak acids*

Acid	K_a / mol dm^{-3}
chloroethanoic	1.30×10^{-3}
benzoic	6.30×10^{-5}
ethanoic	1.70×10^{-5}
hydrocyanic	4.90×10^{-10}

Calculating the pH of weak acids

In Topic 21.2 you calculated the pH of solutions of strong acids, by assuming that they are fully dissociated. For example, in a 1.00 mol dm^{-3} solution of nitric acid, $[H^+] = 1.00$ mol dm^{-3}. In weak acids this is no longer true, and you must use the acid dissociation expression to calculate $[H^+]$.

Calculating the pH of 1.00 mol dm^{-3} ethanoic acid

The concentrations in mol dm^{-3} are:

	$CH_3COOH(aq)$	$\rightleftharpoons CH_3COO^-(aq)$	$+ H^+(aq)$
Before dissociation:	1.00	0	0
At equilibrium:	$1.00 - [CH_3COO^-(aq)]$	$[CH_3COO^-(aq)]$	$[H^+(aq)]$

$$K_a = \frac{[CH_3COO^-(aq)][H^+(aq)]}{[CH_3COOH(aq)]}$$

But as each CH_3COOH molecule that dissociates produces one CH_3COO^- ion and one H^+ ion:

$$[CH_3COO^-(aq)] = [H^+(aq)]$$

Since the degree of dissociation of ethanoic acid is so small (it is a weak acid), $[H^+(aq)]_{eqm}$ is very small and, to a good approximation, $1.00 - [H^+(aq)] \approx 1.00$.

$$K_a = \frac{[H^+(aq)]^2}{1.00}$$

From Table 1, $K_a = 1.70 \times 10^{-5}$ mol dm^{-3}

$$1.70 \times 10^{-5} = [H^+(aq)]^2$$

$$[H^+(aq)] = \sqrt{1.70 \times 10^{-5}}$$

$$[H^+(aq)] = 4.12 \times 10^{-3} \text{ mol dm}^{-3}$$

Taking logs: $\log [H^+(aq)] = -2.385$

$$pH = 2.385 = 2.39 \text{ to 2 d.p.}$$

Calculating the pH of 0.100 mol dm^{-3} ethanoic acid

Using the same method, you get:

$$K_a = \frac{[H^+(aq)]^2}{0.10 - [H^+(aq)]}$$

Again, $0.100 - [H^+(aq)] \approx 0.10$

so: $$1.70 \times 10^{-5} = \frac{[H^+(aq)]^2}{0.10}$$

$$1.70 \times 10^{-6} = [H^+(aq)]^2$$

$$[H^+(aq)] = 1.30 \times 10^{-3} \text{ mol dm}^{-3}$$

$$pH = 2.89 \text{ to 2 d.p.}$$

Study tip

It is acceptable to omit the eqm subscripts unless they are specifically asked for.

Hint

The symbol ≈ means approximately equal to.

pK_a

For a weak acid pK_a is often referred to. This is defined as:

$$pK_a = -\log_{10} K_a$$

Think of 'p' as meaning $-\log_{10}$ of.

pK_a can be useful in calculations (Table 2). It gives a measure of how strong a weak acid is – the smaller the value of pK_a, the stronger the acid.

▼ **Table 2** *Values of K_a and pK_a for some weak acids*

Acid	K_a/mol dm^{-3}	pK_a
chloroethanoic	1.30×10^{-3}	2.88
benzoic	6.30×10^{-5}	4.20
ethanoic	1.70×10^{-5}	4.77
hydrocyanic	4.90×10^{-10}	9.31

Summary questions

1 Which is the strongest acid in Table 2?

2 What can you say about the concentration of H^+ ions compared with the concentration of ethanoate ions in all solutions of pure ethanoic acid?

3 Calculate the pH of the following solutions:

  a 0.100 mol dm^{-3} chloroethanoic acid

  b 0.0100 mol dm^{-3} benzoic acid

21.4 Acid–base titrations

A titration is used to find the concentration of a solution by gradually adding to it a second solution with which it reacts. One of the solutions is of known concentration. To use a titration, you must know the equation for the reaction.

pH changes during acid–base titrations

In an acid–base titration, an acid of known concentration is added from a burette to a measured amount of a solution of a base (an alkali) until an indicator shows that the base has been neutralised. Alternatively, the base is added to the acid until the acid is neutralised. You can then calculate the concentration of the alkali from the volume of acid used.

You can also follow a neutralisation reaction by measuring the pH with a pH meter (Figure 2) in which case you do not need an indicator. The pH meter is calibrated by placing the probe in a buffer solution of a known pH (Topic 21.6).

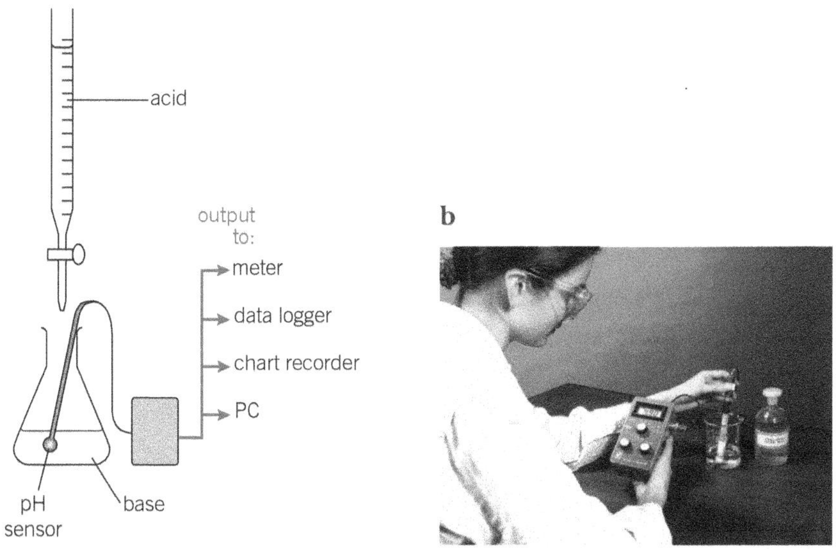

▲ **Figure 2** **a** *Apparatus to investigate pH changes during a titration* **b** *A pH meter*

Titration procedure

The titration curves in Figure 3 were determined using the apparatus shown in Figure 2. Using a pipette, 25 cm^3 acid was placed in the conical flask. The base was added, 1 cm^3 at a time, from a burette and the pH recorded. At areas on the curve where the pH was changing rapidly, the experiment was repeated adding the base 0.1 cm^3 at a time to find the shape of the curve more precisely.

Learning objectives:

- → Describe how pH is determined experimentally.
- → Describe the shapes of the pH curves for acid–base titrations.
- → Define the equivalence point.

Specification reference: 3.1.12

Synoptic link

See Practical 8 on page 526.

▲ **Figure 1** *An acid–base titration, to find the concentration of a base. A volumetric pipette is used to deliver an accurately measured volume of base of unknown concentration into the flask. The acid of known concentration is in the burette*

Titration curves

Figure 3 shows the results obtained for four cases using monoprotic acids. In these cases the *base* is added from the burette and the acid has been accurately measured into a flask. The shape of each titration curve is typical for the type of acid–base titration.

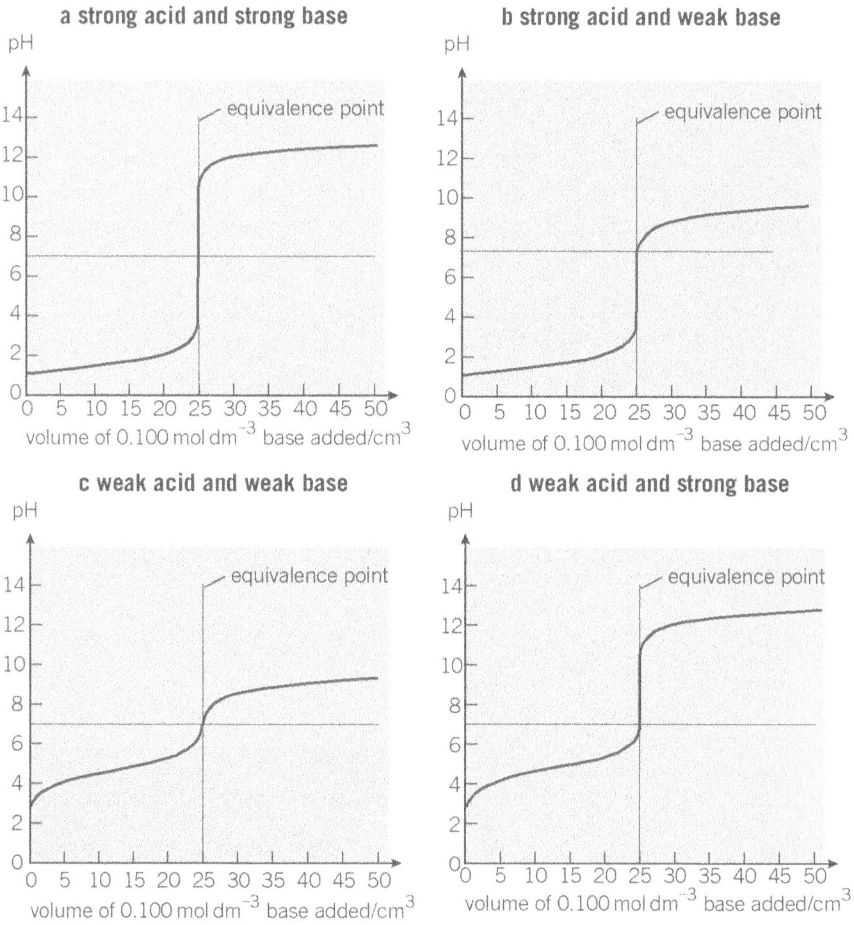

▲ **Figure 3** *Graphs of pH changes for titrations of different acids and bases*

The first thing to notice about these curves is that the pH does not change in a linear manner as the base is added. Each curve has almost horizontal sections where a lot of base can be added without changing the pH much. There is also a very steep portion of each curve, except weak acid–weak base, where a single drop of base changes the pH by several units.

In a titration, the **equivalence point** is the point at which sufficient base has been added to just neutralise the acid (or vice-versa). In each of the titrations in Figure 3 the equivalence point is reached after 25.0 cm^3 of base has been added. However, the pH at the equivalence point is not always exactly 7.

In each case, except the weak acid–weak base titration, there is a large and rapid change of pH at the equivalence point (i.e., the curve is almost vertical) even though this is may not be centred on pH 7.

Study tip

A monoprotic acid has only one acidic hydrogen (i.e., one that can be donated to a base). Hydrochloric acid, HCl, is monoprotic, whereas sulfuric acid, H_2SO_4, is diprotic.

Study tip

The equivalence point is also called the **stoichiometric point.**

Hint

The glass electrode at the tip of a pH meter probe is very delicate and should be treated with care.

Hint

In each case in Figure 3 the base was added to 25 cm^3 of 0.100 mol dm^{-3} acid.

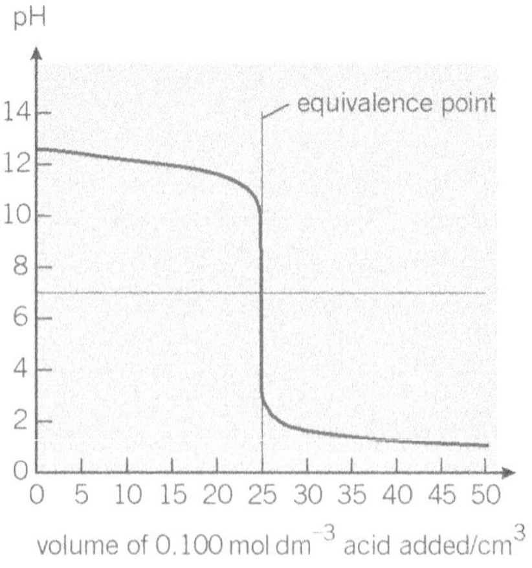

▲ **Figure 4** *Titration of a strong base–strong acid, adding 0.100 mol dm^{-3} HCl(aq) to 25.0 cm^3 of 0.100 mol dm^{-3} NaOH(aq)*

This is relevant to the choice of indicator for a particular titration (see Topic 21.5).

You can add the acid to the base for these pH curves and the shape will be flipped around the pH 7 line. For example, Figure 4 shows a strong acid–strong base curve.

Working out concentrations

The equivalence point can be used to work out the concentration of the unknown acid (or base).

Worked example: A monoprotic acid

In a titration, the equivalence point is reached when 25.00 cm^3 of 0.0150 mol dm^{-3} sodium hydroxide is neutralised by 15.00 cm^3 hydrochloric acid. What is the concentration of the acid?

$$HCl(aq) + NaOH(aq) \rightarrow NaCl(aq) + H_2O(l)$$

The equivalence point shows that 15.0 cm^3 hydrochloric acid of concentration A contains the same number of moles as 25.00 cm^3 of 0.0150 mol dm^{-3} sodium hydroxide.

$$\text{number of moles in solution} = c \times \frac{V}{1000}$$

Where c is concentration in mol dm^{-3} and V is volume in cm^3

From the equation, number of moles HCl = number of moles NaOH

$$25.00 \times \frac{0.0150}{1000} = 15.00 \times \frac{A}{1000}$$

$$A = 0.025$$

So, the concentration of the acid is 0.0250 mol dm^{-3}.

Worked example: A diprotic acid

In a titration, the equivalence point is reached when 20.00 cm^3 of 0.0100 mol dm^{-3} sodium hydroxide is neutralised by 15.00 cm^3 sulfuric acid. What is the concentration of the acid?

$$H_2SO_4(aq) + 2NaOH(aq) \rightarrow Na_2SO_4(aq) + 2H_2O(l)$$

The equivalence point shows that 15.00 cm^3 sulfuric acid of concentration B contains the same number of moles as 20.00 cm^3 of 0.0100 mol dm^{-3} sodium hydroxide.

$$\text{number of moles in solution} = c \times \frac{V}{1000}$$

Where c is concentration in mol dm^{-3} and V is volume in cm^3.

Number of moles of NaOH = $20.00 \times \frac{0.0100}{1000} = \frac{0.2}{1000}$

From the equation, number of moles $H_2SO_4 = \frac{1}{2}$ number of moles NaOH

So number of moles of $H_2SO_4 = \frac{0.1}{1000}$

Number of moles of $H_2SO_4 = 15.00 \times \frac{B}{1000} = \frac{0.1}{1000}$

So, the concentration, B, of the acid is 0.0067 mol dm^{-3}.

Summary questions

1 25.0 cm^3 sodium hydroxide is neutralised by 15.0 cm^3 sulfuric acid, H_2SO_4, of concentration 0.100 mol dm^{-3}.

- **a** Write the equation for this reaction.
- **b** From the equation, how many moles of sulfuric acid will neutralise 1.00 mol of sodium hydroxide?
- **c** How many moles of sulfuric acid are used in the neutralisation?
- **d** What is the concentration of the sodium hydroxide?

2 The graph below shows two titration curves of two acids labelled A and B with a base.

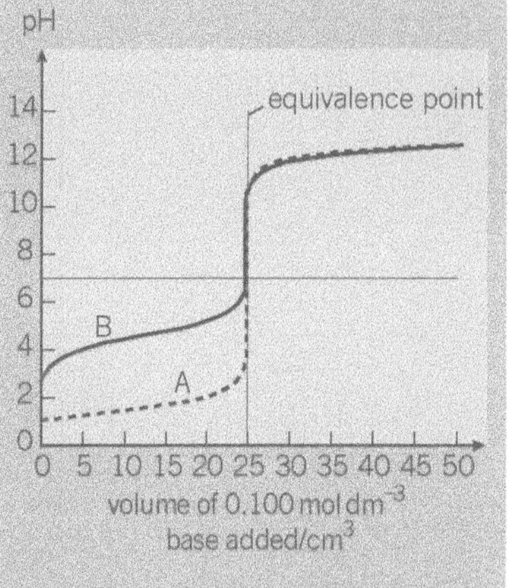

- **a** Which curve represents
  - **i** a strong acid
  - **ii** a weak acid?
- **b** Which one could represent:
  - **i** ethanoic acid, CH_3COOH
  - **ii** hydrochloric acid, HCl?
- **c** Was the base strong or weak? Explain your answer.

21.5 Choice of indicators for titrations

Learning objectives:

- → Describe how pH curves can be used to select a suitable indicator.
- → Describe how the end point can be found from the curves.
- → Explain the significance of the half-neutralisation point on these curves.

Specification reference: 3.1.12

An acid–base titration uses an indicator to find the concentration of a solution of an acid or alkali. The equivalence point is the volume at which exactly the same number of moles of hydrogen ions (or hydroxide ions) has been added as there are moles of hydroxide ions (or hydrogen ions). The **end point** is the volume of alkali or acid added when the indicator just changes colour. Unless you choose the right indicator, the equivalence point and the end point may not always give the same answer.

A suitable indicator for a particular titration needs the following properties:

- The colour change must be sharp rather than gradual at the end point, that is, no more than one drop of acid (or alkali) is needed to give a complete colour change. An indicator that changes colour gradually over several cubic centimetres would be unsuitable and would not give a sharp end point.
- The end point of the titration given by the indicator must be the same as the equivalence point, otherwise the titration will give the wrong answer.
- The indicator should give a distinct colour change. For example, the colourless to pink change of phenolphthalein is easier to see than the red to yellow of methyl orange.

Some common indicators are given in Table 1 with their approximate colour changes. Notice that the colour change of most indicators takes place over a pH range of around two units, centred around the value of pK_a for the indicator. For this reason, not all indicators are suitable for all titrations. Universal indicator is not suitable for any titration because of its gradual colour changes.

▼ **Table 1** *Some common indicators. Universal indicator is a mixture of indicators that change colour at different pHs*

Indicator	pK_a value	Colour change
universal indicator		
methyl orange	3.7	red — change — yellow
bromophenol blue	4.0	yellow — change — blue
methyl red	5.1	red — change — yellow
bromothymol blue	7.0	yellow — change — blue
phenolphthalein	9.3	colourless — change — red

pH scale: 0 1 2 3 4 5 6 7 8 9 10 11 12 13 14 pH

0 = very acidic; 7 = neutral; 14 = very alkaline

The following examples compare the suitability of two common indicators – phenolphthalein and methyl orange – for four different types of acid–base titration. In each case, the base is being added to the acid.

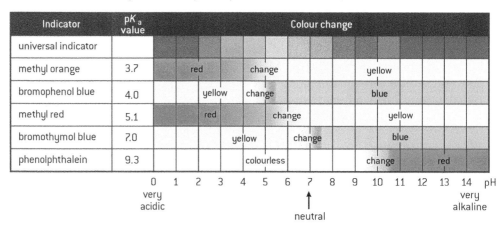

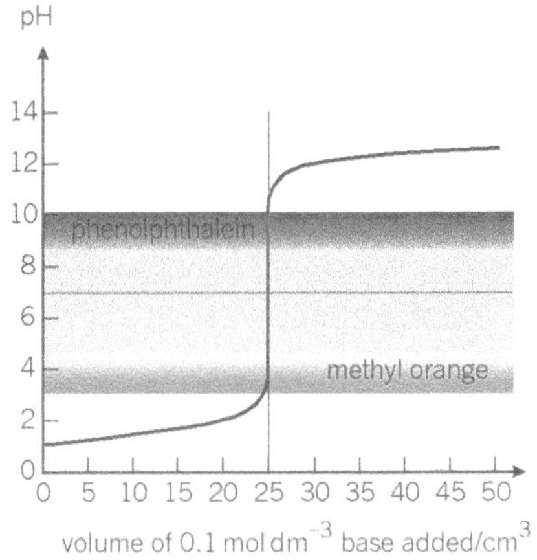

▲ **Figure 1** *Titration of a strong acid–strong base, adding 0.1 mol dm^{-3} NaOH(aq) to 25 cm^3 of 0.1 mol dm^{-3} HCl(aq)*

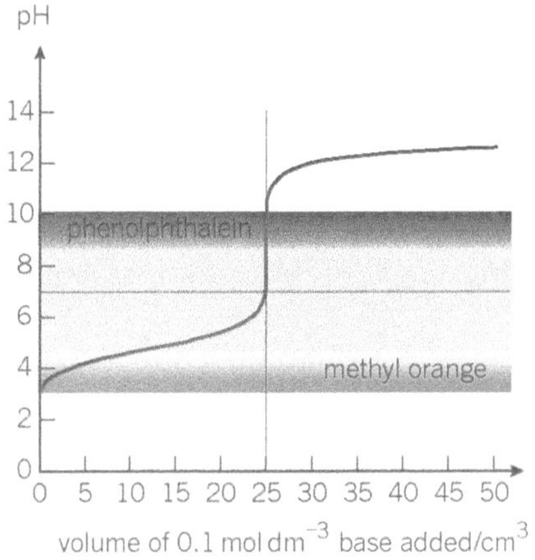

▲ **Figure 2** *Titration of a weak acid–strong base, adding 0.1 mol dm^{-3} NaOH(aq) to 25 cm^3 of 0.1 mol dm^{-3} CH_3COOH(aq)*

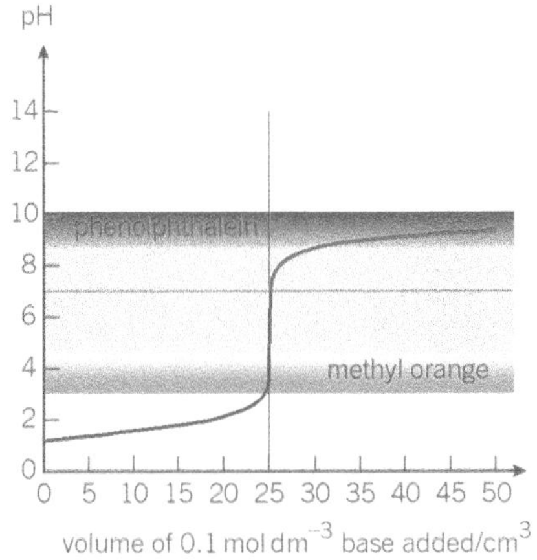

▲ **Figure 3** *Titration of a strong acid–weak base, adding 0.1 mol dm^{-3} NH_3(aq) to 25 cm^3 of 0.1 mol dm^{-3} HCl(aq)*

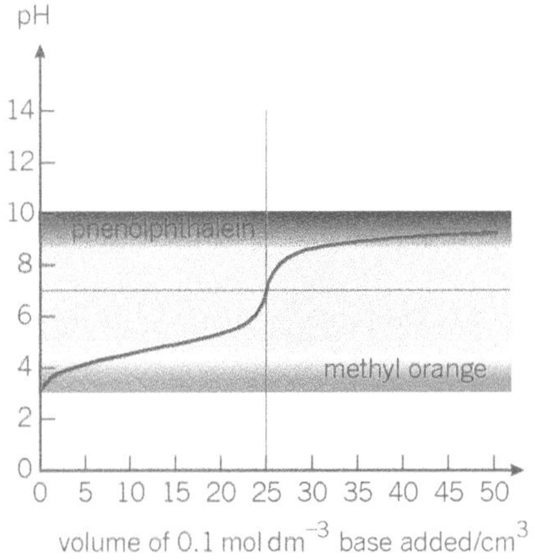

▲ **Figure 4** *Titration of a weak acid–weak base, adding 0.1 mol dm^{-3} NH_3(aq) to 25 cm^3 of 0.1 mol dm^{-3} CH_3COOH(aq)*

Study tip

Practise using pH curves to select a suitable indicator.

1 *Strong acid–strong base, for example, hydrochloric acid and sodium hydroxide*

Figure 1 is the graph of pH against volume of base added. The pH ranges over which two indicators change colour are shown. To fulfil the first two criteria above, the indicator must change within the vertical portion of the pH curve. Here either indicator would be suitable, but phenolphthalein is usually preferred because of its more easily seen colour change.

2 *Weak acid–strong base titration, for example, ethanoic acid and sodium hydroxide*

Methyl orange is not suitable (Figure 2). It does not change in the vertical portion of the curve and will change colour in the 'wrong' place and over the addition of many cubic centimetres of base. Phenolphthalein will change sharply at exactly 25 cm^3, the equivalence point, and would therefore be a good choice.

3 Strong acid–weak base titration, for example, hydrochloric acid and ammonia

Here methyl orange will change sharply at the equivalence point but phenolphthalein would be of no use (Figure 3).

4 Weak acid–weak base, for example, ethanoic acid and ammonia

Here neither indicator is suitable (Figure 4). In fact, no indicator could be suitable as an indicator requires a vertical portion of the curve over two pH units at the equivalence point to give a sharp change.

The half-neutralisation point

If you look at the titration curves you can see that there is always a very gently sloping, almost horizontal, part to the curve before you reach the steep line on which the equivalence point lies. As you add acid (or base), there is very little change to the pH, almost up to the volume of the equivalence point. The point half-way between the zero and the equivalence point is the **half-neutralisation point**. This is significant because the knowledge that you can add acid (or base) up to this point with the certainty that the pH will change very little is relevant to the theory of buffers. It also allows you to find the pK_a of the weak acid.

$$HA + OH^- \rightarrow H_2O + A^-$$

At the half-neutralisation point, half the HA has been converted into A^- and half remains, so:

$$[HA] = [A^-]$$

Therefore $$K_a = \frac{[H^+]\cancel{[A^-]}}{\cancel{[HA]}}$$

$$K_a = [H^+]$$

And $$-\log_{10} K_a = -\log_{10} [H^+]$$

$$pK_a = pH$$

Study tip

This provides a method of determining pK_a values experimentally.

Worked example: to find the pK_a of ethanoic acid

Use the titration curve.

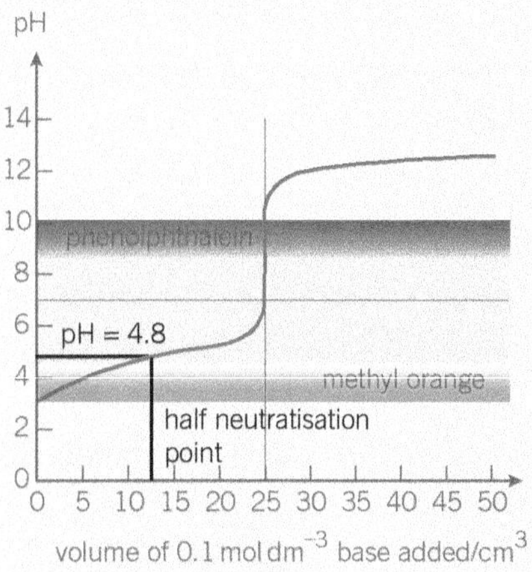

The half-neutralisation point is when 12.5 cm^3 of sodium hydroxide solution has been added.

At this point, the pH is 4.8, so pK_a of ethanoic acid = 4.8

Summary questions

1 The indicator bromocresol purple changes colour between pH 5.2 and 6.8. For which of the following titration types would it be suitable?

a weak acid–weak base

b strong acid–weak base

c weak acid–strong base

d strong acid–strong base

2 For which of the above titrations would bromophenol blue be suitable?

21.6 Buffer solutions

Buffers are solutions that can resist changes of pH when small amounts of acid or alkali are added to them.

How buffers work

Buffers are designed to keep the concentration of hydrogen ions and hydroxide ions in a solution almost unchanged. They are based on an equilibrium reaction which will move in the direction to remove either additional hydrogen ions or hydroxide ions if these are added.

Acidic buffers

Acidic buffers are made from weak acids. They work because the dissociation of a weak acid is an equilibrium reaction.

Consider a weak acid, HA. It will dissociate in solution:

$$HA(aq) \rightleftharpoons H^+(aq) + A^-(aq)$$

From the equation, $[H^+(aq)] = [A^-(aq)]$, and as it is a weak acid, $[H^+(aq)]$ and $[A^-(aq)]$ are both very small because most of the HA is undissociated.

Adding alkali

If a little alkali is added, the OH^- ions from the alkali will react with HA to produce water molecules and A^-:

$$HA(aq) + OH^-(aq) \rightarrow H_2O(aq) + A^-(aq)$$

This removes the added OH^- so the pH tends to remain almost the same.

Adding acid

If H^+ is added, the equilibrium shifts to the left: H^+ ions combining with A^- ions to produce undissociated HA. But, since $[A^-]$ is small, the supply of A^- soon runs out and there is no A^- left to combine the added H^+. So the solution is not a buffer.

However, you can add to the solution a supply of extra A^- by adding a soluble salt of HA, which fully ionises, such as Na^+A^-. This increases the supply of A^- so that more H^+ can be used up. So, there is a way in which both added H^+ and OH^- can be removed.

An acidic buffer is made from a mixture of a weak acid and a soluble salt of that acid. It will maintain a pH of below 7 (acidic).

The function of the weak acid component of a buffer is to act as a source of HA which can remove any added OH^-:

$$HA(aq) + OH^-(aq) \rightarrow A^-(aq) + H_2O(l)$$

The function of the salt component of a buffer is to act as a source of A^- ions which can remove any added H^+ ions:

$$A^-(aq) + H^+(aq) \rightarrow HA(aq)$$

Buffers don't ensure that *no* change in pH occurs. The addition of acid or alkali will still change the pH, but only slightly and by far less than the change that adding the same amount to a non-buffer would cause.

Learning objectives:

- → State the definition of a buffer.
- → Describe how buffers work.
- → Describe how the pH of an acidic buffer solution can be calculated.
- → Describe what buffers are used for.

Specification reference: 3.1.12

Study tip

Practise writing equations to explain how a buffer solution reacts when small amounts of acid or base are added.

Hint

Buffers are important in brewing – the enzymes that control many of the reactions involved work best at specific pHs.

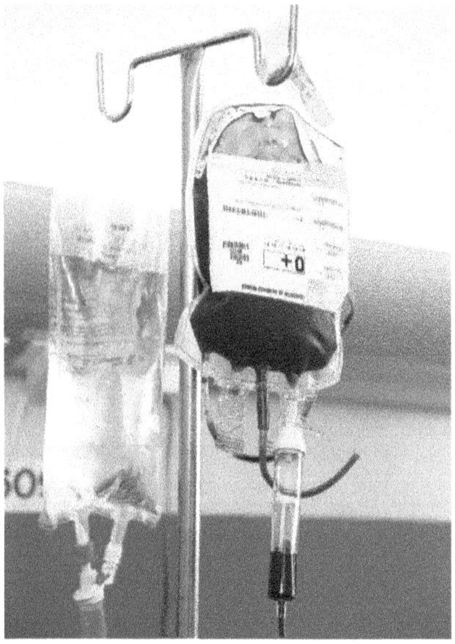

▲ **Figure 1** *Blood is buffered to a pH of 7.4*

It is also possible to saturate a buffer – add so much acid or alkali that all of the available HA or A^- is used up.

Another way of achieving a mixture of weak acid and its salt is by neutralising some of the weak acid with an alkali such as sodium hydroxide. If you neutralise half the acid (Topic 21.5) you end up with a buffer whose pH is equal to the pK_a of the acid as it has an equal supply of HA and A^-.

At half-neutralisation: pH = pK_a

This is a very useful buffer because it is equally efficient at resisting a change in pH whether acid or alkali is added.

Basic buffers

Basic buffers also resist change but maintain a pH at above 7. They are made from a mixture of a weak base and a salt of that base.

A mixture of aqueous ammonia and ammonium chloride, $NH_4^+Cl^-$, acts as a basic buffer. In this case:

- The aqueous ammonia removes added H^+:

  $NH_3(aq) + H^+(aq) \rightarrow NH_4^+(aq)$
- the ammonium ion, NH_4^+, removes added OH^-:

  $NH_4^+(aq) + OH^-(aq) \rightarrow NH_3(aq) + H_2O(l)$

Examples of buffers

An important example of a system involving a buffer is blood, the pH of which is maintained at approximately 7.4. A change of as little as 0.5 of a pH unit may be fatal.

Blood is buffered to a pH of 7.4 by a number of mechanisms. The most important is:

$$H^+(aq) + HCO_3^-(aq) \rightleftharpoons CO_2(aq) + H_2O(l)$$

Addition of extra H^+ ions moves this equilibrium to the right, removing the added H^+. Addition of extra OH^- ions removes H^+ by reacting to form water. The equilibrium above moves to the left releasing more H^+ ions. (The same equilibrium reaction acts to buffer the acidity of soils.)

There are many examples of buffers in everyday products, such as detergents and shampoos. If either of these substances become too acidic or too alkaline, they could damage fabric or skin and hair.

▲ **Figure 2** *Most shampoos are buffered so that they are slightly alkaline*

Calculations on buffers

Different buffers can be made which will maintain different pHs. When a weak acid dissociates:

$$HA(aq) \rightleftharpoons H^+(aq) + A^-(aq)$$

You can write the expression:

$$K_a = \frac{[H^+(aq)][A^-(aq)]}{[HA(aq)]}$$

You can use this expression to calculate the pH of buffers.

Worked example: Calculating pH of a buffer solution 1

A buffer consists of 0.100 mol dm^{-3} ethanoic acid and 0.100 mol dm^{-3} sodium ethanoate (Figure 3). What is the pH of the buffer? (K_a for ethanoic acid is 1.70×10^{-5}, $pK_a = 4.77$.)

Calculate [H$^+$(aq)] from the equation.

$$K_a = \frac{[H^+(aq)][A^-(aq)]}{[HA(aq)]}$$

Sodium ethanoate is fully dissociated, so $[A^-(aq)] = 0.100\,mol\,dm^{-3}$

Ethanoic acid is almost undissociated, so $[HA(aq)] \approx 0.100\,mol\,dm^{-3}$

$$1.70 \times 10^{-5} = [H^+(aq)] \times \frac{0.100}{0.100}$$

$$1.70 \times 10^{-5} = [H^+(aq)] \text{ and } pH = -\log_{10}[H^+(aq)]$$

$$pH = 4.77$$

When you have equal concentrations of acid and salt, pH of the buffer is equal to pK_a of acid used, and this is exactly the same situation as the half-neutralisation point.

Changing the concentration of HA or A$^-$ will affect the pH of the buffer. If you use 0.200 mol dm^{-3} ethanoic acid and 0.100 mol dm^{-3} sodium ethanoate, the pH will be 4.50. Check that you can do this by doing a calculation like the one above.

Worked example: Calculating pH of a buffer solution 2

Calculate the pH of the buffer formed when 500 cm^3 of 0.400 mol dm^{-3} NaOH is added to 500 cm^3 1.00 mol dm^{-3} HA. $K_a = 6.25 \times 10^{-5}$.

Some of the weak acid is neutralised by the sodium hydroxide leaving a solution containing A$^-$ and HA, which will act as a buffer.

$$\text{moles HA} = c \times \frac{V}{1000} = 1.00 \times \frac{500}{1000} = 0.500\,mol$$

$$\text{moles NaOH} = \text{moles OH}^- = c \times \frac{V}{1000} = 0.400 \times \frac{500}{1000} = 0.200\,mol$$

Equation:	HA	+	NaOH	→	H_2O	+	NaA
Initially:	0.500 mol		0.200 mol				0
Finally:	0.300 mol		0 mol				0.200 mol

This leaves 1000 cm^3 of a solution containing 0.300 mol HA and 0.200 mol A$^-$ since all the NaA is dissociated to give A$^-$.

The concentrations are [HA] = 0.300 mol dm^{-3} and [A−] = 0.200 mol dm^{-3}

$$K_a = \frac{[H^+(aq)][A^-(aq)]}{[HA(aq)]}$$

$$6.25 \times 10^{-5} = [H^+(aq)] \times \frac{0.200}{0.300}$$

$$[H^+(aq)] = 6.25 \times 10^{-5} \times \frac{3}{2} = 9.375 \times 10^{-5}$$

$$\text{So pH} = -\log[H^+(aq)] = 4.03$$

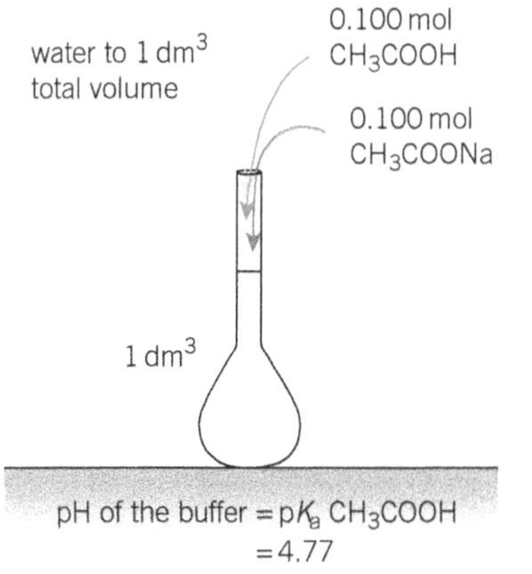

▲ **Figure 3** *Making a buffer*

Study tip

We must calculate the actual quantities to add to make the buffer.

M_r CH_3COONa 82.1, so 0.100 mol has a mass of 8.21 g.

M_r CH_3COOH 60.1, so 0.100 mol has a mass of 6.01 g. This is a liquid, so it is easier to measure its volume. Its density is 1.049 cm^{-3}.

Density = mass / volume

So volume = mass / density.

Volume = 6.01 / 1.049 = 5.72 cm^3.

So we add 6.01 g of solid sodium ethanoate and 5.72 cm^3 of ethanoic acid to the flask. Then we add distilled water to dissolve these and then more distilled water to the mark on the volumetric flask.

Study tip

In a buffer solution $[H^+] \neq [A^-]$ so do *not* use the simplified expression

$$K_a = \frac{[H^+]^2}{[HA]}$$

because it is incorrect. It only applies to a solution containing a weak acid.

The pH change when an acid or a base is added to a buffer

Adding acid

You can calculate how the pH changes when acid is added to a buffer. Suppose you start with 1.00 dm^3 of a buffer solution of ethanoic acid at concentration 0.10 mol dm^{-3} and sodium ethanoate at concentration 0.10 mol dm^{-3}. K_a is 1.70×10^{-5}. This has a pH of 4.77, as shown in the calculation previously.

Now add 10.0 cm^3 of hydrochloric acid of concentration 1.00 mol dm^{-3} to this buffer. Virtually all the added H^+ ions will react with the ethanoate ions, $[A^-]$, to form molecules of ethanoic acid, [HA].

Before adding the acid:

- Number of moles of ethanoic acid = 0.10
- Number of moles of sodium ethanoate = 0.10

Number of moles of hydrochloric acid added is $c \times \frac{V}{1000}$

$$1.00 \times \frac{10.0}{1000} = 0.010$$

After adding the acid, this means:

- The amount of acid is increased by 0.010 mol to 0.110 mol.
- The amount of salt is decreased by 0.010 mol to 0.090 mol.

So, the concentration of acid [HA] is now 0.110 mol dm^{-3}.

And, the concentration of salt $[A^-]$ is now 0.090 mol dm^{-3}.

In calculating these concentrations you have ignored the volume of the added hydrochloric acid, 10 cm^3 in 1000 cm^3, only a 1% change.

$$K_a = \frac{[H^+(aq)][A^-(aq)]}{[HA(aq)]}$$

So $$1.70 \times 10^{-5} = [H^+(aq)]\frac{0.090}{0.110}$$

$$[H^+(aq)] = 1.70 \times 10^{-5} \times \frac{0.110}{0.090} = 2.08 \times 10^{-5}$$

$$pH = 4.68$$

Note how small the pH change is (from 4.77 to 4.68).

Making a buffer solution

You may need to make up buffer solutions of specified pHs, to calibrate a pH meter, for example.

To find suitable concentrations of weak acid and its salt, you use the equation

$$K_a = \frac{[H^+][A^-]}{[HA]}$$

You can rearrange this equation to make $[H^+]$ the subject and then taking logs of both sides. This results in an equation that is easier to use for calculations on buffers. It is called the **Henderson–Hasselbalch equation.**

$$pH = pK_a - \log \left(\frac{[HA]}{[A^-]}\right)$$

It tells you that the pH of a buffer solution depends on the pK_a of the weak acid on which it is based and on the **ratio** of the concentration of the acid to that of its salt.

For example, to make a buffer of pH = 4.50 you first select a weak acid whose pK_a is close to the required pH. Benzoic acid has a pK_a of 4.20, so you could make a buffer from benzoic acid and sodium benzoate.

Then substituting into the Henderson equation:

$$4.50 = 4.20 - \log\left(\frac{[HA]}{[A^-]}\right)$$

$$\log\left(\frac{[HA]}{[A^-]}\right) = -0.30$$

Taking antilogs:

$$\left(\frac{[HA]}{[A^-]}\right) = 0.50$$

This means that the concentration of acid is half the concentration of the salt.

So, for example, a solution that is 0.05 mol dm^{-3} in benzoic acid and 0.10 mol dm^{-3} in sodium benzoate would be a suitable buffer.

The formula of benzoic acid is C_6H_5COOH and that of sodium benzoate is $C_6H_5COO^-Na^+$.

1 Calculate the relative molecular mass of **a** benzoic acid and **b** sodium benzoate.
2 Calculate the mass of **a** benzoic acid and **b** sodium benzoate required to make up 250 cm^3 of a buffer solution that is 0.05 mol dm^{-3} in benzoic acid and 0 .10 mol dm^{-3} in sodium benzoate.
3 Describe the procedure for making up such a solution using a 250 cm^3 graduated flask.

Study tip

Always quote pH values to two decimal places.

Adding base

If you add 10 cm^3 of 0.10 mol dm^{-3} sodium hydroxide to the original buffer, it will react with the H^+ ions and more HA will ionise, so this time you decrease the concentration of the acid [HA] by 0.010 and increase the concentration of ethanoate ions by 0.010.

Using similar steps to those above gives the new pH as 4.89. Check that you agree with this answer.

Note how small the pH change is (from 4.77 to 4.89).

Summary question

1 Find the pH of the following buffers.
   a Using [ethanoic acid] = 0.10 mol dm^{-3}, [sodium ethanoate] = 0.20 mol dm^{-3} (K_a of ethanoic acid = 1.7×10^{-5}).
   b Using [benzoic acid] = 0.10 mol dm^{-3}, [sodium benzoate] = 0.10 mol dm^{-3} (K_a of benzoic acid = 6.3×10^{-5}).

1 In this question give all values of pH to 2 decimal places.
The acid dissociation constant K_a for propanoic acid has the value $1.35 \times 10^{-5}\ mol\,dm^{-3}$ at 25 °C.

$$K_a = \frac{[H^+][CH_3CH_2COO^-]}{[CH_3CH_2COOH]}$$

(a) Calculate the pH of a $0.169\ mol\,dm^{-3}$ solution of propanoic acid.

(3 marks)

(b) A buffer solution contains 0.250 mol of propanoic acid and 0.190 mol of sodium propanoate in $1000\ cm^3$ of solution.
A 0.015 mol sample of solid sodium hydroxide is then added to this buffer solution.
(i) Write an equation for the reaction of propanoic acid with sodium hydroxide.
(ii) Calculate the number of moles of propanoic acid and of propanoate ions present in the buffer solution after the addition of the sodium hydroxide.
(iii) Hence, calculate the pH of the buffer solution after the addition of the sodium hydroxide.

(6 marks)
AQA, 2008

2 In this question, give all values of pH to 2 decimal places.

(a) The ionic product of water has the symbol K_w
(i) Write an expression for the ionic product of water.

(1 mark)

(ii) At 42 °C, the value of K_w is $3.46 \times 10^{-14}\ mol^2\,dm^{-6}$.
Calculate the pH of pure water at this temperature.

(2 marks)

(iii) At 75 °C, a $0.0470\ mol\,dm^{-3}$ solution of sodium hydroxide has a pH of 11.36
Calculate a value for K_w at this temperature.

(2 marks)

(b) Methanoic acid (HCOOH) dissociates slightly in aqueous solution.
(i) Write an equation for this dissociation.

(1 mark)

(ii) Write an expression for the acid dissociation constant K_a for methanoic acid.

(1 mark)

(iii) The value of K_a for methanoic acid is $1.78 \times 10^{-4}\ mol\,dm^{-3}$ at 25 °C. Calculate the pH of a $0.0560\ mol\,dm^{-3}$ solution of methanoic acid.

(3 marks)

(iv) The dissociation of methanoic acid in aqueous solution is endothermic.
Deduce whether the pH of a solution of methanoic acid will increase, decrease, or stay the same if the solution is heated. Explain your answer.

(3 marks)

(c) The value of K_a for methanoic acid is $1.78 \times 10^{-4}\ mol\,dm^{-3}$ at 25 °C.
A buffer solution is prepared containing 2.35×10^{-2} mol of methanoic acid and 1.84×10^{-2} mol of sodium methanoate in $1.00\ dm^3$ of solution.
(i) Calculate the pH of this buffer solution at 25 °C.

(3 marks)

(ii) A $5.00\ cm^3$ sample of $0.100\ mol\,dm^{-3}$ hydrochloric acid is added to the buffer solution in part **(c)** (i).
Calculate the pH of the buffer solution after this addition.

(4 marks)
AQA, 2013

3 In this question give all values of pH to 2 decimal places.

(a) The dissociation of water can be represented by the following equilibrium.

$$H_2O(l) \rightleftharpoons H^+(aq) + OH^-(aq)$$

(i) Write an expression for the ionic product of water, K_w.
(ii) The pH of a sample of pure water is 6.63 at 50 °C.
Calculate the concentration in $mol\,dm^{-3}$ of H^+ ions in this sample of pure water.

(iii) Deduce the concentration in mol dm^{-3} of OH^- ions in this sample of pure water.
(iv) Calculate the value of K_w at this temperature.

(4 marks)

(b) At 25 °C the value of K_w is 1.00×10^{-14} mol^2 dm^{-6}.
Calculate the pH of a 0.136 mol dm^{-3} solution of KOH at 25 °C.

(2 marks)
AQA, 2008

4 Titration curves labelled A, B, C, and D for combinations of different aqueous solutions of acids and bases are shown below.
All solutions have a concentration of 0.1 mol dm^{-3}.

A

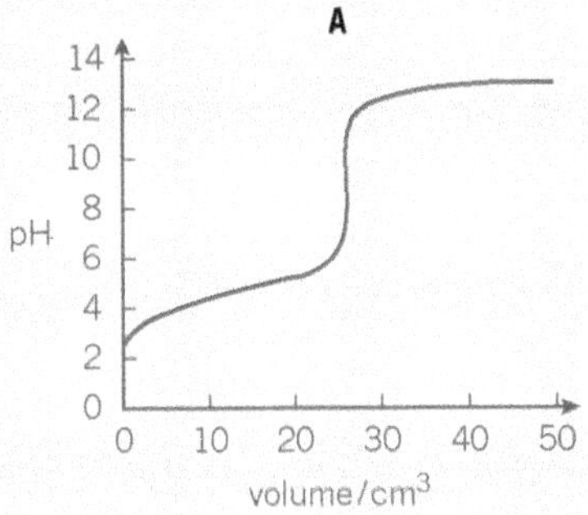

B

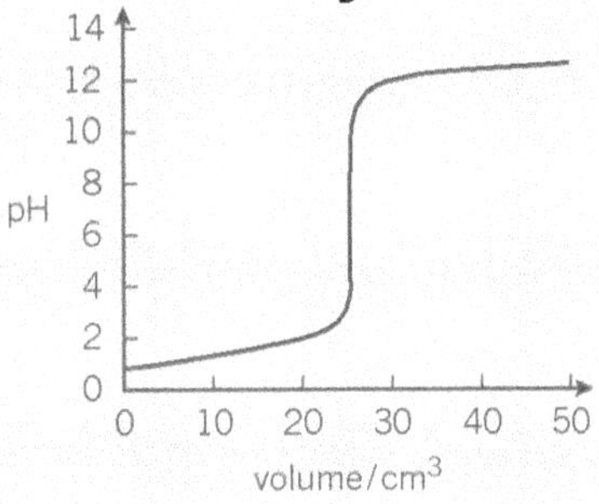

C

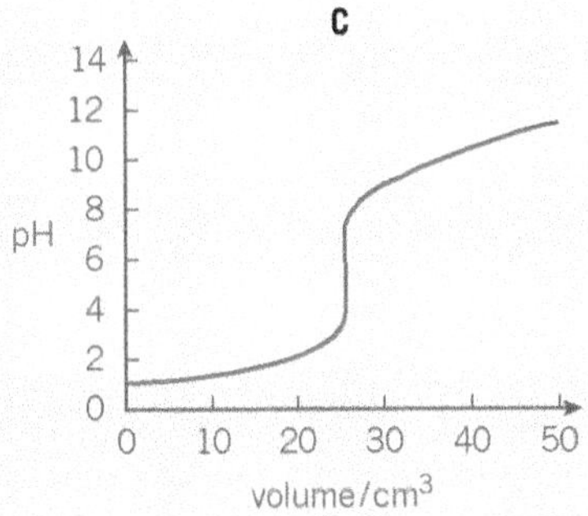

D

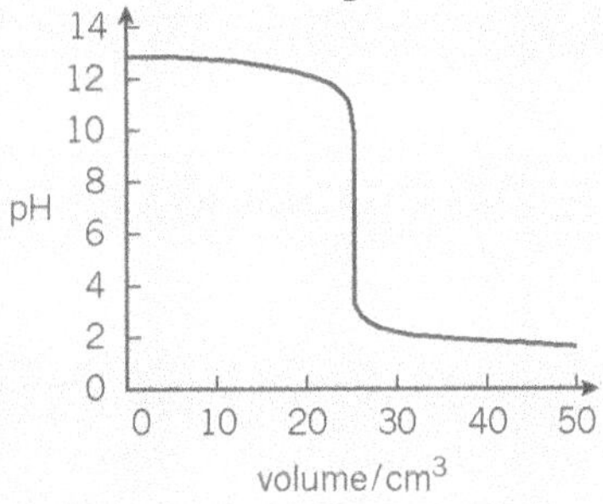

(a) In this part of the question write the appropriate letter.
From the curves **A**, **B**, **C**, and **D**, choose the curve produced by the addition of:
(i) ammonia to 25 cm^3 of hydrochloric acid
(ii) sodium hydroxide to 25 cm^3 of ethanoic acid
(iii) nitric acid to 25 cm^3 of potassium hydroxide.

(3 marks)

(b) A table of acid–base indicators is shown below.
The pH ranges over which the indicators change colour and their colours in acid and alkali are also shown.

Indicator	pH range	Colour in acid	Colour in alkali
Trapaeolin	1.3–3.0	red	yellow
Bromocresol green	3.8–5.4	yellow	blue
Cresol purple	7.6–9.2	yellow	purple
Alizarin yellow	10.1–12.0	yellow	orange

(i) Select from the table an indicator that could be used in the titration that produces curve **B** but **not** in the titration that produces curve **A**.

(1 mark)

(ii) Give the colour change at the end point of the titration that produces curve **D** when cresol purple is used as the indicator.

(1 mark)
AQA, 2011

Answers to the Practice Questions and Section Questions are available at
www.oxfordsecondary.com/oxfordaqaexams-alevel-chemistry

Section 4 practice questions

1 The oxides nitrogen monoxide, NO, and nitrogen dioxide, NO_2, both contribute to atmospheric pollution.
The table gives some data for these oxides and for oxygen.

	$S^\ominus$/JK mol^{-1}	$\Delta_f H$/kJ mol^{-1}
$O_2(g)$	211	0
NO(g)	205	+90
$NO_2(g)$	240	+34

Nitrogen monoxide is formed in internal combustion engines. When nitrogen monoxide comes into contact with air, it reacts with oxygen to form nitrogen dioxide.

$$NO(g) + \frac{1}{2}O_2(g) \rightarrow NO_2(g)$$

(a) Calculate the enthalpy change for this reaction.

(2 marks)

(b) Calculate the entropy change for this reaction.

(2 marks)

(c) Calculate the temperature below which this reaction is spontaneous.

(2 marks)

(d) Suggest **one** reason why nitrogen dioxide is **not** formed by this reaction in an internal combustion engine.

(1 mark)

(e) Write an equation to show how nitrogen monoxide is formed in an internal combustion engine.

(1 mark)

(f) Use your equation from part **(e)** to explain why the free-energy change for the reaction to form nitrogen monoxide stays approximately constant at different temperatures.

(2 marks)

AQA, 2012

2 The balance between enthalpy change and entropy change determines the feasibility of a reaction. The table below contains enthalpy of formation and entropy data for some elements and compounds.

	$N_2(g)$	$O_2(g)$	NO(g)	C(graphite)	C(diamond)
$\Delta_f H^\ominus$/kJ mol^{-1}	0	0	+90.4	0	+1.9
$S^\ominus$/JK^{-1} mol^{-1}	192.2	205.3	211.1	5.7	2.4

(a) Explain why the entropy value for the element nitrogen is much greater than the entropy value for the element carbon (graphite).

(2 marks)

(b) Suggest the condition under which the element carbon (diamond) would have an entropy value of zero.

(1 mark)

(c) Write the equation that shows the relationship between ΔG, ΔH, and ΔS for a reaction.

(1 mark)

(d) State the requirement for a reaction to be feasible.

(1 mark)

(e) Consider the following reaction that can lead to the release of the pollutant NO into the atmosphere.

$$\frac{1}{2}N_2(g) + \frac{1}{2}O_2(g) \rightarrow NO(g)$$

Use data from the table to calculate the minimum temperature above which this reaction is feasible.

(5 marks)

(f) At temperatures below the value calculated in part **(e)**, decomposition of NO into its elements should be spontaneous. However, in car exhausts this decomposition reaction does **not** take place in the absence of a catalyst. Suggest why this spontaneous decomposition does **not** take place.

(1 mark)

(g) A student had an idea to earn money by carrying out the following reaction.

$$C(\text{graphite}) \rightarrow C(\text{diamond})$$

Use the data to calculate values for ΔH and ΔS for this reaction. Use these values to explain why this reaction is **not** feasible under standard pressure at any temperature.

(3 marks)
AQA, 2011

3 Use the table below, where appropriate, to answer the questions which follow.

Standard electrode potentials	E^{\ominus}/V
$2H^+(aq) + 2e^- \rightarrow H_2(g)$	0.00
$Br_2(aq) + 2e^- \rightarrow 2Br^-(aq)$	+1.09
$2BrO_3^-(aq) + 12H^+(aq) + 10e^- \rightarrow Br_2(aq) + 6H_2O(l)$	+1.52

Each of the above reactions can be reversed under suitable conditions.

(a) State the hydrogen ion concentration and the hydrogen gas pressure when, at 298 K, the potential of the hydrogen electrode is 0.00 V.

(2 marks)

(b) A diagram of a cell using platinum electrodes **X** and **Y** is shown below.

solution containing $Br_2(aq)$ and $Br^-(aq)$

solution containing $BrO_3^-(aq)$, $H^+(aq)$ and $Br_2(aq)$

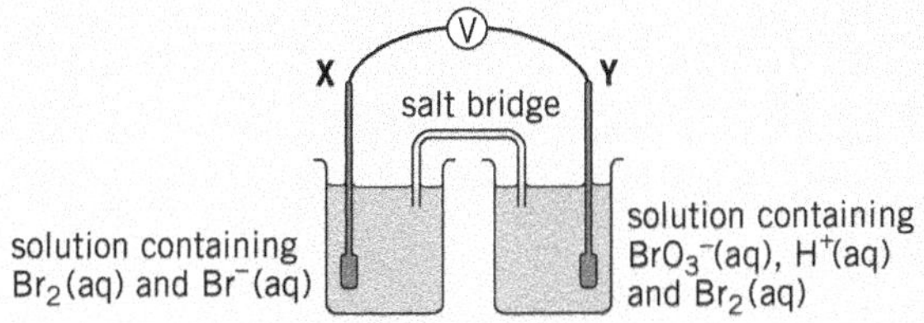

 (i) Use the data in the table above to calculate the emf of the above cell under standard conditions.

 (ii) Write a half equation for the reaction occurring at electrode **X** and an overall equation for the cell reaction which occurs when electrodes **X** and **Y** are connected.

(4 marks)
AQA, 2004

4 Formic acid is a weak acid sometimes used in products to remove limescale from toilet bowls, sinks, and kettles. The structure of formic acid is:

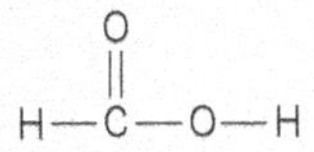

(a) What is the systematic name of formic acid? *(1 mark)*
(b) What is meant by the term *weak acid*? *(1 mark)*
(c) The equation for the dissociation of formic acid is:

$$H-\overset{\overset{\displaystyle O}{\|}}{C}-O-H \ (aq) \rightleftharpoons H-\overset{\overset{\displaystyle O}{\|}}{C}-O^- + H^+$$

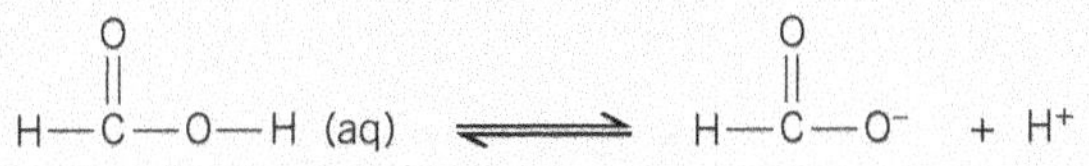

Explain why the hydrogen atom bonded to the oxygen is the one that dissociates rather than that bonded to the carbon atom.

(1 mark)

(d) Write an expression for K_a, the dissociation constant of formic acid. You may use the symbol HA for formic acid and A^- for the formate ion. *(2 marks)*

(e) The values of pK_a for some other weak acids are in the table.

Acid	pK_a
sulfamic	0.99
formic	3.8
ethanoic	4.8

Which is the strongest acid? Explain your answer. *(2 marks)*

(f) Use the expression for K_a to calculate $[H^+]$ in a 1 mol dm^{-3} solution of formic acid and hence the pH of this solution. *(4 marks)*

(g) Limescale contains calcium carbonate, $CaCO_3$. Write an equation for the reaction of calcium carbonate with formic acid. *(2 marks)*

(h) What mass of calcium carbonate would react with 4.5 g of formic acid? *(2 marks)*

(i) Bleach-based toilet cleaners contain chloric(I) acid in which the following equilibrium is set up:

$Cl_2(g) + OH^-(aq) \rightarrow HClO(aq) + Cl^-(aq)$

Suggest what would happen if formic acid is added to such a cleaner. Why would this be potentially dangerous. *(3 marks)*

5 Butadiene dimerises according to the equation

$$2C_4H_6 \rightarrow C_8H_{12}$$

The kinetics of the dimerisation are studied and the graph of the concentration of a sample of butadiene is plotted against time. The graph is shown in the figure.

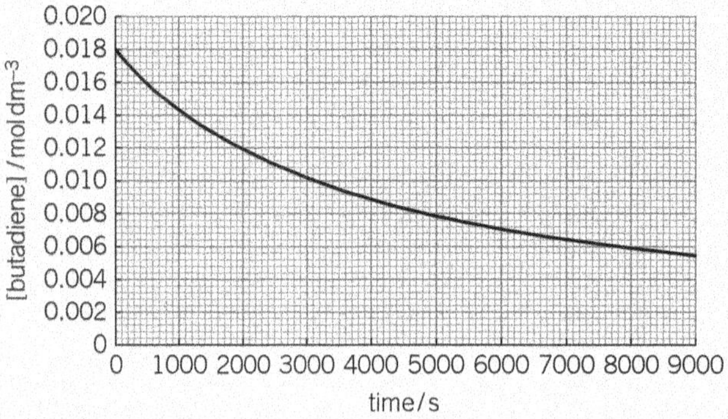

(a) Draw a tangent to the curve when the concentration of butadiene is 0.0120 mol dm^{-3}. *(1 mark)*

(b) The initial rate of reaction in this experiment has the value

$$4.57 \times 10^{-6} \text{ mol dm}^{-3}\text{ s}^{-1}.$$

Use this value, together with a rate obtained from your tangent, to justify that the order of the reaction is 2 with respect to butadiene. *(5 marks)*

6 The human stomach typically contains 1 dm^3 of hydrochloric acid (HCl) which is used to aid digestion of food. The concentration of this acid is approximately 0.01 mol dm^{-3}. Hydrochloric acid is a strong acid.

(a) What is meant by the term *strong acid*? *(1 mark)*

(b) Write an equation for the dissociation of HCl in water. *(1 mark)*

(c) Calculate:

(i) the number of moles of hydrochloric acid in the stomach and

(ii) the pH of the acid in the stomach. *(2 marks)*

Too much acid in the stomach can cause heartburn and many medicines contain antacids – compounds that react to neutralise excess acid. One such medicine contains, in the recommended dose, 0.267 g of sodium hydrogencarbonate ($NaHCO_3$) and 0.160 g of calcium carbonate ($CaCO_3$).

(d) Write equation for the reactions of:

(i) sodium hydrogencarbonate and

(ii) calcium carbonate with hydrochloric acid. *(4 marks)*

(e) Using these equations, calculate how many moles of hydrochloric acid are neutralised by:

(i) the sodium hydrogencarbonate

(ii) the calcium carbonate

(iii) the recommended dose of the antacid. *(5 marks)*

(f) (i) Calculate the number of moles of hydrochloric acid that remain in the stomach.

(ii) Calculate the pH of the stomach acid remaining. (Remember that the acid has a volume of 1 dm^3.) *(2 marks)*

Section 4 summary

Section 1 Physical chemistry

buffers
resists changes in pH where small amount of acid or base are added

acidic buffer
- weak acid
- salt of weak acid

basic buffer
- weak base
- salt of weak base

$pK_a = -\log_{10} K_a$

$K_a = \frac{[H^+] \times [A^-]}{[HA]}$

$K_w = [H^+] \times [OH^-]$

$pH = -\log_{10} [H^+]$

strong acids are fully dissociated
$HA \rightarrow H^+ + A^-$
weak acids are partially dissociated
$HA \rightleftharpoons H^+ + A^-$

acids and bases

acids are proton donors
bases are proton acceptors

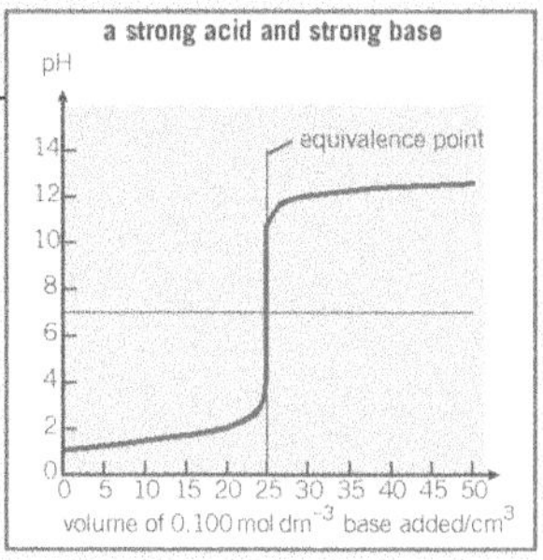

evidence of covalent character in ionic compounds

lattice enthalpy values
$M^+(g) + X^-(g) \rightarrow MX(s)$

Born–Haber cycles

enthalpy charge of solution = – lattice enthalpy + enthalpy change of hydration of ions

Gibbs free energy change ΔG
$\Delta G = \Delta H - T\Delta S$
ΔG must be 0 or negative for reaction to be feasible

Arrhenius constant
$k = Ae^{-\frac{E_a}{RT}}$

rate = $k[A]^m [B]^n$
- [A] = concentration of A
- m = order of reaction with respect to A
- overall order $m + n$

initial rate of reaction gradient of tangent when time = 0

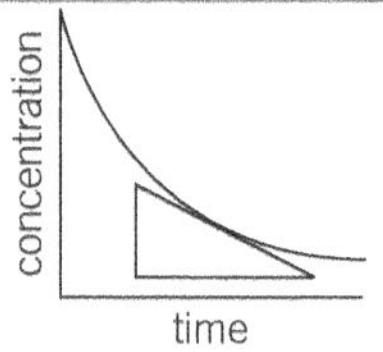

gradient of the tangent gives rate of reaction

zero order

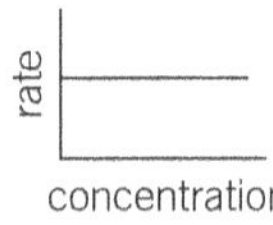

first order

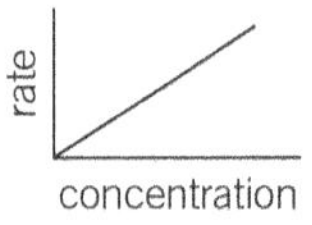

order greater than one

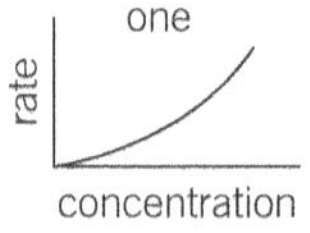

$aA(g) + bB(g) \rightleftharpoons cC(g) + dD(g)$

$K_p = \frac{pC^c \times pD^d}{pA^a \times pB^b}$

standard conditions:
- 298 K
- 100 kPa
- 1.00 mol dm^{-3}

standard hydrogen electrode

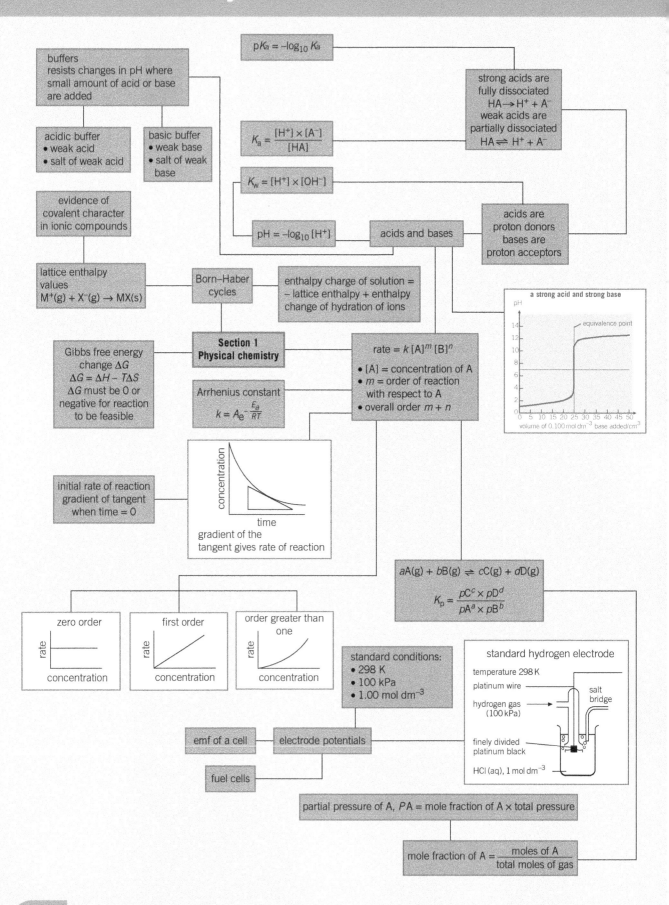

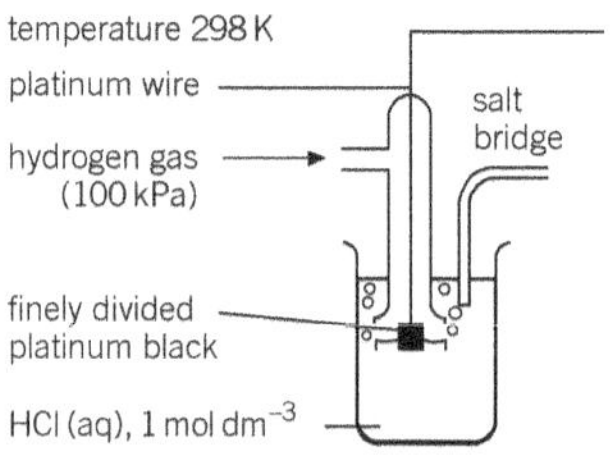

emf of a cell

electrode potentials

fuel cells

partial pressure of A, PA = mole fraction of A × total pressure

mole fraction of A = $\frac{\text{moles of A}}{\text{total moles of gas}}$

Practical skills

In this section you have met the following ideas:

- Finding the entropy change for the vaporisation of water.
- Finding the order of reaction for a reactant.
- Making simple cells and using them to measure an unknown electrode potential.
- Finding how changing conditions effects the emf of a cell.
- Using $E^\ominus$ values to predict the direction of simple redox reactions.
- Finding out the value of K_a for a weak acid.
- Exploring how the pH changes during neutralisation reactions.
- Making and testing a buffer solution.

Maths skills

In this section you have met the following maths skills:

- Using the expression $\Delta G = \Delta H - T\Delta S$ to solve problems.
- Determining the value of ΔS and ΔH from a graph of ΔG versus T.
- Working out the rate equation from given data.
- Using a concentration–time graph to calculate the rate of reaction.
- Calculating the value of the rate constant from the gradient of a zero order concentration–time graph.
- Calculating the partial pressures of reactants and products at equilibrium.
- Finding out the value of K_p
- Calculating the pH of a solution from $[H^+]$ and the value of $[H^+]$ from the pH.
- Finding the pH of solutions.
- Calculating the pH of a buffer solution.

Extension

Produce a report exploring how ideas about entropy and enthalpy can be applied to everyday life.

Suggested resources:

Atkins, P[2010], *The Laws of Thermodynamics: A Very Short Introduction*. Oxford University Press, UK. ISBN 978-0-19-957219-9

Price, G[1998], *Thermodynamics of Chemical Processes: Oxford Chemistry Primers*. Oxford University Press, UK. ISBN 978-0-19-855963-4

Section 5
A2 Inorganic chemistry 2

Chapters in this section

Periodicity introduces the chemical properties of the Period 3 elements, and some of their compounds are described in order to establish patterns and trends in chemical behaviour across a period.

The transition metals have unique chemical structures which give their compounds characteristic (and useful) properties. These include coloured compounds, complex formation, catalytic activity, and variable oxidation states.

Reactions of inorganic compounds in aqueous solution looks at some reactions of some metal ions in solution – including acid–base reactions and ligand substitution reactions – forming metal-aqua ions.

What you already know:

The material in this unit builds on knowledge and understanding that you have built up at AS level. In particular the following:

- Atoms may be held together by covalent, ionic, or metallic bonds.
- Elements in the Periodic Table show patterns in their properties, which are related to their electronic structures.
- The transition metals form a block in the Periodic Table between the s-block elements and the p-block elements. In general they are hard and strong and have typical metallic properties.
- Acid–base reactions involve the transfer of H^+ ions.

22 Periodicity

22.1 Reactions of Period 3 elements

Learning objectives:

→ State how, and under what conditions, sodium and magnesium react with water.

→ State how the elements from sodium to sulfur react with oxygen.

Specification reference: 3.2.4

Synoptic link

Make sure you are confident with the content covered in Topic 8.2, Trends in the properties of elements of Period 3, Topic 8.3, More trends in the properties of the elements of Period 3, Topic 9.1, The physical and chemical properties of Group 2, and Chapter 3, Bonding. You will also need to be able to balance equations, as covered in Topic 2.5, Balanced equations and related calculations.

Hint

Oxidation states may also be called **oxidation numbers.** The two terms mean exactly the same thing.

Synoptic link

You saw in Topic 10.4, Uses of chlorine, that chlorine reacts with water in a reversible reaction to form chloric(I) acid and hydrochloric acid, HCl.

As you move across a period in the Periodic Table from left to right, there are a number of trends in the properties of the elements.

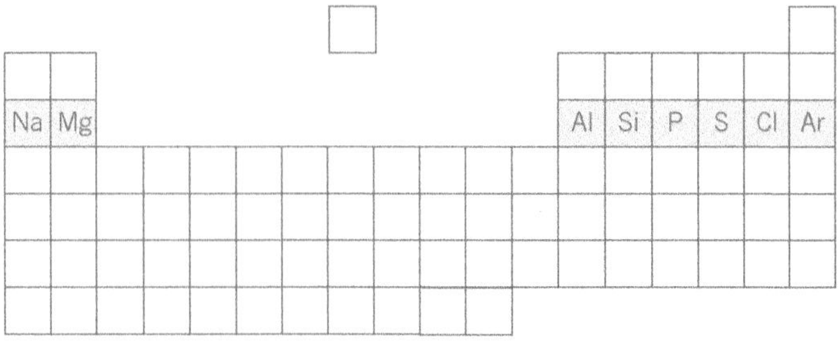

You have already looked at the physical properties of the elements from sodium to argon in Period 3. Here you will examine some of the chemical reactions of these elements.

The elements

The most obvious physical trend in the elements is from metals on the left to non-metals on the right.

- Sodium, magnesium, and aluminium are metallic – they are shiny (when freshly exposed to air), conduct electricity, and react with dilute acids to give hydrogen and salts.
- Silicon is a semi-metal (or metalloid) – it conducts electricity to some extent, a property that is useful in making semiconductor devices.
- Phosphorus, sulfur, and chlorine are typical non-metals – in particular, they do not conduct electricity and have low melting and boiling points.
- Argon is a noble gas. It is chemically unreactive and exists as separate atoms.

The redox reactions of the elements

The reactions of the elements in Period 3 are all redox reactions, since every element starts with an oxidation state of zero, and, after it has reacted, ends up with a positive or a negative oxidation state.

Reactions with water

Sodium and magnesium are the only metal elements in Period 3 that react with cold water. (Chlorine is the only non-metal that reacts with water.)

Sodium

The reaction of sodium with water is vigorous – the sodium floats on the surface of the water and fizzes rapidly, melting because of the heat energy released by the reaction. A strongly alkaline solution of sodium hydroxide is formed (pH 13–14). The oxidation state changes are shown as small numbers above the following symbol equations:

$$2\overset{0}{Na}(s) + 2\overset{+1}{H}_2\overset{-2}{O}(l) \rightarrow 2\overset{+1}{Na}\overset{-2}{O}\overset{+1}{H}(aq) + \overset{0}{H_2}(g)$$

Magnesium

The reaction of magnesium is very slow at room temperature, only a few bubbles of hydrogen are formed after some days. The resulting solution is less alkaline than in the case of sodium because magnesium hydroxide is only sparingly soluble (pH around 10).

$$\overset{0}{Mg}(s) + 2\overset{+1\,-2}{H_2O}(l) \rightarrow \overset{+2\;\,-2+1}{Mg(OH)_2}(aq) + \overset{0}{H_2}(g)$$

The reaction is much faster with heated magnesium and steam and gives magnesium oxide and hydrogen.

$$\overset{0}{Mg}(s) + \overset{+1\,-2}{H_2O}(g) \rightarrow \overset{+2\,-2}{MgO}(s) + \overset{0}{H_2}(g)$$

All these reactions are redox ones, in which the oxidation state of the metal increases and that of some of the hydrogen atoms decreases.

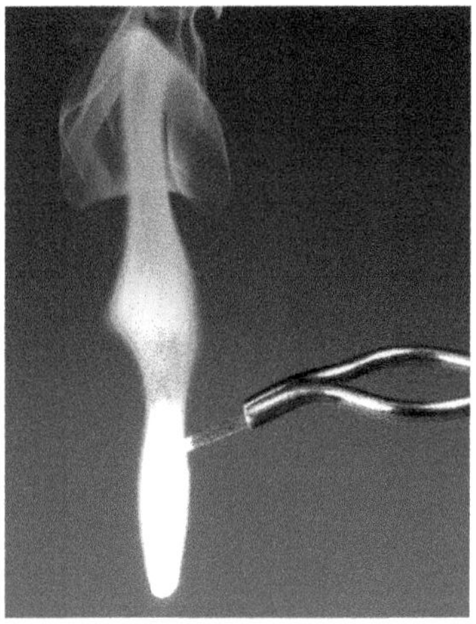

▲ **Figure 1** *Magnesium burning in oxygen from the air*

Reaction with oxygen

All the elements in Period 3 (except for argon) are relatively reactive. Their oxides can all be prepared by direct reaction of the element with oxygen. The reactions are exothermic.

Sodium burns brightly in air (with a characteristic yellow flame) to form white sodium oxide:

$$2\overset{0}{Na}(s) + \frac{1}{2}\overset{0}{O_2}(g) \rightarrow \overset{+1\,-2}{Na_2O}(s)$$

Hint

The sodium oxide formed may have a yellowish appearance due to the production of some sodium peroxide, Na_2O_2.

Magnesium

A strip of magnesium ribbon burns in air with a bright white flame. The white powder that is produced is magnesium oxide. If burning magnesium is lowered into a gas jar of oxygen the flame is even more intense (Figure 1).

magnesium + oxygen → magnesium oxide

$$2\overset{0}{Mg}(s) + \overset{0}{O_2}(g) \rightarrow 2\overset{+2\,-2}{MgO}(s)$$

The oxidation states show how magnesium has been oxidised (its oxidation state has increased) and oxygen has been reduced (its oxidation number has decreased).

Aluminium

When aluminium powder is heated and then lowered into a gas jar of oxygen, it burns brightly to give aluminium oxide – a white powder. Aluminium powder also burns brightly in air (Figure 2).

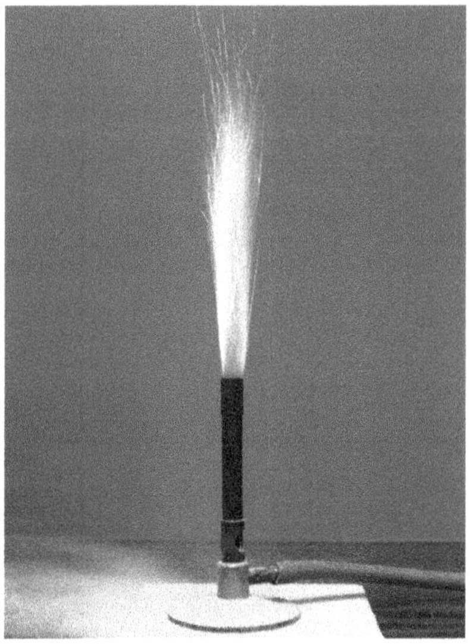

▲ **Figure 2** *Aluminium burning in oxygen from the air. Powdered aluminium is being sprinkled into the flame*

aluminium + oxygen → aluminium oxide

$$4\overset{0}{Al}(s) + 3\overset{0}{O_2}(g) \rightarrow 2\overset{+3\,-2}{Al_2O_3}(s)$$

Aluminium is a reactive metal, but it is always coated with a strongly bonded surface layer of oxide – this protects it from further reaction. So, aluminium appears to be an unreactive metal and is used for many everyday purposes – saucepans, garage doors, window frames, and so on.

Hint

The sum of the oxidation states in Al_2O_3 is zero, as it is in all compounds without a charge:

$(2 \times 3) + (3 \times -2) = 0$

Even if the surface is scratched, the exposed aluminium reacts rapidly with the air and seals off the surface.

Silicon

Silicon will also form the oxide if it is heated strongly in oxygen:

$$\overset{0}{Si}(s) + \overset{0}{O_2}(g) \rightarrow \overset{+4\,-2}{SiO_2}(s)$$

Phosphorus

Red phosphorus must be heated before it will react with oxygen. White phosphorus spontaneously ignites in air and the white smoke of phosphorus pentoxide is given off. Red and white phosphorus are **allotropes** of phosphorus – the same element with the atoms arranged differently.

$$\overset{0}{4P}(s) + \overset{0}{5O_2}(g) \rightarrow \overset{+5\,-2}{P_4O_{10}}(s)$$

If the supply of oxygen is limited, phosphorus trioxide, P_2O_3, is also formed.

Sulfur

When sulfur powder is heated and lowered into a flask of oxygen, it burns with a blue flame to form the colourless gas sulfur dioxide (and a little sulfur trioxide also forms) (Figure 3).

sulfur + oxygen → sulfur dioxide

$$\overset{0}{S}(s) + \overset{0}{O_2}(g) \rightarrow \overset{+4-2}{SO_2}(g)$$

▲ **Figure 3** *Sulfur burning in oxygen*

In all these redox reactions, the oxidation state of the Period 3 element increases and that of the oxygen decreases (from 0 to –2 in each case). The oxidation state changes are shown as small numbers above the symbol equations above. The oxidation number of the Period 3 element in the oxide increases as you move from left to right across the period.

Hint

The empirical formula of phosphorus pentoxide is P_2O_5. In the gas phase it forms molecules of P_4O_{10} and is sometimes referred to as phosphorus(V) oxide.

$$4P + 5O_2 \rightarrow P_4O_{10}$$

Phosphorus and oxygen also form phosphorus trioxide, P_2O_3. Phosphorus trioxide is also called phosphorus(III) oxide.

Summary questions

1 Metals are shiny, conduct electricity, and react with acids to give hydrogen if they are reactive. Give three more properties typical of metals.

2 Non-metals do not conduct electricity. Give two more properties typical of non-metals.

3 a What is the oxidation state of sodium in all its compounds?

b What is the oxidation state of oxygen in sodium peroxide, Na_2O_2? What is unusual about this?

c Show that the sum of the oxidation states in magnesium hydroxide is zero.

4 What is the oxidation state of sulfur in sulfur trioxide?

22.2 The oxides of elements in Period 3

As you move across Period 3 from left to right there are some important trends in the physical properties of the Period 3 compounds. These trends are a result of the change from metal elements on the left of the Periodic Table to non-metal elements on the right. The oxides are representative of these trends.

Learning objectives:

- Describe how the physical properties of the oxides are explained in terms of their structure and bonding.
- State how the oxides react with water.
- Describe how the structures of the oxides explain the trend in their reactions in water.

Specification reference: 3.2.4

The metal oxides

Sodium, magnesium, and aluminium oxides are examples of compounds formed by a metal combined with a non-metal. They form giant ionic lattices where the bonding extends throughout the compound. This results in high melting points.

The bonding in aluminium oxide is ionic but has some covalent character. This is because aluminium forms a very small ion with a large positive charge and so can approach closely to the O^{2-} and distort its electron cloud. So, the bond also has some added covalent character (Figure 1).

It is possible to predict the ionic character of a bond by considering the difference in electronegativities between the two atoms. The bigger the difference, the greater the ionic character of the bond. Caesium oxide, Cs_2O, is about 80% ionic. The electronegativities are Cs = 0.7 and O = 3.5. A bond between two identical atoms, such as that in oxygen, *must* be 100% covalent.

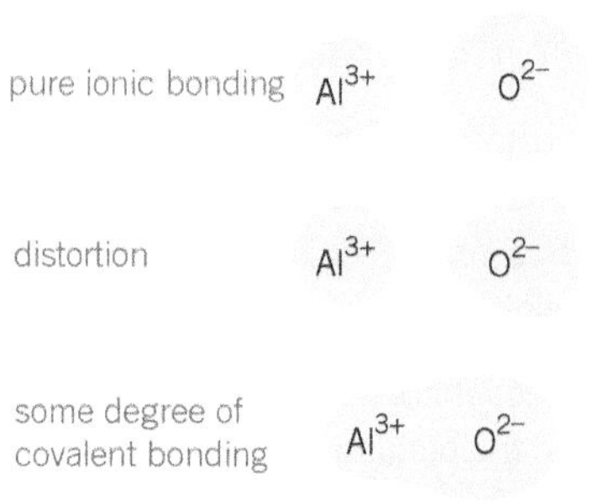

▲ **Figure 1** *The covalent character of the bonding in aluminium oxide*

The non-metal oxides

Silicon oxide has a giant covalent (macromolecular) structure. Again the bonding extends throughout the giant structure, but this time it is covalent (Figure 2). Again you have a compound with a high melting point because many strong covalent bonds must be broken to melt it.

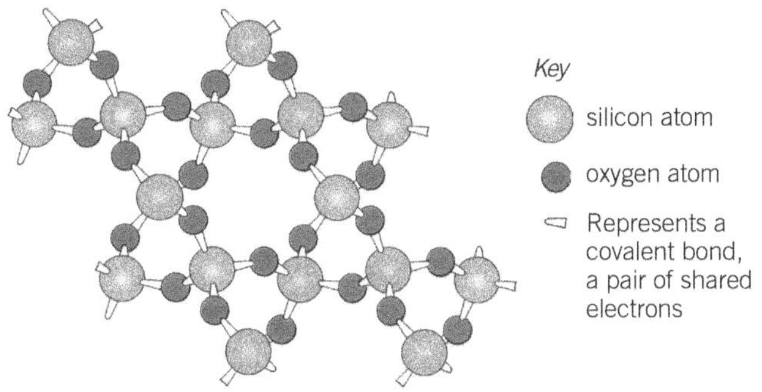

▲ **Figure 2** *Silicon dioxide is a macromolecule*

Synoptic link

In Topic 17.3, More enthalpy changes, you saw that the bonding in zinc selenide was not purely ionic, but had some degree of covalent character.

Phosphorus and sulfur oxides exist as separate covalently bonded molecules. The phosphorus oxides are solids. The intermolecular forces are weak van der Waals and dipole–dipole forces. Their melting points are relatively low. Sulfur dioxide and sulfur trioxide are gases at 298 K.

Synoptic link

Look back at Topic 3.7, Forces acting between molecules, to revise intermolecular forces.

The trends in the physical properties of the oxides are summarised in Table 1.

▼ **Table 1** *The trends in the physical properties of some of the oxides in Period 3*

	Na_2O	MgO	Al_2O_3	SiO_2	P_4O_{10}	SO_3	SO_2
T_m / K	1548	3125	2345	1883	573	290	200
Bonding	ionic	ionic	ionic/covalent	covalent	covalent	covalent	covalent
Structure	giant ionic	giant ionic	giant ionic	giant covalent (macromolecular)	molecular	molecular	molecular

Study tip

It is important that you understand the link between physical properties and bonding.

Note the trend in melting points: $P_4O_{10} > SO_3 > SO_2$

This is related to the increase in intermolecular van der Waals forces between the larger molecules.

The structures of oxo-acids and their anions

Phosphoric(V) acid, H_3PO_4, sulfuric(VI) acid, H_2SO_4, and sulfuric (IV) acid (sulfurous acid, H_2SO_3) are called oxo acids. You can understand the structures of these acids and their anions by drawing dot-and-cross diagrams.

A phosphorus atom has five electrons in its outer shell and the bonding in phosphoric(V) acid is as shown. Note that as phosphorus is in Period 3, it can have more than eight electrons in its outer main shell.

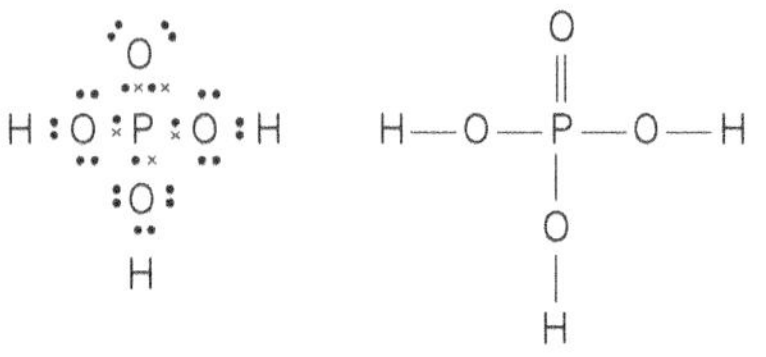

▲ **Figure 3** *Phosphoric(V) acid*

If all three hydrogens are lost as protons, the phosphate(V) ion is formed.

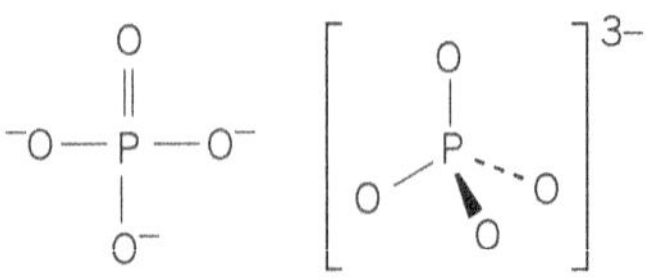

▲ **Figure 4** *Phosphate(V) ion*

At first sight this appears to have one phosphorus–oxygen double bond and three P—O single bonds.

However, it turns out that all three bonds are the same, the bonding electrons are spread out equally over all four bonds, and the shape of the ion is a perfect tetrahedron. This is an example of **delocalisation** which you see in graphite and in benzene.

Sulfur has six electrons in its outer shell and it, too, may have more than eight electrons in its outer main shell. The bonding in sulfuric(VI) acid is as shown.

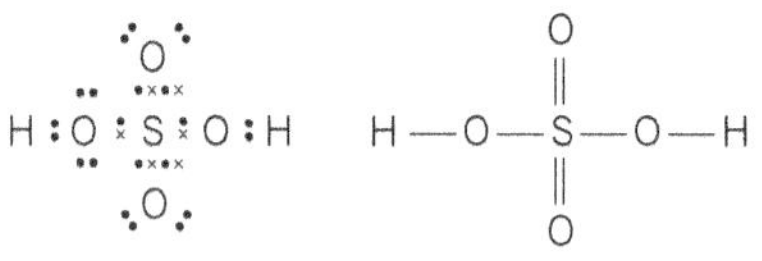

▲ **Figure 5** *Sulfuric(VI) acid*

Like the phosphate(V) ion, the sulfate(VI) anion is a perfect tetrahedron due to delocalisation.

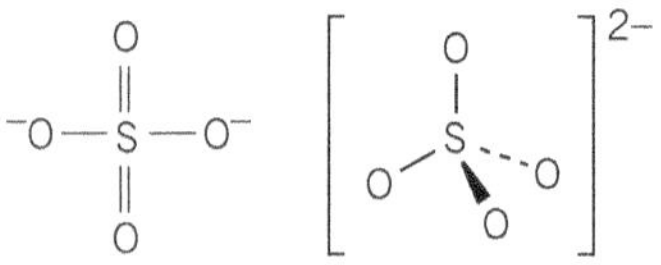

▲ **Figure 6** *Sulfate(VI) anion*

Sulfuric(IV) acid immediately dissociates in aqueous solution. The sulfate(IV) anion's bonding is as shown. Delocalisation makes the three S—O bonds the same length but the bond angle is 106°, smaller than that of a perfect tetrahedron due to the lone pair of electrons.

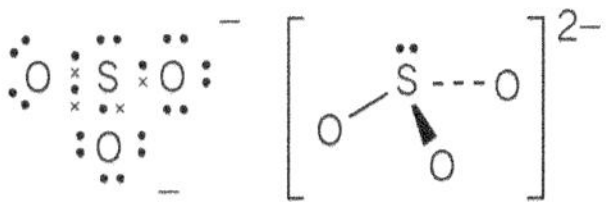

▲ **Figure 7** *Sulfate(IV) anion*

1 What is the angle of a perfect tetrahedron?
2 What can be said about the lengths of the bonds in PO_4^{3-}?
3 How many electrons does phosphorus have in its outer main shell in phosphoric(V) acid?
4 The S—O bond length is 0.157 nm and the S═O bond length is 0.143 nm. Predict the sulfur–oxygen bond length in the sulfate(VI) ion.
5 How many electrons does sulfur have in its outer main shell in the sulfate(IV) anion?

Reaction of oxides with water

Basic oxides

Sodium and magnesium oxides are both bases.

Sodium oxide reacts with water to give sodium hydroxide solution – a strongly alkaline solution:

$$Na_2O(s) + H_2O(l) \rightarrow 2Na^+(aq) + 2OH^-(aq) \text{ pH of solution} \sim 14$$

Magnesium oxide reacts with water to give magnesium hydroxide, which is sparingly soluble in water and produces a somewhat alkaline solution:

$$MgO(s) + H_2O(l) \rightarrow Mg(OH)_2(s) \rightleftharpoons Mg^{2}+(aq) + 2OH^-(aq) \quad \text{pH of solution} \sim 9$$

Insoluble oxides

Aluminium oxide is insoluble in water.
Silicon dioxide is insoluble in water.

Acidic oxides

Non-metals on the right of the Periodic Table typically form acidic oxides. For example, phosphorus pentoxide reacts quite violently with water to produce an acidic solution of phosphoric(V) acid. This ionises, so the solution is acidic.

$$P_4O_{10}(s) + 6H_2O(l) \rightarrow 4H_3PO_4(aq)$$

$H_3PO_4(aq)$ ionises in stages, the first being:

$$H_3PO_4(aq) \rightleftharpoons H^+(aq) + H_2PO_4^-(aq)$$

Sulfur dioxide is fairly soluble in water and reacts with it to give an acidic solution of sulfuric(IV) acid (sulfurous acid). This partially dissociates producing H^+ ions, which cause the acidity of the solution:

$$SO_2(g) + H_2O(l) \rightarrow H_2SO_3(aq)$$

$$H_2SO_3(aq) \rightleftharpoons H^+(aq) + HSO_3^-(aq)$$

Sulfur trioxide reacts violently with water to produce sulfuric acid (sulfuric(VI) acid):

$$SO_3(g) + H_2O(l) \rightarrow H_2SO_4(aq) \rightarrow H^+(aq) + HSO_4^-(aq)$$

Hint

Silicon dioxide is the main constituent of sand, which does not dissolve in water.

The overall pattern is that metal oxides (on the left of the period) form alkaline solutions in water and non-metal oxides (on the right of the period) form acidic ones, whilst those in the middle do not react.

Table 2 summarises these reactions.

▼ **Table 2** *The oxides in water*

Oxide	Bonding	Ions present after reaction with water	Acidity/ alkalinity	Approx. pH (Actual values depend on concentration)
Na_2O	ionic	$Na^+(aq)$, $OH^-(aq)$	strongly alkaline	13–14
MgO	ionic	$Mg^{2+}(aq)$, $OH^-(aq)$	somewhat alkaline	10
Al_2O_3	covalent/ ionic	insoluble, no reaction	—	7
SiO_2	covalent	insoluble, no reaction	—	7
P_4O_{10}	covalent	$H^+(aq) + H_2PO_4^-(aq)$	fairly strong acid	1–2
SO_2	covalent	$H^+(aq) + HSO_3^-(aq)$	weak acid	2–3
SO_3	covalent	$H^+(aq) + HSO_4^-(aq)$	strong acid	0–1

The behaviour of the oxides with water can be understood if you look at their bonding and structure (Table 1).

- Sodium and magnesium oxides, to the left of the Periodic Table, are composed of ions.
- Sodium oxide contains the oxide ion, O^{2-}, which is a very strong base (it strongly attracts protons) and so readily reacts with water to produce hydroxide ions – a strongly alkaline solution.
- Magnesium oxide also contains oxide ions. However, its reaction with water produces a less alkaline solution than sodium oxide because it is less soluble than sodium oxide.
- Aluminium oxide is ionic but the bonding is too strong for the ions to be separated, partly because of the additional covalent bonding it has.
- Silicon dioxide is a giant macromolecule and water will not affect this type of structure.
- Phosphorus oxides and sulfur oxides are covalent molecules and react with water to form acid solutions.

General trend

Solutions of the oxides of the elements go from alkaline to acidic across the period.

Summary questions

1 Write down an equation for the reaction of sodium oxide with water.

  a i State the oxidation number of sodium before and after the reaction.

  ii Has the sodium been oxidised, reduced, or neither?

2 a What ion is responsible for the alkalinity of the solutions formed when sodium oxide and magnesium oxide react with water?

  b What range of pH values represents an alkaline solution?

3 Phosphorus forms another oxide, P_4O_6.

  a Would you expect it to react with water to form a neutral, acidic, or alkaline solution?

  b Explain your answer.

  c Write an equation for its reaction with water.

22.3 The acidic/basic nature of the Period 3 oxides

Learning objectives:

- → Describe how the oxides of the elements in Period 3 react with acids.
- → Describe how the oxides of the elements in Period 3 react with bases.
- → State the equations for these reactions.

Specification reference: 3.2.4

As you saw in Topic 22.2, the general trend is alkalis to acids as you go across the period from left to right. So, you could predict that the alkaline oxides will react with acids and the acidic oxides will react with bases.

Sodium and magnesium oxides

Sodium oxide and magnesium oxide react with acids to give salt and water only.

For example, sodium oxide reacts with sulfuric acid to give sodium sulfate:

$$Na_2O(s) + H_2SO_4(aq) \rightarrow Na_2SO_4(aq) + H_2O(l)$$

Magnesium oxide reacts with hydrochloric acid to give magnesium chloride:

$$MgO(s) + 2HCl(aq) \rightarrow MgCl_2(aq) + H_2O(l)$$

Aluminium oxide

Aluminium oxide reacts *both* with acids and alkalis. It is called an **amphoteric** oxide.

For example, with hydrochloric acid, aluminium chloride is formed:

$$Al_2O_3(s) + 6HCl(aq) \rightarrow 2AlCl_3(aq) + 3H_2O(l)$$

With hot, concentrated sodium hydroxide, sodium aluminate is formed:

$$Al_2O_3(s) + 2NaOH(aq) + 3H_2O(l) \rightarrow 2NaAl(OH)_4(aq)$$

Silicon dioxide

Silicon dioxide will react as a weak acid with strong bases, for example, with hot concentrated sodium hydroxide a colourless solution of sodium silicate is formed:

$$SiO_2(s) + 2NaOH(aq) \rightarrow Na_2SiO_3(aq) + H_2O(l)$$

Study tip

It is important that you know the products of these reactions and that you can write equations for the reactions occurring.

Iron production

In the production of iron, at the high temperatures inside the blast furnace, calcium oxide reacts with the impurity silicon dioxide (sand) to produce a liquid slag, calcium silicate. This is also an example of the acidic silicon dioxide reacting with a base:

$$SiO_2(s) + CaO(l) \rightarrow CaSiO_3(l)$$

Phosphorus pentoxide, P_4O_{10}

The reaction of phosphorus pentoxide with an alkali is really the reaction of phosphoric(V) acid, H_3PO_4, because this is formed when phosphorus pentoxide reacts with water. Phosphoric(V) acid has three –OH groups, and each of these has an acidic hydrogen atom. So it will react with sodium hydroxide in three stages, as each hydrogen in turn reacts with a hydroxide ion and is replaced by a sodium ion:

$$H_3PO_4(aq) + NaOH(aq) \rightarrow NaH_2PO_4(aq) + H_2O(l)$$

$$NaH_2PO_4(aq) + NaOH(aq) \rightarrow Na_2HPO_4(aq) + H_2O(l)$$

$$Na_2HPO_4(aq) + NaOH(aq) \rightarrow Na_3PO_4(aq) + H_2O(l)$$

Overall:

$$3NaOH(aq) + H_3PO_4(aq) \rightarrow Na_3PO_4(aq) + 3H_2O(l)$$

Sulfur dioxide

If you add sodium hydroxide to an aqueous solution of sulfur dioxide, first sodium hydrogensulfate(IV) is formed:

$$SO_2(aq) + NaOH(aq) \rightarrow NaHSO_3(aq)$$

Followed by sodium sulfate(IV):

$$NaHSO_3(aq) + NaOH(aq) \rightarrow Na_2SO_3(aq) + H_2O(l)$$

Calcium sulfite

Sulfur dioxide reacts with the base calcium oxide to form calcium sulfite (calcium sulfate(IV)). This is the first step of one of the methods of removing sulfur dioxide from flue gases in power stations:

$$CaO(s) + SO_2(g) \rightarrow CaSO_3(s)$$

The calcium sulfite is further converted to calcium sulfate, $CaSO_4$, and sold as gypsum for plastering.

Summary questions

1. Write balanced symbol equations for the reactions of:
   - **a** sodium oxide with hydrochloric acid
   - **b** magnesium oxide with sulfuric acid
   - **c** aluminium oxide with nitric acid.
2. $SiO_2(s) + 2NaOH(aq) \rightarrow Na_2SiO_3(aq) + H_2O(l)$
   - **a** Copy the equation above and write the oxidation numbers above each atom.
   - **b** Is the reaction a redox reaction?
   - **c** Explain your answer to **b**.
3. Write a balanced symbol for the reaction of phosphorus pentoxide with water.

22.4 The chlorides of the Period 3 elements

Reactions of the elements with chlorine

All the elements from sodium to sulfur will react directly with chlorine, if heated, to form chlorides as shown in the table. These are all redox reactions in which the Period 3 element is oxidised. Argon, as expected, does not react.

	NaCl	$MgCl_2$	$AlCl_3$	$SiCl_4$	PCl_5	S_2Cl_2
T_m / K	1074	987	463	203	435	193
Bonding	Ionic	Ionic	Covalent	Covalent	Covalent	Covalent
Structure	Giant	Giant	Dimeric molecules, Al_2Cl_6, also form	Molecular	Molecular	Molecular
Notes			In the solid state a largely ionic lattice is formed		Some PCl_3 is also formed	Other chlorides are also formed

The change from ionic to covalent bonding is due to the decrease in electronegativity of the Period 3 element as we go from sodium to sulfur.

Sodium chloride and magnesium chloride are ionic and their formulae are as expected from the positions of the elements in the Periodic Table. The bonding in solid aluminium chloride is largely ionic with a good deal of covalent character. $AlCl_3$ and Al_2Cl_6 form in the gas phase. In $AlCl_3$, the aluminium has only six electrons in its outer main shell and it forms a dimer in which co-ordinate covalent bonds form between lone pairs of electrons on the chlorine atoms and the electron-deficient aluminium atoms.

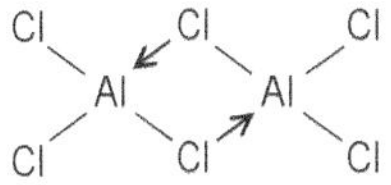

Silicon tetrachloride is covalent as expected. In phosphorus pentachloride, the phosphorus atom has ten electrons in its outer main shell. This is possible as this shell is the main shell three which can hold up to 18 electrons. As expected, the melting point of the chlorides of sodium and magnesium are high as they come from giant ionic lattices. The rest of the chlorides form molecules held together by weaker intermolecular forces only.

Learning objectives:

- → State how and under what conditions the elements of Period 3 react with chlorine.
- → State how and under what conditions the chlorides of the elements of Period 3 react with water.
- → Describe the structures of the acids and the anions formed when the chlorides $SiCl_4$ and PCl_5 react with water.

Specification reference 3.2.4

Hint

The arrow in a co-ordinate bond represents a pair of electron being donated.

The reactions of the chlorides with water

Sodium chloride and magnesium chloride dissolve in water to form neutral solutions containing the ions Na^+ and Cl^- and Mg^{2+} and Cl^- respectively. These ions are hydrated: $Na^+(aq)$; $Mg^{2+}(aq)$, and $Cl^-(aq)$ respectively.

Aluminium chloride reacts rapidly and vigorously with water to give an acidic solution containing H^+ and Cl^- ions. This may be represented simply as

$$AlCl_3(s) + 3H_2O\ (l) \rightarrow Al(OH)_3(s) + 3H^+(aq) + 3Cl^-(aq)$$

In more detail, some hydrated Al^{3+} ions are formed, $[Al(H_2O_6)]^{3+}$.

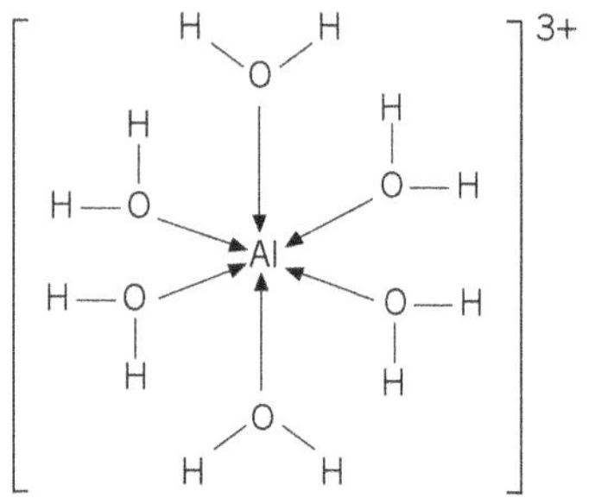

The high charge of the Al^{3+} ion pulls electrons towards it from one of the Al—O bonds resulting in the loss of a H^+ ion which makes the solution acidic. This is the same as the acidity of solutions containing the Fe^{3+} ion which is described in more detail in Topic 24.1, The acid-base chemistry of aqueous transition metal ions.

$$[Al(H_2O_6)]^{3+}(aq) \rightarrow [Al(H_2O_5)(OH)]^{2+}(aq) + H^+(aq)$$

Silicon tetrachloride reacts violently with cold water to form a strongly acidic solution and a colourless 'silica' gel.

$$SiCl_4(l) + 2H_2O(l) \rightarrow SiO_2(s) + 4H^+(aq) + 4Cl^-(aq)$$

Phosphorus pentachloride reacts with boiling water to produce phosphoric acid and hydrochloric acid.

$$PCl_5(s) + 4H_2O(l) \rightarrow H_3PO_4(aq) + 5H^+(aq) + 5Cl^-(aq)$$

Summary questions

1 Write a balanced symbol equation for the reaction of silicon with chlorine. State the change in oxidation number for each element.

2 Is the reaction $AlCl_3(s) + 3H_2O\ (l) \rightarrow Al(OH)_3(s) + 3H^+(aq) + 3Cl^-(aq)$ a redox reaction? Explain your answer.

3 Phosphorus trichloride also reacts with water, forming phosphorous acid, H_3PO_3 and an acid solution. Write an equation for this reaction.

Practice questions

1 **(a)** Suggest why the melting point of magnesium oxide is much higher than the melting point of magnesium chloride.

(2 marks)

(b) Magnesium oxide and sulfur dioxide are added separately to water. In each case describe what happens. Write equations for any reactions which occur and state the approximate pH of any solution formed.

(6 marks)

(c) Write equations for two reactions which together show the amphoteric character of aluminium hydroxide.

(4 marks)
AQA, 2006

2 State what is observed when separate samples of sodium oxide and phosphorus(V) oxide are added to water. Write equations for the reactions which occur and, in each case, state the approximate pH of the solution formed.

(6 marks)
AQA, 2005

3 In the questions below, each of the three elements **X**, **Y**, and **Z** is one of the Period 3 elements Na, Mg, Al, Si, or P.

(a) The oxide of element **X** has a high melting point. The oxide reacts readily with water to form a solution with a high pH.
(i) Deduce the type of bonding present in the oxide of element **X**.
(ii) Identify element **X**.
(iii) Write an equation for the reaction between water and the oxide of element **X**.

(3 marks)

(b) Element **Y** has an oxide which reacts vigorously with water to form a solution containing strong acid.
(i) Deduce the type of bonding present in the oxide of element **Y**.
(ii) Identify element **Y**.
(iii) Identify an acid which is formed when the oxide of element **Y** reacts with water.

(3 marks)

(c) The oxide of element **Z** is a crystalline solid with a very high melting point. This oxide is classified as an acidic oxide but it is not soluble in water.
(i) Deduce the type of crystal shown by the oxide of element **Z**.
(ii) Identify element **Z**.
(iii) Write an equation for a reaction which illustrates the acidic nature of the oxide of element **Z**.

(4 marks)
AQA, 2004

4 Consider the following oxides.
Na_2O, MgO, Al_2O_3, SiO_2, P_4O_{10}, SO_3

(a) Identify one of the oxides from the above which:
(i) can form a solution with a pH less than 3
(ii) can form a solution with a pH greater than 12.

(2 marks)

(b) Write an equation for the reaction between:
(i) MgO and HNO_3
(ii) SiO_2 and $NaOH$
(iii) Na_2O and H_3PO_4.

(3 marks)

(c) Explain, in terms of their type of structure and bonding, why P_4O_{10} can be vaporised by gentle heat but SiO_2 cannot.

(4 marks)
AQA, 2003

5 Write equations for the reactions which occur when the following compounds are added separately to water. In each case, predict the approximate pH of the solution formed when one mole of each compound is added to 1 dm^3 of water.

(a) Sodium oxide

(b) Sulfur dioxide.

(2 marks)
AQA, 2003

6 **(a)** **P** and **Q** are oxides of Period 3 elements.
Oxide **P** is a solid with a high melting point. It does not conduct electricity when solid but does conduct when molten or when dissolved in water. Oxide **P** reacts with water forming a solution with a high pH.
Oxide **Q** is a colourless gas at room temperature. It dissolves in water to give a solution with a low pH.

(i) Identify **P**. State the type of bonding present in **P** and explain its electrical conductivity. Write an equation for the reaction of **P** with water.

(ii) Identify **Q**. State the type of bonding present in **Q** and explain why it is a gas at room temperature. Write an equation for the reaction of **Q** with water.

(9 marks)

(b) **R** is a hydroxide of a Period 3 element. It is insoluble in water but dissolves in both aqueous sodium hydroxide and aqueous sulfuric acid.

(i) Give the name used to describe this behaviour of the hydroxide.

(ii) Write equations for the reactions occurring.

(iii) Suggest why **R** is insoluble in water.

(6 marks)
AQA, 2002

7 This question is about some Period 3 elements and their oxides.

(a) Describe what you would observe when, in the absence of air, magnesium is heated strongly with water vapour at temperatures above 373 K.
Write an equation for the reaction that occurs.

(3 marks)

(b) Explain why magnesium has a higher melting point than sodium.

(2 marks)

(c) State the structure of, and bonding in, silicon dioxide.
Other than a high melting point, give **two** physical properties of silicon dioxide that are characteristic of its structure and bonding.

(4 marks)

(d) Give the formula of the species in a sample of solid phosphorus(V) oxide. State the structure of, and describe fully the bonding in, this oxide.

(4 marks)

(e) Sulfur(IV) oxide reacts with water to form a solution containing ions.
Write an equation for this reaction.

(1 mark)

(f) Write an equation for the reaction between the acidic oxide, phosphorus(V) oxide, and the basic oxide, magnesium oxide.

(1 mark)
AQA, 2014

8 Magnesium oxide, silicon dioxide, and phosphorus(V) oxide are white solids but each oxide has a different type of structure and bonding.

(a) State the type of bonding in magnesium oxide.
Outline a simple experiment to demonstrate that magnesium oxide has this type of bonding.

(3 marks)

(b) By reference to the structure of, and the bonding in, silicon dioxide, suggest why it is insoluble in water.

(3 marks)

(c) State how the melting point of phosphorus(V) oxide compares with that of silicon dioxide. Explain your answer in terms of the structure of, and the bonding in, phosphorus(V) oxide.

(3 marks)

(d) Magnesium oxide is classified as a basic oxide.
Write an equation for a reaction that shows magnesium oxide acting as a base with another reagent.

(2 marks)

(e) Phosphorus(V) oxide is classified as an acidic oxide.
Write an equation for its reaction with sodium hydroxide.

(1 mark)
AQA, 2013

9 White phosphorus, P_4, is a hazardous form of the element. It is stored under water.

(a) Suggest why white phosphorus is stored under water.

(1 mark)

(b) Phosphorus(V) oxide is known as phosphorus pentoxide.
Suggest why it is usually represented by P_4O_{10} rather than by P_2O_5.

(1 mark)

(c) Explain why phosphorus(V) oxide has a higher melting point than sulfur(VI) oxide. *(2 marks)*

(d) Write an equation for the reaction of P_4O_{10} with water to form phosphoric(V) acid. Give the approximate pH of the final solution.

(2 marks)

(e) A waste-water tank was contaminated by P_4O_{10}. The resulting phosphoric(V) acid solution was neutralised using an excess of magnesium oxide. The mixture produced was then disposed of in a lake.

(i) Write an equation for the reaction between phosphoric(V) acid and magnesium oxide. *(1 mark)*

(ii) Explain why an excess of magnesium oxide can be used for this neutralisation. *(1 mark)*

(iii) Explain why the use of an excess of sodium hydroxide to neutralise the phosphoric(V) acid solution might lead to environmental problems in the lake. *(1 mark)*

AQA, 2012

23 The transition metals

23.1 The general properties of transition metals

Learning objectives:

- → Describe the characteristic properties of the elements titanium to copper.
- → Explain these in terms of electronic structure.

Specification reference: 3.2.5

The elements from titanium to copper lie within the d-block elements.

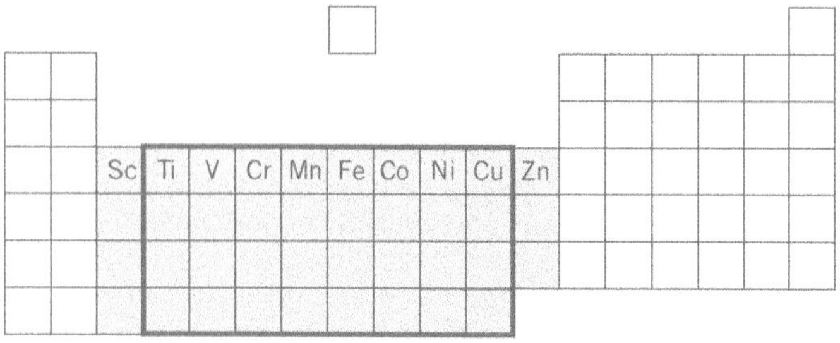

▲ **Figure 1** *The d-block elements (shaded) and the transition metals (outlined)*

Across a period, electrons are being added to a d-sub-shell (3d in the case of titanium to copper). The elements from titanium to copper are metals. They are good conductors of heat and electricity. They are hard, strong, and shiny, and have high melting and boiling points.

These physical properties, together with fairly low chemical reactivity, make these metals extremely useful. Examples include iron (and its alloy steel) for vehicle bodies and to reinforce concrete, copper for water pipes, and titanium for jet engine parts that must withstand high temperatures.

Synoptic link

Look back at Topic 8.1, The Periodic Table, to revise the blocks of the Periodic Table.

Hint

Electron shells are often referred to as 'levels' or 'energy levels'.

Electronic configurations in the d-block elements

Figure 2 shows the electron arrangements for the elements in the first row of the d-block.

In general there are two outer 4s electrons and as you go across the period, electrons are added to the *inner* 3d sub-shell. This explains the overall similarity of these elements. The arrangements of chromium, Cr, and copper, Cu, do not quite fit the pattern. The d-sub-shell is full ($3d^{10}$) in Cu and half full ($3d^5$) in Cr and there is only one electron in the 4s outer shell. It is believed that a half-full d-shell makes the atoms more stable in the same way as a full outer main shell makes the noble gas atoms stable.

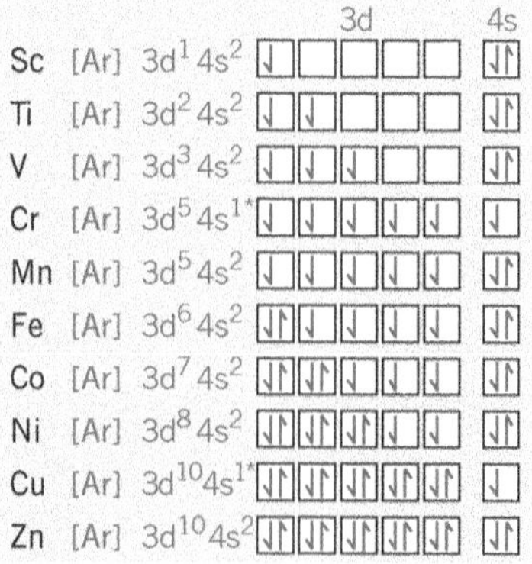

▲ **Figure 2** *Electronic arrangements of the elements in the first d-series. [Ar] represents the electron arrangement of argon – $1s^22s^22p^63s^23p^6$*

Electronic configurations of the ions of d-block elements

To work out the configuration of the ion of an element, first write down the configuration of the element using the Periodic Table, from its atomic number.

Worked example: Electron configuration of V^{2+}

Vanadium, V, has an atomic number of 23. Its electron configuration is:

$$1s^22s^22p^63s^23p^63d^34s^2$$

The vanadium ion V^{2+} has lost the two $4s^2$ electrons and has the electron configuration:

$$1s^22s^22p^63s^23p^63d^3$$

Worked example: Electron configuration of the Cu^{2+} ion

The atomic number of copper is 29. The electron configuration is therefore:

$$1s^2 2s^2 2p^6 3s^2 3p^6 3d^{10} 4s^1$$

The Cu^{2+} ion has lost two electrons, so it has the electron configuration:

$$1s^2 2s^2 2p^6 3s^2 3p^6 3d^9$$

Worked example: Electron configuration of the Cr^{3+} ion

The atomic number of chromium is 24. The electron configuration is therefore:

$$1s^2 2s^2 2p^6 3s^2 3p^6 3d^5 4s^1$$

The Cr^{3+} ion has lost three electrons, so it has the electron configuration:

$$1s^2 2s^2 2p^6 3s^2 3p^6 3d^3$$

In fact, with all transition elements, the 4s electrons are lost first when ions are formed.

The definition of a transition element

The formal definition of a transition element is that it is one that forms at least one stable ion with a *part* full d-shell of electrons. Scandium only forms Sc^{3+} ($3d^0$) in all its compounds, and zinc only forms Zn^{2+} ($3d^{10}$) in all its compounds. They are therefore d-block elements but not transition elements. The transition elements are outlined in red in Figure 1.

Chemical properties of transition metals

The chemistry of transition metals has four main features which are common to all the elements:

- Variable oxidation states: Transition metals have more than one oxidation state in their compounds, for example, Cu(I) and Cu(II). They can therefore take part in many redox reactions.
- Colour: The majority of transition metal ions are coloured, for example, $Cu^{2+}(aq)$ is blue.
- Catalysis: Catalysts affect the rate of reaction without being used up or chemically changed themselves. Many transition metals, and their compounds, show catalytic activity. For example, iron is the catalyst in the Haber process, vanadium(V) oxide in the Contact process, and manganese(IV) oxide in the decomposition of hydrogen peroxide.
- Complex formation: Transition elements form complex ions. A complex ion is formed when a transition metal ion is surrounded by ions or other molecules, collectively called ligands, which are bonded to it by co-ordinate bonds. For example, $[Cu(H_2O)_6]^{2+}$ is a complex ion that is formed when copper sulfate dissolves in water.

Study tip

The electronic configuration of any element can be deduced using the Periodic Table.

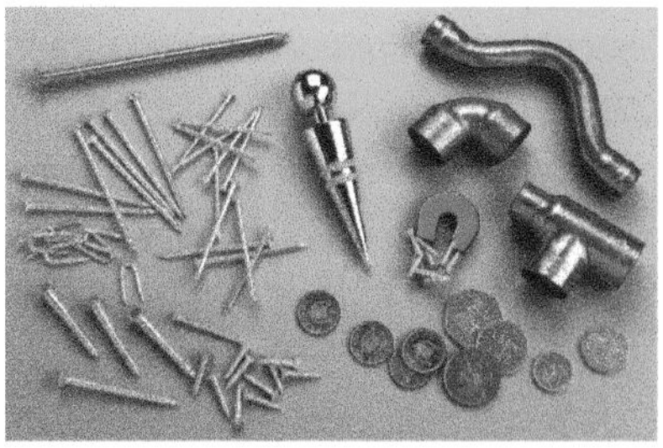

▲ **Figure 3** *Some transition metals in use*

Summary questions

1 The electron arrangement of manganese is:

$1s^2\,2s^2\,2p^6\,3s^2\,3p^6\,3d^5\,4s^2$

Write the electron arrangement of:

a a Mn^{2+} ion

b a Mn^{3+} ion

2 The electron arrangement of iron can be written [Ar] $3d^6\,4s^2$.

a What does the [Ar] represent?

b Which two electrons are lost to form Fe^{2+}?

c Which further electron is lost in forming Fe^{3+}?

23.2 Complex formation and the shape of complex ions

Learning objectives:

- → State what the terms ligand, co-ordinate bond, and co-ordination number mean.
- → Explain what bidentate and multidentate ligands are.
- → Explain how the size of the ligand affects the shape of the complex ion.
- → Explain how ligand charge determines the charge on the complex ion.

Specification reference: 3.2.5

Synoptic link

You will need to understand covalent and co-ordinate bonding, shapes of simple molecules and ions studied in Topic 3.5, The shapes of molecules and ions.

The formation of complex ions

All transition metal ions can form co-ordinate bonds by accepting electron pairs from other ions or molecules. The bonds that are formed are **co-ordinate** (dative) bonds. An ion or molecule with a lone pair of electrons that forms a co-ordinate bond with a transition metal is called a **ligand**. Some examples of ligands are H_2O**:**, **:**NH_3, **:**Cl^-, **:**CN^-.

In some cases, two, four, or six ligands bond to a single transition metal ion. The resulting species is called a complex ion. The number of co-ordinate bonds to ligands that surround the d-block metal ion is called the **co-ordination number**.

- Ions with co-ordination number six are usually octahedral, for example, $[Co(NH_3)_6]^{3+}$.
- Ions with co-ordination number four are usually tetrahedral, for example, $[CoCl_4]^{2-}$.
- Some ions with co-ordination number four are square planar, for example, $[NiCN_4]^{2-}$.

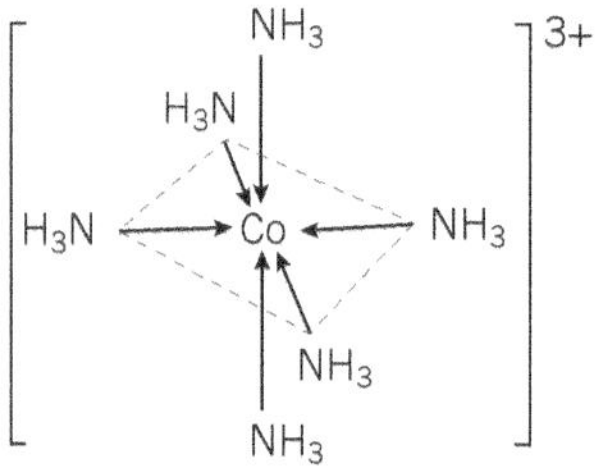

$[Co(NH_3)_6]^{3+}$

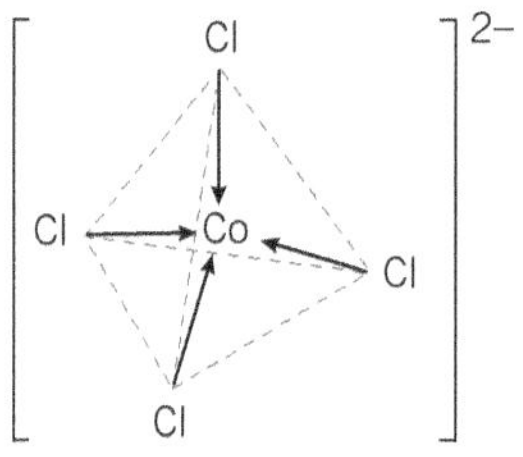

$[CoCl_4]^{2-}$

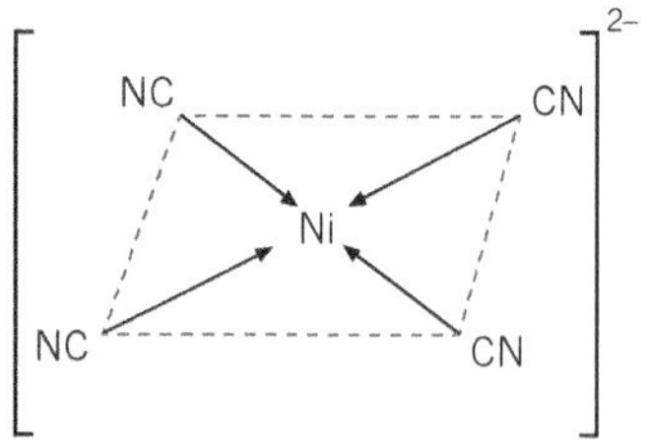

$[Ni(CN)_4]^{2-}$

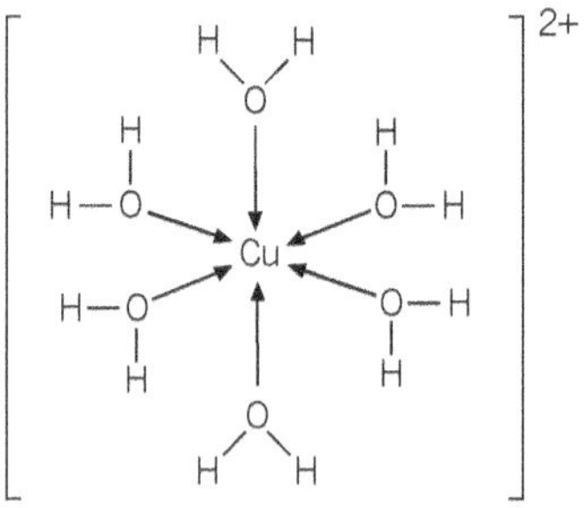

▲ **Figure 1** *Copper(II) ion surrounded by water molecules*

Hint

Some ligands are neutral and others have a negative charge.

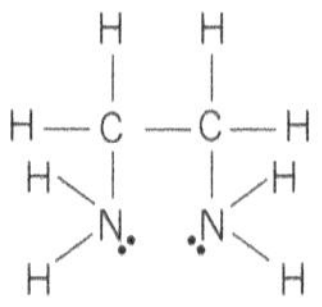

▲ **Figure 2** *Ethane-1,2-diamine*

Aqua ions

If you dissolve the salt of a transition metal in water, for example, copper sulfate, the positively charged metal ion becomes surrounded by water molecules acting as ligands (Figure 1). Normally there are six water molecules in an octahedral arrangement. Such species are called **aqua ions**.

Multidentate ligands – chelation

Some molecules or ions, called **multidentate ligands**, have more than one atom with a lone pair of electrons which can bond to a transition metal ion.

Bidentate ligands include:

- Ethane-1,2-diamine, sometimes called 1,2-diaminoethane or ethylene diamine (Figure 2). Each nitrogen has a lone pair which can form a co-ordinate bond to the metal ion. The name of this ligand is often abbreviated to en, for example, $[Cr(en)_3]^{3+}$. It is a neutral ligand and the chromium ion has a 3+ charge, so the complex ion also has a 3+ charge.
- The ethanedioate (oxalate) ion, $C_2O_4^{2-}$ (Figure 3).
- Benzene-1,2-diol, sometimes called 1,2-dihydroxybenzene, is also a neutral ligand (Figure 4).

An important multidentate ligand is the ion ethylenediaminetetraacetate, called $EDTA^{4-}$ (Figure 5).

This can act as a hexadentate ligand using lone pairs on four oxygen and both nitrogen atoms. Complex ions with polydentate ligands are called **chelates**. Chelates can be used to effectively remove d-block metal ions from solution.

The chelate effect

If you add a hexadentate ligand such as EDTA to a solution of a transition metal salt, the EDTA will replace all six water ligands in the aqua ion $[Cu(H_2O)_6]^{2+}$ as shown:

$$[Cu(H_2O)_6]^{2+}(aq) + EDTA^{4-}(aq) \rightarrow [CuEDTA]^{2-}(aq) + 6H_2O(l)$$

In this equation, two species are replaced by seven. This increase in the number of particles causes a significant increase in entropy which drives the reaction to the right. For this reason chelate complexes with polydentate ligands are favoured over complexes with monodentate ligands. This is called the **chelate effect**.

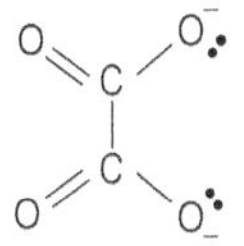

▲ **Figure 3** *Ethanedioate*

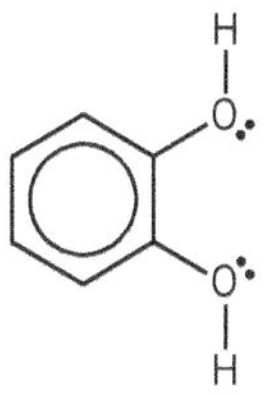

▲ **Figure 4** *Benzene-1,2-diol*

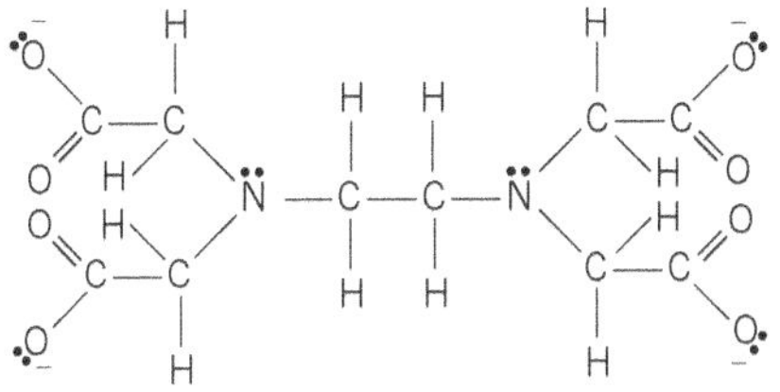

▲ **Figure 5** *EDTA*

Study tip

Both ligand sites of bidentate ligands usually bond to the same metal forming a ring. However, they can act as bridges between two metal ions.

Synoptic link

Look back at Topic 17.4, Why do chemical reactions take place?, to revise entropy.

Haemoglobin

Haemoglobin is the red pigment in blood. It is responsible for carrying oxygen from the lungs to the cells of the body. The molecule consists of an Fe^{2+} ion with a co-ordination number of six. Four of the co-ordination sites are taken up by a ring system called a porphyrin which acts as a tetradentate ligand. This complex is called 'haem'.

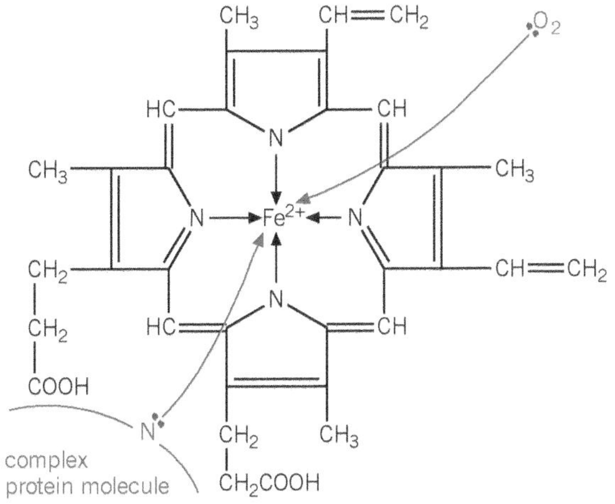

▲ **Figure 6** *Haemoglobin*

Below the plane of this ring is a fifth nitrogen atom acting as a ligand. This atom is part of a complex protein called globin. The sixth site can accept an oxygen molecule as a ligand. The Fe^{2+} to O_2 bond is weak, as $:O_2$ is not a very good ligand, allowing the oxygen molecule to be easily given up to cells.

Better ligands than oxygen can bond irreversibly to the iron and so destroy haemoglobin's oxygen-carrying capacity. This explains the poisonous effect of carbon monoxide, which is a better ligand than oxygen. Carbon monoxide is often formed by incomplete combustion in faulty gas heaters. Because it binds more strongly to the iron than oxygen, it is possible to suffocate in a room with plentiful oxygen.

Anaemia is a condition which may be caused by a shortage of haemoglobin. The body suffers from a lack of oxygen and the symptoms include fatigue, breathlessness, and a pale skin colour. The causes may be loss of blood or deficiency of iron in the diet. The latter may be treated by taking 'iron' tablets which contain iron(II) sulfate.

Study tip

The chlorine atom is in Period 3 and so has an extra shell compared with the nitrogen atom in Period 2.

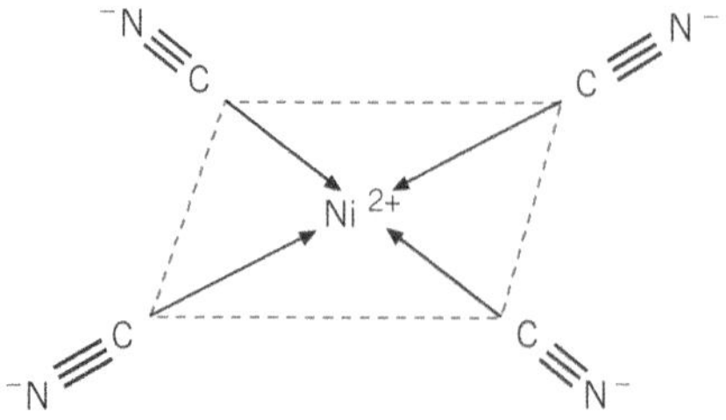

▲ **Figure 7** *A square planar complex*

Synoptic link

You will learn more about using Tollens' reagent to test for aldehydes and ketones in Topic 26.2, Reactions of the carbonyl group in aldehydes and ketones.

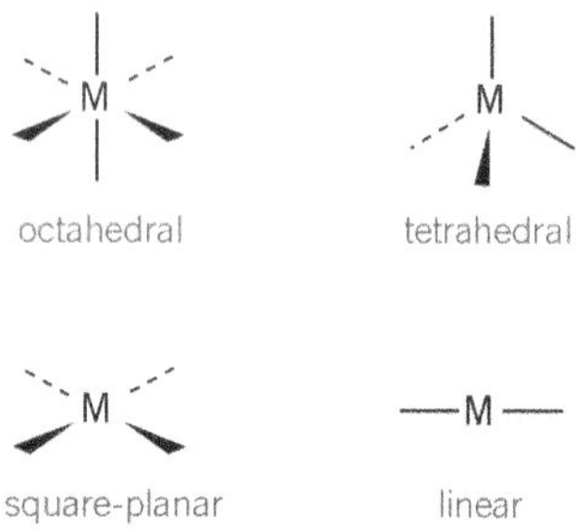

▲ **Figure 8** *The four main shapes of transition metal complexes using wedge and dotted bonds*

Hint

All six ligand positions in an octahedral complex are equivalent as are all the positions in the other three geometries.

Synoptic link

You will learn more about cisplatin in Topic 30.3, The action of anti-cancer drugs.

Shapes of complex ions

As you have seen the $[Co(NH_3)_6]^{3+}$ ion, with six ligands, is an octahedral shape. An octahedron has six points but *eight* faces. The metal ion, Co^{3+}, has a charge of +3 and as the ligands are all neutral, the complex ion has an overall charge of +3.

The $[CoCl_4]^{2-}$ ion, with four ligands, is tetrahedral. The metal ion, Co^{2+}, has a charge of +2 and each of the four ligands $:Cl^-$, has a charge of −1, so the complex ion has an overall charge of −2.

The reason for this difference in shape is that the chloride ion is a larger ligand than the ammonia molecule and so fewer ligands can fit around the central metal ion.

A few complexes of co-ordination number four adopt a square planar geometry (Figure 7).

Some complexes are linear, one example being $[Ag(NH_3)_2]^+$:

$$[H_3N \rightarrow Ag \leftarrow NH_3]^+$$

A solution containing this complex ion is called Tollens' reagent and is used in organic chemistry to distinguish aldehydes from ketones. Aldehydes reduce the $[Ag(NH_3)_2]^+$ to Ag (metallic silver), while ketones do not. The silver forms a mirror on the surface of the test tube, giving the name of the test – the silver mirror test.

Complex ions may have a positive charge or a negative charge.

Representing the shapes of complex ions

Representing three-dimensional shapes on paper can be tricky. Some diagrams in this topic have thin red construction lines to help you to visualise the shapes. These are not bonds. Another way to represent shape is to use wedge bonds and dotted bonds. Wedge bonds come out of the paper and dotted bonds go in (Figure 8).

Isomerism in transition metal complexes

You have met isomerism in organic chemistry. Isomers are compounds with the same molecular formula but with different arrangements of their atoms in space. Transition metal complexes can form both geometrical isomers (*cis-trans*, or *E-Z* isomers) and optical isomers.

Geometrical isomerism

Here ligands differ in their position in space relative to one another. This type of isomerism occurs in octahedral and square planar complexes. Take the octahedral complex ion $[CrCl_2(H_2O)_4]^+$. The Cl^- ligands may be next to each other (the *cis*- or *Z*- form) or on opposite sides of the central chromium ion (the *trans*- or *E*-form) (Figure 9).

In the square planar complex platin, the Cl^- ligands may be next to each other (the *cis*- or *Z*- form) or on opposite sides of the central chromium ion (the *trans*- or *E*-form). A pair of geometrical isomers will have different chemical properties. For example, cisplatin is one of the most successful anti-cancer drugs whilst the *trans*-isomer has no therapeutic effect,

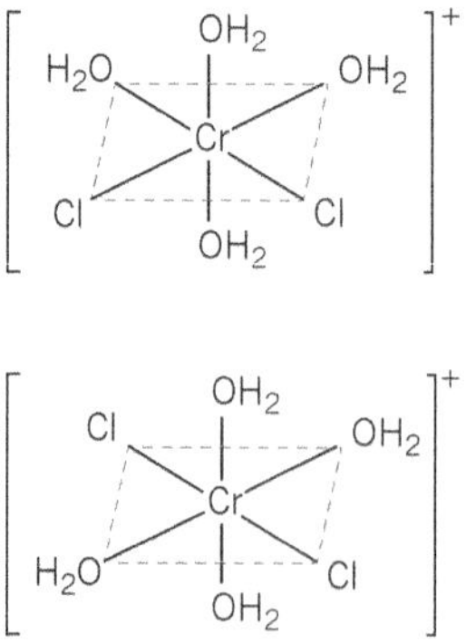

▲ **Figure 9** *The* cis- *or* Z-*isomer (top) and the* trans- *or* E-*isomer (bottom)*

Optical isomerism

Here the two isomers are non-superimposable mirror images of each other. In transition metal complexes this occurs when there are two or more bidentate ligands in a complex (Figure 10).

Look at Figure 10 and imagine rotating one of the complexes around a vertical axis until the two chlorine atoms are in the same position as in the other one. You should be able to see that the positions of the en ligands no longer match. The best way to be sure about this is to use molecular models.

Optical isomers are said to be chiral. They have identical chemical properties but can be distinguished by their effect on polarised light. One isomer will rotate the plane of polarisation of polarised light clockwise and the other anticlockwise.

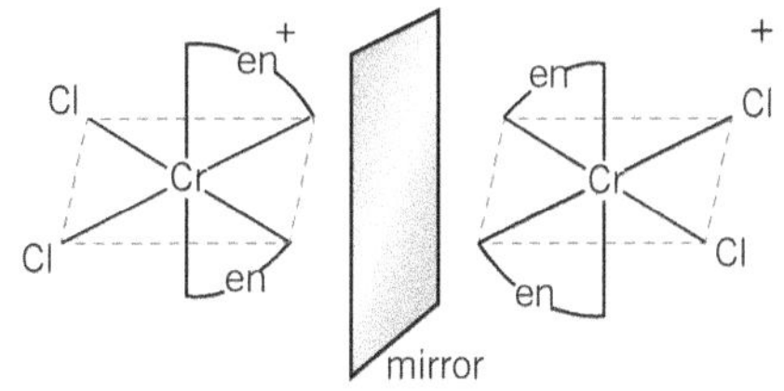

en is an abbreviation for ethane-1,2-diamine:

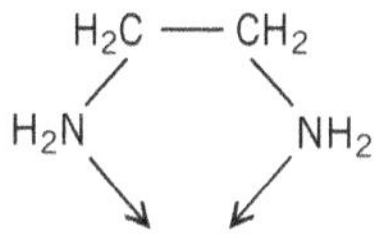

▲ **Figure 10** *Transition metal complexes that are non-identical mirror images of each other (top) and the structure of the ethane-1,2-diamine (en) ligand (bottom)*

Ionisation isomerism

This is a third form of isomerism found in transition metal chemistry. Consider the compound of formula $CrCl_3(H_2O)_6$. Both chloride ions, Cl^-, and water molecules can act as ligands. This compound can exist as three different isomers depending on how many of the chloride ions are bound to the chromium atom as ligands and how many are free as negative ions.

The three isomers are:

Compound 1: $[Cr(H_2O)_6]^{3+} + 3Cl^-$ violet

Compound 2: $[CrCl(H_2O)_5]^{2+} + 2Cl^- + H_2O$ light green

Compound 3: $[CrCl_2(H_2O)_6]^+ + Cl^- + 2H_2O$ dark green

1. What feature of both Cl^- and H_2O allows them to act as ligands?
2. If solutions are made of the same concentration of each compound, which would you expect to conduct electricity best? Explain your answer.
3. Only free chloride ions (and not those bonded as ligands) will react with silver nitrate. Write a balanced equation for the reaction of chloride ions with silver nitrate solution.
4. Which compound would react with most silver nitrate? Explain your answer.
5. State which other type of isomerism is shown by Compound 3.
6. The structural formula of Compound 1 is:

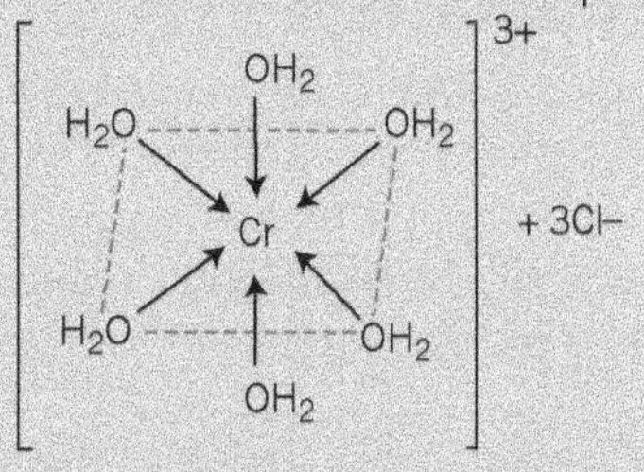

Draw the structural formula of Compound 2 in the same style.

Synoptic link

See Topic 25.2, Optical isomerism, for more detail about optical isomerism in organic chemistry.

Summary questions

1. a Predict the shapes of the following:
   i $[Cu(H_2O)_6]^{2+}$
   ii $[Cu(NH_3)_6]^{2+}$
   iii $[CuCl_4]^{2-}$

   b What is the co-ordination number of the transition metal in each complex in **a**?

   c Explain why the co-ordination numbers are different.
2. Benzene-1,2-dicarboxylate is shown below.

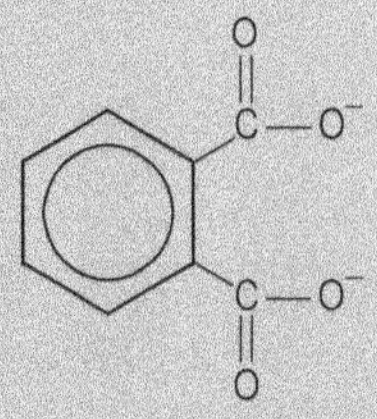

   a Suggest which atoms are likely to form co-ordinate bonds with a metal ion.

   b Mark the lone pairs.

   c Is it likely to be a mono-, bi-, or hexadentate ligand?

23.3 Coloured ions

Learning objectives:

→ Describe the origin of the colour of a transition metal complex ion.

→ Explain what factors determine the colour of a complex ion.

Specification reference: 3.2.5

Study tip

Make sure that you know the factors which give rise to colour changes and are able to illustrate each change by an equation.

Hint

ΔE is also related to wavelength λ by the equation $\Delta E = \frac{hc}{\lambda}$, where c is the velocity of light.

▼ **Table 1** *Colours of four vanadium species*

Oxidation number	Species	Colour
5	$VO_2^+(aq)$	yellow
4	$VO^{2+}(aq)$	blue
3	$V^{3+}(aq)$	green
2	$V^{2+}(aq)$	violet

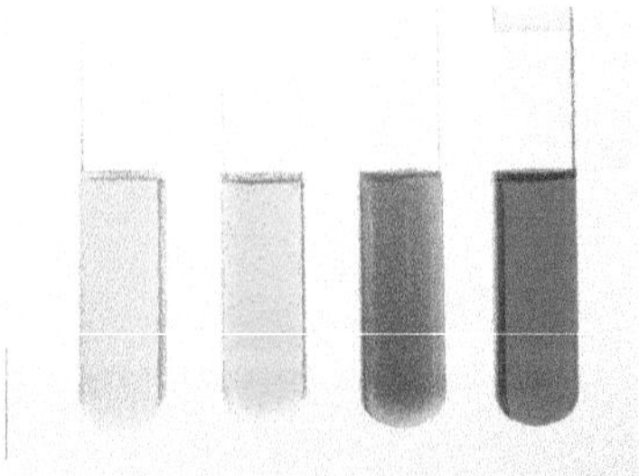

▲ **Figure 2** *Zinc ions in acid solution will reduce vanadium(V) through oxidation states V(IV) and V(III) to V(II). The final flask is normally stoppered because oxygen in the air will rapidly oxidise V(II)*

Most transition metal compounds are coloured. The colour is caused by the compounds absorbing energy that corresponds to light in the visible region of the spectrum. If a solution of a substance looks purple, it is because it absorbs all the light from a beam of white light shone at it except red and blue. The red and blue light passes through and the solution and appears purple (Figure 1).

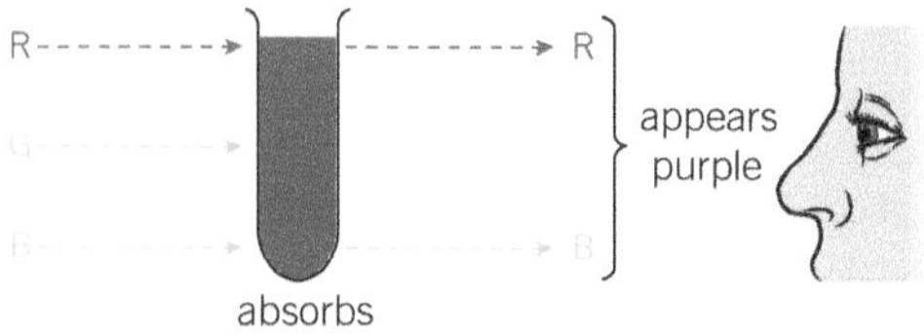

▲ **Figure 1** *Solutions look coloured because they absorb some colours and let others pass through*

Why are transition metal complexes coloured?

This is a simplified explanation, but the general principle is as follows:

- Transition metal compounds are coloured because they have part-filled d-orbitals.
- It is therefore possible for electrons to move from one d-orbital to another.
- In an isolated transition metal atom, all the d-orbitals are of exactly the same energy, but in a compound, the presence of other atoms nearby makes the d-orbitals have slightly different energies.
- When electrons move from one d-orbital to another of a higher energy level (called an excited state), they often absorb energy in the visible region of the spectrum equal to the difference in energy between levels.
- This colour is therefore missing from the spectrum and you see the combination of the colours that are not absorbed.

The frequency of the light is related to the energy difference by the expression $\Delta E = h\nu$, where E is the energy, ν the frequency, and h a constant called Planck's constant. The frequency is related to the colour of light. Violet is of high energy and therefore high frequency and red is of low energy and low frequency.

The colour of a transition metal complex depends on the energy gap ΔE, which in turn depends on the oxidation state of the metal and also on the ligands (and therefore the shape of the complex ion), so different compounds of the same metal will have different colours. For example, Table 1 shows the colours of four vanadium species each with a different oxidation state.

Some more examples of how changing the oxidation state of the metal affects the colour of the complex are given in Table 2.

▼ **Table 2**

Oxidation state of metal	2	3
iron complexes	$[Fe(H_2O)_6]^{2+}$ green	$[Fe(H_2O)_6]^{3+}$ pale brown
chromium complexes	$[Cr(H_2O)_6]^{2+}$ blue	$[Cr(H_2O)_6]^{3+}$ red-violet
cobalt complexes	$[Co(NH_3)_6]^{2+}$ brown	$[Co(NH_3)_6]^{3+}$ yellow

Colorimetry

A simple colorimeter uses a light source and a detector to measure the amount of light of a particular wavelength that passes through a coloured solution. This is also called **spectroscopy**. The more concentrated the solution, the less light transmitted through the solution. A colorimeter is used, with a suitable calibration graph to measure the concentration of solutions of coloured transition metal compounds.

Hint

Usually the experiment is made more sensitive by using a coloured filter in the colorimeter. The filter is chosen by finding out the colour of light that the red solution absorbs most. Red absorbs light in the blue region of the visible spectrum, so a blue filter is used (Figure 3), so that only blue light passes into the sample tube.

Finding the formula of a transition metal complex using colorimetry

A colorimeter can be used to find the ratio of metal ions to ligands in a complex, which gives us the formula of the complex. Two solutions are mixed together, one containing the metal ion and one the ligand, in different proportions. When they are mixed in the same ratio as they are in the complex, there is the maximum concentration of complex in the solution.
So, the solution will absorb most light.

For example, take the blood red complex formed with Fe^{3+} ions and thiocyanate ions, SCN^-.

If potassium thiocyanate is added to a solution of $Fe^{3+}(aq)$, *one* of the water molecules is replaced by a thiocyanate ion and a blood red complex forms:

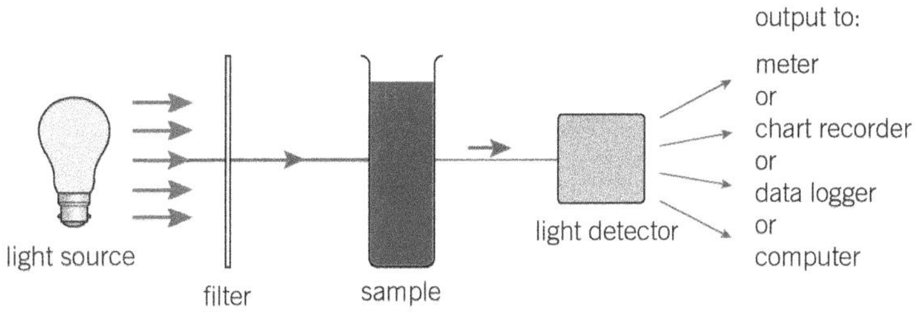

▲ **Figure 3** *Using a colorimeter to find a formula*

$$[Fe(H_2O)_6]^{3+}(aq) + SCN^- \rightarrow [Fe(SCN)(H_2O)_5]^{2+}(aq) + H_2O(l)$$

As the concentration of the red complex increases, less and less light will pass through the solution.

Start with two solutions of the same concentration, one containing $Fe^{3+}(aq)$ ions, for example, iron(III) sulfate, and one containing $SCN^-(aq)$ ions, for example, potassium thiocyanate. Mix them in the proportions shown in Table 3, adding water so that all the tubes have the same total volume of solution.

▼ **Table 3** *The absorbance of different mixtures of $Fe^{3+}(aq)$ and $SCN^-(aq)$*

Tube	1	2	3	4	5	6	7	8
Vol. of $Fe^{3+}(aq)$ solution / cm^3	10.00	10.00	10.00	10.00	10.00	10.00	10.00	10.00
Vol. of $SCN^-(aq)$ solution / cm^3	2.00	4.00	6.00	8.00	10.00	12.00	14.00	16.00
Vol. of water / cm^3	28.00	26.00	24.00	22.00	20.00	28.00	16.00	14.00
Absorbance	0.15	0.33	0.48	0.63	0.70	0.70	0.70	0.70

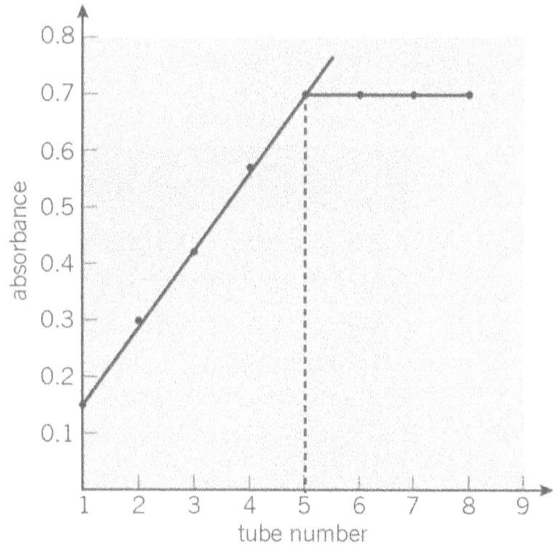

▲ **Figure 4** *A graph of absorbance against tube number*

Each tube is put in the colorimeter and a reading of absorbance taken. Absorbance is a measure of the light absorbed by the solution. A graph of absorbance is plotted against tube number (Figure 4).

From the graph, the maximum absorbance occurs in tube 5 – after this, adding more thiocyanate ions makes no difference. So this shows the ratio of SCN^- ions to Fe^{3+} ions in the complex. From Table 3, tube 5 has equal amounts of SCN^- ions and Fe^{3+} ions so their ratio in the complex must be 1 : 1. So this confirms the formula is $[Fe(SCN)(H_2O)_5]^{2+}$. (The SCN^- has substituted for one of the water molecules in the complex ion $[Fe(H_2O)_6]^{3+}$.)

The colour of gemstones

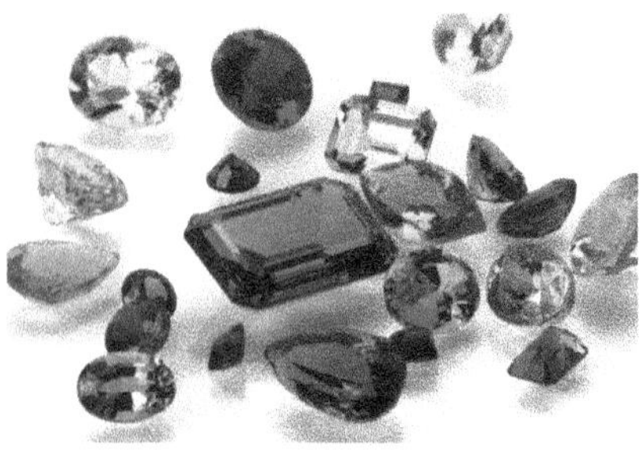

▲ **Figure 5** *The colours of many gemstones are caused by transition metal compounds*

Transition metal ions are responsible for the colours of most gemstones. Rubies are made of aluminium oxide, Al_2O_3, which is colourless – the red colour is caused by trace amounts of Cr^{3+} ions which replace some of the Al^{3+} ions in the crystal lattice. The oxide ions, O^{2-}, are the ligands surrounding the Cr^{3+} ions.

The green colour of emeralds is also caused by Cr^{3+} ions, but in this case, the material from which the gemstone is made is beryllium aluminium silicate, $Be_3Al_2(SiO_3)_6$. The ligand surrounding the Cr^{3+} is silicate, SiO_3^{2-}, in this case. This illustrates the effect of changing the ligand on the colour of the ion.

The red of garnet and the yellow-green colour of peridot are both caused by Fe^{2+} ions – surrounded by eight silicate ligands in the case of garnet and six in peridot.

Other examples include Cu^{2+} which is responsible for the blue-green of turquoise and Mn^{2+} for the pink of tourmaline.

Summary questions

1 a Explain why copper sulfate is coloured, whereas zinc sulfate is colourless.

b A solution of copper sulfate is blue. What colour light passes through this solution?

c What happens to the other colours of the visible spectrum?

2 The graph shown the absorbance of a series of mixtures containing different proportions of two solutions of the same concentration – one containing Ni^{2+} ions and the other containing a ligand called $EDTA^{4-}$ for short. The two solutions react together to form a coloured complex.

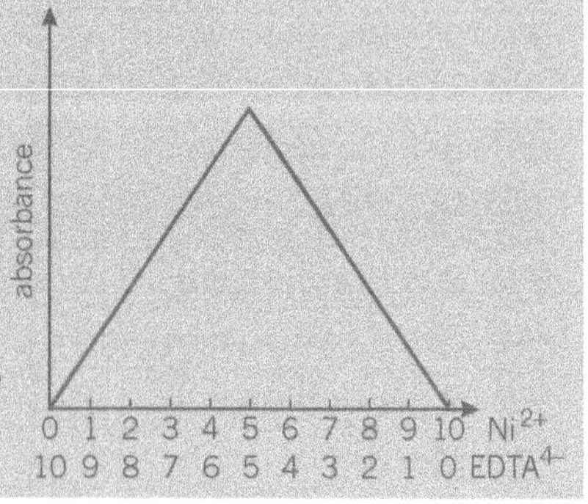

a Which mixture absorbs most light?

b Which mixture contains the highest concentration of the nickel EDTA complex?

c What is the simplest (empirical) formula of the complex?

23.4 Variable oxidation states of transition elements

Learning objective:

→ Describe how the concentration of iron(II) ions in aqueous solution can be found.

Specification reference: 3.2.5

Group 1 metals lose their outer electron to form only +1 ions and Group 2 lose their outer two electrons to form only +2 ions in their compounds. A typical transition metal can use its 3d-electrons as well as its 4s-electrons in bonding, and this means that it can have a greater variety of oxidation states in different compounds. Table 1 shows this for the first d-series. Zinc and scandium are shown as part of the d-series although they are not transition metals.

▼ **Table 1** *Oxidation numbers shown by the elements of the first d-series in their compounds*

Sc	Ti	V	Cr	Mn	Fe	Co	Ni	Cu	Zn
	+I	+I	+I	+I	+I	+I	+I	+I	
	+II	+II	+II	+II	+II	+II	+II	+II	+II
+III	+III	+III	+III	+III	+III	+III	+III	+III	
	+IV	+IV	+IV	+IV	+IV	+IV	+IV		
		+V	+V	+V	+V	+V			
			+VI	+VI	+VI				
				+VII					

The most common oxidation states are shown in red, though they are not all stable.

Except for scandium and zinc all the elements show both the +1 and +2 oxidation states. These are formed by the loss of 4s electrons.

For example, nickel has the electron configuration $1s^2 2s^2 2p^6 3s^2 3p^6 3d^8 4s^2$ and Ni^{2+} is $1s^2 2s^2 2p^6 3s^2 3p^6 3d^8$.

Iron has the electron configuration $1s^2 2s^2 2p^6 3s^2 3p^6 3d^6 4s^2$ and Fe^{2+} is $1s^2 2s^2 2p^6 3s^2 3p^6 3d^6$.

Only the lower oxidation states of transition metals actually exist as simple ions, so that, for example, Mn^{2+} ions exist but Mn^{7+} ions do not. In all Mn(VII) compounds, the manganese is covalently bonded to oxygen in a compound ion as in MnO_4^- (Figure 1).

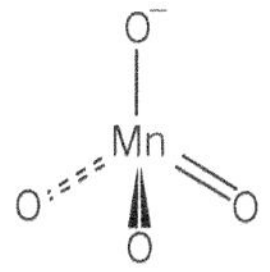

▲ **Figure 1** *Bonding in the $[MnO_4]^-$ ion*

Synoptic link

You will need to understand redox equations and volumetric analysis studied in Topic 7.3, Redox equations, and Topic 2.4, Empirical and molecular formulae.

Redox reactions in transition metal chemistry

Many of the reactions of transition metal compounds are redox reactions, in which the metals are either oxidised or reduced. For example, iron shows two stable oxidation states – Fe^{3+} and Fe^{2+}.

Fe^{2+} is the less stable state – it can be oxidised to Fe^{3+} by the oxygen in the air and also by chlorine. For example:

$$\overset{+2}{2Fe^{2+}(aq)} + \overset{0}{Cl_2(g)} \rightarrow \overset{+3}{2Fe^{3+}(aq)} + \overset{-1}{2Cl^-(aq)}$$

In this reaction, chlorine is the oxidising agent – its oxidation number drops from 0 to −1 (as it gains an electron), whilst that of the iron increases from +2 to +3 (as it loses an electron). Remember the phrase OIL RIG – oxidation is loss, reduction is gain (of electrons).

Using half equations

Potassium manganate(VII) reactions

The technique of using half equations is useful for constructing balanced equations in more complex reactions. Potassium manganate(VII) can act as an oxidising agent in acidic solution (one containing $H^+(aq)$ ions) and will, for example, oxidise Fe^{2+} to Fe^{3+}. During the reaction the oxidation number of the manganese falls from +7 to +2.

First construct the half equation for the reduction of Mn(VII) to Mn(II):

$$MnO_4^-(aq) \rightarrow Mn^{2+}(aq)$$

The oxygen atoms must be balanced using H^+ ions and H_2O molecules:

$$MnO_4^- + 8H^+(aq) \rightarrow Mn^{2+}(aq) + 4H_2O(l)$$

Then balance for charge using electrons:

$$MnO_4^-(aq) + 5e^- + 8H^+(aq) \rightarrow Mn^{2+}(aq) + 4H_2O(l)$$

The half equation for the oxidation of iron(II) to iron(III) is straightforward:

$$Fe^{2+}(aq) \rightarrow Fe^{3+}(aq) + e^-$$

To construct a balanced symbol equation for the reaction of acidified potassium manganate(VII) with $Fe^{2+}(aq)$, first multiply the Fe^{2+} / Fe^{3+} half reaction by five (so that the numbers of electrons in each half reaction are the same) and then add the two half equations:

$$5Fe^{2+}(aq) \rightarrow 5Fe^{3+}(aq) + 5e^-$$

$$MnO_4^-(aq) + 5e^- + 8H^+(aq) \rightarrow Mn^{2+}(aq) + 4H_2O(l)$$

$$5Fe^{2+}(aq) + MnO_4^-(aq) + \cancel{5e^-} + 8H^+(aq) \rightarrow 5Fe^{3+}(aq) + \cancel{5e^-} + Mn^{2+}(aq) + 4H_2O(l)$$

$$5Fe^{2+}(aq) + MnO_4^-(aq) + 8H^-(aq) \rightarrow 5Fe^{3+}(aq) + Mn^{2+}(aq) + 4H_2O(l)$$

This technique makes balancing complex redox reactions much easier.

> **Hint**
>
> H^+ ions are likely to be involved because the reaction takes place in acidic solution. Five electrons are involved because the oxidation state of each manganese atom drops by five.

Redox titrations

You may wish to measure the concentration of an oxidising or a reducing agent. One way of doing this is to do a redox titration. This is similar in principle to an acid–base titration in which you find out how much acid is required to react with a certain volume of base (or vice versa).

One example is in the analysis of iron tablets for quality control purposes. Iron tablets contain iron(II) sulfate and may be taken by patients whose diet is short of iron for some reason.

As you have seen, $Fe^{2+}(aq)$ reacts with manganate(VII) ions (in potassium manganate(VII)) in the ratio 5 : 1. The reaction does not need an indicator, because the colour of the mixture changes as the reaction proceeds (Table 2).

Using a burette, you gradually add potassium manganate(VII) solution (which contains the $MnO_4^-(aq)$ ions) to a solution containing $Fe^{2+}(aq)$ ions, acidified with excess dilute sulfuric acid. The purple colour disappears as the MnO_4^- ions are converted to pale pink $Mn^{2+}(aq)$ ions to leave a virtually colourless solution. Once just enough $MnO_4^-(aq)$ ions have been added to react with all the $Fe^{2+}(aq)$ ions, one more drop of $MnO_4^-(aq)$ ions will turn the solution purple. This is the end point of the titration.

> **Hint**
>
> The body needs iron compounds to make haemoglobin, the compound that carries oxygen in the blood.

▼ **Table 2** *The colours of the ions in the reaction between potassium manganate(VII) and iron(II) sulfate*

Ion	Colour
$Fe^{2+}(aq)$	pale green
$MnO_4^-(aq)$	intense purple
$Fe^{3+}(aq)$	pale violet
$Mn^{2+}(aq)$	pale pink

The apparatus used is shown in Figure 2.

You cannot use hydrochloric acid, as an alternative to sulfuric acid, to supply the $H^+(aq)$ ions in the reaction between potassium manganate(VII) and $Fe^{2+}(aq)$. You can see why this is the case by using $E^\ominus$ values.

Hydrochloric acid contains Cl^- ions. These are oxidised by MnO_4^- ions, as shown by the calculation of emf for the reaction below. This would affect the titration, because the manganate(VII) ions must be used only to oxidise Fe^{2+} ions. Manganate(VII) ions do not oxidise sulfate ions.

The relevant half equations with their values of $E^\ominus$ are:

$$MnO_4^-(aq) + 5e^- + 8H^+(aq) \rightleftharpoons Mn^{2+}(aq) + 4H_2O(l) \quad E^\ominus = +1.51\ V$$

$$\frac{1}{2}Cl_2 + e \rightleftharpoons Cl^- \quad E^\ominus = +1.36\ V$$

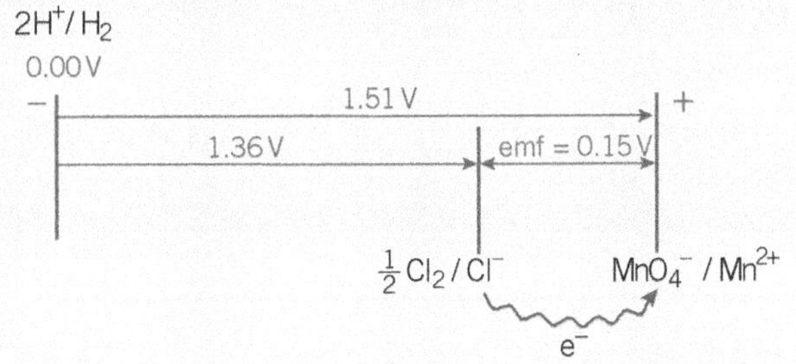

▲ **Figure 3** $E^\ominus$ *values show that acidified* MnO_4^- *ions will oxidise* Cl^- *ions*

Figure 3 shows that the electron flow is from Cl_2 to MnO_4^-, and the emf is 0.15 V, so the half reactions must be as follows:

$$MnO_4^-(aq) + 5e^- + 8H^+(aq) \rightarrow Mn^{2+}(aq) + 4H_2O(l)$$

$$Cl^- \rightarrow \frac{1}{2}Cl_2 + e^-$$

Multiplying the lower half equation by 5, to balance the electrons, and adding the equations together gives:

$$MnO_4^-(aq) + \cancel{5e^-} + 8H^+(aq) + 5Cl^-(aq) \rightleftharpoons Mn^{2+}(aq) + 4H_2O(l)$$

Remember to cancel electrons $\quad + 2\frac{1}{2}Cl_2(aq) + \cancel{5e^-}$

emf = +0.15 V

This reaction is feasible, so MnO_4^- ions will oxidise Cl^- ions and so hydrochloric acid is not suitable for this titration.

Worked example: Iron tablets

A brand of iron tablets has this stated on the pack. 'Each tablet contains 0.200 g of iron(II) sulfate.' The following experiment was done to check this.

One tablet was dissolved in excess sulfuric acid and made up to 250 cm^3 in a volumetric flask. 25.00 cm^3 of this solution was pipetted into a flask and titrated with 0.001 00 mol dm^{-3} potassium manganate(VII) solution until the solution just became purple. Taking an average of several titrations, 26.30 cm^3 of potassium manganate(VII) solution was needed.

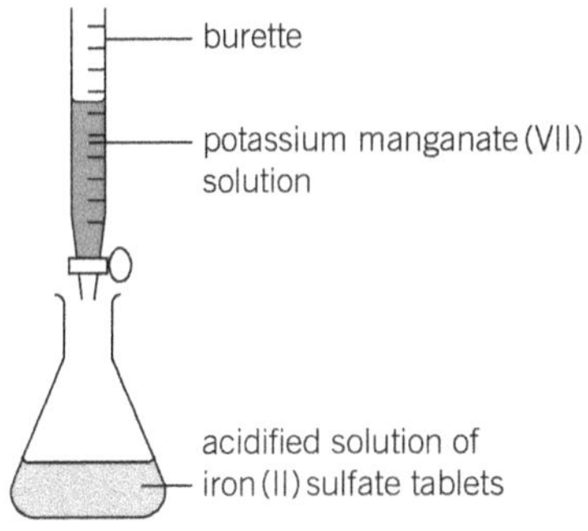

▲ **Figure 2** *Apparatus for a titration*

Synoptic link

Look back at Topic 20.2, Predicting the direction of redox reactions, to revise $E^\ominus$ values.

Hint

The value of $E^\ominus$ for the MnO_4^-/Mn^{2+} half cell will vary with pH.

Hint

The redox potential for reactions involving transition metals ions will also vary according to the ligands in the complex ion.

Study tip

Try to show all your working in calculations to demonstrate your understanding when answers are incomplete or the final answer is wrong.

Number of moles potassium manganate(VII) solution $= c \times \frac{V}{1000}$

where c is the concentration of the solution in mol dm^{-3} and V is the volume of solution used in cm^3.

No. of moles potassium manganate(VII) solution $= 0.001\,00 \times \frac{26.30}{1000}$

$= 2.63 \times 10^{-5}$ mol

$5Fe^{2+}(aq) + MnO_4^-(aq) + 8H^+(aq) \rightarrow 5Fe^{3+}(aq) + Mn^{2+}(aq) + 4H_2O(l)$

From the equation, 5 mol of Fe^{2+} reacts with 1 mol of MnO_4^-:

Number of moles of $Fe^{2+} = 5 \times 2.63 \times 10^{-5}$ mol $= 1.315 \times 10^{-4}$ mol

25.00 cm^3 of solution contained $\frac{1}{10}$ tablet.

So one tablet contains $1.315 \times 10^{-4} \times 10 = 1.315 \times 10^{-3}$ mol Fe^{2+}

Since 1 mol iron(II) sulfate contains 1 mol Fe^{2+}, each tablet contains 1.315×10^{-3} mol $FeSO_4$.

The relative formula mass of $FeSO_4$ is 151.9.

So, each tablet contains $1.315 \times 10^{-3} \times 151.9 = 0.200$ g of iron(II) sulfate as stated on the bottle.

Potassium dichromate(VI) titrations

Acidified potassium dichromate(VI) can also be used in a titration to measure the concentration of Fe^{2+} ions. Here the half equations are:

$$Cr_2O_7^{2-}(aq) + 14H^+(aq) + 6e^- \rightarrow 2Cr^{3+}(aq) + 7H_2O(l)$$

$$Fe^{2+}(aq) \rightarrow Fe^{3+}(aq) + e^-$$

So the second half equation must be multiplied by six before adding and cancelling the electrons.

$$6Fe^{2+}(aq) \rightarrow 6Fe^{3+}(aq) + 6e^-$$

$$Cr_2O_7^{2-}(aq) + 14H^+(aq) + 6e^- \rightarrow 2Cr^{3+}(aq) + 7H_2O(l)$$

$$6Fe^{2+}(aq) + Cr_2O_7^{2-}(aq) + 14H^+(aq) + \cancel{6e^-} \rightarrow 6Fe^{3+}(aq) + 2Cr^{3+}(aq) + 7H_2O(l) + \cancel{6e^-}$$

$$6Fe^{2+}(aq) + Cr_2O_7^{2-}(aq) + 14H^+(aq) \rightarrow 6Fe^{3+}(aq) + 2Cr^{3+}(aq) + 7H_2O(l)$$

Note that although chromium is reduced from +6 to +3, the ion $Cr_2O_7^{2-}$ contains two chromium atoms so six electrons are needed.

As before, the $Fe^{2+}(aq)$ solution is placed in the flask with the dichromate in the burette with excess dilute sulfuric acid to provide the H^+ ions.

As it is not possible to see the colour change when a small volume of orange solution is added to a pale green solution, an indicator must be used – sodium diphenylaminesulfonate, which turns from colourless to purple at the end point.

Oxidation of transition metal ions in alkaline solutions

In both the above examples, a high oxidation states of a metal (Mn(VII) are Cr(VI)) are reduced in acidic solution. Oxidation of lower oxidation states of transition metal ions tends to happen in alkaline solution. This is because in alkaline solution there is a tendency to form negative ions. Since oxidation is electron loss, this is easier from negatively charged species than positively charged or neutral ones.

Typical transition metal species, where M represents a transition metal:

- Acid solution: $M(H_2O)_6{}^{2+}$ positively charged.
- Neutral solution: $M(H_2O)_4(OH)_2$ neutral.
- Alkaline solution: $M(H_2O)_2(OH)_4{}^{2-}$ negatively charged.

Low oxidation states of transition metals, such as Fe^{2+} are often stabilised against oxidation by air by keeping them in acid solution.

To oxidise a transition metal to a high oxidation state, an alkali is often added, followed by an oxidising agent.

Example – some cobalt chemistry

Many M^{2+} ions will be oxidised to M^{3+} in alkaline solution, for example cobalt(II) to cobalt(III):

$$\overset{+2}{2[Co(OH)_6]^{4-}(aq)} + H_2O_2(aq) \rightarrow \overset{+3}{2[Co(OH)_6]^{3-}(aq)} + 2OH^-(aq)$$

In ammoniacal solution, Co^{2+} ions can be oxidised by oxygen in the air.

If you add an excess of ammonia solution to an aqueous solution containing cobalt(II) ions, you get a brownish complex ion formed, $[Co(NH_3)_6]^{2+}$, containing cobalt(II) ions.

The reactions are as follows:

1 First a precipitate is formed by reaction with OH^- ions from the ammonia solution, which is alkaline:

$$[Co(H_2O)_6]^{2+} + 2OH^- \rightarrow Co(H_2O)_4(OH)_2(s) + 2H_2O(l)$$

2 Then the precipitate dissolves in excess ammonia:

$$Co(H_2O)_4(OH)_2(s) + 6NH_3(aq) \rightarrow [Co(NH_3)_6]^{2+} + 2OH^-(aq) + 4H_2O(l)$$

3 The resulting complex ion is oxidised by oxygen in air (or rapidly by hydrogen peroxide solution) to the yellow cobalt(III) ion, $[Co(NH_3)_6]^{3+}$.

You can use half equations to produce a balanced equation for the redox reaction.

The half equations are:

$$[Co(NH_3)_6]^{2+}(aq) \rightarrow [Co(NH_3)_6]^{3+}(aq) + e^-$$

$$O_2(g) + 2H_2O(l) + 4e^- \rightarrow 4OH^-(aq)$$

Multiplying the first equation by four and adding these:

$$4[Co(NH_3)_6]^{2+}(aq) \rightarrow 4[Co(NH_3)_6]^{3+}(aq) + 4e^-$$

$$O_2(g) + 2H_2O(l) + 4e^- \rightarrow 4OH^-(aq)$$

$$4[Co(NH_3)_6]^{2+}(aq) + O_2(g) + 2H_2O(l) + \cancel{4e^-} \rightarrow 4[Co(NH_3)_6]^{3+}(aq) + \cancel{4e^-} + 4OH^-(aq)$$

$$4[Co(NH_3)_6]^{2+}(aq) + O_2(g) + 2H_2O(l) \rightarrow 4[Co(NH_3)_6]^{3+}(aq) + 4OH^-(aq)$$

Synoptic link

Topic 20.2, Predicting the direction of redox reactions, contains the $E^\ominus$ values you will need that are not given here.

Summary questions

1 Zinc will reduce VO_2^+ ions to VO^{2+}; VO^{2+} to V^{3+} and V^{3+} ions to V^{2+} ions. The relevant half equations are:

$Zn(s) \rightarrow Zn^{2+}(aq) + 2e^-$

$VO_2^+(aq) + 2H^+(aq) + e^- \rightarrow H_2O(l) + VO^{2+}(aq)$

$VO^{2+}(aq) + 2H^+(aq) + e^- \rightarrow H_2O(l) + V^{3+}(aq)$

$V^{3+}(aq) + e^- \rightarrow V^{2+}(aq)$

a Write the balanced equation for each of the reduction steps.

b V^{2+} has to be protected from air. Suggest a reason for this.

2 A titration to determine the amount of iron(II) sulfate in an iron tablet was carried out. The tablet was dissolved in excess sulfuric acid and made up to 250 cm^3 in a volumetric flask. 25.00 cm^3 of this solution was pipetted into a flask and titrated with 0.0010 mol dm^{-3} potassium manganate(VII) solution until the solution just became purple. Taking an average of several titrations, 25.00 cm^3 of potassium manganate(VII) solution was needed. How many grams of iron are in this tablet? A_r Fe = 55.8.

3 The $E^\ominus$ value for $Cr_2O_7^{2-}(aq) + 14H^+(aq) + 6e^- \rightleftharpoons 2Cr^{3+}(aq) + 7H_2O(l)$ is +1.33 V.

Use $E^\ominus$ values to show that acidified $Cr_2O_7^{2-}$ ions can be used in a redox titration with Fe^{2+} when Cl^- ions are present, that is, that $Cr_2O_7^{2-}$ ions will not oxidise Cl^-.

4 What are the oxidation states of the metal atoms in these ions?

a MnO_4^- **b** CrO_4^{2-} **c** $Cr_2O_7^{2-}$

5 Classify the reaction as redox or acid–base. Explain your answer.

$2CrO_4^{2-}(aq) + 2H^+(aq) \rightleftharpoons Cr_2O_7^{2-}(aq) + H_2O(l)$

23.5 Catalysis

Catalysts affect the rate of a reaction without being chemically changed themselves at the end of the reaction. Catalysts play an important part in industry because they allow reactions to proceed at lower temperatures and pressures thus saving valuable resources. Modern cars have a catalytic converter in the exhaust system which is based on platinum and rhodium. This catalyses the conversion of carbon monoxide, nitrogen oxides, and unburnt petrol to carbon dioxide, nitrogen, and water.

Many catalysts used in industry are transition metals or their compounds. Catalysts can be divided into two groups:

- heterogeneous
- homogeneous.

Heterogeneous catalysts

Heterogeneous catalysts are present in a reaction in a different phase (solid, liquid, or gas) than the reactants. They are usually present as solids, whilst the reactants may be gases or liquids. Their catalytic action occurs at active sites on the solid surface. The reactants pass over the catalyst surface, which remains in place so the catalyst is not lost and does not need to be separated from the products.

Making heterogenous catalysts more efficient

Catalysts are often expensive, so the more efficiently they work, the more the costs can be minimised. Since their activity takes place on the surface you can:

- Increase their surface area – the larger the surface area, the better the efficiency.
- Spread the catalyst onto an inert support medium, or even impregnate it into one. This increases the surface-to-mass ratio so that a little goes a long way. The more expensive catalysts are often used in this way. For example, the catalytic converter in a car has finely divided rhodium and platinum on a ceramic material.

Catalysts do not last forever.

- Over time, the surfaces may become covered with unwanted impurities. This is called poisoning. The catalytic converters in cars gradually become poisoned by substances used in fuel additives. Until a few years ago, lead-based additives were used in petrol. The lead poisoned the catalysts and so leaded fuel could not be used in cars with converters.
- The finely divided catalyst may gradually be lost from the support medium.

Learning objectives:

→ State what is meant by heterogeneous and homogeneous catalysts.

→ Describe how heterogeneous catalysts can be made more efficient.

→ Explain how a homogeneous catalyst works.

Specification reference: 3.2.5

Synoptic link

Catalysts were first introduced in Topic 5.3, Catalysts.

You will learn more about the use of catalysts in organic chemistry in Topic 26.4, Reactions of carboxylic acids and esters.

Hint

Transition metals have partly full d-orbitals which can be used to form weak chemical bonds with the reactants. This has two effects – weakening bonds within the reactant and holding the reactants close together on the metal surface in the correct orientation for reaction.

Study tip

Try to understand the factors that determine cost and catalyst efficiency.

Some important examples of heterogeneous catalysts

The Haber process

You have already met the Haber process, where ammonia is made by the reaction of nitrogen with hydrogen. The catalyst for the process is iron – present as pea-sized lumps to increase the surface area:

$$N_2(g) + 3H_2(g) \xrightleftharpoons{\text{iron catalyst}} 2NH_3(g)$$

The iron catalyst lasts about five years before it becomes poisoned by impurities in the gas stream such as sulfur compounds, and has to be replaced at considerable cost.

The Contact process

The Contact process produces sulfuric acid – a vital industrial chemical. Around two million tonnes are produced each year in the UK and it is involved in the manufacture of many goods.

It is made from sulfur, oxygen, and water, the key step being:

$$2SO_2 + O_2 \rightleftharpoons 2SO_3$$

This is catalysed by vanadium(V) oxide, V_2O_5, in two steps as follows:

The vanadium(V) oxide oxidises sulfur dioxide to sulfur trioxide and is itself reduced to vanadium(IV) oxide:

$$SO_2 + V_2O_5 \rightarrow SO_3 + V_2O_4$$

The vanadium(IV) oxide is then oxidised back to vanadium(V) oxide by oxygen:

$$2V_2O_4 + O_2 \rightarrow 2V_2O_5$$

The vanadium(V) oxide is regenerated unchanged. Each of the two steps has a lower activation energy than the uncatalysed single step and therefore the reaction goes faster.

This is a good example of how the variability of oxidation states of a transition metal is useful in catalysis.

Hint

Haber's original process used osmium as the catalyst but this was extremely expensive. The chemical engineer Karl Bosch, who scaled up the process, developed the use of iron.

Synoptic link

You first met the Haber process in Topic 5.3, Catalysts.

Hint

Vanadium(V) oxide is also called vanadium pentoxide.

Synoptic link

Look back at Topic 5.1, Collision theory, to revise activation energy.

The manufacture of methanol

Synthesis gas is made from methane, present in natural gas and steam:

$$CH_4(g) + H_2O(g) \rightarrow CO + 3H_2(g)$$

It is a mixture of carbon monoxide and hydrogen and is used to make methanol:

$$\underbrace{CO(g) + 2H_2(g)}_{\text{synthesis gas}} \rightarrow \underbrace{CH_3OH(g)}_{\text{methanol}}$$

This reaction may be catalysed by chromium oxide, Cr_2O_3. Today, the most widely used catalyst is a mixture of copper, zinc oxide, and aluminium oxide.

Methanol is an important industrial chemical (over 30 million tonnes are made each year worldwide) and is used mainly as a starting material for the production of plastics such as Bakelite, Terylene, and Perspex.

Homogeneous catalysts

When the catalyst is in the same phase as the reactant, an intermediate species is formed. For example, in the gas phase chlorine free radicals act as catalysts to destroy the ozone layer. The intermediate here is the ClO• free radical.

Homogeneous catalysis by transition metals

Peroxodisulfate ions, $S_2O_8^{2-}$, oxidise iodide ions to iodine. This reaction is catalysed by Fe^{2+} ions. The overall reaction is:

$$S_2O_8^{2-}(aq) + 2I^-(aq) \rightarrow 2SO_4^{2-}(aq) + I_2(aq)$$

The catalysed reaction takes place in two steps. First the peroxodisulfate ions oxidise iron(II) to iron(III):

$$S_2O_8^{2-}(aq) + 2Fe^{2+}(aq) \rightarrow 2SO_4^{2-}(aq) + 2Fe^{3+}(aq)$$

The Fe^{3+} then oxidises the I^- to I_2, regenerating the Fe^{2+} ions so that none are used up in the reaction:

$$2Fe^{3+}(aq) + 2I^-(aq) \rightarrow 2Fe^{2+}(aq) + I_2(aq)$$

So iron first gives an electron to the peroxodisulfate and later takes one back from the iodide ions.

The uncatalysed reaction takes place between two ions of the same charge (both negative), which repel, therefore giving a high activation energy. Both steps of the catalysed reaction involve reaction between pairs of oppositely charged ions. This helps to explain the increase in rate.

Figure 1 shows the reaction profile. Although there are two steps in the catalysed reaction, the overall activation energy is lower than that for the uncatalysed reaction.

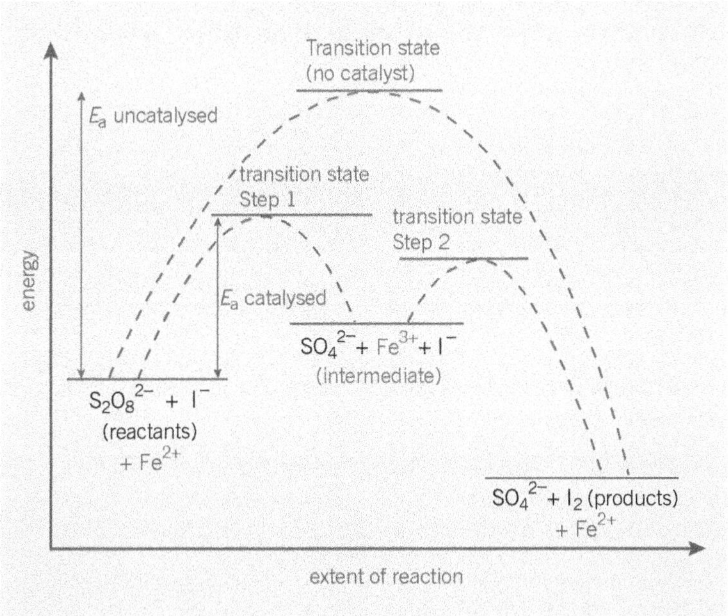

▲ **Figure 1** *Possible reaction profile for the iodine/peroxodisulfate reaction. E_a for the catalysed reaction is the energy gap between the reactants and the higher of the two transition states (transition state Step 1)*

Synoptic link

The destruction of the ozone layer by chlorine free radicals was introduced in Topic 12.5, The formation of halogenoalkanes.

Hint

Do not be put off by unfamiliar chemical names (such as peroxodisulfate ions). Make sure you understand the process that they are used to illustrate.

Study tip

Fe^{3+} can also act as a catalyst for this reaction. When Fe^{3+} is used, I^- is first oxidised to I_2.

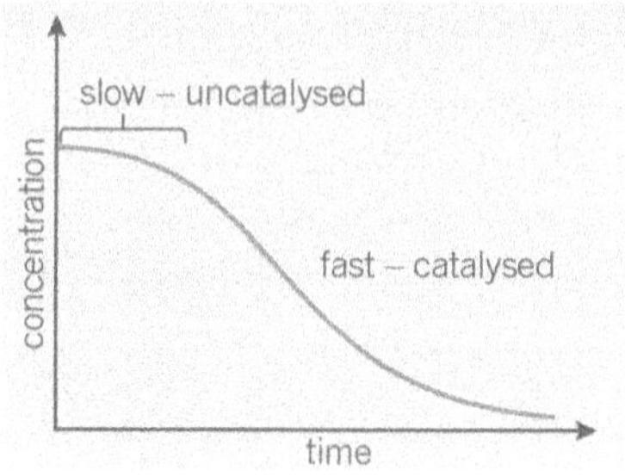

▲ **Figure 2** *A concentration/time graph for an autocatalytic reaction*

Autocatalysis

An interesting example of catalysis occurs when one of the products of the reaction is a catalyst for the reaction. Such a reaction starts slowly at the uncatalysed rate. As the concentration of the product that is also the catalyst builds up, the reaction speeds up to the catalysed rate. From then on it behaves like a normal reaction, gradually slowing down as the reactants are used up. This leads to an odd-looking rate curve (Figure 2).

The oxidation of ethanedioic acid by manganate(VII) ions

One example of an autocatalysed reaction is that between a solution of ethanedioic acid (oxalic acid) and an acidified solution of potassium manganate(VII). It is used as a titration to find the concentration of potassium manganate(VII) solution.

$$2MnO_4^-(aq) + 16H^+(aq) + 5C_2O_4^{2-}(aq) \rightarrow 2Mn^{2+}(aq) + 8H_2O(l) + 10CO_2(g)$$

manganate(VII) ions	hydrogen ions	ethanedioate ions	manganese(II) ions	water	carbon dioxide

The catalyst, Mn^{2+} ions, is not present at the beginning of the reaction. Once a little Mn^{2+} has formed, it can react with MnO_4^- ions to form Mn^{3+} as an intermediate species, which then reacts with $C_2O_4^{2-}$ ions to reform Mn^{2+}:

$$4Mn^{2+}(aq) + MnO_4^-(aq) + 8H^+(aq) \rightarrow 5Mn^{3+}(aq) + 4H_2O(l)$$

$$2Mn^{3+}(aq) + C_2O_4^{2-}(aq) \rightarrow 2CO_2(g) + 2Mn^{2+}(aq)$$

The reaction can easily be followed using a colorimeter to measure the concentration of MnO_4^-, which is purple. The reaction curve looks like the one in Figure 2.

Summary questions

1 a State the difference between a homogeneous and a heterogeneous catalyst.

b Classify each of the examples below as homogeneous or heterogeneous.

i A gauze of platinum and rhodium catalyses the oxidation of ammonia gas to nitrogen monoxide during the manufacture of nitric acid.

ii Nickel catalyses the hydrogenation of vegetable oils.

iii The enzymes in yeast catalyse the production of ethanol from sugar.

2 Why does a catalyst make a reaction go faster? Why is this particularly important for industry?

3 The peroxodisulfate / iodide reaction above is catalysed by Fe^{3+} ions (as well as by Fe^{2+} ions):

$$S_2O_8^{2-}(aq) + 2I^-(aq) \rightarrow 2SO_4^{2-}(aq) + I_2(aq)$$

Write down the two equations that explain this and explain why it is slow in the absence of the catalyst.

4 Show how the overall equation for the autocatalytic reaction between MnO_4^- and $C_2O_4^{2-}$ can be obtained from the equations for the two catalytic steps.

Practice questions

1 Transition metals and their complexes have characteristic properties.

(a) Give the electron configuration of the Zn^{2+} ion.
Use your answer to explain why the Zn^{2+} ion is **not** classified as a transition metal ion.

(2 marks)

(b) In terms of bonding, explain the meaning of the term *complex*.

(2 marks)

(c) Identify **one** species from the following list that does **not** act as a ligand. Explain your answer.

H_2 $\quad$ O^{2-} $\quad$ O_2 $\quad$ CO *(2 marks)*

(d) The element palladium is in the d block of the Periodic Table. Consider the following palladium compound which contains the sulfate ion.

$$[Pd(NH_3)_4]SO_4$$

(i) Give the oxidation state of palladium in this compound. *(1 mark)*

(ii) Give the names of two possible shapes for the complex palladium ion in this compound.

(2 marks)

AQA, 2011

2 This question is about copper chemistry.

(a) Aqueous copper(II) ions $[Cu(H_2O)_6]^{2+}(aq)$ are blue.

(i) With reference to electrons, explain why aqueous copper(II) ions are blue.

(3 marks)

(ii) By reference to aqueous copper(II) ions, state the meaning of each of the **three** terms in the equation $\Delta E = h\nu$.

(3 marks)

(iii) Write an equation for the reaction, in aqueous solution, between $[Cu(H_2O)_6]^{2+}$ and an excess of chloride ions.
State the shape of the complex produced and explain why the shape differs from that of the $[Cu(H_2O)_6]^{2+}$ ion.

(3 marks)

(b) Draw the structure of the ethanedioate ion, $C_2O_4^{2-}$.
Explain how this ion is able to act as a ligand.

(2 marks)

(c) When a dilute aqueous solution containing ethanedioate ions is added to a solution containing aqueous copper(II) ions, a substitution reaction occurs. In this reaction four water molecules are replaced and a new complex is formed.

(i) Write an ionic equation for the reaction. Give the co-ordination number of the complex formed and name its shape.

(4 marks)

(ii) In the complex formed, the two water molecules are opposite each other. Draw a diagram to show how the ethanedioate ions are bonded to a copper ion and give a value for one of the O—Cu—O bond angles. You are **not** required to show the water molecules.

(2 marks)

AQA, 2011

3 **(a)** Octahedral and tetrahedral complex ions are produced by the reaction of transition metal ions with ligands which form co-ordinate bonds with the transition metal ion. Define the term *ligand* and explain what is meant by the term *co-ordinate bond*.

(3 marks)

(b) (i) Some complex ions can undergo a ligand substitution reaction in which both the co-ordination number of the metal and the colour change in the reaction. Write an equation for one such reaction and state the colours of the complex ions involved.

(ii) Bidentate ligands replace unidentate ligands in a metal complex by a ligand substitution reaction.
Write an equation for such a reaction and explain why this reaction occurs.

(8 marks)

AQA, 2005

Answers to the Practice Questions and Section Questions are available at
www.oxfordsecondary.com/oxfordaqaexams-alevel-chemistry

24 Reactions of inorganic compounds in aqueous solutions

24.1 The acid–base chemistry of aqueous transition metal ions

Learning objectives:

- → Describe metal aqua ions.
- → State what determines the acidity of metal aqua ions in aqueous solution.

Specification reference: 3.2.6

If you dissolve a salt of a transition metal such as iron(II) nitrate, $Fe(NO_3)_2$, in water, water molecules cluster around the Fe^{2+} ion so it actually exists as the complex ion $[Fe(H_2O)_6]^{2+}$. Six water molecules act as ligands bonding to the metal ion in an octahedral arrangement. They each use one of their lone pairs of electrons to form a co-ordinate (dative) bond with the metal ion. A similar situation occurs with an iron(III) salt – here the complex formed is $[Fe(H_2O)_6]^{3+}$. These complexes are called **aqua ions**.

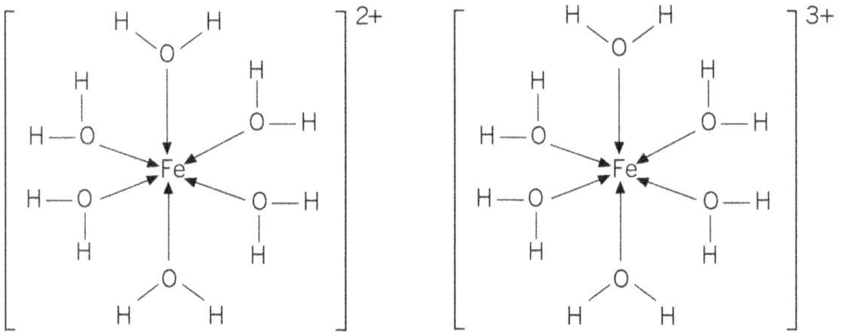

▲ **Figure 1** *$[Fe(H_2O)_6]^{2+}$ (left) and $[Fe(H_2O)_6]^{3+}$ (right)*

However, there is a significant difference in the acidity of these two complexes.

Solutions of Fe^{2+}(aq) are not noticeably acidic, whereas a solution of Fe^{3+}(aq) ($pK_a = 2.2$) is a stronger acid than ethanoic acid ($pK_a = 4.8$). Why is Fe^{3+}(aq) acidic at all and why the difference with Fe^{2+}(aq)? This is because the Fe^{3+} ion is both smaller and more highly charged than Fe^{2+} (it has a higher charge density) making it more strongly polarising. So in the $[Fe(H_2O)_6]^{3+}$(aq) ion the iron strongly attracts electrons from the oxygen atoms of the water ligands, so weakening the O—H bonds in the water molecules. This complex ion will then readily release an H^+ ion making the solution acidic (Figure 2). Fe^{2+} is less polarising and so fewer O—H bonds break in solution.

Hint

When drawing co-ordinate bonds, the arrow → represents the donated pair of electrons.

Synoptic link

pK_a was discussed in Topic 21.3, Weak acids and bases. It is a measure of the strength of an acid. The *smaller* the value of pK_a, the *stronger* the acid.

Hint

In a dilute solution of iron(II) nitrate there will be many times more water molecules than nitrate ions so these are far more likely to act as ligands.

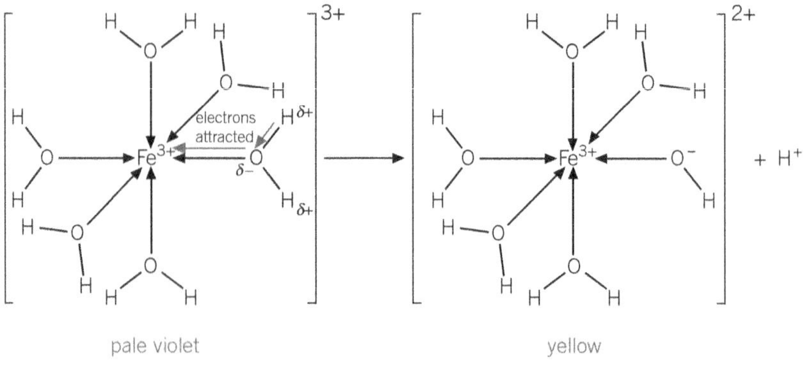

▶ **Figure 2** *The acidity of Fe^{3+}(aq) ions*

Written as an equation:

$$[Fe(H_2O)_6]^{3+}(aq) \rightleftharpoons [Fe(H_2O)_5(OH)]^{2+}(aq) + H^+(aq)$$

With transition metals, there is a general rule that aqua ions of M^{3+} are significantly more acidic than those of M^{2+}. This is because 3+ ions have a higher charge to size ratio, which is also called charge density.

A similar situation occurs in solutions of $Al^{3+}(aq)$, although aluminium is not a transition metal.

Reactions such as the above are often called **hydrolysis** (reaction with water) because they may also be represented as:

$$[Fe(H_2O)_6]^{3+}(aq) + H_2O(l) \rightleftharpoons [Fe(H_2O)_5(OH)]^{2+}(aq) + H_3O^+(aq)$$

This stresses the fact that the $[Fe(H_2O)_6]^{3+}$ ion is donating a proton, H^+, to a water molecule and behaving as a Brønsted–Lowry acid.

Study tip

A hydrolysis reaction is one in which O—H bonds of water are broken and new species are formed.

Synoptic link

You will need to understand Brønsted–Lowry acids and bases studied in Topic 21.1, Defining an acid.

Lewis acids and bases

The Brønsted–Lowry theory of acidity describes acids as proton (H^+ ion) donors, and bases, such as OH^- ions, as proton acceptors.

Another theory (the Lewis theory) is also used to describe acids. This theory defines acids as electron pair acceptors, and bases as electron pair donors in the formation of co-ordinate (dative) covalent bonds. For example:

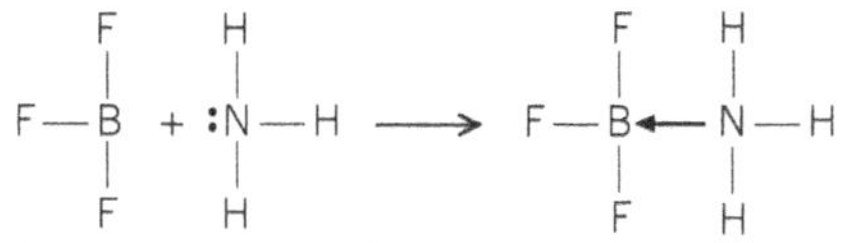

boron trifluoride   ammonia

Here, boron trifluoride is acting as a Lewis acid (electron pair acceptor) and ammonia as a Lewis base (electron pair donor). The Lewis definition of acids is wider than the Brønsted–Lowry one. Boron trifluoride contains no hydrogen and so cannot be an acid under the Brønsted–Lowry definition. H^+ ions have no electrons at all and so can *only* form bonds by accepting an electron pair.

$$H^+ + {}^-{:}O-H \longrightarrow H \leftarrow O-H$$

A water molecule has two lone pairs of electrons and it can use one of these to accept a proton (acting as a Lewis base and as a Brønsted–Lowry base) or, for example, to form a co-ordinate bond with a metal ion (acting as a Lewis base).

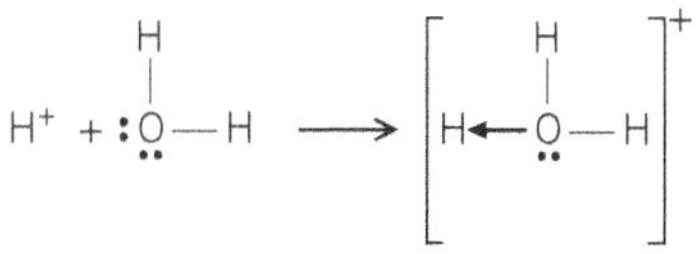

Lewis base and Brønsted–Lowry base

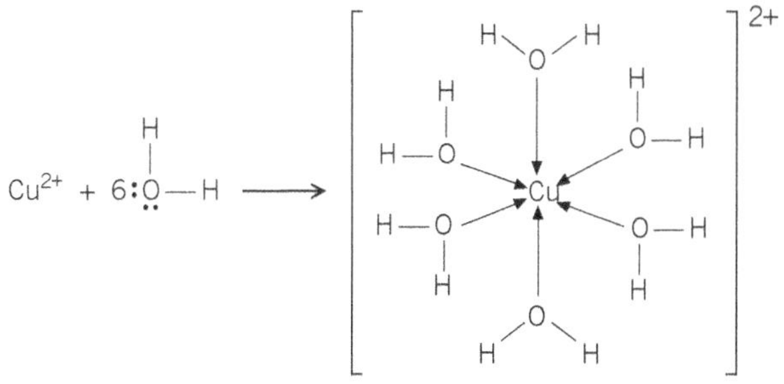

Lewis base

All Brønsted–Lowry acids are also Lewis acids. Ligands which form bonds to transition metal ions using lone pairs are acting as Lewis bases and the metal ions as Lewis acids.

Which of the following can act as Lewis acids and which as Lewis bases?
$AlCl_3$, AlF_3, V^{3+}, Zn^{2+}

Acids: $AlCl_3$, AlF_3 Bases: V^{3+}, Zn^{2+}

Theories of acidity over the years

Acids are a group of compounds with similar properties, for example, neutralising bases, producing hydrogen with the more reactive metals, releasing carbon dioxide from carbonates. They were probably first recognised as a group by their sour taste. Today, of course, no one would dream of tasting a newly synthesised compound before it had been thoroughly tested for toxicity (which could take some time), but in old chemical papers it is not uncommon to find the taste of new compounds reported along with colour, crystal form, melting point, and so on.

Many theories of acidity have been proposed, and these have been discarded or modified as new facts have come along. This is how scientific understanding progresses. Theories of acidity include:

Lavoisier (1777) proposed that all acids contain oxygen. This is fine for many acids, for example, nitric, HNO_3, sulfuric, H_2SO_4, and ethanoic (acetic), CH_3COOH, and was a good working theory. However, once the formula of hydrochloric acid, HCl, was worked out, it became clear that this theory could not be correct.

Davy (1816) suggested that all acids contain hydrogen. This looks better – all the above acids fit and the theory has no problem including HCl. However, it does not explain why the hydrogen is important.

Liebig (1838) defined acids as substances containing hydrogen which could be replaced by a metal. This is an improvement on Davy's theory as it explains why not all hydrogen-containing compounds are acidic, for example, ammonia, NH_3, is not acidic. There must be something special about that hydrogen that makes it replaceable by a metal. This is a theory that is not far from one that could be used today.

Arrhenius (1887) thought of acids as producing hydrogen ions, H^+. This is a development of Liebig's theory. It tells us what exactly is special about the hydrogen – it must be able to become an H^+ ion.

The **Brønsted–Lowry** description of acidity (developed in 1923 by Thomas Lowry and Johannes Brønsted independently) is the most generally useful current theory. This defines an acid as a substance which can donate a proton (an H^+ ion) and a base as a substance which can accept a proton. However, this theory has difficulty with acids that do not contain hydrogen – aluminium chloride, $AlCl_3$, or boron trifluoride, BF_3, for example.

Hint

The boron in boron trifluoride has only six electrons in its outer shell and is therefore able to accept an electron pair from ammonia, for example.

Another theory (the **Lewis theory**) is also used today to describe acids. This theory regards acids as electron pair acceptors and bases as electron pair donors in the formation of co-ordinate covalent bonds.

Acid–base reactions of M^{2+}(aq) and M^{3+}(aq) ions

If you add a base (such as OH^-) it will remove protons from the aqueous complex. This takes place in a series of steps.

In the case of M^{3+}

$$[M(H_2O)_6]^{3+}(aq) + OH^-(aq) \rightarrow [M(H_2O)_5(OH)]^{2+}(aq) + H_2O\ (l)$$

$$[M(H_2O)_5(OH)]^{2+}(aq) + OH^-\ (aq) \rightarrow [M(H_2O)_4(OH)_2]^{+}(aq) + H_2O\ (l)$$

$$[M(H_2O)_4(OH)_2]^{+}(aq) + OH^-\ (aq) \rightarrow M(H_2O)_3(OH)_3(s) + H_2O\ (l)$$

The neutral metal(III) hydroxide, $M(H_2O)_3(OH)_3$, is in effect $M(OH)_3$, which is uncharged and insoluble and forms as a precipitate.

In the case of M^{2+}

$$[M(H_2O)_6]^{2+}(aq) + OH^-\ (aq) \rightarrow [M(H_2O)_5(OH)]^{+}(aq) + H_2O\ (l)$$

$$[M(H_2O)_5(OH)]^{+}(aq) + OH^-\ (aq) \rightarrow M(H_2O)_4(OH)_2(s) + H_2O\ (l)$$

The neutral metal(II) hydroxide, $M(H_2O)_4(OH)_2$, is in effect $M(OH)_2$, which is uncharged and insoluble and forms a precipitate.

Ammonia, which is basic, has the same effect as OH^- ions in removing protons.

$$[M(H_2O)_6]^{3+}(aq) + 3NH_3(aq) \rightarrow M(H_2O)_3(OH)_3\ (s) + 3NH_4^+$$

$$[M(H_2O)_6]^{2+}(aq) + 2NH_3(aq) \rightarrow M(H_2O)_4(OH)_2(s) + 2NH_4^+$$

Reactions with the base CO_3^{2-}, the carbonate ion

The greater acidity of the aqueous Fe^{3+} ion explains why iron(III) carbonate does not exist, but iron(II) carbonate does. The carbonate ion is able to remove protons from $[Fe(H_2O)_6]^{3+}(aq)$ to form hydrated iron(III) hydroxide but cannot do so from $[Fe(H_2O)_6]^{2+}(aq)$.

$$[Fe(H_2O)_6]^{3+}(aq) + 3CO_3^{2-}(aq) \rightleftharpoons Fe(OH)_3(H_2O)_3(s) + 3HCO_3^-(aq)$$

The overall reaction is:

$$2[Fe(H_2O)_6]^{3+}(aq) + 3CO_3^{2-}(aq) \rightarrow 2[Fe(H_2O)_3(OH)_3](aq) + 3CO_2(g) + 3H_2O(l)$$

The reaction can be derived as a combination of the following:

$$2H_3O^+(aq) + CO_3^{2-}\ (aq) \rightarrow 3H_2O(l) + CO_2(g)$$

With the removal of H_3O^+ displacing the hydrolysis equilibrium below to the right.

$$\underset{\text{pale violet}}{[Fe(H_2O)_6]^{3+}(aq)} + 3H_2O(l) \rightleftharpoons \underset{\text{brown}}{Fe(H_2O)_3(OH)_3\ (s)} + 3H_3O^+(aq)$$

In the case of the aqueous Fe^{2+} ion, which is less acidic than $Fe^{3+}(aq)$, insoluble iron(II) carbonate is formed:

$$\underset{\text{pale green}}{[Fe(H_2O)_6]^{2+}(aq)} + CO_3^{2-}\ (aq) \rightarrow \underset{\text{green}}{FeCO_3\ (s)} + 6H_2O(l)$$

In general, carbonates of transition metal ions in oxidation state +2 exist, whilst those of ions in the +3 state do not.

So, for example, Cu^{2+} ions react with sodium carbonate solution to give a plae blue precipitate of copper carbonate.

$$[Cu(H_2O)_6[^{2+}(aq) + CO_3^{2-}(aq) \rightarrow \underset{\text{pale blue}}{CuCO_3(s)} + 6H_2O(l)$$

In contrast, Al^{3+} ions give carbon dioxide gas and a white precipitate of aluminium hydroxide.

Synoptic link

You will need to understand bond polarity studied in Topic 3.6, Electronegativity – bond polarity in covalent bonds, and equilibria in Chapter 6, Equilibria.

Hint

If a solution of sodium carbonate is added to a solution containing Fe^{3+} ions (e.g., $Fe(NO_2)_3$) it will fizz due to carbon dioxide being released. A solution containing Fe^{2+} ions will form a precipitate of the carbonate.

$$[Al(H_2O)_6]^{3+}(aq) + 3CO_3^{2-}(aq) \rightarrow 2[Al(H_2O)_3(OH)_3]\ (s) + 3CO_2(g) + 3H_2O(l)$$

white

Distinguishing iron ions

As you have seen, both Fe^{2+} and Fe^{3+} exist in aqueous solution as octahedral hexa-aqua ions. $[Fe(H_2O)_6]^{2+}$ is pale green and $[Fe(H_2O)_6]^{3+}$ is pale brown, and dilute solutions are hard to tell apart. A simple test to distinguish the two is to add dilute alkali, which precipitates the hydroxides whose colours are more obviously different.

$$[Fe(H_2O)_6]^{3+}(aq) + 3OH^-(aq) \rightarrow Fe(H_2O)_3(OH)_3(s) + 3H_2O(l)$$

pale violet — iron(III) hydroxide (brown)

$$[Fe(H_2O)_6]^{2+}(aq) + 2OH^-(aq) \rightarrow Fe(H_2O)_4(OH)_2(s) + 2H_2O(l)$$

pale green — iron(III) hydroxide (green)

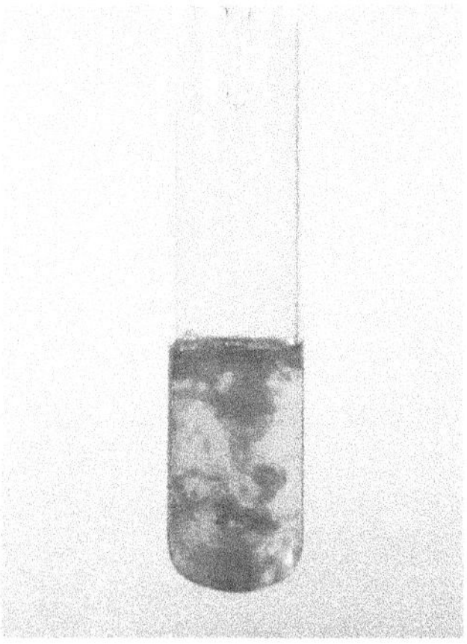

▲ **Figure 3** *Iron(III) hydroxide precipitate*

▲ **Figure 4** *Iron(II) hydroxide precipitate*

Amphoteric hydroxides

Amphoteric means showing both acidic and basic properties. Aluminium hydroxide is an example of this – it will react with both acids and bases. For example:

$$Al(H_2O)_3(OH)_3 + 3HCl \rightarrow Al(H_2O)_6^{3+} + 3Cl^-$$

This is what you would expect from a normal metal hydroxide – it reacts with acid and is therefore basic.

But aluminium hydroxide also shows acidic properties – it will react with the base sodium hydroxide to give a colourless solution of sodium tetrahydroxoaluminate:

$$Al(H_2O)_3(OH)_3 + OH^- \rightarrow [Al(OH)_4]^- + 3H_2O$$

Study tip

An excess of a strong base (e.g., NaOH) is needed to redissolve the $Al(H_2O)_3(OH)_3$ precipitate.

Summary questions

1 Copper chloride forms the ion $[Cu(H_2O)_6]^{2+}$ in aqueous solution.

a What is the oxidation state of the copper in this ion?

b What is the co-ordination number of the copper in this ion?

c Write an equation for the reaction you would expect if excess hydroxide ions are added to a solution of copper chloride.

24.2 Ligand substitution reactions

The water molecules that act as ligands in metal aqua ions can be replaced by other ligands – either because the other ligands form stronger co-ordinate bonds or because they are present in higher concentration and an equilibrium is displaced.

Replacing water as a ligand

There are a number of possibilities:

- The water molecules may be replaced by other neutral ligands, such as ammonia.
- The water molecules may be replaced by negatively charged ligands, such as chloride ions.
- The water molecules may be replaced by bi- or multidentate ligands – this is called **chelation**.
- Replacement of the water ligands may be complete or partial.

Replacement by neutral ligands – no change in co-ordination number

In general for an M^{2+} ion, water molecules may be replaced one at a time by ammonia. Both ligands are uncharged and are of similar size, so there is no change in co-ordination number or charge on the ion:

$$[M(H_2O)_6]^{2+} + NH_3 \rightleftharpoons [M(NH_3)(H_2O)_5]^{2+} + H_2O$$

$$[M(NH_3)(H_2O)_5]^{2+} + NH_3 \rightleftharpoons [M(NH_3)_2(H_2O)_4]^{2+} + H_2O$$

$$[M(NH_3)_2(H_2O)_4]^{2+} + NH_3 \rightleftharpoons [M(NH_3)_3(H_2O)_3]^{2+} + H_2O$$

$$[M(NH_3)_3(H_2O)_3]^{2+} + NH_3 \rightleftharpoons [M(NH_3)_4(H_2O)_2]^{2+} + H_2O$$

$$[M(NH_3)_4(H_2O)_2]^{2+} + NH_3 \rightleftharpoons [M(NH_3)_5(H_2O)]^{2+} + H_2O$$

$$[M(NH_3)_5(H_2O)]^{2+} + NH_3 \rightleftharpoons [M(NH_3)_6]^{2+} + H_2O$$

Overall:

$$[M(H_2O)_6]^{2+} + 6NH_3 \rightleftharpoons [M(NH_3)_6]^{2+} + 6H_2O$$

There is a complication, because ammonia is a base as well as a ligand, and therefore contains OH^- ions, a precipitate may form and then redissolve

$$[M(H_2O)_6]^{2+} + 2OH^-(aq) \rightarrow M(H_2O)_4(OH)_2(s) + 2H_2O(l)$$

$$M(H_2O)_4(OH)_2(s) + 6NH_3(aq) \rightleftharpoons [M(NH_3)_6]^{2+}(aq) + 4H_2O(l) + 2OH^-(aq)$$

Learning objectives:

→ Explain the changes in the co-ordination numbers and charges of complexes when different ligands are substituted.

→ Explain why complexes formed with multidentate ligands are more stable than those with monodentate ligands.

Specification reference: 3.2.6

Synoptic link

Chelation was covered in Topic 23.2, Complex formation and the shape of complex ions.

Synoptic link

You will need to understand free energy change and entropy studied in Topic 17.4, Why do chemical reactions take palce?

Study tip

Because NH_3 and H_2O ligands are similar in size and both are uncharged, ligand exchange occurs without a change in charge or co-ordination number.

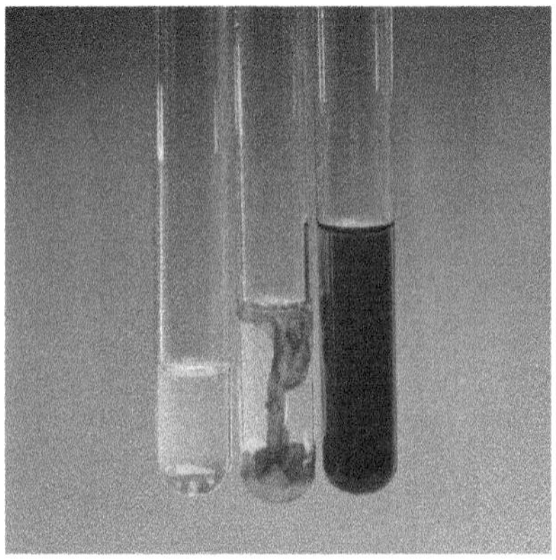

▲ **Figure 1** *Pale blue solution of* $[Cu(H_2O)_6]^{2+}$, the pale blue precipitate of $[Cu(OH)_2(H_2O)_4]$, *and the deep blue solution of* $[Cu(NH_3)_4(H_2O)_2]^{2+}$

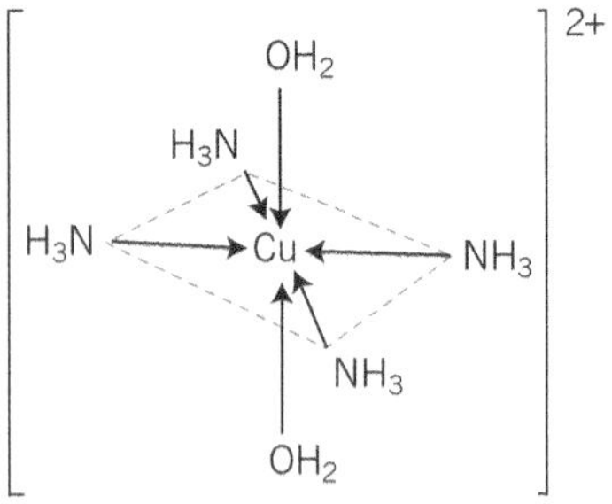

▲ **Figure 2** *The shape of the* $[Cu(NH_3)_4(H_2O)_2]^{2+}$ *ion. The dotted lines are not bonds, they are construction lines to show the square-planar arrangement of the* NH_3 *ligands*

Synoptic link

There are some further examples of ligand substitution reactions of other metal ions, both M^{2+} and M^{3+}, summarised in Topic 24.3, A summary of acid–base and substitution reactions of some metal ions.

Copper(II)

When aqueous copper ions react with ammonia in aqueous solution, ligand replacement is only partial – only four of the water ligands are replaced. The overall reaction is:

$$[Cu(H_2O)_6]^{2+} + 4NH_3 \rightleftharpoons [Cu(NH_3)_4(H_2O)_2]^{2+} + 4H_2O$$

$[Cu(H_2O)_6]^{2+}$ is pale blue whilst $[Cu(NH_3)_4(H_2O)_2]^{2+}$ is a very deep blue.

The steps are similar to those above for Co^{2+}. The ammonia first acts as a base removing protons from two of the water molecules in $[Cu(H_2O)_6]^{2+}$ to form $[Cu(OH)_2(H_2O)_4](s)$. The first thing we see is a pale blue precipitate of copper hydroxide. When more of the concentrated ammonia is added, the precipitate dissolves to form a deep blue solution containing $[Cu(NH_3)_4(H_2O)_2]^{2+}$ (Figure 1). The ammonia has replaced both OH^- ligands, and two of the H_2O ligands:

$$[Cu(OH)_2(H_2O)_4](s) + 4NH_3(aq) \rightleftharpoons [Cu(NH_3)_4(H_2O)_2]^{2+}(aq) + 2H_2O(l) + 2OH^-(aq)$$

The shape of the $[Cu(NH_3)_4(H_2O)_2]^{2+}$ *ion*

The $[Cu(NH_3)_4(H_2O)_2]^{2+}$ is octahedral, as expected for a six co-ordinate ion. The four ammonia molecules exist in a square-planar arrangement around the metal ion with the two water molecules above and below the plane (Figure 2).

The Cu—O bonds are longer (and therefore weaker) than the Cu—N bonds, as would be expected because water is a poorer ligand than ammonia. The octahedron is slightly distorted.

Replacement by chloride ions – change in co-ordination number

When aqueous copper ions react with concentrated hydrochloric acid there is a change in both charge and co-ordination number. Concentrated hydrochloric acid provides a high concentration of Cl^- ligands:

$$[Cu(H_2O)_6]^{2+} + 4Cl^- \rightleftharpoons [CuCl_4]^{2-} + 6H_2O$$

The pale blue colour of the $[Cu(H_2O)_6]^{2+}$ ion is replaced by the yellow $[CuCl_4]^{2-}$ ion. (Although the solution may look green as some $[Cu(H_2O)_6]^{2+}$ will remain.) Again, the actual replacement takes place in steps. The co-ordination number of the ion is four and the ion is tetrahedral.

$[Cu(H_2O)_6]^{2+}$ is six co-ordinate and $[CuCl_4]^{2-}$ is four co-ordinate (Figure 3), because Cl^- is larger than H_2O and fewer ligands can physically fit around the central copper ion.

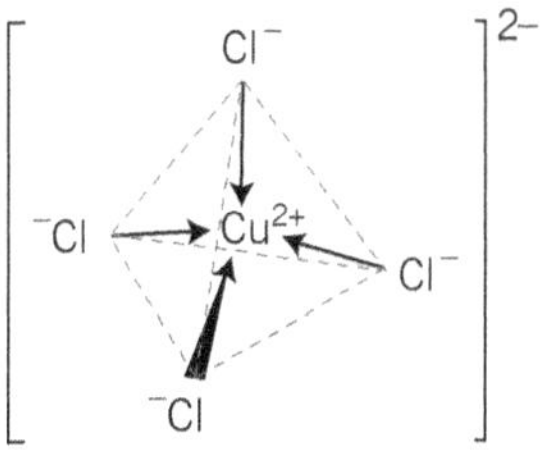

▲ **Figure 3** *The shape of the* $[CuCl_4]^{2-}$ *ion*

Chelation

Chelation is the formation of complexes with multidentate ligands. These are ligands with more than one lone pair so they can form more than one co-ordinate bond. Examples include ethylene diamine, benzene-1,2-diol, and $EDTA^{4-}$. These complexes are usually more stable than those with monodentate ligands. This increased stability is mainly due to the entropy change of the reaction.

Ethylene diamine (often represented as en for short) is a bidentate ligand and can be thought of as two ammonia ligands linked by a short hydrocarbon chain. Each en can replace two water molecules:

$$\underset{\text{four entities}}{[Cu(H_2O)_6]^{2+}(aq) + 3en} \rightarrow \underset{\text{seven entities}}{[Cu(en)_3]^{2+}(aq) + 6H_2O(l)}$$

In this reaction, *three* molecules of ethylene diamine release *six* of water. The larger number of entities on the right in this reaction means that there is a significant entropy increase as the reaction goes from left to right. This favours the reaction.

A single hexadentate ligand $EDTA^{4-}$ can displace all six water ligands from $[M(H_2O)_6]^{2+}$. For example:

$$\underset{\text{two entities}}{[Cu(H_2O)_6]^{2+}(aq) + EDTA^{4-}(aq)} \rightarrow \underset{\text{seven entities}}{[CuEDTA]^{2-}(aq) + 6H_2O(l)}$$

In this reaction, *one* ion of $EDTA^{4-}$ releases six water ligands. The larger number of entities on the right in this reaction means that there is a significant entropy increase as the reaction goes from left to right. This entropy increase favours the formation of chelates (complexes with polydentate ligands) over complexes with monodentate ligands.

Study tip

Practise writing equations for the step-by-step replacement of one ligand with another. They are straightforward but can be tedious, and it is easy to make mistakes with charges. Make sure each step balances for both ligands and charge.

Hint

Ethylene diamine (en) is also called ethane-1,2-diamine and 1,2-diaminoethane. Its formula is $H_2NCH_2CH_2NH_2$. Each nitrogen atom has a lone pair which makes it a bidentate ligand.

Synoptic link

Entropy is a measure of the degree of disorder (Topic 17.4, Why do chemical reactions take place?) so the more molecules that are released and are free to move around, the greater the entropy.

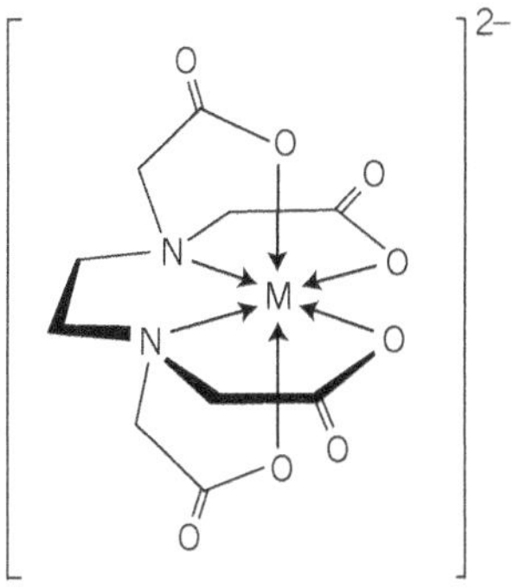

▲ **Figure 4** *A metal(II)–EDTA complex*

Summary questions

1 In the stepwise conversion of $[Cu(H_2O)_6]^{2+}$ to $[CuCl_4]^{2-}$, one of the species formed is neutral. Suggest two possible formulae that it could have.

2 **a** Draw the shape of $[Cu(H_2O)_6]^{2+}$ and **b** predict the shape of $[CuBr_4]^{2-}$. Explain your answer.

3 Write equations for the step-by-step replacement of all the water ligands in $[Cu(H_2O)_6]^{2+}$ by en. How many entities are there on each side of each equation? Predict the likely sign of the entropy change for each reaction.

4 When concentrated hydrochloric acid is added to an aqueous solution containing Co(II) ions, the following change takes place:

$[Co(H_2O)_6]^{2+} \rightarrow [CoCl_4]^{2-}$ and the colour changes from pink to blue.

a Is there any change in the oxidation state of the cobalt?

b Give the shapes of the two ions concerned.

c Suggest two possible reasons for the colour change.

24.3 A summary of acid–base and substitution reactions of some metal ions

Learning objective:

→ Describe products of the reactions between bases and metal aqua ions.

Specification reference: 3.2.6

Synoptic link

You will need to understand co-ordinate bonding studied in Topic 3.2, Covalent bonding.

Synoptic link

See Practical 9 on page 527.

The products of many of the reactions of transition metal compounds can be identified by their colours.

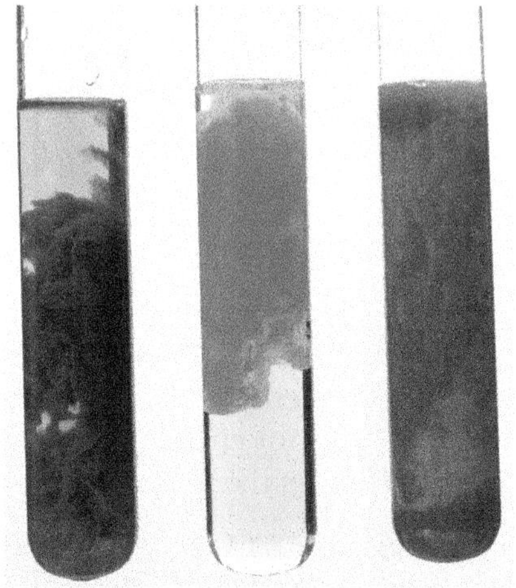

▲ **Figure 1** *Precipitates of (left to right) iron (III) hydroxide (left), copper (II) hydroxide (centre) and a suspension of chromium (III) hydroxide (right).*

Table 1 and Table 2 show a number of examples of reactions involving different bases. You should be able to rationalise these observations using the principles explained in Topic 24.2. In each case the effect of adding the base shown on the left to solutions of aqua ions is shown in the following tables.

▼ **Table 1** *Adding a base to $M^{2+}(aq)$ complexes*

	$[Fe(H_2O)_6]^{2+}(aq)$ pale green	$[Cu(H_2O)_6]^{2+}(aq)$ pale blue
OH^- little	green gelatinous ppt of $[Fe(H_2O)_4(OH)_2]$*	pale blue ppt of $[Cu(H_2O)_4(OH)_2]$
OH^- excess	green gelatinous ppt of $[Fe(H_2O)_4(OH)_2]$*	pale blue ppt of $[Cu(H_2O)_4(OH)_2]$
NH_3 little	green gelatinous ppt of $[Fe(H_2O)_4(OH)_2]$*	pale blue ppt of $[Cu(H_2O)_4(OH)_2]$
NH_3 excess	green gelatinous ppt of $[Fe(H_2O)_4(OH)_2]$*	deep blue solution of $[Cu(NH_3)_4(H_2O)_2]^{2+}$
CO_3^{2-}	green ppt of $FeCO_3$	blue-green ppt of $CuCO_3$

*Pale green $[Fe(H_2O)_4(OH)_2]$ is soon oxidised by air to brown $[Fe(H_2O)_3(OH)_3]$

▼ **Table 2** *Adding a base to $M^{3+}(aq)$ complexes*

	$[Fe(H_2O)_6]^{3+}(aq)$ purple/yellow/brown	$[Al(H_2O)_6]^{3+}(aq)$ colourless
OH^- little	brown gelatinous ppt of $[Fe(H_2O)_3(OH)_3]$	white ppt of $[Al(H_2O)_3(OH)_3]$
OH^- excess	brown gelatinous ppt of $[Fe(H_2O)_3(OH)_3]$	colourless solution of $[Al(OH)_4]^-]$
NH_3 little	brown gelatinous ppt of $[Fe(H_2O)_3(OH)_3]$	white ppt of $[Al(H_2O)_3(OH)_3]$
NH_3 excess	brown gelatinous ppt of $[Fe(H_2O)_3(OH)_3]$	white ppt of $[Al(H_2O)_3(OH)_3]$
CO_3^{2-}	brown gelatinous ppt of $[Fe(H_2O)_3(OH)_3]$ and bubbles of CO_2	white ppt of $[Al(H_2O)_3(OH)_3]$ and bubbles of CO_2

In the case of the M^{2+} ions, precipitates of the metal carbonates form when carbonate ions are added. In the case of the M^{3+} ions, bubbles of carbon dioxide are produced instead. This is a reflection of the greater acidity of $[M(H_2O)_6]^{3+}$ compared with $[M(H_2O)_6]^{2+}$, see Topic 24.1, The acid-base chemistry of aqueous transition metal ions–.

Summary questions

1 Why are M^{3+} aqua ions more acidic than M^{2+} aqua ions?

2 Explain why all the compounds of aluminium are colourless.

3 Explain why $[Co(H_2O)_6]^{2+}$ and $[Co(NH_3)_6]^{2+}$ both have a co-ordination number of six and have the same charge.

4 a Write the equations for the reactions of:

  i $[Fe(H_2O)_6]^{3+}(aq)$ with sodium hydroxide solution

  ii $[Cu(H_2O)_6]^{2+}(aq)$ with excess ammonia.

 b What colour changes would you expect to see in **a i** and **ii**?

Practice questions

1 Consider the reaction scheme below and answer the questions which follow.

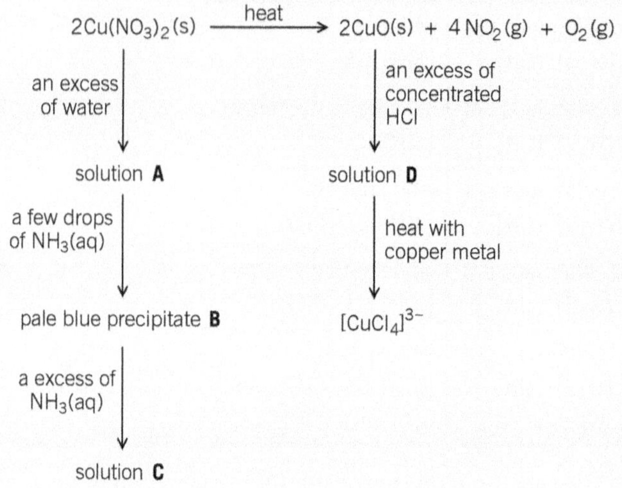

(a) A redox reaction occurs when $Cu(NO_3)_2$ is decomposed by heat. Deduce the oxidation state of nitrogen in $Cu(NO_3)_2$ and in NO_2 and identify the product formed by oxidation in this decomposition.

(3 marks)

(b) Identify and state the shape of the copper-containing species present in solution **A**.

(2 marks)

(c) Identify the pale blue precipitate **B** and write an equation, or equations, to show how **B** is formed from the copper-containing species in solution **A**.

(2 marks)

(d) Identify the copper-containing species present in solution **C**. State the colour of this copper-containing species and write an equation for its formation from precipitate **B**.

(3 marks)

(e) Identify the copper-containing species present in solution **D**. State the colour and shape of this copper-containing species.

(3 marks)

(f) The oxidation state of copper in $[CuCl_4]^{3-}$ is +1.

(i) Give the electron arrangement of a Cu^+ ion.

(ii) Deduce the role of copper metal in the formation of $[CuCl_4]^{3-}$ from the copper-containing species in solution **D**.

(2 marks)

AQA, 2005

2 Consider the following reaction scheme that starts from aqueous $[Cu(H_2O)_6]^{2+}$ ions.

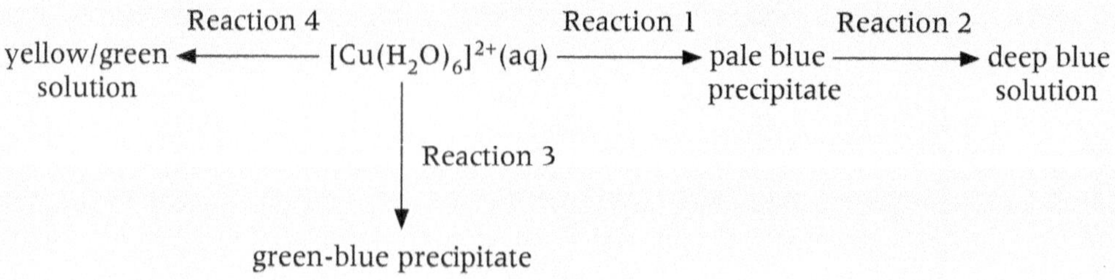

For each of the reactions 1 to 4, identify a suitable reagent, give the formula of the copper-containing species formed and write an equation for the reaction.

(a) Reaction 1 *(3 marks)*

(b) Reaction 2 *(3 marks)*

(c) Reaction 3 *(3 marks)*

(d) Reaction 4 *(3 marks)*

AQA, 2014

3 The scheme below shows some reactions of copper(II) ions in aqueous solution. **W**, **X**, **Y**, and **Z** are all copper-containing species.

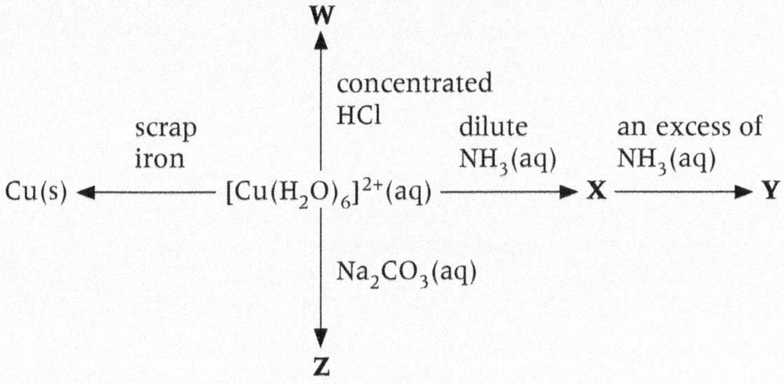

(a) Identify ion **W**. Describe its appearance and write an equation for its formation from $[Cu(H_2O)_6]^{2+}(aq)$ ions. *(3 marks)*

(b) Identify compound **X**. Describe its appearance and write an equation for its formation from $[Cu(H_2O)_6]^{2+}(aq)$ ions. *(3 marks)*

(c) Identify ion **Y**. Describe its appearance and write an equation for its formation from **X**. *(3 marks)*

(d) Identify compound **Z**. Describe its appearance and write an equation for its formation from $[Cu(H_2O)_6]^{2+}(aq)$ ions. *(3 marks)*

(e) Copper metal can be extracted from a dilute aqueous solution containing copper(II) ions using scrap iron.

(i) Write an equation for this reaction and give the colours of the initial and final aqueous solutions. *(3 marks)*

(ii) This method of copper extraction uses scrap iron. Give **two** other reasons why this method of copper extraction is more environmentally friendly than reduction of copper oxide by carbon. *(2 marks)*

AQA, 2010

Answers to the Practice Questions and Section Questions are available at www.oxfordsecondary.com/oxfordaqaexams-alevel-chemistry

Section 5 practice questions

1 Due to their electron arrangements, transition metals have characteristic properties including catalytic action and the formation of complexes with different shapes.

(a) Give **two other** characteristic properties of transition metals. For each property, illustrate your answer with a transition metal of your choice.

(4 marks)

(b) Other than octahedral, there are several different shapes shown by transition metal complexes. Name **three** of these shapes and for each one give the formula of a complex with that shape.

(6 marks)

(c) It is possible for Group 2 metal ions to form complexes. For example, the $[Ca(H_2O)_6]^{2+}$ ion in hard water reacts with $EDTA^{4-}$ ions to form a complex ion in a similar manner to hydrated transition metal ions. This reaction can be used in a titration to measure the concentration of calcium ions in hard water.

(i) Write an equation for the equilibrium that is established when hydrated calcium ions react with $EDTA^{4-}$ ions.

(1 mark)

(ii) Explain why the equilibrium in part **(c)**(i) is displaced almost completely to the right to form the EDTA complex.

(3 marks)

(iii) In a titration, 6.25 cm^3 of a 0.0532 mol dm^{-3} solution of EDTA reacted completely with the calcium ions in a 150 cm^3 sample of a saturated solution of calcium hydroxide. Calculate the mass of calcium hydroxide that was dissolved in 1.00 dm^3 of the calcium hydroxide solution.

(3 marks)

AQA, 2012

2 Iron is an important element in living systems. It is involved in redox and in acid–base reactions.

(a) Explain how and why iron ions catalyse the reaction between iodide ions and $S_2O_8^{2-}$ ions. Write equations for the reactions that occur.

(5 marks)

(b) Iron(II) compounds are used as moss killers because iron(II) ions are oxidised in air to form iron(III) ions that lower the pH of soil.

(i) Explain, with the aid of an equation, why iron(III) ions are more acidic than iron(II) ions in aqueous solution.

(3 marks)

(ii) In a titration, 0.321 g of a moss killer reacted with 23.60 cm^3 of acidified 0.0218 mol dm^{-3} $K_2Cr_2O_7$ solution.
Calculate the percentage by mass of iron in the moss killer. Assume that all of the iron in the moss killer is in the form of iron(II).

(5 marks)

(c) Some sodium carbonate solution was added to a solution containing iron(III) ions. Describe what you would observe and write an equation for the reaction that occurs.

(3 marks)

AQA, 2011

3 **(a)** State what is meant by the term *homogeneous* as applied to a catalyst.

(b) (i) State what is meant by the term *autocatalysis*. *(1 mark)*

(ii) Identify the species which acts as an autocatalyst in the reaction between ethanedioate ions and manganate(VII) ions in acidic solution.

(2 marks)

(c) When petrol is burned in a car engine, carbon monoxide, carbon dioxide, oxides of nitrogen and water are produced. Catalytic converters are used as part of car exhaust systems so that the emission of toxic gases is greatly reduced.

(i) Write an equation for a reaction which occurs in a catalytic converter between two of the toxic gases. Identify the reducing agent in this reaction.

(ii) Identify a transition metal used in catalytic converters and state how the converter is constructed to maximise the effect of the catalyst.

(5 marks)

AQA, 2004

4 **(a)** Using complex ions formed by Co^{2+} with ligands selected from H_2O, NH_3, Cl^-, $C_2O_4^{2-}$ and $EDTA^{4-}$, give an equation for each of the following.

(i) A ligand substitution reaction which occurs with no change in either the co-ordination number or in the charge on the complex ion.

(ii) A ligand substitution reaction which occurs with both a change in the co-ordination number and in the charge on the complex ion.

(iii) A ligand substitution reaction which occurs with no change in the co-ordination number but a change in the charge on the complex ion.

(iv) A ligand substitution reaction in which there is a large change in entropy.

(8 marks)

(b) An aqueous solution of iron(II) sulfate is a pale-green colour. When aqueous sodium hydroxide is added to this solution a green precipitate is formed. On standing in air, the green precipitate slowly turns brown.

(i) Give the formula of the complex ion responsible for the pale-green colour.

(ii) Give the formula of the green precipitate.

(iii) Suggest an explanation for the change in the colour of the precipitate.

(4 marks)

AQA, 2004

5 A co-ordinate bond is formed when a transition metal ion reacts with a ligand.

(a) Explain how this co-ordinate bond is formed.

(2 marks)

(b) Describe what you would observe when dilute aqueous ammonia is added dropwise, to excess, to an aqueous solution containing copper(II) ions.
Write equations for the reactions that occur.

(4 marks)

(c) When the complex ion $[Cu(NH_3)_4(H_2O)_2]^{2+}$ reacts with 1,2-diaminoethane, the ammonia molecules but not the water molecules are replaced.
Write an equation for this reaction.

(1 mark)

(d) Suggest why the enthalpy change for the reaction in part **(c)** is approximately zero.

(2 marks)

(e) Explain why the reaction in part **(c)** occurs despite having an enthalpy change that is approximately zero.

(2 marks)

AQA, Specimen paper 1

6 **Table 1** shows observations of changes from some test-tube reactions of aqueous solutions of compounds **Q, R,** and **S** with five different aqueous reagents. The initial colours of the solutions are not given.

▼ **Table 1**

	$BaCl_2$ + HCl	$AgNO_3$ + HNO_3	NaOH	Na_2CO_3	HCl (conc)
Q	no change observed	pale cream precipitate	white precipitate	white precipitate	no change observed
R	no change observed	white precipitate	white precipitate, dissolves in excess of NaOH	white precipitate, bubbles of a gas	no change observed
S	white precipitate	no change observed	brown precipitate	brown precipitate, bubbles of a gas	yellow solution

(a) Identify each of compounds **Q**, **R**, and **S**.
You are **not** required to explain your answers.

(6 marks)

(b) Write ionic equations for each of the positive observations with **S**.

(4 marks)

AQA, Specimen paper 1

7 **(a)** Define the term transition metal.

(1 mark)

(b) Explain why scandium, Sc, is classified as a d-block element but not as a transition metal element. *(2 marks)*

8 **(a)** Explain what is meant by the terms *complex ion* and *ligand*.

(2 marks)

(b) Complete the electron configuration of:
(i) Cu atom: $1s^2$, $2s^2$

(1 mark)

(ii) Cu^{2+} ion: $1s^2$, $2s^2$

(1 mark)

(c) Consider the hexaaquachromium(III), $[Cr(H_2O)_6]^{3+}$, complex.
(i) Draw the shape of this complex.

(1 mark)

(ii) Name the shape of this complex.

(1 mark)

Answers to the Practice Questions and Section Questions are available at
www.oxfordsecondary.com/oxfordaqaexams-alevel-chemistry

Section 5 summary

monodentate ligands
- donate one pair of electrons to form a new co-ordinate bond
- H_2O, NH_3, Cl^-

bidentate ligands
- donate two pairs of electrons to form new co-ordinate bonds
- en, $C_2O_4^{2-}$

multidentate ligands
- $EDTA^{4-}$

ligand substitution

chelate effect

cis-trans (E-Z) isomerism

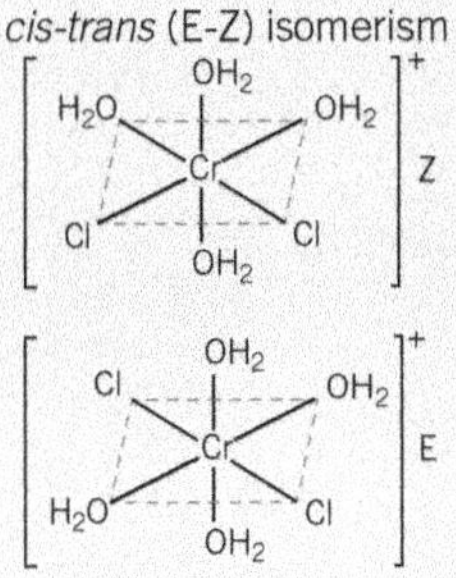

optical isomerism

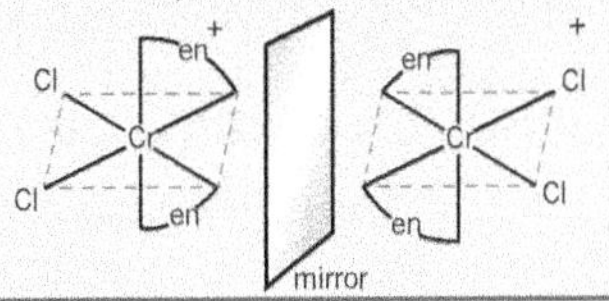

co-ordination number: 6

octahedral

co-ordination number: 4

tetrahedral

co-ordination number: 4

square-planar

cisplatin

co-ordination number: 2

—M—

linear

co-ordination number

form complexes

ligands donate a pair of electrons to form a co-ordinate bond

form coloured compounds

transition metals

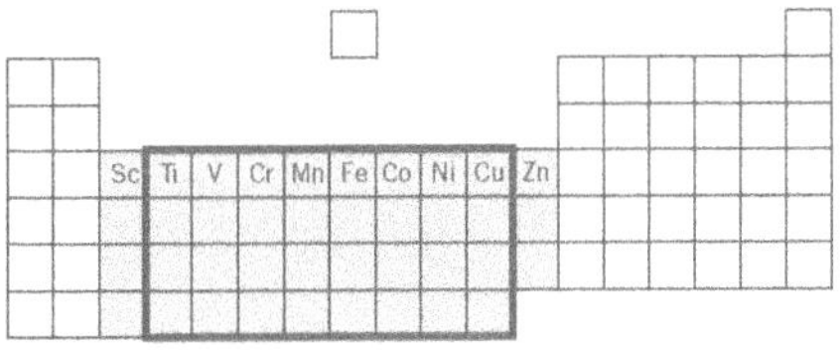

Section 2 Inorganic chemistry

form ions with an incomplete d-sub-shell

Period 3

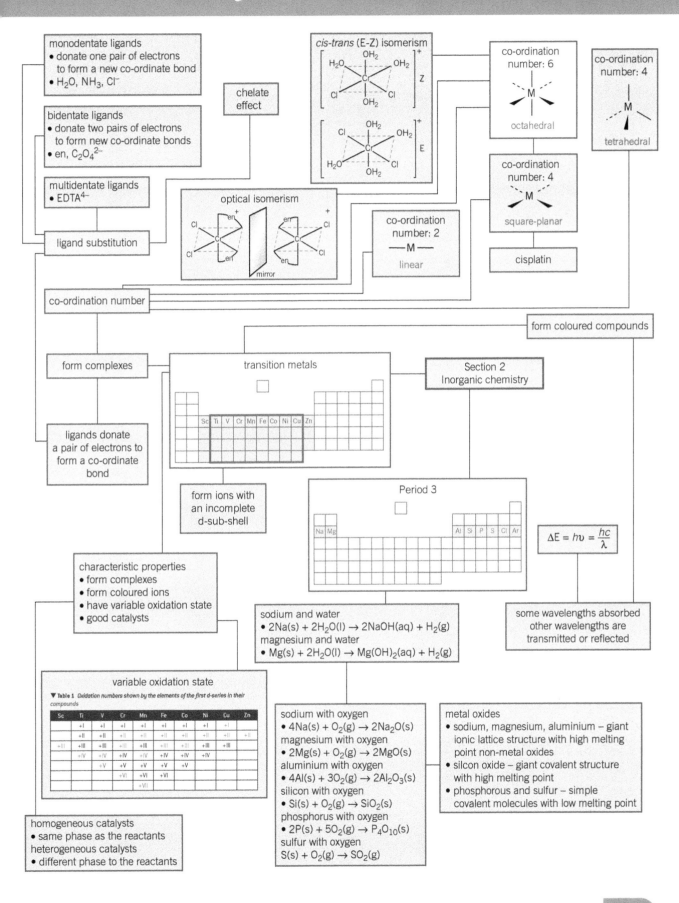

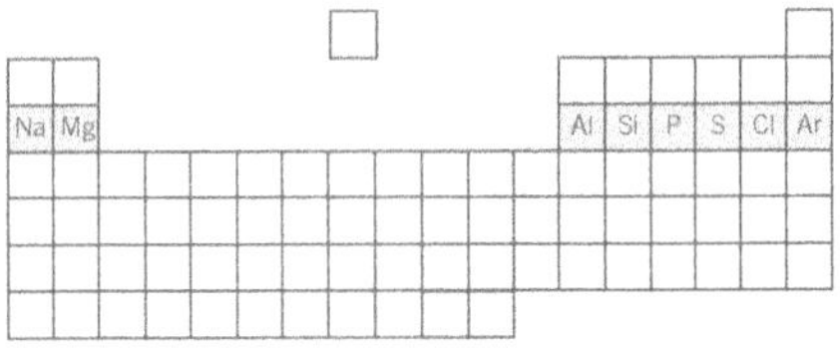

$$\Delta E = h\upsilon = \frac{hc}{\lambda}$$

some wavelengths absorbed other wavelengths are transmitted or reflected

characteristic properties
- form complexes
- form coloured ions
- have variable oxidation state
- good catalysts

sodium and water
- $2Na(s) + 2H_2O(l) \rightarrow 2NaOH(aq) + H_2(g)$

magnesium and water
- $Mg(s) + 2H_2O(l) \rightarrow Mg(OH)_2(aq) + H_2(g)$

variable oxidation state

▼ **Table 1** *Oxidation numbers shown by the elements of the first d-series in their compounds*

Sc	Ti	V	Cr	Mn	Fe	Co	Ni	Cu	Zn
	+I	+I	+I	+I	+I	+I	+I	+I	
	+II	+II	+II	+II	+II	+II	+II	+II	+II
+III	+III	+III	+III	+III	+III	+III	+III	+III	
	+IV	+IV	+IV	+IV	+IV	+IV	+IV		
		+V	+V	+V	+V	+V			
			+VI	+VI	+VI				
				+VII					

sodium with oxygen
- $4Na(s) + O_2(g) \rightarrow 2Na_2O(s)$

magnesium with oxygen
- $2Mg(s) + O_2(g) \rightarrow 2MgO(s)$

aluminium with oxygen
- $4Al(s) + 3O_2(g) \rightarrow 2Al_2O_3(s)$

silicon with oxygen
- $Si(s) + O_2(g) \rightarrow SiO_2(s)$

phosphorus with oxygen
- $2P(s) + 5O_2(g) \rightarrow P_4O_{10}(s)$

sulfur with oxygen
$S(s) + O_2(g) \rightarrow SO_2(g)$

metal oxides
- sodium, magnesium, aluminium – giant ionic lattice structure with high melting point non-metal oxides
- silcon oxide – giant covalent structure with high melting point
- phosphorous and sulfur – simple covalent molecules with low melting point

homogeneous catalysts
- same phase as the reactants

heterogeneous catalysts
- different phase to the reactants

Practical skills

In this section you have met the following ideas:

- Investigating the reactions of Period 3 oxides.
- Investigating ligand substitution reactions.
- Finding the concentration of a solution using colorimetry.
- Investigating the reduction of vanadate(V) ions.
- Investigating redox titrations.
- Finding out about autocatalysis.
- Investigating metal-aqua reactions.
- Identifying positive and negative ions and finding the identity of unknown substances.

Maths skills

In this section you have met the following maths skills:

- Using information about ligands to draw the shapes of complex ions.
- Working out how to draw *cis* and *trans* and optical isomers of complexes.
- Calculating the concentration of a solution from a graph of absorption versus concentration.

Extension

Produce a report exploring how the electronic configuration of transition metals affects their reactivity and properties.

Suggested resources:

Winter, M[2015], *d-Block Chemistry: Oxford Chemistry Primers.* Oxford University Press, UK. ISBN 978-0-19-870096-8

McCleverty, J[1999], *Chemistry of the First Row Transition Metals. Oxford Chemistry Primers.* Oxford University Press, UK. ISBN 978-0-19-850151-0

Section 6

A2 Organic chemistry 2

Nomenclature and isomerism revisits the IUPAC naming system introduced earlier and applies it to further families of organic compounds. A further type of isomerism, optical isomerism based on mirror image molecules, is introduced.

Compounds containing the carbonyl group introduces the chemistry of aldehydes, ketones, carboxylic acids, and esters, all of which contain the carbonyl group, C=O.

Aromatic chemistry looks at the chemistry of compounds based on the benzene ring, which have unexpected properties due to their system of electrons delocalised over a hexagonal ring of carbon atoms.

Amines are organic compounds based on the ammonia molecule. Their nitrogen atom has a lone pair of electrons which explains their reactivity as bases and nucleophiles.

Polymerisation looks at two types of long chain molecules based on smaller repeating units – condensation and addition polymers. It describes their synthesis and uses and also their biodegradability (or lack of it).

Amino acids and proteins are two groups of biologically important groups of polymers which are vital for life. The chapter looks at how proteins are built up from amino acids. It shows how some simple test tube reactions can help to identify organic chemicals.

Organic synthesis and analysis shows how a series of organic reactions can be linked together to make a target molecule from a given starting material.

Structure determination explains the techniques of proton nuclear magnetic resonance (NMR) and carbon-13 NMR and shows how these techniques can be used to help deduce the structures of organic compounds.

Chromatography describes a group of techniques used for separating mixtures of organic compounds and shows how they can be linked with mass spectrometry to help identify the components.

Chapters in this section

What you already know:

The material in this unit builds on knowledge and understanding that you have built up at AS level. In particular the following:

- ☐ Organic compounds are based on chains and rings of carbon atoms along with hydrogen.
- ☐ Organic compound exist in families called homologous series.
- ☐ Different families of organic compounds have different functional groups.
- ☐ There is a systematic naming system for organic compounds.
- ☐ Organic compounds can exist as isomers with the same molecular formula but different arrangement of atoms.
- ☐ Some important groups of organic compound include alkanes, alkenes, halogenoalkanes, and alcohols.
- ☐ Characteristic reactions and infrared spectroscopy can be used to help identify organic compounds.
- ☐ The principles of mass spectrometry.

25 Nomenclature and isomerism

25.1 Naming organic compounds

The IUPAC systematic naming system for organic chemistry is based on a root, which describes the length of the carbon chain. Functional groups have names with numbers to show where they occur on the chain. You can use Table 1 to revise this system.

Table 1 *Examples of systematic naming of organic compounds*

Structural formula	Name	Notes
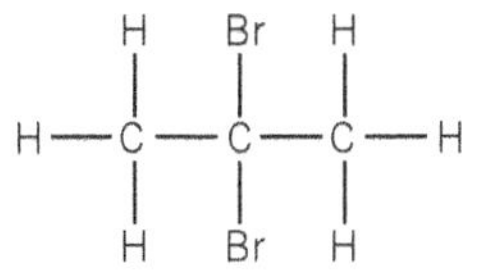	2,2-dibromopropane	The di tells you there are two bromine atoms. The 2,2 says where the two bromine atoms are on the chain.
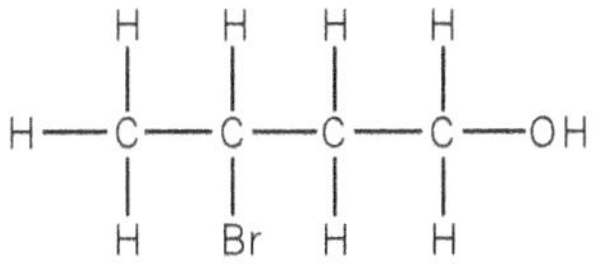	3-bromobutan-1-ol	The suffix '-ol' defines the compound as an alcohol. The –OH group defines the end that the chain is counted from, so the bromine is attached to carbon 3.
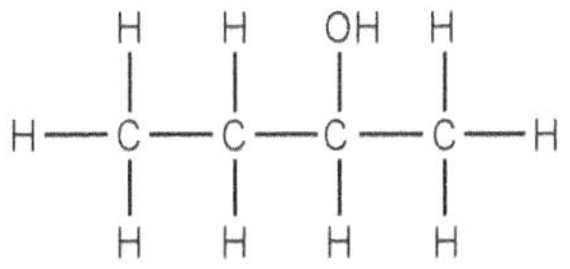	butan-2-ol	Not butan-3-ol as the smallest locant possible is used.
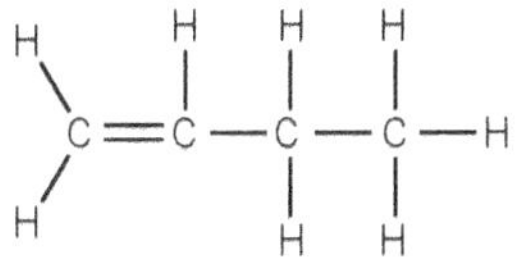	but-1-ene	Not but-2-ene, but-3-ene, or but-4-ene as the smallest locant possible is used.
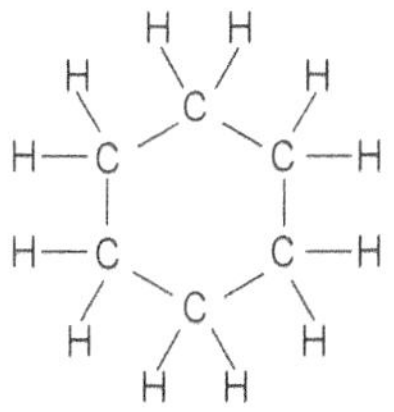	cyclohexane	'Cyclo-' is used to indicate a ring.
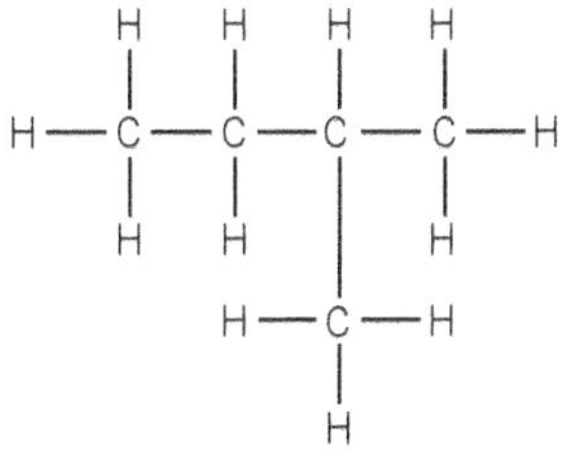	methylbutane	There is no need to use a number to locate the side chain because it must be on carbon number 2.

Learning objective:

→ Describe how IUPAC rules are used for naming organic compounds.

Specification reference: 3.3.1

Synoptic link

You will need to know the nomenclature and isomerism studied in Topic 11.1, Carbon compounds, and Topic 11.2, Nomenclature - naming organic compounds, and the shapes and bond angles in simple molecules in Topic 3.5, The shapes of molecules and ions.

Structural formula	Name	Notes
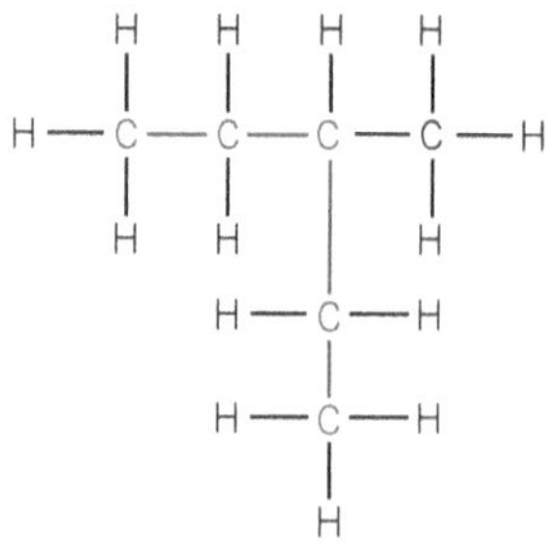	3-methylpentane	This is not 2-ethylbutane. The rule is to base the name on the longest unbranched chain, in this case pentane (picked out in red). Remember the bond angles are 109.5°, not 90°.
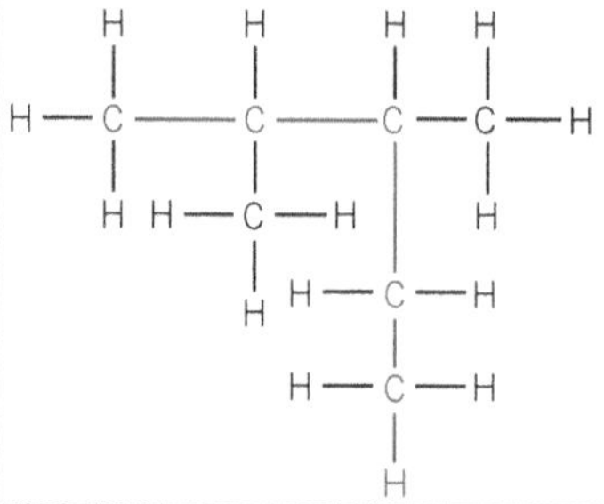	2,3-dimethylpentane	Again remember the root is based on the longest unbranched chain.

More functional groups

Table 2 shows how to name organic compound and includes the functional groups that you will meet in the following chapters, as well as those you have already met.

▼ **Table 2** *The suffixes and prefixes of some functional groups*

Family	Formula	Suffix	Prefix	Example
alkenes	$RCH{=}CH_2$	-ene		propene, CH_3CHCH_2
alkynes	$RC{\equiv}CH$	-yne		propyne, CH_3CCH
halogenoalkanes	R—X (X is F, Cl, Br, or I)		halo- (fluoro-, chloro-, bromo-, iodo-)	chloromethane, CH_3Cl
carboxylic acids	RCOOH	-oic acid		ethanoic acid, CH_3COOH
anhydrides	RCOOCOR′	-anhydride		ethanoic anhydride, $CH_3COOCOCH_3$
esters	RCOOR′	-oate (Esters are named from their parent alcohol and acid, so propyl ethanoate is derived from propanol and ethanoic acid.)		propyl ethanoate, $CH_3COOC_3H_7$
acyl chlorides	RCOCl	-oyl chloride		ethanoyl chloride, CH_3COCl
amides	$RCONH_2$	-amide		ethanamide, CH_3CONH_2
nitriles	$RC{\equiv}N$	-nitrile		ethanenitrile, $CH_3C{\equiv}N$
aldehydes	RCHO	-al		ethanal, CH_3CHO
ketones	RCOR′	-one		propanone, CH_3COCH_3
alcohols	ROH	-ol	hydroxy-	ethanol, C_2H_5OH 2-hydroxyethanal, $HOCH_2CHO$
amines	RNH_2	-amine	amino-	ethylamine, $CH_3CH_2NH_2$
arenes	C_6H_5R			methylbenzene, $C_6H_5CH_3$

Some functional groups may be identified by either a prefix or a suffix. For example, alcohols have the suffix '-ol' and the prefix 'hydroxy-'. The suffix '-ol' is used if it is the only functional group. When there are two (or more) functional groups, the IUPAC rules have a comprehensive system of priority. In Table 2 the higher in the list is a suffix and the lower a prefix. So the amino acid alanine (see below), which is both a carboxylic acid and an amine, has the systematic name 2-aminopropanoic acid.

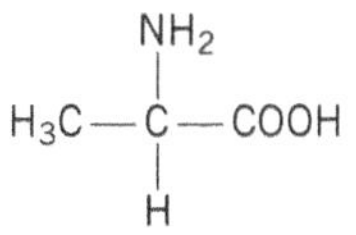

The International Union of Pure and Applied Chemistry

According to its website, the International Union of Pure and Applied Chemistry (IUPAC) 'serves to advance the worldwide aspects of the chemical sciences and to contribute to the application of chemistry in the service of mankind. As a scientific, international, non-governmental and objective body, IUPAC can address many global issues involving the chemical sciences.'

One of its services to the world of chemistry is to have developed a systematic naming system for organic chemicals, the full rules for which are held in a publication known as the Red Book. There is a companion Blue Book for inorganic chemistry.

This means that any organic chemical can be given a name which can be recognised and used by chemists throughout the world. This can obviously reduce confusion. For example, one well-respected database lists a total of 28 names that are in use for the compound with the IUPAC name butanone, a relatively simple compound:

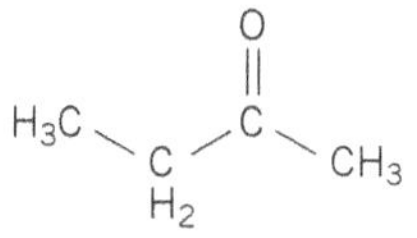

butanone

IUPAC also rules on the naming of newly discovered elements. An element 'can be named after a mythological concept, a mineral, a place or country, a property, or a scientist' and the discoverer has the right to propose a name and symbol.

Summary questions

1 Draw the displayed formula of:

a 3-ethyl-3-methylhexane

b 2,4-dimethylpentane.

2 What is the name of:

a $CH_3CH_2CH_2OH$

b $CH_3CH(Cl)CH_3$

c $CH_3CH_2CH{=}CHCH_2CH_3$

d CH_3CH_2OH

e C_4H_9COOH?

Study tip

In chemical names, strings of numbers are separated by commas. A hyphen is placed between words and numbers.

Hint

Sometimes hydrocarbon molecules form rings. These are indicated by the prefix 'cyclo-', for example, cyclohexane:

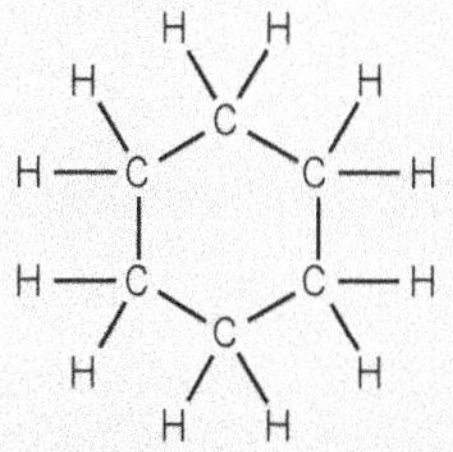

The molecular formula of cyclohexane is C_6H_{12}. It is different from the molecular formula for hexane, C_6H_{14}, as there are no $-CH_3$ groups at the end of the carbon chain as the carbons are joined in a ring.

Study tip

Compounds are named according to the longest hydrocarbon chain. When the molecule is branched, look at it carefully and count the longest chain before you start numbering the carbon atoms.

25.2 Optical isomerism

Learning objective:

→ Describe what type of molecules show optical isomerism.

Specification reference: 3.3.7

Isomers are compounds with the same molecular formula but which have different molecular structures or a different arrangement of atoms in space. Organic chemistry provides many examples of isomerism. Structural isomers:

- have different functional groups (Figure 1a) or
- have functional groups attached to the main chain at different points (Figure 1b) or
- have a different arrangement of carbon atoms in the skeleton of the molecule (Figure 1c).

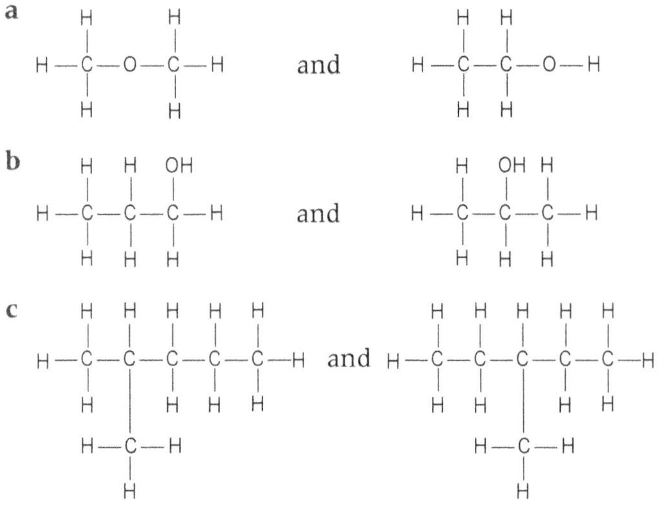

▲ **Figure 1** *Pairs of the different types of structural isomers*

Stereoisomerism

Stereoisomerism is where two (or more) compounds have the same structural formula. They differ in the *arrangement* of the bonds in space. There are two types:

- *E-Z* isomerism
- optical isomerism.

Optical isomerism

Optical isomers occur when there are four different substituents attached to one carbon atom. This results in two isomers that are non-superimposable mirror images of one another, but are not identical. For example, bromochlorofluoromethane exists as two mirror image forms:

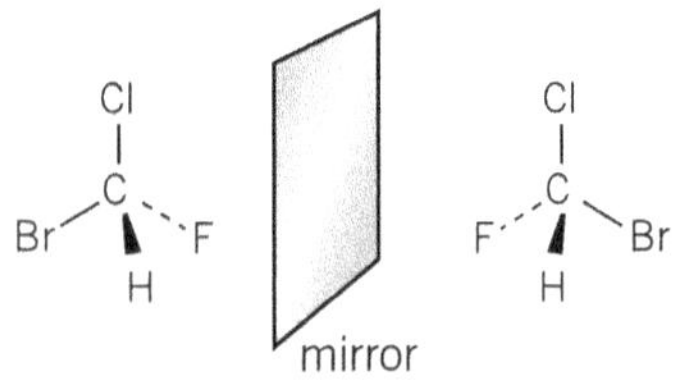

Link

You met with structural isomerism and *E-Z* isomerism in Topic 11.3, Isomerism. *E-Z* involves the arrangement of atoms around double bonds.

The ball and stick models of bromochlorofluoromethane in Figure 2 may help you to see that these are not identical.

Imagine rotating one of the molecules about the C—Cl bond (pointing upwards) until the two bromine atoms (in red) are in the same position.

Hint

It really helps to use models to understand optical isomerism. Even matchsticks and Plasticine will do.

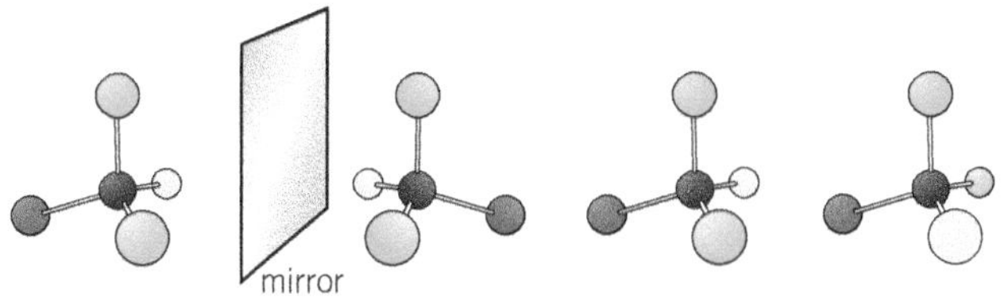

▲ **Figure 2** *Bromochlorofluoromethane has a pair of mirror isomers which are not identical*

Study tip

Drawn structures of optical isomers should show a tetrahedral arrangement of groups around a central carbon atom.

The positions of the hydrogen (blue) and fluorine atoms (yellow) will not match – you cannot superimpose one molecule onto the other.

This is just like a pair of shoes. A left shoe and a right shoe are mirror images, but they are not identical, that is, they cannot be superimposed.

Pairs of molecules like this are called **optical isomers** because they differ in the way they rotate the plane of polarisation of polarised light – either clockwise ((+)-isomer) or anticlockwise ((−)-isomer).

Chirality

Optical isomers are said to be **chiral** and the two isomers are called a pair of **enantiomers**. The carbon bonded to the four different groups is called the **chiral centre** or the **asymmetric carbon atom**, and is often indicated on formulae by *. You can easily pick out a chiral molecule because it contains at least one carbon atom that has four different groups attached to it.

- All α-amino acids, except aminoethanoic acid (glycine) the simplest one (Figure 3), have a chiral centre. For example, the chiral centre of α-aminopropanoic acid (2-aminopropanoic acid) is:

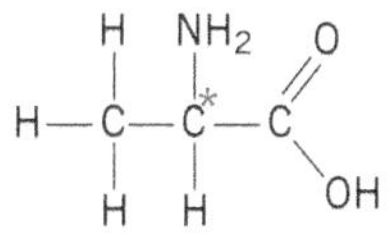

α-aminopropanoic acid

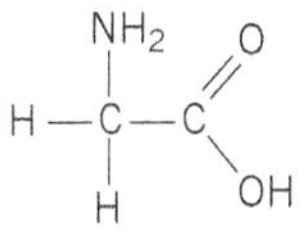

glycine

▲ **Figure 3** *Aminoethanoic acid (glycine)*

- 2-hydroxypropanoic acid (non-systematic name lactic acid) is also chiral. Although the chiral carbon is bonded to two other carbon atoms, these carbons are part of different groups and you must count the whole group.

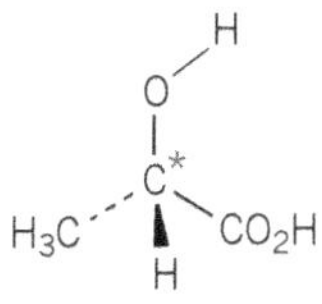

2-hydroxypropanoic acid

Synoptic link

There is more about α-amino acids in Topic 30.1, Introduction to amino acids.

Optical isomerism happens because the isomers have three-dimensional structures so it can only be shown by three-dimensional representations or by models.

Optical activity

Light consists of vibrating electric and magnetic fields. You can think of it as waves with vibrations occurring in all directions at right angles to the direction of motion of the light wave. If the light passes through a special filter, called a **polaroid** (as in polaroid sunglasses) all the vibrations are cut out except those in one plane, for example, the vertical plane (Figure 4).

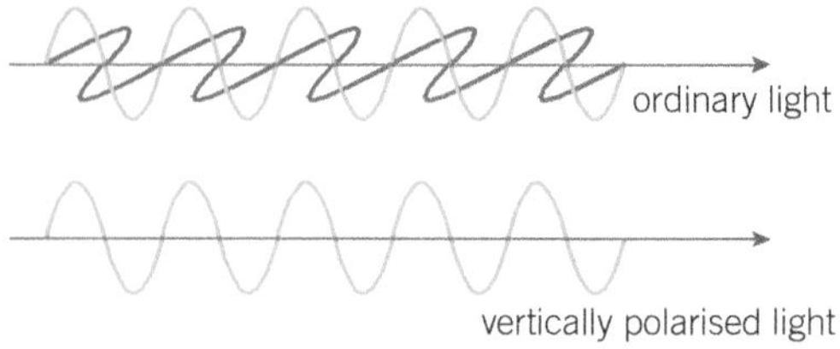

▲ **Figure 4** *Polarised light*

The light is now vertically polarised and it will be affected differently by different optical isomers of the same substance.

Optical rotation can be measured using a **polarimeter** (Figure 5).

1 Polarised light is passed through two solutions of the same concentration, each containing a different optical isomer of the same substance.

2 One solution will rotate the plane of polarisation through a particular angle, clockwise. This is the (+)-isomer.

3 The other will rotate the plane of polarisation by the same angle, anticlockwise. We call this the (−)-isomer.

(There are several other systems in use for distinguishing pairs of isomers, as well as (+) and (−) you may see R and S, D and L, or d and l.)

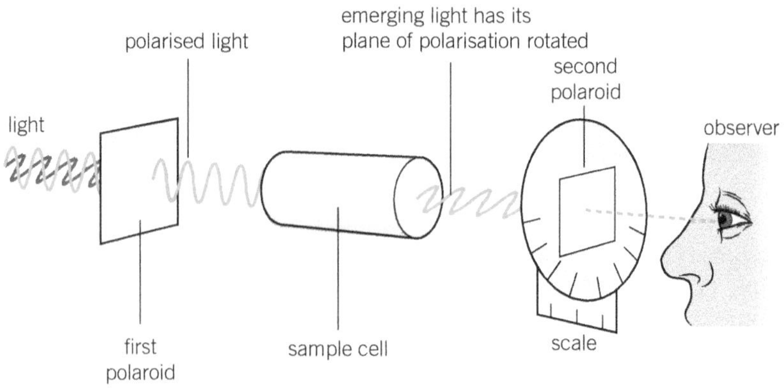

▲ **Figure 5** *A polarimeter measures rotation of the plane of polarisation of polarised light*

Summary questions

1 Which of the following compounds show optical isomerism?

a 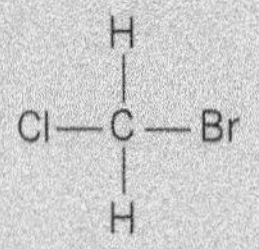

b 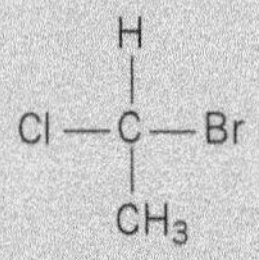

c 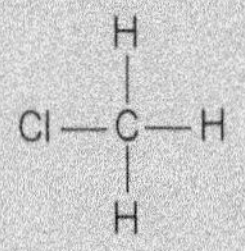

d 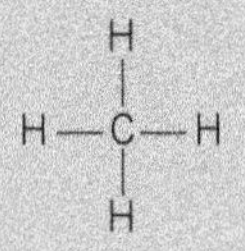

2 Mark the chiral centre on this molecule with a *:

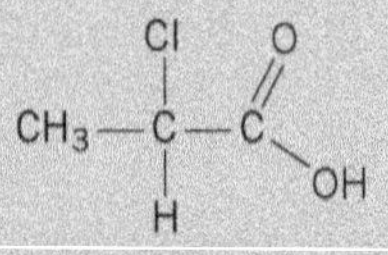

3 The molecule below has two chiral centres. Mark them both.

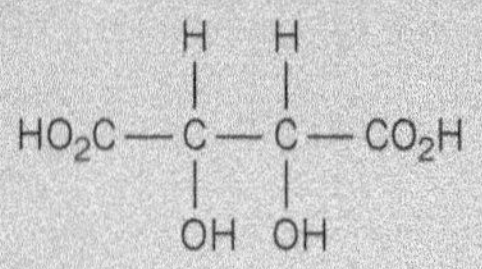

25.3 Synthesis of optically active compounds

A great many of the reactions that are used in organic synthesis to produce optically active compounds actually produce a 50:50 mixture of two optical isomers. This is called a racemic mixture or **racemate** and is not optically active because the effects of the two isomers cancel out.

Learning objectives:

- → State what a racemate is.
- → Describe how a racemate is formed by synthesis.

Specification reference: 3.3.7

The synthesis of 2-hydroxypropanoic acid (lactic acid)

2-hydroxypropanoic acid (lactic acid) has a chiral centre, marked by * in the structure.

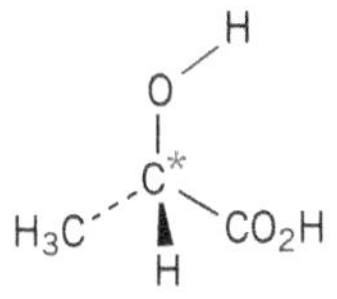

2-hydroxypropanoic acid

The synthesis below produces a mixture of optical isomers. It can be made in two stages from ethanal.

Stage 1

Hydrogen cyanide is added across the C=O bond to form 2-hydroxypropanenitrile.

This is a nucleophilic addition reaction in which the nucleophile is the CN^- ion. It takes place as follows:

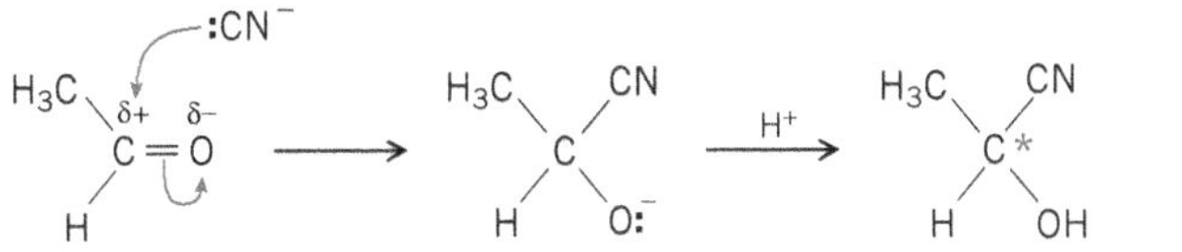

ethanal 2-hydroxypropanenitrile

2-hydroxypropanenitrile has a chiral centre, the starred carbon, which has four different groups ($-CH_3$, –H, –OH, and –CN). The reaction used does not favour one of these isomers over the other (i.e., the –CN group could add on with equal probability from above or below the CH_3CHO which is planar) so you get a racemic mixture.

Stage 2

The nitrile group is converted into a carboxylic acid group.

The 2-hydroxypropanenitrile is reacted with water acidified with a dilute solution of hydrochloric acid. This is a hydrolysis reaction:

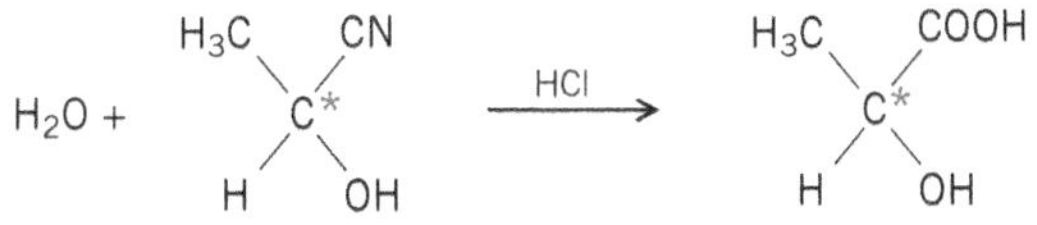

2-hydroxypropanoic acid

Study tip

If a mixture of equal amounts of the two enantiomers is formed, this is optically inactive as their effects cancel out.

Synoptic link

You will cover nucleophilic addition reactions in Topic 26.2, Reactions of the carbonyl group in aldehydes and ketones.

Hint

The carbon chain length of the product is one greater than that of the starting material. You started with *ethanal* (two carbon atoms) and ended with 2-hydroxy*propane*nitrile (three carbon atoms). This type of reaction is important in synthesis, whenever a carbon chain needs to be lengthened.

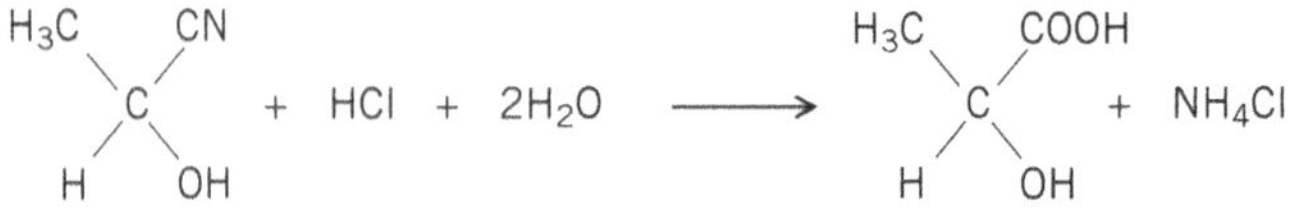

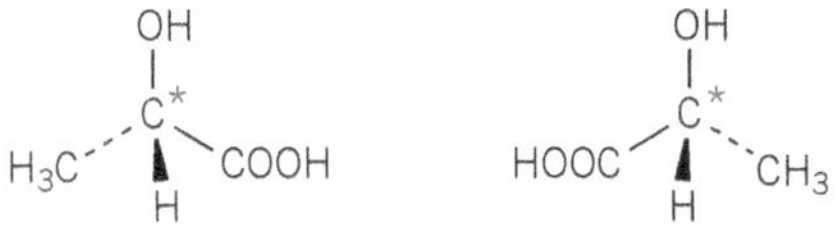

▲ **Figure 2** *Lactic acid is produced naturally in sour milk. In muscle tissue a build-up of lactic acid causes cramp. The two situations produce different optical isomers*

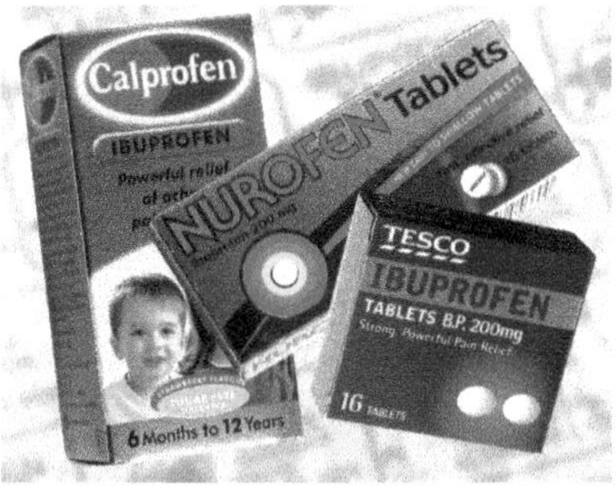

▲ **Figure 3** *Nurofen contains ibuprofen, which is a mixture of two optical isomers*

The balanced equations for the two steps are shown below:

Step 1

$$H_3C\text{--}CH\text{=}O + HCN \longrightarrow H_3C\text{--}C(CN)(OH)\text{--}H$$

Step 2

$$H_3C\text{--}C(CN)(OH)\text{--}H + HCl + 2H_2O \longrightarrow H_3C\text{--}C(COOH)(OH)\text{--}H + NH_4Cl$$

The 2-hydroxypropanoic acid that is produced still has a chiral centre – this has not been affected by the hydrolysis reaction, which only involves the –CN group. So you still have a racemic mixture of two optical isomers (Figure 1).

▲ **Figure 1** *The optical isomers of 2-hydroxypropanoic acid*

It is often the case that a molecule with a chiral centre that is made synthetically ends up as a mixture of optical isomers. However, the same molecule produced naturally in living systems will often be present as only one optical isomer. Amino acids are a good example of this. All naturally occurring amino acids (except aminoethanoic acid, glycine) are chiral, but in every case only one of the isomers is formed in nature. This is because most naturally occurring molecules are made using enzyme catalysts, which only produce one of the possible optical isomers.

Optical isomers in the drug industry

Some drugs are optically active molecules. For some purposes, a racemic mixture of the two optical isomers will do. For other uses, only one isomer is required. Many drugs work by a molecule of the active ingredient fitting an area of a cell (called a receptor) or an enzyme's active site like a piece in a jigsaw puzzle. Because receptors have a three-dimensional structure, only one of a pair of optical isomers will fit. In some cases, one optical isomer is an effective drug and the other is inactive. This is a problem and there are three options:

1 Separate the two isomers – this may be difficult and expensive as optical isomers have very similar properties.
2 Sell the mixture as a drug – this is wasteful because half of it is inactive.
3 Design an alternative synthesis of the drug that makes only the required isomer.

The over-the-counter painkiller and anti-inflammatory drug ibuprofen (sold as Nurofen and Calprofen) is an example.

The structure of ibuprofen is:

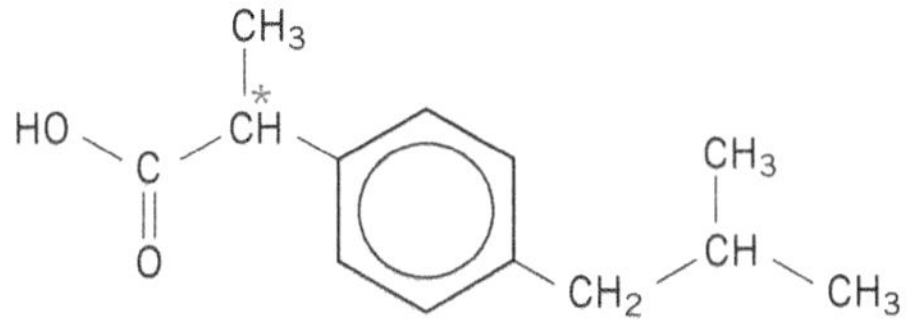

The starred carbon is the chiral centre. At present, most ibruprofen is made and sold as a racemate.

In some cases one of the optical isomers is an effective drug and the other is toxic or has unacceptable side effects. For example, naproxen has one isomer that is used to treat arthritic pain, whilst the other causes liver poisoning. In this case it is vital that only the effective optical isomer is sold.

The structure of ibuprofen

Ibuprofen is a popular remedy for mild pain and inflammation that is available over-the-counter under a variety of trade names such as Nurofen and Cuprofen. The skeletal formula of ibuprofen in shown in Figure 4.

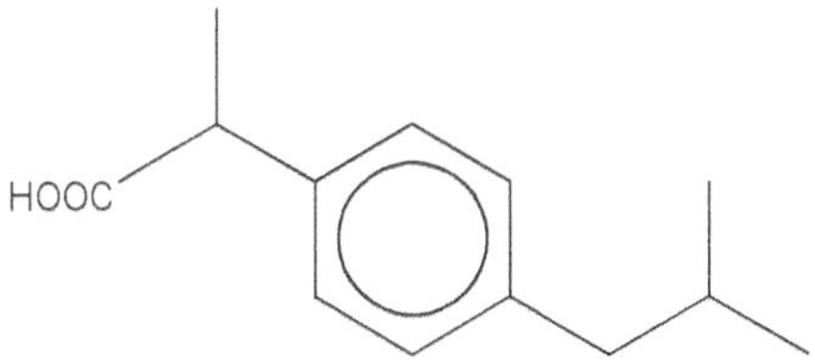

▲ **Figure 4** *Skeletal formula of ibuprofen*

Optical activity of ibuprofen

Ibuprofen can exist as a pair of optical isomers that are mirror images of each other. These mirror images are non-superimposable. This mirror image property occurs in molecules that have a carbon atom to which four different groups are bonded. The two optical isomers of ibuprofen are identified by the prefixes *R*– and *S*+,

Mirror image isomers are identical in many properties such as solubility, melting point, and boiling point. They can be distinguished by the fact that they rotate the plane of polarisation of polarised light in different directions– the (+)-isomer clockwise as the observer looks at the light, and the (–)-isomer anticlockwise. The symbols *R* and *S* refer to the 3-D arrangement of the atoms in space. The two isomers do, however, behave differently when they interact with other 'handed' molecules such as the prostaglandins, which are involved in the process of inflammation. Of the two optical isomers of ibuprofen it is the *S*+ form which has the anti-inflammatory and pain-killing effect.

However, it has been found that there is an enzyme in the body that converts the *R*– form into the *S*+. In fact, 60% of the *R*– form is converted into *S*+. This means that in a typical dose of ibuprofen of 400 mg, 200 mg is *S*+ and 200 mg *R*–. Of the 200 mg of *R*–, 60% (i.e., 120 mg) is converted into the active *S*+ form, giving a total of active form of 320 mg. Therefore there is little point in going to the trouble of synthesising the *S*+ form only, and ibuprofen is sold as a racemic mixture (one initially containing equal amounts of both optical isomers). However, a synthesis is possible that produces a pure sample of just one of the isomers.

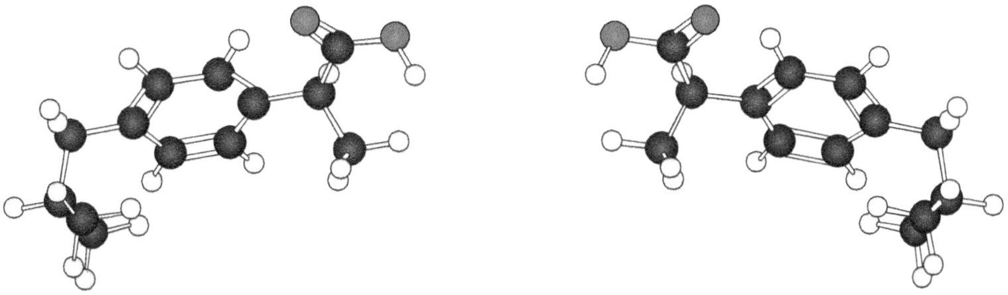

▲ **Figure 5** *S+ibuprofen (left) and R–ibuprofen (right) showing their mirror image relationship*

Identify the functional group in ibuprofen.

carboxylic acid

The thalidomide tragedy

Around the late 1950s there was a spate of incidents of children born with serious and unusual birth defects – missing, shortened, or deformed limbs. There were over 10 000 of these worldwide, almost 500 of them in the UK. Eventually it was realised that the common factor was that their mothers had all taken a drug called thalidomide in early pregnancy. The drug had been prescribed to relieve the symptoms of morning sickness. It had been tested on animals and considered safe (although the testing regime was much more relaxed than it would be today) but, crucially, there had been no tests on pregnant animals. In 1961 the drug was withdrawn.

Thalidomide exists as a pair of optical isomers. They differ in how they interact with other chiral molecules, which are common in living things. The isomers are extremely hard to separate and thalidomide was supplied as a racemic mixture produced when the the drug was synthesised. Apparently no one thought to test the two isomers separately.

Figure 6 shows one of the two enantiomers of thalidomide with the chiral carbon marked. The other isomer would have the positions of the C—H and the C—N bonds reversed, that is, the C—H going into the paper and the C—N coming out.

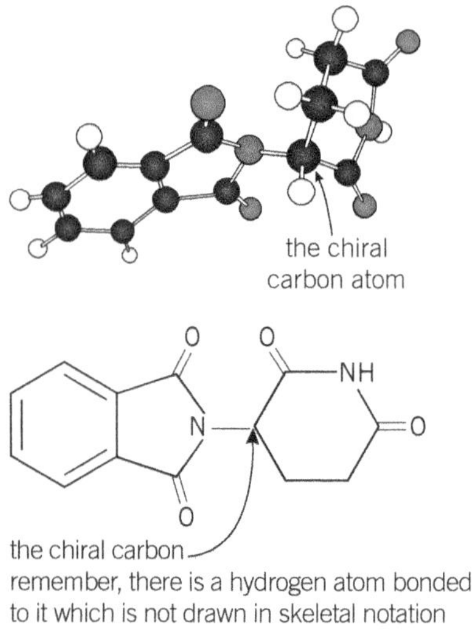

▲ **Figure 6** *One of the enantiomers of thalidomide with the chiral carbon atom marked.*
Key: black = carbon, red = oxygen, pale blue = hydrogen, dark blue = nitrogen

One of the enantiomers, called the *S*-form, is the one that caused the birth defects whilst the other, the *R*-form, is a safe sedative. It has been suggested that if just the *R*-form had been used, the tragedy would have been averted. However, there is evidence that in the human body, *R*-thalidomide is converted into *S*-thalidomide and so even if the pure *R*-form had been produced and taken, patients would have ended up with some of the *S*-form in their bodies.

Even after the ban in the early 1960s, pharmacologists continued to work with thalidomide. It appears that so long as it is not given in pregnancy, it is a safe and potentially useful drug and it is now being investigated as a treatment for a number of conditions including leprosy. Chemists have also produced a number of related compounds that are up to 4000 times more effective and have fewer side effects.

Study tip

Appreciate that different optical isomers can have very different drug action.

Synoptic link

The symbol ⌬ represents an aryl group – see Topic 27.1, Introduction to arenes.

Summary questions

1 a What would be the product of the reaction of propanal with hydrogen cyanide followed by reaction with dilute hydrochloric acid?

  b Does this molecule have a chiral centre? Explain your answer.

2 a What would be the product of the reaction of propanone with hydrogen cyanide followed by reaction with dilute hydrochloric acid?

  b Does this molecule have a chiral centre? Explain your answer.

3 Explain why the carbon marked ** in the formula of ibuprofen is *not* a chiral centre.

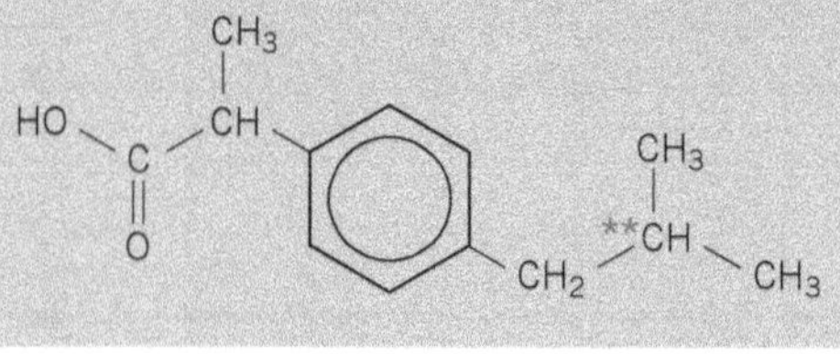

Practice questions

1 The amino acid alanine is shown below.

$$H_2N-\underset{\underset{H}{|}}{\overset{\overset{CH_3}{|}}{C}}-COOH$$

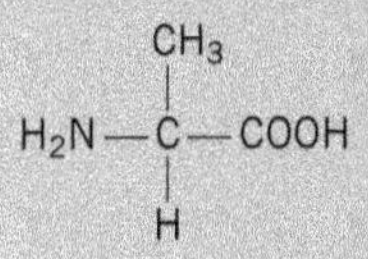

Give the systematic name for alanine.

(1 mark)
AQA, 2007

2 Phenylethanone, $C_6H_5COCH_3$, reacts with HCN according to the equation below.

$$C_6H_5COCH_3 + HCN \longrightarrow C_6H_5-\underset{\underset{CN}{|}}{\overset{\overset{OH}{|}}{C}}-CH_3$$

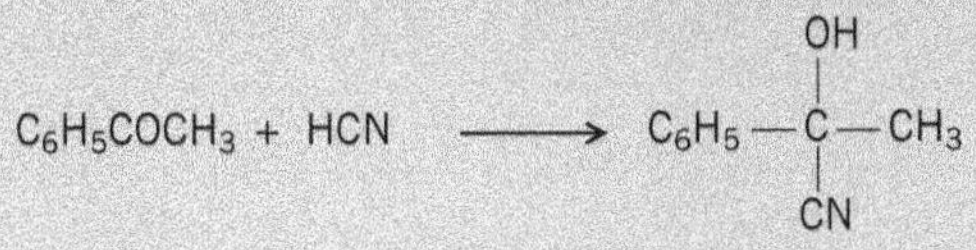

The product formed exists as a racemic mixture. State the meaning of the term *racemic mixture* and explain why such a mixture is formed in this reaction.

(3 marks)
AQA, 2007

3 The reaction of but-2-ene with hydrogen chloride forms a racemic mixture of the stereoisomers of 2-chlorobutane.
 (a) Name the type of stereoisomerism shown by 2-chlorobutane and give the meaning of the term *racemic mixture*. State how separate samples of the stereoisomers could be distinguished.

(4 marks)

 (b) By considering the shape of the reactive intermediate involved in the mechanism of this reaction, explain how a racemic mixture of the two stereoisomers of 2-chlorobutane is formed.

(3 marks)
AQA, 2006

4 Consider the reactions shown below.

$$CH_3CH_2-\underset{\underset{CH_3}{|}}{\overset{\overset{H}{|}}{C}}-CH_2OH \quad \textbf{J}$$

Reaction 1 → $\underset{\underset{CH_3}{}}{\overset{\overset{CH_3CH_2}{}}{C}}=CH_2$ **K**

Reaction 2 → $CH_3CH_2-\underset{\underset{CH_3}{|}}{\overset{\overset{H}{|}}{C}}-CHO$ **L**

Reaction 3 → $CH_3CH_2-\underset{\underset{CH_3}{|}}{\overset{\overset{H}{|}}{C}}-COOH$ **M**

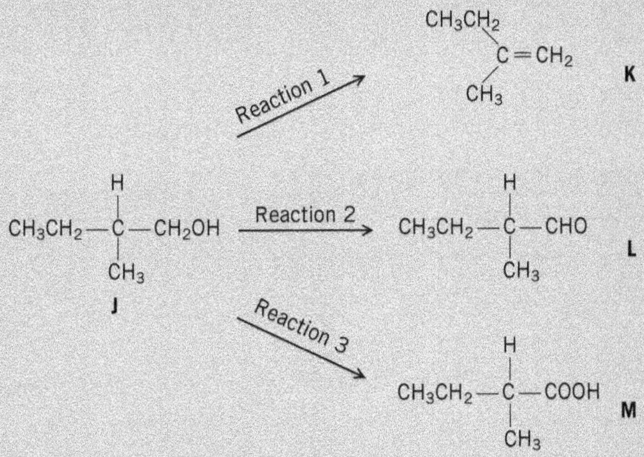

 (a) Name compound **J**.
 (b) Compound **J** exists as a pair of stereoisomers. Name this type of stereoisomerism.
 (c) Draw the structure of an isomer of **K** which shows stereoisomerism.

(3 marks)
AQA, 2007

5 (a) State the meaning of the term *stereoisomerism*.

(2 marks)

 (b) Draw the structure of an isomer of C_5H_{10} which shows *E–Z* isomerism and explain how this type of isomerism arises. Name the structure you have drawn.

(3 marks)

 (c) Name the structure below and state the type of isomerism it shows.

(2 marks)

$$H-\underset{\underset{OH}{|}}{\overset{\overset{COOH}{|}}{C}}-CH_3$$

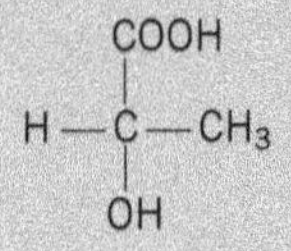

 (d) State how the different isomers of this structure can be distinguished from each other.

(2 marks)

6 (a) Define the term *stereoisomer*.

(1 mark)

 (b) (i) Draw the displayed formula of but-2-ene.

(1 mark)

 (ii) What type of isomerism is shown by but-2-ene?

(1 mark)

 (iii) What are the conditions necessary for this type of isomerism?

(2 mark)

7 Consider molecule A which is optically active.

$$H_5C_2{\overset{\overset{OH}{|}}{\underset{\underset{H}{|}}{C}}}COOH$$

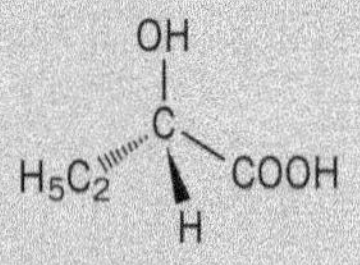

Molecule A

 (a) Define the term *optically active*.

(1 mark)

 (b) Draw the optical isomer of molecule A.

(1 mark)

 (c) (i) What is a racemic mixture?

(1mark)

 (ii) Why is a racemic mixture not optically active?

(1 mark)

8 The display formula of α-aminopropanoic acid is shown below.

$$H-\underset{\underset{H}{|}}{\overset{\overset{H}{|}}{C}}-\underset{\underset{H}{|}}{\overset{\overset{NH_2}{|}}{C}}-C\overset{O}{\underset{OH}{<}}$$

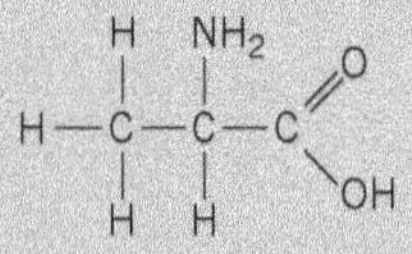

 (a) Circle the chiral centre.

(1 mark)

 (b) Use the display formula above to explain why this molecule is optically active.

(1 mark)

Answers to the Practice Questions and Section Questions are available at
www.oxfordsecondary.com/oxfordaqaexams-alevel-chemistry

26 Compounds containing the carbonyl group

26.1 Introduction to aldehydes and ketones

The carbonyl group consists of a carbon–oxygen double bond: 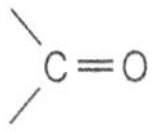

The group is present in **aldehydes** and **ketones**.

In aldehydes, the carbon bonded to the oxygen (the carbonyl carbon) has at least one hydrogen atom bonded to it, so the general formula of an aldehyde is:

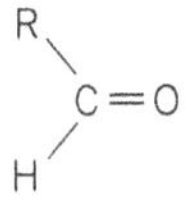

This is sometimes written as RCHO.

In ketones, the carbonyl carbon has two organic groups, which can be represented by R and R′, so the formula of a ketone is:

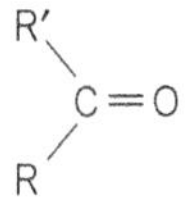

The R groups in both aldehydes and ketones may be alkyl or aryl.

How to name aldehydes and ketones

Aldehydes are named using the suffix '-al'. The carbon of the aldehyde functional group is counted as part of the carbon chain of the root. So:

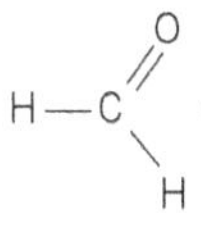 or HCHO is methanal and 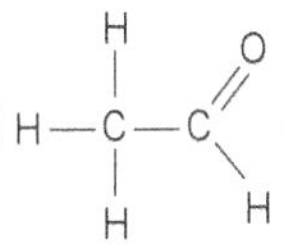 or CH_3CHO is ethanal.

The aldehyde group can *only* occur at the end of a chain, so a numbering system is not needed to show its location.

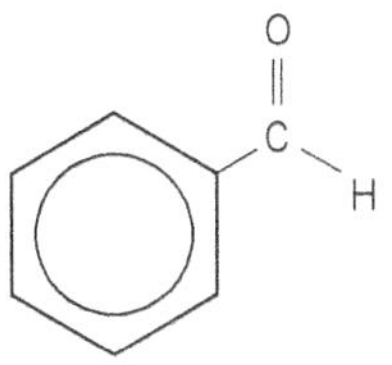

or C_6H_5CHO, is counted as a derivative of benzene (not of methylbenzene) and is called benzenecarbaldehyde. It is often still called by the old name 'benzaldehyde'.

Ketones are named using the suffix '-one'. In the same way as aldehydes, the carbon atom of the ketone functional group is counted as part of the root. So the simplest ketone:

Learning objectives:

→ Describe aldehydes and ketones.

→ State how aldehydes and ketones are named.

Specification reference: 3.3.8

Synoptic link

Alkyl means based on a saturated hydrocarbon group. *Aryl* means based on an aromatic system, see Topic 27.1, Introduction to arenes.

Synoptic link

You will need to know the oxidation of alcohols studied in Topic 15.2, The reactions of alcohols, and bond polarity studied in Topic 3.6, Electronegativity – bond polarity in covalent bonds.

Hint

When an aldehyde group is a substituent on a benzene ring, the suffix '-carbaldehyde' is used and the carbon is not counted as part of the root name.

Synoptic link

Look back at Topic 3.7, Forces acting between molecules, to remind yourself about hydrogen bonding.

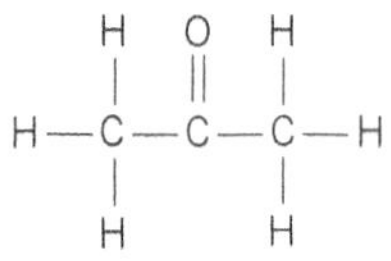 or CH_3COCH_3, is called propanone.

No ketone with fewer than three carbon atoms is possible.

You do not need to number the carbon in propanone or in butanone:

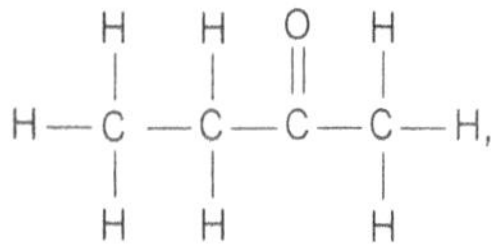

$C_2H_5COCH_3$, because the carbonyl group can only be in one position. With larger numbers of carbon atoms, numbers are needed to locate the carbonyl group on the chain, for example, pentanone could be pentan-3-one, $CH_3CH_3COCH_2CH_3$, or pentan-2-one, $CH_3COCH_2CH_2CH_3$.

Physical properties of carbonyl compounds

The carbonyl group is strongly polar, $C^{\delta+}{=}O^{\delta-}$, so there are permanent dipole–dipole forces between the molecules. These forces mean that boiling points are higher than those of alkanes of comparable relative molecular mass but not as high as those of alcohols, where hydrogen bonding can occur between the molecules (Table 1).

▼ **Table 1** *Boiling point data*

Name	Formula	M_r	T_b / K
butane	$CH_3CH_2CH_2CH_3$	60	273
propanone	CH_3COCH_3	58	359
propan-1-ol	$CH_3CH_2CH_2OH$	60	370

Solubility in water

Shorter chain aldehydes and ketones mix completely with water because hydrogen bonds form between the oxygen of the carbonyl compound and water (Figure 1). As the length of the carbon chain increases, carbonyl compounds become less soluble in water.

Methanal, HCHO, is a gas at room temperature. Other short chain aldehydes and ketones are liquids, with characteristic smells (propanone, sometimes known as acetone, is found in many brands of nail varnish remover). Benzenecarbaldehyde smells of almonds and is used to scent soaps and flavour food.

The reactivity of carbonyl compounds

The C=O bond in carbonyl compounds is strong (Table 2) and you might think that the C=O bond would be the least reactive bond. However, almost all reactions of carbonyl compounds involve the C=O bond.

This is because the big difference in electronegativity between carbon and oxygen makes the $C^{\delta+}{=}O^{\delta-}$ strongly polar. So, nucleophilic reagents can attack the $C^{\delta+}$. Also, since they contain a double bond, carbonyl compounds are unsaturated and addition reactions are possible.

In fact, the most typical reactions of the carbonyl group are nucleophilic additions.

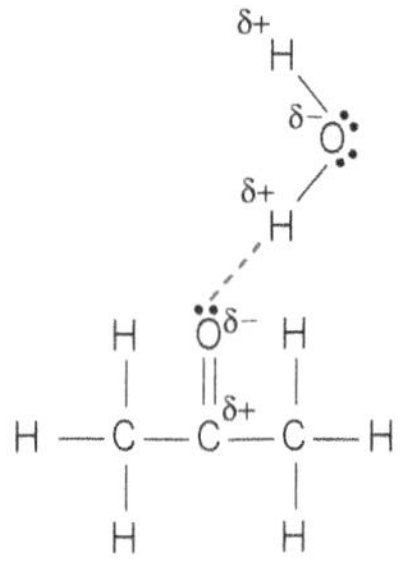

▲ **Figure 1** *Hydrogen bonding between propanone and water*

▼ **Table 2** *Comparison of bond strengths*

Bond	Mean bond enthalpy / kJ mol^{-1}
C=O	740
C=C	612
C—O	358
C—C	347

Summary questions

1. Name the following compounds:

   a 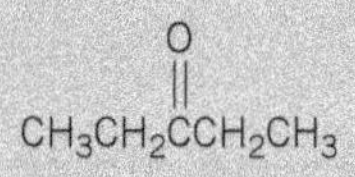

   b 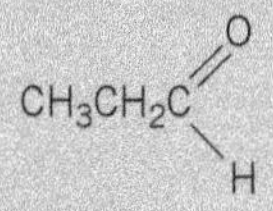

2. Explain why:

   a No ketone with fewer than three carbons is possible.

   b No numbering system is needed in the ketone butanone.

   c No numbering is ever needed to locate the position of the C=O group when naming aldehydes.

3. Explain why there are no hydrogen bonds between propanone molecules.

4. Explain why hydrogen bonds can form between propanone and water molecules.

26.2 Reactions of the carbonyl group in aldehydes and ketones

Learning objectives:

- → Describe the mechanism of nucleophilic addition reactions of carbonyl compounds.
- → Describe how these compounds react when oxidised or reduced.

Specification reference: 3.3.8

Many of the reactions of carbonyl compounds are nucleophilic addition reactions.

They also undergo redox reactions.

Nucleophilic addition reactions

By representing the nucleophile as :Nu⁻, the general reaction is:

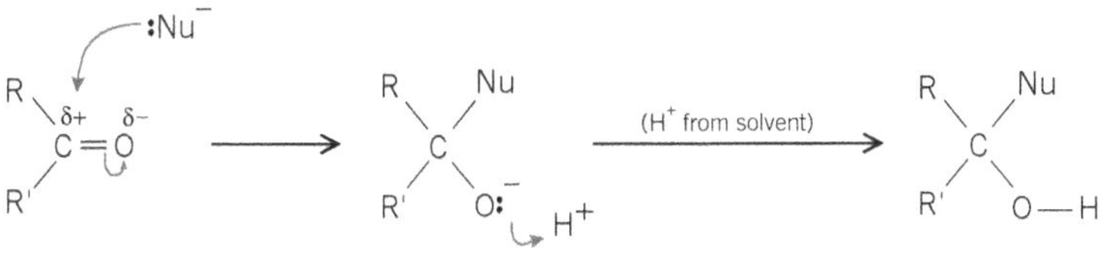

The addition of hydrogen cyanide is a good example of a nucleophilic addition.

Addition of hydrogen cyanide

Sodium cyanide or potassium cyanide is used as a source of cyanide ions followed by the addition of dilute hydrochloric acid. You will not carry out this reaction in the laboratory because of the toxic nature of the CN^- ion. The products are called **hydroxynitriles**. Hydroxynitriles are useful in synthesis because both the –OH and –CN groups are reactive and can be converted into other functional groups. Here the nucleophile is $:CN^-$

With a ketone:

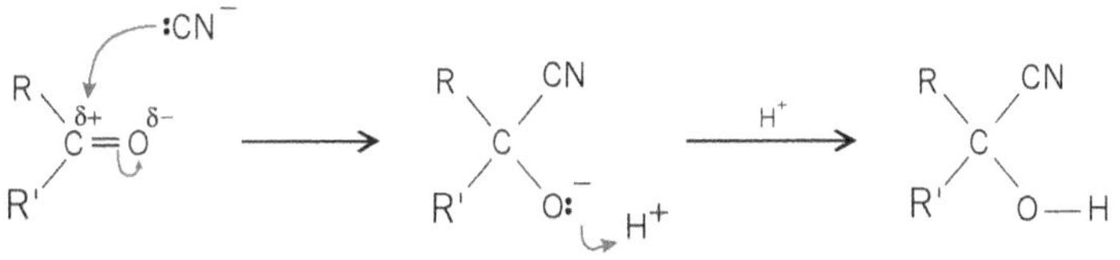

Or with an aldehyde:

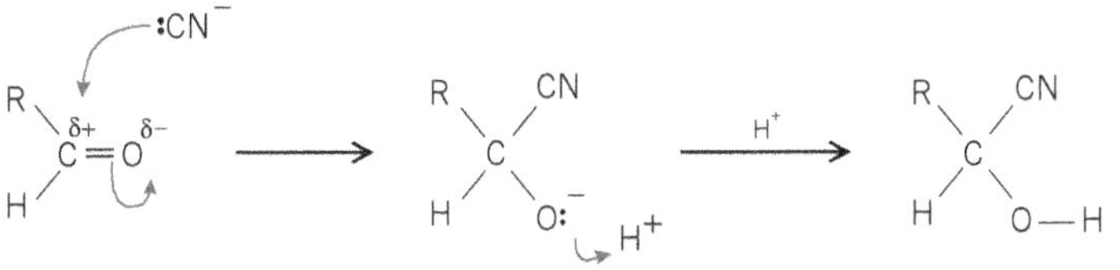

The overall balanced equation for the reaction with an aldehyde is:

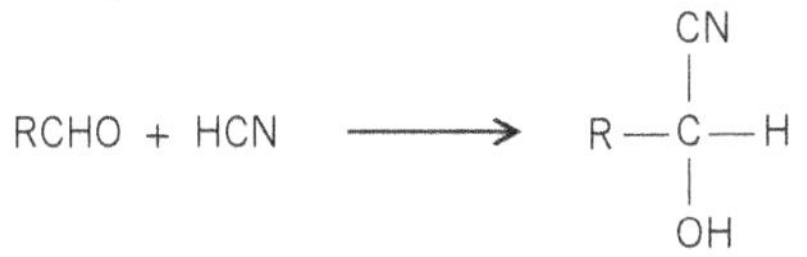

This reaction is important in organic synthesis because it increases the length of the carbon chain by one carbon. The products are called hydroxynitriles. (This is an example where the –OH group is named using the prefix 'hydroxy-' rather than the suffix '-ol'.)

This reaction will produce a racemic mixture of two optical isomers (enantiomers) when carried out with an aldehyde or an unsymmetrical ketone, because there is an equal probability that the $:CN^-$ ion may attack from above or below the flat C=O group (Figure 1).

Study tip

Nucleophiles have a lone pair to attack the $C^{\delta+}$. Some nucleophiles are negatively charged, others use the negative end of a dipole to attack $C^{\delta+}$.

Study tip

In theory, hydrogen cyanide could be used as the nucleophile. However, this is toxic and, being a gas, it is hard to stop it escaping into the laboratory.

Synoptic link

Optical isomerism and racemic mixtures were covered in Topic 25.3, Synthesis of optically active compounds.

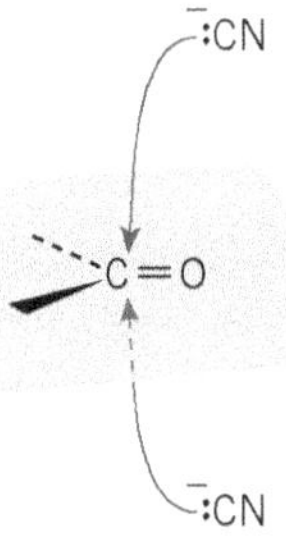

▲ **Figure 1** *The $:CN^-$ ion may attack from above or below the C=O group*

Redox reactions

Oxidation

Aldehydes can be oxidised to carboxylic acids. Remember that [O] is used to represent the oxidising agent.

One oxidising agent commonly used is acidified (with dilute sulfuric acid) potassium dichromate(VI), $K_2Cr_2O_7/H^+$.

Ketones *cannot* be oxidised easily to carboxylic acids because, unlike aldehydes, a C—C bond must be broken. Stronger oxidising agents break the hydrocarbon chain of the ketone molecule resulting in a shorter chain molecule, carbon dioxide, and water.

Distinguishing aldehydes from ketones

Weak oxidising agents can oxidise aldehydes but not ketones. This is the basis of two tests to distinguish between them.

Fehling's test

Fehling's solution is made from a mixture of two solutions – Fehling's A which contains the Cu^{2+} ion and is therefore coloured blue, and Fehling's B which contains an alkali and a complexing agent.

- When an aldehyde is warmed with Fehling's solution, a brick red precipitate of copper(I) oxide is produced as the copper(II) oxidises the aldehyde to a carboxylic acid, and is itself reduced to copper(I).
- Ketones give no reaction to this test.

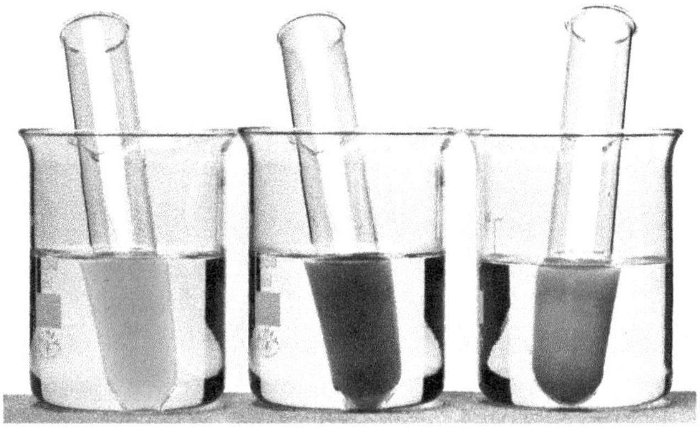

▲ **Figure 2** *When an aldehyde is warmed with Fehling's solution, the blue colour will turn green then a brick-red precipitate forms*

The silver mirror test

Tollens' reagent contains the complex ion $[Ag(NH_3)_2]^+$ which is formed when aqueous ammonia is added to an aqueous solution of silver nitrate.

- When an aldehyde is warmed with Tollens' reagent, metallic silver is formed. Aldehydes are oxidised to carboxylic acids by Tollens' reagent. The Ag^+ is reduced to metallic silver. A silver mirror will be formed on the inside of the test tube (which has to be spotlessly clean).

  $RCHO + [O] \rightarrow RCOOH$ The aldehyde is oxidised.

  $[Ag(NH_3)_2]^+ + e^- \rightarrow Ag + 2NH_3$ The silver is reduced.

- Ketones give no reaction to this test.

Synoptic link

You met with aldehydes and ketones in Topic 15.2, The reactions of alcohols. They are formed from the oxidation of primary and secondary alcohols, respectively.

Synoptic link

Remember what was covered on the Fehling's test in Topic 15.2, The reactions of alcohols.

Synoptic link

A complexing agent can form co-ordinate (dative bonds) with metal atoms or ions, see Topic 23.2, Complex formation and the shape of complex ions.

Hint

Benedict's solution is similar to Fehling's solution but is more convenient as it does not have to be prepared by mixing. It also contains Cu^{2+} ions but has a different complexing agent.

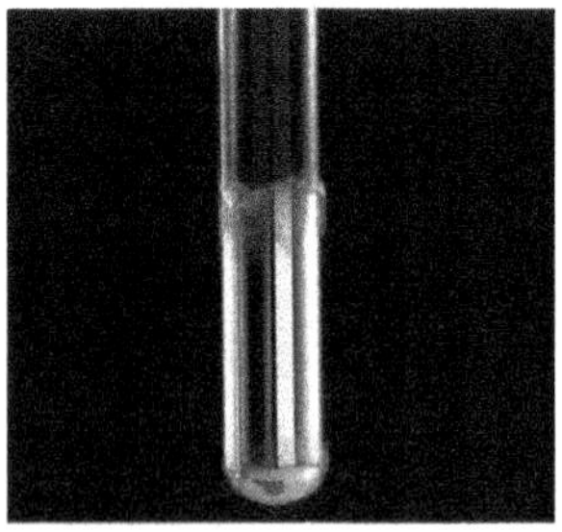

▲ **Figure 3** *The reaction of aldehydes with Ag^+ ions was once used as a method of silvering mirrors.*

Study tip

Compounds containing the carbonyl group have a strong absorption in the infrared spectrum at around 1700 cm^{-1}. This can be used to show the presence of this bond.

Reduction

Many reducing agents will reduce both aldehydes and ketones to alcohols. One such reducing agent is sodium tetrahydridoborate(III), $NaBH_4$, in aqueous solution. This generates the nucleophile $:H^-$, the hydride ion.

This reduces $C^{\delta+}{=}O^{\delta-}$ but not C=C as $:H^-$ is repelled by the high electron density in the C=C bond, but is attracted to the $C^{\delta+}$ of the C=O bond.

Reducing an aldehyde

Aldehydes are reduced to primary alcohols by the following mechanism in which H^- acts as a nucleophile:

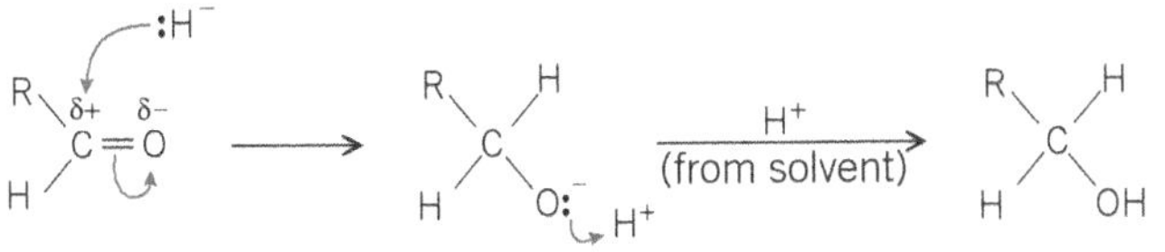

primary alcohol

[H] is used to represent reduction in equations.

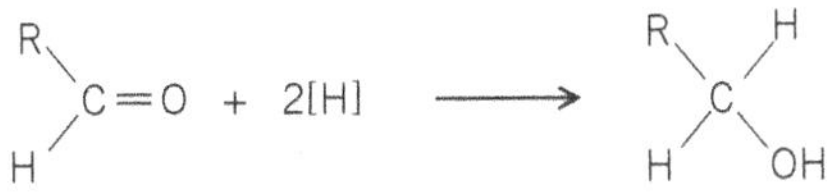

Reducing a ketone

Ketones are reduced to secondary alcohols in a similar way.

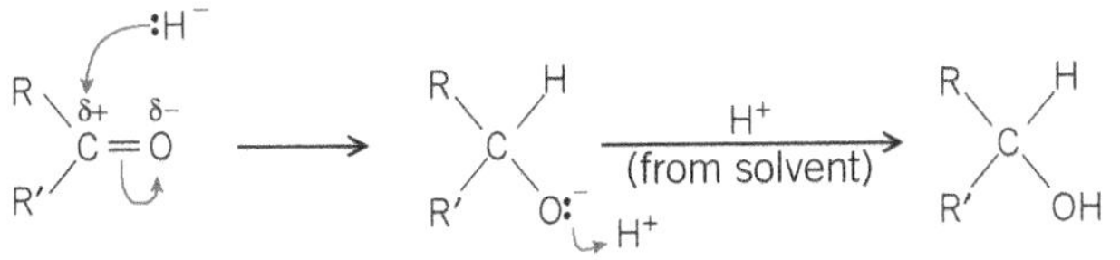

secondary alcohol

Using [H]:

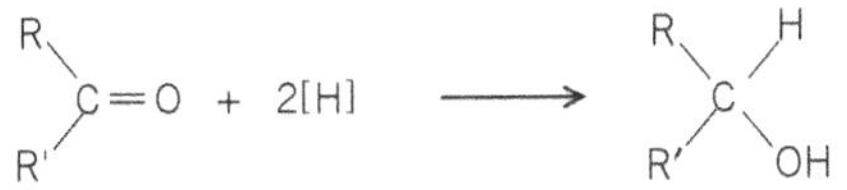

These reactions are **nucleophilic addition reactions** (because the H^- ion is a nucleophile).

Hint

Sodium tetrahydridoborate(III) is sometimes called sodium borohydride.

Summary questions

1. Which of the following is a nucleophile?
   H^+, Cl^-, Cl•
2. Sodium tetrahydridoborate(III) generates the nucleophile $:H^-$ and converts aldehydes and ketones to alcohols.
   a. Would you expect this reagent to reduce

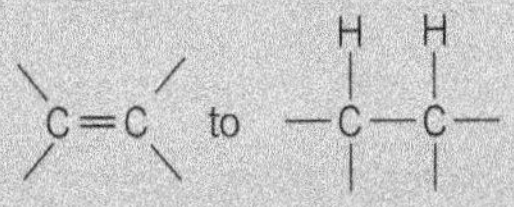

   b. Explain your answer.
   c. Predict the product when sodium tetrahydridoborate(III) reacts with:

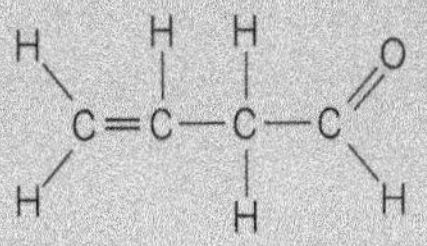

3. Hydrogen with a suitable catalyst will add on to C=C bonds as well as reducing the carbonyl group to an alcohol. Predict the product when hydrogen reacts with the compound in question **2c** in the presence of a suitable catalyst.
4. Explain why the reaction of CH_3CHO with HCN forms a racemic mixture, whilst that with CH_3COCH_3 forms a single compound.

26.3 Carboxylic acids and esters

Learning objectives:

- → Describe carboxylic acids and esters.
- → State how they are named.

Specification reference: 3.3.9

Synoptic link

You will need to know: the principles of the IUPAC naming system covered in Topic 11.2, Nomenclature – naming organic compounds, and Topic 25.1, Naming organic compounds.

Carboxylic acids

The carboxylic acid functional group is 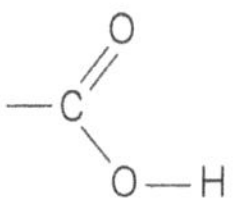

This is sometimes written as –COOH or as $-CO_2H$. This group can only be at the end of a carbon chain.

Carboxylic acids have two functional groups that you have seen before:

- the carbonyl group, 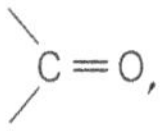, found in aldehydes and ketones
- the hydroxy group, –OH, found in alcohols.

Having two groups on the same carbon atom changes the properties of each group. The most obvious difference is that the –OH group in carboxylic acids is much more acidic than the –OH group in alcohols.

The most familiar carboxylic acid is ethanoic acid (acetic acid), which is the acid in vinegar.

How to name carboxylic acids

Carboxylic acids are named using the suffix '-oic acid'. The carbon atom of the functional group is counted as part of the carbon chain of the root. So, HCOOH is methanoic acid, CH_3COOH is ethanoic acid, and so on.

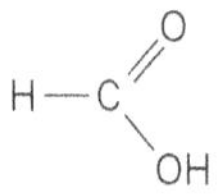

methanoic acid

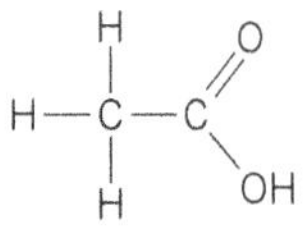

ethanoic acid

Where there are substituents or side chains on the carbon chain, they are numbered, counting from the carbon of the carboxylic acid as carbon number one. So, $CH_3CHBrCOOH$ is 2-bromopropanoic acid and $CH_3CH(CH_3)CH_2COOH$ is 3-methylbutanoic acid.

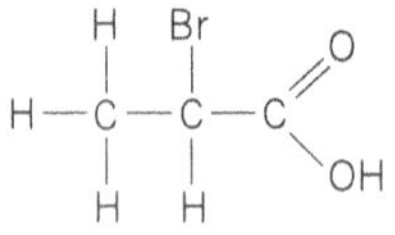

2-bromopropanoic acid

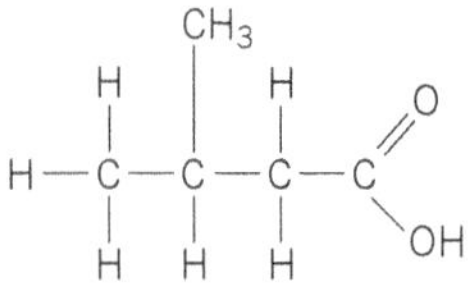

3-methylbutanoic acid

When the functional group is attached to a benzene ring, the suffix '-carboxylic acid' is used and the carbon of the functional group is *not* counted as part of the root. So, C_6H_5COOH is benzenecarboxylic acid.

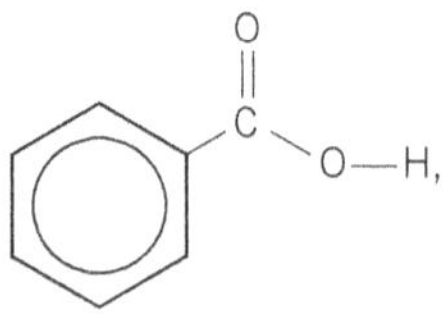

benzenecarboxylic acid (this is still often called benzoic acid)

Physical properties of carboxylic acids

The carboxylic acid group can form hydrogen bonds with water molecules (Figure 1). For this reason carboxylic acids up to, and including, four carbons (butanoic acid) are completely soluble in water.

The acids also form hydrogen bonds with one another in the solid state (Figure 2). They therefore have much higher melting points than the alkanes of similar relative molecular mass. Ethanoic acid ($M_r = 60$) melts at 290 K whilst butane ($M_r = 58$) melts at 135 K.

One way of identifying a carboxylic acid is to measure its melting point and compare it with tables of known melting points. A Thiele tube may be used (Figure 3), or the melting point can be found electrically (Figure 4).

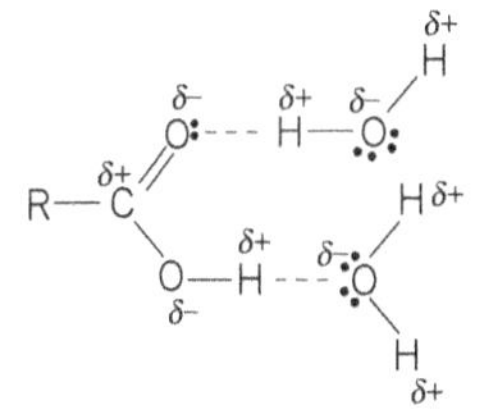

▲ **Figure 1** *A molecule of a carboxylic acid forming hydrogen bonds with water*

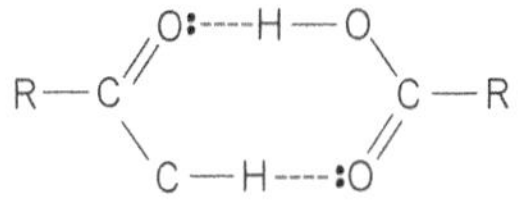

▲ **Figure 2** *Two carboxylic acid molecules can hydrogen bond together to form a pair called a* ***dimer***

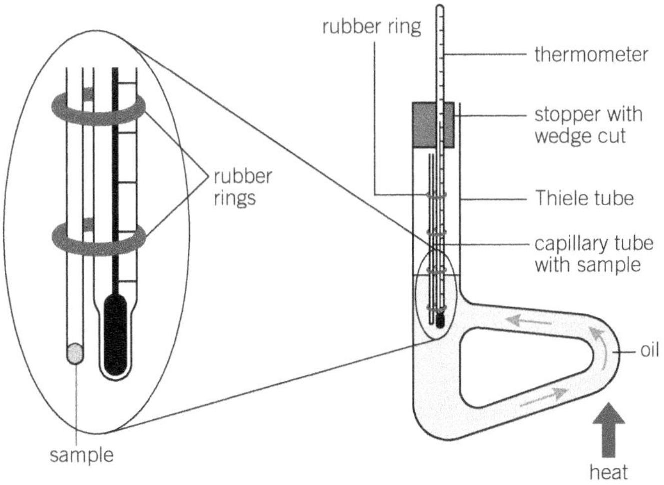

▲ **Figure 3** *A Thiele tube may be used to measure a melting point*

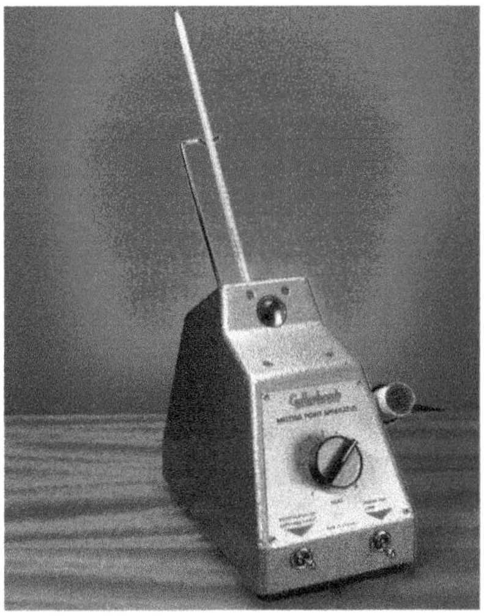

▲ **Figure 4** *Modern electrical melting point apparatus*

Pure ethanoic acid is sometimes called **glacial** ethanoic acid because it may freeze on a cold day – its freezing point is 13 °C (260 K).

The acids have characteristic smells. You will recognise the smell of ethanoic acid as vinegar, whilst butanoic acid causes the smell of rancid butter.

The non-systematic names of hexanoic and octanoic acids are caproic and capryllic acid respectively, from the same derivation as Capricorn the goat. They are present in goat fat and cause its unpleasant smell.

Esters

Esters are derived from carboxylic acids. The hydrogen (from the –OH group of the acid) is replaced by a hydrocarbon group – an alkyl or aryl group (OH is replaced by OR). So the general formula is:

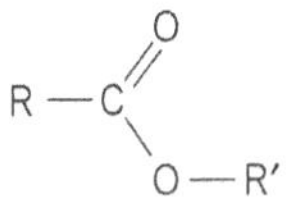

or RCOOR′.

Study tip

Take care with the names of esters. It is easy to get them the wrong way round. The part of the name relating to the acid comes last. Also, remember that the acid is named from the number of carbon atoms – including the carbon of the functional group.

Study tip

Esters such as ethyl ethanoate are ingredients of many brands of nail varnish remover. They dissolve the nitrocellulose-based polymer in the varnish.

Hint

Plasticisers are added to plastics such as PVC to make them softer and more flexible. The small molecules of the plasticiser get in between the long chain molecules and allow them to slide across one another more easily. Without plasticiser, PVC is rigid and is used for drain pipes, for example. With plasticiser, it is flexible and can be used as a waterproof fabric in tablecloths and aprons, for example.

How to name esters

The names of esters are based on that of the parent acid, for example, all esters from ethanoic acid are called ethanoates. But, the name always begins with the alkyl (or aryl) group that has replaced the hydrogen of the acid, rather than the name of the acid.

For example:

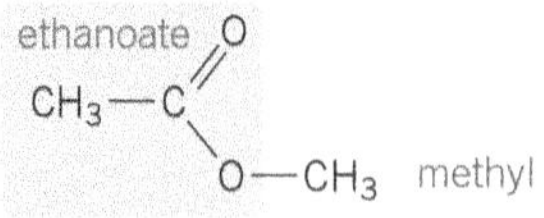

is called methyl ethanoate.

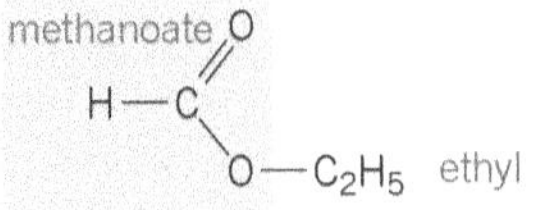

is called ethyl methanoate.

Short chain esters are fairly volatile and have pleasant fruity smells, so that they are often used in flavourings and perfumes. For example, 3-methylbutyl ethanoate has the smell of pear drops, octyl ethanoate is orange flavoured, whilst pentyl pentanoate smells and tastes of apples. They are also used as solvents and plasticisers. Fats and oils are esters with longer carbon chains.

Summary questions

1 Give the name of:

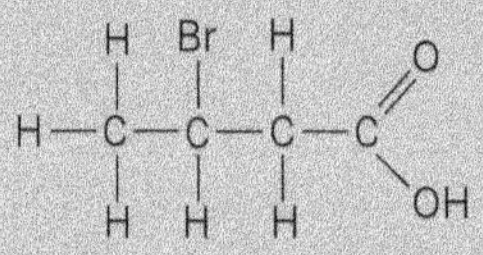

2 Write the displayed formula for 3-chloropropanoic acid.

3 Why is it not necessary to call propanoic acid the name 1-propanoic acid?

4 Give the names of the following esters:

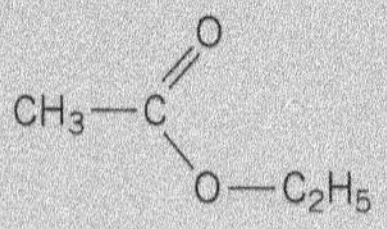

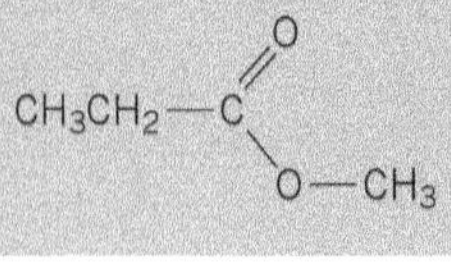

26.4 Reactions of carboxylic acids and esters

Learning objectives:

- → Describe how carboxylic acids react.
- → State how esters are formed from carboxylic acids.
- → Describe how esters are hydrolysed.
- → Describe how esters are used.

Specification reference: 3.3.9

Reactivity of carboxylic acids

The carboxylic acid group is polarised as shown:

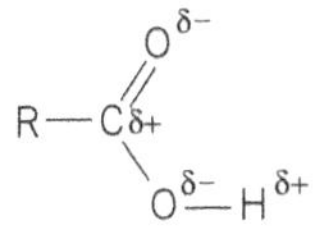

- The $C^{\delta+}$ is open to attack from nucleophiles.
- The $O^{\delta-}$ of the C=O may be attacked by positively charged species (like H^+, in which case you say it has been protonated).
- The $H^{\delta+}$ may be lost as H^+, in which case the compound is behaving as an acid.

Loss of a proton

If the hydrogen of the –OH group is lost, a negative ion – a carboxylate ion – is left.

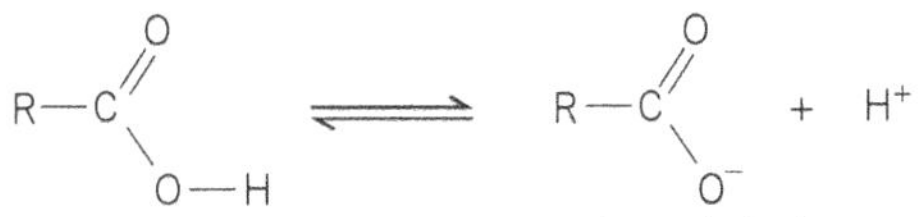

a carboxylate ion

The negative charge is shared over the whole of the carboxylate group.

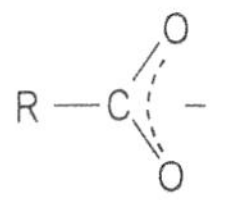

This **delocalisation** makes the resulting ion more stable.

Carboxylic acids are weak acids, so the equilibrium is well over to the left:

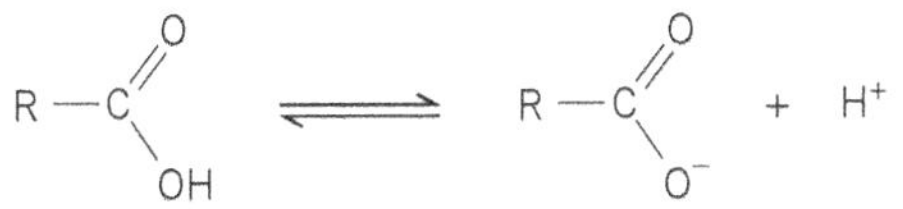

Even so, they are strong enough to react with sodium hydrogencarbonate, $NaHCO_3$, to release carbon dioxide. This distinguishes them from other organic compounds that contain the –OH group, such as alcohols.

$$CH_3COOH(aq) + NaHCO_3(aq) \rightarrow CH_3COONa(aq) + H_2O(l) + CO_2(g)$$

ethanoic acid, sodium hydrogencarbonate, sodium ethanoate, water, carbon dioxide

Reactions of acids

Carboxylic acids are proton donors and show the typical reactions of acids.

They form ionic salts with the more reactive metals, alkalis, metal oxides, or metal carbonates in the usual way. The salts that are formed have the general name **carboxylates**, and are named from the particular acid. Methanoic acid gives **methanoates**, ethanoic acid gives **ethanoates**, propanoic acid gives **propanoates**, and so on.

Study tip

The carboxylic acid group contains both the carbonyl group and the alcohol group. However, the two groups react differently when they are next to each other in a molecule.

Study tip

The stability of the carboxylate ion is what allows the H^+ ion to be released and makes the molecules acidic.

Study tip

Carboxylic acids give CO_2 with $NaHCO_3(aq)$, solid Na_2CO_3, and solid $NaHCO_3$.

▲ **Figure 1** *Carboxylic acids fizz with sodium carbonate*

For example, ethanoic acid reacts with aqueous sodium hydroxide:

$$\underset{\text{ethanoic acid}}{CH_3COOH(aq)} + \underset{\text{sodium hydroxide}}{NaOH(aq)} \rightarrow \underset{\text{sodium ethanoate}}{CH_3COONa(aq)} + \underset{\text{water}}{H_2O(l)}$$

Ethanoic acid reacts with aqueous sodium carbonate:

$$\underset{\text{ethanoic acid}}{2CH_3COOH(aq)} + \underset{\text{sodium carbonate}}{Na_2CO_3(aq)} \rightarrow \underset{\text{sodium ethanoate}}{2CH_3COONa(aq)} + \underset{\text{water}}{H_2O(l)} + \underset{\text{carbon dioxide}}{CO_2(g)}$$

Esters

Formation of esters

Esters, general formula RCOOR′, are **acid derivatives**.

Carboxylic acids react with alcohols to form esters. This reaction is speeded up by a strong acid catalyst. This is a reversible reaction and forms an equilibrium mixture of reactants and products. For example:

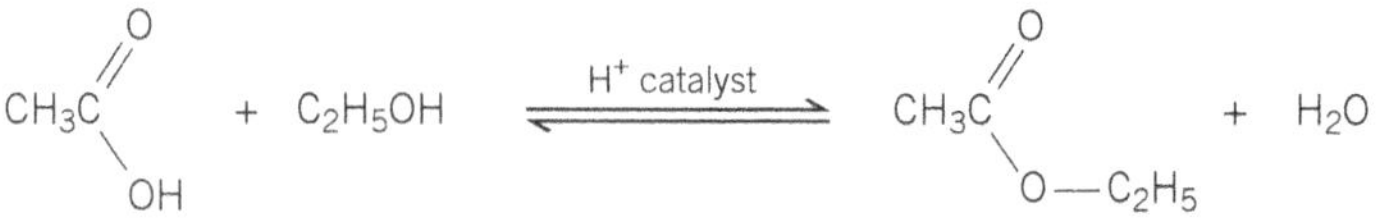

ethanoic acid ethanol ethyl ethanoate water

Hydrolysis of esters

The carbonyl carbon atom of an ester has a δ+ charge and is therefore attacked by water acting as a weak nucleophile. This is the reverse of the reaction above. The equation is:

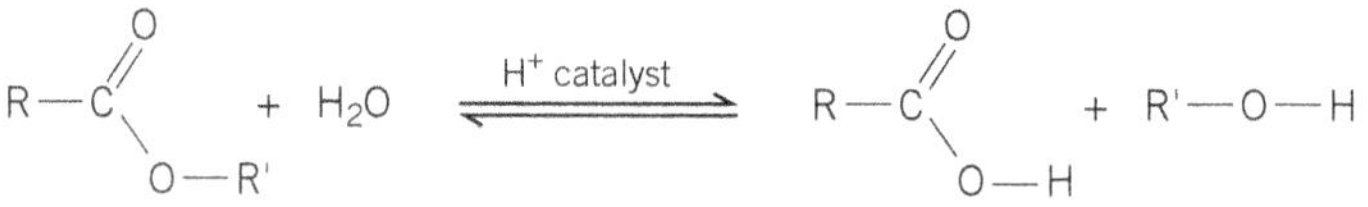

ester carboxylic acid alcohol

The hydrolysis (reaction with water) of esters does not go to completion. It produces an equilibrium mixture containing the ester, water, acid, and alcohol. The acid is a catalyst so it affects only the rate at which equilibrium is reached, not the composition of the equilibrium mixture.

An ester can be hydolysed at room temperature when a strong acid catalyst is used. The balanced equation for the acid catalysed hydrolysis of ethyl ethanoate is:

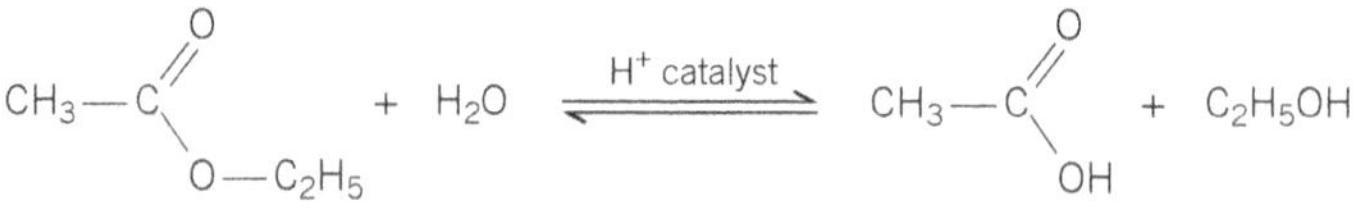

Bases also catalyse hydrolysis of esters. In this case, the salt of the acid is produced rather than the acid itself. This removes the acid from the reaction mixture so an equilibrium is not established and the reaction goes to completion, so there is more product in the mixture.

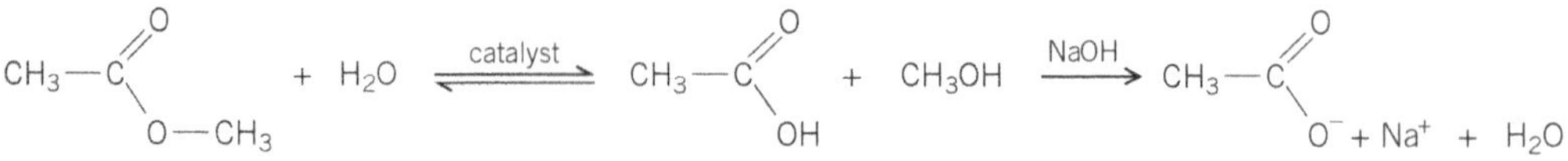

sodium ethanoate

Esters

In the laboratory, esters are made by warming the appropriate acid and alcohol with concentrated sulfuric acid. The ester will be more volatile than the original alcohol and carboxylic acid and may be distilled off the reaction mixture.

Synoptic link

Equilibrium is covered in Topic 6.3, The equilibrium constant K_c.

Study tip

When an acid catalyst is used in the hydrolysis of an ester an equilibrium mixture of reactants and products is obtained.

The mechanism of base hydrolysis

The mechanism of base hydrolysis of esters can be explained using 'curly arrows' to show the movement of electron pairs:

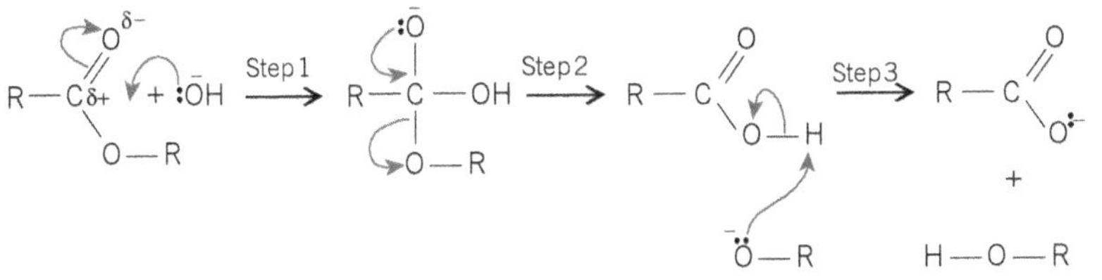

1 Describe Step 1.
2 What is the leaving group in Step 2.
3 How is RO^- acting in Step 3.

1 nucleophilic attack 2 RO^- 3 As a base

Hint

Esters are also used as perfumes, food flavourings (e.g., peardrop sweets), solvents, and plasticisers to modify the properties of polymers.

Uses of esters

Animal and vegetable oils and fats are the esters of the alcohol propane-1,2,3-triol (non-systematic name is glycerol). The only difference between a fat and an oil is that oils are liquid at room temperature, whilst fats are solid. Oils and fats contain three molecules of long chain (around 12–18 carbons) carboxylic acids called **fatty acids**. Since they are based on glycerol, fats and oils are referred to as **triglycerides** (Figure 2).

Fats and oils can be hydrolysed in acid conditions to give a mixture of glycerol and the component fatty acids.

They can also be hydrolysed by boiling with sodium hydroxide. Both the products are useful – glycerol and a mixture of sodium salts of the three acids which formed part of the ester. These salts are soaps. Soap can be a mixture containing many different salts. The type of soap depends on the fatty acids initially present in the ester.

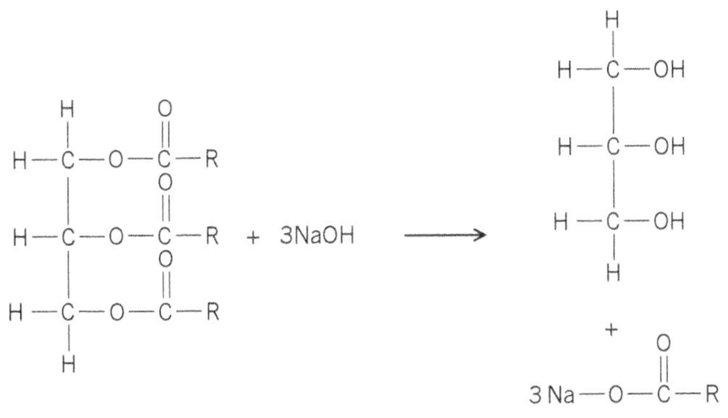

These sodium salts are ionic and dissociate to form Na^+ and $RCOO^-$. $RCOO^-$ has two distinct ends: a long hydrocarbon chain which is non-polar and the COO^- group which is polar and ionic. The hydrocarbon will mix with grease, while the COO^- mixes with water (Figure 3). So, these tadpole-shaped molecules allow grease and water to mix and therefore are used as cleaning agents.

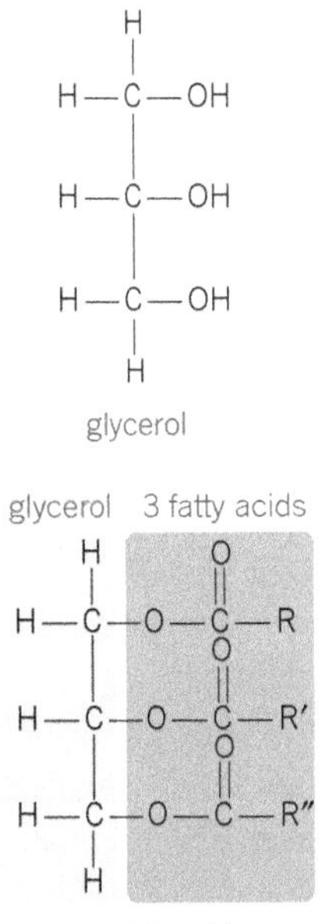

▲ **Figure 2** *Glycerol and a triglyceride*

▼ **Table 1** *Some common fatty acids*

Name	Formula	Details
stearic acid	$CH_3(CH_2)_{16}CO_2H$	present in most animal fats
palmitic acid	$CH_3(CH_2)_{14}CO_2H$	used in making soaps
oleic acid	$CH_3(CH_2)_7CH{=}CH(CH_2)_7CO_2H$	monounsaturated – it has one double bond, present in most fats and in olive oil
linoleic acid	$CH_3(CH_2)_4(CH{=}CHCH_2)_2(CH_2)_6CO_2H$	polyunsaturated, present in many vegetable oils

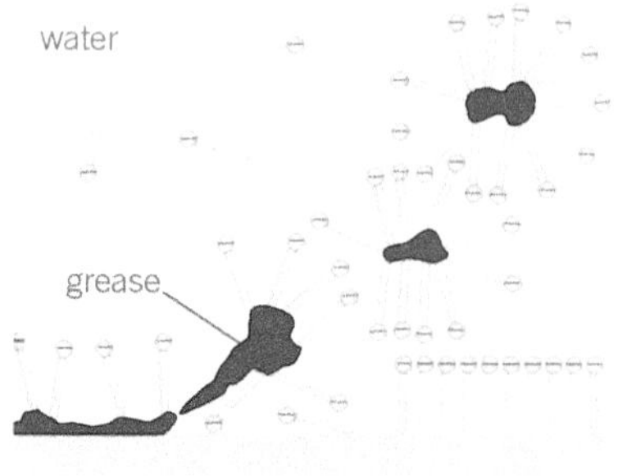

▲ **Figure 3** *The action of soap. The hydrocarbon ends of the ions, in yellow, mix with grease and the COO⁻ ends, in blue, lift it into aqueous solution*

▲ **Figure 4** *Many vehicles now run off fuels which contain biodiesel*

Summary questions

1. Carboxylic acids, being acidic, will react with the reactive metals. Give three other reactions that are typical of acids.
2. Name the acid and the alcohol that would react together to give the ester methyl ethanoate.
3. Name the acid and the alcohol that would react together to give the ester ethyl methanoate.
4. Methyl ethanoate and ethyl methanoate are a pair of isomers. Explain what this means.

Propane-1,2,3-triol (glycerol)

Glycerol has three O—H bonds, so it readily forms hydrogen bonds and is very soluble in water. It is a very important chemical in many industries and has a really wide range of uses.

- It is used extensively in many pharmaceutical and cosmetic preparations. Because it attracts water, it is used to prevent ointments and creams from drying out.
- It is used as a solvent in many medicines, and is present in toothpastes.
- It is used as a solvent in the food industry, for example, for food colourings.
- It is used to plasticise various materials like sheets and gaskets, cellophane, and special quality papers. Plasticisers are introduced between the molecules of the polymer which makes up the material and by allowing the molecules to slip over each other, the material becomes flexible and smooth. PVC may contain up to 50% plasticiser, such as esters of hexanedioic acid. Over time, the plasticiser leaks away, leaving the plastic brittle and inflexible.

Biodiesel

One possible solution to the reliance on crude oil as a source of fuel for motor vehicles is **biodiesel**. This is a renewable fuel, as it is made from oils derived from crops such as rape seed. Rape seed oil is a triglyceride ester. To make biodiesel the oil is reacted with methanol (with a strong alkali as a catalyst) to form a mixture of methyl esters, which can be used as a fuel in diesel vehicles with little or no modification. This process is being introduced commercially, but as the chemistry is relatively simple, some people are making their own biodiesel at home starting with used chip-shop oil, for example. Germany has thousands of filling stations supplying biodiesel, and it is cheaper there than ordinary diesel fuel. All fossil diesel fuel sold in France contains between 2% and 5% biodiesel.

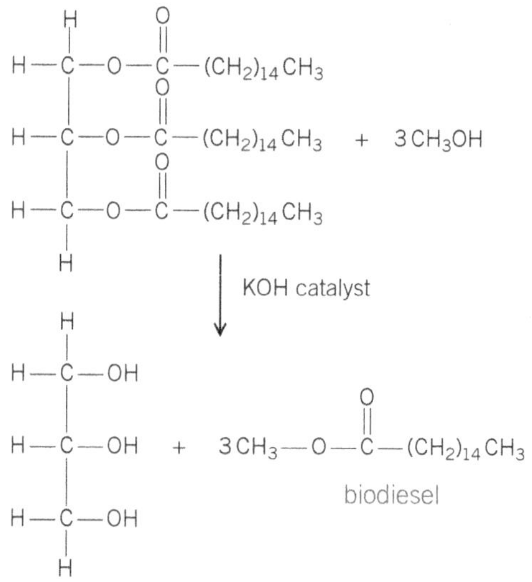

26.5 Acylation

Acylation is the process by which the acyl group is introduced into another molecule. The acyl group is:

$$R-C{\overset{O}{\underset{}{\diagdown}}}$$

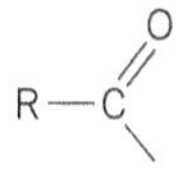

There is a group of compounds called **acid derivatives**, which all have the acyl group as part of their structure. Two important acid derivatives are acid chlorides and acid anhydrides. Acid derivatives are derived from carboxylic acids and have the general formula:

$$R-C{\overset{O}{\diagup}\atop{\diagdown Z}}$$

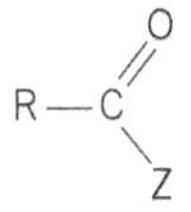

Z may be a variety of groups (Table 1).

If R is CH_3, the group is called **ethanoyl**.

▼ **Table 1** *Some acid derivatives*

−Z	Name of acid derivative	General formula	Example
−OR′	ester	RCOOR′	ethyl ethanoate, $CH_3COOC_2H_5$
−Cl	acid chloride	RCOCl	ethanoyl chloride, CH_3COCl
−OCOR′	acid anhydride	RCOOCOR′	ethanoic anhydride, $CH_3COOCOCH_3$

The carbonyl group of an acid derivative is polarised as shown:

$$R\overset{\delta+}{-}C{\overset{O^{\delta-}}{\diagup}\atop{\diagdown Z}}$$

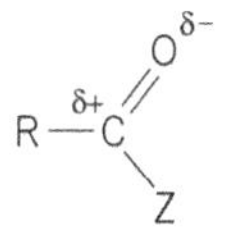

It is attacked by nucleophiles at the $C^{\delta+}$ and, in the process, the nucleophile replaces Z and the nucleophile therefore acquires an acyl group. So the nucleophile has been acylated.

The general reaction is:

$$R-\underset{Z}{\overset{\overset{\delta-}{O}}{C}}{}^{\delta+} + :Nu^- \longrightarrow R-\underset{Nu}{\overset{O}{C}} + :Z^-$$

How readily the reaction occurs depends on three factors:

1 The magnitude of the δ+ charge on the carbonyl carbon, which in turn depends on the electron-releasing or attracting power of Z.
2 How easily Z is lost. (Z is called the leaving group.)
3 How good the nucleophile is.

Learning objectives:

→ Describe acylation reactions.
→ Explain the nucleophilic addition–elimination mechanism for acylation reactions.

Specification reference: 3.3.9

Synoptic link

You will need to know the bond polarity and shapes of molecules studied in Topic 3.6, Electronegativity – bond polarity in covalent bonds, and Topic 3.5, The shapes of molecules and ions, and nucleophilic substitution reactions of halogenoalkanes studied in Topic 13.2, Nucleophilic substitution in halogenoalkanes

Study tip

A nucleophile has a lone pair of electrons and attacks positively charged carbon atoms.

Factors 1 and 2 tend to be linked – groups which strongly attract electrons tend to form stable negative ions, Z^-, and are good leaving groups.

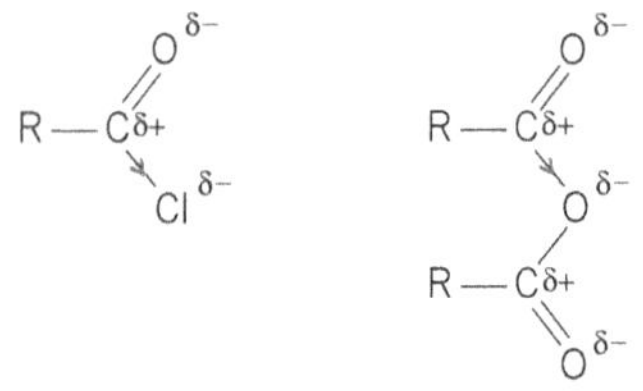

The Z groups of acyl chlorides and acid anhydrides *withdraw* electrons from the carbonyl carbon. This makes the carbon more positive and makes these compounds reactive towards nucleophiles. So, acyl chlorides and acid anhydrides are both good acylating agents. Acyl chlorides are somewhat more reactive than acid anhydrides.

Nucleophiles

Nucleophiles must have a lone pair of electrons which they use to attack an electron-deficient carbon, $C^{\delta+}$. The best nucleophiles are the ones that are best at donating their lone pair.

Acyl chlorides and acid anhydrides will both react with all the following nucleophiles, listed in order of reactivity:

The products of the reactions of these nucleophiles with acyl chlorides and acid anhydrides are shown in Table 2. These reactions are called **addition–elimination reactions**. These nucleophiles are all neutral so they must lose a hydrogen ion during the reaction.

One way of looking at these reactions is that a hydrogen atom of the nucleophile ('the active hydrogen') has been replaced by an acyl group.

- If the nucleophile is ammonia, the product is an amide.

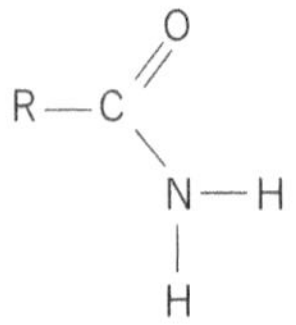

- If the nucleophile is a primary amine, the product is an *N*-substituted amide.

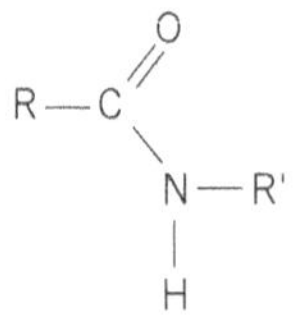

- If the nucleophile is the OH group of an alcohol, the product is an ester.
- If the nucleophile is OH from water, the product is a carboxylic acid.

▼ **Table 2** *The products of the reactions of acid derivatives with nucleophiles. All reactions take place at room temperature*

← increasing reactivity

acid derivative / nucleophile (↑ increasing reactivity)	R—C(=O)—Cl **acid chloride**	R—C(=O)—O—C(=O)—R **anhydride**
NH_3 ammonia	R—C(=O)—NH_2 amide	R—C(=O)—NH_2 amide
R′ – NH_2 amine	R—C(=O)—NHR′ *N*-substituted amide	R—C(=O)—NHR′ *N*-substituted amide
R′ – OH alcohol	R—C(=O)—O—R′ ester	R—C(=O)—O—R′ ester
H_2O water	R—C(=O)—OH carboxylic acid	R—C(=O)—OH carboxylic acid

The mechanism of the reactions

The mechanism of these reactions follows the same pattern, shown below.

1 Ethanoyl chloride and water (called hydrolysis).

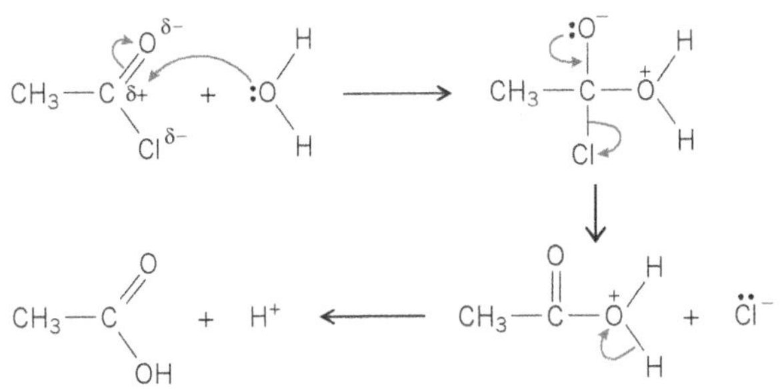

The overall equation may be written:

$$CH_3COCl + H_2O \rightarrow CH_3COOH + HCl$$

Study tip

A curly arrow shows movement of a pair of electrons.

Study tip

Many students find that it is best to work out the products of these reactions from the mechanism rather than to remember them.

Study tip

A second NH_3 removes a proton from the intermediate ion to give the final product.

2 Ethanoyl chloride and ethanol.

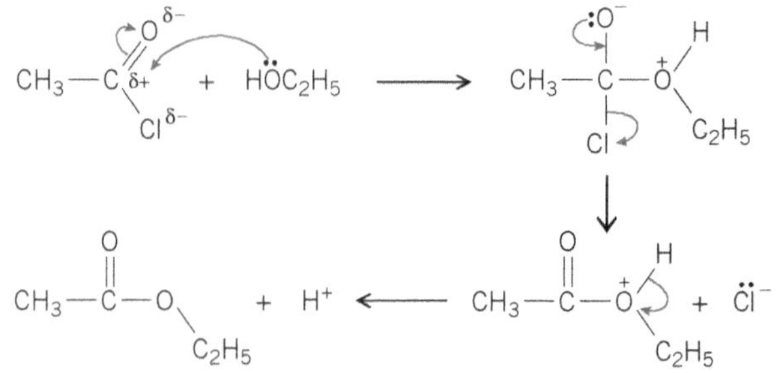

The overall equation may be written:

$$CH_3COCl + C_2H_5OH \rightarrow CH_3COOC_2H_5 + HCl$$

3 Ethanoyl chloride and ammonia. (The H^+ ion that is lost then reacts with a second molecule of ammonia to form NH_4^+.)

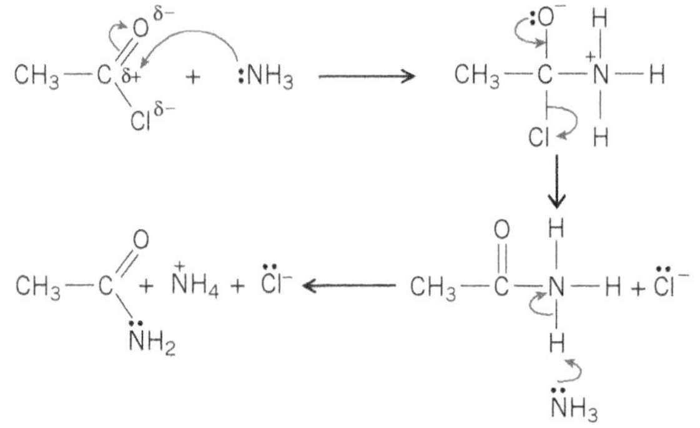

The overall equation may be written:

$$CH_3COCl + 2NH_3 \rightarrow CH_3CONH_2 + NH_4Cl$$

4 Ethanoyl chloride and methylamine.

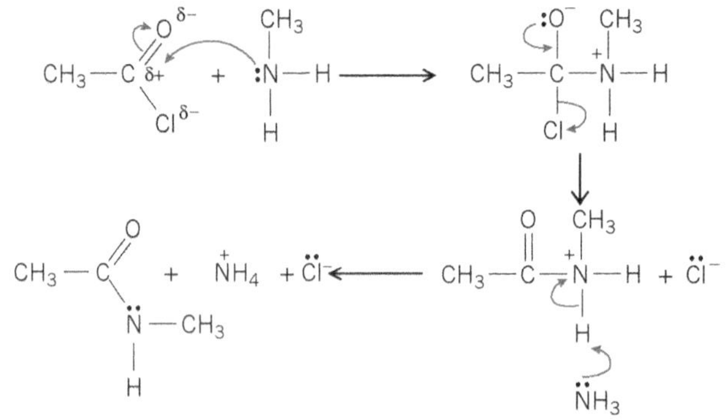

The overall equation may be written:

$$CH_3COCl + CH_3NH_2 \rightarrow CH_3CONHCH_3 + HCl$$

Uses of acylation reactions

Ethanoic anhydride is manufactured on a large scale. Its advantages over ethanoyl chloride as an acylating agent are:

- it is cheaper
- it is less corrosive
- it does not react with water as readily
- it is safer, as the by-product of its reaction is ethanoic acid rather than hydrogen chloride.

One important use is in the production of aspirin.

Aspirin

Aspirin (systematic name 2-ethanoyloxybenzenecarboxylic acid) is probably the most used medicine of all time – it must have been used to treat millions of headaches. It is often thought of as an over-the-counter remedy for moderate pain, which also reduces fever. However, it has more recently been shown to have many other effects, such as reducing the risk of heart attacks and some cancers. It is not without risks itself, for example, it can cause intestinal bleeding, and it has been suggested that if it were to be introduced as a new drug today it would be prescription-only.

Aspirin has a long history. Compounds related to it were originally extracted from willow bark. One old theory held that cures to diseases could be found near the cause, and willow, which grows in damp places, was suggested as a cure for rheumatism, which is made worse by dampness.

In 1890, the German chemist Felix Hofmann produced the ethanoyl (or acyl) derivative of salicylic acid (2-hydroxybenzenecarboxylic acid), from willow bark extract, and used it to treat his father's rheumatism. This derivative is what is now used for aspirin.

The reagent used is ethanoyl anhydride:

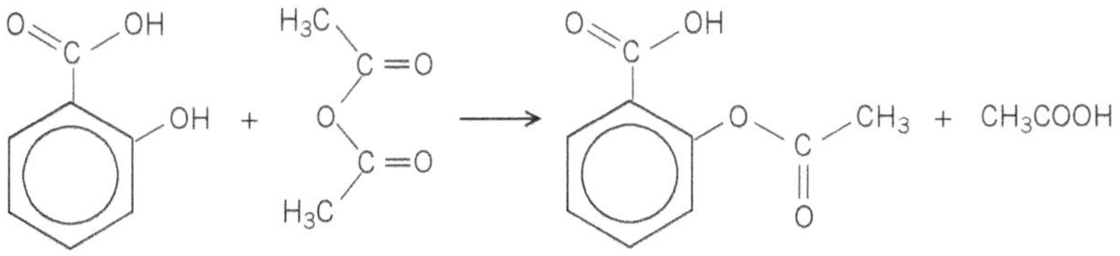

Aspirin

The by-product is ethanoic acid.

1 Work out the atom economy for the reaction in which aspirin is formed. You will probably need to draw out the displayed formulae first.
2 An alternative acylating agent for this reaction would be ethanoyl chloride.
   a Write the equation for the reaction of salicylic acid (2-hydroxybenzenecarboxylic acid) with ethanoyl chloride.
   b What is the by-product in this case?
   c Work out the atom economy for the reaction.

1 (180 ÷ 240) × 100 = 75.0% 2 a $HOOCC_6H_4OH + CH_3COCl \rightarrow HOOCC_6H_4OOCCH_3 + HCl$ b HCl c (180 ÷ 216.5) × 100 = 83.1%

Synoptic link

See Topic 2.6, Balanced equations, atom economies, and percentage yields.

Summary questions

1 Why is ethanoyl chloride a good acylating agent?
2 Which of the following could be acylated? **a** NH_4^+ **b** OH^- **c** CH_4? Explain your answers.
3 Why is acylation called an addition–elimination reaction?
4 Write down the equation for the formation of propanamide from the reaction between ammonia and propanoyl chloride. Give the mechanism for this reaction.

Practice questions

1 (a) Write an equation for the formation of methyl propanoate, $CH_3CH_2COOCH_3$, from methanol and propanoic acid.

(1 mark)

(b) Name and outline a mechanism for the reaction between methanol and propanoyl chloride to form methyl propanoate.

(5 marks)

(c) Propanoic anhydride could be used instead of propanoyl chloride in the preparation of methyl propanoate from methanol. Draw the structure of propanoic anhydride.

(1 mark)

(d) (i) Give **one** advantage of the use of propanoyl chloride instead of propanoic acid in the laboratory preparation of methyl propanoate from methanol.

(ii) Give **one** advantage of the use of propanoic anhydride instead of propanoyl chloride in the industrial manufacture of methyl propanoate from methanol.

(2 marks)

AQA, 2006

2 Consider the sequence of reactions below:

$$CH_3CH_2CHO \xrightarrow[HCN]{\text{Reaction 1}} \underset{\underset{OH}{|}}{\overset{\overset{H}{|}}{CH_3CH_2-C-CN}} \xrightarrow{\text{Reaction 2}} \underset{\underset{OH}{|}}{\overset{\overset{H}{|}}{CH_3CH_2-C-COOH}}$$

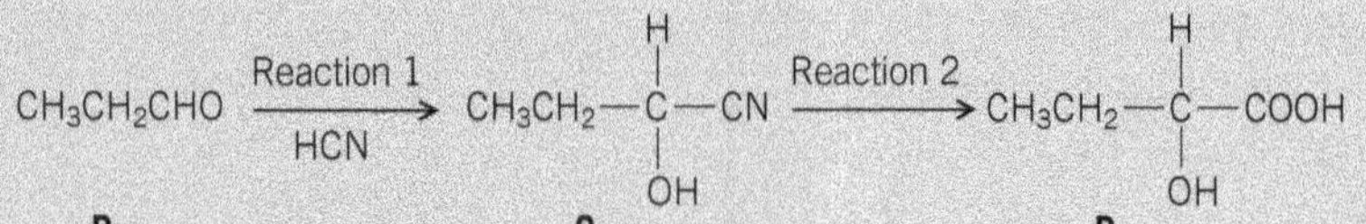

P · Q · R

(a) Name and outline a mechanism for Reaction 1.

(5 marks)

(b) (i) Name compound **Q**.

(ii) The molecular formula of **Q** is C_4H_7NO. Draw the structure of the isomer of **Q** which shows geometrical isomerism and is formed by the reaction of ammonia with an acyl chloride.

(3 marks)

(c) Draw the structure of the main organic product formed in each case when **R** reacts separately with the following substances:

(i) methanol in the presence of a few drops of concentrated sulfuric acid

(ii) acidified potassium dichromate(VI)

(iii) concentrated sulfuric acid in an elimination reaction.

(3 marks)

AQA, 2006

3 (a) Name and outline a mechanism for the reaction between propanoyl chloride, CH_3CH_2COCl, and methylamine, CH_3NH_2.

(5 marks)

(b) Draw the structure of the organic product.

(1 mark)

AQA, 2005

4 A naturally-occurring triester, shown below, was heated under reflux with an excess of aqueous sodium hydroxide and the mixture produced was then distilled. One of the products distilled off and the other was left in the distillation flask.

$$CH_3(CH_2)_{16}COOCH_2$$
$$|$$
$$CH_3(CH_2)_{16}COOCH$$
$$|$$
$$CH_3(CH_2)_{16}COOCH_2$$

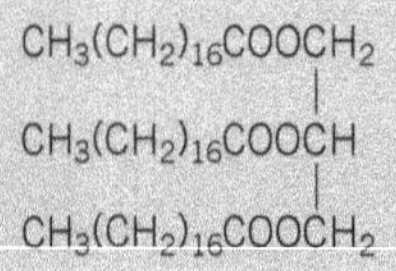

(a) Draw the structure of the product distilled off and give its name.

(2 marks)

(b) Give the formula of the product left in the distillation flask and give a use for it.

(2 marks)

AQA, 2005

Answers to the Practice Questions and Section Questions are available at
www.oxfordsecondary.com/oxfordaqaexams-alevel-chemistry

27 Aromatic chemistry

27.1 Introduction to arenes

Arenes are hydrocarbons based on benzene, C_6H_6, which is the simplest one. Although benzene is an unsaturated molecule, it is very stable. It has a hexagonal (six-sided) ring structure with a special type of bonding. Arenes were first isolated from sweet-smelling oils, such as balsam, and this gave them the name aromatic compounds. Arenes are still called aromatic compounds, but this now refers to their structures rather than their aromas. Benzene and other arenes have characteristic properties.

Benzene is given the special symbol:

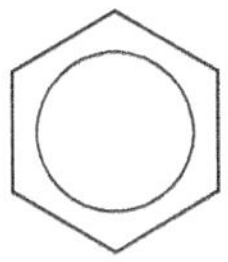

This is a skeletal formula, which does not show the carbon or hydrogen atoms. There is one carbon atom and one hydrogen atom at each point of the hexagon.

An arene can have other functional groups (substituents) replacing one or more of the hydrogen atoms in its structure.

Learning objectives:

→ Explain why the benzene ring is a planar hexagon.

→ Explain why the carbon–carbon bonds are of equal length.

Specification reference: 3.3.10

Hint

A linear six-carbon alkane would have the formula C_6H_{14} – more than twice the number of hydrogens compared with benzene.

Synoptic link

You will need to know the covalent bonding studied in Topic 3.2, Covalent bonding, and bonding in alkenes studied in Topic 14.1, Alkenes.

Bonding and structure of benzene

The bonding and structure of benzene, C_6H_6, was a puzzle for a long time to organic chemists, because:

- in spite of being unsaturated, it does not readily undergo addition reactions
- all the carbon atoms were equivalent, which implied that all the carbon–carbon bonds are the same.

Benzene consists of a flat, regular hexagon of carbon atoms, each of which is bonded to a single hydrogen atom. The geometry of benzene is shown in Figure 1. Notice the difference between the flat benzene ring (Figure 2) and the puckered cyclohexane ring (Figure 3).

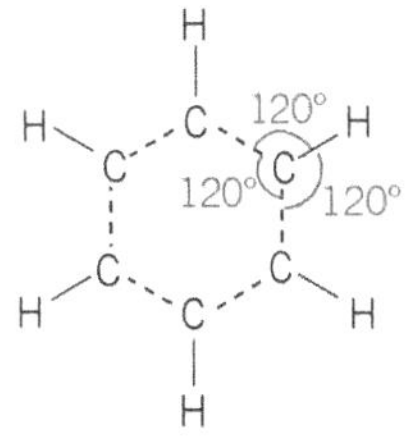

▲ **Figure 1** *The geometry of benzene (the dashed lines show the shape and do not represent single bonds)*

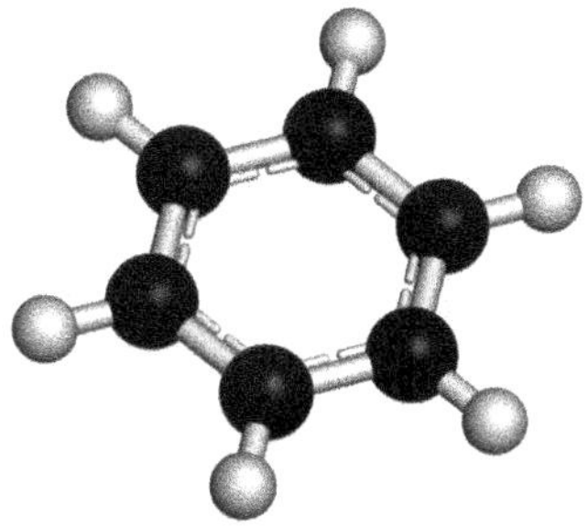

▲ **Figure 2** *The flat benzene ring*

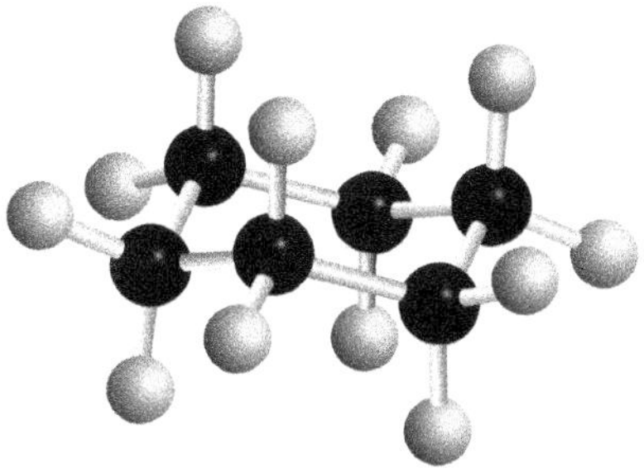

▲ **Figure 3** *The puckered cyclohexane ring*

The C—C bond lengths in benzene are intermediate between those expected for a carbon–carbon single bond and a carbon–carbon double bond (Table 1). So, each bond is intermediate between a single and a double bond.

▼ **Table 1** *Carbon–carbon bond lengths*

Bond	Length / nm
C—C	0.154
C⋯C (in benzene)	0.140
C=C	0.134
C≡C	0.120

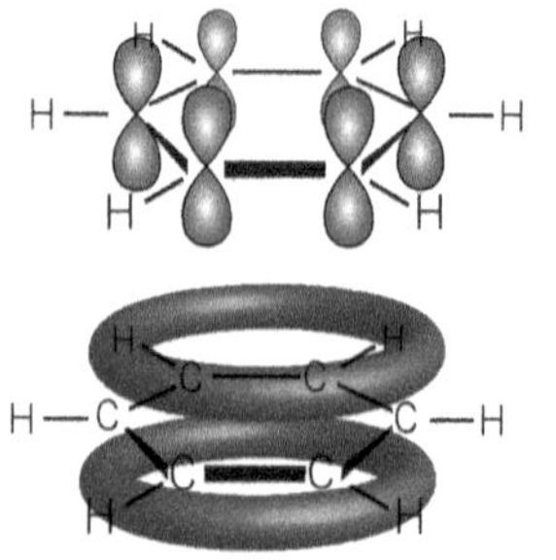

▲ **Figure 4** *Delocalisation of p-electrons to form areas of electron density above and below the ring*

Study tip

Benzene is more stable than the hypothetical molecule cyclohexa-1,3,5-triene because of delocalisation.

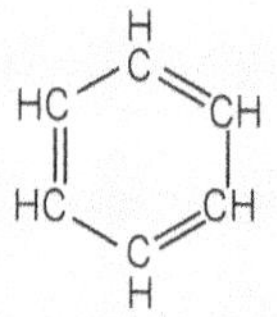

The symbol 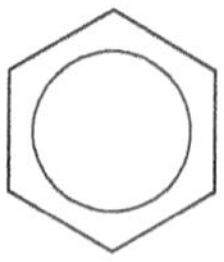 is used to represent this.

This can be explained by using the idea that some of the electrons are delocalised. Delocalisation means that electrons are spread over more than two atoms – in this case the six carbon atoms that form the ring.

Each carbon has three covalent bonds – one to a hydrogen atom and the other two to carbon atoms. The fourth electron of each carbon atom is in a p-orbital, and there are six of these – one on each carbon atom. The p-orbitals overlap and the electrons in them are **delocalised**. They form a region of electron density above and below the ring (Figure 4).

Overall, each carbon–carbon bond is intermediate between a single and a double bond. The delocalised system is very important in the chemistry of benzene and its derivatives. It makes benzene unusually stable. This is sometimes called **aromatic stability**.

The thermochemical evidence for stability

The enthalpy change for the hydrogenation of cyclohexene is:

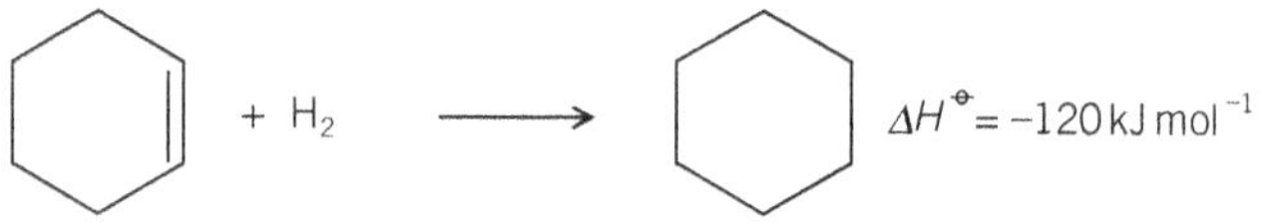

cyclohexene cyclohexane

So the hydrogenation of a ring with alternate double bonds would be expected to be three times this:

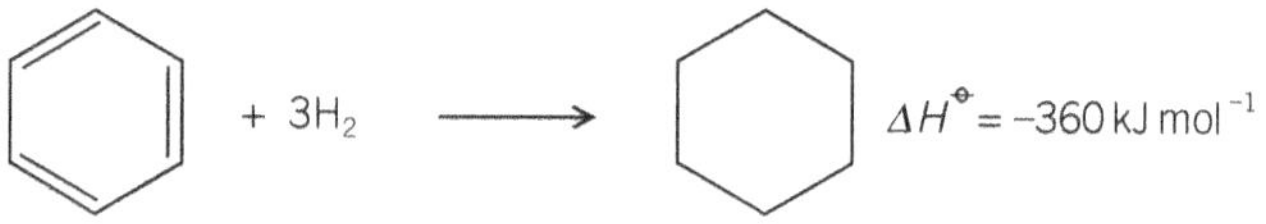

hypothetical non-delocalised benzene cyclohexane

The enthalpy change for benzene is in fact:

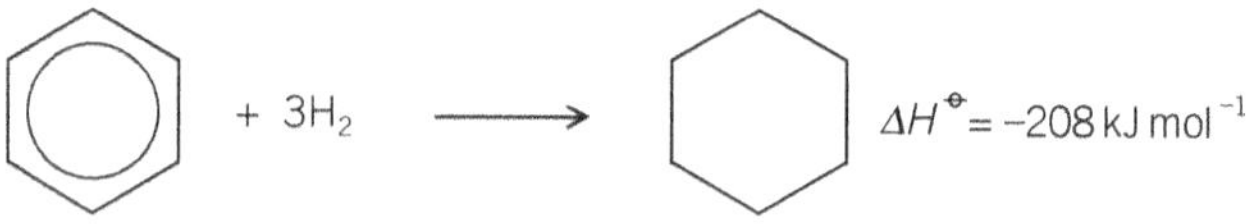

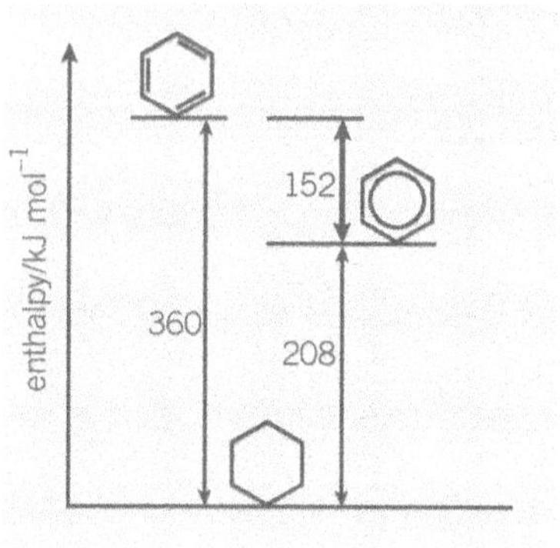

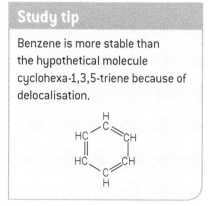

▲ **Figure 5** *Enthalpy diagram for the hydrogenation of cyclohexa-1,3,5-triene, cyclohexane, and benzene*

If these values are put on an enthalpy diagram (Figure 5), you can see that benzene is 152 kJ mol^{-1} more stable than the unsaturated ring structure.

The most important dream in history?

This is how Friedrich August von Kekulé's insight into a chemical mystery – the structure of benzene – has been described. Benzene, C_6H_6, had been discovered by Michael Faraday but its structure was a puzzle, as the proportion of carbon to hydrogen seemed to be too great for conventional theories. In 1865, the Belgian chemist Friedrich August von Kekulé published a paper in which he suggested that benzene's structure was based on a ring of carbon atoms with alternating double and single bonds.

His idea resulted from a dream of whirling snakes.

'I turned my chair to the fire [after having worked on the problem for some time] and dozed. Again the atoms were gambolling before my eyes. This time the smaller groups kept modestly to the background. My mental eye, rendered more acute by repeated visions of this kind, could now distinguish larger structures, of manifold conformation – long rows, sometimes more closely fitted together – all twining and twisting in snakelike motion. But look! What was that? One of the snakes had seized hold of its own tail, and the form whirled mockingly before my eyes. As if by a flash of lighting I awoke.'

However, even this insight left a number of problems:

- A cyclic triene should show addition reactions, which benzene rarely does.

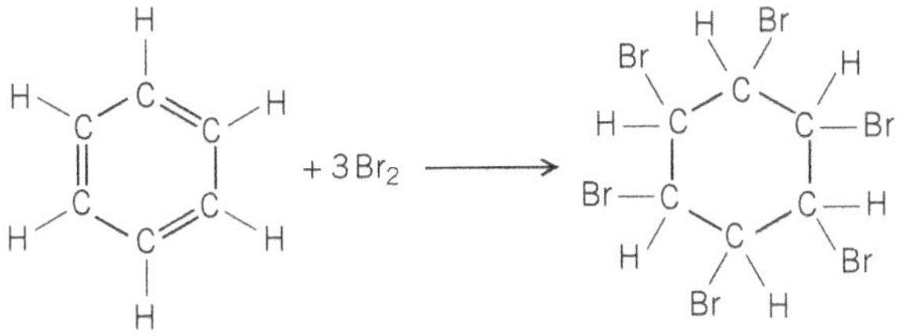

- Kekulé's structure should give rise to two isomeric di-substituted compounds as shown, using skeletal notation:

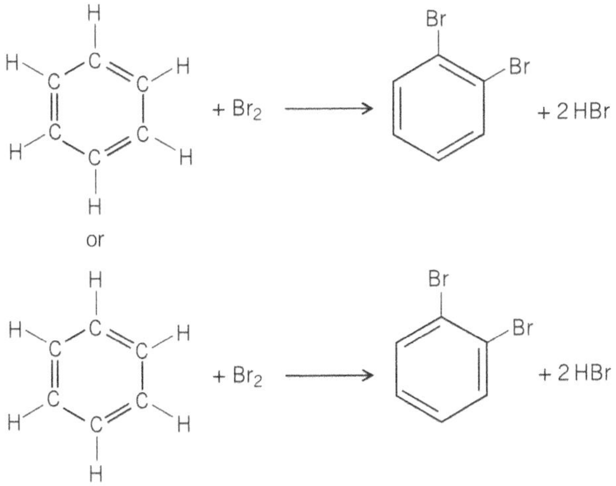

- The hexagon would not be symmetrical – double bonds are shorter than single bonds (Figure 6).

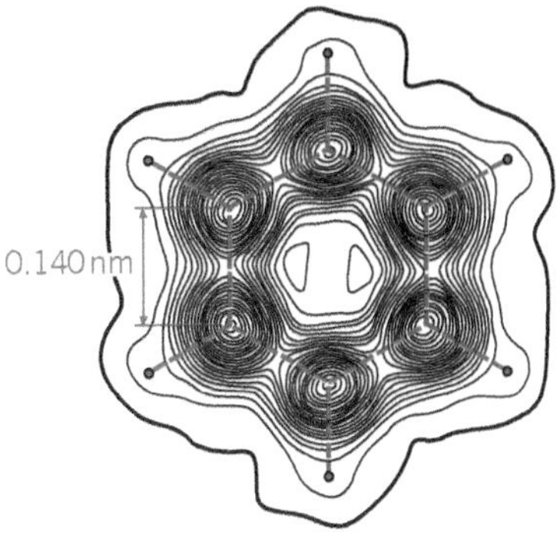

▲ **Figure 6** *A technique called X-ray diffraction shows a contour map of the electron density in an individual benzene molecule. This shows that the benzene molecule is a perfect hexagon and each carbon–carbon bond length is 0.140 nm*

Kekulé himself suggested a solution to the second dilemma by proposing that benzene consisted of structures in rapid equilibrium:

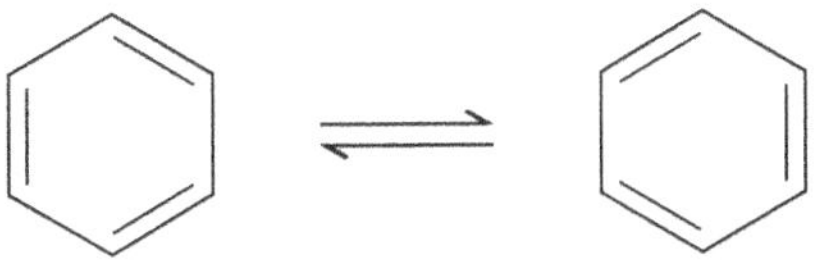

Later this rapid alternation of two structures evolved into the idea of **resonance** between two structures, both of which contribute to the actual structure. The actual structure was thought to be a hybrid (a sort of average) of the two. Such **resonance hybrids** were believed to be more stable than either of the separate structures.

Summary questions

1 What is the empirical formula of benzene?

2 How many molecules of hydrogen, H_2, would need to be added onto a benzene molecule to give a fully saturated product cyclohexane?

3 Explain what is meant by *delocalisation* of electrons in the benzene ring.

4 Look at the two di-substituted compounds formed with the bromination of Kekulé's proposed structure of benzene. Which of the two hypothetical di-substituted compounds would have the shorter bond between the two carbon atoms bonded to the bromine atoms?

27.2 Arenes – physical properties, naming, and reactivity

Learning objectives:

- → Describe how substituted arenes are named.
- → State how the arene ring affects reactivity.

Specification reference: 3.3.10

Hint

Freezing point and melting point are exactly the same temperature.

Hint

Benzene itself is **carcinogenic** (may cause cancer), so in your practical work in school you are likely to use related compounds that are safer, such as methylbenzene.

Physical properties of arenes

Benzene is a colourless liquid at room temperature. It boils at 353 K and freezes at 279 K. Its boiling point is comparable with that of hexane (354 K) but its melting point is much higher than hexane's (178 K). This is because benzene's flat, hexagonal molecules (Figure 1) pack together very well in the solid state. They are therefore harder to separate and this must happen for the solid to melt.

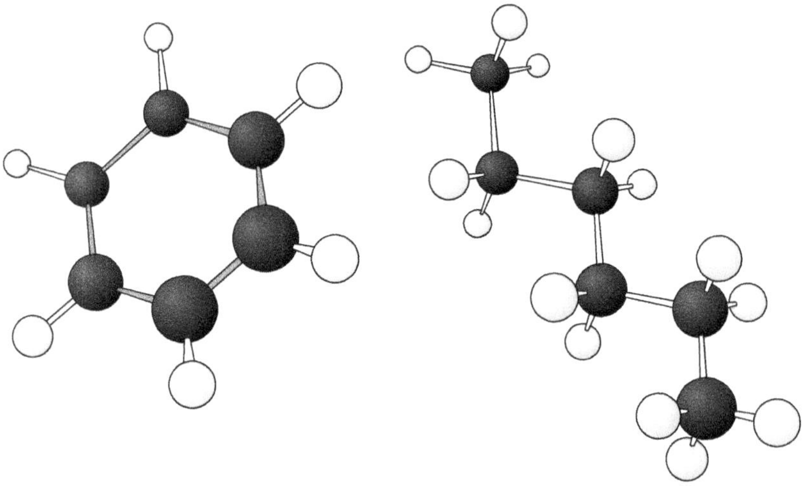

▲ **Figure 1** *Benzene molecules (left) can pack together better than hexane molecules (right) so benzene has a higher melting point than hexane*

Like other hydrocarbons that are non-polar, arenes do not mix with water but mix with other hydrocarbons and non-polar solvents.

Naming aromatic compounds

Substituted arenes are generally named as derivatives of benzene, so benzene forms the root of the name.

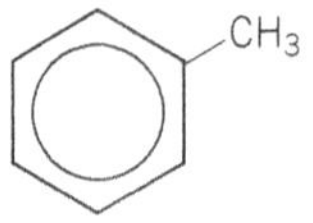

$C_6H_5CH_3$, is called methylbenzene.

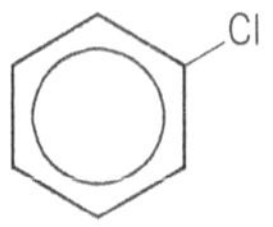

C_6H_5Cl, is called chlorobenzene, and so on.

If there is more than one substituent, the ring is numbered:

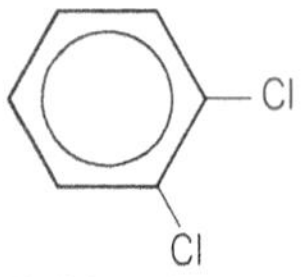

1,2-dichlorobenzene

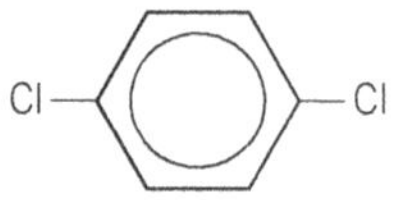

1,4-dichlorobenzene

Examples

You can test yourself by covering the names or the structures.

ethylbenzene	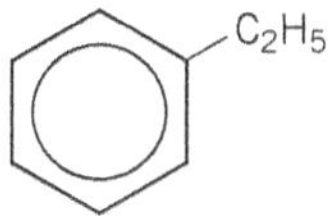	$C_6H_5C_2H_5$
nitrobenzene	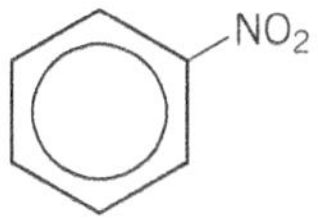	$C_6H_5NO_2$
1,2-dimethylbenzene	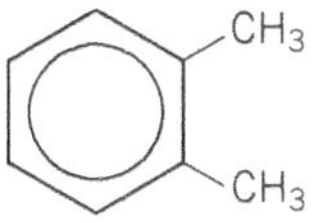	$C_6H_4(CH_3)_2$

The reactivity of aromatic compounds

Two factors are important to the reactivity of aromatic compounds:

- The ring is an area of high electron density, because of the delocalised bonding (Topic 27.1) and is therefore attacked by electrophiles.
- The aromatic ring is very stable. It needs energy to be put in to break the ring before the system can be destroyed. This is called the delocalisation energy. It means that the ring almost always remains intact in the reactions of arenes.

The above two points mean that most of the reactions of aromatic systems are **electrophilic substitution** reactions.

Summary questions

1 What intermolecular forces act between non-polar molecules?

2 Name:

a 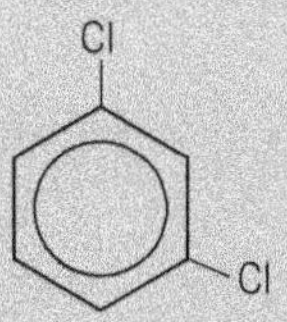

b 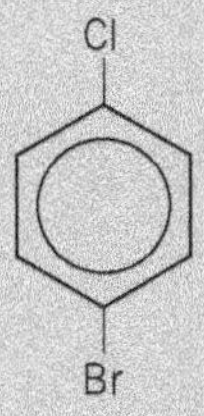

3 Draw the structure of:

a 1,4-dimethylbenzene

b 1,2-dimethylbenzene

4 Which of the following is an electrophile?

R^+ $:NH_3$ NO_2 Cl^-

Hint

Non-systematic names are still used for many derivatives of benzene. In particular, benzenecarboxylic acid is nearly always called benzoic acid and benzenecarbaldehyde is usually called benzaldehyde.

Study tip

An electrophile has a positive charge – either as a positive ion or the positive end of a dipole.

Synoptic link

You will need to know about enthalpy change studied in Topic 4.2, Enthalpy.

27.3 Reactions of arenes

Learning objectives:

- → Explain why arenes react by electrophilic substitution.
- → Describe the mechanism of nitration.
- → Describe the mechanism of acylation.

Specification reference: 3.3.10

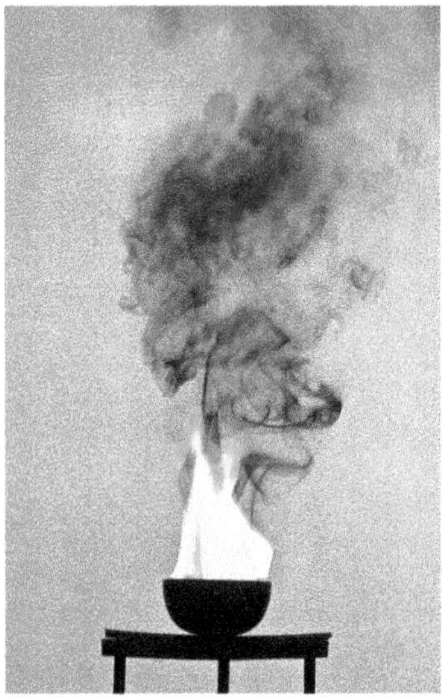

▲ **Figure 1** *Arenes burn with a smoky flame*

Hint

Compare the C : H ratio of benzene, C_6H_6, (1 : 1) to that of cyclohexene, C_6H_{12} (1 : 2).

Synoptic link

See Practical 10 on page 527.

Synoptic link

You will need to know the electrophilic addtion to alkenes studied in Topic 14.2, Reactions of alkenes.

Combustion

Arenes burn in air with flames that are noticeably smoky. This is because they have a high carbon : hydrogen ratio compared with alkanes. There is usually unburnt carbon remaining when they burn in air and this produces soot. A smoky flame suggests an aromatic compound.

Electrophilic substitution reactions

Although benzene is unsaturated it does not react like an alkene.

The most typical reaction is an electrophilic substitution that leaves the aromatic system unchanged, rather than addition which would require the input of the delocalisation energy to destroy the aromatic system.

The mechanism of electrophilic substitutions

The delocalised system of the aromatic ring has a high electron density that attracts electrophiles. At the same time the electrons are attracted towards the electrophile, El^+.

A bond forms between one of the carbon atoms and the electrophile. But, to do this, the carbon must use electrons from the delocalised system. This destroys the aromatic system. To get back the stability of the aromatic system, the carbon loses an H^+ ion with the electron in the C—H bond returning to the delocalised system. The sum of these reactions is the substitution of H^+ by El^+.

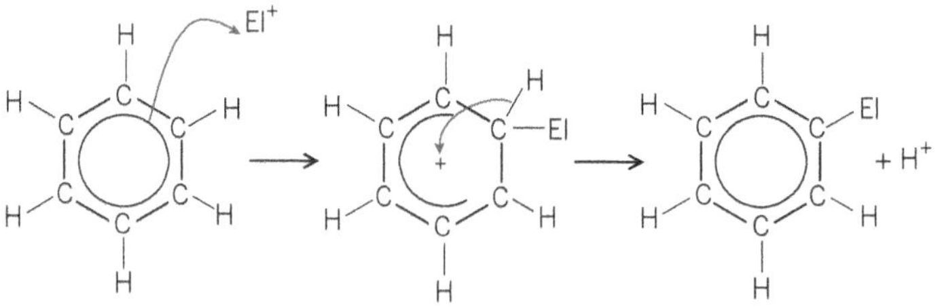

The same overall process occurs in, for example, nitration and Friedel–Crafts acylation reactions.

Nitration

Nitration is the substitution of an NO_2 group for one of the hydrogen atoms on an arene ring. The electrophile NO_2^+ is generated in the reaction mixture of concentrated nitric and concentrated sulfuric acids:

$$H_2SO_4 + HNO_3 \rightarrow H_2NO_3^+ + HSO_4^-$$

Sulfuric acid is a stronger acid than nitric acid and donates a proton, H^+, to HNO_3.

$H_2NO_3^+$ then loses a molecule of water to give NO_2^+, which is called the **nitronium ion** or **nitryl cation**.

$$H_2NO_3^+ \rightarrow NO_2^+ + H_2O$$

The overall equation for the generation of the NO_2^+ electrophile is:

$$H_2SO_4 + NHO_3 \rightarrow NO_2^+ + HSO_4^- + H_2O$$

NO_2^+ is an electrophile and the following mechanism occurs:

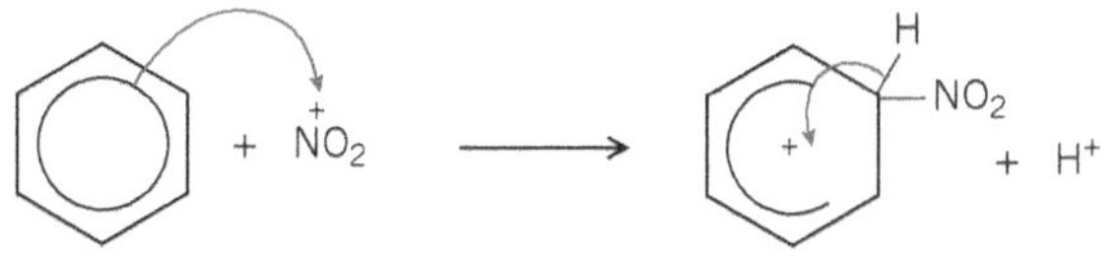

The overall product of the reaction of the nitronium ion, NO_2^+, with benzene is nitrobenzene:

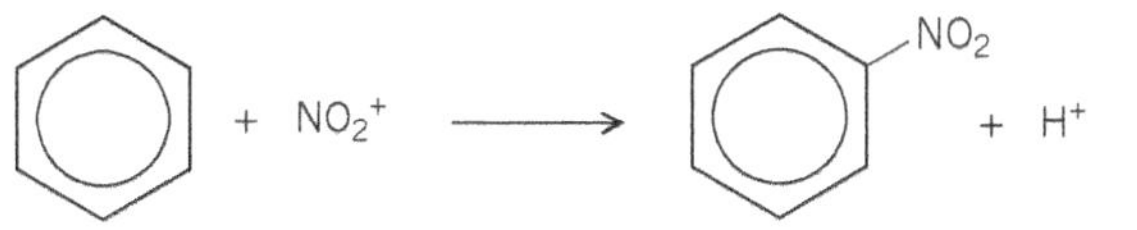

nitrobenzene

The H^+ then reacts with the HSO_4^- to regenerate H_2SO_4, making sulfuric acid a catalyst. The balanced equation is:

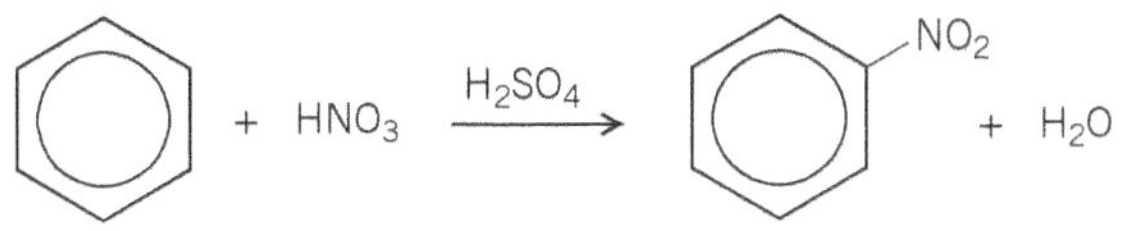

The uses of nitrated arenes

Nitration is an important step in the production of explosives like TNT. Nitration is the first step in making aromatic amines, and these in turn are used to make industrial dyes.

Study tip

In organic chemistry, curly arrows are used to indicate the movement of a pair of electrons. They run from areas of high electron density to more positively charged areas.

Synoptic link

You will learn about making aromatic amines in Topic 28.1, Introduction to amines.

Synoptic link

You will need to know the co-ordinate bonding and bond polarity studied in Topic 3.2, Covalent bonding, and Topic 3.6, Electronegativity – bond polarity in covalent bonds, and shapes of molecules studied in Topic 3.5, The shapes of molecules and ions.

Further substitution of arenes

An atom or group of atoms already on a benzene ring will affect further substitution reactions in two ways:

1. It may release electrons onto the benzene ring and therefore make it more susceptible to further electrophilic substitution reactions (i.e., these reactions will go faster and there may be more than one substituent). Or it will withdraw electrons from the ring, making it less susceptible to further electrophilic substitution.
2. It will direct further substitution to particular positions on the ring. Electron-releasing groups direct further substitution to the 2, 4, and 6 positions. Electron-withdrawing groups direct further substitution to the 3 and 5 positions.

Electron-releasing groups include $-CH_3$, $-OCH_3$, $-OH$, and $-NH_2$.

Electron withdrawing groups include $-NO_2$, $-COCl$.

Halogens are exceptions to the rule – they withdraw electrons but direct substitution to the 2 and 4 positions.

These rules explain why phenol can be nitrated to 2, 4, 6-trinitrophenol because $-OH$ is electron-releasing.

1. Predict the likely products of single nitration of chlorobenzene.
2. The nitration of methylbenzene can be done as a school practical exercise. Explain why there is no danger of the formation of the explosive 2, 4, 6-trinitromethylbenzene (trinitrotoluene, TNT).

1 2-nitrochlorobenzene 4-nitrochlorobenzene
2 Nitro groups withdraw electrons and make further substitution unlikely.

Friedel–Crafts acylation reactions

These reactions use aluminium chloride as a catalyst. The method of doing this was discovered by Charles Friedel and James Crafts.

The mechanism for acylation is a substitution, with RCO substituting for a hydrogen on the aromatic ring.

Acyl chlorides provide the RCO group. They react with $AlCl_3$ to form $AlCl_4^-$ and RCO^+.

$$RCOCl + AlCl_3 \rightarrow RCO^+ + AlCl_4^-$$

This reaction takes place because the aluminium atom in aluminium chloride has only six electrons in its outer main shell and readily accepts a lone pair from the chlorine atom of RCOCl.

RCO^+ is a good electrophile that is attacked by the benzene ring to form substitution products.

The aluminium chloride is a catalyst – it is reformed by reaction of the $AlCl_4^-$ ion with H^+ from the benzene ring:

$$AlCl_4^- + H^+ \rightarrow AlCl_3 + HCl$$

The mechanism for the reaction is:

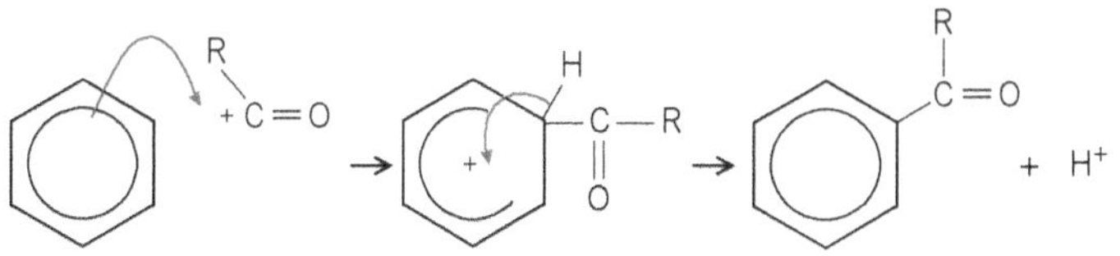

The products are acyl-substituted arenes. The overall reactions are:

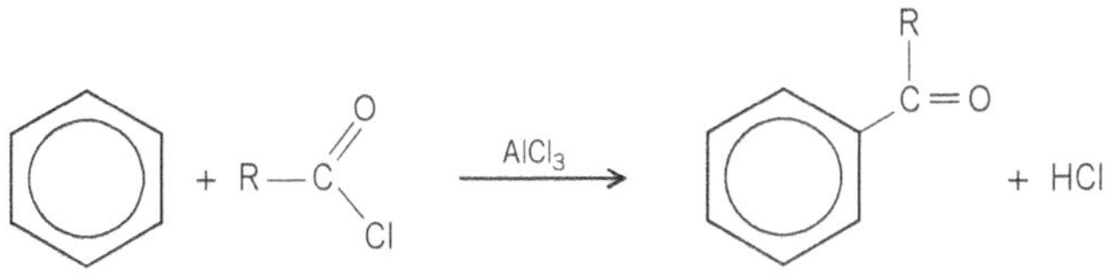

For example, ethanoyl chloride reacts with benzene to form:

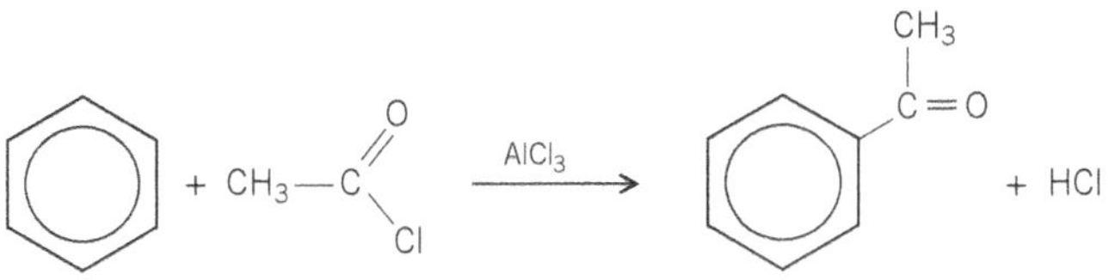

Acylation is a useful step in the synthesis of new substituted aromatic compounds.

Summary questions

1 Classify **a** nitration **b** Friedel– Crafts reactions as:

A electrophilic substitution

B nucleophilic substitution

C electrophilic addition

D free-radical addition

E free radical substitution.

2 Name the two isomers of 1,3-dinitrobenzene.

3 Explain why most of the reactions of benzene are substitutions rather than additions.

4 Write the equation for the reaction between propanoyl chloride with benzene. What species attacks the benzene ring?

27.4 Further reactions of arenes

Learning objectives

- → Describe the mechanisms of sulfonation and alkylation of arenes
- → Describe the free radical chlorination of arenes
- → Describe how chlorine-substituted arenes react

Sulfonation

Sulfonation is another example of electrophilic substitution. Here the electrophile is sulfur trioxide, SO_3. This is present in concentrated sulfuric acid due to the following equilibrium:

$$H_2SO_4 \rightleftharpoons SO_3 + H_2O$$

It is also present in greater concentration in fuming sulfuric acid, $H_2S_2O_7$, which is essentially a solution of sulfur trioxide in concentrated sulfuric acid.

Because oxygen is more electronegative than sulfur, SO_3 is polarised

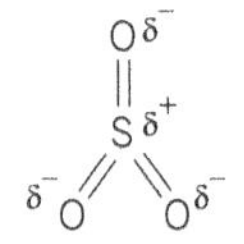

When benzene is sulfonated the product is called benzenesulfonic acid. The reaction is:

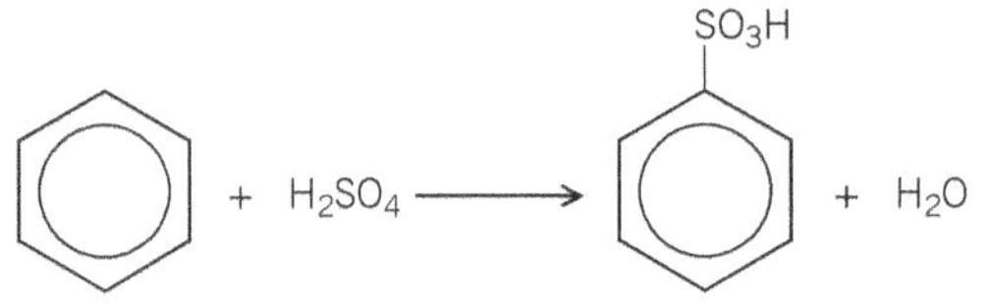

The mechanism is:

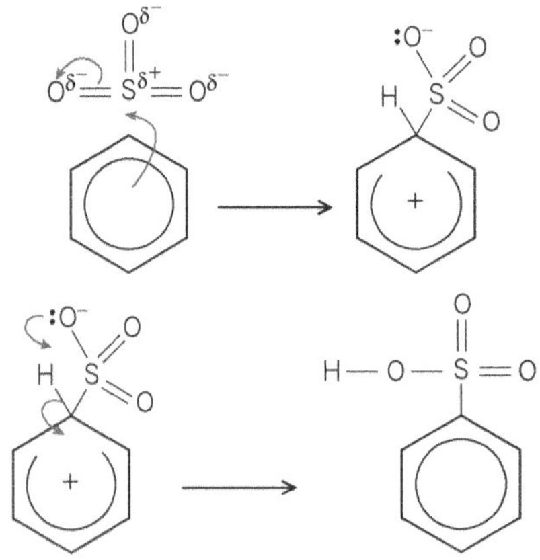

▲ **Figure 1** *Compounds such as benzene dodecylsulfonate are used in shampoos*

Benzenesulfonic acid can be converted to phenol, C_6H_5OH by reaction with water. Derivatives of benzenesulfonic acid with long alkyl side chains on the benzene ring are used as detergents. The sulfonate end of the molecule is polar and will dissolve in water, while the alkane side chain will mix with grease. Benzenesulfonates are also used in the manufacture of sulfonamide drugs.

Alkylation

This is a variant of the Friedel Crafts acylation reaction, see Topic 27.3, which uses a chloroalkane, rather than an acid chloride as the reagent with a catalyst of aluminium chloride used to generate an electrophile. For example:

$$CH_3Cl + AlCl_3 \rightarrow CH_3^+ + AlCl_4^-$$

This is followed by an electrophilic substitution to form an alkyl benzene. So for example benzene reacts to form methylbenzene

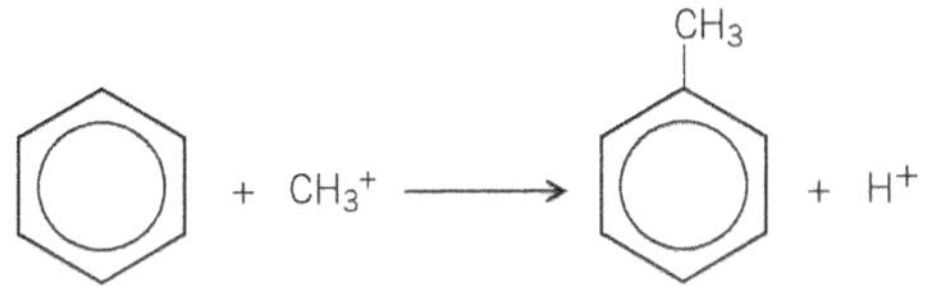

methylbenzene methylbenzene

Methylbenzene

Methylbenzene (non-systematic name toluene) reacts similarly to benzene but is more reactive than benzene towards electrophilic substitution because the methyl group releases electrons onto the benzene ring, making it more susceptible to electrophilic attack.

Nitration of methylbenzene

When nitrated with a mixture of nitric and sulfuric acids it produces a mixture of 2-nitrobenzene and 4-nitrobenzene.

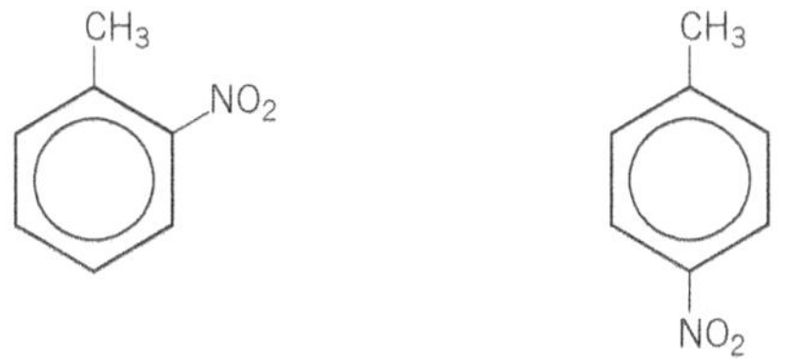

methyl-2-nitrobenzene methyl-4-nitrobenzene

The two isomers can be separated by thin layer chromatography, see Topic 33.1. With more vigorous conditions, more nitro-groups can be substituted onto the benzene ring, resulting in the explosive, trinitrotoluene (TNT).

TNT

▲ **Figure 2** *Filling TNT shells (1940)*

TNT is short for trinitrotoluene. It is made by nitrating methylbenzene, commonly called toluene. TNT is an important high explosive with both military and peaceful applications. It is a solid of low melting point. This property is used both in filling shells (Figure 2) and by bomb disposal teams who can steam TNT out of unexploded bombs.

The explosion of TNT is shown in the following equation:

2 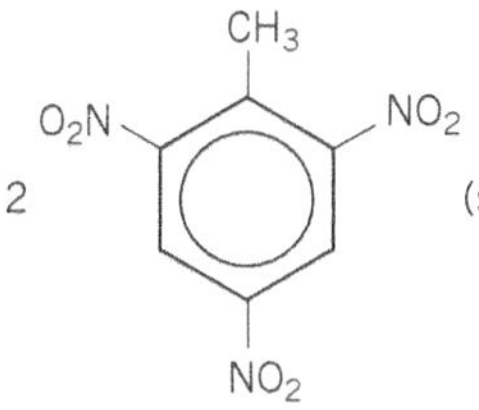 $(s) + 10\frac{1}{2}O_2(g) \longrightarrow 14CO_2(g) + 3N_2(g) + 5H_2O(g)$

The reaction is strongly exothermic. The rapid formation of a lot of gas as well as heat produces the destructive effect.

Many other compounds with several nitrogen atoms in the molecule are explosive. Another example is 2,4,6-trinitrophenol, which can explode on impact and is therefore useful as a detonator to set off other explosives.

What is the systematic name for TNT? The methyl group is at position 1.

1-methyl-2,4,6-trinitrobenzene

Sulfonation of methylbenzene also produces a mixture of the 2- and 4- substituted products.

Free radical reactions

The reaction of benzene with chlorine when exposed to ultra-violet (UV) light is unusual in that it is an *addition* reaction which involves chlorine free radicals produced by the UV light which breaks the Cl—Cl bond.

Cl—Cl → 2Cl•

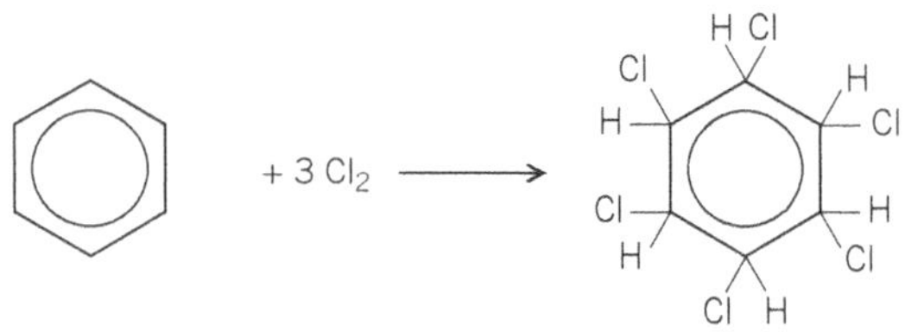

The final product is called 1, 2, 3, 4, 5, 6-hexachlorocyclohexane which exists as a number of isomers.

This reaction contrasts with the reactions of alkanes with chlorine in UV light which produces *substitution* products, see Topic 12.5.

Study tip

You will need to recall the free radical substitution reactions of alkanes covered in Topic 12.5, The formation of halogenoalkanes.

Methylbenzene, which is both an alkane and an arene, produces substitution products of the methyl group rather than addition to the ring when boiled in the presence of chlorine and exposed to UV light. That is, methylbenzene behaves like an alkane in this situation. The hydrogen atoms of the methyl group are substituted by chlorine in a series of steps. Each step produces a molecule of HCl.

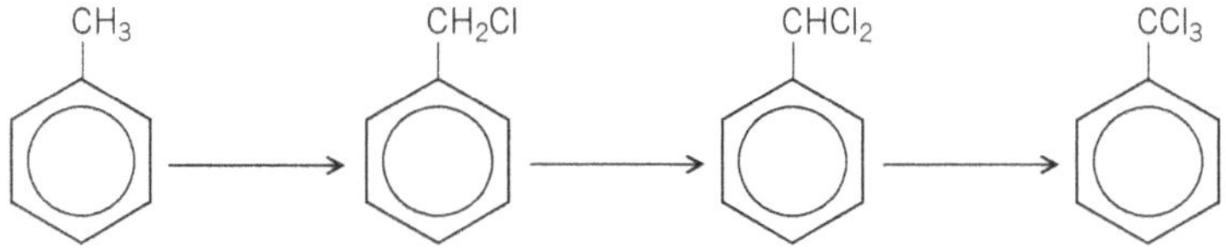

Reactions of halogen-substituted arenes

Alkylbenzenes such as methylbenzene may have halogen substituents on the ring or on the alkyl group. The reactivities of these compounds differ considerably according to where the halogen atom is.

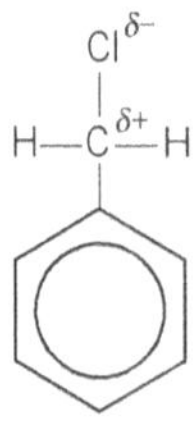

is called phenylchloromethane. It reacts in a similar way to chloromethane – the Cl atom may be substituted by an OH group by heating with an alkaline solution containing $^-$OH ions. This is a nucleophilic substitution reaction typical of halogeno*alkanes*, see Topic 13.2.

The isomer 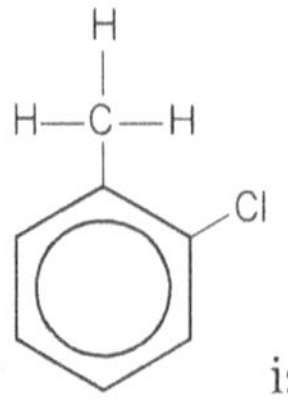

is called 1-chloromethybenzene. The Cl atom can only be substituted under very vigorous conditions. The molecule behaves like an arene which does not easily react with nucleophiles because of the high electron density on the benzene ring.

Methylbenzene

Substituents affect the reactivity of the benzene ring towards electrophiles. They do this by releasing electrons onto the ring or by withdrawing them. The methyl group releases electrons so methylbenzene is more reactive than benzene towards electrophiles. Further substituents go to either the 2- or 4- positions on the ring.

Benzene is a carcinogen so school experiments to show the properties of aromatic compounds often use methylbenzene rather than benzene itself.

Comparing reactivity

The C—Cl bond in 1-chloromethybenzene is stronger than that in phenylchloromethane which explains why it is harder to substitute the chlorine atom. This is because electrons in filled p-orbitals on the Cl atom interact with the delocalised electrons on the benzene ring.

Summary questions

1 What substitution product is formed by heating 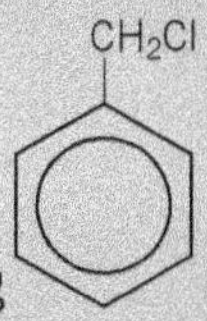

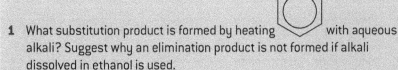

with aqueous alkali? Suggest why an elimination product is not formed if alkali dissolved in ethanol is used.

2 Suggest what reagent and catalyst would be required to make propylbenzene from benzene.

3 In the sufonation of benzene, explain why the new bond is formed with the sulfur atom rather than the oxygen atom.

4 Write a balanced equation for the first step of the free radical reaction of chlorine with methylbenzene.

1 Give reagents and conditions and write equations to show the formation of nitrobenzene from benzene.
Name and outline a mechanism for this reaction of benzene.

(8 marks)
AQA, 2007

2 A possible synthesis of phenylethene (*styrene*) is outlined below.

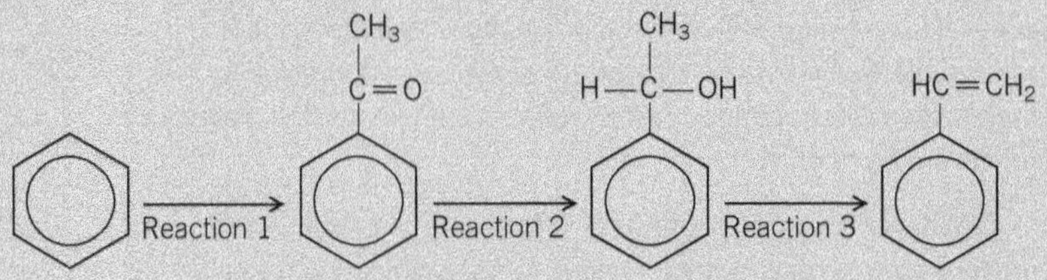

In Reaction 1, ethanoyl chloride and aluminium chloride are used to form a reactive species which then reacts with benzene.
Write an equation to show the formation of the reactive species.
Name and outline the mechanism by which this reactive species reacts with benzene.

(6 marks)
AQA, 2006

3 An acylium ion has the structure $R—\overset{+}{C}═O$ where R is any alkyl group.
In the conversion of benzene into phenylethanone, $C_6H_5COCH_3$, an acylium ion, $CH_3\overset{+}{C}O$, reacts with a benzene molecule.
Write an equation to show the formation of this acylium ion from ethanoyl chloride and one other substance.
Name and outline the mechanism of the reaction of this acylium ion with benzene.

(6 marks)
AQA, 2007

4 An equation for the formation of phenylethanone is shown below. In this reaction a reactive intermediate is formed from ethanoyl chloride. This intermediate then reacts with benzene.

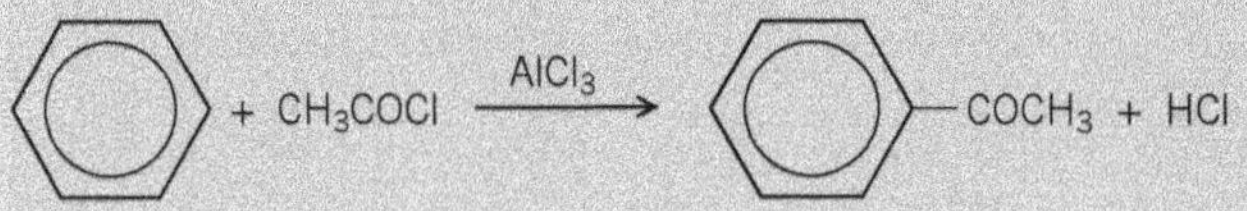

(a) Give the formula of the reactive intermediate.
(b) Outline a mechanism for the reaction of this intermediate with benzene to form phenylethanone.

(4 marks)

5 Consider compound **P** shown below that is formed by the reaction of benzene with an electrophile.

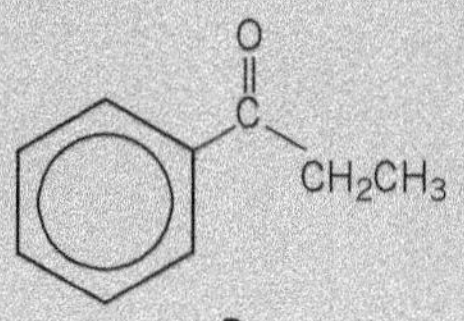

P

(a) Give the **two** substances that react together to form the electrophile and write an equation to show the formation of this electrophile. *(3 marks)*

(b) Outline a mechanism for the reaction of this electrophile with benzene to form **P**. *(3 marks)*

(c) Compound **Q** is an isomer of **P** that shows optical isomerism. **Q** forms a silver mirror when added to a suitable reagent.
Identify this reagent and suggest a structure for **Q**. *(2 marks)*
AQA, 2010

6 The hydrocarbons benzene and cyclohexene are both unsaturated compounds. Benzene normally undergoes substitution reactions, but cyclohexene normally undergoes addition reactions.

(a) The molecule cyclohexatriene does not exist and is described as hypothetical. Use the following data to state and explain the stability of benzene compared with the hypothetical cyclohexatriene.

$$\Delta H^{\ominus} = -120 \, \text{kJ mol}^{-1}$$

$$\Delta H^{\ominus} = -208 \, \text{kJ mol}^{-1}$$

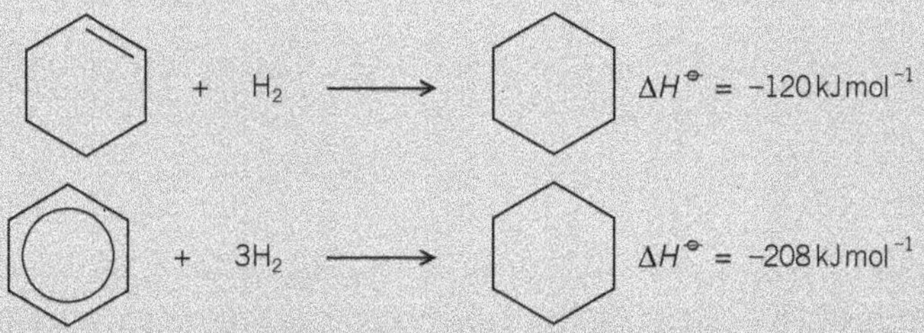

(4 marks)

(b) Benzene can be converted into amine **U** by the two-step synthesis shown below.

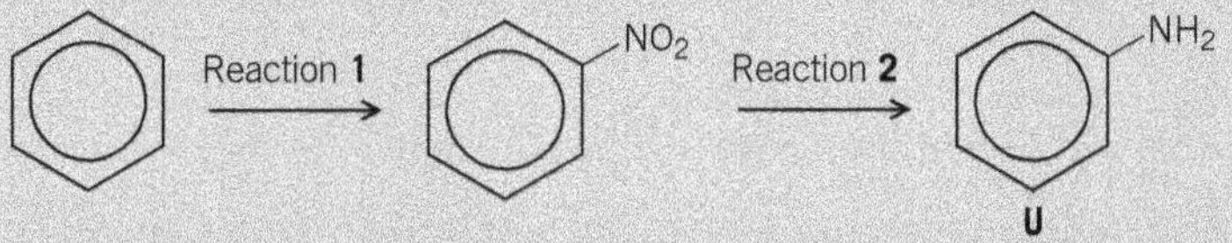

The mechanism of Reaction **1** involves attack by an electrophile.

Give the reagents used to produce the electrophile needed in Reaction **1**.

Write an equation showing the formation of this electrophile.

Outline a mechanism for the reaction of this electrophile with benzene.

(6 marks)
AQA, 2011

7 Propanoyl chloride can be used, together with a catalyst, in the synthesis of 1-phenylpropene from benzene. The first step in the reaction is shown below.

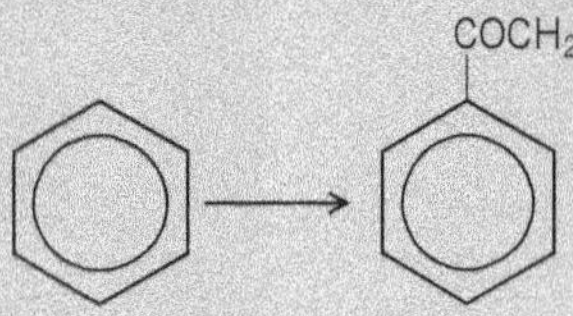

The mechanism of this reaction is an electrophilic substitution. Write an equation to show the formation of the electrophile from propanoyl chloride. Outline the mechanism of the reaction of this electrophile with benzene.

(5 marks)
AQA, 2004

8 5-amino-2-methylbenzenesulfonic acid can be prepared from methylbenzene in a three-step synthesis:

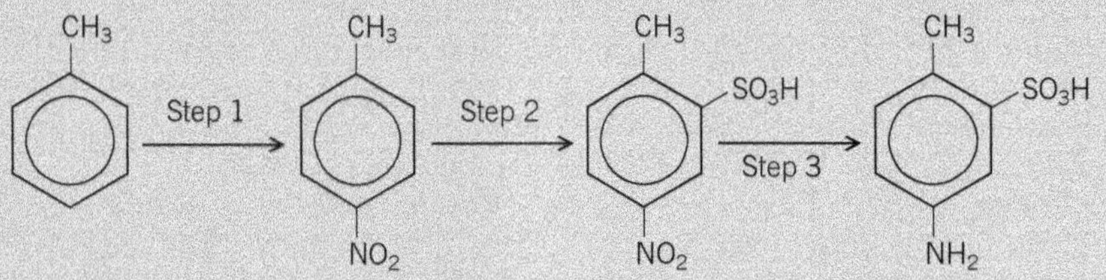

(a) State the type of reaction taking place in Step 1 and give suitable reagent(s) for this step.

(3 marks)

(b) Write an equation for the formation of the reactive inorganic species involved in the mechanism for Step 1.

(1 mark)

(c) Identify the reactive inorganic species involved in the mechanism in Step 2 and outline the mechanism.

(5 marks)
Oxford International AQA specimen papers 2016

9 The nitration of benzene is an important industrial reaction.

(a) State the reagents required for the nitration of benzene.

(1 mark)

(b) Name an important material whose manufacture involves the nitration of benzene.

(1 mark)

(c) (i) Write a balanced equation for the nitration of benzene.

(2 marks)

(ii) Explain why the NO_2^+ ion is described as an electrophile.

(1 mark)

(iii) Name the type of mechanism involved in the nitration of benzene.

(1 mark)

28 Amines

28.1 Introduction to amines

Learning objectives:

- → Describe what amines are.
- → Describe how they are named.
- → Describe how they react.

Specification reference: 3.3.11

This chapter is about a group of compounds called amines. Amines can be thought of as derivatives of ammonia in which one or more of the hydrogen atoms in the ammonia molecule have been replaced by alkyl or aryl groups.

ammonia   primary amine   secondary amine   tertiary amine

Amines are very reactive compounds, so they are useful as intermediates in **synthesis** – the making of new molecules.

The terms primary, secondary, and tertiary are used for amines slightly differently from the way they are used with alcohols. In amines, 1°, 2°, and 3° refer to the number of substituents (R-groups) on the *nitrogen* atom. (In alcohols, 1°, 2°, and 3° refer to the number of substituents on the *carbon* atom bonded to the –OH group.)

Synoptic link

Primary, secondary, and tertiary alcohols were covered in Topic 15.1, Alcohols – introduction.

Hint

1°, 2°, and 3° are shorthand for primary, secondary, and tertiary, respectively.

How to name amines

Primary amines have the general formula RNH_2, where the R can be an alkyl or aryl group. Amines are named using the suffix -amine, for example:

$CH_3—NH_2$ is methylamine
$C_2H_5—NH_2$ is ethylamine.

$C_6H_5NH_2$ is phenylamine.

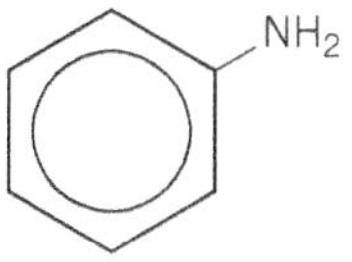

Secondary amines have the general formula RR′NH, for example:

$(CH_3)_2NH$ is dimethylamine.

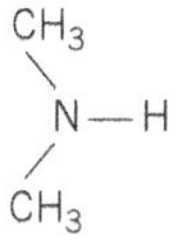

Tertiary amines have the general formula RR′R″N, for example:

$(C_2H_5)_3NH$ is triethylamine.

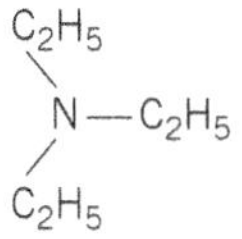

Different substituents are written in alphabetical order:

$CH_3(C_3H_7)NH$ is N–methylpropylamine.

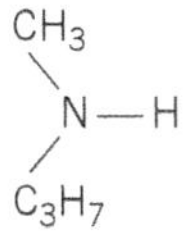

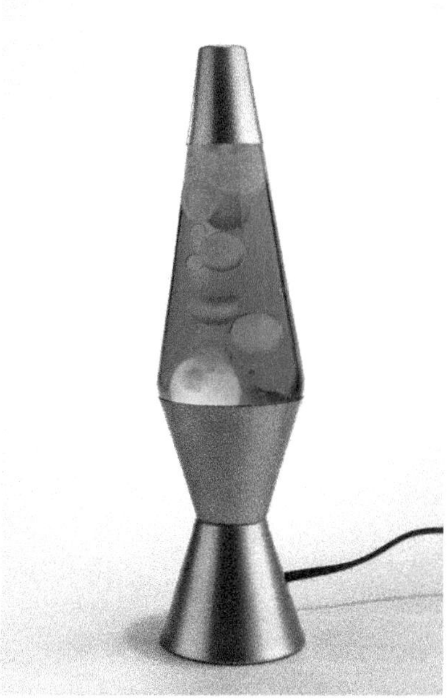

▲ **Figure 1** *Phenylamine has almost the same density as water and is not soluble in it. Heat from a bulb at the base of the lava lamp changes the density enough for the phenylamine (here dyed red) to float when hot and sink when cool*

Synoptic link

You can remind yourself of the effect of lone pairs on the shapes of molecules in Topic 3.5, The shapes of molecules and ions.

The properties of primary amines

Shape

Ammonia is a pyramidal molecule with bond angles of approximately 107°. The angles of a perfect tetrahedron are 109.5°. The difference is caused by the lone pair, which repels more than the bonding pairs of electrons in the N—H bonds. Amines keep this basic shape (Figure 2).

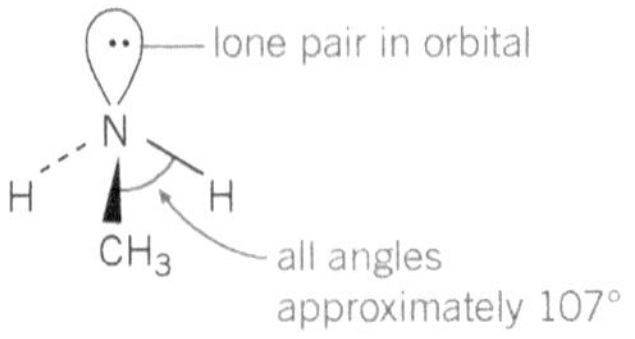

▲ **Figure 2** *The shape of the methylamine molecule*

Boiling points

Amines are polar:

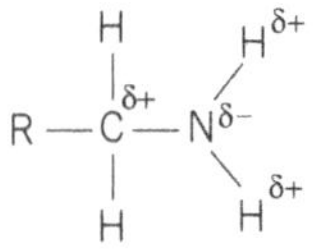

Primary amines can hydrogen bond to one another using their $-NH_2$ groups (in the same way as alcohols with their –OH groups). However, as nitrogen is less electronegative than oxygen (electronegativities – O = 3.5, N = 3.0), the hydrogen bonds are not as strong as those in alcohols. The boiling points of amines are lower than those of comparable alcohols:

methylamine, M_r = 31, CH_3—NH_2, boiling point = 267 K

methanol, M_r = 32, CH_3—OH, boiling point = 338 K

Hint

A lower boiling point means the molecules are easier to separate.

Shorter chain amines such as methylamine and ethylamine are gases at room temperature, and those with slightly longer chains are volatile liquids. They have fishy smells. Rotting fish and rotting animal flesh smell of di- and triamines, produced when proteins decompose (Figure 3).

▲ **Figure 3** *Rotting fish smell of di- and triamines*

Solubility

Primary amines with chain lengths up to about four carbon atoms are very soluble both in water and in alcohols because they form hydrogen bonds with these solvents. Most amines are also soluble in less polar solvents. Phenylamine, $C_6H_5NH_2$, is not very soluble in water due to the benzene ring, which cannot form hydrogen bonds.

The reactivity of amines

Amines have a lone pair of electrons and this is important in the way they react. The lone pair may be used to form a bond with:

- a H^+ ion, when the amine is acting as a **base**
- an electron-deficient carbon atom, when the amine is acting as a **nucleophile**.

Synoptic link

An amine is a proton acceptor so it is a Brønsted–Lowry base, see Chapter 21, Acids, bases, and buffers. An amine is also a Lewis base as it is a lone-pair donor, see Topic 24.1, The acid–base chemistry of aqueous transition metal ions.

Summary questions

1. Classify C_2H_5—N(H)—C_3H_7 as primary, secondary, or tertiary.
2. Name the compound in **1**.
3. Write the structural formula of trimethylamine.
4. Predict whether dimethylamine will be a solid, liquid, or gas at room temperature.
5. Explain your answer to **4**.

28.2 The properties of amines as bases

Amines as bases

Amines can accept a proton (an H^+ ion) so they are Brønsted–Lowry bases.

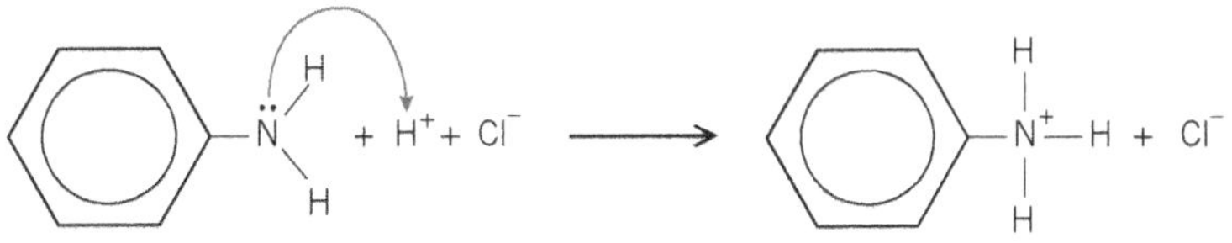

phenylamine — phenylammonium chloride, a water-soluble, ionic salt

Reaction as bases

Amines react with acids to form salts. For example, ethylamine, a soluble alkylamine, reacts with dilute hydrochloric acid:

$$C_2H_5\ddot{N}H_2 + H^+ + Cl^- \longrightarrow C_2H_5NH_3^+ + Cl^-$$

ethylamine — ethylammonium chloride

The products are ionic compounds that will crystallise as the water evaporates.

Phenylamine, an arylamine, is relatively insoluble, but it will dissolve in excess hydrochloric acid because it forms the soluble ionic salt.

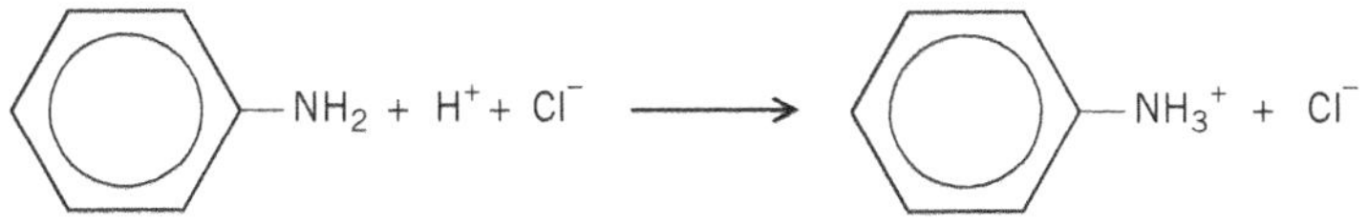

phenylamine — phenylammonium chloride, a water-soluble, ionic salt

If a strong base like sodium hydroxide is added, it removes the proton from the salt and regenerates the insoluble amine.

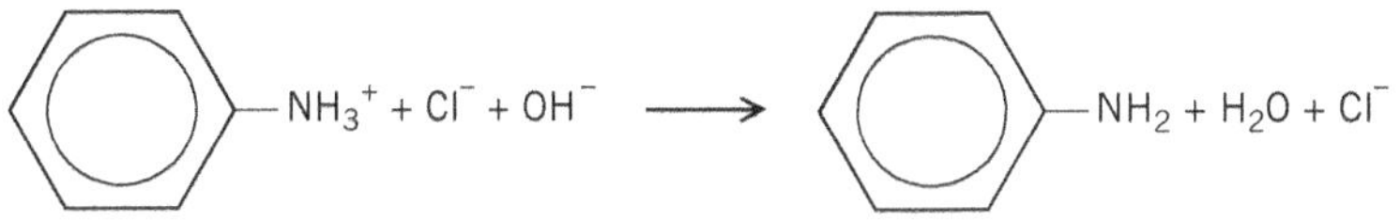

phenylamine

Comparing base strengths

The strength of a base depends on how readily it will accept a proton, H^+. Both ammonia and amines have a lone pair of electrons that attract a proton. So amines are weak bases.

Alkyl groups *release* electrons away from the alkyl group and towards the nitrogen atom. This is called the **inductive effect** and is shown by an arrow (Figure 1).

The inductive effect of the alkyl group increases the electron density on the nitrogen atom and therefore makes it a better electron pair donor (i.e., more attractive to protons). So, primary alkylamines are stronger bases than ammonia.

Learning objectives:

- → Explain why amines behave as Brønsted–Lowry bases.
- → Explain why the base strengths of amines differ from each other and from ammonia.

Specification reference: 3.3.11

Hint

The salts of amines are sometimes named as the hydrochloride of the parent amine. So ethylammorium chloride is also called ethylamine hydrochloride.

Hint

The smell of a solution of an amine disappears when an acid is added due to the formation of the ionic (and therefore involatile) salt. The smell returns if a strong base is then added.

$$R \rightarrow \ddot{N}H_2$$

▲ **Figure 1** *A primary amine. The arrow shows that R releases electrons. This is called the inductive effect*

Solubility of drugs

A number of medicinal drugs are amines, for example, the nasal decongestant Sudafed, active ingredient pseudoephedrine. Longer chain amines are relatively insoluble in water so when they are used in medicines they are often supplied as hydrochlorides to make them more soluble in the bloodstream.

Pseudoephedrine has the formula:

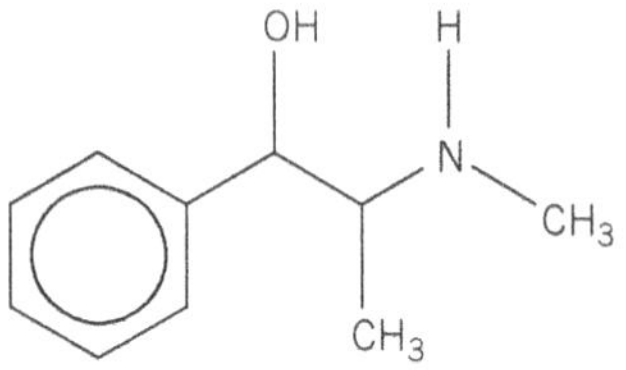

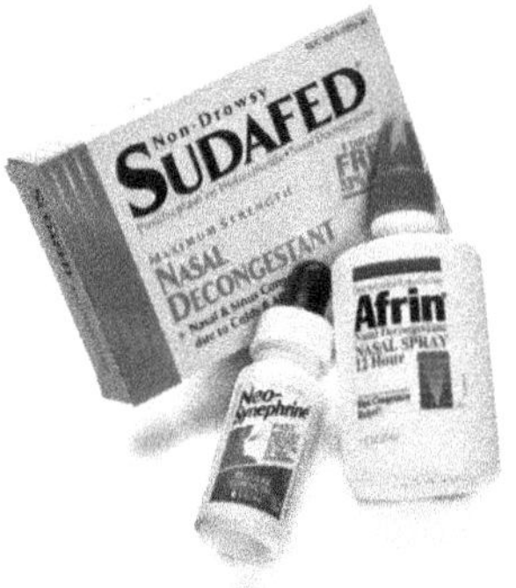

▲ **Figure 2** *Nasal decongestant sprays reduce swelling in the blood vessels inside your nose, helping to relieve breathing issues caused by colds or hayfever*

1. Is pseudoephedrine a primary, secondary, or tertiary amine?
2. Draw the formula of pseudoephedrine hydrochloride.
3. Explain why pseudoephedrine hydrochloride is more soluble in water than pseudoephedrine.
4. Is pseudoephedrine hydrochloride likely to be a solid, liquid, or gas? Explain your answer.
5. As well as having an amine group, pseudoephedrine has two other functional groups. Name them.
6. Pseudoephedrine has two chiral centres in its molecule. Mark them with a * on your formula of pseudoephedrine hydrochloride. Hint, it may help if you draw in the hydrogen atoms that are not marked on the skeletal formula.
7. What problems might this chirality have in pseudoephedrine's use as a drug?

Secondary alkylamines have two inductive effects and are therefore stronger bases than primary alkylamines. However, tertiary alkylamines are not stronger bases than secondary ones because they are less soluble in water.

Aryl groups *withdraw* electrons from the nitrogen atom because the lone pair of electrons on nitrogen atom overlaps with the delocalised system on the benzene ring, as shown for phenylamine.

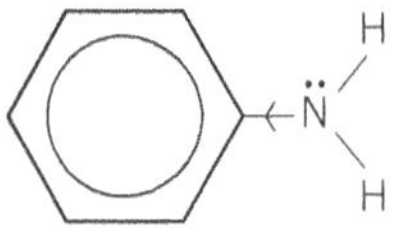

The nitrogen is a weaker electron pair donor and therefore less attractive to protons, so arylamines are weaker bases than ammonia.

ethylamine > ammonia > phenylamine

strongest ⟶ weakest

Summary questions

1. **a** Write the equation for dimethylamine reacting with hydrochloric acid.
   b Name the product.
2. Phenylamine is not very soluble in water. It forms oily drops that float in the water. Predict what you would see if you:
   a add concentrated hydrochloric acid to a mixture of phenylamine and water.
   b then add sodium hydroxide solution to the resulting solution.
3. Suggest whether dimethylamine will be a weaker or stronger base than ethylamine. Explain your answer.

28.3 Amines as nucleophiles and their synthesis

The lone pair of electrons from an amine will attack positively charged carbon atoms. So amines, like ammonia, will act as nucleophiles.

Reactions of ammonia with halogenoalkanes

Primary alkyl amines are produced when halogenoalkanes are reacted with ammonia. There is nucleophilic substitution of the halide by NH_2.

$$NH_3 + RX \rightarrow RNH_3^+\,X^-$$

$$RNH_3^+X^- + NH_3 \rightarrow RNH_2 + NH_4^+\,X^-$$

primary amine

However, the primary amine produced is also a nucleophile and this will react with the halogenoalkane to produce a secondary amine:

$$RNH_2 + RX \rightarrow R_2NH_2^+X^-$$

$$R_2NH_2 + X^- + NH_3 \rightarrow R_2NH + NH_4^+X^-$$

secondary amine

The secondary amine will react to give a tertiary amine:

$$R_2NH + RX \rightarrow R_3NH^+X^-$$

$$R_3NH^+X^- + NH_3 \rightarrow R_3N + NH_4^+X^-$$

tertiary amine

This in turn will react to a produce a quarternary ammonium salt:

$$R_3N + RX \rightarrow R_4N^+X^-$$

So a mixture of primary, secondary, and tertiary amines and a quarternary ammonium salt is produced. This means that this is not a very efficient way of preparing an amine, though the products may be separated by fractional distillation. A large excess of ammonia gives a better yield of primary amine.

The mechanism of the reaction

For all the above reactions the mechanism is essentially the same:

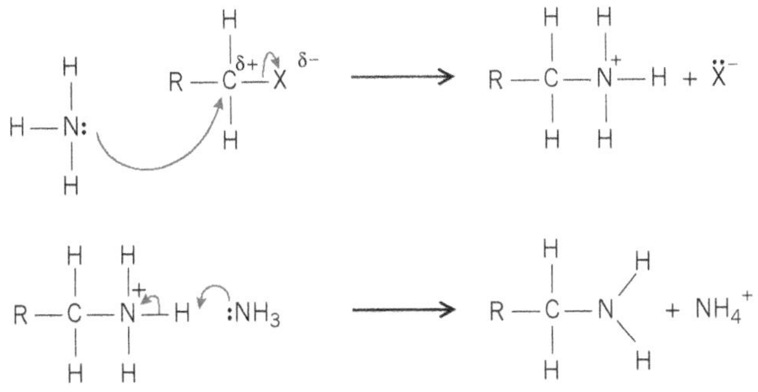

Initially ammonia acts as a nucleophile. In the second stage, it acts as a base.

Learning objectives:

- → Explain why ammonia and amines act as nucleophiles.
- → State how halogenoalkanes react with ammonia and amines.
- → State how amines are prepared from nitriles.
- → State how aromatic amines are synthesised from benzene.

Specification reference: 3.3.11

Synoptic link

You will need to know the nucleophilic substitution reactions in halogenoalkanes studied in Topic 13.2, Nucleophilic substitution in halogenoalkanes, and redox reactions and oxidation states studied in Topic 7.2, Oxidation states, and Topic 7.3, Redox equations.

Study tip

Remember that in these reactions a proton is removed from the initial substitution intermediate so two moles of ammonia or amine are required for each mole of halogenoalkane.

Synoptic link

You will cover the formation of nylon in Topic 29.1, Condensation polymers.

Preparation of amines

Primary amines

Reduction of nitriles

Primary aliphatic amines can be prepared from halogenoalkanes in a two-step process:

Study tip

Nitriles cannot be reduced to amines by $NaBH_4$.

Step 1: Halogenoalkanes react with the cyanide ion in aqueous ethanol. The cyanide ion replaces the halide ion by nucleophilic substitution to form a nitrile:

$$RBr + CN^- \rightarrow R{-}C{\equiv}N + Br^-$$

Step 2: Nitriles contain the functional group $-C{\equiv}N$. They can be reduced to primary amines, for example, with a nickel / hydrogen catalyst:

$$R{-}C{\equiv}N + 2H_2 \rightarrow R{-}CH_2NH_2$$

This gives a purer product than a bromoalkane and ammonia because only the primary amine can be formed. The carbon chain of the product is one carbon atom longer than in the starting material.

Hint

This preparation would not be attempted in a school laboratory as benzene is a carcinogen.

Phenylamine

Phenylamine is the simplest arylamine. It is the starting point for making many other chemicals and is made in industry using benzene produced from crude oil.

▲ **Figure 1** *Aromatic amines are used in the manufacture of dyestuffs*

Making phenylamine

Phenylamine can be made from benzene.

Step 1: Benzene is reacted with a mixture of concentrated nitric and concentrated sulfuric acid. This produces nitrobenzene:

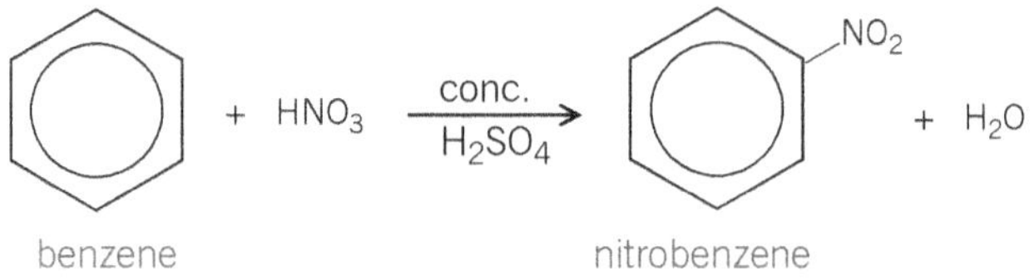

Synoptic link

Remember a mixture of nitric and sulfuric acids generates the electrophile NO_2^+ – see Topic 27.3, Reactions of arenes. So this reaction is an electrophilic substitution.

Step 2: Nitrobenzene is reduced to phenylamine, using tin and hydrochloric acid as the reducing agent.

The tin and hydrochloric acid react to form hydrogen, which reduces the nitrobenzene by removing oxygen atoms of the NO_2 group and replacing them with hydrogen atoms.

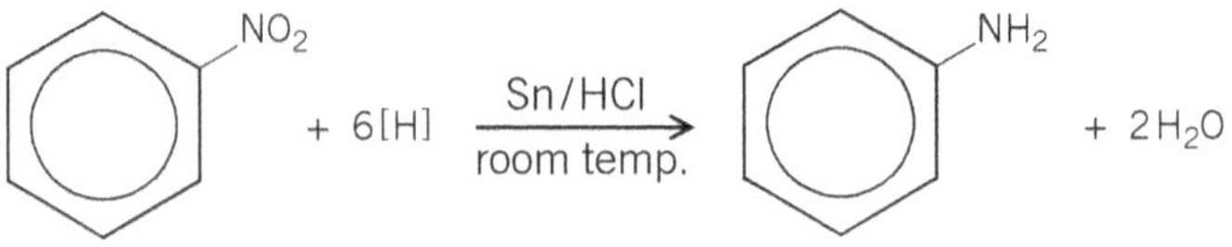

This could also be written:

$$C_6H_5NO_2 + 6[H] \rightarrow C_6H_5NH_2 + 2H_2O$$

Since the reaction is carried out in hydrochloric acid, the salt $C_6H_5NH_3{}^+Cl^-$ is formed and sodium hydroxide is added to liberate the free amine:

$$C_6H_5NH_3^+Cl^- + NaOH \rightarrow C_6H_5NH_2 + H_2O + NaCl$$

Study tip

It is acceptable to use [H] when writing equations for reduction, but the equation *must* balance.

The formation of amides

Amines will react with acid chlorides and acid anhydrides. These are nucleophilic substitution reactions (sometimes called addition – elimination reactions) and the products are N-substituted amides.

The equations are:

$$CH_3-C(=O)Cl \ + \ CH_3-N(H)H \longrightarrow CH_3-C(=O)N(H)-CH_3 \ + \ HCl$$

$$CH_3-C(=O)-O-C(=O)-CH_3 \ + \ CH_2-N(H)H \longrightarrow CH_3-C(=O)N(H)-CH_3 \ + \ CH_3-C(=O)OH$$

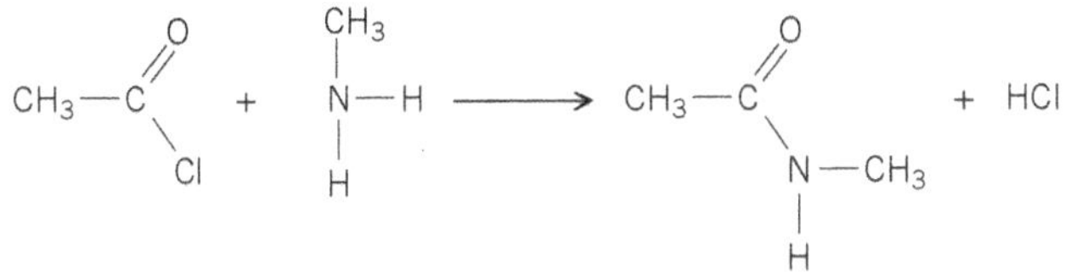

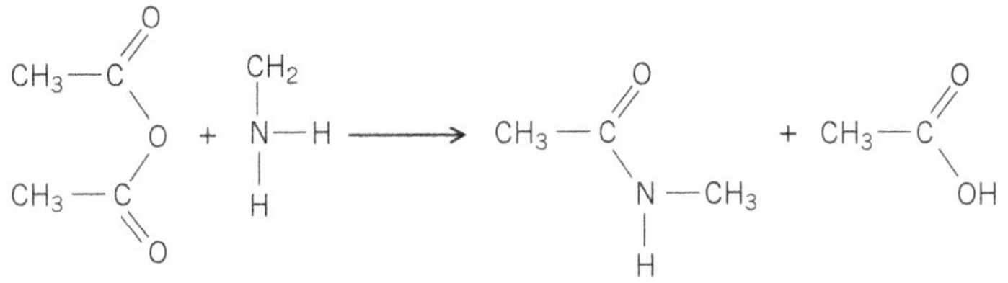

The amine adds on to the acid chloride and then HCl is eliminated.

This reaction is useful in forming polymers such as nylon.

The economic importance of amines

Amines are used in the manufacture of synthetic materials such as nylon and polyurethane, dyes, and drugs.

Quaternary ammonium compounds are used industrially in the manufacture of hair and fabric conditioners. They have a long hydrocarbon chain and a positively charged organic group, so they form cations:

$$\text{(long chain)}-N^+(CH_3)(CH_3)-CH_3 \quad Br^-$$

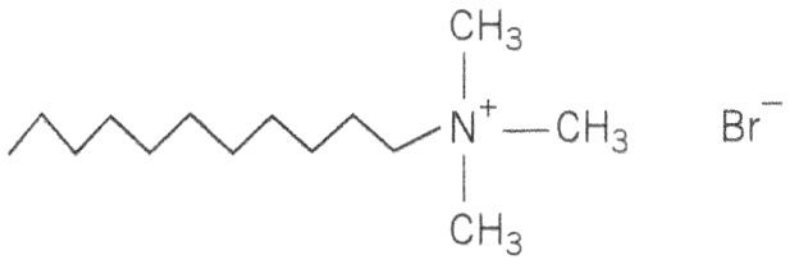

Both wet fabric and wet hair pick up negative charges on their surfaces. So the positive charges of the cations attract them to the wet surface and form a coating that prevents the build-up of static electricity. This keeps the surface of the fabric smooth (in fabric conditioner) and prevents flyaway hair in hair conditioners.

They are called cationic surfactants because in aqueous solution the ions cluster with their charged ends in the water and their hydrocarbon tails on the surface.

Synoptic link

You were first introduced to amines reacting with acid chlorides in Topic 26.5, Acylation.

Study tip

Remember that cations are named because they move towards the negatively charged cathode (i.e., they are positively charged).

▲ **Figure 1** Hair conditioner

Sulfa drugs

The story of the antibiotic penicillin is well known. It was the result of a chance observation of mould on a discarded Petri dish by Alexander Fleming, and was developed by Howard Flory and Ernst Chain (and a massive industrial effort) into a drug that saved thousands of lives in the Second World War and since. However, it was not the first anti-bacterial drug. Another class of drugs, the sulfanilamides, were already in use before penicillin and may also have had an effect on the course of the war – by saving the life of Prime Minister Winston Churchill.

Towards the end of the nineteenth century, it had been noticed that some dyes used to stain bacteria to make them visible under the microscope could also kill them. Since these dyes were absorbed by the bacteria rather than their surroundings, they might be expected to kill the bacteria but not their host. Eventually the dye Prontosil Rubrum began to be used in medicine to fight bacterial infections.

By the early 1940s it had been established that Prontosil was converted in the body into the compound sulfanilamide which was the active ingredient.

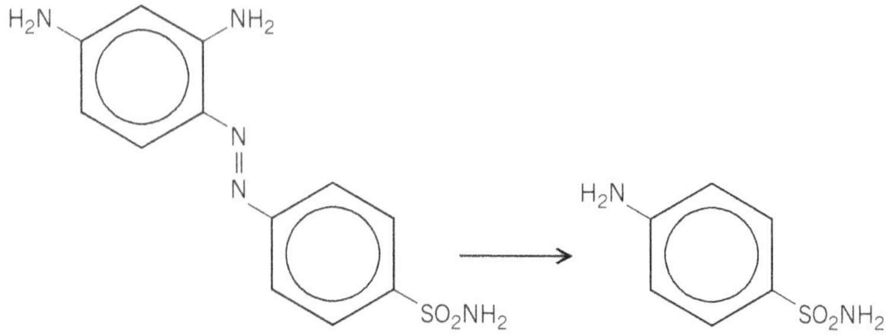

Prontosil Rubrum   sulfanilamide

The drug worked by preventing the bacteria making folic acid, which they need to synthesise DNA and therefore replicate. Bacteria make folic acid from a compound called *para*-aminobenzoic acid (PABA). The sulfanilamide molecule is of a similar shape to PABA and the bacteria try to use it to make folic acid but without success, as it is the wrong molecule. Humans do not need to synthesise folic acid – they get it from their food – and so sulfanilamide kills bacteria but is harmless to humans.

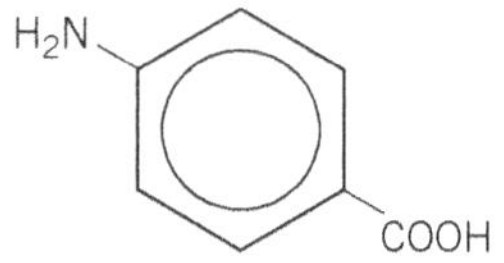

PABA

Since the 1940s over 5000 variations on the sulfanilamide molecule have been synthesised by chemists in an effort to find molecules that are more effective, have fewer side effects, are absorbed at a different rate, and so on. This is one of the main methods used to discover new drugs – to take a molecule with a known beneficial effect and make variations on it in the hope of maintaining or enhancing its activity but reducing any disadvantages that it might have. Nowadays, this process can be speeded up by the technique of combinatorial chemistry.

Although less common than they once were, sulfa drugs are still used today. The one that cured Winston Churchill's pneumonia in 1943 was sulfapyridine, known at the time as M & B 693, after the makers the May & Baker Company. May & Baker is still in business and supplies chemicals for school laboratories as well as making drugs. Look for their labels in your school preparation room.

What is the systematic name of PABA?

4-aminobenzenecarboxylic acid (or 4-aminobenzoic acid)

Hint

Para- is part of an older naming system for locating substituents on aromatic rings. It means opposite the original substituent, (i.e., in the 4 position).
Ortho- means adjacent (the 2 position) and *meta*- the 3 position.

Synoptic link

The use of robots in synthesis is described in Topic 31.1, Synthetic routes.

Summary questions

1 Why is nucleophilic substitution of a halogenolkane not a good method for preparing a primary amine?

2 a Write the equation for the reaction of chloroethane with an excess of ammonia. Give the reaction mechanism.

b What are the other possible products of this reaction?

Practice questions

1 (a) Name the compound $(CH_3)_2NH$.

(1 mark)

(b) $(CH_3)_2NH$ can be formed by the reaction of an excess of CH_3NH_2 with CH_3Br. Name and outline a mechanism for this reaction.

(5 marks)

(c) Name the type of compound produced when a large excess of CH_3Br reacts with CH_3NH_2. Give a use for this type of compound.

(2 marks)

AQA, 2006

2 Consider the following reaction sequence.

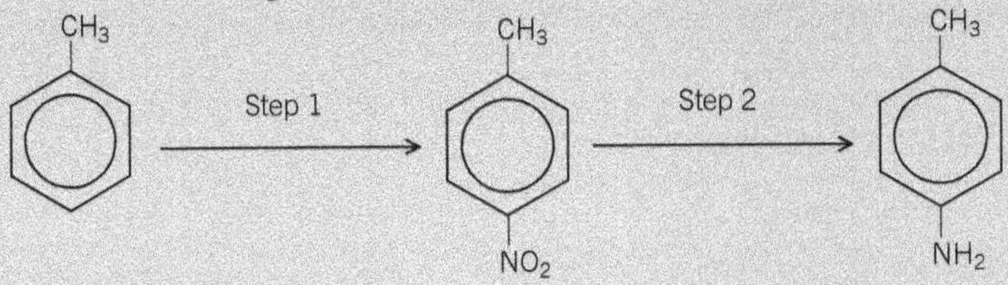

methylbenzene **E** **F**

(a) For Step 2, give a reagent or combination of reagents. Write the equation for this reaction using [H] to represent the reducing agent.

(2 marks)

(b) Draw the structure of the species formed by **F** in an excess of hydrochloric acid.

(2 marks)

(c) Compounds **G** and **H** are both monosubstituted benzenes and both are isomers of **F**. **G** is a primary amine and **H** is a secondary amine. Draw the structures of **G** and **H**.

(2 marks)

AQA, 2005

3 (a) Name and outline a mechanism for the formation of butylamine, $CH_3CH_2CH_2CH_2NH_2$, by the reaction of ammonia with 1-bromobutane, $CH_3CH_2CH_2CH_2Br$.

(5 marks)

(b) Butylamine can also be prepared in a two-step synthesis starting from 1-bromopropane, $CH_3CH_2CH_2Br$. Write an equation for each of the two steps in this synthesis.

(3 marks)

(c) Explain why butylamine is a stronger base than ammonia.

(2 marks)

(d) Draw the structure of a tertiary amine which is an isomer of butylamine.

(1 mark)

AQA, 2004

4 Propylamine, $CH_3CH_2CH_2NH_2$, can be formed either by nucleophilic substitution or by reduction.

(a) Draw the structure of a compound which can undergo nucleophilic substitution to form propylamine.

(1 mark)

(b) Draw the structure of the nitrile which can be reduced to form propylamine.

(1 mark)

(c) State and explain which of the two routes to propylamine, by nucleophilic substitution or by reduction, gives the less pure product. Draw the structure of a compound formed as an impurity.

(3 marks)

AQA, 2006

5 This question is about the primary amine $CH_3CH_2CH_2NH_2$

(a) The amine $CH_3CH_2CH_2NH_2$ reacts with CH_3COCl.
Name and outline a mechanism for this reaction.
Give the IUPAC name of the organic product.

(6 marks)

(b) Isomers of $CH_3CH_2CH_2NH_2$ include another primary amine, a secondary amine, and a tertiary amine.

(i) Draw the structures of these **three** isomers.
Label each structure as primary, secondary, or tertiary.

(3 marks)

6 **(a)** Name and outline a mechanism for the reaction of $CH_3CH_2NH_2$ with CH_3CH_2COCl.
Name the amide formed.

(6 marks)

(b) Halogenoalkanes such as CH_3Cl are used in organic synthesis.
Outline a three-step synthesis of $CH_3CH_2NH_2$ starting from methane. Your first step should involve the formation of CH_3Cl.
In your answer, identify the product of the second step and give the reagents and conditions for each step.
Equations and mechanisms are **not** required.

(6 marks)
AQA, 2013

7 Consider the reaction shown below.

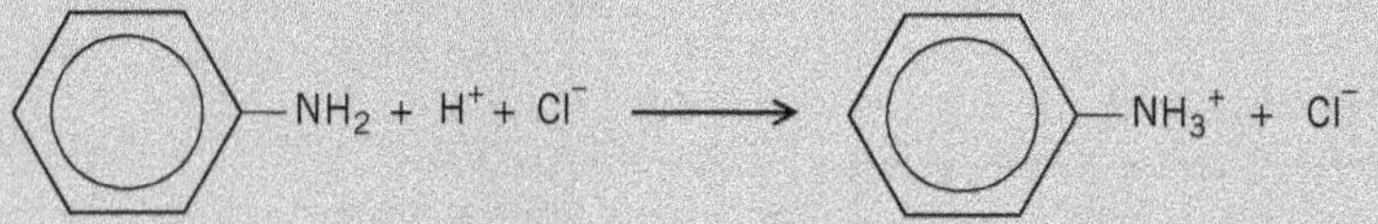

In this reaction phenylamine reacts with hydrochloric acid to form phenylammonium chloride.

(a) Explain how this reaction shows that phenylamine is a Brønsted-Lowry base.

(1 mark)

(b) Explain why phenylammonium chloride is soluble in water.

(1 mark)

Answers to the Practice Questions and Section Questions are available at www.oxfordsecondary.com/oxfordaqaexams-alevel-chemistry

29 Polymerisation
29.1 Condensation polymers

A condensation reaction occurs when two molecules react together and a small molecule, often water or hydrogen chloride, is eliminated. For example, esters are formed when carboxylic acids and alcohols react together. This is a condensation reaction because water, H_2O, is eliminated – hydrogen from the alcohol and an –OH group from the carboxylic acid.

$$R-C \overset{O}{\underset{OH}{\big|}} \quad + \quad HO-R' \quad \longrightarrow \quad R-C \overset{O}{\underset{O-R'}{\big|}} \quad + \quad H_2O$$

carboxylic acid alcohol ester water

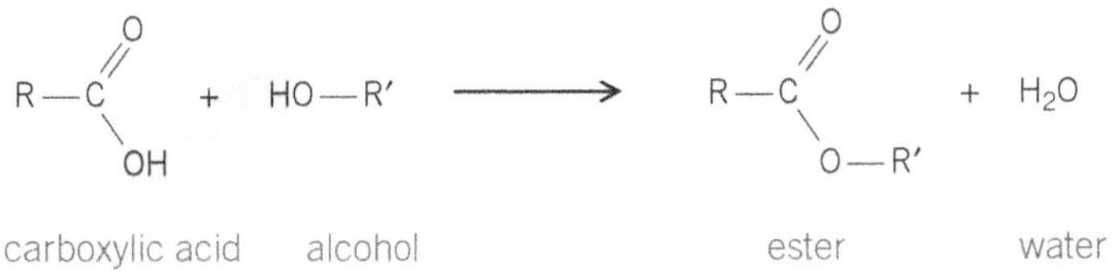

Condensation polymers are normally made from two different monomers, each of which has *two* functional groups. Both functional groups can react, forming long-chain polymers.

Polyesters, polyamides, and polypeptides are all examples of condensation polymers (Figure 1).

Polyesters

A *poly*ester has the ester linkage –COO– repeated over and over again.

To make a polyester diols are used, which have two –OH groups, and dicarboxylic acids, which have two carboxylic acid, –COOH, groups:

$$HO-A-OH \qquad\qquad \overset{O \qquad\qquad O}{\underset{HO \qquad\qquad OH}{C-B-C}}$$

diol dicarboxylic acid

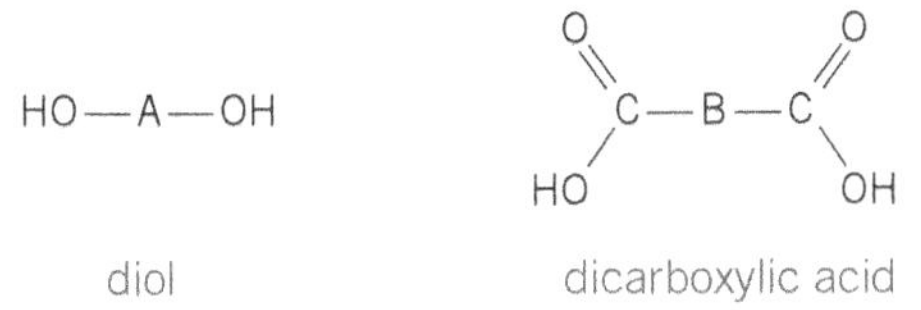

A and B represent unspecified organic groups, often $+CH_2\!+_n$. The functional groups on the ends of each molecule react to form a chain. For example, diols and dicarboxylic acids react together to give a polyester by eliminating molecules of water (Figure 2).

$$HO-A-OH \quad HO-\overset{O}{\overset{\|}{C}}-B-\overset{O}{\overset{\|}{C}}-OH \quad HO-A-OH \; HO-\overset{O}{\overset{\|}{C}}-B-\overset{O}{\overset{\|}{C}}-OH$$

diol dicarboxylic acid diol

H_2O H_2O H_2O

$$-O-A-O-\overset{O}{\overset{\|}{C}}-B-\overset{O}{\overset{\|}{C}}-O-A-O-\overset{O}{\overset{\|}{C}}-B-\overset{O}{\overset{\|}{C}}-$$

polyester

▲ **Figure 2** *Making a polyester*

Learning objectives:
→ Describe a condensation polymer.
→ Explain what sorts of molecules react to form condensation polymers.

Specification reference: 3.3.12

▲ **Figure 1** *Examples of polyesters and polyamides (nylon)*

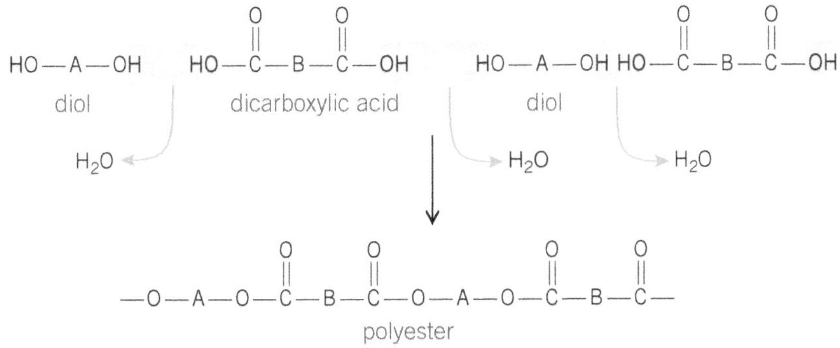

The fibre Terylene is a polyester made from benzene-1,4-dicarboxylic acid and ethane-1,2-diol (Figure 3).

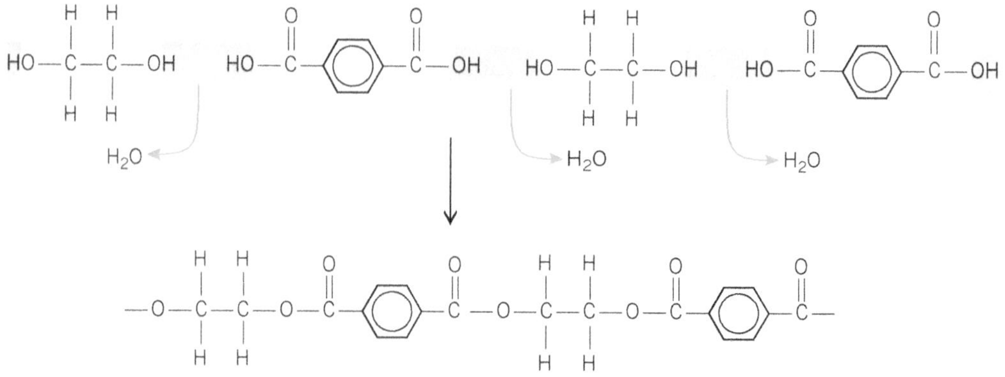

▲ **Figure 3** *Terylene is a polyester. Notice how the C—O is alternately to the left and to the right of the C═O*

> **Study tip**
>
> It is important you understand that, unlike polyalkenes, polyester and polyamides can be hydrolysed and are biodegradable.

Polyamides

An amide is formed when a carboxylic acid and an amine react together:

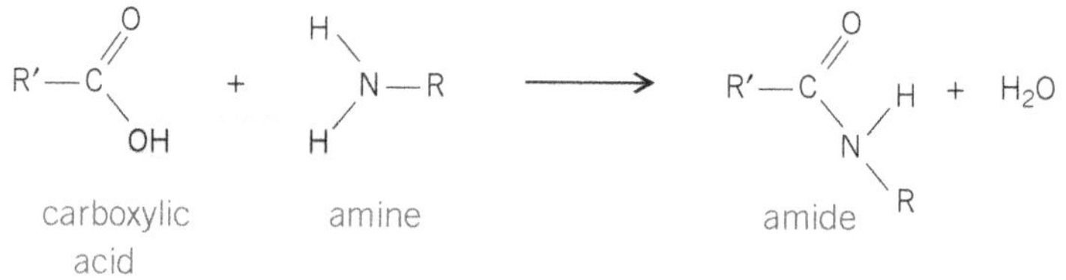

*Poly*amides have the amide linkage –CONH– repeated over and over again. To make polyamides from two different monomers, a diaminoalkane (which has two amine groups) reacts with a dicarboxylic acid (which has two carboxylic acid groups) (Figure 4).

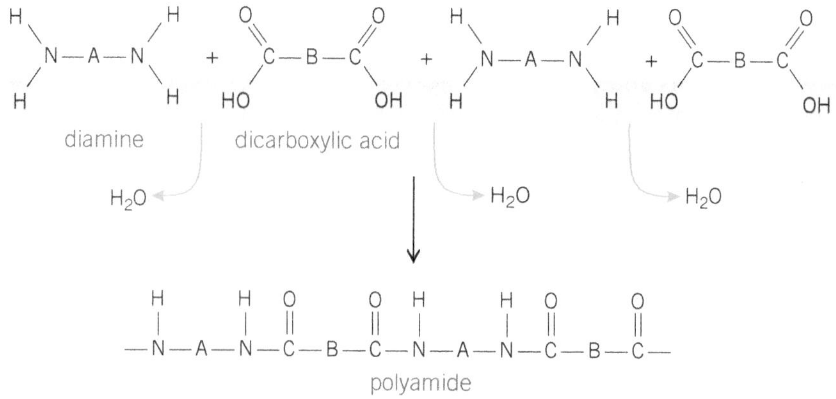

▲ **Figure 4** *The general equation for making a polyamide, such as Nylon-6,6 or Kevlar*

Both Nylon and Kevlar are condensation polymers.

Nylon

Industrially, Nylon-6,6 is made from 1,6-diaminohexane and hexane-1,6-dicarboxylic acid:

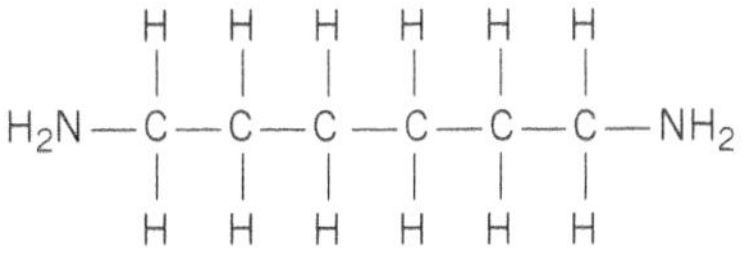

1,6-diaminohexane

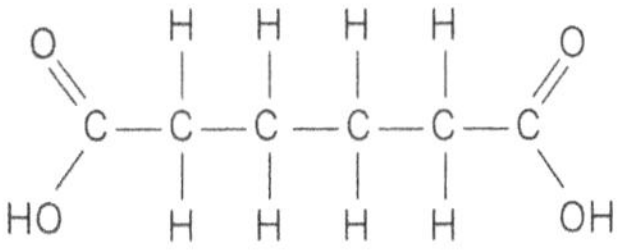

hexane-1,6-dicarboxylic acid

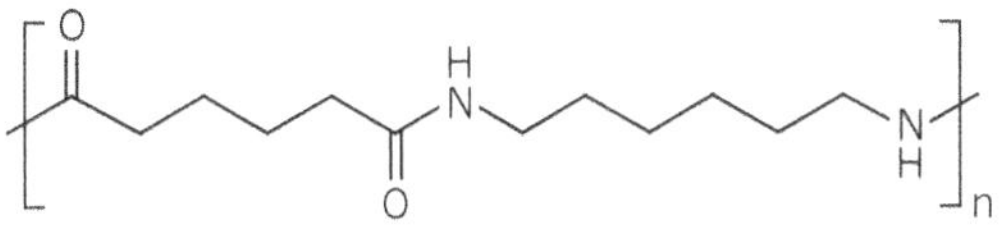

Nylon-6,6

In the laboratory, the reaction goes faster if a diacid chloride is used rather than the dicarboxylic acid, and in this case hydrogen chloride is eliminated. Nylon-6,10 is made from 1,6-diaminohexane and decane-1,10-dicarboxylic acid. Many other nylons are made each with slightly different properties.

▲ **Figure 5** *When 1-6-diaminohexane and hexane-1,6-dioyldichloride meet, Nylon-6,6 is formed at the interface. This demonstration was first performed by Stephanie Kwolek who developed Kevlar*

Kevlar

Kevlar is made from benzene-1,4-diamine and benzene-1,4-dicarboxylic acid (Figure 6).

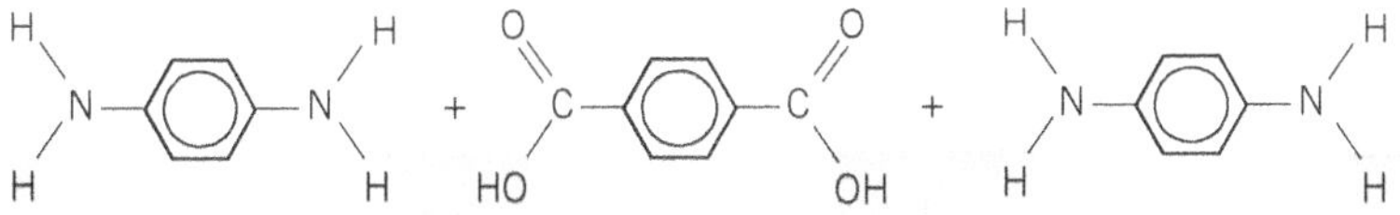

benzene-1,4-diamine benzene-1,4-dicarboxylic acid

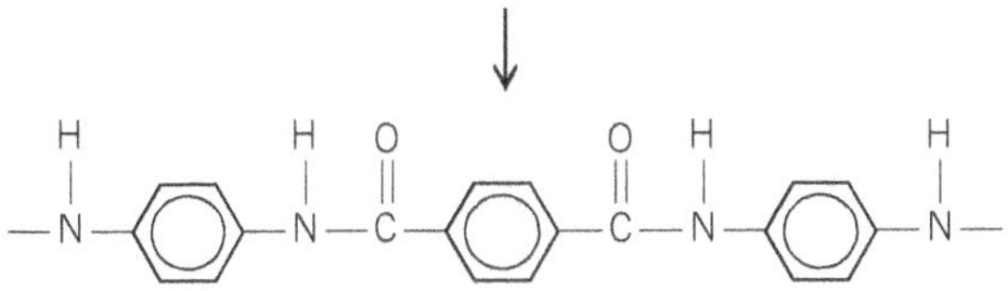

Kevlar

▲ **Figure 6** *Kevlar is a polyamide. Because the amide groups are linking rigid benzene rings, Kevlar has very different properties to nylon*

▲ **Figure 7** *Formula 1 drivers' helmets need to be lightweight as the less weight it adds to a driver's head, the smaller the risk of whiplash injuries under the extreme G-forces experienced in accelerating and braking. Helmets worn by racing drivers contain Kevlar, which is five times stronger than steel, weight for weight*

Kevlar's strength is due to the rigid chains and the ability of the flat aromatic rings to pack together held by strong intermolecular forces. The polymer, developed in the 1960s by Stephanie Kwolek of the DuPont company, is credited with saving some 3000 lives because of its use in bullet proof vests and anti-stab clothing as worn by the police. You may have Kevlar oven gloves at home.

Polypeptides and proteins

Polypeptides are also polyamides. They may be made from a single amino acid monomer, or many different ones.

In a polypeptide, *each* amino acid has both an amine group and a carboxylic acid group. So the amine group of one amino acid can react with the carboxylic acid group of another. A molecule of water is eliminated and a condensation polymer can begin to form:

Hint

Once a dipeptide is formed, tri-, tetra, and polypeptides can form by further reaction at each end of the molecule.

Hint

When an amide linkage is formed between amino acids, it is often called a **peptide link.**

Synoptic link

Amino acids are covered in Topic 30.1, Introduction to amino acids.

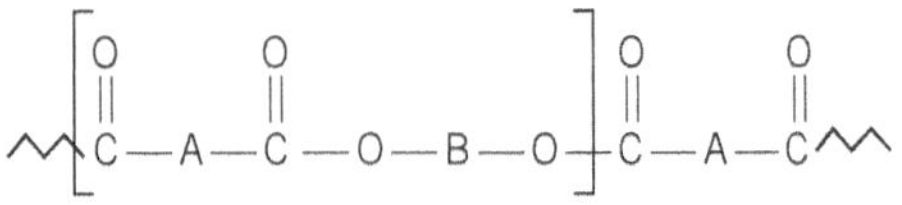

▲ **Figure 8** *The repeat unit is in brackets*

Study tip

It is important that you can deduce the monomers from which a polymer is formed and deduce the repeating unit in each type of polymer.

Intermolecular forces in condensation polymers

The intermolecular forces between condensation polymers can be quite strong. In nylon, for example, hydrogen bonds can form between the C=O and N–H groups as shown. This allows the polymer chains to align and attract one another thus forming strong fibres.

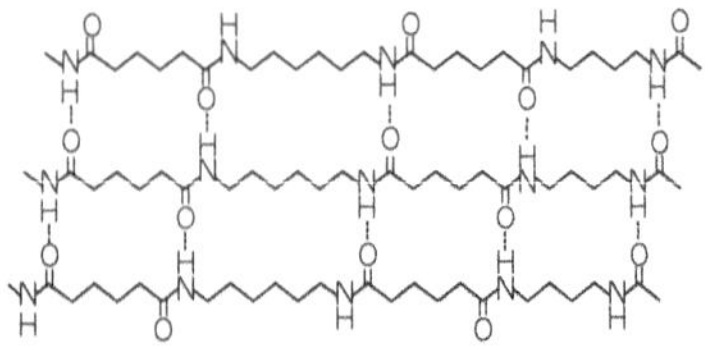

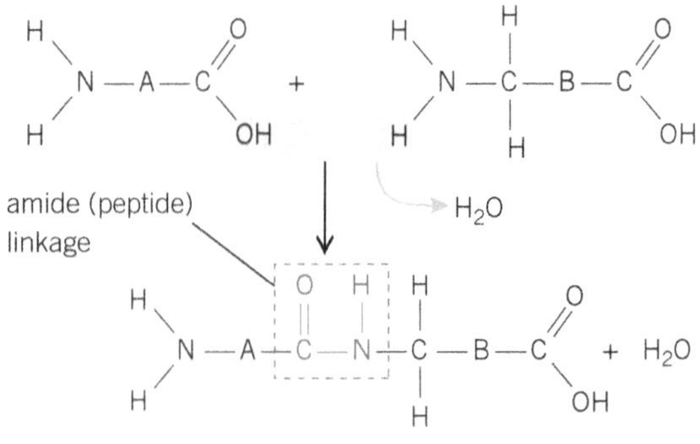

There is a difference between a polymer like Nylon-6,6 (where there are two monomers (one a diamine, $H_2N—X—NH_2$ and one a dicarboxylic acid, HOOC—Y—COOH)) and a polypeptide (where each amino acid monomer has one $–NH_2$ group and one –COOH group, $H_2NCHRCOOH$). There are 20 naturally occurring varieties of amino acids.

Identifying the repeat unit of a condensation polymer

The repeat unit of a condensation polymer is found by starting at any point in the polymer and stopping when the same pattern of atoms begins again (Figure 8).

Identifying the monomer(s) of a condensation polymer

The best way to work out the monomer(s) in a condensation polymer is to try and recognise the links formed by familiar functional groups (Table 1).

▼ **Table 1** *Condensation polymers – the repeat unit is inside the bracket*

Monomer 1	Monomer 2	Polymer
 dicarboxylic acid	HO—B—OH diol	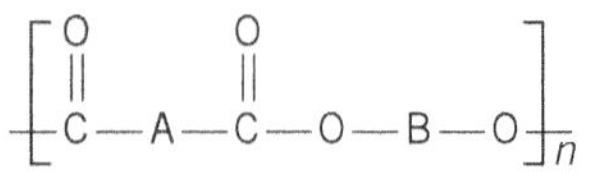
 dicarboxylic acid	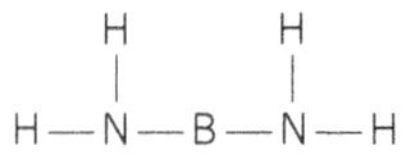 diamine	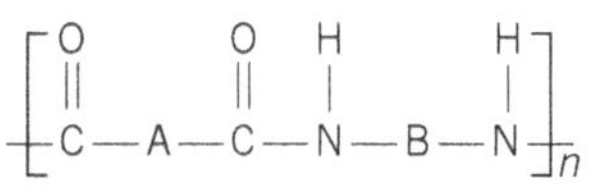
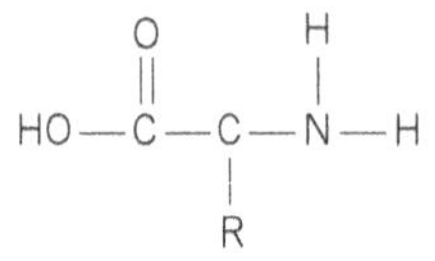 amino acid		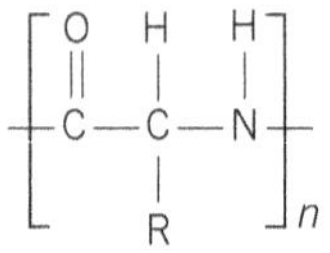

1 Start with the repeat unit.
2 Break the linkage (at the C—O for a polyester or C—N for a polyamide).
3 Add back the components of water for each ester or amide link.

For example:

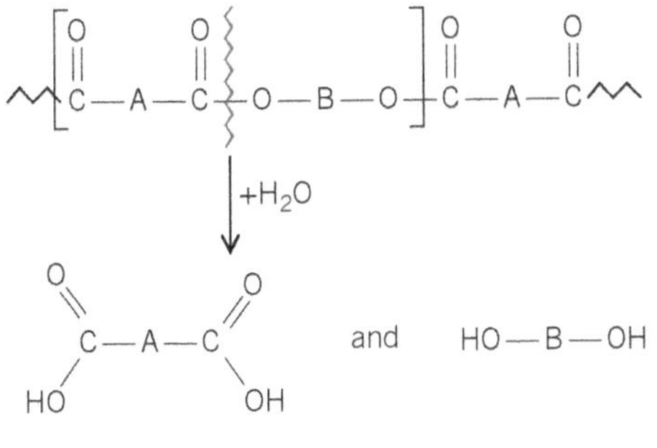

This is exactly the same process that occurs when condensation polymers are hydrolysed.

Disposal of polymers

Poly(ethene) and poly(propene) are not **biodegradable** because they are basically long-chain alkane molecules. Alkanes are unreactive because they have only strong, non-polar C—H and C—C bonds. There is nothing in the natural environment that will easily break them down and they persist for many years. They are usually disposed of in landfill sites, along with other rubbish, or by incineration. Some may be melted down and remoulded.

Poly(alkenes) can be burnt to carbon dioxide and water to produce energy, although poisonous carbon monoxide may be released into the atmosphere if combustion is incomplete (when there is a shortage of oxygen).

Burning poly(alkenes) does add to the problem of increasing the level of carbon dioxide in the atmosphere:

$$\text{+}CH_2\text{+}_n + 1\frac{1}{2}nO_2 \rightarrow nCO_2 + nH_2O$$

Other addition polymers, such as polystyrene, may release toxic products on burning. Complete combustion of polystyrene (a hydrocarbon) would produce carbon dioxide and water only. However, under certain conditions the polymer may depolymerise to produce toxic styrene vapour. Incomplete combustion produces carbon monoxide and unburnt carbon particles – black smoke.

Condensation polymers like polyesters and polyamides can be broken down by hydrolysis and are potentially biodegradable by the reverse of the polymerisation reaction by which they were formed.

The reaction below shows the hydrolysis of a polyamide such as nylon. However, this reaction is so slow under everyday conditions that you do not need to worry about your nylon umbrella depolymerising in the rain.

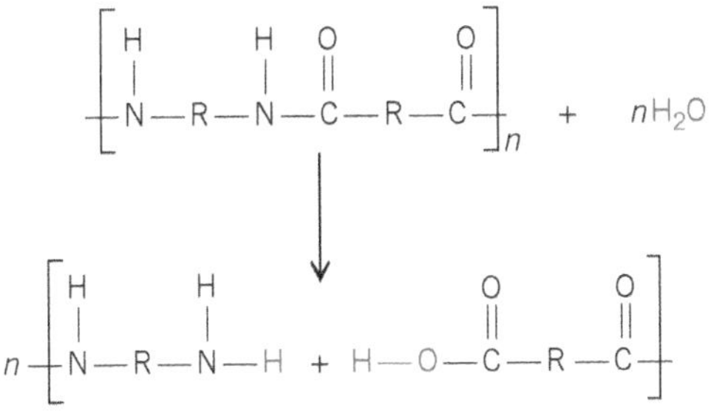

Throughout the polymer, the N—C bond is broken.

Synoptic link

You will need to know bond polarity studied in Topic 3.6, Electronegativity – bond polarity in covalent bonds

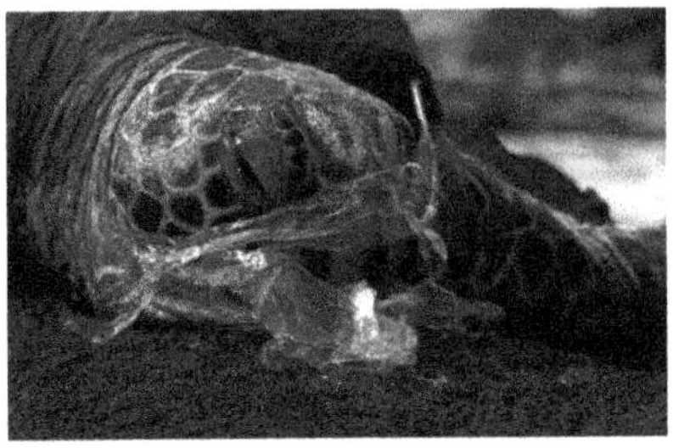

▲ **Figure 9** *Undecomposed poly(ethene) and poly(propene) cause problems for wildlife*

Recycling plastics

Many polyester materials are now recycled. They are being collected, sorted, and then melted and reformed. Fleece garments may well be made from recycled soft drink bottles. With all recycling, the costs and benefits have to be balanced. Melting and reforming of plastics can only be done a limited number of times, as during the process the polymer chains tend to break and shorten, thus degrading the properties of the polymer.

Advantages of recycling

Almost all plastics are derived from crude oil. Recycling saves this expensive and ever diminishing resource, as well as the energy used in refining it.

If plastics are not recycled they mostly end up in landfill sites.

Disadvantages of recycling

The plastics need to be collected, transported, and sorted, which uses energy and manpower and is therefore expensive.

Hermann Staudinger

Hermann Staudinger is considered to be the father of polymer chemistry, and he received the 1953 Nobel Prize for chemistry for his discoveries in this field – work which started in the 1920s.

Today, the idea of giant molecules (macromolecules) made up of chains of smaller ones is universally accepted. However, in the 1920s this idea was at odds with the established theory, and molecules with relative molecular masses (then called molecular weights) of over 5000 or so – such as rubber, starch, proteins, and so on – were considered to be made up of small molecules held together by some unknown force. Staudinger was already an established academic chemist (he had a reaction named after him) and put his reputation on the line by taking up the study of rubber. One distinguished colleague, with ill-disguised contempt, advised him to: 'Drop the idea of large molecules – organic molecules with a molecular weight higher than 5000 do not exist. Purify your rubber, then it will crystallise.'

Staudinger proved that polymers were indeed giant molecules made up of monomers by linking together molecules of methanal (formaldehyde, HCHO) one at a time to make successively bigger molecules, CH_2O, $(CH_2O)_2$, $(CH_2O)_3$, and so on until he produced the high molecular weight substance paraldehyde. He showed that the properties of these molecules gradually changed from those typical of small molecules to those of very large ones. So the very large molecule was simply a chain of small molecules held together by normal covalent bonds – no unknown force was required. A few years later, X-ray diffraction was able to confirm the structures of polymers.

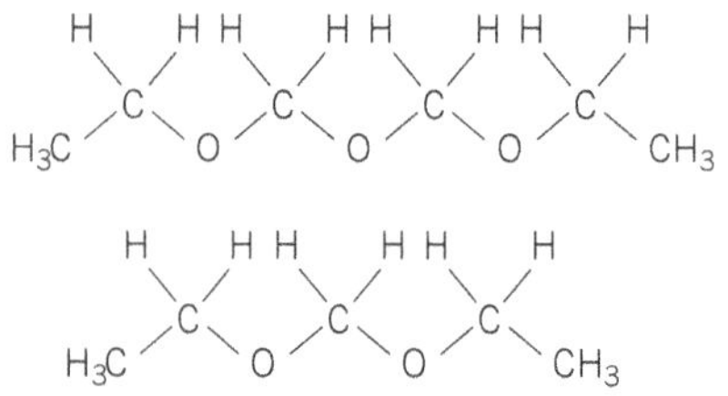

▲ **Figure 10** *Two of Staudinger's molecules – the CH_3 groups are the ends of the chains*

Staudinger's work led to modern synthetic polymers such as polythene (poly(ethene)) and nylon (a polyamide) and to an understanding of the structures of natural ones such as proteins, starch, and of course, rubber – which is poly(isoprene). Staudinger actually predicted artificial fibres – nylon was produced by Wallace Carothers in the late 1930s.

1 This is the structural formula of isoprene, the monomer from which rubber is made. What is its systematic (IUPAC) name?

2 Is isoprene likely to form an addition or a condensation polymer? Explain your answer.

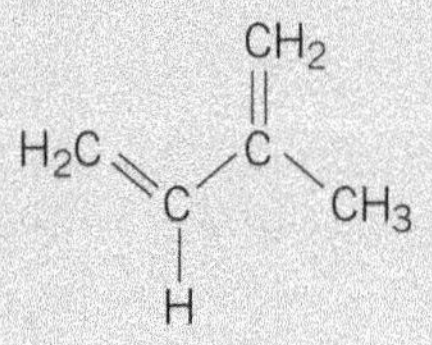

Summary questions

1 There are a number of different types of nylon made from two monomers – a dicarboxylic acid and a diamine.

   a The one made from hexane-1,6-dicarboxylic acid and 1,6-diaminohexane is called Nylon-6,6. Suggest where the numbers come from.

   b Nylon-6,10 is made from the same dicarboxylic acid as Nylon-6,6. What is the other monomer? Give its name and its formula.

2 Nylons are polyamides. Explain why proteins and peptides are also called polyamides.

3 Terylene is a polyester made from benzene-1,4-dicarboxylic acid and ethane-1,2-diol. Suggest another diol that would react with this acid to make a different polyester.

4 What are the linkages called in the the two polymers below?

   a 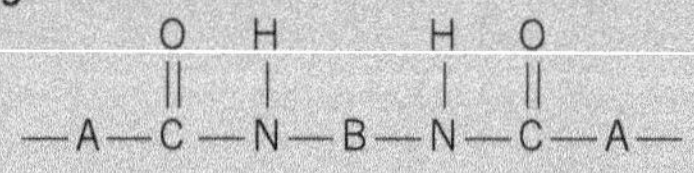

   b 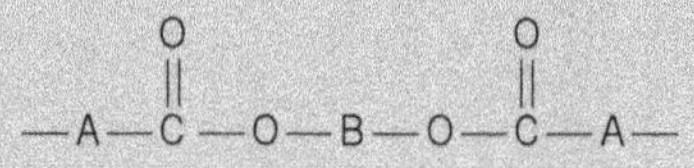

5 Write an equation for the hydrolysis of a polyester.

Practice questions

1 The repeating units of two polymers, **P** and **Q**, are shown below.

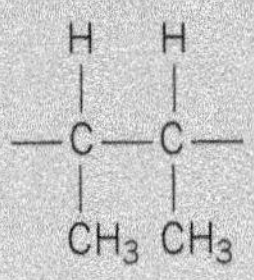

P

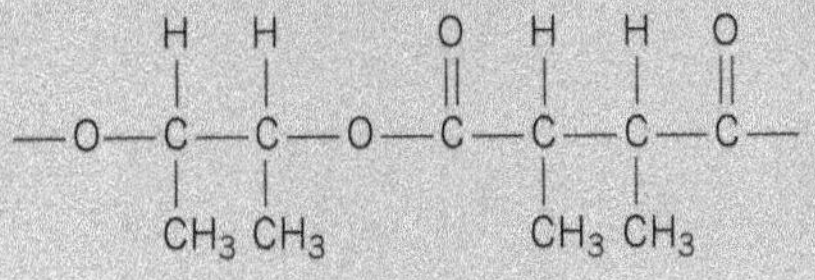

Q

(a) Draw the structure of the monomer used to form polymer **P**. Name the type of polymerisation involved.

(2 marks)

(b) Draw the structures of **two** compounds which react together to form polymer **Q**. Name these **two** compounds and name the type of polymerisation involved.

(5 marks)

(c) Identify a compound which, in aqueous solution, will break down polymer **Q** but not polymer **P**.

(1 mark)

AQA, 2006

2 The structure below shows the repeating unit of a polymer.

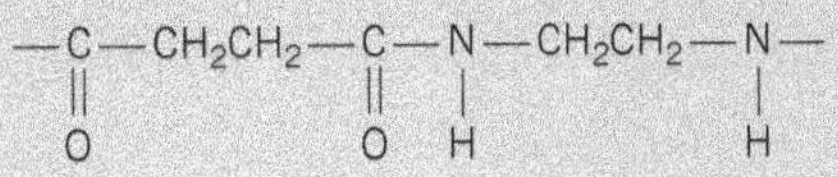

By considering the functional group formed during polymerisation, name this type of polymer and the type of polymerisation involved in its formation.

(2 marks)

AQA, 2006

3 **(a)** The compound $H_2C{=}CHCN$ is used in the formation of acrylic polymers.

(i) Draw the repeating unit of the polymer formed from this compound.

(ii) Name the type of polymerisation involved in the formation of this polymer.

(2 marks)

(b) The repeating unit of a polyester is shown below.

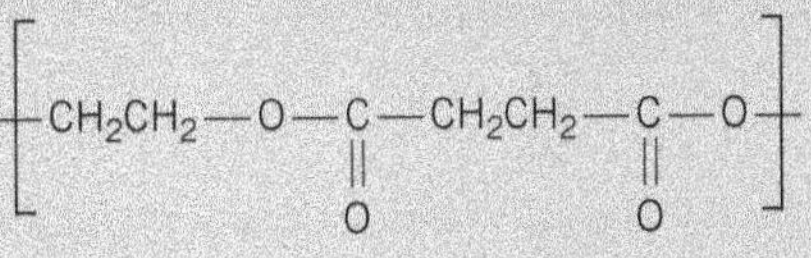

(i) Deduce the empirical formula of the repeating unit of this polyester.

(ii) Draw the structure of the acid which could be used in the preparation of this polyester and give the name of this acid.

(iii) Give **one** reason why the polyester is biodegradable.

(4 marks)

AQA, 2004

4 Consider the hydrocarbon **G**, $(CH_3)_2C{=}CHCH_3$, which can be polymerised.

(a) Name the type of polymerisation involved and draw the repeating unit of the polymer.

(2 marks)

(b) Draw the structure of an isomer of **G** which shows geometrical isomerism.

(1 mark)

(c) Draw the structure of an isomer of **G** which does not react with bromine water.

(1 mark)

AQA, 2004

5 (a) The hydrocarbon **M** has the structure shown below.

$$CH_3CH_2 - C = CH_2$$
$$|$$
$$CH_3$$

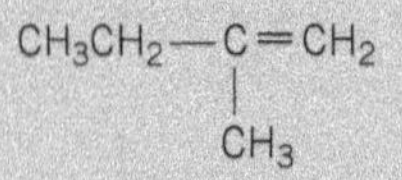

(i) Name hydrocarbon **M**.
(ii) Draw the repeating unit of the polymer which can be formed from **M**. State the type of polymerisation occurring in this reaction.

(3 marks)

(b) Draw the repeating unit of the polymer formed by the reaction between butanedioic acid and hexane-1,6-diamine. State the type of polymerisation occurring in this reaction and give a name for the linkage between the monomer units in this polymer.

(4 marks)
AQA, 2003

6 (a) Synthetic polyamides are produced by the reaction of dicarboxylic acids with compounds such as $H_2N(CH_2)_6NH_2$
(i) Name the compound $H_2N(CH_2)_6NH_2$
(ii) Give the repeating unit in the polyamide Nylon 6,6.

(2 marks)

(b) Synthetic polyamides have structures similar to those found in proteins.
(i) Draw the structure of 2-aminopropanoic acid.
(ii) Draw the organic product formed by the condensation of two molecules of 2-aminopropanoic acid.

(2 marks)
AQA, 2002

7 (a) Explain why polyalkenes are chemically inert.

(2 marks)

(b) Explain why polyesters and polyamides are biodegradeable.

(2 marks)

(c) Discuss the advantages of recycling polymers.

(2 marks)

8 The displayed formula of two organic compounds are shown below.

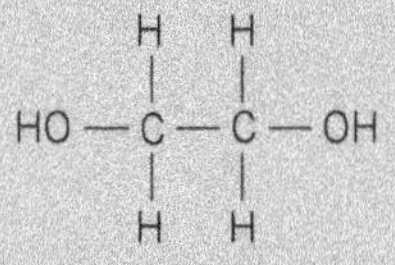

monomer A monomer B

(a) (i) Monomer A is diol. Name compound A.

(1 mark)

(ii) What type of compound is monomer B?

(1 mark)

(b) Monomer A and monomer B can react together to form a useful new substance named compound C.
(i) Draw a repeat unit of the new substance compound C.

(1 mark)

(ii) Circle the ester linkage in compound C.

(1 mark)

(iii) Name the non-organic product of this reaction.

(1 mark)

(iv) State the type of reaction that has taken place.

(1 mark)

(v) Suggest why a lab coat made from compound C may be damaged if concentrated sodium hydroxide was accidentally spilt on it.

(1 mark)

Answers to the Practice Questions and Section Questions are available at
www.oxfordsecondary.com/oxfordaqaexams-alevel-chemistry

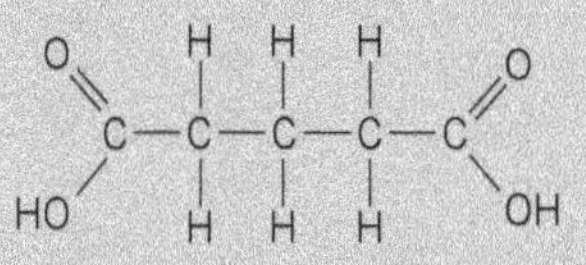

Amino acids are the building blocks of proteins, which in turn are vital components of all living systems.

Amino acids have two functional groups – a carboxylic acid and a primary amine. There are 20 important naturally occurring amino acids and they are all α-amino acids (also called 2-amino acids), which means that the amine group is on the carbon next to the $-CO_2H$ group (Figure 1).

α-amino acids have the general formula:

$$H_2N-\overset{\overset{\displaystyle R}{|}}{\underset{\underset{\displaystyle H}{|}}{C^*}}-\overset{\overset{\displaystyle O}{\parallel}}{C}-O-H$$

This structure has a carbon bonded to four different groups. The molecule is therefore chiral. Almost all naturally occurring amino acids exist as the (−) enantiomer.

Acid and base properties

Amino acids have both an acidic and a basic functional group.

* The carboxylic acid group has a tendency to lose a proton (act as an acid):

$$-\overset{\overset{\displaystyle O}{\parallel}}{C}-OH \rightleftharpoons -\overset{\overset{\displaystyle O}{\parallel}}{C}-O^- + H^+$$

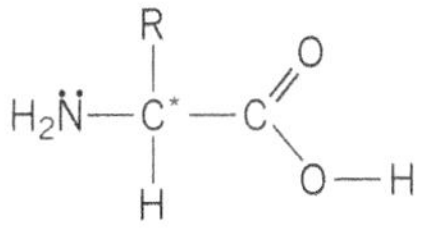

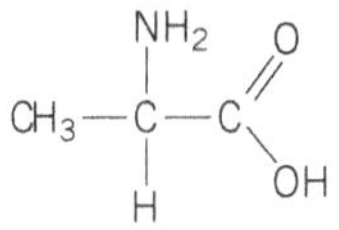

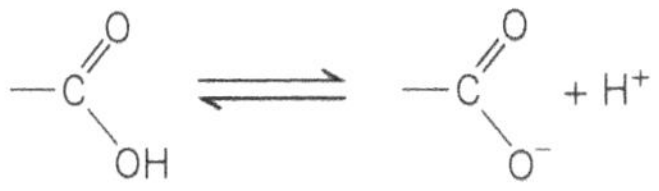

* The amine group has a tendency to accept a proton (act as a base):

$$H^+ + H-\overset{\overset{\displaystyle \cdot\cdot}{}}{\underset{\underset{\displaystyle H}{|}}{N}}- \rightleftharpoons H-\overset{\overset{\displaystyle H}{|}}{\underset{\underset{\displaystyle H}{|}}{\overset{+}{N}}}-$$

Amino acids exist as **zwitterions**. Ions like these have both a permanent positive charge and a permanent negative charge, though the compound is neutral overall (Figure 2).

Because they are ionic, amino acids have high melting points and dissolve well in water but poorly in non-polar solvents. A typical amino acid is a white solid at room temperature and behaves very much like an ionic salt.

In strongly acidic conditions the lone pair of the H_2N-group accepts a proton to form the positive ion (Figure 3).

The amino group has gained a hydrogen ion – it is protonated.

In strongly alkaline solutions, the –OH group loses a proton to form the negative ion (Figure 4).

The carboxylic acid group has lost a hydrogen ion – it is **deprotonated**.

Table 1 shows some naturally occurring amino acids. Each of these is usually referred to by its non-systematic name (the IUPAC names can be complex) and also by a three-letter abbreviation, which is useful when describing the sequences of amino acids in proteins, see Topic 30.2.

Learning objectives:

→ State what amino acids are.

→ Describe why they have both acidic and basic properties.

Specification reference: 3.3.13

▲ **Figure 1** *α-aminopropanoic acid, also called alanine, written in shorthand as $CH_3CH(NH_2)COOH$*

Synoptic link

Look back at Topic 25.2, Optical isomerism.

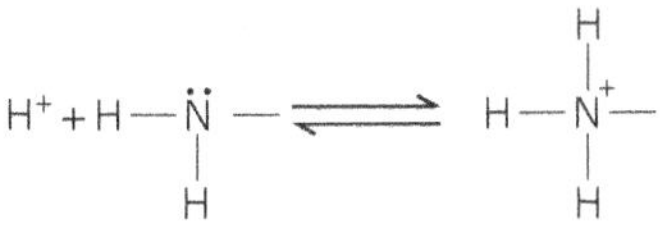

Hint

Compounds with two functional groups are called **bifunctional compounds**.

Synoptic link

You will need to know the nature of ionic bonding and states of matter studied in Topic 3.4, Bonding and physical properties.

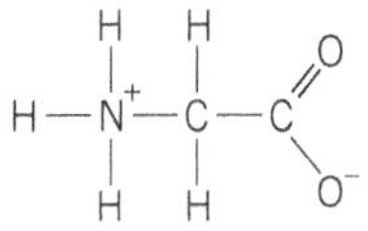

▲ **Figure 2** *A zwitterion*

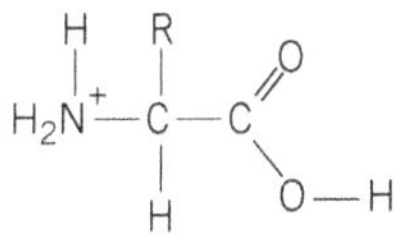

▲ **Figure 3** *A protonated amino acid*

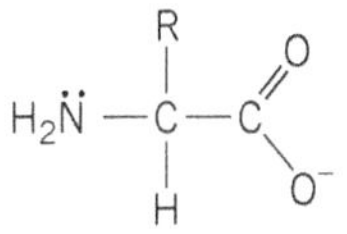

▲ **Figure 4** *A deprotonated amino acid*

▼ **Table 1** *20 naturally occurring amino acids*

Formula	Name and abbreviation	Formula	Name and abbreviation
H_2NCHCO_2H H	glycine (Gly)	H_2NCHCO_2H $CHOH$ CH_3	threonine (Thr)
H_2NCHCO_2H CH_3	alanine (Ala)	H_2NCHCO_2H CH_2SH	cysteine (Cys)
H_2NCHCO_2H $CHCH_3$ CH_3	valine (Val)	H_2NCHCO_2H CH_2 $CONH_2$	asparagine (Asn)
H_2NCHCO_2H CH_2 $CH_3(CH_3)_2$	leucine (Leu)	H_2NCHCO_2H CH_2 CH_2CONH_2	glutamine (Gln)
H_2NCHCO_2H CHC_2H_5 CH_3	isoleucine (Ile)	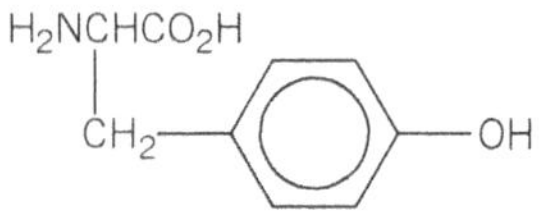	tyrosine (Tyr)
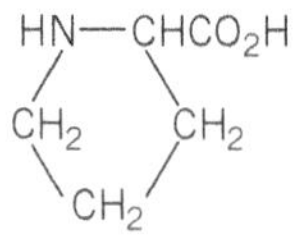	proline (Pro) (proline is a secondary amine)	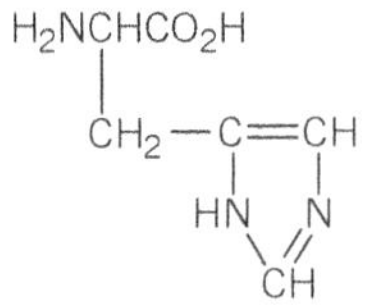	histidine (His)
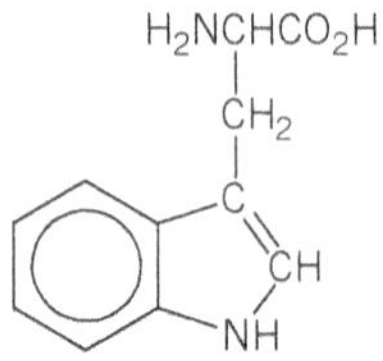	tryptophan (Try)	H_2NCHCO_2H $(CH_2)_3$ NH $NH{=}C{-}NH_2$	arginine (Arg)
H_2NCHCO_2H CH_2 CH_2SCH_3	methionine (Met)	H_2NCHCO_2H $(CH_2)_3$ CH_2NH_2	lysine (Lys)
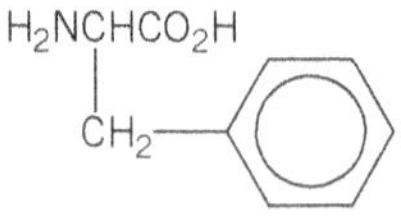	phenylalanine (Phe)	H_2NCHCO_2H CH_2CO_2H	aspartic acid (Asp)
H_2NCHCO_2H CH_2OH	serine (Ser)	H_2NCHCO_2H CH_2 CH_2CO_2H	glutamic acid (Glu)

Summary questions

1. The systematic name of glycine is 2-aminoethanoic acid. What is the systematic name of alanine (Table 1)?
2. Explain why alanine is chiral whereas glycine is not.

30.2 Peptides, polypeptides, and proteins

Amino acids link together to form peptides. Molecules containing up to about 50 amino acids are referred to as polypeptides. When there are more than 50 amino acids they are called proteins. Naturally occurring proteins are everywhere – enzymes, wool, hair, and muscles are all examples.

Amino acids and the peptide link

An amide has the functional group $—CONH_2$ or 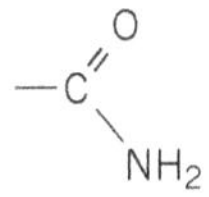

The amine group of one amino acid can react with the carboxylic acid group of another to form an **amide linkage** –CONH–.

This linkage is shown by shading in Figure 1.

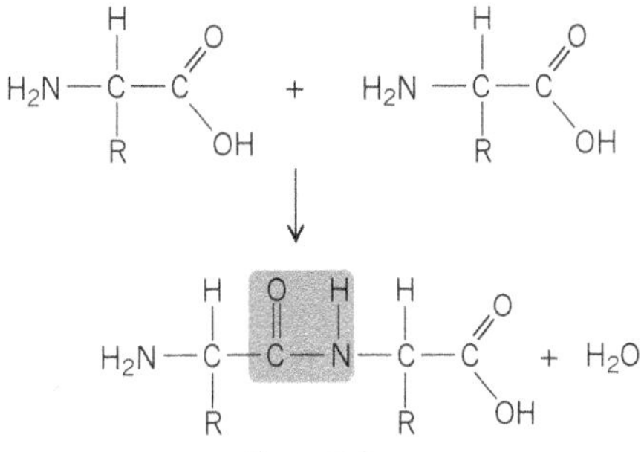

▲ **Figure 1** *Formation of a dipeptide*

Compounds formed by the linkage of amino acids are called **peptides**, and the amide linkage is called a peptide linkage in this context. A peptide with two amino acids is called a **dipeptide**. The dipeptide still retains $–NH_2$ and $–CO_2H$ groups and so can react further to give tri- and tetrapeptides, and so on (Figure 2).

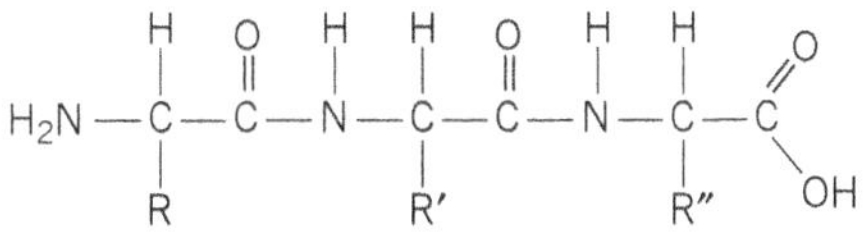

▲ **Figure 2** *A tripeptide – R, R′, and R″ may be the same or different*

A particular protein will have a fixed sequence of amino acids in its chain. This is called the **primary structure** of the protein. For example, just one short sequence of the protein insulin (the hormone controlling sugar metabolism) runs:

-ala-glu-ala-leu-tyr-

Polypeptides and proteins are condensation polymers because a small molecule (in this case water) is eliminated as each link of the chain forms.

Learning objectives:

- → State what peptides are.
- → Describe how amino acids form proteins.
- → Describe the primary, secondary, and tertiary structures of proteins.
- → State what bonds hold protein molecules in their particular shapes.
- → Describe how proteins can be broken down.

Specification reference: 3.3.13

Hint

Table 1 in Topic 30.1 gives the names, formula, and the three letter abbreviations of 20 naturally occurring amino acids.

Synoptic link

You covered condensation polymers in Topic 29.1, Condensation polymers.

Study tip

Hydrolysis is a reaction with water (often boiling) that may be catalysed by an acid, an alkali, or an enzyme. As they can be hydrolysed, proteins are biodegradable.

Hydrolysis

When a protein or a peptide is boiled with hydrochloric acid of concentration 6 mol dm^{-3} for about 24 hours, it breaks down to a mixture of all the amino acids that made up the original protein or peptide. All the peptide linkages are hydrolysed by the acid (Figure 3).

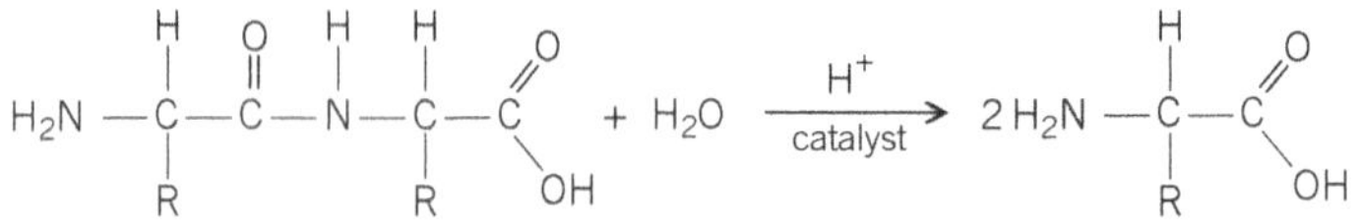

▲ **Figure 3** *The hydrolysis of the peptide link*

The structure of proteins

Proteins have complex shapes that are held in position by hydrogen bonds and other intermolecular forces as well as sulfur–sulfur bonds. The shapes of proteins are vital to their functions, for example, as enzymes and structural materials in living things. Many proteins are helical (spiral). Hydrogen bonding holds the helix in shape (Figures 4 and 5).

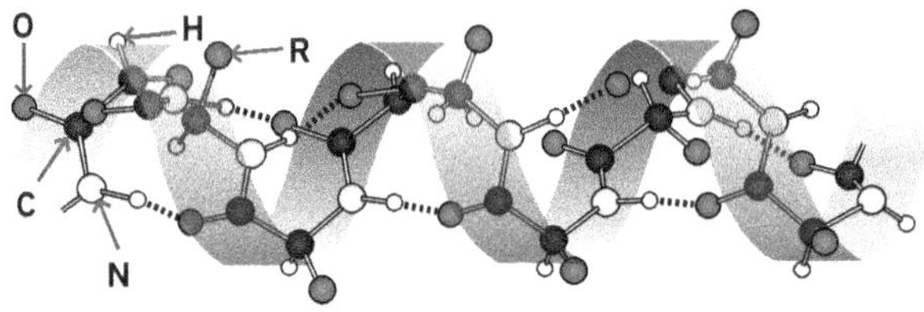

▲ **Figure 4** *The helical structure of a protein. Hydrogen bonds are shown as dotted lines*

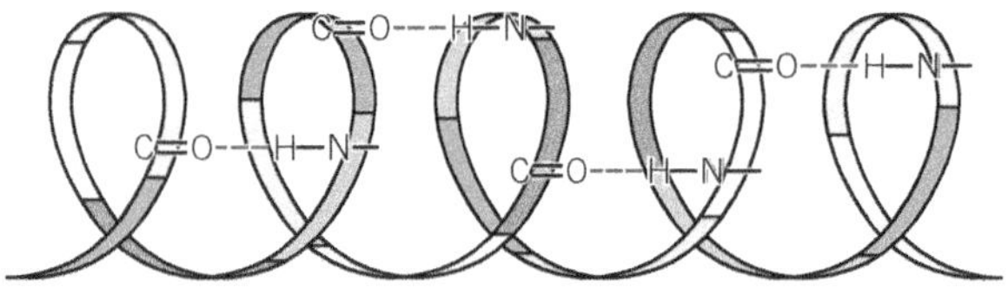

▲ **Figure 5** *The helix of a protein is held together by hydrogen bonding. The coloured strips represent amino acids (there are 18 to every 5 turns of the helix)*

Synoptic link

Bond polarity and intermolecular forces were covered in Topic 3.6, Electronegativity – bond polarity in covalent bonds, and Topic 3.7, Forces acting between molecules.

Another arrangement of a protein is called pleating and the protein ends up as a pleated sheet. The hydrogen bonding is shown in Figure 6.

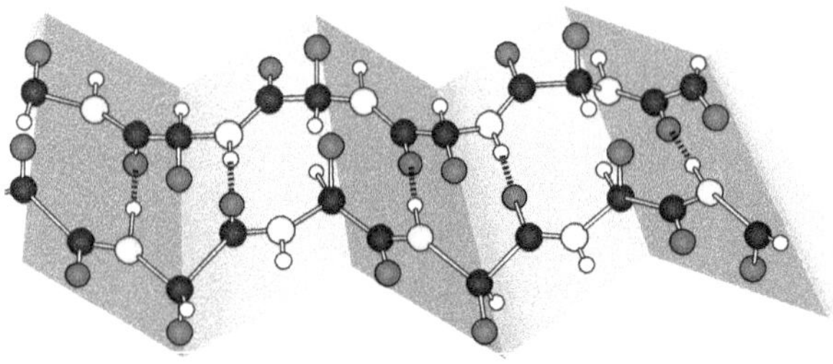

▲ **Figure 6** *A pleated sheet protein showing the hydrogen bonds as dotted lines*

Hydrolysis by enzymes

Certain enzymes will partially hydrolyse specific proteins. For example, the enzyme trypsin will only break the peptide bonds formed by lysine and arginine. Detective work, based on this and other techniques, enables chemists to find the sequence of amino acids in different proteins. The first protein to be fully sequenced was insulin. Fred Sanger won the 1958 Nobel Prize for chemistry for this achievement. He also won the Prize in 1980 for sequencing of DNA. He is the only person ever to win two Nobel Prizes for chemistry.

The stretchiness of wool

Wool is a protein fibre with a helix which is, as usual, held together by hydrogen bonds (Figure 7). When wool is gently stretched, the hydrogen bonds stretch (Figure 8) and the fibre extends. Releasing the tension allows the hydrogen bonds to return to their normal length and the fibre returns to its original shape. However, washing at high temperatures can permanently break the hydrogen bonds and a garment may permanently lose its shape.

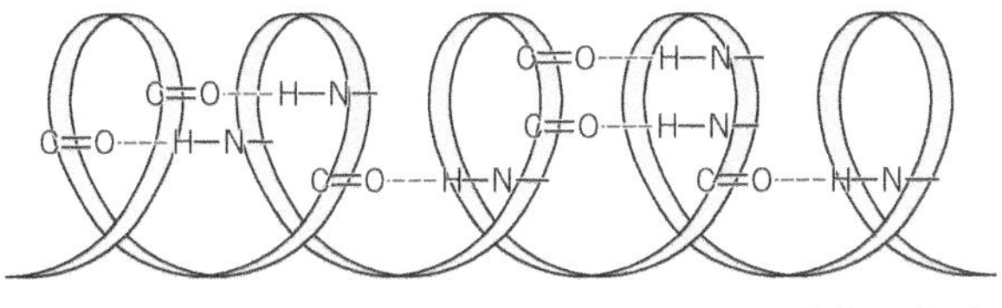

▲ **Figure 7** *Hydrogen bonds in wool*

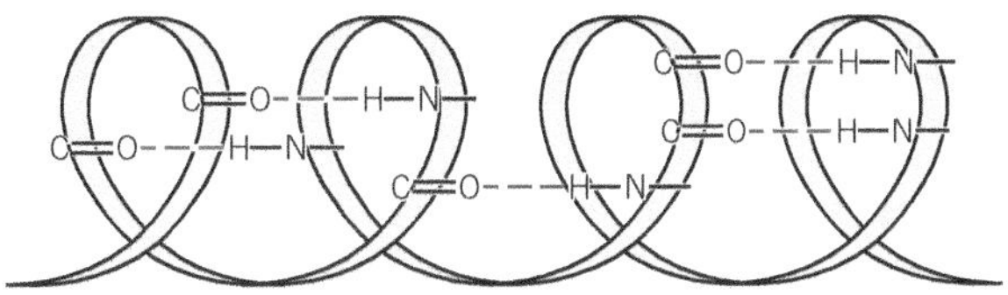

▲ **Figure 8** *The wool is gently stretched*

Bonding between amino acids

The amino acids in a protein chain can bond together in a number of ways.

- Hydrogen bonding between, for example, C=O groups and –N—H groups as shown C=O···H—N.
- Ionic attractions between groups on the side chains of amino acids such as $-COO^-$ (for example, on glutamic acid) and $-NH_3^+$ (for example, on lysine).
- Sulfur–sulfur bonds. The amino acid cysteine has a side chain with an $-CH_2SH$ group. Under suitable oxidising conditions, two cysteine molecules may react together to form a sulfur–sulfur bond that forms a bridge between the two molecules and creates a double amino acid called cystine. This is called sulfur–sulfur bridging or a disulfide bridge.

$$-CH_2SH + HSCH_2- + [O] \rightarrow -CH_2S-SCH_2- + H_2O$$

Hint

Remember the use of [O] in equations to represent an oxidising agent.

Levels of protein structure

All proteins have three (and sometimes more) levels of structure – primary, secondary, and tertiary.

Primary structure

The sequence of amino acids along a protein chain is called its primary structure. It can be represented simply by the sequence of three-letter names of the relevant amino acids, for example, gly-ala-ala-val-leu, and so on. This structure is held together by covalent bonding. Therefore it is relatively stable – it requires harsh conditions such as boiling with 6 mol dm^{-3} hydrochloric acid to break the amino acids apart.

Secondary structure

A protein chain may form a helix (the α-helix) or a folded sheet (called a β-pleated sheet). This is called the secondary structure and is held in place by hydrogen bonding between, for example, C=O groups and –N–H groups. Hydrogen bonds are much weaker than covalent bonds and this level of structure can relatively easily be disrupted by gentle heating or changes in pH.

Synoptic link

Make sure that you are confident about the different types of intermolecular forces which are described in Topic 3.7, Forces acting between molecules.

Tertiary structure

The α-helix or β-pleated sheet can itself be folded into a three-dimensional shape – this is called the tertiary structure and is held in place by a mixture of hydrogen bonding, ionic interactions, and sulfur–sulfur bonds (as well as van der Waals forces which exist between all molecules). Figure 9 shows an example of part of the tertiary structure and the bonds that hold it in place.

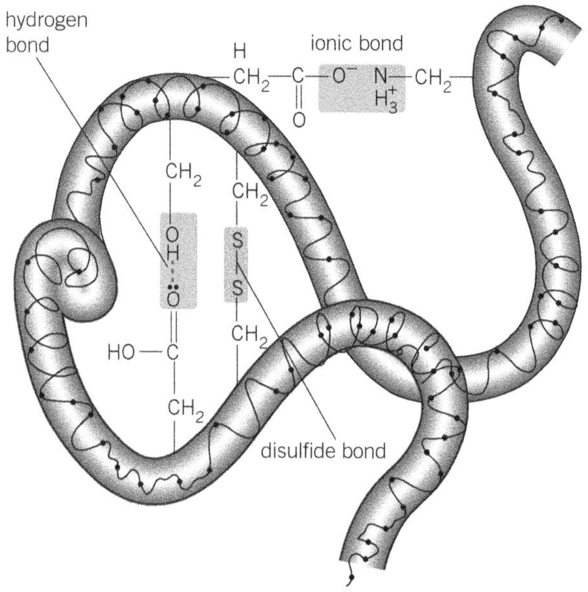

▲ **Figure 9** *An example of part of the tertiary structure of a helical protein*

Many proteins fold into globular shapes. The shapes of protein molecules are vital to their function – especially as enzymes.

Finding the structure of proteins

The shapes of proteins are of great importance. Many techniques are used to determine the secondary and tertiary structures of proteins including X-ray diffraction, which can locate the actual positions of atoms in space. However, these are beyond the scope of this book. The first step in determining the *primary* structure is to find out the number of each type of amino acid present in the protein. To begin this process, the protein is refluxed with 6 mol dm^{-3} hydrochloric acid. This process is called **hydrolysis**. It breaks the amide bonds between the amino acids and results in a mixture containing all the individual amino acids in the original protein.

Thin-layer chromatography

After hydrolysis, the amino acids can then be separated and identified by a technique called thin-layer chromatography (TLC). TLC is similar to paper chromatography but the paper is replaced by a chromatography plate consisting of a thin, flexible plastic sheet coated with a thin layer of silica (Figure 10). The IUPAC name of silica is silicon dioxide, SiO_2. This white powder is called the stationary phase.

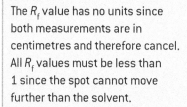

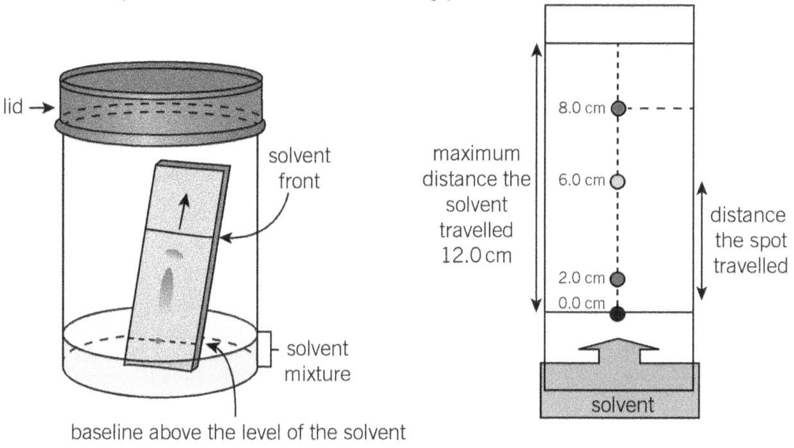

▲ **Figure 10** *A thin-layer chromatography experiment and the chromatogram that results*

1. A small spot containing the mixture of amino acids to be separated is placed on a line about 1 cm up the plate and the plate is placed in a tank containing a suitable solvent to a depth of about $\frac{1}{2}$ cm. The starting line must be above the initial level of the solvent. The solvent (or mixture of solvents) is called the mobile phase (or eluent).
2. A lid is placed on the tank so that the inside of it is saturated with solvent vapour and the solvent is allowed to rise up the plate (Figure 10). As it does so, it carries the amino acids with it. Each amino acid lags behind the solvent front to an extent that depends on its affinity for the solvent compared with its affinity for the stationary phase. This depends on the intermolecular forces that act between the amino acid and the solvent – the stronger they are, the closer the amino acid is to the solvent front.
3. When the solvent has almost reached the top of the plate, the plate is removed from the tank and the position to which the solvent front has moved is marked. Amino acids are colourless, so the positions they have reached have to be made visible. This is done by spraying the plate with a developing agent such as ninhydrin, which reacts with amino acids to form a purple compound, or by shining ultraviolet light on the plate. If the solvent is suitable, the amino acids will be completely separated.

R_f values are then calculated for each amino acid spot.

$$R_f = \frac{\text{distance moved by the spot}}{\text{distance moved by the solvent}}$$

So the R_f value for the red spot (8.0 cm) in Figure 10 is $\frac{8.0\text{ cm}}{12.0\text{ cm}} = 0.67$

This allows each amino acid in the mixture to be identified by comparing the R_f value of each spot with the values obtained by known pure amino acids run in the same solvent mixture.

Hint

TLC plates may use glass or aluminium rather than plastic sheet. There are alternative materials to silica for the stationary phase. Plastic sheets are convenient because they can be cut to the required size with scissors.

Study tip

The R_f value has no units since both measurements are in centimetres and therefore cancel. All R_f values must be less than 1 since the spot cannot move further than the solvent.

Synoptic link

You will learn more about other chromatography techniques in Topic 33.1, Chromatography.

2-dimensional TLC

Often, two amino acids have very similar R_f values in a particular solvent. This makes it hard to distinguish them. One solution to this problem is to use 2-dimensional TLC. Here a square piece of TLC film is used. The plate is spotted in one corner and a chromatogram is run in the usual way so that the spots are separated along one side of the plate. The plate is then turned through 90° and the chromatogram is run again with a different solvent (Figure 11). This makes it easier to see the separation between the spots and gives two R_f values (one for each solvent). If both these values match those for a known amino acid, you can be more confident in your identification.

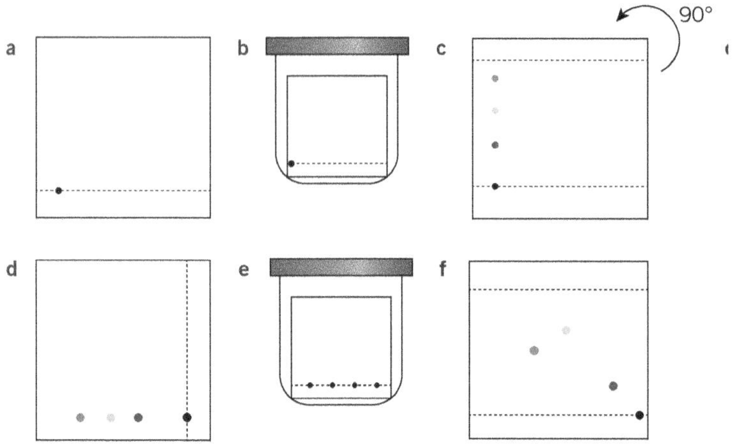

▲ **Figure 11** *The principle of 2-D TLC*

Calculate the R_f values of each of the amino acids represented by the orange and blue spots in Figure 11(f). You will need to use a ruler to measure the distances travelled.

Summary questions

1 a What are the functional groups in an amino acid?

b Which group is acidic and which basic?

2 How many amide (peptide) linkages are there in a tripeptide?

3 In what form will amino acid residues exist after a protein has been hydrolysed with 6 mol dm^{-3} hydrochloric acid. Draw the structural formula of an alanine residue.

4 Draw the formulae of the three amino acids that would be formed by the hydrolysis of the tripeptide shown below.

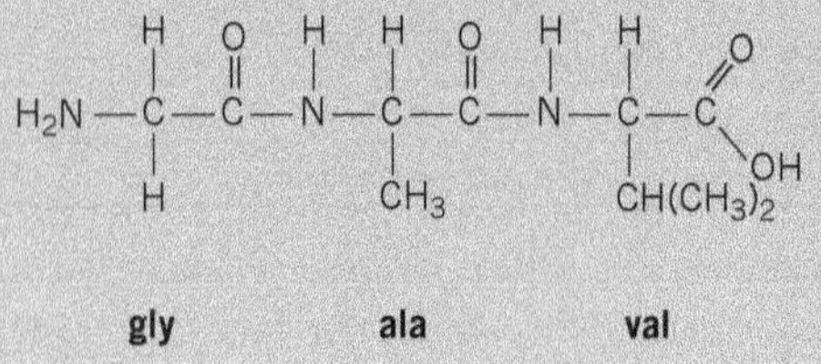

30.3 The action of anti-cancer drugs

Cisplatin, an anti-cancer drug

Cancer is not one disease but many. What cancers have in common is 'rogue' cells which have lost control over their growth and replication and grow much faster than normal cells. Cisplatin was discovered in 1965 and is one of the most successful cancer treatments, for example, giving survival rates of up to 90% in testicular cancer.

Cisplatin is square planar and has the formula:

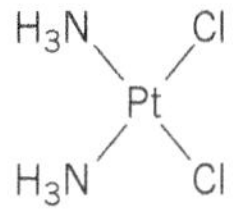

It works by bonding to strands of DNA, distorting their shape and preventing replication of the cells. The molecule bonds to nitrogen atoms on two adjacent guanine bases on a strand of DNA.

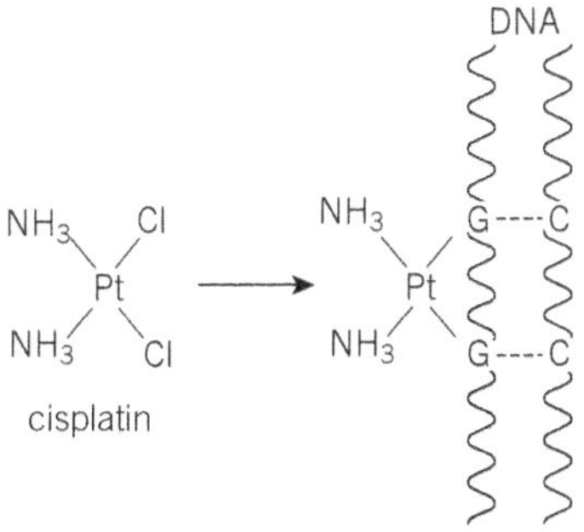

This works because the nitrogen atoms of the guanine molecules have lone pairs of electrons which form dative covalent bonds with the platinum. They displace the chloride ions because they are better ligands. This is an example of a **ligand substitution** reaction.

Like all drugs, cisplatin has side effects – it will bond to DNA in healthy cells as well as in cancerous ones but cancer cells are replicating faster than healthy cells, and so the effect of the drug is greater on cancer cell, than on normal cells. However, healthy cells that replicate quickly, such as hair follicles, are significantly affected and this is why patients undergoing chemotherapy (drug treatment for cancer) often lose their hair. Work is underway to find drugs and delivery systems that can better discriminate between healthy and cancerous cells.

Summary question

1 Draw the formula of transplatin and suggest why it is not an effective anti-cancer drug.

Learning objective:

→ Describe how the anti-cancer drug cisplatin works.

Specification reference 3.3.13 and 3.2.5

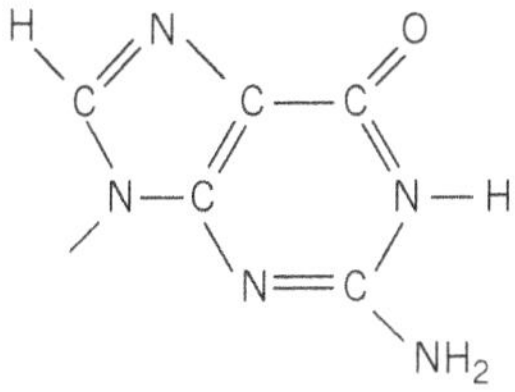

▲ **Figure 1** *The structure of the guanine base*

Synoptic link

Ligand substitution reactions were covered in Topic 24.2, Ligand substitution reactions.

Practice questions

1 (a) The structure of the amino acid alanine is shown below.

$$H_2N-\underset{\underset{H}{|}}{\overset{\overset{CH_3}{|}}{C}}-COOH$$

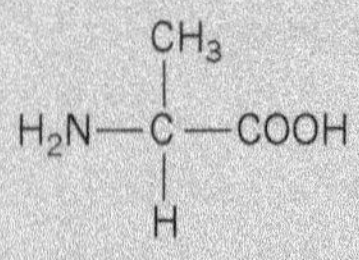

 (i) Draw the structure of the zwitterion formed by alanine.
 (ii) Draw the structure of the organic product formed in each case from alanine
 when it reacts with:
 • CH_3OH, in the presence of a small amount of concentrated sulfuric acid.
 • Na_2CO_3
 • CH_3Cl in a 1 : 1 mole ratio.

(4 marks)

 (b) The amino acid lysine is shown below.

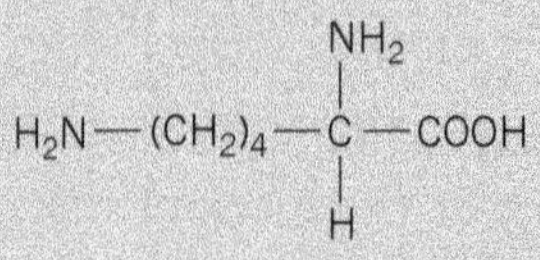

$$H_2N-(CH_2)_4-\underset{\underset{H}{|}}{\overset{\overset{NH_2}{|}}{C}}-COOH$$

Draw the structure of the lysine species present in a solution at low pH.

(1 mark)

 (c) The amino acid proline is shown below.

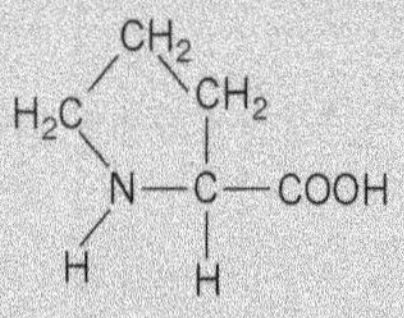

$$\begin{array}{c} CH_2 \\ H_2C \qquad CH_2 \\ N-C-COOH \\ | \qquad | \\ H \qquad H \end{array}$$

Draw the structure of the dipeptide formed from two proline molecules.

(1 mark)
AQA, 2007

2 Draw the structures of the **two** dipeptides which can form when one of the amino acids
 shown below reacts with the other.

$$H_2N-\underset{\underset{H}{|}}{\overset{\overset{CH_3}{|}}{C}}-COOH \qquad\qquad H_2N-\underset{\underset{H}{|}}{\overset{\overset{CH_2OH}{|}}{C}}-COOH$$

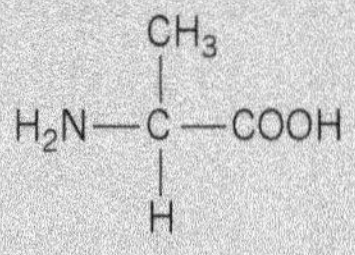

 structure 1 structure 2

(2 marks)
AQA, 2006

3 Consider the following amino acid.

$$H_2N-\underset{\underset{CH(CH_3)_2}{|}}{\overset{\overset{H}{|}}{C}}-COOH$$

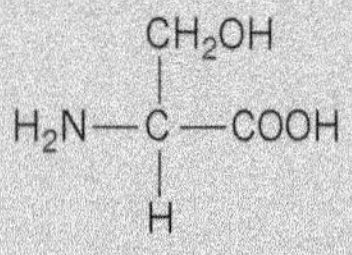

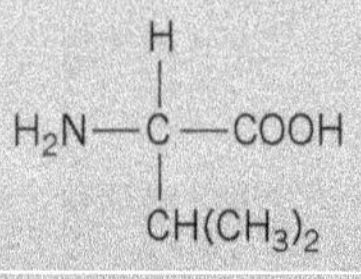

 (a) Draw the structure of the amino acid present in the solution at pH 12.
 (b) Draw the structure of the dipeptide formed from two molecules of this amino acid.
 (c) Protein chains are often arranged in the shape of a helix. Name the type of
 interaction that is responsible for holding the protein chain in this shape.

(3 marks)
AQA, 2004

4 The structures of the amino acids alanine and glycine are shown below.

$$H_2N-\underset{\underset{H}{|}}{\overset{\overset{CH_3}{|}}{C}}-COOH \qquad\qquad H_2N-\underset{\underset{H}{|}}{\overset{\overset{H}{|}}{C}}-COOH$$

alanine glycine

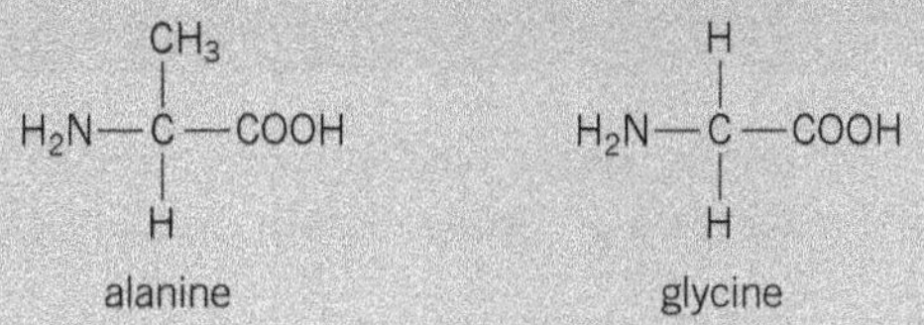

Alanine exists as a pair of stereoisomers.
(a) Explain the meaning of the term *stereoisomers*.
(b) State how you could distinguish between the stereoisomers.

(4 marks)
AQA, 2003

Answers to the Practice Questions and Section Questions are available at
www.oxfordsecondary.com/oxfordaqaexams-alevel-chemistry

31 Organic synthesis and analysis

31.1 Synthetic routes

Learning objective:

→ Describe how organic reactions can be used to synthesise target molecules.

Specification reference: 3.3.14

Synoptic link

You will need to know all the organic chemistry studied previously.

Study tip

It is important to know the reactions of these functional groups:

- alkanes
- alkenes
- halogenoalkanes
- alcohols
- amines
- aldehydes
- ketones
- acid derivatives
- arenes.

This chapter is about working out a series of reactions for making (synthesising) a given molecule, usually called the **target molecule**.

Synthesis of a target molecule is a common problem in industries like drug or pesticide manufacture. Suppose a molecule is found to have a particular effect, for example, as an antibiotic. Drug companies may synthesise, on a small scale, a number of compounds of similar structures. These will be screened for possible antibiotic properties. Any promising compounds may then be made in larger quantities for thorough investigation of their effectiveness, safety, side effects, and so on, before the final step goes ahead – producing them commercially.

Using the organic reactions you have already met, you can work out a reaction scheme to convert a starting material into a target molecule.

Working out a scheme

Start by writing down the formula of the starting molecule, A, and that of the target molecule, X.

One way of working out what route to take is to write down all the compounds which can be made from A and all the ways in which X can be prepared (Figure 1).

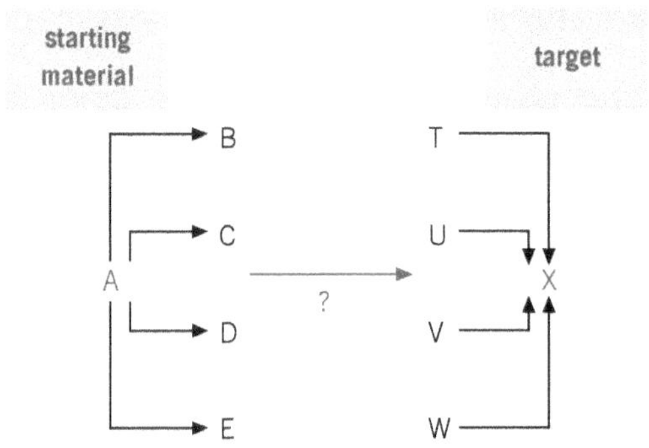

▲ **Figure 1** *Devising a synthesis of compound X from compound A*

Synoptic link

Yield is a measure of the efficiency of the conversion of a starting material into a product. See Topic 2.6, Balanced equations, atom economies, and percentage yields.

You may then see how B, C, D, or E can be converted, in one or more steps to T, U, V, W, or direct to X. It is important to keep the number of steps as small as possible to maximise the yield of the target.

Sometimes you will be able to see straight away that a particular reaction will be needed. For example, if the target molecule has one more carbon atom than the starting material, it is probable that the reaction of cyanide ions with a halogenoalkane will be needed at some stage, as this reaction increases the length of the carbon chain by one, for example:

$$CH_3Br + CN^- \rightarrow CH_3C{\equiv}N + Br^-$$

bromomethane → ethanenitrile

How the functional groups are connected

The inter-relationships between the functional groups you should know are shown in Figure 2. Make sure you can recall the reagents and conditions for each conversion.

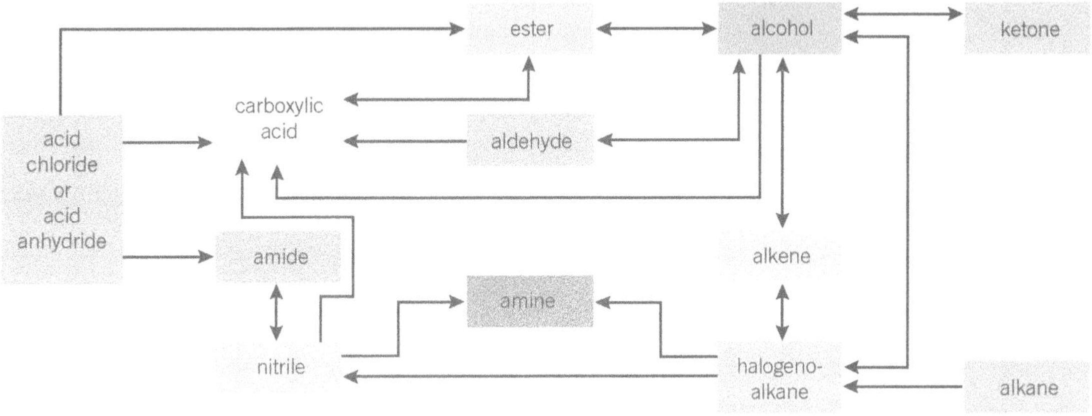

▲ **Figure 2** *Inter-relationships between functional groups. You can use this chart to revise your knowledge of organic reactions*

Synthetic robots

Routine chemical synthesis in the pharmaceutical industry is now often done by robots – not androids but arrays of reaction tubes along with computer-controlled syringes to measure out and mix the reactants. This produces a 'library' of related compounds. For example, you could oxidise several alcohols of different chain lengths to produce a library of aldehydes. The target compound can then be tested to see if any of them have any potential for use as medicines. A chemist is still needed to work out the reaction and program the computer.

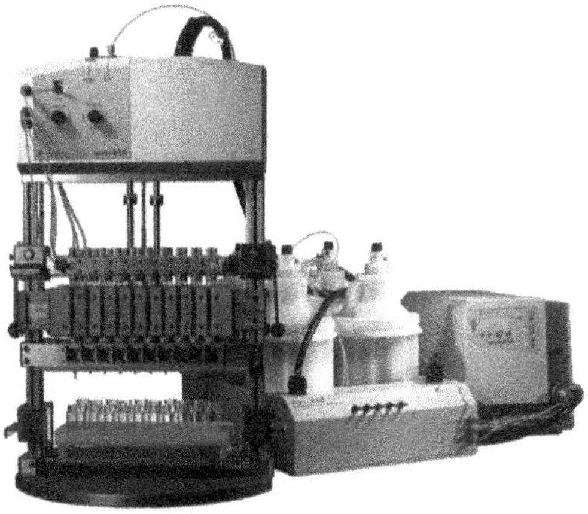

Hint

You need to be able to recall the reactions of all the functional groups you have met, including conditions such as heating, refluxing, use of acidic or alkaline conditions, and catalysts.

Reagents used in organic chemistry

Oxidising agents

Potassium dichromate(VI), $K_2Cr_2O_7$, acidified with dilute sulfuric acid will oxidise primary alcohols to aldehydes, and aldehydes to carboxylic acids. Secondary alcohols are oxidised to ketones.

Reducing agents

Different reducing reagents have different capabilities:

- Sodium tetrahydridoborate(III), $NaBH_4$, will reduce C═O but not C═C. It can be used in aqueous solution. This reducing agent will reduce polar unsaturated groups, such as $C^{\delta+}═O^{\delta-}$, but not non-polar ones, such as C═C. This is because it generates the nucleophile $:H^-$ which attacks $C^{\delta+}$ but is repelled by the electron-rich C═C.
- Hydrogen with a nickel catalyst, H_2/Ni, is used to reduce C═C but not C═O.
- Tin and hydrochloric acid, Sn/H^+, may be used to reduce $R—NO_2$ to $R—NH_2$.

Dehydrating agents

Alcohols can be converted to alkenes by passing their vapours over heated aluminium oxide or by acid-catalysed elimination reactions.

Examples of reaction schemes

1 How can propanoic acid be synthesised from 1-bromopropane?

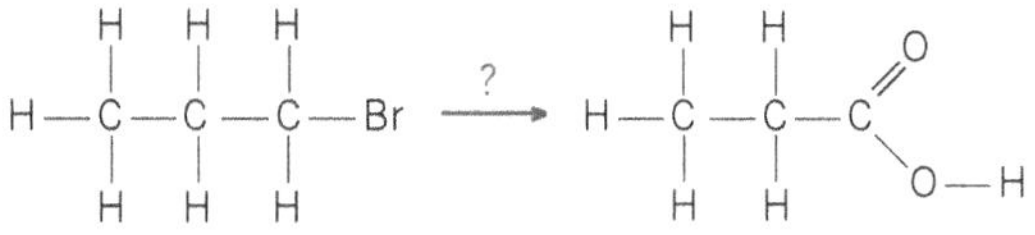

Both the starting material and the target have the same number of carbon atoms, so no alteration to the carbon skeleton is needed.

Write down all the compounds which can be made in one step from 1-bromopropane and all those from which propanoic acid can be made in one step as shown in Figure 3. You may use Figure 2 to help you.

In this case two of the compounds are the same – the ones in red.

Study tip

Cover the list of compounds in the centre and try to remember them. It is also important to recall the reagents and conditions for each reaction.

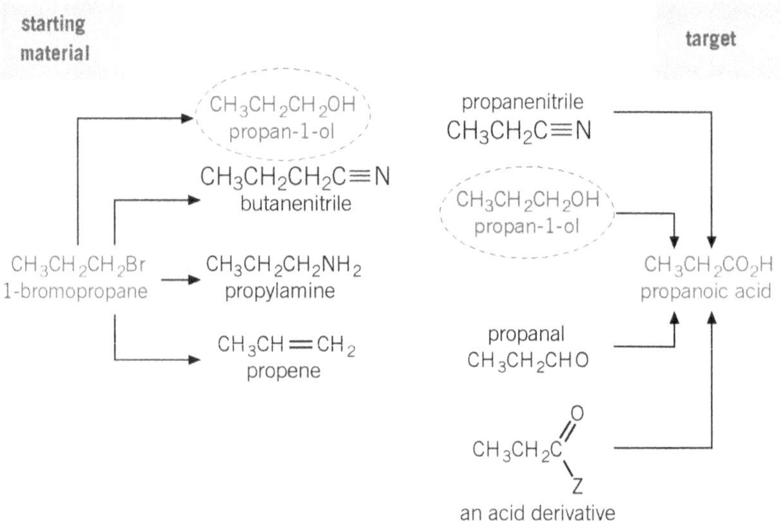

▲ **Figure 3** *Devising a synthesis of propanoic acid from 1-bromopropane*

So, 1-bromopropane can be converted into propan-1-ol which can be converted into propanoic acid. The conversion required can be done in two steps:

Step 1 $CH_3CH_2CH_2Br \xrightarrow{\text{reflux with NaOH(aq)}} CH_3CH_2CH_2OH$
1-bromopropane → propan-1-ol

Step 2 $CH_3CH_2CH_2OH \xrightarrow{\text{reflux with } K_2Cr_2O_7/H^+} CH_3CH_2CO_2H$
propan-1-ol → propanoic acid

Both these reactions have a good yield.

2 How can propylamine by synthesised from ethene?

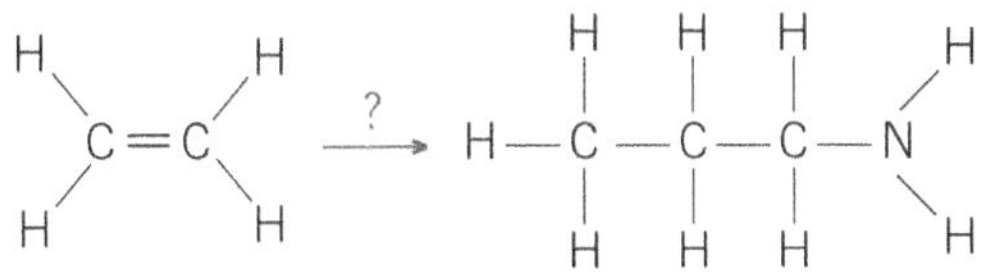

Propylamine has one more carbon atom than ethene. This suggests that the formation of a nitrile is involved at some stage.

Write down all the compounds that can be made from ethene and all the compounds from which the propylamine can be made (Figure 4).

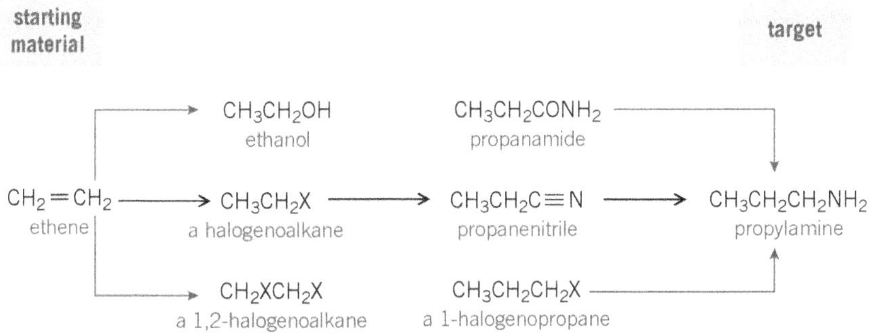

▲ **Figure 4** *Devising a synthesis of propylamine from ethene*

In this instance, no compound that can be made in one step from the starting material, can be converted into the product, so more than two steps must be required. You already know that the formation of a nitrile is required. A halogenoethane can be converted into propanenitrile so the synthesis can be completed in three steps:

Step 1 $CH_2CH_2 \xrightarrow{HBr} CH_3CH_2Br$
ethene → bromoethane

Step 2 $CH_3CH_2Br \xrightarrow{\text{KCN/dil. } H_2SO_4} CH_3CH_2C{\equiv}N$
bromoethane → propanenitrile

Step 3 $CH_3CH_2C{\equiv}N \xrightarrow{Ni/H_2} CH_3CH_2CH_2NH_2$
propanenitrile → propylamine

Chloroethane or iodoethane could have been chosen instead of bromoethane.

Propylamine, a primary amine, is itself a reactive molecule. It could be converted to a secondary amine, methyl propylamine for example, by reaction with a halogenoalkane such as bromomethane.

$$CH_3CH_2CH_2NH_2 + CH_3Br \rightarrow CH_3CH_2CH_2NHCH_3 + HBr$$

'Green' chemistry

Green or sustainable chemistry is a movement that focuses on designing chemical products and processes that:

- minimise the use and generation of hazardous substances
- minimise waste
- reduce pollution
- reduce the consumption of non-renewable resources
- reduce the energy demands of chemical processes.

This can be done by:

- selecting reactions with a high atom economy and yield (see Topic 2.6, Balanced equations, atom economies, and percentage yields) to minimise waste
- devising processes with the fewest possible number of steps
- devising processes that do not require solvents, as these need energy to remove them at the end of a reaction
- use reactions that take place at room temperature and pressure where possible.

Aromatic reactions

Figure 5 summarises some of the important reactions of aromatic compounds using benzene as the starting material.

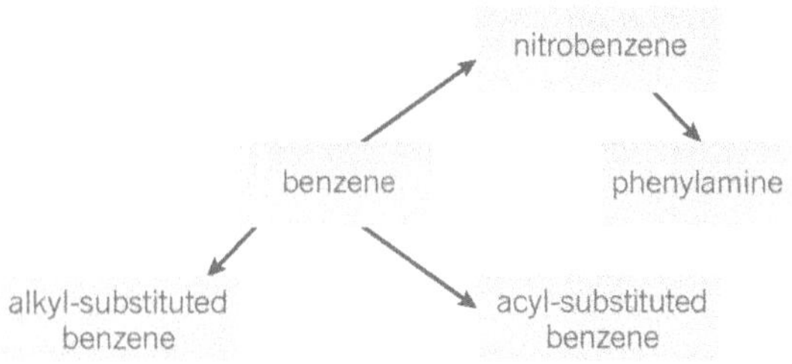

▲ **Figure 5** *Some inter-relationships between functional groups in aromatic compounds. Make sure you can recall the reagents and conditions for each conversion*

Summary questions

1 Give a one step reaction to convert:

a 1-bromobutane to pentanenitrile

b ethanoic acid to methyl ethanoate

c but-1-ene to butan-2-ol

d cyclohexanol to cyclohexene.

2 Give a two step reaction to convert:

a ethene to ethanoic acid

b propanone to 2-bromopropane.

3 For each step, name the type of reaction taking place and the reagents required.

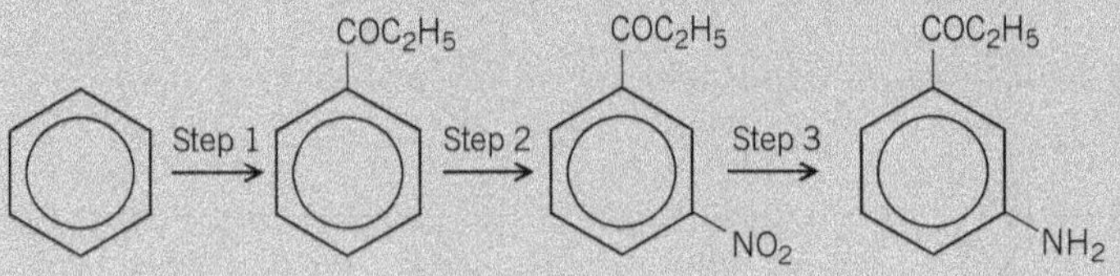

31.2 Organic analysis

This topic builds on the corresponding chapter in the AS/1st year A-level (chapter 16).

Remember, when identifying an organic compound, you need to know the functional groups present.

Chemical reactions

Some tests are very straightforward:

- Is the compound acidic (suggests carboxylic acid)?
- Is the compound solid (suggests long carbon chain or ionic bonding), liquid (suggests medium length carbon chain or polar or hydrogen bonding), or gas (suggests short carbon chain, little or no polarity)?
- Does the compound dissolve in water (suggests polar groups) or not (suggests no polar groups)?
- Does the compound react with water (suggests acid chloride or acid anhydride)?
- Does the compound burn with a smoky flame (suggests high C:H ratio, possibly aromatic) or non-smoky flame (suggests low C:H ratio, probably non-aromatic)?

Some specific chemical tests are listed in Table 1.

▼ **Table 1** *Chemical tests for functional groups*

Functional group	Test	Result
alkene –C=C–	shake with bromine water	red-brown colour disappears
halogenoalkane R—X	1. add NaOH(aq) and warm 2. acidify with HNO_3 3. add $AgNO_3$(aq)	precipitate of AgX
alcohol R—OH	add acidified $K_2Cr_2O_7$	orange colour turns green with primary or secondary alcohols (also with aldehydes)
aldehyde R—CHO	warm with Fehling's solution or warm with Tollens' solution or add acidified $K_2Cr_2O_7$	blue colour turns to red precipitate silver mirror forms orange colour turns green
carboxylic acid R—COOH	add $NaHCO_3$(aq)	bubbles observed as carbon dioxide given off

Learning objective:

→ Describe how organic groups can be identified.

Specification reference: 3.3.6

Synoptic link

You will need to know all the organic chemistry studied in your A Level course including the reactions from AS Level, which are recapped here.

Synoptic link

Look back at Topic 10.3, Reactions of halide ions, for more detail on how the silver precipitate, AgX, can be used to identify the halogen.

Hint

You cannot use the silver nitrate test to identify a fluoroalkane as silver(I) fluoride, AgF, is soluble in water.

Hint

Benedict's solution may be used instead of a mixture of Fehling's 1 and 2.

Summary questions

1 A student carried out the following procedure to test whether substance A was a halogenoalkane. He warmed it with sodium hydroxide, acidified the resulting solution with dilute hydrochloric acid and then added silver nitrate solution.

a What is the reason for warming with sodium hydroxide? Write an equation for what would happen with a halogenoalkane RX.

b What mistake has he made that would invalidate the test? Explain your answer. What should he have done?

2 A student wishes to distinguish between the three isomers B, C and D.

B 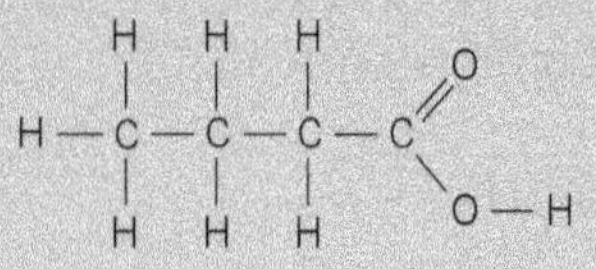

C 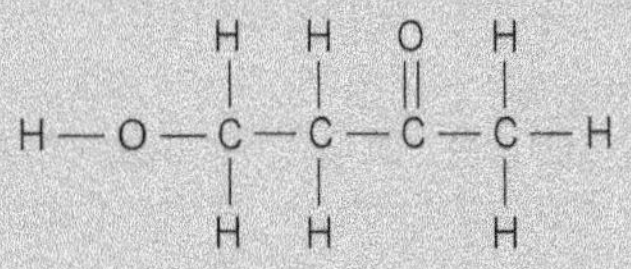

D 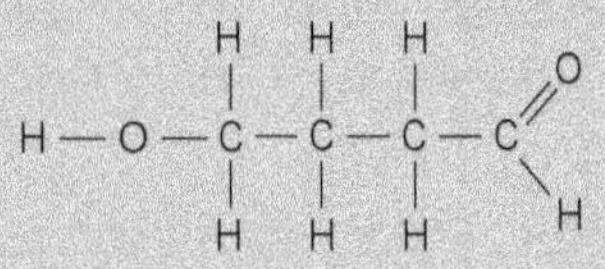

a What are isomers?

b What functional group(s) are present in B, C, and D?

c Suggest simple test-tube reactions that would enable her to distinguish the three isomers.

Practice questions

1 Describe how you could distinguish between the compounds in the following pairs using **one** simple test-tube reaction in each case.
For each pair, identify a reagent and state what you would observe when both compounds are tested separately with this reagent.

(a)

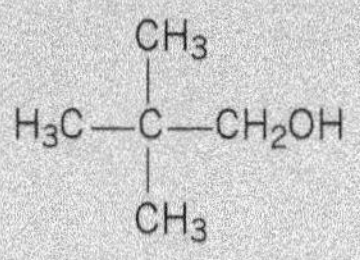

$H_3C-\underset{\underset{CH_3}{|}}{\overset{\overset{CH_3}{|}}{C}}-CH_2OH$ $H_3C-\underset{\underset{OH}{|}}{\overset{\overset{CH_3}{|}}{C}}-CH_2CH_3$

 R **S**

(3 marks)

(b)

$O=\underset{\underset{OCH_2CH_3}{\diagdown}}{\overset{\overset{CH_3}{\diagup}}{C}}$ $O=\underset{\underset{CH_2CH_3}{\diagdown}}{\overset{\overset{OH}{\diagup}}{C}}$

 T **U**

(3 marks)

(c) $H_3C-\underset{\underset{O}{\|}}{C}-CH_2-\underset{\underset{O}{\|}}{C}-CH_3$ $H-\underset{\underset{O}{\|}}{C}-CH_2-\underset{\underset{O}{\|}}{C}-H$

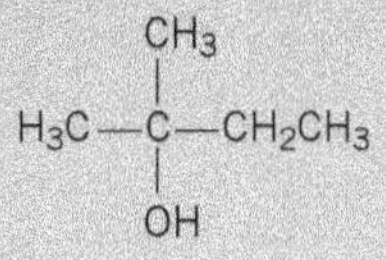

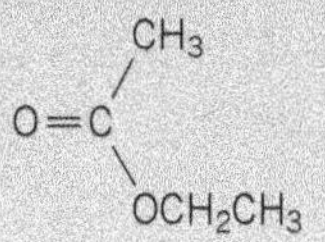

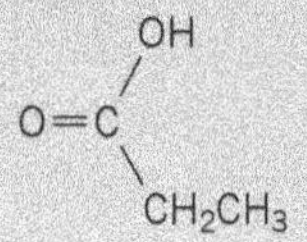

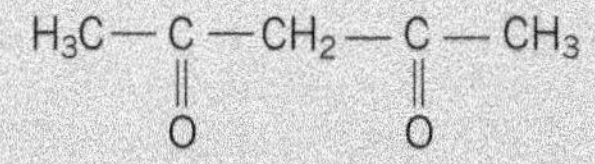

 V **W**

(3 marks)
AQA, 2013

2 (a) A chemist discovered four unlabelled bottles of liquid, each of which contained a different pure organic compound. The compounds were known to be propan-1-ol, propanal, propanoic acid, and 1-chloropropane.
Describe four **different** test-tube reactions, one for each compound that could be used to identify the four organic compounds.
Your answer should include the name of the organic compound, the reagent(s) used, and the expected observation for each test.

(8 marks)
AQA, 2012

3 Chemists have to design synthetic routes to convert one organic compound into another. Propanone can be converted into 2-bromopropane by a three-step synthesis.
Step 1: propanone is reduced to compound **L**.
Step 2: compound **L** is converted into compound **M**.
Step 3: compound **M** reacts to form 2-bromopropane.
Deduce the structure of compounds **L** and **M**.
For each of the three steps, suggest a reagent that could be used and name the mechanism.
Equations and curly arrow mechanisms are **not** required.

(8 marks)
AQA, 2012

4 (a) Copy and complete the diagram by giving the structural formula of the product in each of the boxes provided.

Reaction 1

$C_6H_6 \xrightarrow{\text{conc } H_2SO_4 \text{ and conc } HNO_3}$ i)

Reaction 2

$CH_3(CH)_2CH_3 \xrightarrow[\text{Ni}]{H_2}$ ii)

Reaction 3

$CH_3CHO \xrightarrow[\text{dilute}]{K_2Cr_2O_7}$ iii)

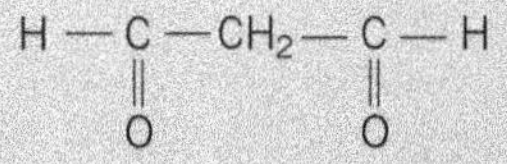

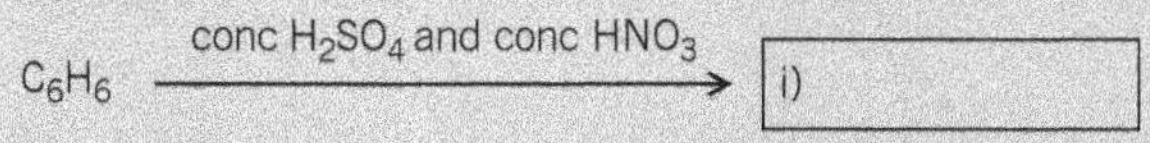

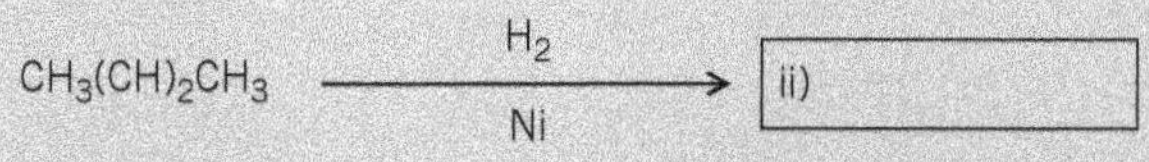

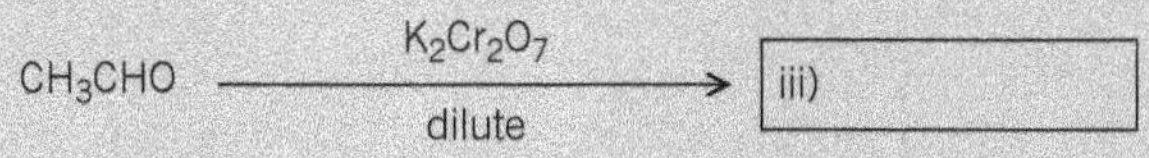

(3 marks)

(b) (i) State the role of the concentrated sulfuric acid in **Reaction 1**.

(1 mark)

(ii) State the role of the nickel in **Reaction 2**.

(1 mark)

(iii) Why is potassium dichromate(VI) used in **Reaction 3**?

(1 mark)

5 A chemist is given a sample of a halogenoalkane labelled compound A. Explain how the chemist could test to see if compound A was a chloroalkane. Describe the test the chemist could carry out and how they could use the results of the test to confirm whether or not compound A is a chloroalkane.

(4 marks)

6 One mole of compound X has a mass of 58.0 g. A chemist tests the compound by warming a sample of X with Fehling's solution. The chemist observes that the Fehling's solution turns from a blue solution to a red precipitate.

(a) What type of substance is compound X?

(1 mark)

(b) Name compound X.

(1 mark)

7 Describe how a chemist could test for the presence of the alkene functional group. Describe the how to carry out the test and how to interpret the results of the test.

(2 marks)

Answers to the Practice Questions and Section Questions are available at
www.oxfordsecondary.com/oxfordaqaexams-alevel-chemistry

32 Structure determination

32.1 Nuclear magnetic resonance (NMR) spectroscopy

Nuclear magnetic resonance spectroscopy (NMR) is used particularly in organic chemistry. It is a powerful technique that can help find the structures of even quite complex molecules.

A magnetic field is applied to a sample, which is surrounded by a source of radio waves and a radio receiver. This generates an energy change in the nuclei of atoms in the sample that can be detected. Electromagnetic energy is emitted, which can then be interpreted by a computer.

Learning objectives:

- → Explain the principles of NMR.
- → Describe the ^{13}C NMR spectrum.
- → Explain the chemical shift.
- → Describe what information a ^{13}C NMR spectrum gives.

Specification reference: 3.3.15

A brief theory of NMR

Although you will only be examined on *interpreting* NMR spectra, this background reading may help you to understand how NMR works, although in some respects it is an over-simplification.

Many nuclei with odd mass numbers, such as ^{1}H, ^{13}C, ^{15}N, ^{19}F and ^{31}P, have the property of *spin* (as do electrons). This gives them a magnetic field like that of a bar magnet.

If bar magnets are placed in an external magnetic field, they will line up parallel to the field (Figure 1a).

It is also possible that the bar magnets could line up anti-parallel to the field, as in (Figure 1b), but this orientation has a higher energy as the bar magnets have to be forced into position against the repulsion of the external magnetic field. The stronger the external magnetic field and the stronger the bar magnets, the larger the energy gap between the parallel and anti-parallel states.

Something similar applies to nuclei with spin, such as ^{1}H and ^{13}C. There will be some of the nuclei in each energy state but more of them will be in the lower (parallel) one. If electromagnetic energy just equal in energy to the difference between the two positions (ΔE in Figure 2) is supplied, some nuclei will flip between the parallel and anti-parallel positions. This is called **resonance**. The energy required to cause this is in the radio region of the electromagnetic spectrum. It is supplied by a radio frequency source, and the resonances are detected by a radio receiver (Figure 3). The frequency of the radio waves required to cause flipping for a particular magnetic field is called the **resonant frequency** of that atomic nucleus. A higher frequency corresponds to a larger energy gap between the two states. If the magnetic field is kept constant and the radio frequency gradually increases, different atomic nuclei will come into resonance at different frequencies depending on the strength of their atomic magnets.

In fact, modern instruments use pulses of radio waves of a range of frequencies all at once and analyse the response by a computer technique called Fourier transformation, but the principle remains of finding the frequencies at which different nuclei resonate.

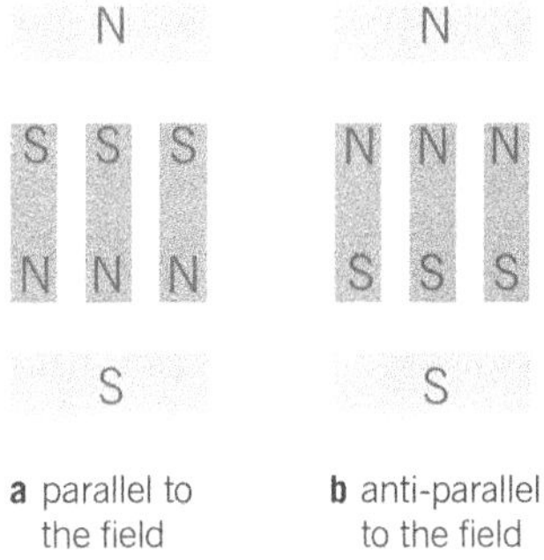

◀ **Figure 1** *The two possible orientations of bar magnets in a magnetic field*

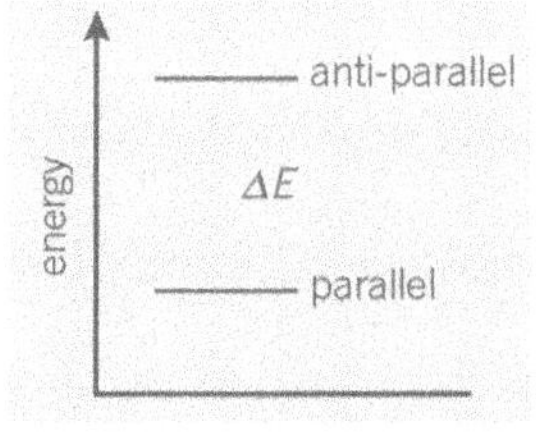

◀ **Figure 2** *Energy level diagram of the two orientations of bar magnets in a magnetic field*

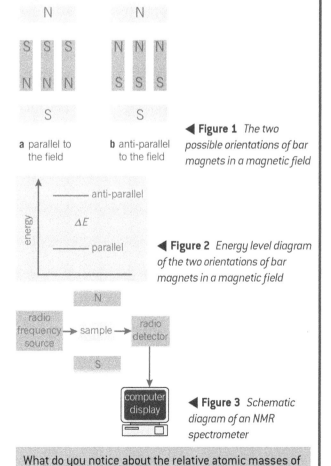

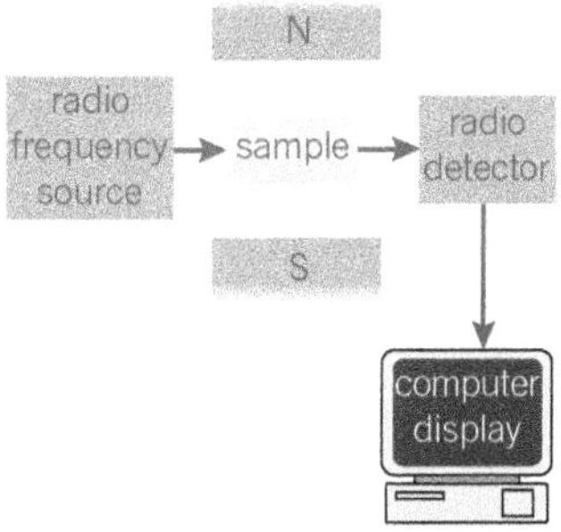

◀ **Figure 3** *Schematic diagram of an NMR spectrometer*

What do you notice about the relative atomic masses of the nuclei with spin?

Carbon-13, ^{13}C, NMR

NMR is most often used with organic compounds. Although carbon-12, ^{12}C, has no nuclear spin, carbon-13, ^{13}C, does have one. Whilst only 1% of carbon atoms are carbon-13, modern instruments are sensitive enough to obtain a carbon-13 spectrum.

Not all the carbon-13 atoms in a molecule resonate at exactly the same magnetic field strength. Carbon atoms in different functional groups feel the magnetic field differently. This is because all nuclei are **shielded** from the external magnetic field by the electrons that surround them. Nuclei with more electrons around them are better shielded. The greater the electron density around a carbon-13 atom, the smaller the magnetic field felt by the nucleus and the lower the frequency at which it resonates. The NMR instrument produces a graph of energy absorbed (from the radio signal) vertically against a quantity called **chemical shift** (which is related to the resonant frequency) horizontally.

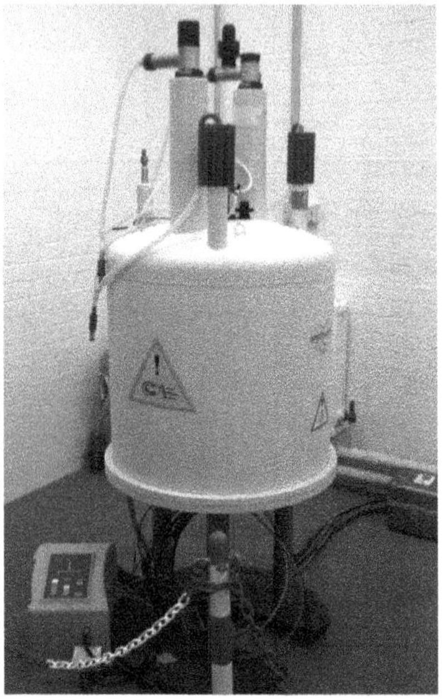

▲ **Figure 4** *Modern NMR instruments use electromagnets with superconducting coils to produce the strong magnetic fields required. The large white tank holds a jacket of liquid nitrogen surrounding an inner jacket of liquid helium which cools the magnet coils to 4 K*

The chemical shift

Chemical shift δ is measured in units called parts per million (ppm) from a defined zero related to a compound called tetramethylsilane, TMS (see Topic 32.2). Chemical shift is related to the difference in frequency between the resonating nucleus and that of TMS. In ^{13}C, NMR values of δ range from 0 to around 200 ppm.

The main point about ^{13}C NMR is that carbon atoms in different environments will give different chemical shift values. Figure 5 shows the ^{13}C NMR spectrum of ethanol. It has two peaks, one for each carbon, because the carbon atoms are in different environments – one is further from the oxygen atom than the other. The oxygen atom, being electronegative, draws electrons away from the carbon atom to which it is directly bonded.

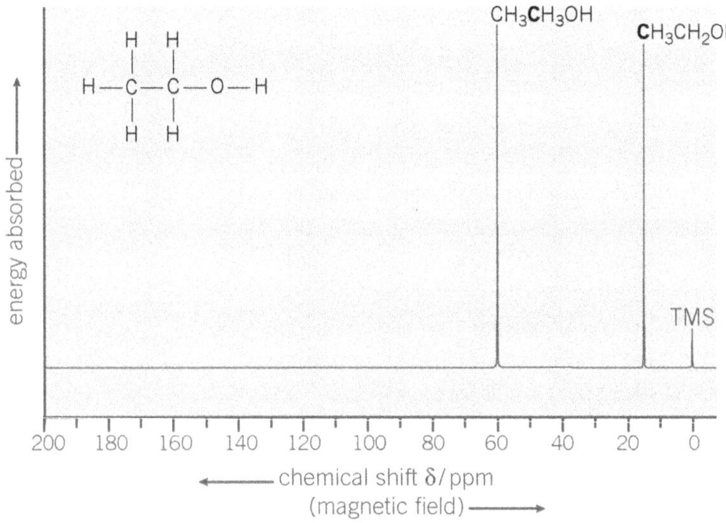

▲ **Figure 5** *Carbon-13 spectrum of ethanol*

Hint

On NMR spectra the chemical shift *increases* from right to left.

Table 1 shows values of ^{13}C chemical shifts for carbon atoms in a variety of environments. The carbon atom at δ = 60 ppm in the ethanol spectrum is the carbon bonded to the oxygen ($CH_3\mathbf{CH_2}OH$), whilst that at δ = 15 ppm is the other carbon ($\mathbf{CH_3}CH_2OH$).

This is because the electronegative oxygen atom draws electrons away from the carbon bonded to CH_3**CH_2**OH. It is deshielded and feels a greater magnetic field and so resonates at a higher frequency and therefore has a *greater* δ value than the other carbon. The other carbon **CH_3**CH_2OH is surrounded by more electrons and therefore shielded and has a *smaller* δ value.

More examples of ^{13}C NMR spectra

Figures 6 and 7 show the ^{13}C NMR spectra of the isomers propanone, CH_3COCH_3, and propanal, CH_3CH_2CHO. In propanone, there are just two different environments for the carbon atoms – the two CH_3 groups and the C=O. The spectrum shows two peaks:

- At δ = 205 ppm due to the C=O.
- At δ = 30 ppm due to the CH_3 groups.

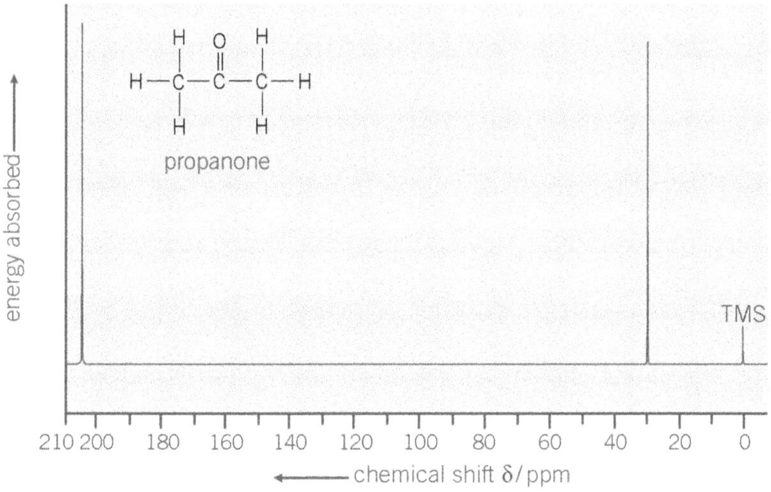

▲ **Figure 6** *^{13}C NMR spectrum of propanone*

Propanal has three different carbon environments and so shows three peaks:

- The CH_3 group at δ = 5 ppm.
- The CH_2 at δ = 37.
- The CHO group at δ = 205 ppm.

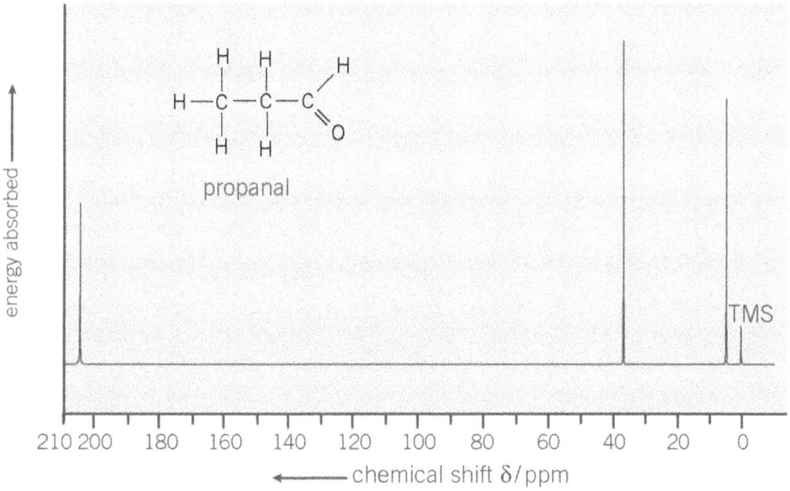

▲ **Figure 7** *^{13}C NMR spectrum of propanal*

Study tip

Tables of chemical shift data are provided by Oxford International AQA Examinations. See page 536.

▼ **Table 1** *^{13}C chemical shift values*

Type of carbon	δ / ppm
–**C**–C–	5–40
R–**C**–Cl or Br	10–70
R–C(=O)–**C**–	20–50
R–**C**–N<	25–60
–**C**–O– alcohols, ethers, or esters	50–90
>**C=C**<	90–150
R–**C**≡N	110–125
benzene ring	110–160
R–**C**(=O)– esters or acids	160–185
R–**C**(=O)– aldehydes or ketones	190–220

Study tip

The heights of the peaks in ^{13}C NMR spectra are not significant, it is their δ values that are important in interpreting spectra.

Summary questions

1 The ^{13}C NMR spectrum of ethanol is discussed above and has two peaks. Methoxymethane is an isomer of ethanol:

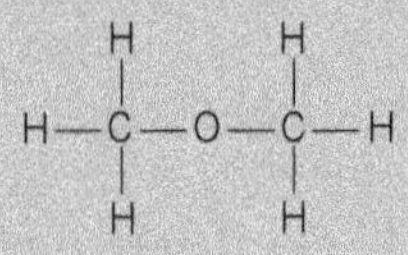

methoxymethane

a How many peaks would you expect to find in its ^{13}C NMR spectrum?

b Explain your answer.

2 The ^{13}C NMR spectra of propan-1-ol and propan-2-ol are given below. State which is which and explain your answer.

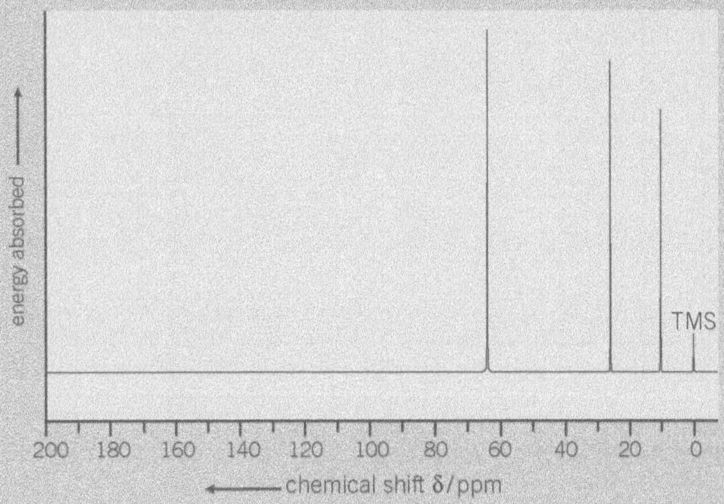

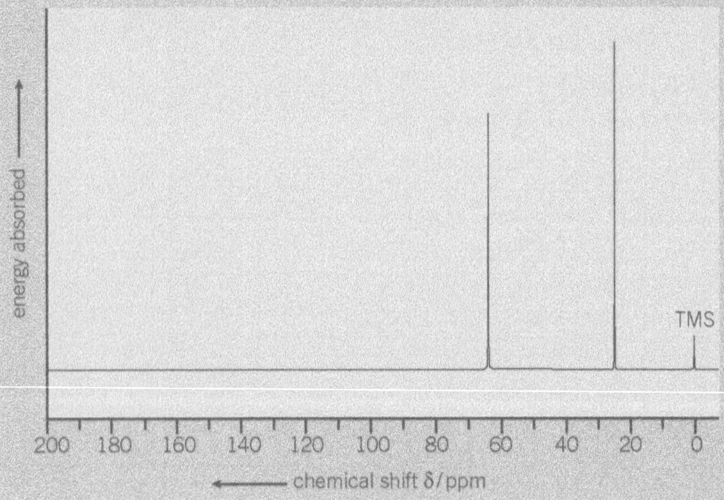

Hint

It will help to draw the displayed or structural formula of the two isomers.

32.2 Proton NMR

Learning objectives:

- → Explain a ^1H NMR spectrum.
- → Describe what information a ^1H NMR spectrum gives.
- → Explain what the integration trace shows.

Specification reference: 3.3.15

In proton NMR, it is the ^1H nucleus that is being examined. Nearly all hydrogen atoms are ^1H so it is easier to get an NMR spectrum for ^1H than for ^{13}C.

Here it is hydrogen atoms attached to different functional groups that feel the magnetic field differently, because all nuclei are shielded from the external magnetic field by the electrons that surround them. Nuclei with more electrons around them are better shielded. The greater the electron density around a hydrogen atom, the smaller the chemical shift δ. The values of chemical shift in proton NMR are smaller than those for ^{13}C NMR, most are between 0 and 10 ppm.

If all the hydrogen nuclei in an organic compound are in identical environments, you get only one chemical shift value. For example, all the hydrogen atoms in methane, CH_4, are in the same environment and have the same chemical shift:

But, in a molecule like methanol, there are hydrogen atoms in two different environments – the three on the carbon atom, and the one on the oxygen atom. The NMR spectrum will show the two environments (Figure 1).

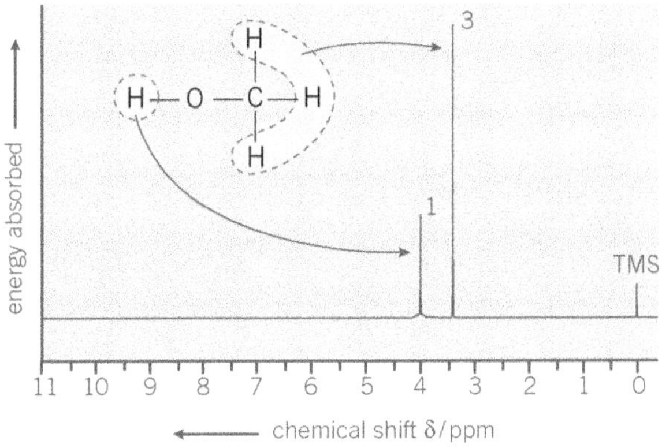

▲ **Figure 1** *The NMR spectrum of methanol – the peak areas are in the ratio 1 : 3*

In general, the further away a hydrogen atom is from an electronegative atom (such as oxygen) the smaller its chemical shift. In ethanol, CH_3CH_2OH, there are three values of δ.

In ^1H NMR, the areas under the peaks (shown here by the numbers next to them) are proportional to the number of hydrogen atoms of each type – in this case three and one.

The integration trace

In proton NMR spectra, the area of each peak is related to the number of hydrogen atoms producing it. So, in the spectrum of methanol, CH_3OH, the CH_3 peak is three times the area of the OH peak. This can be difficult to evaluate by eye, so the instrument produces a line called the integration trace, shown in red in Figure 2. The relative heights of the steps of this trace give the relative number of each type of hydrogen, 3 : 1 in this case.

▼ **Table 1** *Chemical shift values for proton, ^1H, NMR*

Type of proton	δ / ppm
RO**H**	0.5–5.0
RC**H**$_3$	0.7–1.2
RN**H**$_2$	1.0–4.5
R_2C**H**$_2$	1.2–1.4
R_3C**H**	1.4–1.6
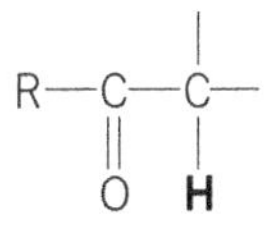	2.1–2.6
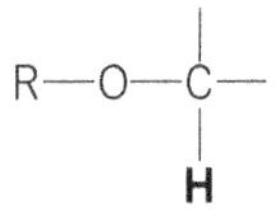	3.1–3.9
RC**H**$_2$Cl or Br	3.1–4.2
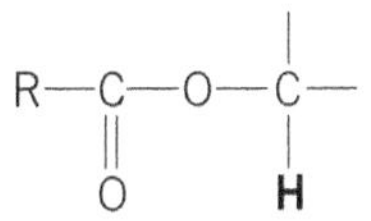	3.7–4.1
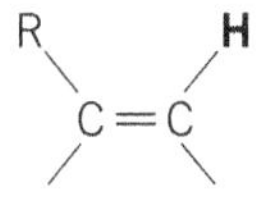	4.5–6.0
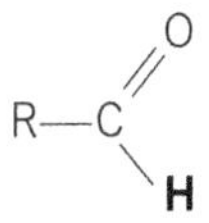	9.0–10.0
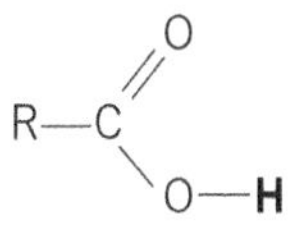	10.0–12.0

Hint

For simplicity, the integrated trace has been omitted from NMR spectra in this book, and the relative number of hydrogen atoms that each peak represents is given.

Summary questions

1 This question is about the isomers propan-1-ol and propan-2-ol.

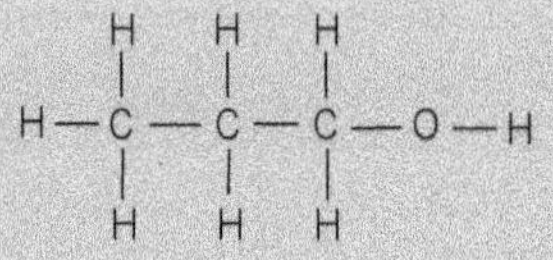

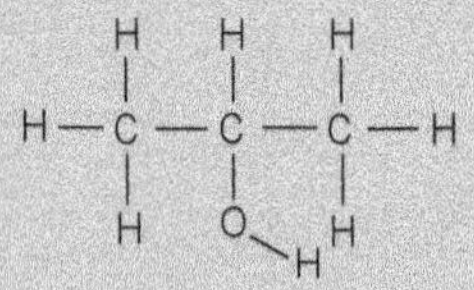

a What is meant by the term *isomer*?

b Write down the formulae of propan-1-ol and propan- 2-ol and mark each of the hydrogen atoms A, B, and so on, to show which are in different environments.

c How many different environments are there for the hydrogen atoms in:

i propan-1-ol

ii propan-2-ol?

d How many hydrogen atoms in each of the different environments, A, B, and so on, are there in:

i propan-1-ol

ii propan-2-ol?

e Predict the order of the chemical shift for each atom in:

i propan-1-ol

ii propan-2-ol.

The chemical shift value at which the peak representing each type of proton appears tells you about its environment – the type of functional group of which it is a part.

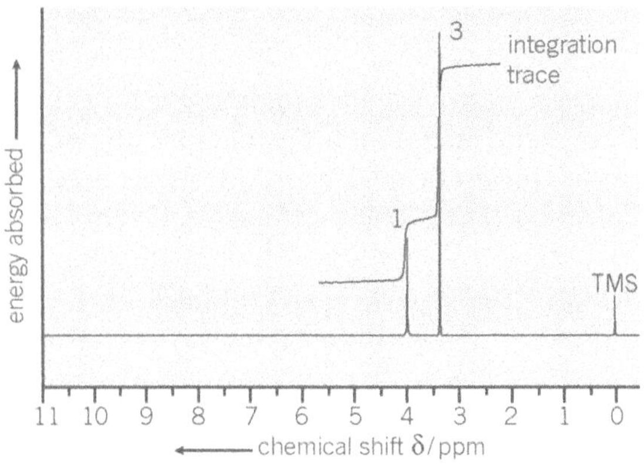

▲ **Figure 2** *The NMR spectrum of methanol showing the integration trace in red*

Chemical shift values

Hydrogen atom(s) in any functional group have a particular chemical shift value (Table 1).

Tetramethylsilane

The δ values of chemical shifts are measured by reference to a standard – the chemical shift of the hydrogen atoms in the compound tetramethylsilane, $Si(CH_4)_4$, TMS (Figure 3).

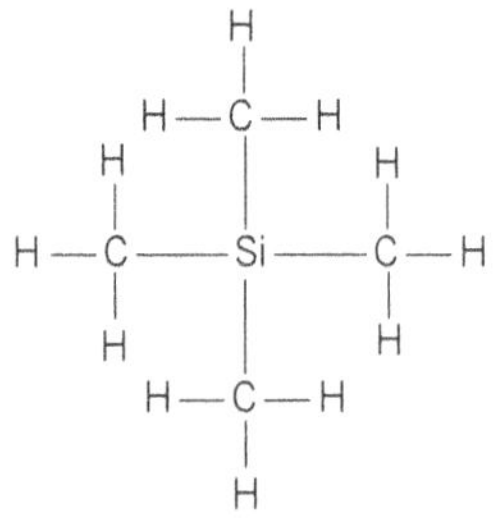

▲ **Figure 3** *Tetramethylsilane (TMS) – all 12 hydrogen atoms are in exactly the same environment, so they produce a single ^1H NMR signal*

The chemical shift value of these hydrogen atoms is zero by definition. A little TMS, which is a liquid, may be added to samples before their NMR spectra are run, and gives a peak at a δ value of exactly zero ppm to calibrate the spectrum (although modern techniques do not require this). All the spectra in this book show a TMS peak at δ = 0.

Other reasons for using TMS are that it is inert, non-toxic, and easy to remove from the sample.

32.3 Interpreting proton, ^{1}H, NMR spectra

If you are presented with a spectrum of an organic compound, such as in Figure 1, you can find out a lot about its structure.

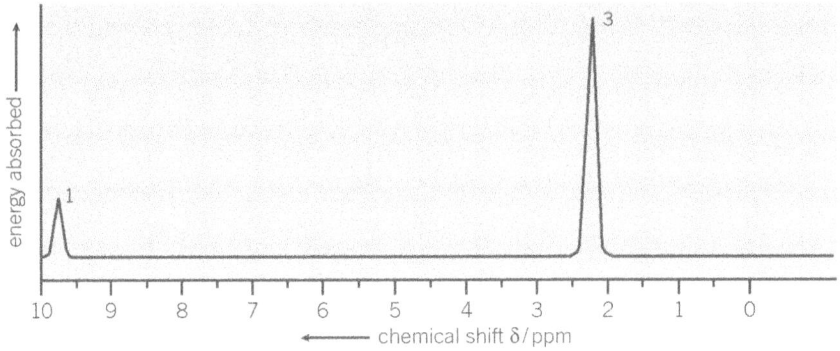

▲ **Figure 1** *The NMR spectrum of an organic compound*

The chemical shift values in Table 1 in Topic 32.2 tell you that the single hydrogen at δ 9.7 is the hydrogen from a –CHO (aldehyde) group and the three hydrogens at δ 2.2 are those of a $-COCH_3$ group. (This peak could also be caused by $-COCH_2R$, but since there are three hydrogens it must be $-COCH_3$.)

So the compound is likely to be ethanal, CH_3CHO (Figure 2).

Spin–spin coupling

If you zoom in on most NMR peaks, they are split into particular patterns – this is called **spin–spin coupling** (also called spin–spin splitting). It happens because the applied magnetic field felt by any hydrogen atom is affected by the magnetic field of the hydrogen atoms on the neighbouring carbon atoms. This spin–spin splitting gives information about the neighbouring hydrogen atoms, which can be very helpful when working out structure.

Figure 3 shows the spin–spin splitting patterns.

The *n* + 1 rule

If there is one hydrogen atom on an adjacent carbon, this will split the NMR signal of a particular hydrogen into two peaks each of the same height.

If there are two hydrogen atoms on an adjacent carbon, this will split the NMR signal of a particular hydrogen into three peaks with the height ratio 1:2:1.

Three adjacent hydrogen atoms will split the NMR signal of a particular hydrogen into four peaks with the height ratio 1:3:3:1.

This is called the *n* + 1 rule:

***n* hydrogens on an adjacent carbon atom will split a peak into *n* + 1 smaller peaks.**

Learning objectives:

- → Explain what causes spin–spin coupling.
- → Describe the *n* + 1 rule.
- → Explain how ^{1}H NMR spectra can be interpreted.

Specification reference: 3.3.15

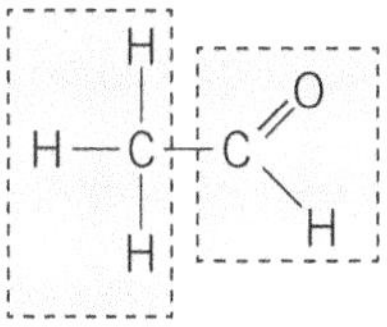

▲ **Figure 2** *The two groups that make up ethanal*

Study tip

Spin–spin coupling is not usually seen in ^{13}C NMR spectra due to the low abundance of ^{13}C. This is one reason why ^{13}C spectra are simpler than ^{1}H spectra.

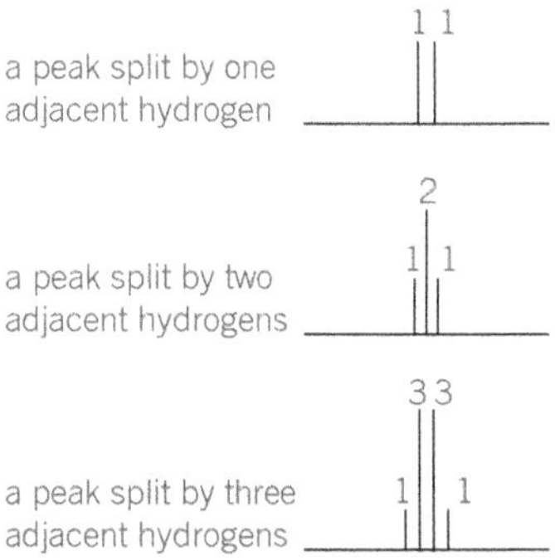

▲ **Figure 3** *NMR splitting patterns*

Some examples of interpreting ¹H NMR spectra

Ethanal

If you zoom in on the peaks in the spectrum of ethanal shown in Figure 1, you will see spin–spin splitting (Figure 4).

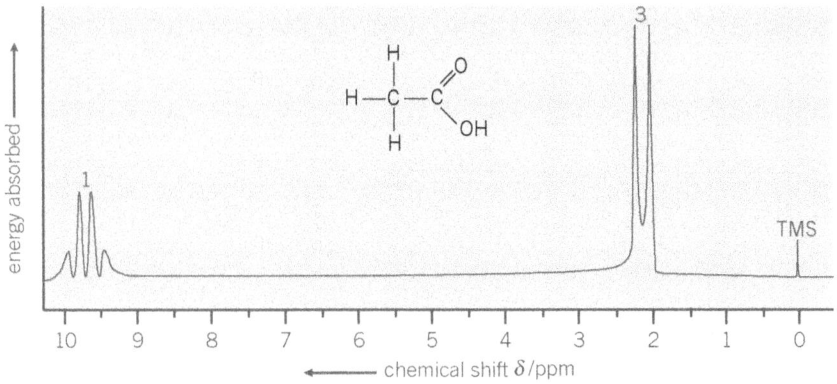

▲ **Figure 4** *The NMR spectrum of ethanal, CH_3CHO*

There are two types of hydrogen environments:

- A single peak of δ 9.7. This is the hydrogen of a –CHO group. This peak is split into four (height ratios 1 : 3 : 3 : 1) by the three hydrogens of the adjacent $–CH_3$ group.
- The peak with δ 2.2 is caused by three hydrogens of a $–CH_3$ group. This peak is split into two (height ratios 1 : 1) by the one hydrogen of the adjacent –CHO group.

Propanoic acid

Figure 5 shows the NMR spectrum of propanoic acid.

It is useful to make a table (Table 1) of the chemical shift of the peaks and what group they could correspond to by reference to Table 1 in Topic 32.2.

From the chemical shift value alone, the peak at 2.4 could be caused by either $–COCH_2R$ or $–COCH_3$. However, the fact that there are just two hydrogens means that it must correspond to $–COCH_2R$.

▼ **Table 1** *Chemical shift of the peaks of the ¹H NMR spectrum of propanoic acid, and what groups they could correspond to*

Chemical shift δ	Type of hydrogen	Number of hydrogens
11.7	–COOH	1
2.4	$–COCH_2R$ or $–COCH_3$	2
1.1	RCH_3	3

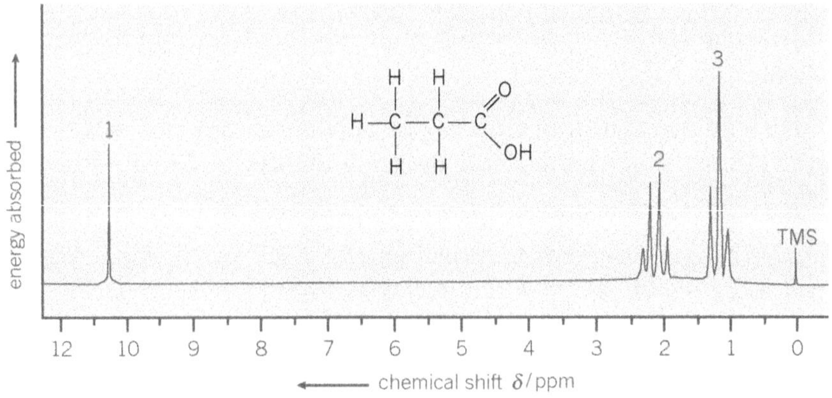

▲ **Figure 5** *The NMR spectrum of propanoic acid, CH_3CH_2COOH*

Looking at the spin–spin splitting:

- The peak at 11.7 is not split. This is because the adjacent carbon has no hydrogens bonded to it, –COOH.
- The peak at 2.4 is split into four. This indicates that the adjacent carbon has three hydrogens bonded to it. So, the R in $-COCH_2R$ must be $-CH_3$.
- The peak at 1.1 is split into three. This indicates that the adjacent carbon has two hydrogens bonded to it. So, the R in RCH_3 must be $-CH_2$.

So, if you put these groups together you make propanoic acid:

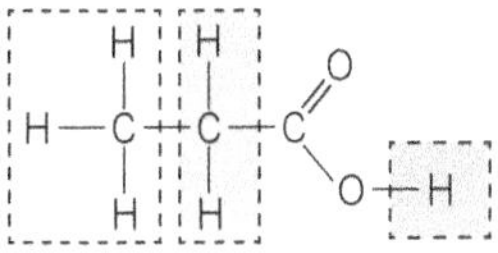

Study tip

Spin–spin coupling is not seen when equivalent hydrogens are on adjacent carbon atoms, for example, $HOCH_2CH_2OH$ has only two single peaks in its NMR spectrum.

Solvents for ^{1}H NMR

NMR spectra are normally run in solution. The solvent must not contain any hydrogen atoms, otherwise the signal from the hydrogen atoms in the solution would swamp the signals from hydrogen atoms in the sample, because there are vastly more of them.

One solvent commonly used is tetrachloromethane, CCl_4, which has no hydrogen atoms. Other solvents contain deuterium, which is an isotope of hydrogen and has the symbol D. Deuterium does not produce an NMR signal in the same range as hydrogen, though it has the same chemical properties. Some examples of deuterium-based solvents are deuterotrichloromethane, $CDCl_3$, deuterium oxide, D_2O, and perdeuterobenzene, C_6D_6.

More examples of interpreting and predicting NMR spectra

Propanone

The NMR spectrum of propanone (Figure 6) has just one peak. This means that all the hydrogen atoms in the molecule are in identical environments. The chemical shift value of 2.1 indicates that this corresponds to $-COCH_3$ or $-COCH_2R$.

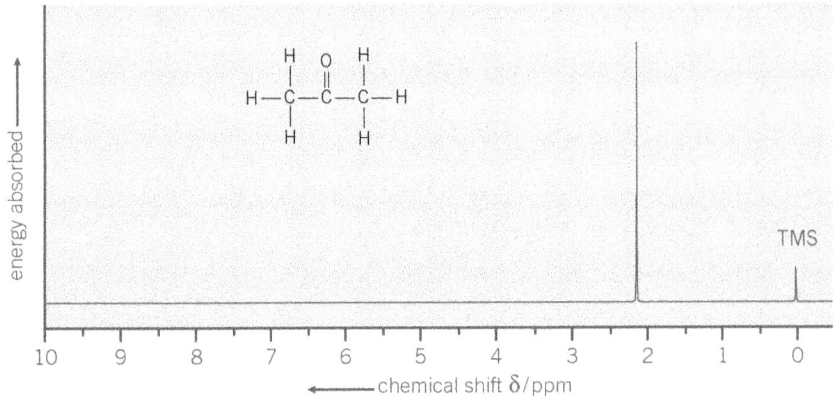

▲ **Figure 6** *The NMR spectrum of propanone*

Study tip

There are six hydrogen atoms in a molecule of propanone, all of which are in an identical environment. However, you cannot tell the number from the spectrum because the peak areas in NMR spectra are only relative. This means that you can tell that in methanol, for example, the area under the $-CH_3$ peak is three times that of the –OH peak but not the absolute number of each type of hydrogen (the absolute numbers could be two and six, for example).

Predicting NMR spectra

Chemists making new compounds may predict the spectrum of a compound they are making and compare their prediction with that of the compound they actually produce, to check that their reaction has gone as intended.

Ethyl ethanoate

There are three sets of hydrogen atoms in different environments. The values of chemical shift are predicted using Table 1 in Topic 32.2.

You can predict the spectrum shown in Figure 7 by dividing up the molecule as shown:

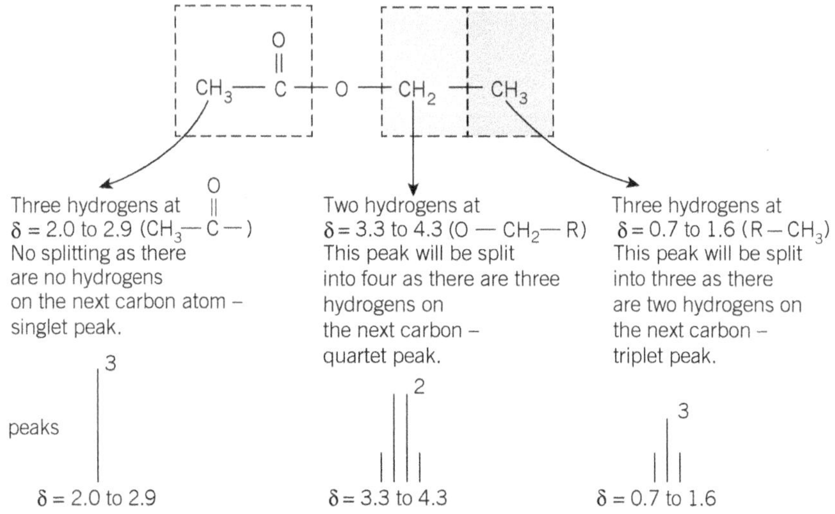

Three hydrogens at δ = 2.0 to 2.9 (CH3—C(=O)—)
No splitting as there are no hydrogens on the next carbon atom – singlet peak.
Two hydrogens at δ = 3.3 to 4.3 (O — CH2— R)
This peak will be split into four as there are three hydrogens on the next carbon – quartet peak.
Three hydrogens at δ = 0.7 to 1.6 (R — CH3)
This peak will be split into three as there are two hydrogens on the next carbon – triplet peak.
peaks
3
2
3
δ = 2.0 to 2.9
δ = 3.3 to 4.3
δ = 0.7 to 1.6

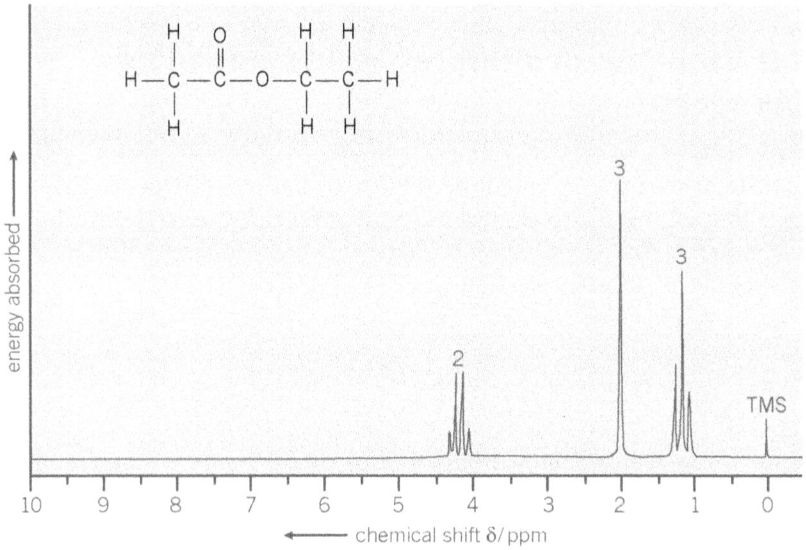

energy absorbed
3
3
2
TMS
10
9
8
7
6
5
4
3
2
1
0
chemical shift δ/ppm

▲ **Figure 7** *The NMR spectrum of ethyl ethanoate*

The birth of NMR

NMR is probably the most important analytical technique used by organic chemists today. Indeed, one Nobel Prize-winning chemist has been quoted as saying 'when the NMR goes down, the organic chemists go home'.

However, the chemical usefulness of the technique was discovered almost by accident. The effect began to be investigated by physicists just before and after the Second World War, and it appears that the researchers were helped in building their apparatus by the availability of cheap electronic components from surplus wartime radar equipment. The aim of the experiment was to measure the magnetic properties of atomic nuclei (their magnetic moments to be precise). They succeeded in their measurements, but were frustrated to find that the same atomic nucleus in different chemicals gave different results. For example, the two nitrogen atoms in ammonium nitrate, NH_4NO_3, gave different values. They realised that this was because the nitrogen nuclei were being shielded from the magnetic field by the electrons that surrounded them, and that as the two nitrogen atoms were in different chemical environments they were shielded to different extents.

What was a frustration to the physicists trying to investigate the nucleus was a gift to chemists whose prime interest was what was happening to the electrons. NMR could tell chemists about the degree to which electrons were surrounding atoms, so it could distinguish between the hydrogen atoms in the CH_3, CH_2, and OH groups in ethanol, CH_3CH_2OH, for example.

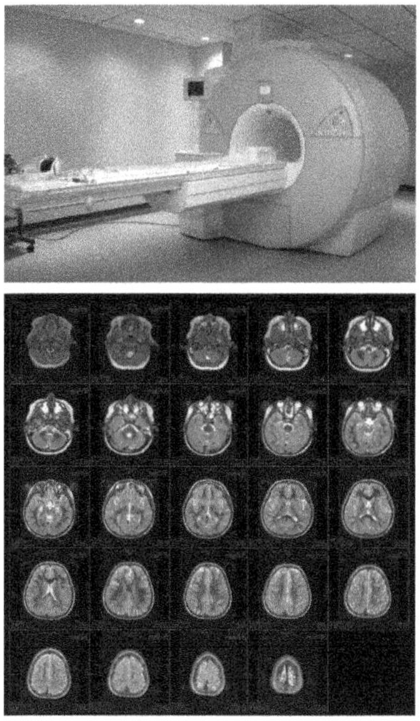

▲ **Figure 8** *MRI scanner and a scan of a female child's brain obtained by using this technique*

Manipulating the data

NMR is a technique that generates a lot of information and it has benefited enormously from the development of computers to process and present the data that it generates. Back in the early days of the 1950s and early 1960s, the data was produced from the instrument on paper tape and had to be manually transferred to punched cards which had to be *posted* to a computer centre to be put onto magnetic tape and processed. (In those days, a powerful computer might be the size of a house – there was no PC in every home and lab then.) The results would be posted back to the researchers, maybe a week later, provided that no one dropped the cards or tore the paper tape. Later, instruments used mechanical chart recorders. Nowadays, a researcher will drop off a compound at the department's NMR facility and expect to have the spectrum up on their networked PC almost before they are back at the lab.

Magnetic resonance imaging (MRI)

NMR can be used to investigate the human body – this was first realised by Felix Bloch, who found he got a strong signal by placing his finger in an NMR spectrometer. This signal was coming from protons in the water molecules that make up a large proportion of the human body. Water in different parts of the body (e.g., normal cells and cancer cells) gives slightly different NMR signals. MRI scanning of parts of the body, to help diagnose medical conditions, is now routine. The patient passes through a scanner where the magnetic field varies across the body. This, along with sophisticated computer processing of the NMR signal, allows a three-dimensional image of the body to be built up. The technique is harmless as, unlike X-rays, neither the radio waves nor the magnetic field can damage cells. However, the name 'magnetic resonance imaging' is used rather than 'nuclear magnetic resonance' because of the association of the word 'nuclear' with radioactivity in the mind of the public.

Summary questions

1 The ^{1}H NMR spectra shown are those of ethanol and of methoxymethane.

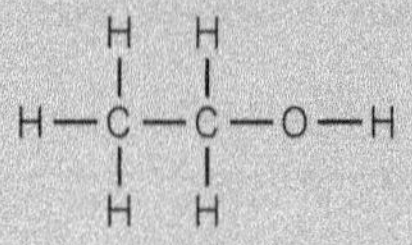

ethanol

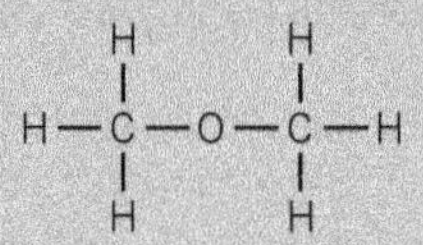

methoxymethane

a

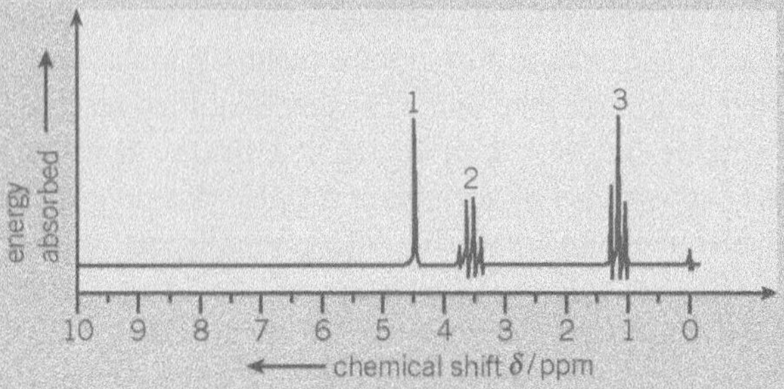

b

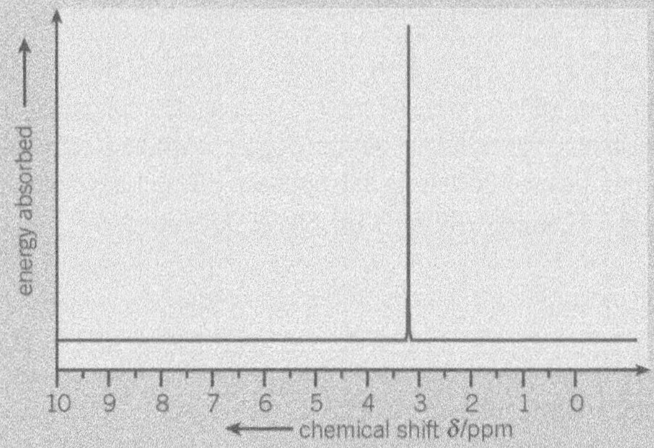

a Work out which spectrum represents which compound.

b Say what type of hydrogen each peak represents.

2 Predict the NMR spectrum of methyl ethanoate, CH_3COOCH_3, using the same procedure as for ethyl ethanoate above.

1 NMR spectroscopy can be used to study the structures of organic compounds.

(a) Compound **J** was studied using ^1H NMR spectroscopy.

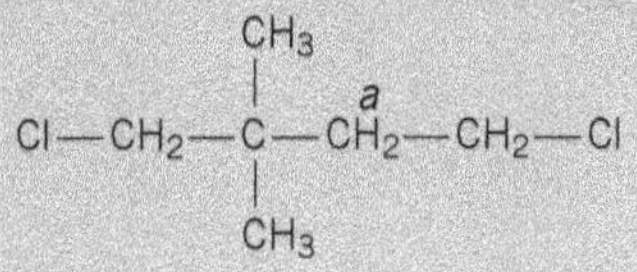

J

(i) Identify a solvent in which J can be dissolved before obtaining its ^1H NMR spectrum.

(1 mark)

(ii) Give the number of peaks in the ^1H NMR spectrum of **J**.

(1 mark)

(iii) Give the splitting pattern of the protons labelled *a*.

(1 mark)

(iv) Give the IUPAC name of **J**.

(1 mark)

(b) Compound **K** was studied using ^{13}C NMR spectroscopy.

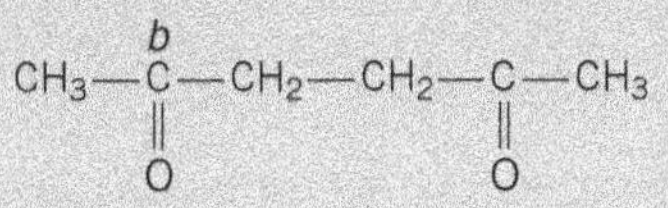

K

(i) Give the number of peaks in the ^{13}C NMR spectrum of **K**.

(1 mark)

(ii) Use Table 1 in Topic 32.1 to suggest a δ value of the peak for the carbon labelled *b*.

(1 mark)

(iii) Give the IUPAC name of **K**.

(1 mark)

AQA, 2013

2 Atenolol is an example of the type of medicine called a beta blocker. These medicines are used to lower blood pressure by slowing the heart rate. The structure of atenolol is shown below.

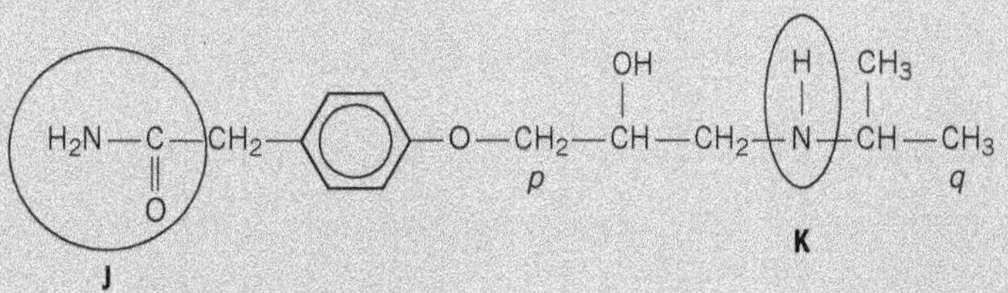

(a) Give the name of each of the circled functional groups labelled **J** and **K** on the structure of atenolol shown above.

(2 marks)

(b) The ^1H NMR spectrum of atenolol was recorded. One of the peaks in the ^1H NMR spectrum is produced by the CH_2 group labelled p in the structure of atenolol. Use Table 1 in Topic 32.2 to suggest a range of δ values for this peak. Name the splitting pattern of this peak.

(2 marks)

(c) NMR spectra are recorded using samples in solution. The ^1H NMR spectrum was recorded using a solution of atenolol in $CDCl_3$.

(i) Suggest why $CDCl_3$ and **not** $CHCl_3$ was used as the solvent.

(1 mark)

(ii) Suggest why $CDCl_3$ is a more effective solvent than CCl_4 for polar molecules such as atenolol.

(1 mark)

(d) The ^{13}C NMR spectrum of atenolol was also recorded.
Use the structure of atenolol given to deduce the total number of peaks in the ^{13}C NMR spectrum of atenolol.

(1 mark)

(e) Part of the ^{13}C NMR spectrum of atenolol is shown below. Use this spectrum and Table 1 in Topic 32.1 where appropriate, to answer the questions which follow.

δ/ppm

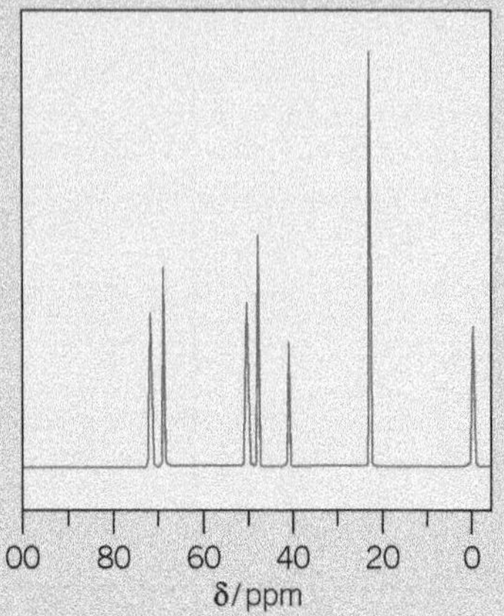

(i) Give the formula of the compound that is used as a standard and produces the peak at $\delta = 0$ ppm in the spectrum.

(1 mark)

(ii) One of the peaks in the ^{13}C NMR spectrum above is produced by the CH_3 group labelled q in the structure of atenolol.
Identify this peak in the spectrum by stating its δ value.

(1 mark)

(iii) There are three CH_2 groups in the structure of atenolol.
One of these CH_2 groups produces the peak at $\delta = 71$ in the ^{13}C NMR spectrum above.
Draw a circle around this CH_2 group in the structure of atenolol shown below.

$$H_2N-\underset{\underset{O}{\|}}{C}-CH_2-\bigcirc-O-CH_2-\underset{\overset{|}{OH}}{CH}-CH_2-\underset{\overset{|}{H}}{N}-\underset{\overset{|}{CH_3}}{CH}-CH_3$$

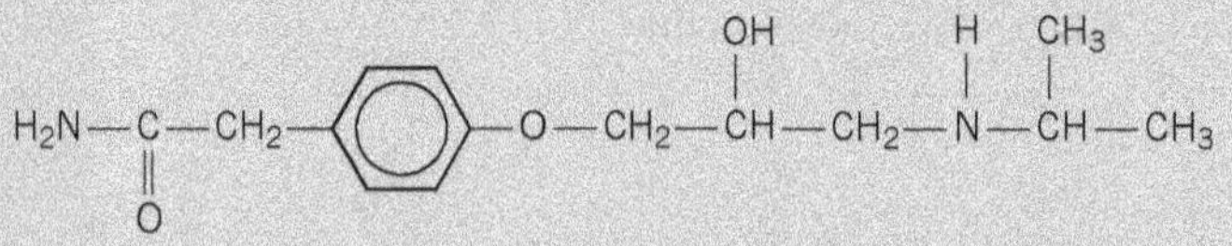

(1 mark)

(f) Atenolol is produced industrially as a racemate (an equimolar mixture of two enantiomers) by reduction of a ketone. Both enantiomers are able to lower blood pressure. However, recent research has shown that one enantiomer is preferred in medicines.

(i) Suggest a reducing agent that could reduce a ketone to form atenolol.

(1 mark)

(ii) Draw a circle around the carbon atom in the structure of atenolol shown above.

(1 mark)

(iii) Suggest how you could show that the atenolol produced by reduction of a ketone was a racemate and **not** a single enantiomer.

(2 marks)

(iv) Suggest **one** advantage and **one** disadvantage of using a racemate rather than a single enantiomer in medicines.

(2 marks)

AQA, 2011

Answers to the Practice Questions and Section Questions are available at
www.oxfordsecondary.com/oxfordaqaexams-alevel-chemistry

33 Chromatography
33.1 Chromatography

Learning objectives:
- → State how similar substances can be separated using chromatography.
- → Describe column chromatography.
- → Describe gas–liquid chromatography.
- → Describe thin-layer chromatography.
- → Describe gas chromatography mass spectrometry (GCMS).

Specification reference: 3.3.16

You will be familiar with paper chromatography, which is often used to separate the dyes in, for example, felt-tip pens.

Chromatography describes a whole family of separation techniques. They all depend on the principle that a mixture can be separated if it is dissolved in a solvent and then the resulting solution (now called the mobile phase) moves over a solid (the stationary phase).

- The moving or **mobile phase** carries the soluble components of the mixture with it. The more soluble the component in the mobile phase, the faster it moves. The solvent in the moving phase is often called the eluent (in column chromatography).
- The **stationary phase** will hold back the components in the mixture that are attracted to it. The more affinity a component in the mixture being separated has for the stationary phase, the slower it moves with the solvent.

So, if suitable moving and stationary phases are chosen, a mixture of similar substances can be separated completely, because every component of the mixture has a unique balance between its affinity for the stationary and for the mobile phase. In fact, chromatography is often the only way that very similar components of a mixture can be separated.

▲ **Figure 1** *The cellulose of the paper holds many trapped water molecules (the stationary phase). Here, ethanol is the mobile phase, or eluent*

Thin-layer chromatography

Thin-layer chromatography (TLC) is a development of paper chromatography. The filter paper is replaced by a glass, metal, or plastic sheet coated with a thin layer of silica gel (silicon dioxide, SiO_2) or alumina (aluminium oxide, Al_2O_3) which acts as the stationary phase. These are often called plates. Plastic- and metal-backed sheets can be cut to size with scissors.

TLC has several advantages over paper chromatography:

- it runs faster
- smaller amounts of mixtures can be separated
- the spots usually spread out less
- the plates are more robust than paper.

When the chromatogram has run, the position of colourless spots may have to be located by shining ultraviolet light on the plate, or chemically by spraying the plate with a locating agent which reacts with the components of the mixture to give coloured compounds.

After the plate has been run, an R_f value is calculated of each component using:

$$R_f = \frac{\text{distance moved by spot}}{\text{distance moved by solvent}}$$

The R_f values can be used to help identify each component.

Synoptic link

TLC is often used for separating mixtures of amino acids, and the procedure for running a thin-layer chromatogram has been described in Topic 30.2, Peptides, polypeptides, and proteins.

Hint

All R_f values must be less than 1.

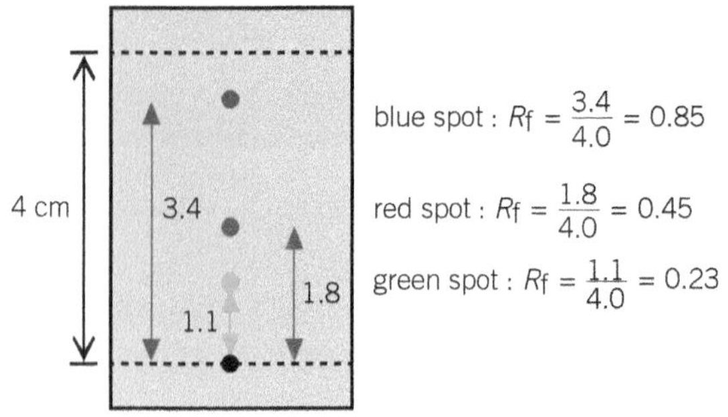

4 cm
3.4
1.1
1.8
blue spot : $Rf = \frac{3.4}{4.0} = 0.85$
red spot : $Rf = \frac{1.8}{4.0} = 0.45$
green spot : $Rf = \frac{1.1}{4.0} = 0.23$

▲ **Figure 2** *Calculating R_f values from thin-layer chromatograms*

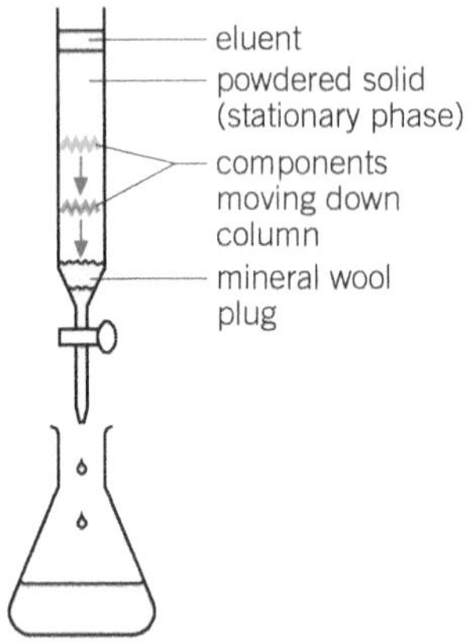

▲ **Figure 3** *Column chromatography*

Column chromatography

Column chromatography uses a powder, such as silica, aluminium oxide, or a resin, as the stationary phase. This is packed into a narrow tube – the column – and a solvent (the eluent) is added at the top (Figure 3). As the eluent runs down the column, the components of the mixture move at different rates and can be collected separately in flasks at the bottom. More than one eluent may be used to get a better separation. This method has the advantage that fairly large amounts can be separated and collected. For example, a mixture of amino acids can be separated into its pure components by this method.

Gas–liquid chromatography (GC)

This technique is one of the most important modern analytical techniques. The basic apparatus is shown in Figure 4.

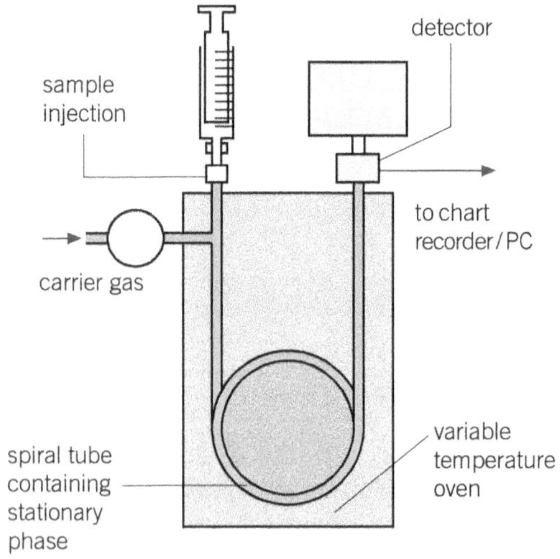

▲ **Figure 4** *Gas–liquid chromatography (GC)*

The stationary phase is a powder, coated with oil. It is either packed into or coated onto the inside of a long capillary tube, up to 100 m long and less than $\frac{1}{2}$ mm in diameter coiled up and placed in an oven whose temperature can be varied. The mobile phase is usually an unreactive gas, such as nitrogen or helium. After injection, the sample is carried along by the gas and the mixture separates as some of the components move along with the gas and some are retained by the oil, each to a

Hint

Gas–liquid chromatography is often simply called **gas chromatography**.

different degree. This means that the components leave the column at different times after injection – they have different **retention times**.

Various types of detectors are used, including ones that measure the thermal conductivity of the emerging gas. The results may be presented on a graph (Figure 5). The area under each peak is proportional to the amount of that component.

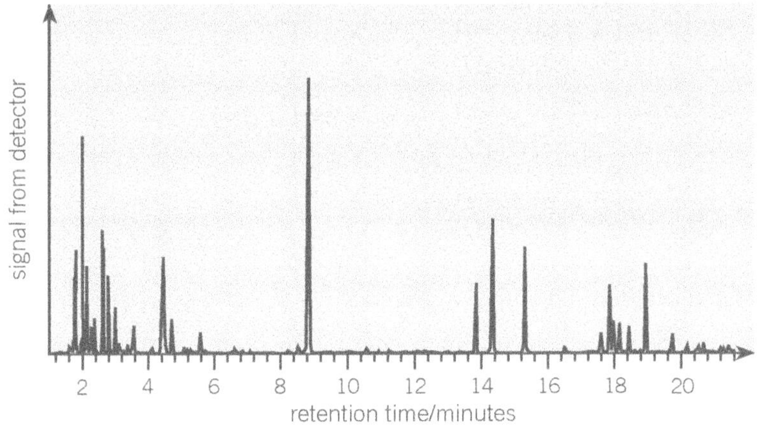

▲ **Figure 5** *Typical GC trace – each peak represents a different component*

In some instruments the components are fed directly into a mass spectrometer, infrared spectrometer, or NMR spectrometer for identification. Today, the whole process is automated and computer controlled (Figure 6).

As an analytical method for separating mixtures, GC is extremely sensitive. It can separate minute traces of substances in foodstuffs, and even link crude oil pollution found on beaches with its tanker of origin by comparing oil samples. One of its best-known uses is for testing athletes' blood or urine for drug taking.

The identification of a component is done by matching its retention time with that of a known substance under the same conditions. This is then confirmed by comparing the mass spectra of the two substances.

GCMS

GCMS stands for Gas Chromatography–Mass Spectrometry. It is essentially two techniques in one. A mass spectrometer is used as the detector for a gas chromatography system. As each component of a mixture comes out of the gas chromatography column, the time it has taken to pass through the column (its retention time) is noted. Each component is fed automatically into mass spectrometer which enables the compound to be identified either by its fragmentation pattern or by measuring its accurate mass.

Study tip

Separation depends on the balance between solubility in the moving phase and retention in the stationary phase.

▲ **Figure 6** *A gas chromatography instrument*

Synoptic link

The analytical use of mass spectrometry was covered in Topic 16.2, Mass spectrometry.

Hint

GCMS can also be written Mass Spectrometry–Gas Chromatography (MSGC)

HPLC

HPLC can be taken to stand for High Pressure Liquid Chromatography or High Performance Liquid Chromatography. Both names are appropriate. Here the mixture to be separated is forced through a column containing the stationary phase by a solvent driven by a high pressure pump. It is similar to column chromatography except that the pump drives the solvent (the eluent) rather than gravity. A variety of materials can be used as the stationary phase, including chiral ones that can separate optical isomers. A variety of detection methods can be used, for example, the absorption of ultraviolet light.

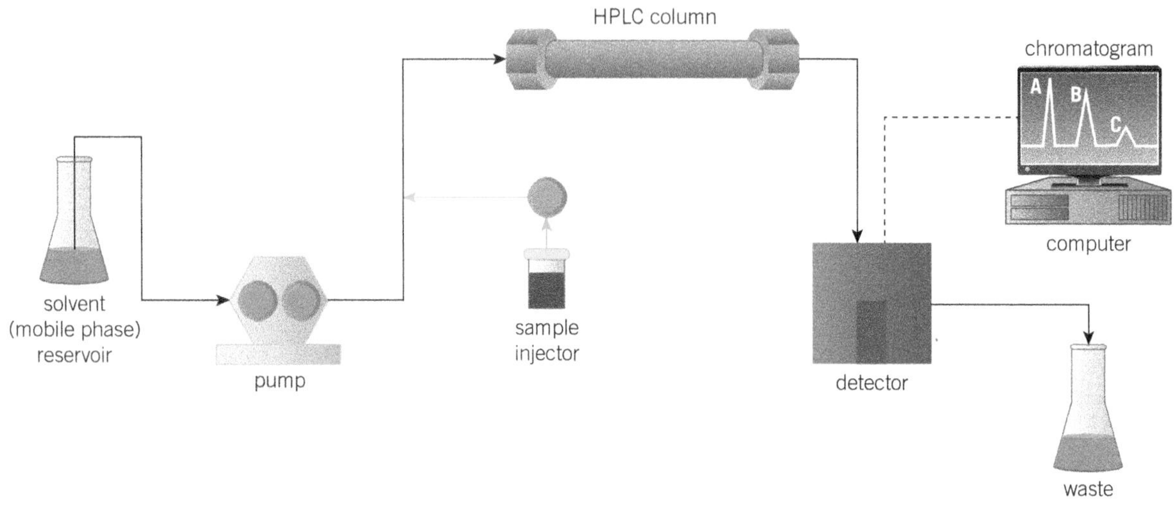

▲ **Figure 7** *Schematic diagram of an HPLC system*

Synoptic link

Optical isomers were covered in Topic 25.2, Optical isomerism.

Summary questions

1. What is the difference between column chromatography and gas–liquid chromatography?
2. Why is GC so important in forensic detective work? Give a possible example not in the text.
3. From the GC in Figure 7 above, identify from A, B, and C:
   a. the most abundant component in the mixture
   b. the one with the greatest affinity for the solid phase
   c. the one with the greatest affinity for the gas phase
   d. the one with the greatest retention time.

Practice questions

1 A peptide is hydrolysed to form a solution containing a mixture of amino acids. This mixture is then analysed by silica gel thin-layer chromatography (TLC) using a developing solvent. The individual amino acids are identified from their R_f values.
Part of the practical procedure is given below.

1. **Wearing plastic gloves to hold a TLC plate**, draw a pencil line 1.5 cm from the bottom of the plate.
2. Use a capillary tube to apply a very small drop of the solution of amino acids to the mid-point of the pencil line.
3. Allow the spot to dry completely.
4. In the developing tank, add the developing solvent to **a depth of not more than 1 cm**.
5. Place your TLC plate in the developing tank **and seal the tank with a lid**.
6. Allow the developing solvent to rise up the plate to at least $\frac{3}{4}$ of its height.
7. Remove the plate and quickly mark the position of the solvent front with a pencil.
8. Allow the plate to dry **in a fume cupboard**.

(a) Parts of the procedure are in bold text.
Explain why these parts of the procedure are essential. *(4 marks)*

(b) Outline the steps needed to locate the positions of the amino acids on the TLC plate and to determine their R_f values. *(4 marks)*

(c) Explain why different amino acids have different R_f values. *(2 marks)*

AQA, Specimen paper 3

2 The figure shows a chromatogram used to separate some amino acids by paper chromatography, using solvent X – a mixture of ethanoic acid and butan-1-ol and water.

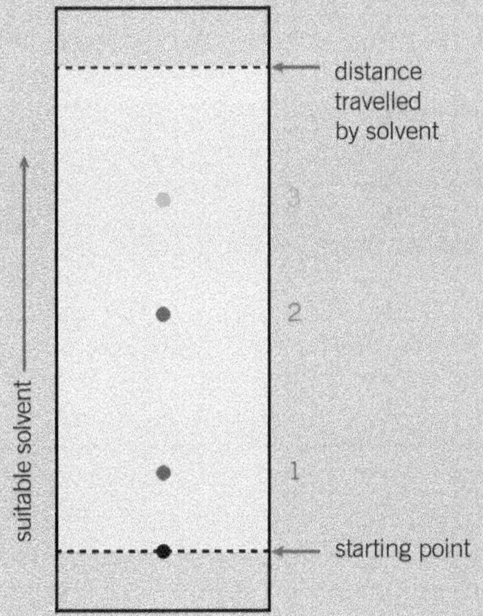

(a) Identify the amino acids using the table below. R_f values of some amino acids using solvent X. *(3 marks)*

alanine	0.38
arginine	0.16
glycine	0.26
leucine	0.73
tyrosine	0.50
valine	0.60

(b) Why is it essential to know the solvent used in the process? *(1 mark)*

3 Two-way paper chromatography was used to separate a mixture. The results are shown below.

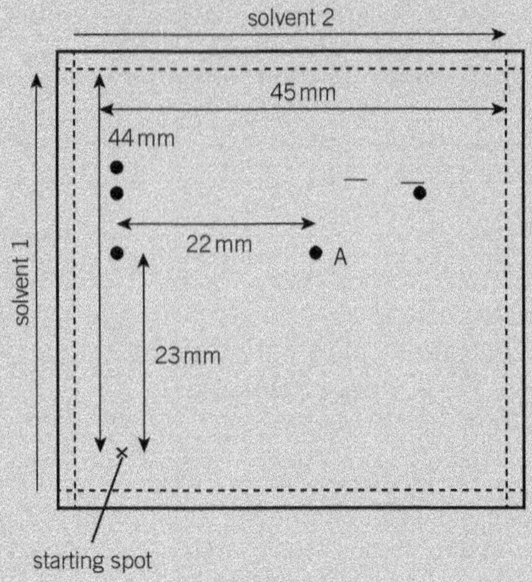

(a) Describe briefly the method of doing this. *(4 marks)*

(b) Why does two-way chromatography makes identification of the components of the mixture more certain. *(1 mark)*

(c) Find the R_f values of:

(i) after the first run in solvent 1

(ii) after the second run in solvent 2. *(2 marks)*

4 **(a)** A chemist discovered four unlabelled bottles of liquid, each of which contained a different pure organic compound. The compounds were known to be propan-1-ol, propanal, propanoic acid, and 1-chloropropane.
Describe four **different** test-tube reactions, one for each compound that could be used to identify the four organic compounds.
Your answer should include the name of the organic compound, the reagent(s) used and the expected observation for each test.

(8 marks)

(b) A fifth bottle was discovered labelled propan-2-ol. The chemist showed, using infrared spectroscopy, that the propan-2-ol was contaminated with propanone.
The chemist separated the two compounds using column chromatography. The column contained silica gel, a polar stationary phase.
The contaminated propan-2-ol was dissolved in hexane and poured into the column. Pure hexane was added slowly to the top of the column. Samples of the eluent (the solution leaving the bottom of the column) were collected.

- Suggest the chemical process that would cause a sample of propan-2-ol to become contaminated with propanone.
- State how the infrared spectrum showed the presence of propanone.
- Suggest why propanone was present in samples of the eluent collected first (those with shorter retention times), whereas samples containing propan-2-ol were collected later.

(4 marks)
AQA, 2012

Section 6 practice questions

1 Kevlar is a polymer used in protective clothing.
 The repeating unit within the polymer chains of Kevlar is shown.

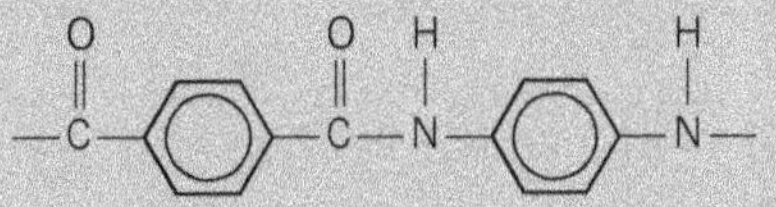

 (a) Name the strongest type of interaction between polymer chains of Kevlar. *(1 mark)*
 (b) One of the monomers used in the synthesis of Kevlar is:

 H_2N—⬡—NH_2

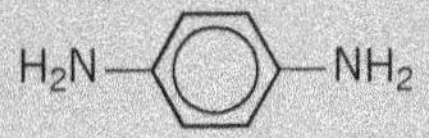

 An industrial synthesis of this monomer uses the following two-stage process starting
 from compound **X**.

 Stage **1**

 Cl—⬡—NO_2 + $2NH_3$ ⟶ H_2N—⬡—NO_2 + NH_4Cl

 X

 Stage **2**

 H_2N—⬡—NO_2 ⟶ H_2N—⬡—NH_2

 (i) Suggest why the reaction of ammonia with **X** in Stage **1** might be considered
 unexpected. *(2 marks)*
 (ii) Suggest a combination of reagents for the reaction in Stage **2**. *(1 mark)*
 (iii) Compound **X** can be produced by nitration of chlorobenzene.
 Give the combination of reagents for this nitration of chlorobenzene.
 Write an equation or equations to show the formation of a reactive intermediate
 from these reagents. *(3 marks)*
 (iv) Name and outline a mechanism for the formation of **X** from chlorobenzene
 and the reactive intermediate in part (iii). *(4 marks)*

 AQA, 2014

2 Each of the following conversions involves reduction of the starting material.
 (a) Consider the following conversion.

 O_2N—⬡—NO_2 ⟶ H_2N—⬡—NH_2

 Identify a reducing agent for this conversion.
 Write a balanced equation for the reaction using molecular formulae for the
 nitrogen-containing compounds and [H] for the reducing agent.
 Draw the repeating unit of the polymer formed by the product of this reaction
 with benzene 1,4-dicarboxylic acid. *(5 marks)*
 (b) Consider the following conversion.

 ⬡ ⟶ ⬡

 Identify a reducing agent for this conversion.
 State the empirical formula of the product.
 State the bond angle between the carbon atoms in the starting material and
 the bond angle between the carbon atoms in the product. *(4 marks)*
 (c) The reducing agent in the following conversion is $NaBH_4$.

 H_3C—C—CH_2CH_3 ⟶ H_3C—CH—CH_2CH_3
 ‖ |
 O OH

(i) Name and outline a mechanism for the reaction. *(5 marks)*
(ii) By considering the mechanism of this reaction, explain why the product formed is optically inactive. *(3 marks)*

AQA, 2013

3 Organic chemists use a variety of methods to identify unknown compounds. When the molecular formula of a compound is known, spectroscopic and other analytical techniques are used to distinguish between possible structural isomers. Use your knowledge of such techniques to identify the compounds described below.
Use spectral data where appropriate.
Each part below concerns a different pair of structural isomers.
Draw **one** possible structure for each of the compounds **A** to **J**, described below.

(a) Compounds **A** and **B** have the molecular formula C_3H_6O.
A has an absorption at 1715 cm^{-1} in its infrared spectrum and has only one peak in its ^1H NMR spectrum.
B has absorptions at 3300 cm^{-1} and at 1645 cm^{-1} in its infrared spectrum and does **not** show *E–Z* isomerism. *(2 marks)*

(b) Compounds **C** and **D** have the molecular formula C_5H_{12}
In their ^1H NMR spectra, **C** has three peaks and **D** has only one. *(2 marks)*

(c) Compounds **E** and **F** are both esters with the molecular formula $C_4H_8O_2$
In their ^1H NMR spectra, **E** has a quartet at $\delta = 2.3$ ppm and **F** has a quartet at $\delta = 4.1$ ppm. *(2 marks)*

(d) Compounds **G** and **H** have the molecular formula $C_6H_{12}O$
Each exists as a pair of optical isomers and each has an absorption at about 1700 cm^{-1} in its infrared spectrum. **G** forms a silver mirror with Tollens' reagent but **H** does not. *(2 marks)*

(e) Compounds **I** and **J** have the molecular formula $C_4H_{11}N$ and both are secondary amines. In their ^{13}C NMR spectra, **I** has two peaks and **J** has three. *(2 marks)*

AQA, 2010

4 In 2008, some food products containing pork were withdrawn from sale because tests showed that they contained amounts of compounds called dioxins many times greater than the recommended safe levels.
Dioxins can be formed during the combustion of chlorine-containing compounds in waste incinerators. Dioxins are very unreactive compounds and can therefore remain in the environment and enter the food chain.
Many dioxins are polychlorinated compounds such as tetrachlorodibenzodioxin (TCDD) shown below.

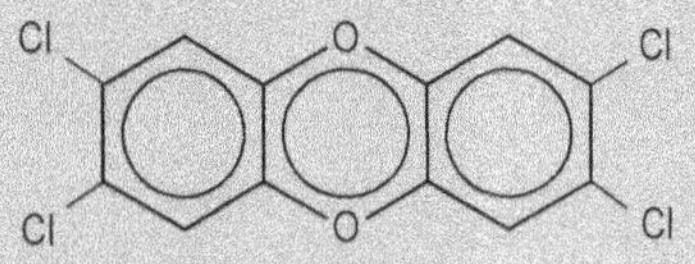

In a study of the properties of dioxins, TCDD and other similar compounds were synthesised. The mixture of chlorinated compounds was then separated before each compound was identified by mass spectrometry.

(a) Fractional distillation is **not** a suitable method to separate the mixture of chlorinated compounds before identification by mass spectrometry.
Suggest how the mixture could be separated. *(1 mark)*

(b) The molecular formula of TCDD is $C_{12}H_4O_2Cl_4$
Chlorine exists as two isotopes, ^{35}Cl (75%) and ^{37}Cl (25%).
Deduce the number of molecular ion peaks in the mass spectrum of TCDD and calculate the *m/z* value of the most abundant molecular ion peak. *(2 marks)*

(c) Suggest **one** operating condition in an incinerator that would minimise the formation of dioxins. *(1 mark)*

(d) TCDD can also be analysed using ^{13}C NMR.
- (i) Give the formula of the compound used as the standard when recording a ^{13}C spectrum. *(1 mark)*
- (ii) Deduce the number of peaks in the ^{13}C NMR spectrum of TCDD. *(1 mark)*

AQA, 2010

5 (a) Acyl chlorides and acid anhydrides are important compounds in organic synthesis. Outline a mechanism for the reaction of CH_3CH_2COCl with CH_3OH and name the organic product formed.

Mechanism
Name of organic product *(5 marks)*

(b) A polyester was produced by reacting a diol with a diacyl chloride. The repeating unit of the polymer is shown below.

$$-O-\underset{\underset{O}{\parallel}}{C}-CH_2CH_2-\underset{\underset{O}{\parallel}}{C}-O-CH_2CH_2CH_2CH_2CH_2-$$

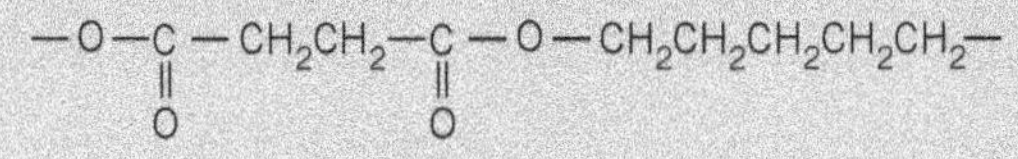

- (i) Name the diol used. *(1 mark)*
- (ii) Draw the displayed formula of the diacyl chloride used. *(1 mark)*

5 (b) (iii) A shirt was made from this polyester. A student wearing the shirt accidentally splashed aqueous sodium hydroxide on a sleeve. Holes later appeared in the sleeve where the sodium hydroxide had been.
Name the type of reaction that occurred between the polyester and the aqueous sodium hydroxide. Explain why the aqueous sodium hydroxide reacted with the polyester. *(3 marks)*

(c) (i) Complete the following equation for the preparation of aspirin using ethanoic anhydride by writing the structural formula of the missing product.

aspirin

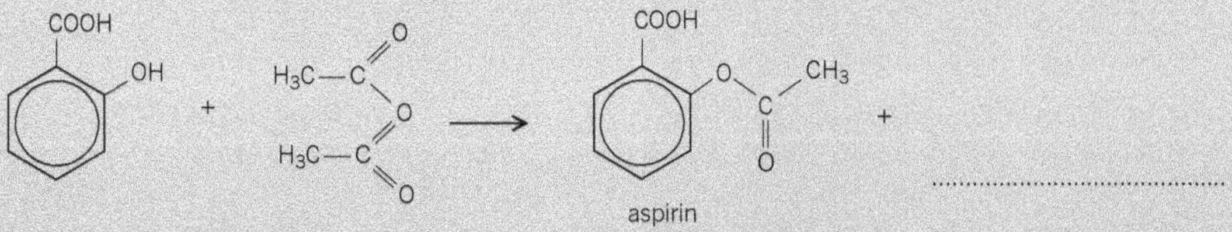

(1 mark)
- (ii) Suggest a name for the mechanism for the reaction in part (i). *(1 mark)*
- (iii) Give **two** industrial advantages, other than cost, of using ethanoic anhydride rather than ethanoyl chloride in the production of aspirin. *(2 marks)*

(d) Complete the following equation for the reaction of one molecule of benzene 1,2-dicarboxylic anhydride (phthalic anhydride) with one molecule of methanol by drawing the structural formula of the single product.

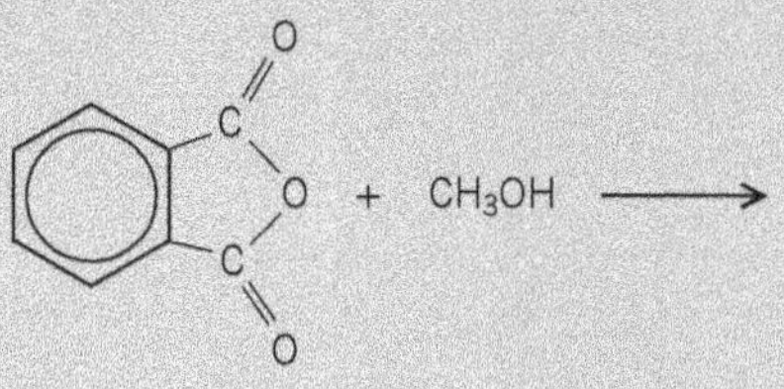

(1 mark)

(e) The indicator phenolphthalein is synthesised by reacting phthalic anhydride with phenol as shown in the following equation.

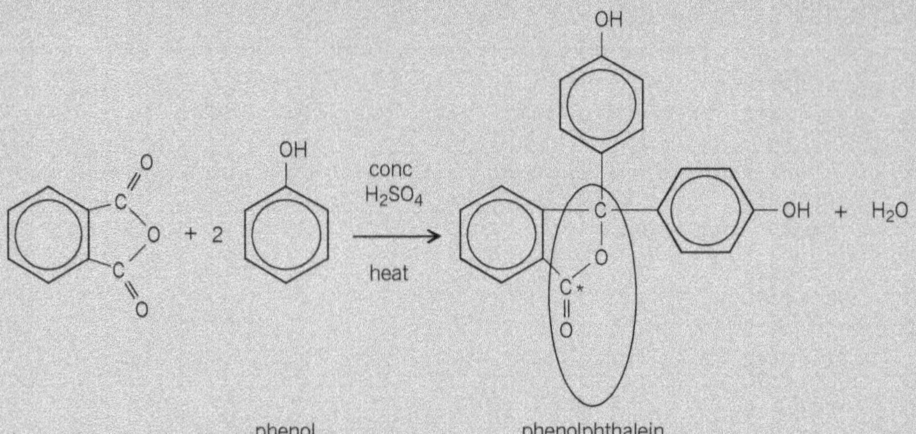

(i) Name the functional group ringed in the structure of phenolphthalein. *(1 mark)*
(ii) Deduce the number of peaks in the ^{13}C NMR spectrum of phenolphthalein. *(1 mark)*
(iii) One of the carbon atoms in the structure of phenolphthalein shown above is labelled with an asterisk (*).
Use Table 1 in Topic 32.1 to suggest a range of δ values for the peak due to this carbon atom in the ^{13}C NMR spectrum of phenolphthalein. *(1 mark)*

AQA, 2012

6 This question is about the over-the counter painkiller and anti-inflammatory agent ibuprofen, whose skeletal formula is shown below.

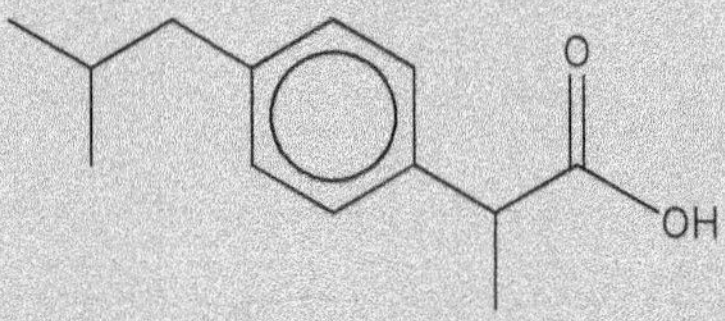

(a) What two functional groups does ibuprofen have? *(2 marks)*
(b) Ibuprofen is optically active. Identify the asymmetric carbon atom in the structure of ibuprofen and explain your reasoning. *(3 marks)*
(c) What potential problem does the fact that ibuprofen is optically active have on the use of ibuprofen as a drug? *(4 marks)*
(d) In order to work quickly, a drug should be soluble in water so that it can quickly get into the bloodstream.
(e) (i) Explain why ibuprofen is not very soluble in water. *(1 mark)*
(ii) Ibuprofen can be made more water-soluble by reacting it with the amino acid lysine whose structure is shown below.

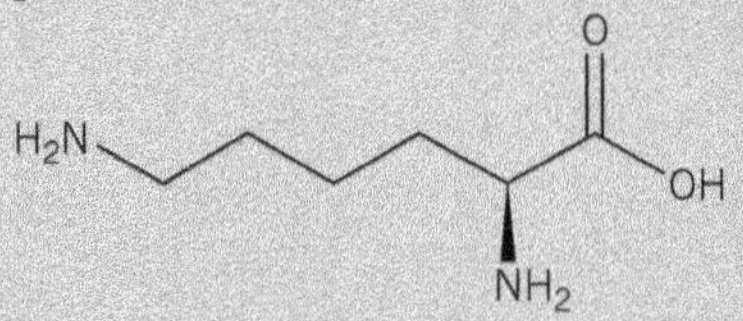

Part of the reason that lysine is water-soluble is that it can exist as a zwitterion. Explain the term zwitterions and show how it makes lysine water-soluble. *(2 marks)*
(iii) Ibuprofen and lysine react together to form a soluble salt. Suggest how this happens. *(2 marks)*
(iv) Why can we be confident that lysine is non-toxic? *(1 mark)*
(f) Ibuprofen is synthesised from a 2-methylpropylbenzene which is derived from crude oil.
(i) Draw the skeletal formula of 2-methylpropylbenzene. *(2 marks)*
(ii) What type of reactions will 2-methylpropylbenzene be most likely to undergo? Select the type of reagent from nucleophilic, electrophilic, and free radical. Select the type of reaction from substitution, addition, and elimination. Explain your answer. *(4 marks)*

Answers to the Practice Questions and Section Questions are available at www.oxfordsecondary.com/oxfordaqaexams-alevel-chemistry

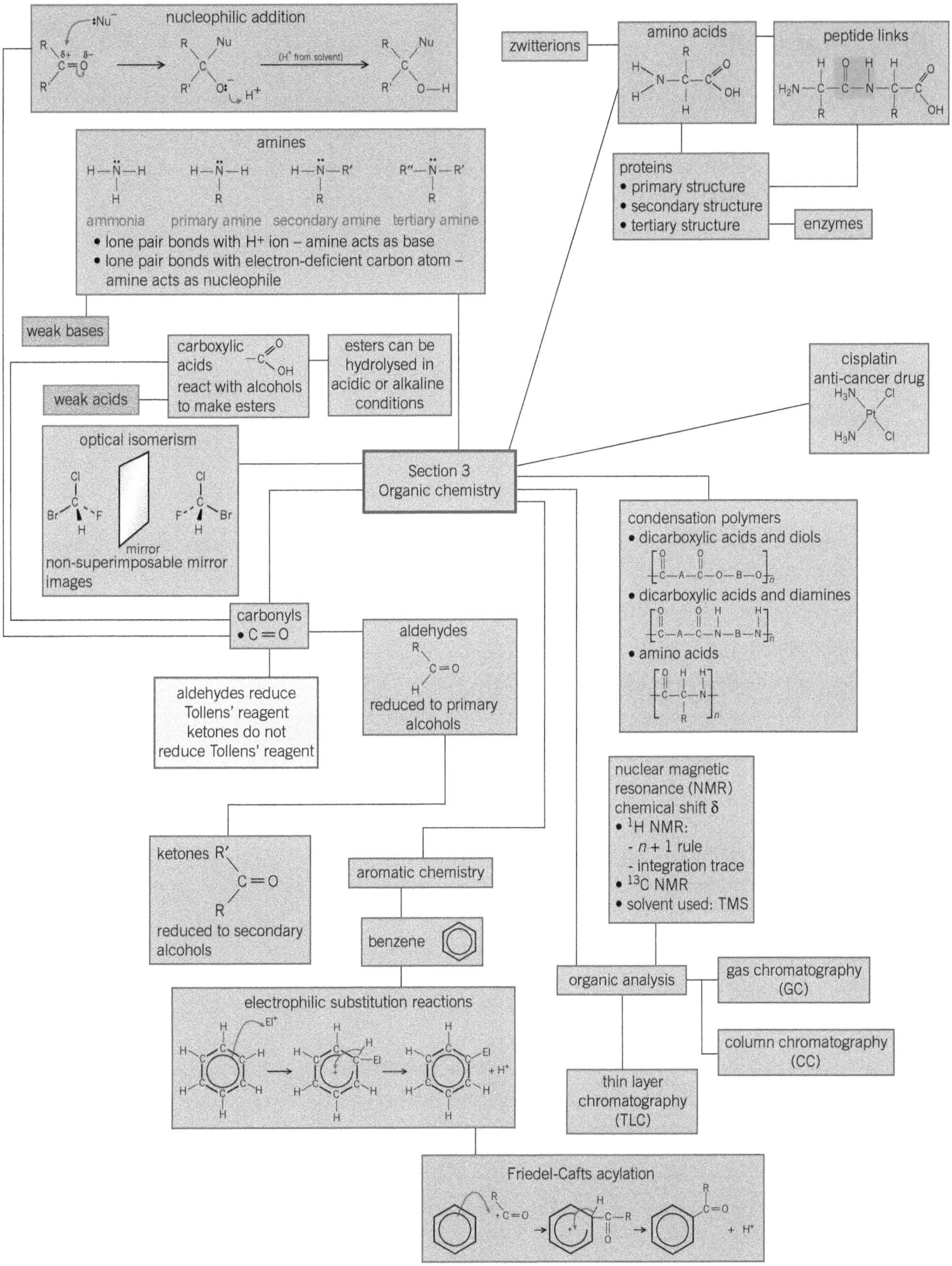
nucleophilic addition
(H+ from solvent)
amines
ammonia
primary amine
secondary amine
tertiary amine
• lone pair bonds with H+ ion – amine acts as base
• lone pair bonds with electron-deficient carbon atom – amine acts as nucleophile
weak bases
carboxylic acids
react with alcohols to make esters
weak acids
esters can be hydrolysed in acidic or alkaline conditions
optical isomerism
mirror
non-superimposable mirror images
Section 3 Organic chemistry
carbonyls
• C=O
aldehydes reduce Tollens' reagent ketones do not reduce Tollens' reagent
aldehydes
reduced to primary alcohols
ketones
reduced to secondary alcohols
aromatic chemistry
benzene
electrophilic substitution reactions
Friedel-Cafts acylation
zwitterions
amino acids
peptide links
proteins
• primary structure
• secondary structure
• tertiary structure
enzymes
cisplatin anti-cancer drug
condensation polymers
• dicarboxylic acids and diols
• dicarboxylic acids and diamines
• amino acids
nuclear magnetic resonance (NMR) chemical shift δ
• 1H NMR:
- n + 1 rule
- integration trace
• 13C NMR
• solvent used: TMS
organic analysis
gas chromatography (GC)
column chromatography (CC)
thin layer chromatography (TLC)

Practical skills

In this section you have met the following ideas:

- Investigating the effect of optical isomers on polarised light.
- Finding out how to distinguish between aldehydes and ketones.
- Finding out how to make and identify esters.
- Finding out how to make soap and biodiesel.
- Investigating the reactions of acid anhydrides and acyl chlorides.
- Finding out how to make aspirin.
- Investigating nitration reactions.
- Using melting points to identify compounds.
- Finding out how to make nylon.

Maths skills

In this section you have met the following maths skills:

- Working out how to draw optical isomers.

Extension

Produce a report explaining the principles of nuclear magnetic resonance and exploring how it can be used.

Suggested resources:

- Hore, P[2015], *Nuclear Magnetic Resonance: Oxford Chemistry Primers*. Oxford University Press, UK. ISBN 978-0-19-870341-9.
- Hore, P[2015], *NMR: The Toolkit: Oxford Chemistry Primers*. Oxford University Press, UK. ISBN 978-0-19-870342-6.

A level additional practice questions

1 **(a)** Use electron pair repulsion theory to state and explain the shape of an ammonia (NH_3) molecule. Draw an NH_3 molecule and include the bond angle. *(5 marks)*

(b) Chromium is a transition metal. Other than their catalytic activity, state three characteristic properties of transition metals. *(3 marks)*

(c) State the full electron configuration of:

(i) Cr

(ii) Cr^{3+} *(2 marks)*

(d) $[Cr(H_2O)_4(NH_3)_2]^{2+}$ is a complex ion. Define the terms:

(i) complex ion

(ii) ligand *(2 marks)*

(e) (i) Predict the shape of the $[Cr(H_2O)_4(NH_3)_2]^{2+}$ ion.

(ii) Deduce the co-ordination number of the $[Cr(H_2O)_4(NH_3)_2]^{2+}$ ion *(3 marks)*

2 **(a)** Describe and explain the trend in atomic radius across Period 3 of the Periodic Table. *(3 marks)*

(b) Describe the bonding and structure in magnesium. Include a diagram in your answer. *(4 marks)*

(c) Explain why magnesium, Mg, has a higher melting point than sodium, Na. *(2 marks)*

(d) Explain why the melting point of phosphorus, P_4, is greater than the melting point of chlorine, Cl_2. *(2 marks)*

(e) (i) Magnesium reacts with oxygen to form magnesium oxide. Write the equation for the reaction. Include state symbols. *(1 mark)*

(ii) Magnesium oxide forms an alkaline solution when it reacts with water. Explain why and include an equation in your answer. *(2 marks)*

3 Isopropyl alcohol is used as an industrial solvent. The skeletal formula of isopropyl alcohol is given below.

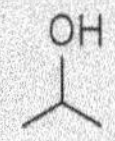

(a) (i) State the molecular formula of isopropyl alcohol.

(ii) Give the IUPAC name for isopropyl alcohol. *(2 marks)*

(b) A chemist analysed a sample of isopropyl alcohol using ^{13}C NMR. Deduce the number of peaks in the NMR spectra of isopropyl alcohol. Choose one answer.

A 1

B 2

C 3

D 4 *(1 mark)*

(c) Deduce the relative molecular mass of isopropyl alcohol. Choose one answer.

A 29

B 32

C 36

D 60 *(1 mark)*

(d) The sample was then analysed using proton NMR. Deduce the number of peaks in the proton NMR of isopropyl alcohol. Choose one answer.

A 2

B 3

C 4

D 9 *(1 mark)*

(e) (i) A student oxidised a sample of isopropyl alcohol using acidified potassium dichromate(VI). Using the structural formula $(CH_3)_2CHOH$ to represent isopropyl alcohol and [O] to represent the oxidising agent, write an equation for this reaction.

(ii) State the IUPAC name for the organic product of the reaction.

(iii) Describe and explain the colour change that the student would observe during the oxidation of isopropyl alcohol by acidified potassium dichromate(VI). *(4 marks)*

(f) The student wanted to use the sample of isopropyl alcohol to produce a carbonyl compound. Explain why the student did not have to distil off the product of the reaction. *(1 mark)*

(g) Draw the displayed formula and give the IUPAC name of the structural isomer of isopropyl alcohol. *(1 mark)*

4 Values for lattice enthalpy can be calculated indirectly using Born–Haber cycles.

Letter	Enthalpy change	Energy/ kJ mol^{-1}
A	Formation of calcium oxide	−635
B	First electron affinity of oxygen	−141
C	Second electron affinity of oxygen	+790
D	First ionisation energy of calcium	+590
E	Second ionisation energy of calcium	+1145
F	Atomisation of oxygen	+249
G	Atomisation of calcium	+178
H	Lattice enthalpy of calcium oxide	

(a) Give the equation for the formation of one mole of calcium oxide from its constituent elements. Include state symbols. *(2 marks)*

(b) Copy and complete the diagram below by stating the correct letter from the table of enthalpy changes above. *(2 marks)*

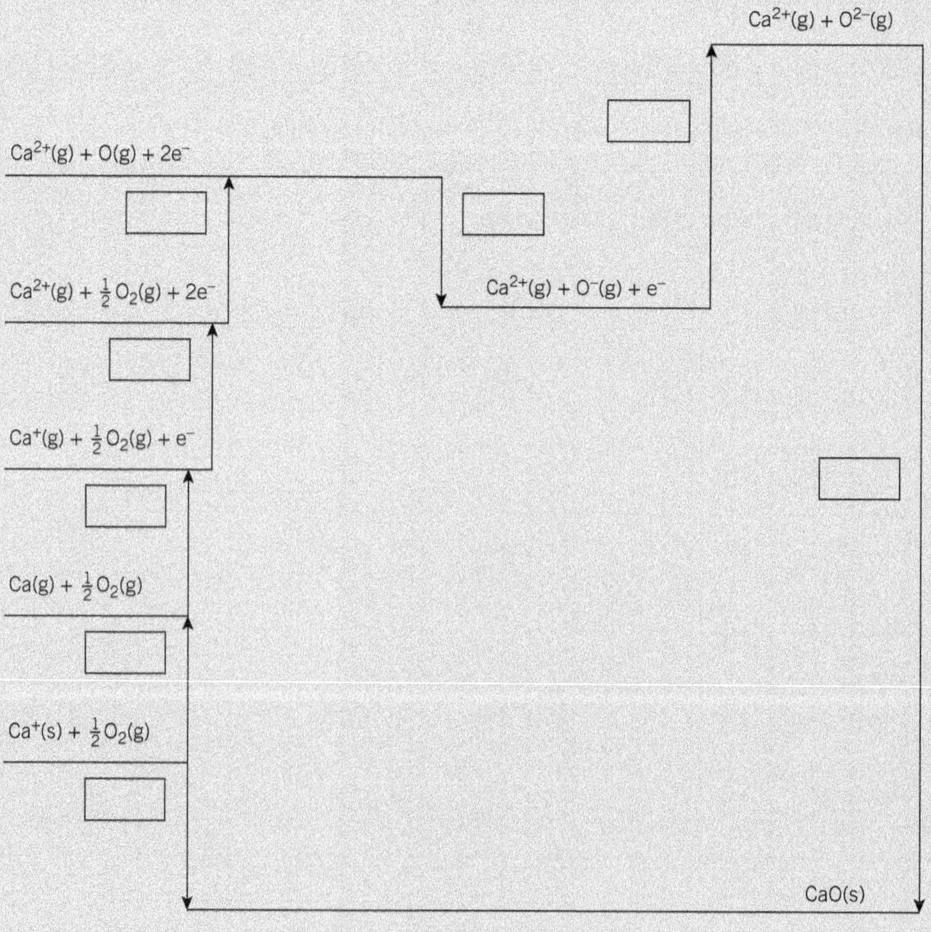

(c) Calculate the lattice enthalpy of calcium oxide. *(2 marks)*

(d) Predict whether the lattice enthalpy of calcium oxide or barium oxide would be the most exothermic. Explain your answer. *(3 marks)*

5 Compound B is a secondary alcohol. The displayed formula of compound B is shown below.

$$C_2H_5 \overset{\overset{\displaystyle OH}{|}}{\underset{\underset{\displaystyle H}{|}}{C}} CH_3$$

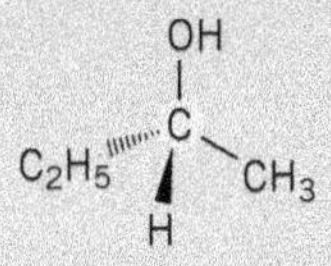

(a) Give the IUPAC name for compound B. *(1 mark)*

(b) Draw the skeletal formula of compound B. *(1 mark)*

(c) Give the IUPAC name of the structural isomer of compound B. *(1 mark)*

(d) Draw the optical isomer of compound B. *(1 mark)*

6 Alanine is an α-amino acid

$$CH_3 - \overset{\overset{\displaystyle NH_2}{|}}{\underset{\underset{\displaystyle H}{|}}{C}} - COOH$$

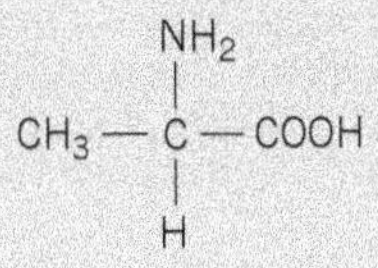

(a) Define the term α-amino acid *(1 mark)*

(b) Draw the zwitterion of alanine. *(1 mark)*

(c) Amino acids are crystalline solids which have surprisingly high melting points. Explain why zwitterions have relatively high melting points. *(2 marks)*

(d) State the IUPAC name of alanine. *(1 mark)*

7 A student carried out a titration between hydrochloric acid and sodium hydroxide and recorded the results below.

Experiment	1	2	3
Final burette reading/ cm^3	32.80	32.40	32.70
Initial burette reading/cm^3	50.00	50.00	50.00
Titre/cm^3			

(a) Complete the table. *(1 mark)*

(b) Name the piece of apparatus used to add sodium hydroxide to the hydrochloric acid. *(1 mark)*

(c) Calculate the mean titre. Give your answer to 3 significant figures. *(1 mark)*

(d) The error in the titre for experiment 1 is +/− 0.10 cm^3. Calculate the percentage error in Experiment 1. Give your answer to an appropriate number of significant figures. Choose one answer. *(1 mark)*

A 0.58%

B 0.6%

C 5.81%

D 0.0058%

(e) The student repeated the experiment using the same volume and concentration of hydrochloric acid and the same equipment. Suggest what the student could do to reduce the percentage error in their results. *(1 mark)*

8 **(a)** Define the term pH. *(1 mark)*

(b) Lactic acid has the structural formula $CH_3CHOHCOOH$. Give the IUPAC name of lactic acid. *(1 mark)*

(c) Lactic acid has a K_a value of 1.4×10^{-4}.

(i) Give the expression for K_a.

(ii) Calculate the pH of a $0.10\,mol\,dm^{-3}$ solution of lactic acid. Give your answer to two decimal places. *(4 marks)*

9 A student added sodium hydroxide solution dropwise to a test tube containing a solution of iron(II) sulfate. The student then left the test tube for several hours.

(a) Describe what the student would see:

(i) When sodium hydroxide was added to the iron(II) sulfate solution.

(ii) When the test tube was left for several hours. *(2 marks)*

(b) Write ionic equations for the reactions that occur when:

(i) The sodium hydroxide was added

(ii) When the test tube was left for several hours. Include state symbols. *(2 marks)*

10 A chemist investigated the equilibrium system below:

$$2NO(g) + 2CO(g) \rightleftharpoons 2CO_2(g) + N_2(g) \qquad \Delta H = -788\,kJ\,mol^{-1}$$

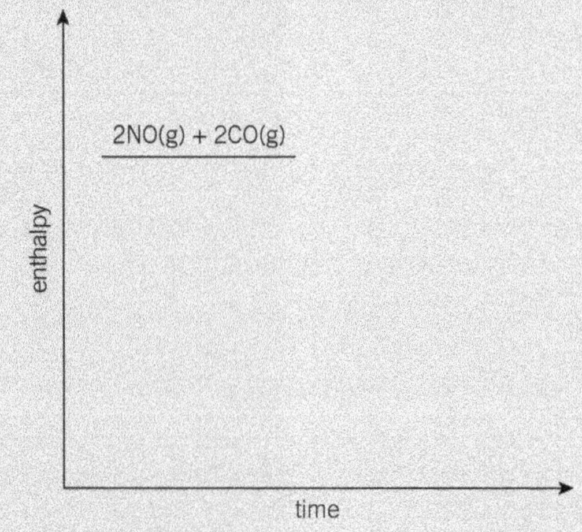

(a) Complete the enthalpy level diagram for this reaction. *(2 marks)*

(b) The chemist mixed 0.76 mol of CO with 0.45 mol of NO. The mixture was left at a constant temperature to reach a dynamic equilibrium.

(i) Define the term *dynamic equilibrium*.

(ii) The chemist analysed the equilibrium mixture and found that 0.30 moles of NO remained. The total volume of the equilibrium mixture was $2.00\,dm^3$. Write the expression for K_c for this reaction. *(2 marks)*

(c) Deduce the units for K_c. Choose one answer.

A $mol\,dm^{-3}$

B $mol^{-1}\,dm^{3}$

C $mol^{-2}\,dm^{6}$

D $mol^{3}\,dm^{-9}$ *(1 mark)*

(d) Calculate the value of K_c for this equilibrium mixture. Show your working. Give your answer to 3 significant figures. *(4 marks)*

11 Ammonia, NH_3, can be produced industrially using the Haber process at a temperature of between 400 and 500 °C.

$$N_2(g) + 3H_2(g) \rightleftharpoons 2NH_3(g) \qquad \Delta H = -92\,kJ\,mol^{-1}$$

(a) State the catalyst used in the Haber process. *(1 mark)*

(b) The diagram below shows a Maxwell-Boltzmann distribution. Explain how the catalyst used in the Haber process will increase the rate of reaction of formation of ammonia.

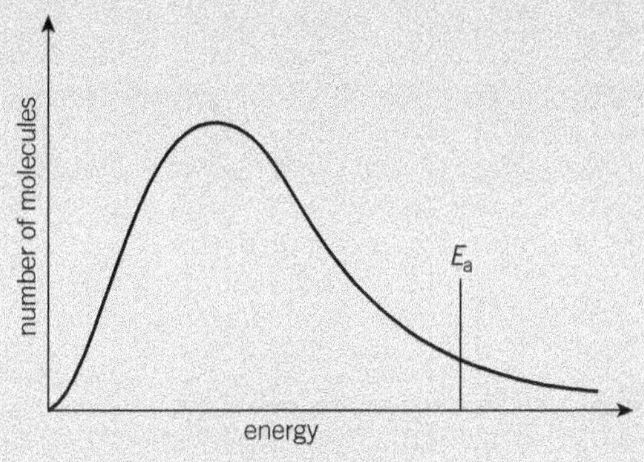

(2 marks)

(a) Ammonia has a higher boiling point than might be expected from other Group 6 hydrides. State the type of intermolecular attraction between ammonia molecules which causes ammonia to have a higher boiling point than expected. Use a diagram to explain your answer. *(5 mark)*

(b) Consider the complex cisplatin, $[Pt(NH_3)_2Cl_2]$, which contains ammonia ligands.

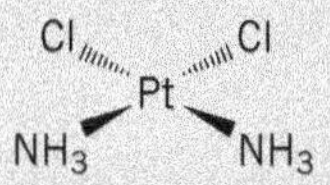

(i) Cisplatin is a neutral complex. Explain why.

(ii) Draw and name the stereoisomer of cisplatin $[Pt(NH_3)_2Cl_2]$. *(3 marks)*

(c) The table below shows the entropy of hydrogen, nitrogen, and ammonia.

Substance	$H_2(g)$	$N_2(g)$	$NH_3(g)$
Entropy $S^\ominus$ / $JK^{-1}mol^{-1}$	131	191	192

(i) Consider the equations for five processes shown in the table below. For each process predict the sign of the entropy change ΔS.

Process	Sign of ΔS
$H_2O(s) \rightarrow H_2O(l)$	
$NH_3(g) \rightarrow NH_3(l)$	
$NaCl(s) + aq \rightarrow Na^+(aq) + Cl^-(aq)$	
$CaCO_3(s) \rightarrow CaO(s) + CO_2(g)$	
$C_2H_5OH(l) + 3O_2(g) \rightarrow 2CO_2(g) + 3H_2O(g)$	

(5 marks)

(ii) Calculate the entropy change for the formation of ammonia from nitrogen and hydrogen.

$N_2(g) + 3H_2(g) \rightleftharpoons 2NH_3(g)$

(iii) Give the expression for the Gibbs free energy change.

(iv) Calculate the Gibbs free energy change for the formation of ammonia from its constituent elements at 25 °C. *(8 marks)*

12 Propanoic acid is a weak acid.

(a) Select the molecular formula of propanoic acid. Choose one answer.

A $C_3H_6O_2$
B CH_3CH_2COOH
C $CH_3CH_2CH_2COOH$
D $C_{1.5}H_3O$ *(1 mark)*

(b) Define the term *weak acid.* *(2 marks)*

(c) A 0.20 mol dm^{-3} solution of propanoic acid has a pH of 2.79.
Calculate the K_a value of propanoic acid. Give your answer to two significant figures. *(3 marks)*

(d) A student placed 25.0 cm^3 of the 0.100 mol dm^{-3} solution of propanoic acid into a flask.

(i) Name the piece of apparatus the student should use to measure exactly 25.0 cm^3 of propanoic acid.
(ii) Calculate the amount, in mol, of propanoic acid in the flask.
(iii) The propanoic acid was titrated with an aqueous solution of calcium hydroxide. 22.0 cm^3 of calcium hydroxide was required for complete neutralisation.
Deduce the number of moles of calcium hydroxide required.
(iv) Calculate the concentration of the calcium hydroxide solution. Give your answer to three significant figures.
(v) State the expression for the ionic product of water K_w.
(vi) Calculate the pH of the calcium hydroxide solution at 25 °C. Give your answer to two decimal places. *(8 marks)*

13 A student analysed a multivitamin and mineral tablet. The tablet contained a small amount of an iron(II) salt. The tablet was powdered and 0.525 g was dissolved in water. A small amount of dilute sulfuric acid was added. The solution was then titrated with potassium manganate(VII).
15.50 cm^3 of 0.002 00 mol dm^{-3} potassium manganate(VII) was required.

(a) State the colour that the student would observe when exactly the right amount of potassium manganate(VII) was added. *(1 mark)*

(b) Consider the two half equations below.

$Fe^{2+}(aq) \rightarrow Fe^{3+}(aq) + e^-$

$MnO_4^-(aq) + 8H^+(aq) + 5e^- \rightarrow Mn^{2+}(aq) + 4H_2O(l)$

Combine the two half equations to give the overall equation for the reaction. *(1 mark)*

(c) (i) Calculate the amount, in mol, of potassium manganate(VII) used.
(ii) Calculate the amount, in mol, of Fe^{2+} used.
(iii) Calculate the percentage by mass of iron in the tablet. Give your answer to three significant figures. *(5 marks)*

Answers to the Practice Questions and Section Questions are available at
www.oxfordsecondary.com/oxfordaqaexams-alevel-chemistry

Section 7
Practical skills

Practical work is firmly at the heart of any chemistry course. It helps you to understand new ideas and difficult concepts, and helps you to appreciate the importance of experiments in testing and developing scientific theories. In addition, practical work develops the skills that scientists use in their everyday work. Such skills involve planning, researching, making and processing measurements, analysing, and evaluating experimental results, as well as the ability to use a variety of apparatus and chemicals safely.

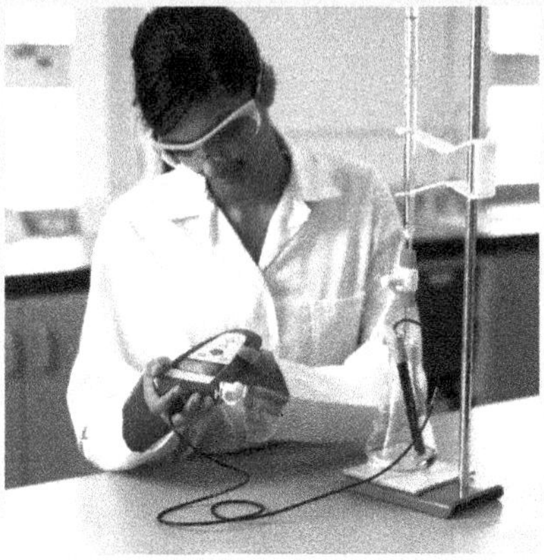

During this course you will be carrying out a number of practicals using a range of apparatus and techniques. Table 1 lists some of the practical activities you will do in your AS/1st year A level course, and questions may be set on these in the written exam. In carrying out these activities, you should become proficient in all the practical skills assessed directly and indirectly in your AS/1st year A-level course. For each activity, the table references the relevant topic or topics in this book. (In addition there will be a teacher assessed pass/fail endorsement of your practical skills.) Table 2 (p525) lists the practicals related to the A2 course.

▼ **Table 1** *The practical activities you will cover as part of your AS Level course*

	Practical	Topic
1	Making up a volumetric solution and carrying out a titration	2.5, Balanced equations and related calculations
2	Measurement of an enthalpy change	4.3, Measuring enthalpy changes–calorimetry 4.4, Hess's law
3	Carrying out test-tube reactions to identify cations and anions in aqueous solutions	9.1, The physical and chemical properties of Group 2 10.3, Reactions of halide ions
4	Distillation of a product from a reaction	15.2, The reactions of alcohols
5	Tests for alcohol, aldehyde, alkene, and carboxylic acid	15.2, The reaction of alcohols

Practice questions

The following questions will give you some practice – you may not have done the actual experiment mentioned but you should think about similar practical work that you have done.

Practical 1

1 Describe how to make 250 cm^3 of an accurately known 0.10 mol dm^{-3} solution of blue, hydrated copper sulfate, $CuSO_4.5H_2O$.
A_r Cu = 63.5, S = 32.1, H = 1.0, O = 16.0

2 a Describe carefully the steps needed to find the concentration of a sodium hydroxide solution, using 0.10 mol dm^{-3} hydrochloric acid and phenolphthalein indicator, which is pink in alkali and colourless in acid. Choose the apparatus used for each step explaining any measures you take to ensure safety and accuracy Select from the following apparatus

conical flask funnel goggles burette

25.0 cm^3 pipette with filler dropping pipette white tile

b 25.00 cm^3 of sodium hydroxide solution were neutralised by 30.00 cm^3 of acid.

i Write the equation for the reaction.

ii Was the sodium hydroxide solution more or less concentrated than the acid? Explain your answer.

iii Find the concentration of the sodium hydroxide solution.

c Why is phenolphthalein a better indicator than universal indicator for titrations?

Practical 2

Determining an enthalpy change that cannot be measured directly

Anhydrous (white) copper sulfate reacts with water to form hydrated (blue) copper sulfate:

$$CuSO_4(s) + 5H_2O(l) \rightarrow CuSO_4.5H_2O(s) \quad \Delta H_3 = ?$$

The enthalpy change cannot be measured directly, but it can be determined using the Hess's law cycle below and measuring the enthalpy changes ΔH_1 and ΔH_2, the enthalpies of solution of the two copper sulfates.

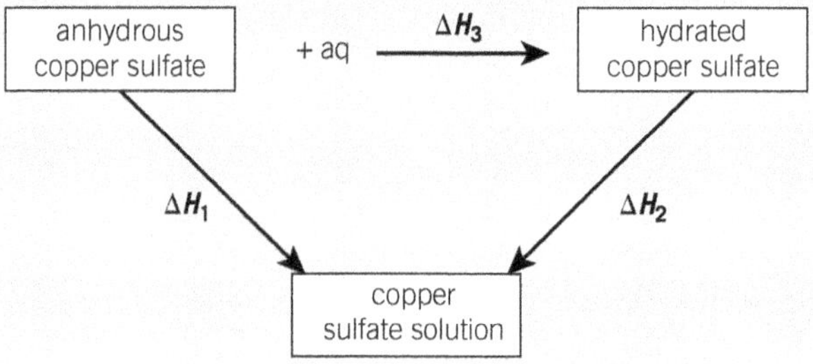

A student weighed out 4.00 g of anhydrous copper sulfate on a top-pan balance that read to 0.01 g. She measured out 50.0 cm^3 water into an expanded polystyrene beaker using a measuring cylinder that measured to 0.5 cm^3. The water had been standing in the laboratory for over 1 hour. She recorded the temperature of the water. She quickly stirred the mixture until all the solid had dissolved and recorded the highest temperature attained using a thermometer that read to 0.1 °C.

She then repeated the procedure using 6.25 g of hydrated copper sulfate and 47.5 cm^3 water (this allows for the water present in the hydrated copper sulfate). This time the temperature dropped.

Results

Anhydrous copper sulfate

Initial temperature of water = 18.7 °C;
Highest temperature attained = 25.9 °C

Hydrated copper sulfate

Initial temperature of water = 18.7 °C;
Lowest temperature attained = 17.5 °C

Calculation

1 Calculate the number of moles of
  a anhydrous copper sulfate
  b hydrated copper sulfate that were used.
  Use A_rCu = 63.5, S = 32.1, O = 16.0, H = 1.0

2 How many moles of water does this amount of hydrated copper sulfate contain?

3 Calculate the temperature change in each experiment.

4 Calculate the heat produced by the anhydrous copper sulfate using $q = m \times c \times \Delta T$. Remember that the mass of the solution includes the mass of the copper sulfate added. Use a value of 4.18 J g^{-1} °C^{-1} for c, the specific heat of a dilute aqueous solution.

5 Calculate the heat absorbed by the hydrated copper sulfate using $q = m \times c \times \Delta t$

6 Calculate the enthalpy change of solution in each case in kJ mol^{-1} giving the correct sign for each.
  What is ΔH_3?

Questions

1 Explain why the heat of hydration of anhydrous copper sulfate cannot be measured directly.

2 What is the percentage error in
  a weighing the anhydrous copper sulfate
  b weighing the hydrated copper sulfate
  c measuring the 50.0 cm^3 water
  d measuring the temperature rise
  e measuring the temperature drop?

3 Why had the water been left to stand for one hour before the experiment?

4 Why were the experiments carried out in expanded polystyrene beakers?

5 Outline how a cooling curve could be used to allow for heat loss – see Topic 4.3, Measuring enthalpy changes – calorimetry.

6 Bearing in mind the number of significant figures in each of the measurements, how many significant figures can be given in the value for ΔH_3?

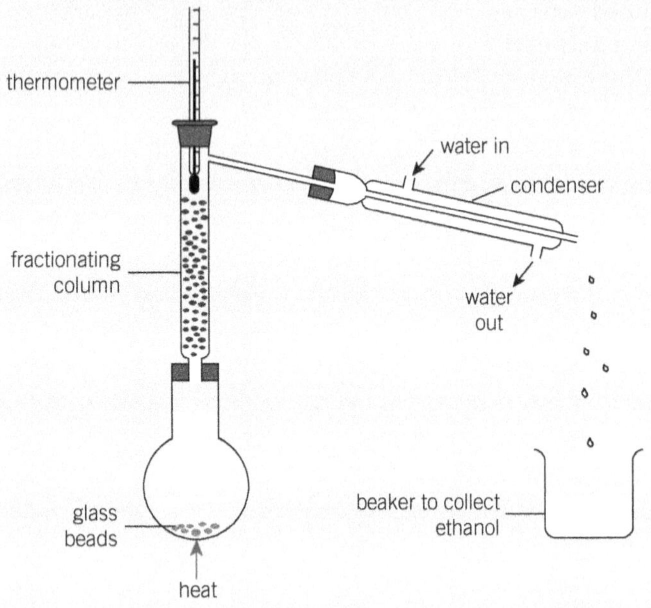

Practical 3

An unlabelled bottle of white powder is found in a school chemical store. Close by is a label 'magnesium bromide'. What tests would you carry out to confirm that the label belongs to the bottle?

Practical 4
Distillation of ethanol

The diagram shows a distillation apparatus set up to produce pure ethanol (boiling temperature 78 °C) from a mixture resulting from the fermentation of a solution containing sugar (sucrose) and yeast.

Look at the diagram and answer the following questions.

1 **There are four faults in the experimental set up.** Point out each one and explain why it is a problem and what should be done to correct it.
2 What is the purpose of the ground glass beads in the boiling flask?
3 Over which of these temperature ranges should the product be collected?
   a 76–80 °C,
   b 70–85 °C,
   c all the vapour should be collected
4 Describe how you would heat the flask.

Practical 5

You have to identify three liquids and two gases. You know you have an alcohol, an alkene, an aldehyde, a carboxylic acid, and air.

These are the tests you decide to do:

1 Add a few drops of bromine water to each gas to identify the gases.
2 Add a sample of each liquid to water and test the pH.
3 Add Benedict's solution to the remaining two liquids and warm in a water bath.
4 Warm with acidified potassium dichromate solution.

Explain how this would work.

▼ **Table 2** *Some practical activities you will cover as part of your A2 course*

Practical		Topic
6	Measuring the rate of a reaction	18.1 The rate of chemical reactions
7	Measuring $E^{\ominus}$ for an electrochemical cell	20.1 Electrode potentials and the electrochemical series
8	Investigating pH change during an acid–base titration	21.4 Acid–base titrations
9	Identifying transition metal ions in solution	24.3 A summary of acid–base and substitution reactions of some metal ions
10	Preparing an organic solid	27.3 Reactions of arenes

Practical 6
Measuring the rate of a reaction

A student was investigating the factors that affect the rate of a reaction between a metal and hydrochloric acid which gives off hydrogen gas. The graphs below show her results for two different concentrations of acid.

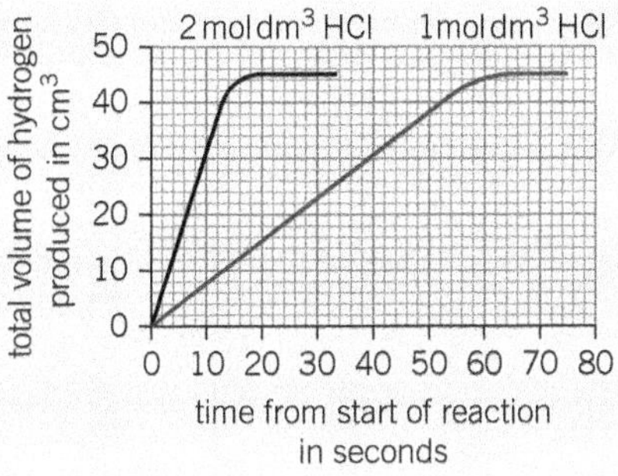

1 Suggest two experimental methods for following this reaction.
2 Use the graphs to calculate the initial rate of the reaction in each experiment.
3 What is important about the *initial* rate of a reaction?
4 What do these results suggest about the order of the reaction with respect to the concentration of hydrochloric acid?
5 What other factors must be kept constant if this conclusion is to be valid?

Practical 7

Measuring $E^\ominus$ for an electrochemical cell

A student set up the apparatus below to measure the value of $E^\ominus$ for $Zn^{2+}(aq) \rightleftharpoons Zn(s)$

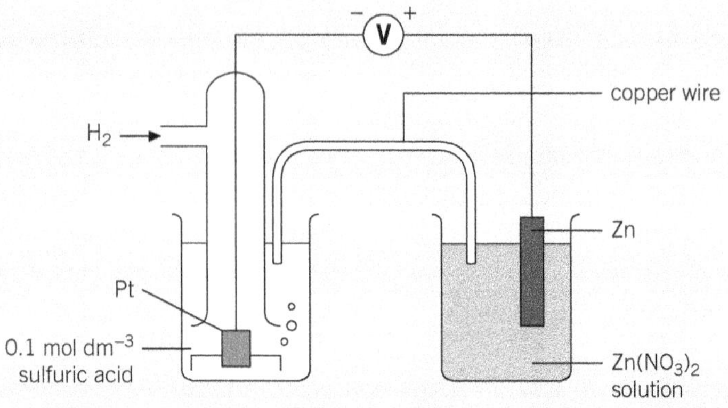

1 Point out **three significant errors** in the experimental setup.
2 State three experimental conditions not mentioned on the diagram that must be controlled if the potential difference measured is to be $E^\ominus$.

Practical 8

Investigating pH change during an acid–base titration

A student obtained the following figures when she added a solution of a base to a solution of an acid of the same concentration.

Volume of base added / cm^3	pH
0	1.00
5	1.13
10	2.24
15	1.31
20	1.41
25	7.01
30	12.00
35	12.11
40	12.24
45	12.34
50	12.41

1 Plot a graph of pH (vertically) against Volume of base added (horizontally)
2 Which value is clearly wrong?
3 The student's teacher suggested that she should have made measurement every 1 cm^3 rather than every 5 cm^3 between 20 cm^3 of base added and 30 cm^3 of base added. Explain why this would be a good idea.
4 Name the instrument used to measure the pH values.
5 Why would universal indicator *not* be suitable for measuring the pH?

6 Which of the following best represents the acid and base used: strong acid/strong base; strong acid/weak base; weak acid/strong base/; weak acid/weak base?

Practical 9
Identifying transition metal ions in solution

You have two very dilute solutions A and B that you want to identify and you know they are solutions of transition metal nitrates.

You have access to sodium hydroxide solution, ammonia solution, and sodium carbonate solution.

You first add a little sodium hydroxide solution to a sample of each and this produces a brown gelatinous precipitate in A and a pale blue precipitate in B, so you think A might contain Fe^{3+} ions and B might contain Cu^{2+} ions.

1 What reaction has taken place to produce the results above?
2 Suggest how you could use the ammonia solution and sodium carbonate solution to confirm that A might contain Fe^{3+} ions and B might contain Cu^{2+} ions explaining clearly the results you would expect.

Practical 10
Preparing an organic solid

A student prepared a sample of methyl 3-nitrobenzoate by the nitration of methyl benzoate. Her notes are summarised below.

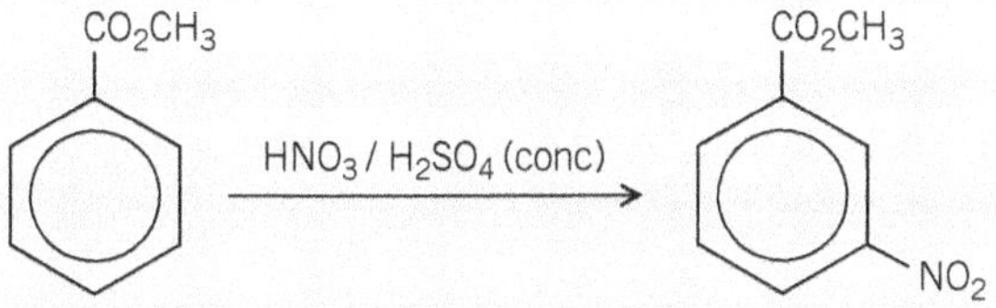

The mixture of acids was slowly added to the methyl benzoate (a liquid) in a conical flask, cooling the flask in an ice-bath to keep the reaction mixture below room temperature. When the reaction was complete, the mixture was poured onto crushed ice to precipitate the solid product. After the ice has melted, the product was separated by suction filtration. It was then purified by **recrystallisation** – the product was dissolved in the minimum volume of boiling ethanol which was then cooled in an ice bath until crystals of the product appeared. These were again separated by suction filtration.

The purity of the product was checked by thin layer chromatography (TLC).

1 Suggest why it was important to keep the reaction mixture cool.
2 What side-product of the reaction is possible?
3 How would the TLC plate appear if the product was
   a pure,
   b contaminated by a side-product?
4 Suggest another method of checking the purity (or otherwise) of the product.
5 Name and briefly describe the method of purifying a solid organic compound.

Section 8
Mathematical skills

Units in calculations

You are expected to use the correct units in calculations.

Units

You still describe the speed of a car in miles per hour. The units miles per hour could be written miles/hour or miles hour^{-1} where the superscript '−1' is just a way of expressing per something. In science you use the metric system of units, and speed has the unit metres per second, written m s^{-1}. In each case, you can think of per as meaning divided by.

Units can be surprisingly useful

A mile is a unit of distance and an hour a unit of time, so the unit miles per hour reminds you that speed is distance divided by time.

In the same way, if you know the units of density are grams per cubic centimetre, usually written g cm^{-3}, where cm^{-3} means per cubic centimetre, you can remember that density is mass divided by volume.

Multiplying and dividing units

When you are doing calculations, units cancel and multiply just like numbers. This can be a guide to whether you have used the right method.

For example:

The density of a liquid is 0.8 g cm^{-3}. What is the volume of a mass of 1.6 g of it?

Density = mass / volume

So volume = mass / density

Putting in the values and the units:

volume = 1.6 g / 0.8 g cm^{-3}

Cancelling the gs

volume = 2.0 / cm^{-3}

volume = 2.0 cm^{3}

If you had started with the wrong equation, such as

volume = density/mass or

volume = mass × density, you would not have ended up with the correct units for volume.

Units to learn

It is a good idea to learn the units of some basic quantities by heart.

	Unit	Comment
volume	dm^3	1 dm^3 is 1 litre (l) which is 1000 cm^3
concentration	mol dm^{-3}	
pressure	pascals, Pa = N m^{-2}	N m^{-2} are newtons per square metre
enthalpy	kJ mol^{-1}	kJ is kilojoule. Occasionally J mol^{-1} is used
entropy	JK^{-1} mol^{-1}	joules per kelvin per mole

Standard form

This is a way of writing very large and very small numbers in a way that makes calculations and comparisons easier.

The number is written as number multiplied by ten raised to a power. The decimal point is put to the right of the first digit of the number.

For example:

22 000 is written 2.2×10^4.

0.000 002 2 is written 2.2×10^{-6}.

How to work out the power to which ten must be raised

Count the number of places you must move the decimal point in order to have one digit before the decimal point.

For example:

$0.0005\,1 = 5.1 \times 10^{-4}$

$51\,000 = 5.1 \times 10^{4}$

Moving the decimal point to the right gives a negative index (numbers less than 1), and to the left a positive index (numbers greater than one).

(The number 1 itself is 10^0, so the numbers 1–9 are followed by $\times 10^0$ when written in standard form. Can you see why?)

Multiplying and dividing

When *multiplying* numbers expressed in this way, *add* the powers (called indices) and when *dividing*, *subtract* them.

Worked examples

Calculate

a $2 \times 10^5 \times 4 \times 10^6$

b $\dfrac{8 \times 10^3}{4 \times 10^2}$

c $\dfrac{5 \times 10^8}{2 \times 10^{-6}}$

Answer

a $2 \times 10^5 \times 4 \times 10^6$

Multiply $2 \times 4 = 8$. Add the indices to give 10^{11}

Answer = 8×10^{11}

b $\dfrac{8 \times 10^3}{4 \times 10^2}$

Divide 8 by 4 = 2. Subtract the indices to give 10^1

Answer = $2 \times 10^1 = 20$

c $\dfrac{5 \times 10^8}{2 \times 10^{-6}}$

Divide 5 by 2 = 2.5. Subtract the indices $(8 - (-6))$ to give 10^{14}

Answer = 2.5×10^{14}

A handy hint for non-mathematicians

Non-mathematicians sometimes lose confidence when using small numbers such as 0.002. If you are not sure whether to multiply or divide, then do a similar calculation with numbers that you are happy with, because the rule will be the same.

Example:

How many moles of water in 0.000 1 g? A mole of water has a mass of 18 g.

Do you divide 18 by 0.000 1 or 0.000 1 by 18?

If you have any doubts about how to do this, then in your head change 0.000 1 g into a more familiar number such as 100 g.

How many moles of water in 100 g? A mole of water has a mass of 18 g.

Now you can see that you must divide 100 by 18. So in the same way you must divide 0.000 1 by 18 in the original problem.

$$\frac{0.0001}{18} = 5.6 \times 10^{-6}$$

Prefixes and suffixes

In chemistry you will often encounter very large numbers (such as the number of atoms in a mole) or very small numbers (such as the size of an atom).

Prefixes and suffixes are often used with units to help express these numbers. You will come across the following which multiply the number by a factor of 10^n. The red ones are the ones you are most likely to use.

Prefix	Conversion Factor	Symbol
pico	10^{-12}	p
nano	10^{-9}	n
micro	10^{-6}	μ
milli	10^{-3}	m
centi	10^{-2}	c
deci	10^{-1}	d
kilo	10^3	k
mega	10^6	M

So 5400 g = 5.4×10^3 g = 5.4 kg

Converting to base units

If you want to convert a number expressed with a prefix to one expressed in the base unit, multiply by the conversion factor. If you have a very small or very large number (and have to handle several zeros) the easiest way is to first convert the number to standard form.

Worked example

Convert a) 2 cm and b) 100 000 000 mm to metres

a 2 cm = 2×10^{-2} m = 0.02 m

b 100 000 000 mm = 1×10^8 mm
= $1 \times 10^8 \times 10^{-3}$ m = 1×10^5 m

Base units

The SI system is founded on base units. The ones you will meet in chemistry are:

Unit	Symbol	Used for
metre	m	length
kilogram	kg	mass
second	s	time
ampere (amp)	A	electric current
kelvin	K	temperature
mole	mol	amount of substance

Handling data

Sorting out significant figures

Many of the numbers used in chemistry are measurements – for example, the volume of a liquid, the mass of a solid, the temperature of a reaction vessel – and no measurement can be exact. When you make a measurement, you can indicate how uncertain it is by the way you write it. For example, a length of 5.0 cm means that you have used a measuring device capable of reading to 0.1 cm, a value of 5.00 cm means that you have measured to the nearest 0.01 cm and so on. So the numbers 5, 5.0, and 5.00 are different, you say they have different numbers of *significant figures*.

What exactly is a significant figure?

In a number that has been found or worked out from a measurement, the significant figures are all the digits known for certain, *plus the first uncertain one* (which may be a zero). The last digit is the uncertain one and is at the limit of the apparatus used for measuring it (Figure 1).

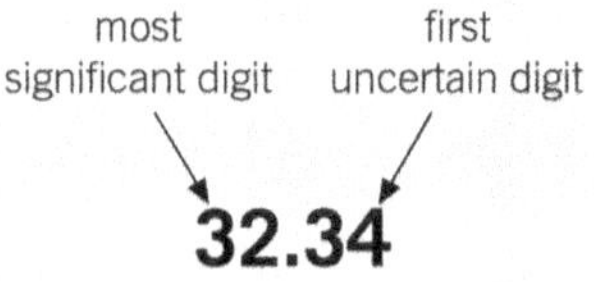

▲ **Figure 1** *A number with four significant figures*

For example, if you say a substance has a mass of 4.56 grams it means that you are certain about the 4 and the 5 but not the 6 as you are approaching the limit of accuracy of our measuring device (you will have seen the last figure on a top-pan balance fluctuate). The number 4.56 has three significant figures (s.f.).

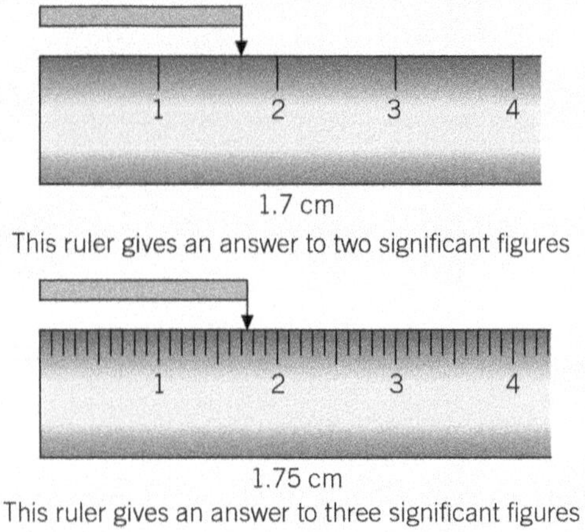

▲ **Figure 2** *Rulers with different precision*

When a number contains zeros, the rules for working out the number of significant figures are given below.

- Zeros between digits are significant.
- Zeros to the left of the first non-zero digit are not significant (even when there is a decimal point in the number).
- When a number with a decimal point ends in zeros to the right of the decimal point, these zeros are significant.
- When a number with no decimal point ends in several zeros, these zeros may or may not be significant. The number of significant figures should ideally be stated. For example, 20 000 (to 3 s.f.) means that the number has been measured to the nearest 100 but 20 000 (to 4 s.f.) means that the number has been measured to the nearest 10.

The following examples should help you to work out the number of significant figures in your data.

Worked examples

What is the number of significant figures in each of the following?

a 11.23

Answer

4 s.f. all non-zero digits are significant.

b 1100

Answer

2 s.f. (but it could be 2, 3, or 4 significant figures). The number has no decimal point so the zeros may or may not be significant. With numbers with zeros at the end it is best to state the number of significant figures.

c 1100.0

Answer

5 s.f. the decimal point implies a different accuracy of measurement to example (b).

d 0.025

Answer

2 s.f. zeros to the left of the decimal point only fix the position of the decimal point. They are not significant.

Question

1 How many significant figures?

a 40 000

b 1.030

c 0.22

d 22.00

Using significant figures in answers

When doing a calculation, it is important that you don't just copy down the display of your calculator, because this may have a far greater number of significant figures than the data in the question justifies. Your answer cannot be more certain than the least certain of the information that you used to calculate it. So your answer should contain the same number of significant figures as the measurement that has the smallest number of them.

Worked example

81.0 g (3 s.f.) of iron has a volume of 10.16 cm^3 (4 s.f.). What is its density?

Answer

$$\text{Density} = \frac{\text{mass}}{\text{volume}} = \frac{81.0\text{ g}}{10.16\text{ cm}^3}$$

$= 7.972\ 440\ 94\text{ g cm}^{-3}$ (this number has 9 s.f.)

Since our least certain measurement was to 3 s.f., our answer should have 3 s.f., i.e., 7.97 g cm^{-3}

If our answer had been 7.976 440 94, you would have rounded it up to 7.98 because the fourth significant figure (6) is five or greater.

The other point to be careful about is *when* to round up. This is best left to the very end of the calculation. Don't round up as you go along as this could make a difference to your final answer.

Decimal places and significant figures

The apparatus you use in the laboratory usually reads to a given number of decimal places (for example, hundredths or thousandths of a gram). For example, the top-pan balances in most schools and colleges usually weigh to 0.01 g, which is to two decimal places.

The number of significant figures of a measurement obtained by using the balance depends on the mass you are finding. A mass of 10.38 g has 4 s.f. but a mass of 0.08 has only 1 s.f. Check this with the rules above.

Hint

Calculator displays usually show numbers in standard form in a particular way. For example, 2.6×10^{-4} may appear as 2.6 – 04, a shorthand form which is not acceptable as a way of writing an answer.

Algebra

Equations

You can write an equation if you can show a connection between sets of measurements (variables).

For example, at a *fixed* volume, if you double the temperature (in kelvin) of a gas, the pressure doubles too.

Mathematically speaking, the pressure P is directly proportional to the temperature T.

$P \propto T$

The symbol $\propto$ means is proportional to.

This is shown in the data in Table1.

▼ **Table 1**

Temperature/K	Pressure/Pa
100	1000
150	1500
200	2000
250	2500

This also means that the pressure, P, is equal to some constant, k, multiplied by the temperature:

$P = kT$

In this case, the constant is 10 and if you multiply the temperature in kelvin by 10, you get the pressure in Pa.

Pressure and volume of a gas also vary. At constant temperature, as the pressure of a gas goes up, its volume goes down. More precisely, if you double the pressure, you halve the volume. Mathematically speaking, volume V is *inversely* proportional to pressure P.

$$V \propto \frac{1}{P}$$

So $$V = k \times \frac{1}{P}$$

Or simply $$V = \frac{k}{P}$$

This is shown by the data in Table 2.

In this case the constant is 24. If you multiply $\frac{1}{P}$ by 24, you get the volume.

▼ **Table 2**

Pressure/Pa	Volume/l	1/P =1/Pa
1	24	1.00
2	12	0.50
3	8	0.33
4	6	0.25

Mathematical symbols

Symbol	Meaning
$\rightleftharpoons$	equilibrium or reversible reaction
$<$	less than
$\ll$	much less than
$>$	greater than
$\gg$	much greater than
$\approx$ or	approximately equal to
$\propto$	proportional to

Question

2a If two variables, x and y, are directly proportional to each other, what happens to one if you quadruple the other?

b Write an expression that means x is inversely proportional to y.

c What happens to the volume of a gas if you triple the pressure at constant temperature?

Handling equations

Changing the subject of an equation

If you can confidently do the next exercise, go straight to the section *Substituting into equations* and try that. Otherwise work through *Rearranging equations*.

Question

3 The equation that connects the pressure P, volume V, and temperature T of a mole of gas is $PV = RT$

where P, V, and T are variables and R is a constant called the gas constant.

Rearrange the equation to find:

a P in terms of V, R, and T

b V in terms of P, R, and T

c T in terms of P, V, and R

d R in terms of P, V, and T

Rearranging equations

Start with a simple relation because the rules are the same however complicated the equation.

$$a = \frac{b}{c}$$

where $b = 10$ and $c = 5$,

It is easy to see when substituting these values into the expression

$$a = \frac{10}{5} = 2.$$

But what do you do if you need to find b or c from this equation?

you need to rearrange the equation so that b (or c) appears on its own on the left-hand side of the equation like this

$b = ?$

$a = \frac{b}{c}$ means $a = b \div c$

Step 1: Multiply both sides of the equation by c, because b is *divided* by c, so to get b on its own you must *multiply* by c.

Remember that to keep an equation valid whatever you do to one side you must do to the other – think of it as a see-saw, with the = sign as the pivot.

So now $c \times a = \frac{b \times c}{c}$

usually written $ca = \frac{bc}{c}$

Now cancel the cs on the right since b is being both multiplied and divided by c

Which leaves $c \times a = b$

Or $b = c \times a$ usually written $b = ca$

You can now rearrange this equation in the same way to find c.

Step 2: $b = c \times a$. Divide both sides by a

$$\frac{b}{a} = \frac{c \times a}{a}$$

Now cancel the as on the right

So $\frac{b}{a} = c$

Notice that because c started on the bottom, a two-step process was necessary. You found an expression for b first and then found one for c.

Question

4 Find the variable in brackets in terms of the others.

a $p = \frac{q}{r}$ (q)

b $n = mt$ (m)

c $g = \frac{fe}{h}$ (h)

d $\frac{pr}{e} = s$ (r)

Substituting into equations

When you are asked to substitute numerical values into an equation, it is essential that you carry out the mathematical operations in the right order.

A useful aid to remembering the order is the word **BIDMAS**:

Brackets

Indices

Division

Multiplication

Addition

Subtraction

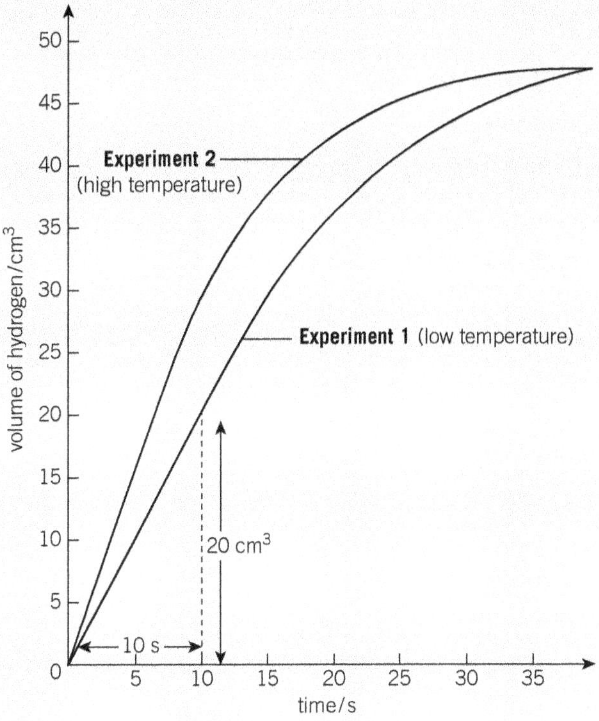

▲ **Figure 3** *Reaction rate graph*

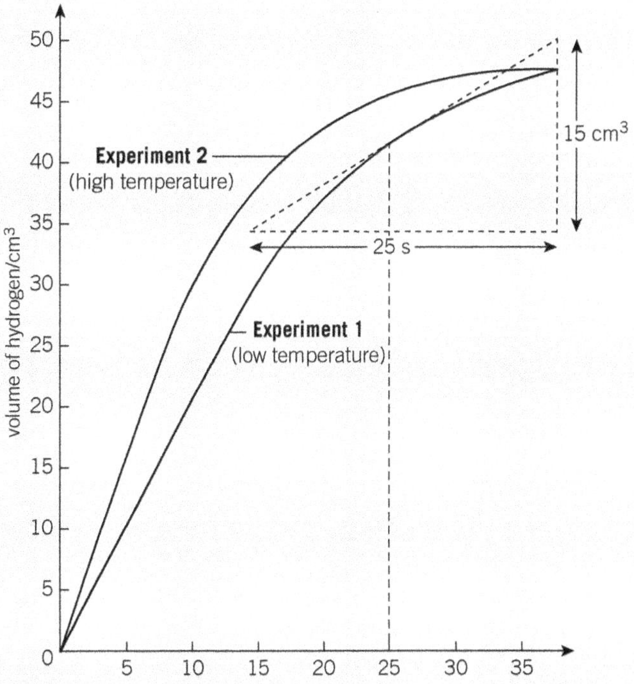

▲ **Figure 4** *Reaction rate graph*

Graphs

The graph in Figures 3 and 4 show the rate of reaction between hydrochloric acid and magnesium to produce hydrogen gas.

The gradient of a straight line section of a graph is found by dividing the length of the line A (vertical) by the length of line B (horizontal).

In the case of Experiment 1 above this tells us the rate at which hydrogen is produced, between 0 ands 10 seconds. It will have units:

Rate = $20\,cm^3/10\,s = 2\,cm^3/s$

or $2\,cm^3\,s^{-1}$

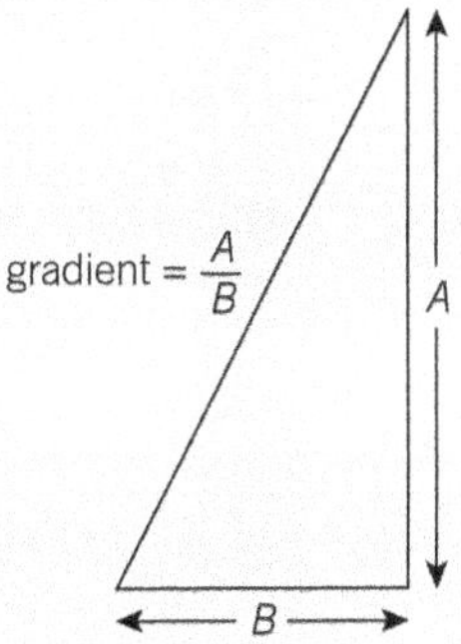

The steeper the line is, the greater the rate.

Question

5 Find the rate for Experiment 2 over the first 10 seconds.

Tangents

When you get to the part of the graph where the line starts to curve, the best you can do is to take a tangent at the point you are investigating and find the gradient of the tangent.

A tangent is a line drawn so that it just touches the graph line at one particular point.

The rate in Experiment 1, after 25 seconds is $\frac{15\,cm^3}{25\,s} = 0.6\ cm^3\,s^{-1}$

Question

6 Find the rate after 30 s in Experiment 2.

Logarithms

A logarithm, or log for short, is a mathematical function – the log of a number represents the power to which a base number (often ten) has to be raised to give the number. This is easy to do for numbers that are multiples of ten such as 100 or 10 000. 100 is 10^2, so the log to the base ten (written $\log_{10}$ or just log) of 100 is 2 and $\log_{10}$ of 10 000 is 4. Logs can also be negative numbers. The log of $\frac{1}{1000}$ is −3 as $\frac{1}{1000}$ is 10^{-3}. With other numbers you must use a calculator to find the log. Log_{10} 72.33 is 1.8593.

Make sure that you are confident using your calculator to find logs.

Question

7 What is the $\log_{10}$ of:

a 1000

b $\frac{1}{100}$

c 0.0001

d 48.2

e 0.037

You will need to use a calculator for parts d and e. You can go back from the log to the original number by using the antilog, or inverse log function, of the calculator. The antilog of 27 is 10^{27} (which shouldn't need a calculator) and the antilog of 4.33 is 21 379.6.

Make sure that you are confident to use your calculator to find antilogs (inverse logs).

Question

8 What is the antilog (inverse log) of:

a 3

b −2

c 14

d 8.2

e 0.37

You will need to use a calculator for parts d and e. The log function turns very large or very small numbers into more manageable numbers without losing the original number (which you can recover using the antilog function). So $\log_{10} 6\times10^{23} = 23.778$ and $\log_{10} 1.6\times10^{-19} = -18.80$.

This can be very useful in plotting graphs as it can allow numbers with a wide range in magnitudes to be fitted onto a reasonable size of graph. For example, the successive ionisation energies of sodium range from 496 to 159 079, whereas their logs range from just 2.695 to 5.201.

Another important use of logs in chemistry is the pH scale, which measures acidity, and depends on the concentration of hydrogen ions (H^+) in a solution. This can vary from around 5 $mol\,dm^{-3}$ to around $5\times10^{-15}\,mol\,dm^{-3}$ – an enormous range. If you use a log scale, this becomes 0.6989 to −14.301, which is a much more manageable range.

When multiplying numbers you add their logs and when dividing you subtract them. This is easy to see with numbers that are multiples of ten.

100 × 10 000 = 1000 000, that is, $10^2\times10^4 = 10^6$ and $\frac{10^4}{10^2} = 10^2$, but the same rules apply for more awkward numbers.

Geometry and trigonometry

Simple molecules adopt a variety of shapes. The most important of these are shown in Figure 5 with the relevant angles. When drawing representations of three-dimensional shapes, the convention is to show bonds coming out of the paper as wedges which get thicker as they come towards you. Bonds going into the paper are usually drawn as dotted lines or reverse wedges.

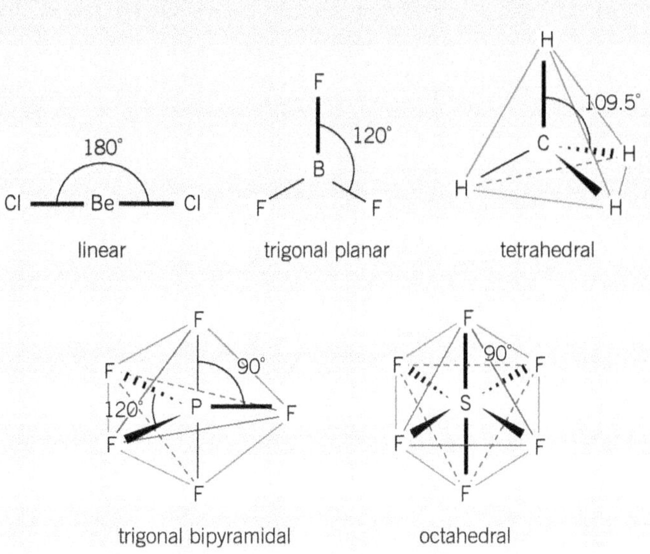

▲ **Figure 5** *Three-dimensional drawings of molecular shapes*

The three-dimensional shapes are based on geometrical solid figures as shown.

Table A Infrared

Bond	Wavenumber / cm^{-1}
C—H	2850–3300
C—C	750–1100
C=C	1620–1680
C=O	1680–1750
C—O	1000–1300
O—H (alcohols)	3230–3550
O—H (acids)	2500–3000
N—H	3300–3500

Table B ^1H NMR chemical shift data

Type of proton	δ/ppm
ROH	0.5–5.0
RCH$_3$	0.7–1.2
RNH$_2$	1.0–4.5
R$_2$CH$_2$	1.2–1.4
R$_3$CH	1.4–1.6
R—C(=O)—C—H	2.1–2.6
R—O—C—H	3.1–3.9
RCH$_2$Cl or Br	3.1–4.2
R—C(=O)—O—C—H	3.7–4.1
R,H C=C	4.5–6.0
R—C(=O)H	9.0–10.0
R—C(=O)O—H	10.0–12.0

Table C ^{13}C NMR chemical shift data

Type of carbon	δ/ppm
—C—C—	5–40
R—C—Cl or Br	10–70
R—C—C(=O)	20–50
R—C—N	25–60
—C—O—	50–90
C=C	90–150
R—C≡N	110–125
(benzene ring)	110–160
R—C(=O)	160–185
R—C(=O)	190–220

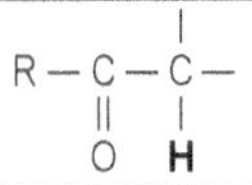

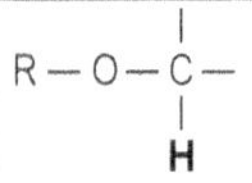

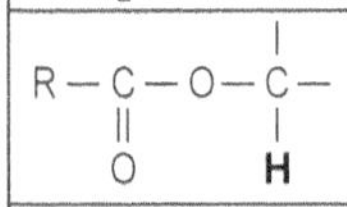

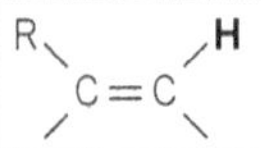

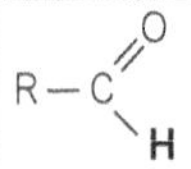

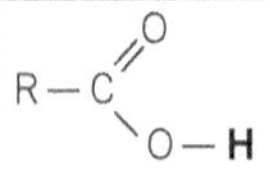

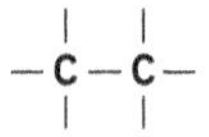

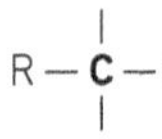

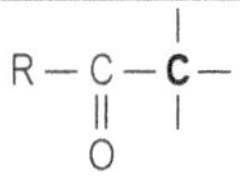

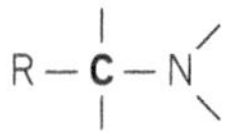

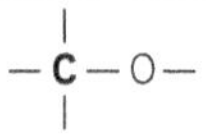

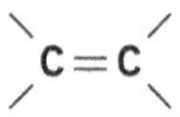

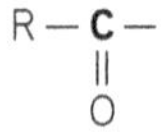

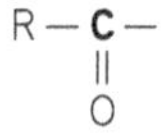

The Periodic Table of the elements

Key

relative atomic mass
atomic symbol
name
atomic (proton) number

1 (1)	2 (2)	(3)	(4)	(5)	(6)	(7)	(8)	(9)	(10)	(11)	(12)	3 (13)	4 (14)	5 (15)	6 (16)	7 (17)	0 (18)
							1.0 **H** hydrogen 1										4.0 **He** helium 2
6.9 **Li** lithium 3	9.0 **Be** beryllium 4											10.8 **B** boron 5	12.0 **C** carbon 6	14.0 **N** nitrogen 7	16.0 **O** oxygen 8	19.0 **F** fluorine 9	20.2 **Ne** neon 10
23.0 **Na** sodium 11	24.3 **Mg** magnesium 12											27.0 **Al** aluminium 13	28.1 **Si** silicon 14	31.0 **P** phosphorus 15	32.1 **S** sulfur 16	35.5 **Cl** chlorine 17	39.9 **Ar** argon 18
39.1 **K** potassium 19	40.1 **Ca** calcium 20	45.0 **Sc** scandium 21	47.9 **Ti** titanium 22	50.9 **V** vanadium 23	52.0 **Cr** chromium 24	54.9 **Mn** manganese 25	55.8 **Fe** iron 26	58.9 **Co** cobalt 27	58.7 **Ni** nickel 28	63.5 **Cu** copper 29	65.4 **Zn** zinc 30	69.7 **Ga** gallium 31	72.6 **Ge** germanium 32	74.9 **As** arsenic 33	79.0 **Se** selenium 34	79.9 **Br** bromine 35	83.8 **Kr** krypton 36
85.5 **Rb** rubidium 37	87.6 **Sr** strontium 38	88.9 **Y** yttrium 39	91.2 **Zr** zirconium 40	92.9 **Nb** niobium 41	95.9 **Mo** molybdenum 42	[98] **Tc** technetium 43	101.1 **Ru** ruthenium 44	102.9 **Rh** rhodium 45	106.4 **Pd** palladium 46	107.9 **Ag** silver 47	112.4 **Cd** cadmium 48	114.8 **In** indium 49	118.7 **Sn** tin 50	121.8 **Sb** antimony 51	127.6 **Te** tellurium 52	126.9 **I** iodine 53	131.3 **Xe** xenon 54
132.9 **Cs** caesium 55	137.3 **Ba** barium 56	138.9 **La*** lanthanum 57	178.5 **Hf** hafnium 72	180.9 **Ta** tantalum 73	183.8 **W** tungsten 74	186.2 **Re** rhenium 75	190.2 **Os** osmium 76	192.2 **Ir** iridium 77	195.1 **Pt** platinum 78	197.0 **Au** gold 79	200.6 **Hg** mercury 80	204.4 **Tl** thallium 81	207.2 **Pb** lead 82	209.0 **Bi** bismuth 83	[209] **Po** polonium 84	[210] **At** astatine 85	[222] **Rn** radon 86
[223] **Fr** francium 87	[226] **Ra** radium 88	[227] **Ac †** actinium 89	[261] **Rf** rutherfordium 104	[262] **Db** dubnium 105	[266] **Sg** seaborgium 106	[264] **Bh** bohrium 107	[277] **Hs** hassium 108	[268] **Mt** meitnerium 109	[271] **Ds** darmstadtium 110	[272] **Rg** roentgenium 111	[285] **Cn** copernicium 112	[286] **Uut** ununtrium 113	[289] **Fl** flerovium 114	[293] **Uup** ununpentum 115	[293] **Lv** Livermorium 116	[294] **Uus** ununseptium 117	[294] **Uuo** ununoctium 118

* **58 – 71** Lanthanides	140.1 **Ce** cerium 58	140.9 **Pr** praseodymium 59	144.2 **Nd** neodymium 60	144.9 **Pm** promethium 61	150.4 **Sm** samarium 62	152.0 **Eu** europium 63	157.3 **Gd** gadolinium 64	158.9 **Tb** terbium 65	162.5 **Dy** dysprosium 66	164.9 **Ho** holmium 67	167.3 **Er** erbium 68	168.9 **Tm** thulium 69	173.0 **Yb** ytterbium 70	175.0 **Lu** lutetium 71
† **90 – 103** Actinides	232.0 **Th** thorium 90	231.0 **Pa** protactinium 91	238.0 **U** uranium 92	237.0 **Np** neptunium 93	239.1 **Pu** plutonium 94	243.1 **Am** americium 95	247.1 **Cm** curium 96	247.1 **Bk** berkelium 97	252.1 **Cf** californium 98	[252] **Es** einsteinium 99	[257] **Fm** fermium 100	[258] **Md** mendelevium 101	[259] **No** nobelium 102	[260] **Lr** lawrencium 103

Glossary

A

Acid Brønsted–Lowry: a proton donor; Lewis: an electron pair acceptor.

Acid derivative An organic compound related to a carboxylic acid of formula RCOZ, where Z = —Cl, —NHR, —OR or —OCOR.

Activation energy The minimum energy that a particle needs in order to react; the energy (enthalpy) difference between the reactants and the transition state.

Aldehyde An organic compound with the general formula RCHO.

Alkaline earth metals The metals in Group 2 of the Periodic Table.

Alkane A hydrocarbon with C—C and C—H single bonds only, with the general formula C_nH_{2n+2}.

Allotropes Pure elements which can exist in different physical forms in which their atoms are arranged differently. For example, diamond, graphite, and buckminsterfullerene are allotropes of carbon.

Anaerobic respiration The process by which energy is released and new compounds formed in living things in the absence of oxygen.

Atom economy This describes the efficiency of a chemical reaction by comparing the total number of atoms in the product with the total number of atoms in the starting materials. It is defined by:

$$\%\ \text{Atom economy} = \frac{\text{mass of desired product}}{\text{total mass of reactants}} \times 100$$

Atomic orbital A region of space around an atomic nucleus where there is a high probability of finding an electron.

Avogadro constant The total number of particles in a mole of substance. Also called the **Avogadro number**. It is numerically equal to 6.022×10^{23}.

B

Base Brønsted–Lowry: a proton acceptor; Lewis: an electron pair donor.

Base peak The peak representing the ion of greatest abundance (the tallest peak) in a mass spectrum.

Bond dissociation enthalpy The enthalpy change required to break a covalent bond with all species in the gaseous state.

Buffer A solution that resists change of pH when small amounts of acid or base are added or on dilution.

C

Calorimeter An instrument for measuring the heat changes that accompany chemical reactions.

Carbocation An organic ion in which one of the carbon atoms has a positive charge.

Carbon-neutral A process, or series of processes, in which as much carbon dioxide is absorbed from the air as is given out.

Catalyst A substance that alters the rate of a chemical reaction but is not used up in the reaction.

Catalytic cracking The breaking, with the aid of a catalyst, of long-chain alkane molecules (obtained from crude oil) into shorter chain hydrocarbons (some of which are alkenes).

Chelation The process by which a multidentate ligand replaces a monodentate ligand in forming co-ordinate (dative) bonds to a transition metal ion.

Chemical feedstock The starting materials in an industrial chemical process.

Chiral This means 'handed'. A chiral molecule exists in two mirror image forms that are not superimposable.

Chiral centre An atom to which four different atoms or groups are bonded. The presence of such an atom causes the parent molecule to exist as a pair of nonsuperimposable mirror images.

Co-ordinate bond A covalent bond in which both the electrons in the bond come from one of the atoms forming the bond. (Also called a dative bond.)

Co-ordination number The number of ligand molecules bonded to a metal ion.

Covalent bonding Describes a chemical bond in a pair of electrons are shared between two atoms.

D

Dative covalent bonding Covalent bonding in which both the electrons in the bond come from one of the atoms in the bond. (Also called co-ordinate bonding.)

Delocalisation Describes the process by which electrons are spread over several atoms and help bond them together.

Delocalised Describes electrons that are spread over several atoms and help to bond them together.

Dipole–dipole force An intermolecular force that results from the attraction between molecules with permanent dipoles.

Displacement reaction A chemical reaction in which one atom or group of atoms replaces another in a compound, for example, Zn + CuO → ZnO + Cu.

Displayed formula The formula of a compound drawn out so that each atom and each bond is shown.

Disproportionation Describes a redox reaction in which the oxidation number of some atoms of a particular element increases and that of other atoms of the same element decreases.

Dynamic equilibrium A situation in which the composition of a constant concentration reaction mixture does not change because both forward and backward reactions are proceeding at the same rate.

E

Electron density The probability of electrons being found in a particular volume of space.

Electron pair repulsion theory A theory which explains the shapes of simple molecules by assuming that pairs of electrons around a central atom repel each other and thus take up positions as far away as possible from each other in space.

Electronegativity The power of an atom to attract the electrons in a covalent bond.

Electrophile An electron-deficient atom, ion, or molecule that takes part in an organic reaction by attacking areas of high electron density in another reactant. It is an electron pair acceptor (electron liking)

Electrophilic addition A reaction in which a carbon–carbon double bond is saturated, by the carbon–carbon double bond attacking an electrophile.

Electrostatic forces The forces of attraction and repulsion between electrically charged particles.

Elimination A reaction in which an atom or group of atoms is removed from a reactant.

Empirical formula The simplest whole number ratio of atoms of each element in a compound.

Enantiomer One of a pair of nonsuperimposable mirror image isomers.

Endothermic Describes a reaction in which heat energy is taken in as the reactants change to products, the temperature therefore drops.

End point The point in a titration when the volume of reactant added just causes the colour of the indicator to change.

Enthalpy change A measure of heat energy given out or taken in when a chemical or physical change occurs at constant pressure.

Enthalpy diagrams Diagrams in which the enthalpies (energies) of the reactants and products of a chemical reaction are plotted on a vertical scale to show their relative levels.

Entropy A numerical measure of disorder in a chemical system.

Equilibrium mixture The mixture of reactants and products formed when a reversible reaction is allowed to proceed in a closed container until no further change occurs. The forward and backward reactions are still proceeding but at the same rate.

Equivalence point The point in a titration at which the reaction is just complete.

Exothermic Describes a reaction in which heat energy is given out as the reactants change to products – the temperature therefore rises.

F

Fatty acid A long-chain carboxylic acid.

Fingerprint region The area of an infrared spectrum below about $1500\,cm^{-1}$. It is caused by complex vibrations of the whole molecule and is characteristic of a particular molecule.

Fraction A mixture of hydrocarbons collected over a particular range of boiling points during the fractional distillation of crude oil.

Free radical A chemical species with an unpaired electron – usually highly reactive.

Functional group An atom or group of atoms in an organic molecule which is responsible for the characteristic reactions of that molecule.

G

Group A vertical column of elements in the Periodic Table. The elements have similar properties because they have the same outer electron arrangement.

H

Half equation An equation for a redox reaction which considers just one of the species involved and shows explicitly the electrons transferred to or from it.

Homologous series A set of organic compounds with the same functional group. The compounds differ in the length of their hydrocarbon chains.

Hydration A reaction in which water is added.

Hydrogen bonding A type of intermolecular force in which a hydrogen atom ($H^{\delta+}$) interacts with a more electronegative atom with a $\delta-$ charge.

Hydrolysis A reaction of a compound or ion with water.

I

Incomplete combustion A combustion reaction in which there is insufficient oxygen for all the carbon in the fuel to burn to carbon dioxide. Carbon monoxide and/or carbon (soot) are formed.

Inductive effect The electron-releasing effect of alkyl groups such as $—CH_3$ or $—C_2H_5$.

Ionic bonding Describes a chemical bond in which an electron or electrons are transferred from one atom to another, resulting in the formation of oppositely charged ions with electrostatic forces of attraction between them.

Ionisation energy The energy required to remove a mole of electrons from a mole of isolated gaseous atoms or ions.

Isomer One of two (or more) compounds with the same molecular formula but different arrangement of atoms in space.

K

Ketone An organic compound with the general formula R_2CO in which there is a C=O double bond.

L

Lattice A regular three-dimensional arrangement of atoms, ions, or molecules.

Leaving group In an organic substitution reaction, the leaving group is an atom or group of atoms that is ejected from the starting material, normally taking with it an electron pair and forming a negative ion.

Ligand An atom, ion, or molecule that forms a co-ordinate (dative) bond with a transition metal ion using a lone pair of electrons.

Lone pair A pair of electrons in the outer shell of an atom. The pair is not involved in bonding.

M

Maxwell–Boltzmann distribution The distribution of energies (and therefore speeds) of the molecules in a gas or liquid.

Mean bond enthalpy The average value of the bond dissociation enthalpy for a given type of bond taken from a range of different compounds.

Metallic bonding Describes a chemical bond in which outer electrons are delocalised within the lattice of metal ions.

Mole A quantity of a substance that contains the Avogadro number (6.022×10^{23}) of particles (e.g., atoms, molecules or ions).

Molecular formula A formula that tells us the actual numbers of atoms of each different element that make up a molecule of a compound.

Molecular ion In mass spectrometry this is a molecule of the sample which has been ionised but which has not broken up during its flight through the instrument.

Monomer A small molecule that combines with many other monomers to form a polymer.

N

Nucleons Protons and neutrons – the sub-atomic particles found in the nuclei of atoms.

Nucleophile A negative ion or molecule that is able to donate a pair of electrons and takes part in an organic reaction by attacking an electron-deficient area in another reactant.

Nucleophilic substitution An organic reaction in which a molecule with a partially positively charged carbon atom is attacked by a reagent with a negative charge or partially negatively charged area (a nucleophile). It results in the replacement of one of the groups or atoms on the original molecule by the nucleophile.

Nucleus The tiny, positively charged centre of an atom composed of protons and neutrons.

O

Optical isomer Pairs of molecules that are non-superimposable mirror images.

Order of reaction In the rate expression, this is the sum of the powers to which the concentrations of all the species involved in the reaction are raised. If rate = $k[A]^a[B]^b$, the overall order of the reaction is $a + b$.

Oxidation A reaction in which an atom or group of atoms loses electrons.

Oxidation state The number of electrons lost or gained by an atom in a compound compared to the uncombined atom. It forms the basis of a way of keeping track of redox (electron transfer) reactions. Also called oxidation number.

Oxidising agent A reagent that oxidises (removes electrons from) another species.

P

Percentage yield In a chemical reaction this is the actual amount of product produced divided by the theoretical amount (predicted from the chemical equation) expressed as a percentage.

Period A horizontal row of elements in the Periodic Table. There are trends in the properties of the elements as we cross a period.

Periodicity The regular recurrence of the properties of elements when they are arranged in atomic number order as in the Periodic Table.

pH A scale for measuring acidity and alkalinity. $pH = -\log_{10} [H^+]$ in a solution.

Polar Describes a molecule in which the charge is not symmetrically distributed so that one area is slightly positively charged and another slightly negatively charged.

Polarised This describes an atom or ion where the distribution of charge around it is distorted from the spherical.

Positive inductive effect Describes the tendency of some atoms or groups of atoms to release electrons via a covalent bond.

Proton number The number of protons in the nucleus of an atom; the same as the atomic number.

Protonated Describes an atom, molecule, or ion to which a proton (an H^+ ion) has been added.

R

Racemate A mixture of equal amounts of two optical isomers of a chiral compound. It is optically inactive.

Rate constant The constant of proportionality in the rate expression.

Rate-determining step The slowest step in the reaction mechanism. It governs the rate of the overall reaction.

Rate expression A mathematical expression showing how the rate of a chemical reaction depends on the concentrations of various chemical species involved.

Reaction mechanism The series of simple steps that lead from reactants to products in a chemical reaction.

Redox reaction Short for reduction–oxidation reaction, it describes reactions in which electrons are transferred from one species to another.

Reducing agent A reagent that reduces (adds electrons to) another species.

Reduction A reaction in which an atom or group of atoms gain electrons.

Relative atomic mass, A_r

$$A_r = \frac{\text{average mass of 1 atom of an element}}{\frac{1}{12}\text{mass of 1 atom of } ^{12}\text{C}}$$

Relative formula mass M_r

$$M_r = \frac{\text{average mass of an entity}}{\frac{1}{12}\text{mass of 1 atom of } ^{12}\text{C}}$$

Relative molecular mass M_r

$$M_r = \frac{\text{average mass of a molecule}}{\frac{1}{12}\text{mass of 1 atom of } ^{12}\text{C}}$$

S

Saturated hydrocarbon A compound containing only hydrogen and carbon with only C—C and C—H single bonds, i.e., one to which no more hydrogen can be added.

Specific heat capacity c The amount of heat needed to raise the temperature of 1 g of substance by 1 K.

Spectator ions Ions that are unchanged during a chemical reaction, that is, they take no part in the reaction.

Standard molar enthalpy change of combustion $\Delta_c H^\ominus$ The enthalpy change when 1 mole of a substance is completely burned in oxygen with all reactants and products in their standard states (298 K and 100 kPa).

Standard molar enthalpy change of formation $\Delta_f H^\ominus$ The enthalpy change when 1 mole of substance is formed from its elements with all reactants and products in their standard states (298 K and 100 kPa).

Stereoisomer Isomers with the same molecular formula and the same structure, but a different position of atoms in space.

Stoichiometry Describes the simple whole number ratios in which chemical species react.

Strong acid An acid that is fully dissociated into ions in solution.

Strong nuclear force The force that holds protons and neutrons together within the nucleus of the atom.

Structural formula A way of writing the formula of an organic compound in which bonds are not shown but each carbon atom is written separately with the atoms or groups of atoms attached to it.

Structural isomer Isomers with the same molecular formula but a different structure.

T

Thermochemical cycle A sequence of chemical reactions (with their enthalpy changes) that convert a reactant into a product. The total enthalpy change of the sequence of reactions will be the same as that for the conversion of the reactant to the product directly (or by any other route).

Triglyceride An ester formed between glycerol (propane-1,2,3-triol) and three fatty acid molecules.

V

van der Waals force A type of intermolecular force of attraction that is caused by instantaneous dipoles and acts between all atoms and molecules.

W

Weak acid An acid that is only slightly dissociated into ions in solution.

Answers to summary questions

1.1

1 a i proton, neutron ii proton, neutron
iii proton, electron iv neutron v electron
b Because they have opposite charges of the same size and the atom is neutral.

1.2

1 a 1 proton, 1 neutron, 1 electron
b 1 proton, 2 neutrons, 1 electron
2 X and Z
3

Element	W	X	Y	Z
a Number of protons	15	7	8	7
b Mass number	31	14	16	15
c Number of neutrons	16	7	8	8

Carbon dating

1 17 190 years (three half lives)
2 Not necessarily – it tells us when the tree from which the wood of bowl was made died. The bowl may have been made later than this and would therefore not be so old.

1.3

1 Because they have lost one (or more) electrons (which have a negative charge)
2 They are attracted by a negatively charged plate.
3 The ions pass through a series of holes or slits.
4 Ions with the smallest m/z.
5 72.63
6 63.6

Relative abundance

Probability of chlorine molecule being
$^{35}Cl—^{35}Cl$ ($m/z = 70$) is $\frac{3}{4} \times \frac{3}{4} = \frac{9}{16}$

Probability of chlorine molecule being
$^{37}Cl—^{37}Cl$ ($m/z = 74$) is $\frac{1}{4} \times \frac{1}{4} = \frac{1}{16}$

Probability of chlorine molecule being
$^{35}Cl—^{37}Cl$ ($m/z = 72$) is $\frac{3}{4} \times \frac{1}{4} = \frac{3}{16}$

Probability of chlorine molecule being
$^{37}Cl—^{35}Cl$ ($m/z = 72$) is $\frac{1}{4} \times \frac{3}{4} = \frac{3}{16}$

Probability of ions of $m/z = 72$ can be added together:
$\frac{3}{16} + \frac{3}{16} = \frac{6}{16}$

1.4

1

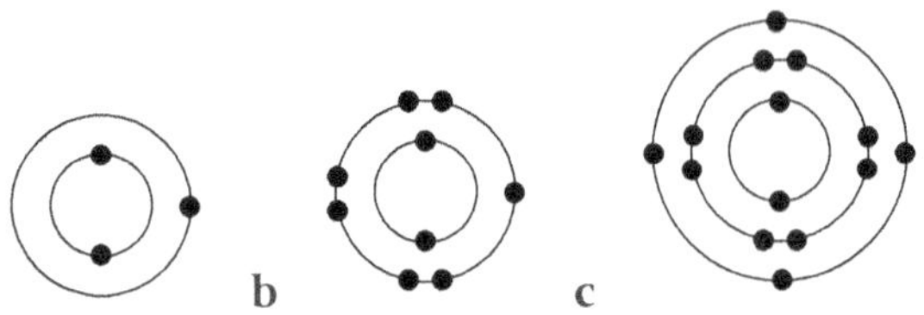

2 a 2,2 b 2,8,3 c 2,8,8
3 A^{2+}, C^-, E^+; B and D are atoms
A: Mg, B: He, C: Cl, D: Ne, E: Li

1.5

1 a $1s^2\ 2s^2\ 2p^6\ 3s^2\ 3p^3$ b [Ne] $3s^2\ 3p^3$
2 a i $1s^2\ 2s^2\ 2p^6\ 3s^2\ 3p^6$ ii $1s^2\ 2s^2\ 2p^6$
b [Ar], [Ne]

1.6

1 The second electron is removed from a positively charged ion whilst the first is removed from a neutral atom. More energy is needed to overcome the additional attractive force and so the second ionisation energy is higher.

2

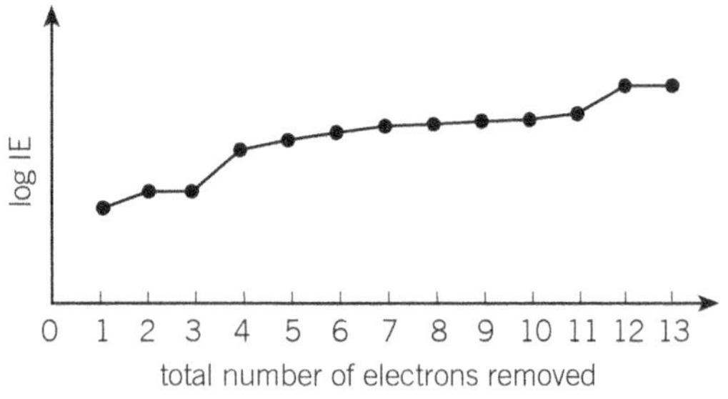

3 a Group 4
b The large jump in ionisation energy comes after the removal of the fourth electron showing that there are four electrons in the outer shell.

2.1

1 a 16.0 b 106.0 c 58.3 d 132.1
2 Many answers possible such that the relative atomic masses add up to 16. For example, 16 hydrogen atoms; or one carbon atom and four hydrogen atoms.
3 a 2 b 0.05 c 0.1
4 4g O_2 5 11g CO_2

Answers to the Practice Questions and Section Questions are available at
www.oxfordsecondary.com/oxfordaqaexams-alevel-chemistry

2.2

1 a 1 mol dm^{-3} b 0.125 mol dm^{-3}
c 10 mol dm^{-3}

2 a 0.002 b 0.025 c 0.05

3 a 58.5 b 0.004 c 0.016

2.3

1 a approximately 8.75×10^6
b The same as part **a**.

2 a 50 360 cm^3 b 113 000 Pa

3 1.94

4 The same as in question **3**. The same number of moles of any gas has the same volume under the same conditions of temperature and pressure.

2.4

1 a H_2SO_4 sulfuric acid
b $Ca(OH)_2$ calcium hydroxide
c $MgCl_2$ magnesium chloride

2 a 0.16 mol Mg, 0.16 mol O b MgO

3 a CH_2 b CHCl c CH

4 $C_2H_6O_2$ 5 C_3H_6O

6 a CH b C_6H_6

Finding the empirical formula of copper oxide

1 The flame burns off excess hydrogen to prevent it entering the laboratory.
2 The green colour is characteristic of copper ions.
3 Water droplets form due to the reaction of oxygen from the copper oxide reacting with the hydrogen. They condense while the tube is still cool.
4 This prevents oxygen from the air entering the tube and reacting with the copper while it is still hot and converting some of it back to copper oxide.

Erroneous results

1 $\frac{0.635}{63.5} = 0.01$ moles copper
0.735 − 0.635 = 0.100 g oxygen
$\frac{0.100}{16.0} = 0.006\,25$ moles oxygen
This gives the formula $Cu_{1.6}O$

2 The student had not allowed the reaction to go to completion, or they had allowed air back into the reduction tube while the copper was still hot.

Another oxide of copper

1 mass of copper = 1.27 g
mass of oxygen = 0.16 g

2 moles of copper = 0.02
moles of oxygen = 0.01

3 Cu_2O

2.5

1 a $2Mg + O_2 \rightarrow 2MgO$
b $Ca(OH)_2 + 2HCl \rightarrow CaCl_2 + 2H_2O$
c $Na_2O + 2HNO_3 \rightarrow 2NaNO_3 + H_2O$

2 0.25 mol dm^{-3}

3 a Yes, there are 0.107 mol Mg. This would be enough to react with 0.0.214 mol HCl, but there is only 0.100 mol HCl.
b 1238 cm^3

4 a i $H_2SO_4(aq) + 2NaOH(aq) \rightarrow Na_2SO_4(aq) + 2H_2O(l)$
ii $2H^+ + SO_4^{2-} + 2Na^+ + 2OH^- \rightarrow 2Na^+ + SO_4^{2-} + 2H_2O$
b Na^+ and SO_4^{2-}

2.6

1 CaCOOO → CaO + COO 56.0%

2 79.8%

3 100% All the reactants are incorporated into the desired product.

4 a 1 mol b 5.6 g c 64.3%

3.1

1 **b** and **c**, they are both metal/non-metal compounds

2 Because they have strong electrostatic attraction between the ions that extends through the whole structure.

3 When they are molten or in aqueous solution.

4 a

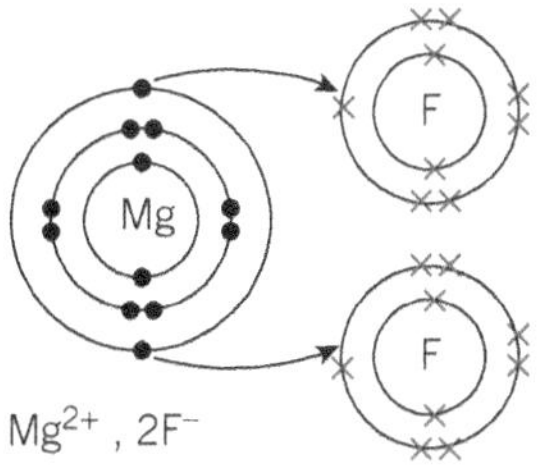

b

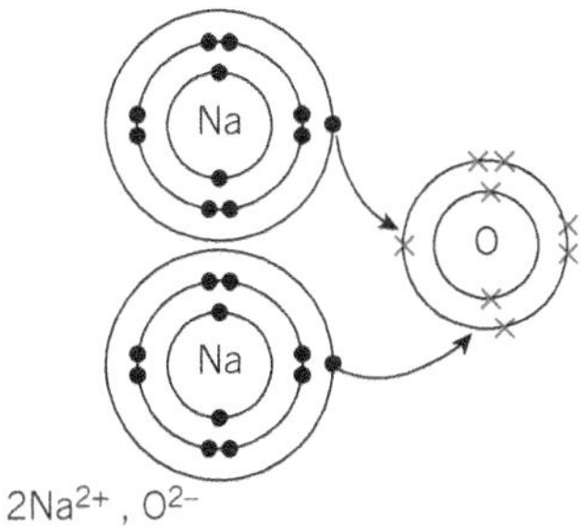

5 MgF_2; Na_2O 6 neon

Noble gas compounds

In xenon, outer electrons are further from the nucleus and more sub-shells means the outer electrons experience more shielding. Less energy required to remove an outer electron.

3.2

1 A pair of electrons shared between two non-metal atoms (usually) that holds the atoms together.

2 **b** and **d**, they are both non-metal/non-metal compounds

3

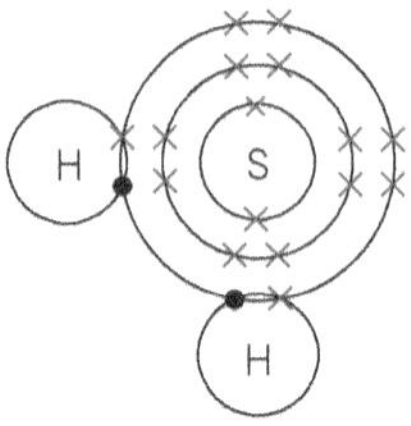

4

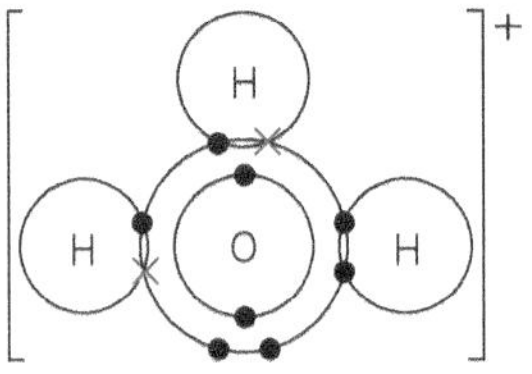

3.3

1 For example: metals conduct heat and electricity well, non-metals do not; metals are shiny, malleable and ductile, non-metals are not.

2 $1s^2\ 2s^2\ 2p^6\ 3s^2\ 3p^6\ 4s^2$

3 The two 4s electrons. 4 2

5 **a** Sodium would have a lower melting temperature because there are fewer electrons in the delocalised system and the charge on the ions is smaller.

b Magnesium would be stronger as there are more electrons in the delocalised system and the charge on the ions is greater.

3.4

1 **a** In a molecular crystal, there is strong covalent bonding between the atoms within the molecules but weaker intermolecular forces between the molecules. In a macromolecular crystal, all the atoms within the crystal are covalently bonded.

b Macromolecular crystals have higher melting and boiling temperatures.

2 The layers of carbon atoms are held together by weak van der Waals forces which allow the layers to slide over one another. They may also allow other molecules such as oxygen to penetrate between the layers.

3 Electricity is conducted via the delocalised electrons that spread along the layers of carbon atoms. Graphite conducts well along the layers but poorly at right angles to them. Metals conduct well in all directions.

4 Both have giant structures in which covalent bonding occurs between many atoms.

5 **a** A, C, D **b** B **c** A **d** B, D
e C **f** D

3.5

1

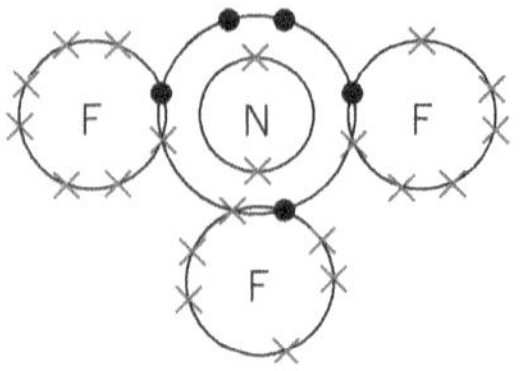

Eight electrons gives four pairs of electrons. This is a triangular pyramid.

2 BF_3 has three electron pairs in its outer shell and is trigonal planar. NF_3 has four electron pairs in its outer shell and its shape is based on that of a tetrahedron (the bond angle is 'squeezed down' by a couple of degrees because one of the electron pairs is a lone pair).

3

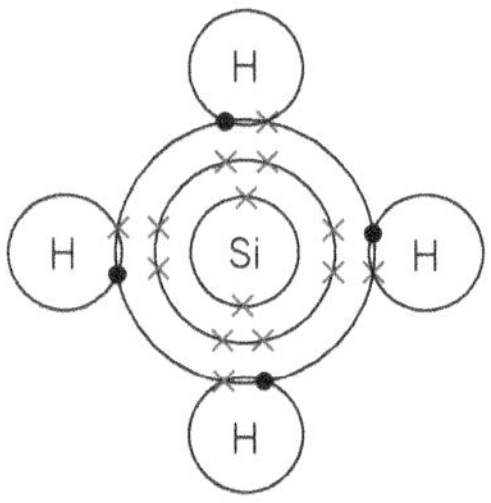

The shape is perfectly tetrahedral.

4 109.5°

5 Angular. It is similar to a water molecule.

3.6

1 Fluorine is a smaller atom and when it forms a covalent bond, the shared electrons are closer to the nucleus.

2 $H^{\delta+}\text{–}Cl^{\delta-}$

3 **a** and **b** In both cases, the two atoms in the molecule are the same and therefore the electrons in the bond are equally shared.

4 **a** H—N < H—O < H—F

b The order of the polarity is the same as the order of electronegativity of the second atom.

3.7

1 He, Ne, Ar, Kr. The van der Waals forces increase as the number of electrons in the atom increases.

2 H_2. It cannot have a permanent dipole because both atoms are the same.

3 Hexane (C_6H_{14}) is a larger molecule than butane (C_4H_{10}) and so has more electrons. This means that there are larger van der Waals forces between the molecules.

4 Because covalent bonds are localised between the two atoms that they bond and there is little attraction between the individual molecules.

5
$$\overset{\delta+}{H}\!-\!\overset{\delta-}{Br}\!\longleftarrow\!\cdots\!\longrightarrow\!\overset{\delta+}{H}\!-\!\overset{\delta-}{Br}$$
 attraction

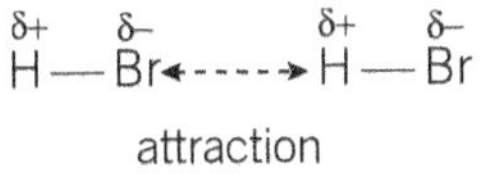

6 HBr

7 a There is no sufficiently electronegative atom.

 b There is no hydrogen atom.

8

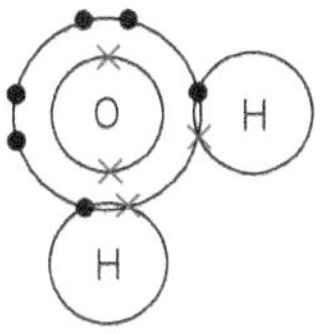

 a 2

 b 2

 c A hydrogen bond requires both a lone pair of electrons on an atom of N, O, or F and a hydrogen atom. In water there are two lone pairs and two hydrogen atoms, allowing the formation of two hydrogen bonds. In ammonia, although there are three hydrogen atoms, there is only one lone pair of electrons on the N. This means that only one hydrogen bond can form.

4.1

1 445 kJ **2** endothermic

3 It is the reverse of the reaction in question **1**.

4 1.6 g

The energy values of fuels

1 $CH_3OH + 2O_2 \rightarrow CO_2 + 1\tfrac{1}{2}H_2O$

2 No carbon dioxide formed.

3 Carbon dioxide is a greenhouse gas.

4.2

1 a change **b** enthalpy (heat energy)

 c That the enthalpy change is measured at 298 K.

 d That heat is given out. **e** exothermic

 f

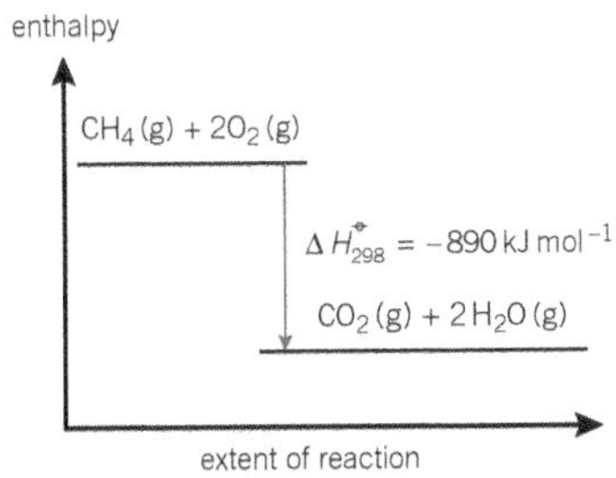

4.3

1 −672 kJ mol⁻¹

2 a −46.2 kJ mol⁻¹

b It will be numerically smaller (less negative).

c This is caused by heat loss.

3 a The enthalpy change.

 b The mass of water (or other substance in which the heat is collected).

 c The specific heat capacity of the water (or other substance).

 d The temperature change.

4.4

1 a −70 kJ mol⁻¹ **b** −217 kJ mol⁻¹

 c −97 kJ mol⁻¹ **d** −191 kJ mol⁻¹

 e +301 kJ mol⁻¹

4.5

1 a via $\Delta_f H^\ominus$ −85 kJ mol⁻¹

 b via $\Delta_c H^\ominus$ −85 kJ mol⁻¹

4.6

1 a −70 kJ mol⁻¹ **b** 217 kJ mol⁻¹

 c −97 kJ mol⁻¹ **d** −191 kJ mol⁻¹

 e +301 kJ mol⁻¹

4.7

1 a

$$H\!-\!\underset{\underset{H}{|}}{\overset{\overset{H}{|}}{C}}\!-\!\underset{\underset{H}{|}}{\overset{\overset{H}{|}}{C}}\!-\!H + Br\!-\!Br \longrightarrow H\!-\!\underset{\underset{H}{|}}{\overset{\overset{H}{|}}{C}}\!-\!\underset{\underset{H}{|}}{\overset{\overset{H}{|}}{C}}\!-\!Br + H\!-\!Br$$

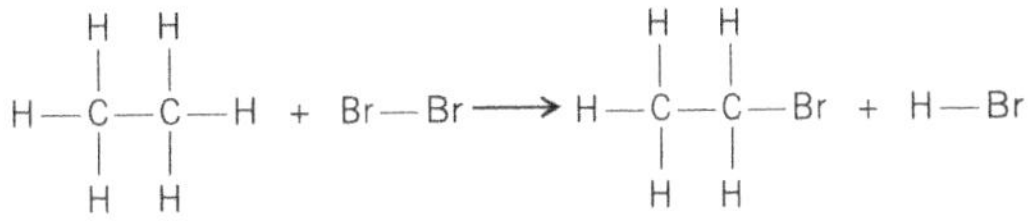

2 a 1 × C—C, 6 × C—H, 1 × Br—Br

 b 3018 kJ mol⁻¹

3 a 1 × C—C, 5 × C—H, 1 × C—Br, 1 × H—Br

 b 3063 kJ mol⁻¹

4 45 kJ mol⁻¹

5 a −45 kJ mol⁻¹ **b** exothermic

5.1

1 temperature, concentration of reactants, surface area of solid reactants, pressure of gaseous reactants, catalyst

2 a reactants **b** products

 c transition state or activated complex

 d activation energy

3 a exothermic

b The products have less energy (enthalpy) than the reactants.

5.2

1 a Fraction of particles with energy *E*.

b energy *E*

c The number of particles with enough energy to react.

d moves to the right e no change

5.3

1 a A = enthalpy, B = extent of reaction, C = transition state with catalyst, D = transition state without catalyst, R = reactants, P = products

b The activation energies without and with catalyst respectively

c Exothermic

6.1

1 a true b false c true d false

2 They are the same.

6.2

1 a Yes. This is a gas phase reaction with different numbers of particles on each side of the arrow.

b No. This is not a gas phase reaction.

c No. This is a gas phase reaction with the same number of particles on each side of the arrow.

2 a move to the left b no change

c Equilibrium would be set up more quickly but the final position would be unchanged.

d High pressure. It would force the equilibrium in the direction of fewest particles.

6.3

1 a $K_c = \frac{[C]_{eqm}}{[A]_{eqm}[B]_{eqm}}$ b $K_c = \frac{[C]_{eqm}}{[A]^2_{eqm}[B]_{eqm}}$

c $K_c = \frac{[C]^2_{eqm}}{[A]^2_{eqm}[B]^2_{eqm}}$

2 a $mol^{-1}\,dm^3$ b $mol^{-2}\,dm^6$ c $mol^{-2}\,dm^6$

3 a $K_c = 3.46$ (no units)

b it cancels out

c further to the left

6.4

1 a 1.01 mol b 1.067 mol c 2.067 mol

6.5

1 a increase b no change c no change

2 More water (and ethyl ethanoate) will be produced to oppose the change so that the equilibrium is maintained and K_c remains constant.

3 a increases b does not change

c increases d does not change

e does not change

7.1

1 a bromine b calcium

c calcium d bromine

e $Ca \rightarrow Ca^{2+} + 2e^-$ $Br_2 + 2e^- \rightarrow 2Br^-$

f bromine g calcium

7.2

1 a Pb +2, Cl −1 b C +4, Cl −1

c Na +1, N +5, O −2

2 −2 before and after

3 before 0, after −2

4 before +2, after +3

5 a +5 b +5 c −3

7.3

1 a +2 0 +3 −1

$Fe^{2+} + \frac{1}{2}Cl_2 \rightarrow Fe^{3+} + Cl^-$

b Iron, because its oxidation number has increased.

c Chlorine, because its oxidation number has decreased.

d $Fe^{2+} \rightarrow Fe^{3+} + e^-$ $\frac{1}{2}Cl_2 + e^- \rightarrow Cl^-$

2 a i $3Cl_2 + 6NaOH \rightarrow NaClO_3 + 5NaCl + 3H_2O$

ii $Sn + 4HNO_3 \rightarrow SnO_2 + 4NO_2 + 2H_2O$

b i $2\frac{1}{2}Cl_2 + 5e^- \rightarrow 5Cl^-$

ii $\frac{1}{2}Cl_2 + 6OH^- \rightarrow ClO_3^- + 5e^- + 3H_2O$

ii $Sn + 2H_2O \rightarrow SnO_2 + 4H^+ + 4e^-$

$4HNO_3 + 4H^+ + 4e^- \rightarrow 4NO_2 + 4H_2O$

8.1

1 a i any two from Br, K, Fe

ii Br and Cl or K and Cs

iii Br, Cl

2 b i Fe ii K or Cs

3 a Xe b Ge c Sr d Ge or Xe e W

8.2

1 a left b right

2 a 0 (8 or 18) b 4

3 A: Na, B: Si

8.3

1 decrease 2 increase 3 increase

4 Because they have the highest nuclear charge.

The discovery of argon

1 5

2 A_r of argon is 50% greater than that of nitrogen and 1 mole of any gas occupies approximately the same volume.

3 Pass the air over a heated metal, e.g., copper.

4 42

5 It did not fit into the Periodic Table as it was understood then.

8.4

1 a $1s^22s^2$ b $1s^22s^22p^1$

2 a 2s b 2p

3 Level 2p is of higher energy than 2s.

9.1

1 a +2 b They all lose their two outer electrons when they form compounds.

2 The outer electrons become further from the nucleus and are thus more easily lost.

3 Electrons are transferred from calcium to chlorine.

4 $\overset{0}{Ca}(s) + 2\overset{+1\,-2}{H_2O}(l) \rightarrow \overset{+2\,-2\,+1}{Ca(OH)_2}(aq) + \overset{0}{H_2}(g)$

5 a more vigorous

b Less vigorous. The Group 2 metals become more reactive as we descend the group due, in part, to the fact that the outer electrons become further from the nucleus and are thus more easily lost.

6 a most soluble

b Least soluble. These are the trends found in the rest of the group.

Lime kilns

1 14 tonnes

2 Carbon dioxide produced in the reaction and by burning the fuel.

3 To increase its surface area.

4 11.9 tonnes

5 a 1 b 1

6 a 56 g b 74 g

7 Lime is more efficient.

8 cost, safety, etc.

The extraction of titanium

1 $2Fe_2O_3(s) + 3C(s) \rightarrow 3CO_2(g) + 4Fe(s)$

2 It is liquid – suggests $TiCl_4$ exists as covalently bonded molecules.

3 $TiO_2(s) + 2C(s) + 2Cl^2(g) \rightarrow Ti\,Cl_4(g) + 2CO_2(g)$

4 Carbon has been oxidised (0 to +4). Chlorine has been reduced (0 to −1)

10.1

1 a solid, very dark colour

b largest atom

c least electronegative

2 These properties can be predicted by extrapolating the trends observed with the halogens from F to I.

3 a approximately 600 K

b It has the most electrons and therefore the strongest van der Waals forces.

10.2

1 a mixture ii only

b Chlorine is a better oxidising agent than iodine and bromine. Therefore it can displace iodine from iodide salt, but bromine cannot displace chlorine from chloride salts.

c $Cl_2(aq) + 2NaI(aq) \rightarrow I_2(s) + 2NaCl(aq)$

Extraction of iodine from kelp

1 It is an oxidation – the oxidation state of the iodine atom goes from −1 to 0.

2 It is added to prevent thyroid problems which can be caused by lack of iodine in the diet.

10.3

1 a It has the least reducing power.

b The F^- ion is the smallest halide ion which means that it is hardest for it to lose an electron (which comes from an outer shell close to the nucleus).

c $NaF(s) + H_2SO_4(l) \rightarrow NaHSO_4(s) + HF(g)$

d No, the oxidation state of the fluorine remains as −1.

2 a Formation of a cream precipitate.

b $AgNO_3(aq) + NaBr(aq) \rightarrow NaNO_3(aq) + AgBr(s)$

c The precipitate would dissolve.

d To remove ions such as carbonate and hydroxide which would also produce a precipitate.

e Chloride ions (from HCl) and sulfate ions (from H_2SO_4) would also form precipitates with silver ions.

f Silver fluoride is soluble in water and does not form a precipitate.

10.4

1 a $Br_2(g) + H_2O(l) \rightarrow HBrO(aq) + HBr(aq)$

b $Br_2(g) + 2NaOH(aq) \rightarrow NaBrO\ (aq) + NaBr(aq) + H_2O(l)$

2 To kill microorganisms and make the water safe to drink.

3 a hydrochloric acid and oxygen

b 0 +1 −2 +1 −1 0

$2Cl_2(g) + 2H_2O(l) \rightarrow 4HCl(aq) + O_2(g)$

c oxygen has been oxidised

d chlorine has been reduced

e chlorine

f oxygen

11.1

1 a 0.4 mol b 1.0 mol c C_2H_5 d C_4H_{10}

e $CH_3CH_2CH_2CH_3$

f

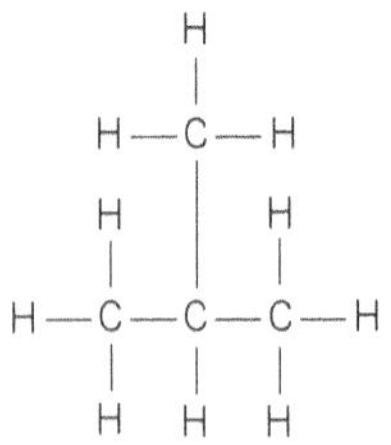

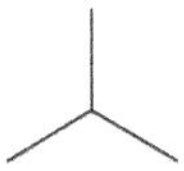

From inorganic to organic

1 A reaction in which the reactant and product have the same molecular formula but a different arrangement of atoms.

2 100% – there is just one starting material and one product.

11.2

1 a 1-chloropropane b pentane

c pent-2-ene d 2-methylpentane

2 a

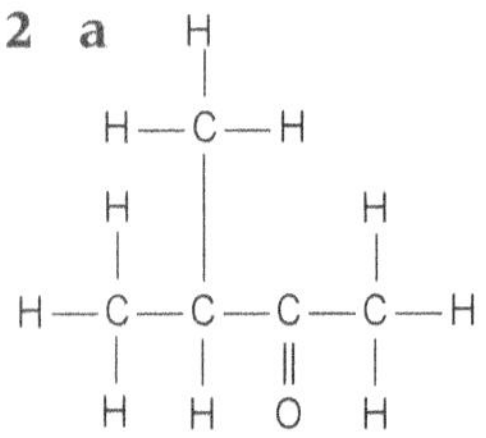

b

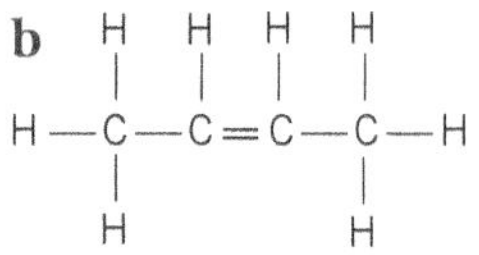

c

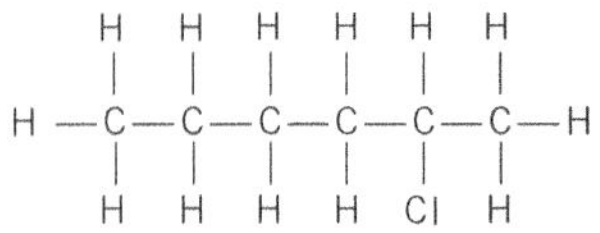

d

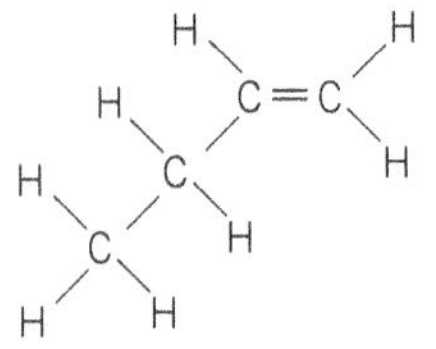

What's in a name?

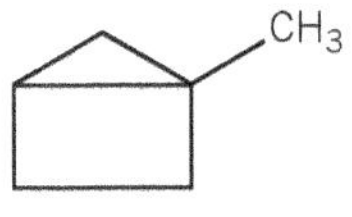

eave-methylhousane

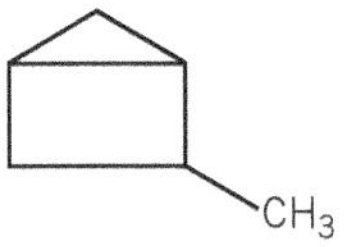

floor-methylhousane

11.3

1 a B b A c C

2 a

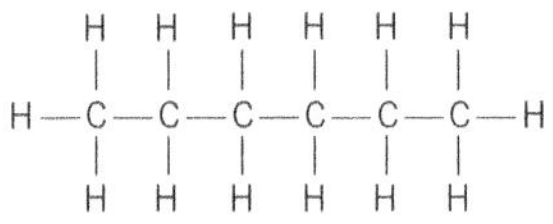

$CH_3CH_2CH_2CH_2CH_2CH_3$

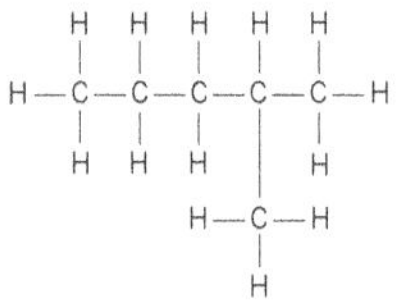

$CH_3CH_2CH_2CH(CH_3)CH_3$

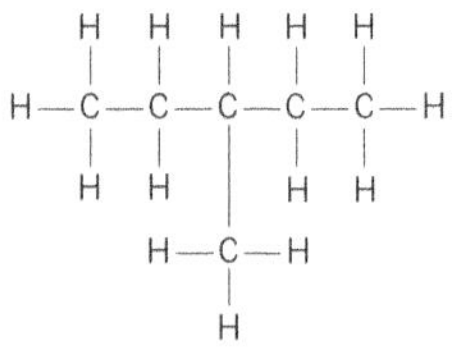

$CH_3CH_2CH(CH_3)CH_2CH_3$

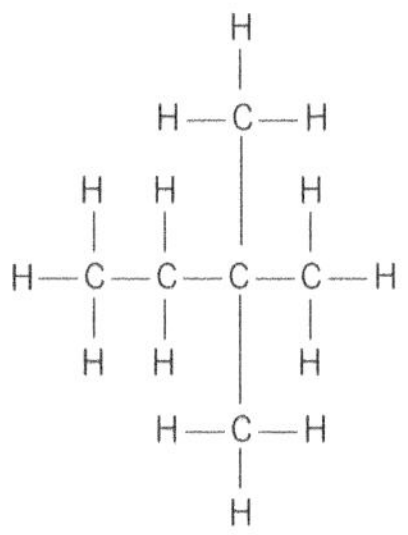

$CH_3CH_2C(CH_3)_2CH_3$

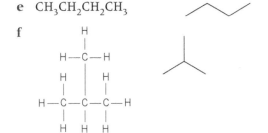

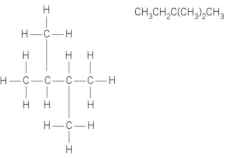

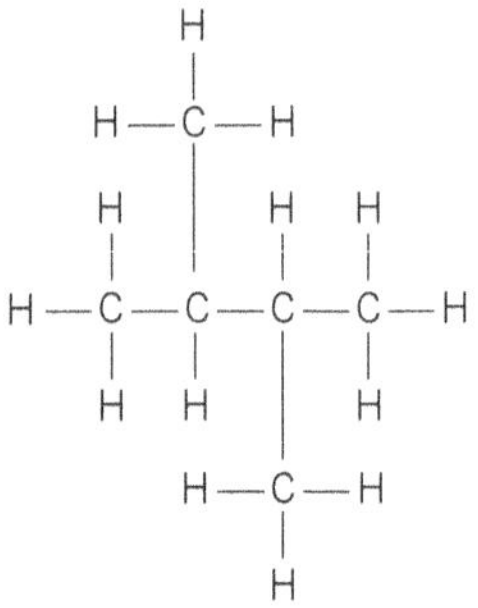

$CH_3CH(CH_3)CH(CH_3)CH_3$

b hexane 2-methylpentane 3-methylpentane 2,2-dimethylbutane 2,3-dimethylbutane

3 D

4 a *E*-pent-2-ene

b *Z*-pent-2-ene

12.1

1 methylbutane

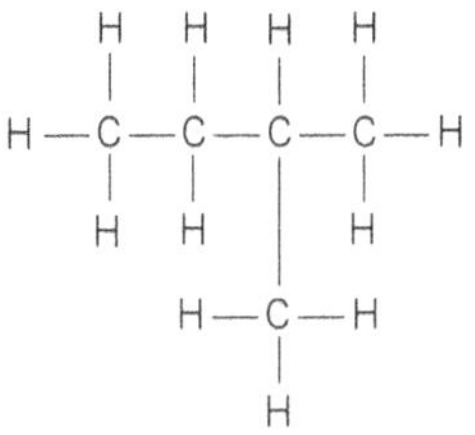

2

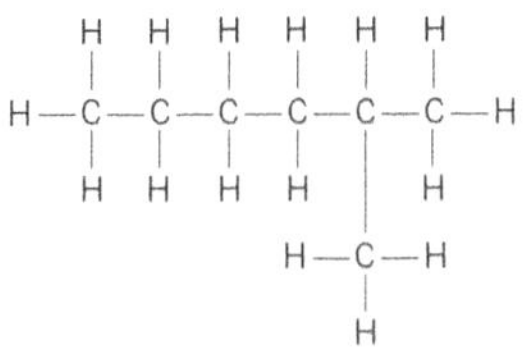

$CH_3CH_2CH_2CH_2CH(CH_3)CH_3$

3 heptane

4 Heptane will have a higher melting temperature because its straight chains will pack together more closely.

12.2

1

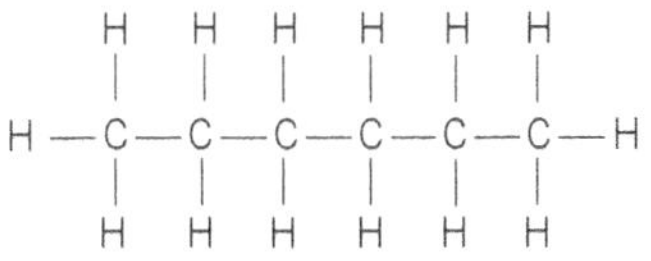

$CH_3CH_2CH_2CH_2CH_2CH_3$

2 petrol

3 Fractional distillation separates a mixture into several components with different ranges of boiling temperatures whereas distillation simply separates all the volatile components of a mixture from the non-volatile ones.

4 e.g., methane, ethane, propane, butane

12.3

1 decane → octane + ethene

2 Many of the products are gases rather than liquids.

3 Octane itself has a short enough chain length to be in demand.

4 By using a catalyst

5 Short chain products are in greater demand than long chain ones. Alkenes are more useful than alkanes as starting materials for further chemical reactions.

12.4

1 a propane + oxygen → carbon dioxide + water

$C_3H_8(g) + 5O_2(g) \rightarrow 3CO_2(g) + 4H_2O(l)$

b butane + oxygen → carbon monoxide + water

$C_4H_{10}(g) + 4\frac{1}{2}O_2 \rightarrow 4CO(g) + 5H_2O(l)$

2 a They produce carbon dioxide (a greenhouse gas) when they burn. They may produce poisonous carbon monoxide when burnt in a restricted supply of oxygen. They are in general non-renewable resources. They may produce nitrogen oxides and sulfur oxides when burnt. Cancer-causing carbon particulates may be produced and unburnt hydrocarbons (which contribute to photochemical smog) may be released into the atmosphere.

b Burn as little fuel as possible and/or offset the CO_2 produced by planting trees, for example. Ensure that burners are serviced and adjusted to burn the fuel completely. Remove sulfur from the fuel before burning or remove SO_2 from the combustion products (by reacting with calcium oxide, for example).

3 Possibilities include wind power, wave power, and nuclear power.

12.5

1 a termination b propagation

c propagation d initiation

2 a $Cl\bullet + O_3 \rightarrow ClO\bullet + O_2$ and $ClO\bullet + O_3 \rightarrow 2O_2 + Cl\bullet$

b $Cl\bullet + Cl\bullet$ $ClO\bullet + ClO\bullet$ $Cl\bullet + ClO\bullet$

13.1

1 a

i

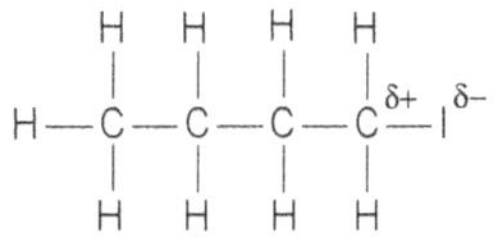

iii

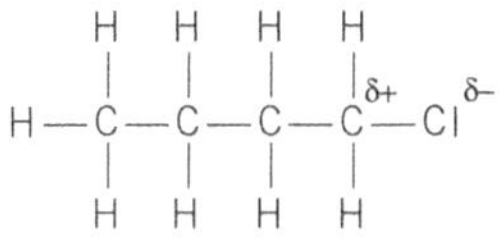

ii

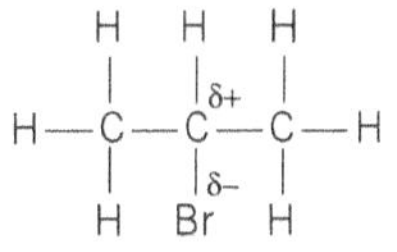

iv

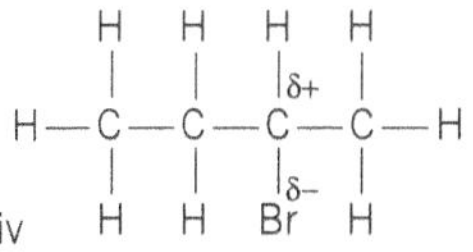

b i 1-iodobutane iii 1-chlorobutane

ii 2-bromopropane iv 2-bromobutane

c i, because it has the highest M_r and therefore most electrons and highest van der Waals forces.

2 Because the C—X bond becomes stronger

13.2

1 a Because haloalkanes do not dissolve in aqueous solutions.

b OH^-

c Because the OH group replaces the halogen atom.

d X^-

e RI

2 a CN^-

b

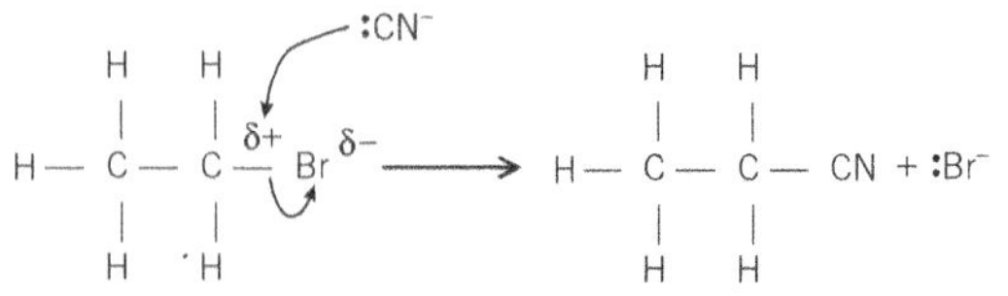

c Propanenitrile

13.3

1 A

2 D

3 a Propan-2-ol, propene

b Show that it decolourises a solution of bromine.

c

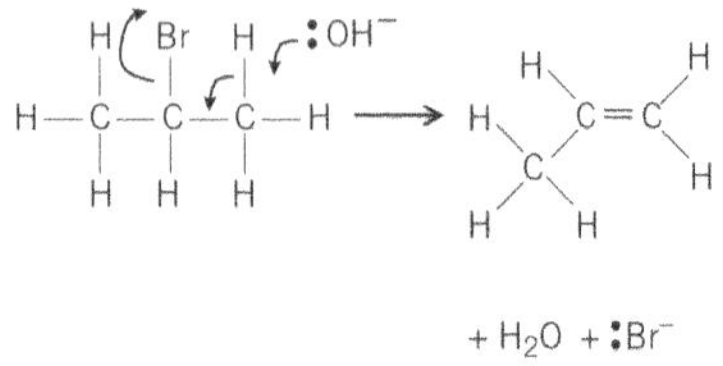

CFCs

1

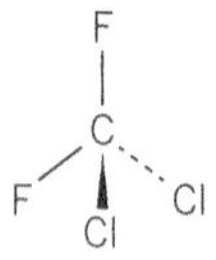

Tetrahedral with all atoms positioned around the carbon atom equivalent. No isomers.

2 a structural **b**

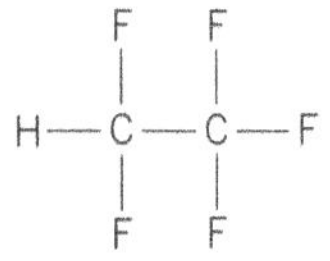

c C_2HF_5

14.1

1 hex-2-ene

2 $CH_3CH_2CH_2CH_2CH{=}CH_2$

3

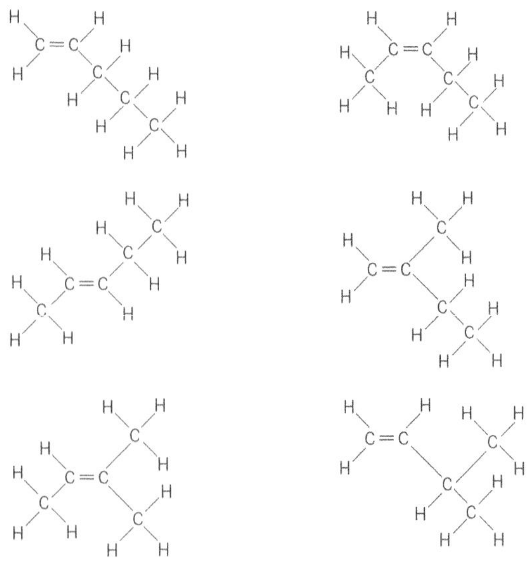

4 a electrophiles

5 a electron-rich

Bond energies

1 $612 - 347 = 265$ kJ mol^{-1}

2 The electron density in the σ-orbital is concentrated between the nuclei and holds them together better than the electron density of the π-orbital which is above and below the plane of the molecule.

14.2

1 $CH_2{=}CHCH_3 + 4\frac{1}{2}O_2 \rightarrow 3CO_2 + 3H_2O$

2 a electrophilic additions

3 a 1-bromopropane and 2-bromopropane

b 2-bromopropane

c It is formed from the more stable of the two intermediate carbocations.

4 chloroethane

5 c Bromine solution is decolourised.

14.3

1 A, B, and D

2 a

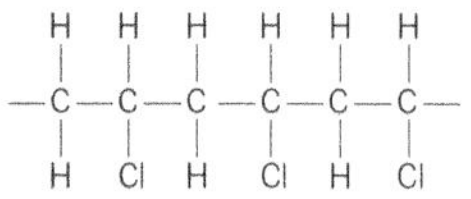

b vinyl chloride

c poly(chloroethene)

3 $CF_2{=}CF_2$

4 $CH_2{=}CHCl$

14.4

1 100%; it is an addition reaction.

2

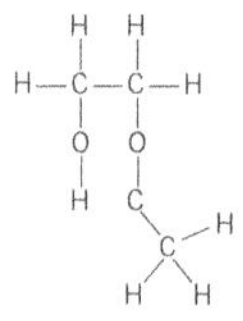

3. They each have one (or more) lone pairs of electrons. The N or O atom has a partial negative charge.

4. Restrict the supply of oxygen.

15.1

1

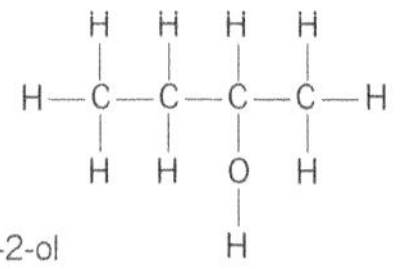

butan-2-ol

2 primary: methanol
secondary: butan-2-ol
tertiary: 2-methylpentan-2-ol

3 Because the oxygen atom has two lone pairs which repel more than bonding pairs.

Antifreeze

1 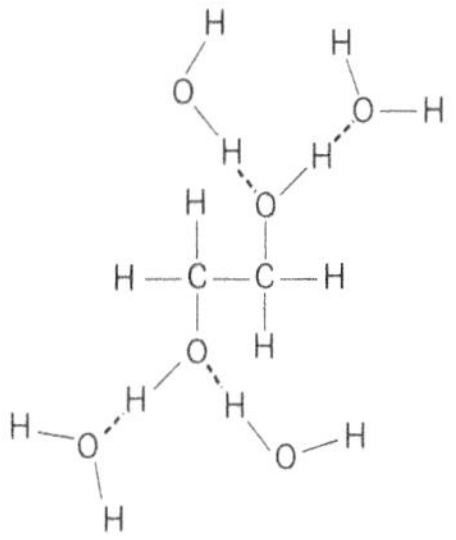

2 The ethane-1,2-diol molecules can form hydrogen bonds with water but cannot fit into the regular three-dimensional structure of ice, see Topic 3.5. So these solutions remain liquid at lower temperatures than pure water.

The reactivity of alcohols

1 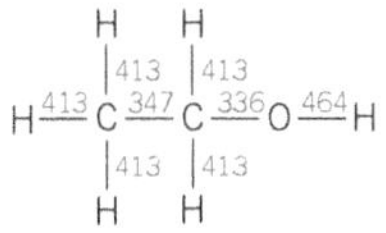

2 The C—C and C—H bonds are strong and relatively non-polar. The weakest bond is C—O which is also polarised $^{\delta+}C—O^{\delta-}$ so the $C^{\delta+}$ can be attached by nucleophiles.

15.2

1 a A carboxylic acid is formed.
   b A ketone is formed.

2 This would require a C—C bond to break.

3 In distillation, the vapour is removed from the original flask and condensed in a different one. In refluxing, the vapour is condensed and returned to the original flask.

4 Gently oxidise the alcohols. In the case of the primary alcohol, an aldehyde will be formed that will give a positive silver mirror or Fehling's test. In the case of the secondary alcohol, a ketone will be formed that will not give a positive silver mirror or Fehling's test.

5 $CH_3CH_2OH \rightarrow CH_2{=}CH_2 + H_2O$ ethene

6 Pent-1-ene and pent-2-ene (*E*- and *Z*-isomers)

16.1

1 Test the solubility of the precipitate in ammonia – see Topic 10.3.

2 a To remove any CO_3^{2-} or OH^- ions present which would also form a precipitate.
   b This would form a precipitate of AgCl.

3 a alkane and carboxylic acid
   b $CH_2{=}CHCO_2H$
   c $CH_2{=}CHCO_2H + Br_2 \rightarrow CH_2Br—CHBrCO_2H$
   $CH_2{=}CHCO_2H + NaHCO_3 \rightarrow CH_2CHCO_2Na + H_2O + CO_2$

16.2

1 a A solution of the molecules passes through a positively charged hollow needle.
   b positive

2 $C_{10}H_{16}$

16.3

1 a or b

2 This IR peak is caused by C=O which is present in both **a** and **b** but not **c**.

3 b or c

4 This IR peak is caused by O—H which is present in **b** and **c** but not **a**.

5 b

6 This compound has both C=O and O—H.

Greenhouse gases

1 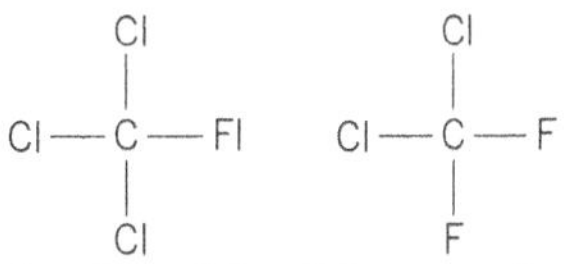

2 C—Cl and C—F; because both molecules have the same bonds.

3 Temperature will vary as well as other climatic conditions.

4 CO_2: 1 × 350 = 350
   CH_4: 30 × 1.7 = 51

17.1

1 a +788 kJ mol^{-1}
   b This is the reverse of the equation for lattice enthalpy formation, so the sign of ΔH is changed.
   c The enthalpy change of lattice dissociation.

2 a The electron is attracted by the sodium nucleus, so energy must be put in to remove it.
   b The electron is attracted by the chlorine nucleus so energy is given out during this process.

3 a i $Al(g) \rightarrow Al^+(g) + e^-$
      ii $Al^+(g) \rightarrow Al^{2+}(g) + e^-$
   b The sum of the enthalpies for **ai** and **ii**.

17.2

1 a

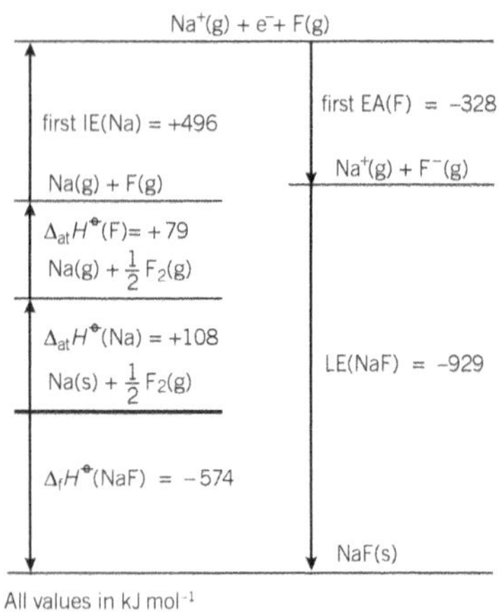

b $-929\ kJ\ mol^{-1}$

The first noble gas compound

1 $O(g) \rightarrow O^+(g) + e^-$

2 Xenon and krypton have lower ionisation energies than helium and neon, therefore it will be easier to form positive ions Xe^+ and Kr^+.

3 It is a compound of a metal and a non-metal – these are usually ionic and therefore expected to have high boiling points.

4 Xe is +1 as it has a single positive charge. F is always −1 in compounds and there are six of them. So the Pt must be +5 so that the sum of the oxidation numbers in the neutral compound is zero.

17.3

1

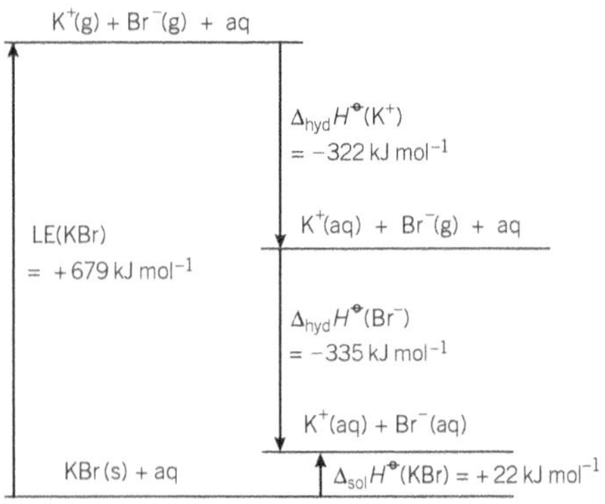

$+22\ kJ\ mol^{-1}$.
Small because the energy put in to break the lattice is of similar size to that given out when the ions are hydrated.

2 They are small and highly charged positive ions so they strongly polarise negative ions. The calculated value would be greater because there is extra covalent bonding.

17.4

1 a i Approximately zero, two solids produce two solids.

ii Significantly positive, a solid produces several moles of gases.

iii Significantly negative, a gas turns into a solid.

iv Significantly positive, a liquid turns into a gas.

b i $-7.8\ J\ K^{-1}\ mol^{-1}$

ii $+876.4\ J\ K^{-1}\ mol^{-1}$

iii $-174.8\ J\ K^{-1}\ mol^{-1}$

iv $+119\ J\ K^{-1}\ mol^{-1}$

The predictions are upheld.

2 a i $+493\ kJ\ mol^{-1}$ **ii** $-52\ kJ\ mol^{-1}$

iii It is feasible at 6000 K

b 5523 K

3 $-284\ J\ K^{-1}\ mol^{-1}$

Determining an entropy change

1 240 kJ   2 $M_r = 18.0$ 5.55 mol

3 43.2 kJ   43 200 J   4 $115.8\ J K^{-1} mol^{-1}$

5 2 s.f. as 2.4 kW has the smallest number of sf in the data, so $\Delta_{vap}S = 120\ J K^{-1} mol^{-1}$ to 2 s.f. (We cannot be sure of the number of s.f. for the two values of 100 g and 100 but they are probably 3 s.f.

6 Heat loss from the sides of the kettle. The kettle should be insulated.

7 1%

8 There are hydrogen bonds between the molecules in water in the liquid state. This makes the liquid state more ordered that for non-hydrogen bonding liquids.

18.1

1 A reactant – its concentration decreases with time.

2 $1.3 \times 10^{-3}\ mol\ dm^{-3}\ s^{-1}$ to 2 s.f.

3 The rate of reaction after 300 seconds.

4 That at time 0 seconds, the gradient will be steeper than that at time 600 seconds.

5 The rate of reaction at the beginning is greatest as the concentration of reactants is greatest here and then decreases as reactants are used up.

Fast reactions

There are several possibilities, for example, two could combine to form an O_2 molecule or one could recombine with a ClO• radical in the reverse of the original reaction, and so on.

18.2

1 rate = $k[A][B][C]^2$

2 a i 1 ii 1 iii 2
b i double ii double iii quadruple
c i 1 ii 5 iii 6
iv 3 iv 3
d $dm^9\ mol^{-3}\ s^{-1}$

3 a the rate constant
b i 2 ii 0 iii 0
iv 1
c 3
d $dm^6\ mol^{-2}\ s^{-1}$
e a catalyst

4 D It is impossible to tell without experimental data.

18.3

1 At 300 K: 36 in 10^{19}
At 310 K: 13 in 10^{18}

2 103.8 $kJmol^{-1}$

18.4

1 a i 1 ii 2
b 3 c 27 $mol\,dm^{-3}\ s^{-1}$ d rate = $k[A][B]^2$
e No, other species might be involved that have not been investigated.

The iodine clock reaction

1 Iodide ions (oxidation state –1) are oxidised to iodine (oxidation state 0)

2 Hydrogen peroxide

3 The rate doubles

4 First order

5 Nothing

6 By making a stock solution and diluting it with water.

7 Temperature, the concentration of potassium iodide, the concentration of H^+ ions.

18.5

1 a B b F and D c Step (ii)

19.1

1 a i move right ii move left
b i move left ii move right
c i no effect ii move right

2 a increase b no change c no change

3 a i move left ii move left
iii no change unit Pa^{-1}
b i move right ii no change
iii no change no unit
c i move right ii move right
iii no change unit Pa

20.1

1 a $Ni(s)|Ni^{2+}(aq)\ ||\ Ag^+(aq)|Ag(s)$
$E = +\ 1.05$ V
b Electrons would flow from the nickel electrode to the silver electrode.
$Ni(s) \rightarrow Ni^{2+}(aq) + 2e^-$
$Ag^+(aq) + e^- \rightarrow Ag(s)$
Overall:
$Ni(s) + 2Ag^+(aq) \rightarrow Ni^{2+}(aq) + 2Ag(s)$

2 a +0.63 V b –0.63 V

20.2

1 0.90 V zinc electrode positive

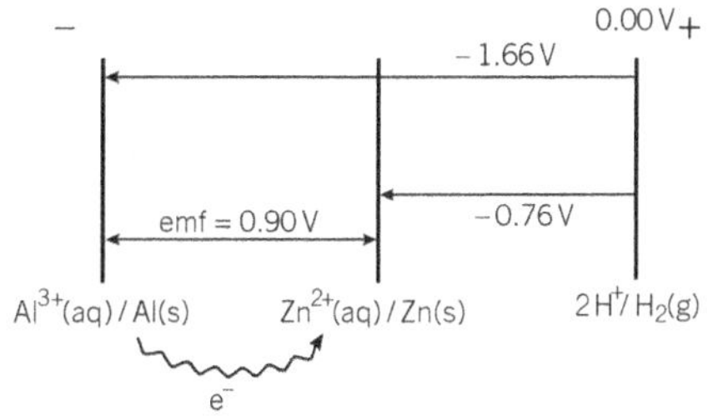

2 a +0.93 V b –0.53 V c +0.97 V d +0.13 V

3 Chlorine and bromine, but not iodine.

4 a no b no

20.3

Answer from main text.

21.1

1 a acid: HNO_3
base: OH^-
b acid: CH_3COOH
base: H_2O

2 $1 \times 10^{-10}\ mol\,dm^{-3}$

3 a H_2O b NH_4^+ c H_3O^+ d HCl

21.2

1 2.00 2 1×10^{-6} mol dm^{-3} 3 1×10^{-5} mol dm^{-3}
4 1.70 5 13.30

Mixing bathroom cleaners

1 $CaCO_3(s) + 2HCl(aq) \rightarrow CaCl_2(aq) + CO_2(g) + H_2O(l)$

2 5×10^{-2} mol HCl 3 5×10^{-2} mol Cl_2 4 1200 cm^3

5 Some chlorine would remain dissolved in the water in the toilet bowl.

21.3

1 chloroethanoic acid 2 They are the same.

3 a 1.94 b 3.10

21.4

1 a $2NaOH(aq) + H_2SO_4(aq) \rightarrow Na_2SO_4(aq) + 2H_2O(l)$

b 0.50 c 1.5×10^{-3} d 0.120 mol dm^{-3}

2 a i A ii B

b i B iv A

c Strong, as there is a rapid pH change in the alkaline region at the equivalence point.

21.5

1 b and d 2 b and d

21.6

1 a 4.47 b 4.20

Making a buffer solution

1 a 122.1 b 144.1

2 a 1.53 g benzoic acid b 3.60 g

3 Accurately weigh out the two compounds into separate weighing boats. Using a funnel in the mouth of the flask, transfer all of each solid into the flask using a wash bottle containing distilled (deionised) water. Fill the flask with water to a few centimetres below the graduation mark. Stopper the flask and shake until all the solid has dissolved. Now carefully fill the flask with water using the wash bottle until the meniscus of the solution sits on the graduation line. Stopper the flask and shake again to ensure that the concentration of the buffer solution is uniform.

22.1

1 They are malleable (can be beaten into sheets) and ductile (can be drawn into wires). They are good conductors of heat.

2 They tend to be brittle – they are poor conductors of heat.

3 a +1

b −1. Oxidation state of oxygen is usually −2.

c +2 −2 +1
$Mg(OH)_2 = +2 + 2(-2 + 1) = 0$

4 +6

22.2

1 a $Na_2O(s) + H_2O(l) \rightarrow 2NaOH(aq)$

b i before +1, after +1 ii neither

2 a OH^- ions b greater than 7

3 a acidic

b Phosphorus is a non-metal.

c $P_4O_6(s) + 6H_2O(l) \longrightarrow 4H_3PO_3(aq)$

The structures of oxo-acids and their anions

1 109.5°

2 They are all the same (and intermediate between the P═O and P—O lengths).

3 10

4 0.150 nm (the average of S—O and S═O)

5 10

22.3

1 a $Na_2O(s) + 2HCl(aq) \rightarrow 2NaCl(aq) + H_2O(l)$

b $MgO(s) + H_2SO_4(aq) \rightarrow MgSO_4(aq) + H_2O(l)$

c $Al_2O_3(s) + 6HNO_3(aq) \rightarrow 3H_2O(l) + 2Al(NO_3)_3(aq)$

2 a
$$\overset{+4\ -2}{SiO_2}(s) + 2\overset{+1\ -2\ +1}{NaOH}(aq) \rightarrow \overset{+1\ +4\ -2}{Na_2SiO_3}(aq) + \overset{+1\ -2}{H_2O}(l)$$

b No

c No change in oxidation number of any of the elements occurs.

3 $P_4O_{10}(s) + 6H_2O(l) \rightarrow 4H_3PO_4(aq)$

22.4

1 $Si(s) + 2Cl_2(g) \rightarrow SiCl_4(l)$

2 No, there is no change in the oxidation number of any element.

3 $PCl_3(p) + 3H_2O(l) \rightarrow H_3PO_3(aq) + 3H^+(aq) + 3Cl^-(aq)$

23.1

1 a $1s^2\ 2s^2\ 2p^6\ 3s^2\ 3p^6\ 3d^5$

b $1s^2\ 2s^2\ 2p^6\ 3s^2\ 3p^6\ 3d^4$

2 a $1s^2\ 2s^2\ 2p^6\ 3s^2\ 3p^6$

b the two 4s electrons

c one of the 3d electrons

23.2

1 a i octahedral ii octahedral iii tetrahedral

b 6, 6, 4

c Cl^- is a larger ligand than either ammonia or water, so fewer ligands can fit around the metal ion.

2 a The two negatively charged oxygen atoms, because they have lone pairs of electrons.

b

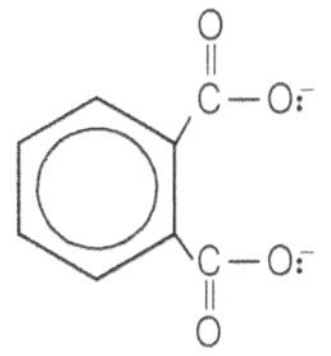

c bidentate

Ionisation isomerism

1 They both a have lone pairs of electrons.

2 Compound 1 as it has most ions.

3 $AgNO_3(aq) + Cl^-(aq) \rightarrow AgCl(s) + NO_3^-(aq)$

4 Compound 1 as it has most free chloride ions.

5 *Cis trans (E–Z)* isomerism

6

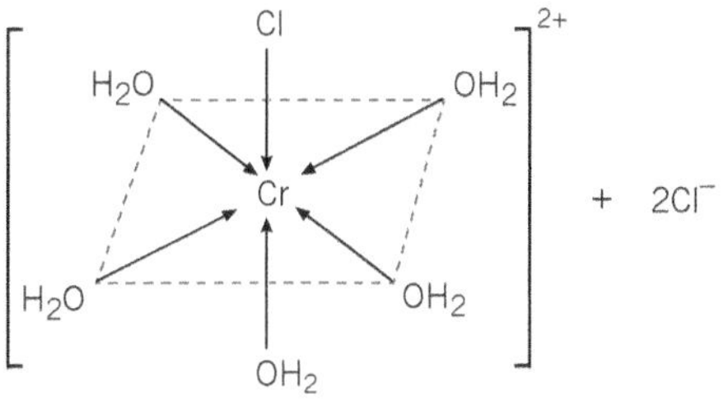

23.3

1 a The copper ion has part-filled d-orbitals, so electrons can move from one d-orbital to another and absorb light. Zinc has full d-orbitals.

b blue c They are absorbed.

2 a 5 : 5 b 5 : 5 c $[NiEDTA]^{2-}$

23.4

1 a $Zn(s) + 2VO_2^+ + 4H^+ \rightarrow 2H_2O + 2VO^{2+} + Zn^{2+}(aq)$

$Zn(s) + 2VO^{2+}(aq) + 4H^+(aq) \rightarrow Zn^{2+}(aq) + 2H_2O(l) + 2V^{3+}(aq)$

$Zn(s) + 2V^{3+}(aq) \rightarrow Zn^{2+}(aq) + 2V^{2+}(aq)$

b This low oxidation state can be oxidised by air.

2 0.0698 g

3 $E^\ominus$ for the reaction:

$Cl_2(g) + 2e^- \rightarrow 2Cl^-$

so

is + 1.36 V, $E^\ominus$ for the reaction of $Cr_2O_7^{2-}$ and Cl^- is –0.03 V. Therefore the reaction is not feasible. Hence potassium dichromate(VI) will not oxidise chloride ions to chlorine.

4 a +7 b +6 c +6

5 acid base, H^+ ions are donated to O atoms.

23.5

1 a Homogeneous catalysts are in the same phase as the reactants, heterogeneous catalysts are in a different phase from the reactants.

b i heterogeneous ii heterogeneous

iii homogeneous

2 It lowers the activation energy. This means that reactions can be carried out at a lower temperature than without the catalyst, so saving energy and money.

3 First, iron(III) oxidises iodide to iodine, itself being reduced to iron(II):

$2Fe^{3+}(aq) + 2I^-(aq) \rightarrow 2Fe^{2+}(aq) + I_2(aq)$

Then the peroxodisulfate ions oxidise the iron(II) back to iron(III):

$S_2O_8^{2-}(aq) + 2Fe^{2+}(aq) \rightarrow 2SO_4^{2-}(aq) + 2Fe^{3+}(aq)$

Catalyst is needed since the reaction is slow. The reaction is slow since both reactants are negative ions. These ions will repel each other, reducing the chance of effective collisions.

4 $4Mn^{2+}(aq) + MnO_4^-(aq) + 8H^+(aq) \rightarrow 5Mn^{3+}(aq) + 4H_2O(l)$

$2Mn^{3+}(aq) + C_2O_4^{2-}(aq) \rightarrow 2CO_2(g) + 2Mn^{2+}(aq)$

Multiply the top equation by 2 and the lower one by 5 to give the same number of Mn^{3+} ions:

$8Mn^{2+}(aq) + 2MnO_4^-(aq) + 16H^+(aq) \rightarrow 10Mn^{3+}(aq) + 8H_2O(l)$

$10Mn^{3+}(aq) + 5C_2O_4^{2-}(aq) \rightarrow 10CO_2(g) + 10Mn^{2+}(aq)$

Then combine the two equations and cancel any species that appear on both sides:

$2MnO_4^-(aq) + 16H^+(aq) + 5C_2O_4^{2-}(aq) \rightarrow 2Mn^{2+}(aq) + 8H_2O(l) + 10CO_2(g)$

24.1

1 2+

2 6

3 $[Cu(H_2O)_6]_2+ (aq) + 2OH^-(aq) \rightarrow Cu(H_2O)_4(OH)_2(s) + 2H_2O(l)$

24.2

1 $Cu(H_2O)_4Cl_2$ or $Cu(H_2O)_2Cl_2$

2 a 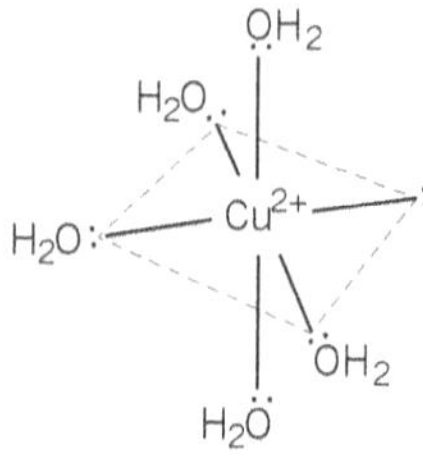

octahedral

b Bromide ions are bigger than water molecules, so fewer can fit around the copper ion. The shape is tetrahedral.

3 $[Cu(H_2O)_6]^{2+}(aq) + en \rightarrow [Cu(H_2O)_4(en)]^{2+}(aq) + 2H_2O(l)$

$[CuH_2O_4(en)]^{2+}(aq) + en \rightarrow [Cu(H_2O)_2(en)_2]^{2+}(aq) + 2H_2O(l)$

$[Cu(H_2O)_2(en)_2]^{2+}(aq) + en \rightarrow [Cu(en)_3]^{2+}(aq) + 2H_2O(l)$

At each step two entities produce three, therefore the entropy change of each step is likely to be positive.

4 a No, it remains at +2.

b $[Co(H_2O)_6]^{2+}$ is octahedral and $[CoCl_4]^{2-}$ is tetrahedral.

c The ligands have changed as has the co-ordination number.

24.3

1 They are smaller and more highly charged and therefore more strongly polarising. They can thus weaken one of the O—H bonds in one of the water molecules that surround them, so releasing a H^+ ion.

2 Aluminium is not a transition metal. It has no part-filled d-orbitals. Most transition metal compounds are coloured because of the electrons moving between part-filled d-orbitals and absorbing light.

3 Both NH_3 and H_2O are of similar size and are neutral ligands.

4 a i $[Fe(H_2O)_6]^{3+}(aq) + 3OH^-(aq) \rightarrow Fe(H_2O)_3(OH)_3(s) + 3H_2O(l)$

ii $[Cu(H_2O)_6]^{2+}(aq) + 4NH_3(aq) \rightarrow [Cu(NH_3)_4(H_2O)_2]^{2+}(aq) + 4H_2O(l)$

b i The orange brown solution changes to a reddish brown solid.

ii The pale blue solution turns to dark blue.

25.1

1 a 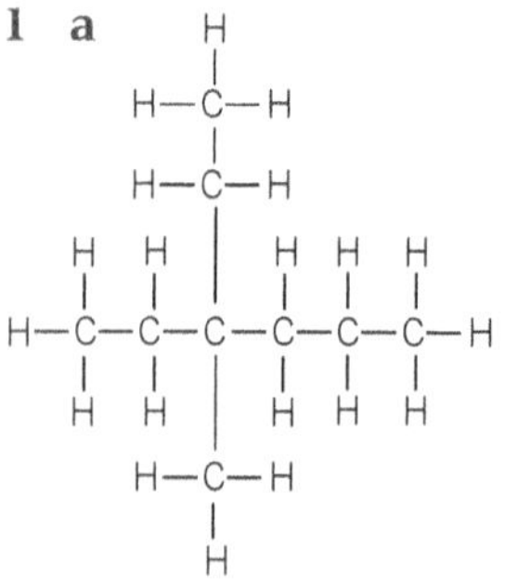

b 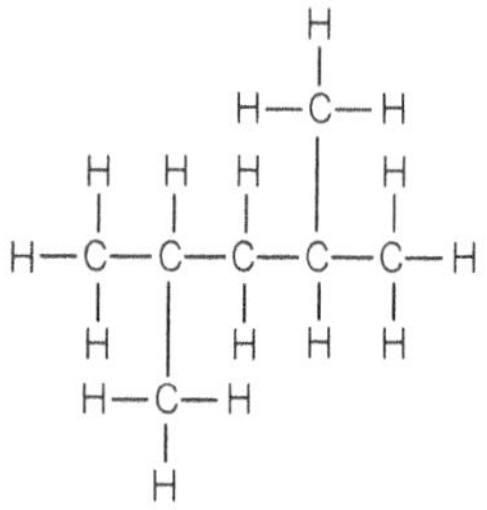

2 a propan-1-ol **b** 2-chloropropane

c hex-3-ene **d** ethanol

e pentanoic acid

25.2

1 b

2 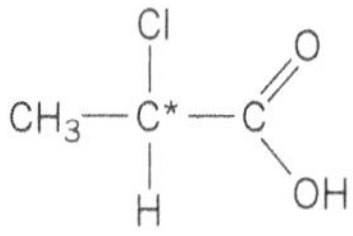

3 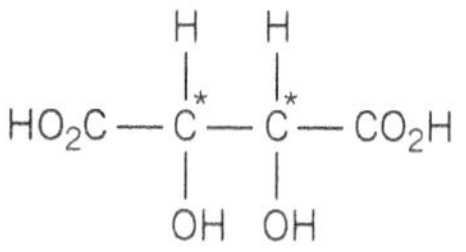

25.3

1 a 2-hydroxybutanoic acid

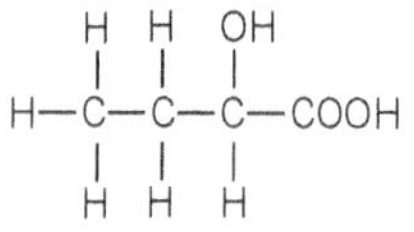

b Yes, the carbon to which the –OH group is bonded has four different groups attached to it.

2 a 2-methyl-2-hydroxypropanoic acid

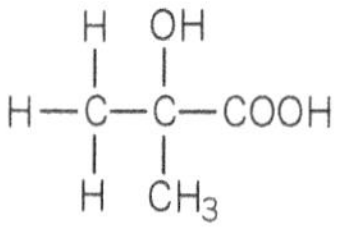

b No, it has no carbon to which four different groups are attached.

3 This carbon atom does not have four different groups attached to it.

26.1

1 a pentan-3-one **b** propanal

2 a A ketone must have a C═O group with two carbon-containing groups attached to it, so at least three carbon atoms are required.

b The C═O group can only be on carbon 2, otherwise the compound would be an aldehyde.

c The C═O group must always be on the end of the chain.

3 A hydrogen bond requires a molecule with a hydrogen atom covalently bonded to an atom of fluorine, oxygen, or nitrogen. There is no such carbon atom in propanone.

4 The hydrogen atom in an –OH group in a water molecule can hydrogen bond with the oxygen atom in the C=O bond of propanone.

26.2

1 Cl^- 2 a no

b A negatively charged nucleophile will be repelled by the high electron density in the C=C.

c

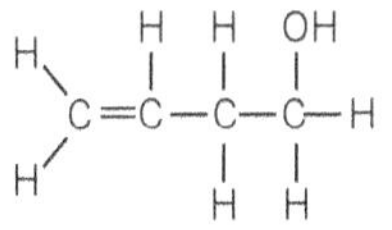

3

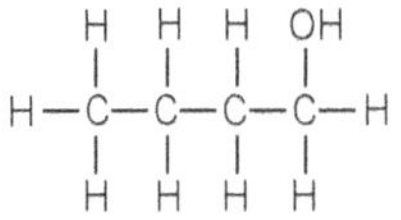

4 The CN^- ions can attack the planar C=O group from above or below. In the case of CH_3CHO this will produce a compound with a chiral centre (four different groups attached to it). This will be a pair of optical isomers. As attack from above or below is equally likely it will be a 50:50 mixture (racemic). In the case of CH_3COCH_3, the product will not have a carbon atom with four groups attached to it. It will not therefore be chiral and will be a single compound.

26.3

1 3-bromobutanoic acid 2

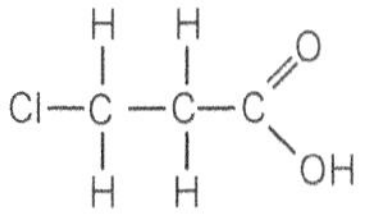

3 The carboxylic acid must be at the end of a chain and therefore in the 1 position.

4 ethyl ethanoate; methyl propanoate

26.4

1 They react with carbonates and hydrogencarbonates to give carbon dioxide, metal oxides to give salts, and alkalis to give salts.

2 methanol and ethanoic acid

3 ethanol and methanoic acid

4 They have the same molecular formula but have a different arrangement of their atoms in space.

26.5

1 The carbon of the C=O group is strongly $C^{\delta+}$ because, as well as oxygen, it is also bonded to an electronegative chlorine atom which draws electrons away from it.

2 OH^-, because it is negatively charged and contains a lone pair of electrons.

3 Because the nucleophile and the acylating agent join together and then a small molecule is lost.

4

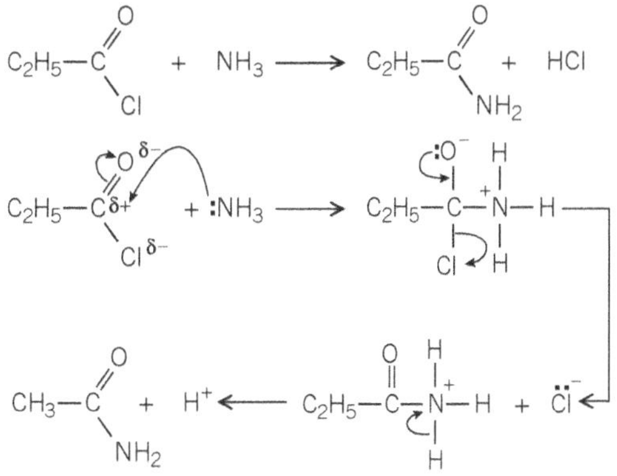

27.1

1 CH 2 3

3 Six electrons are spread out over all six carbon atoms in the ring rather than being localised in three distinct double bonds.

4

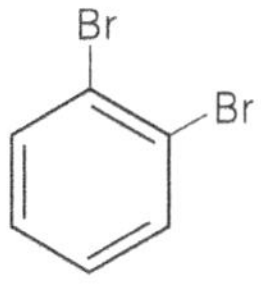

27.2

1 van der Waals forces

2 a 1,3-dichlorobenzene

b 1-bromo-4-chlorobenzene

3 a

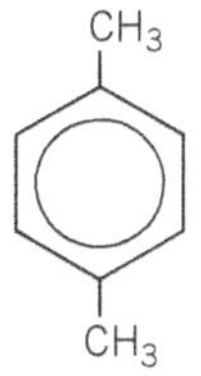

b

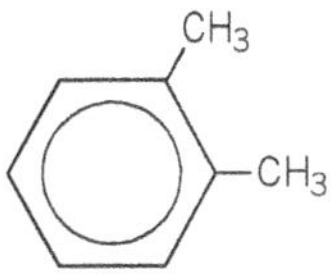

4 R^+

27.3

1 a A electrophilic substitution

b A electrophilic substitution

2 1,2-dinitrobenzene and 1,4-dinitrobenzene

3 Addition reactions would involve the loss of aromatic stability.

4

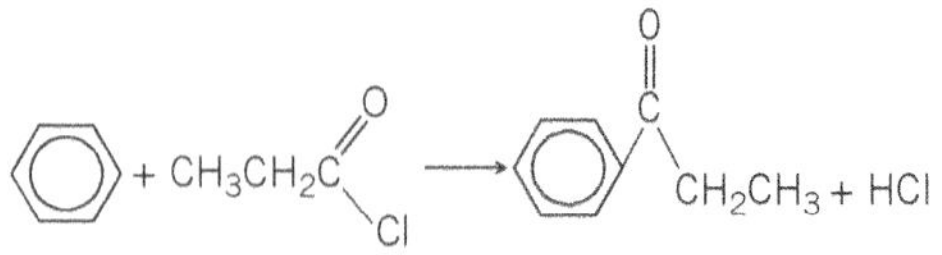

$CH_3CH_2CO^+$

27.4

1

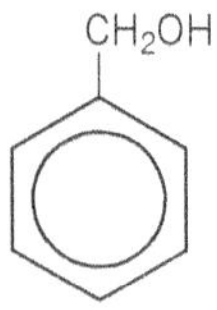

An elimination product would involve removing an H atom from the ring, thus destroying the aromatic system.

2 $CH_3CH_2CH_2Cl$ and $AlCl_3$.

3 The S atom is polarised δ + and will attack the electron-rich benzene ring while the O atoms are polarised δ–.

4

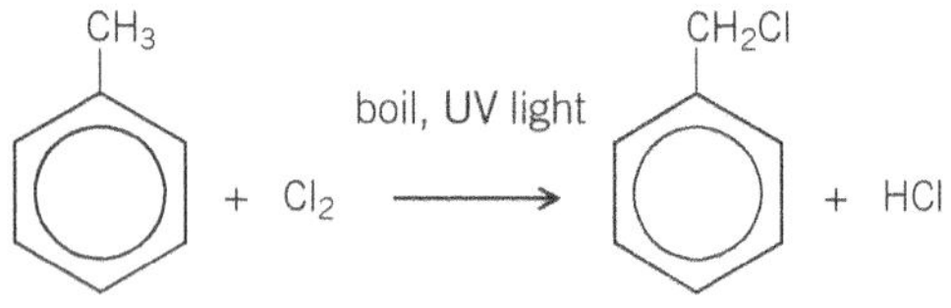

28.1

1 secondary

2 ethylpropylamine

3

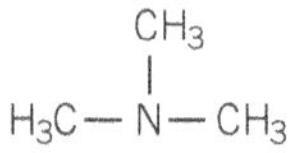

4 a gas

5 It has the same value of M_r as ethylamine, which is a gas.

28.2

1 a $(CH_3)_2NH + HCl \rightarrow (CH_3)_2NH_2^+ + Cl^-$

b dimethylammonium chloride or dimethylamine hydrochloride.

2 a It will dissolve, that is, the oily drops will disappear to give a colourless solution.

b The oily drops will re-appear.

3 Stronger, as it has two electron-releasing alkyl groups.

Solubility of drugs

1 secondary

2

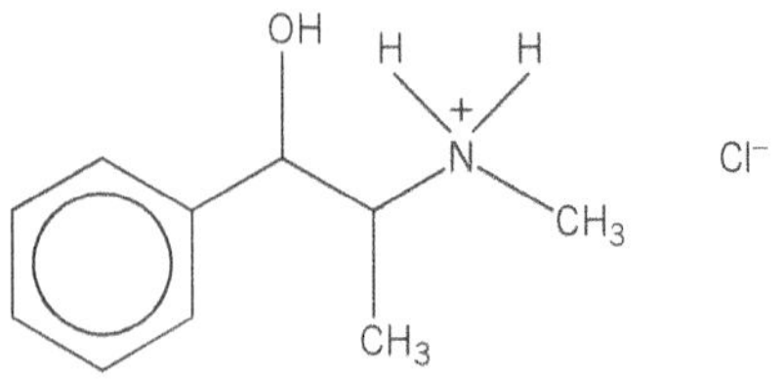

3 It is an ionic compound.

4 Solid, because it is ionically bonded.

5 –OH, alcohol, and an aromatic ring.

6

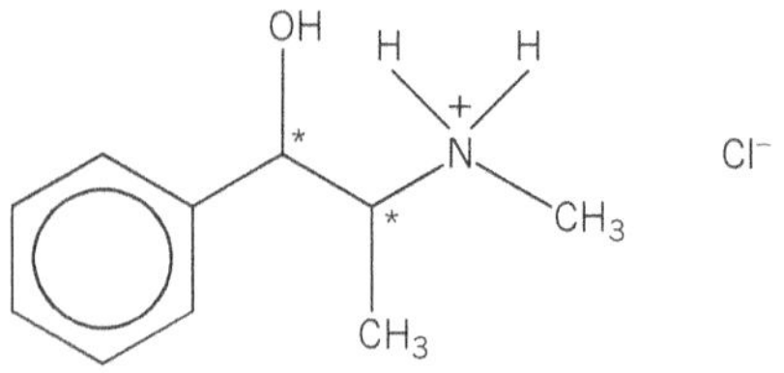

7 It is possible that only one of the optical isomers is active as a drug.

28.3

1 Because secondary and tertiary amines (and quaternary ammonium salts) may be formed as well.

2 a $CH_3CH_2Cl + 2NH_3 \rightarrow CH_3CH_2NH_2 + NH_4Cl$

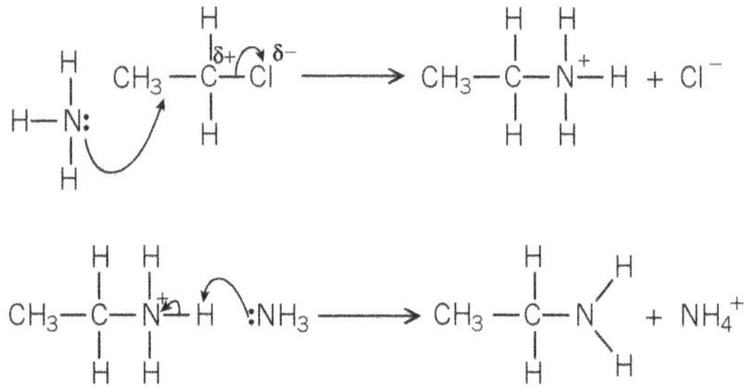

b $(C_2H_5)_2NH$; $(C_2H_5)_3N$; $(C_2H_5)_4N^+Cl^-$

29.1

1 a the number of carbon atoms in each monomer

b 1,10-diaminodecane

$NH_2(CH_2)_{10}NH_2$

2 They also have CONH linkages.

3 any diol, for example, propane-1,3,-diol

4 a amide b ester

5 $\text{+CO–R–CO–O–R'–O+}_n + nH_2O \rightarrow$

$n\text{+CO–R–COOH + HO–R'–O+}$

Hermann Staudinger

1 2-methylbuta-1,3-diene

2 Addition since it contains a —C=C— in the structure.

30.1

1 2-aminopropanoic acid

2 Carbon number 2 in alanine has four different groups attached to it.

30.2

1 a amine and carboxylic acid

b The carboxylic acid is acidic; the amine is basic.

2 two

3 They will be protonated, i.e., there will be a H^+ ion attached to the $-NH_2$. $^+H_3NCH(CH_3)\ CO_2H$

4 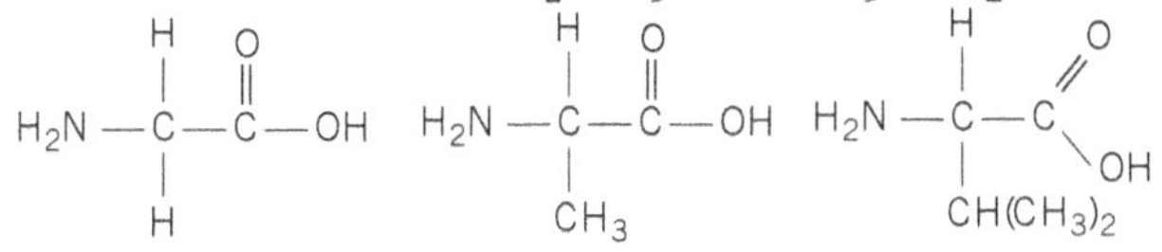

2-dimensional TLC

(approximate values) Solvent 1: orange = 0.85 blue = 0.60

Solvent 2: orange = 0.52 blue = 0.65

30.3

1 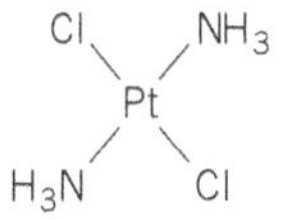

The two chloride ligands are too far apart to effectively bond to two adjacent guanine bases.

31.1

1 a React with HCN.

b React with methanol (with an acid catalyst).

c Addition of water (by reaction with concentrated sulfuric acid followed by water).

d Dehydrate with P_2O_5, for example.

2 a Add water (by reaction with concentrated sulfuric acid followed by water) to produce ethanol, then oxidise with $K_2Cr_2O_7/H_2SO_4$.

b Reduce (using $NaBH_4$, for example) to propan-2-ol, then react with KBr/H_2SO_4.

3 Step 1: Friedel–Crafts acylation using ethanoyl chloride and aluminium chloride.

Step 2: Nitration-electrophilic substitution using a mixture of concentrated nitric and sulfuric acids.

Step 3: Reduction (hydrogenation) using Sn/ concentrated HCl.

31.2

1 a This will release the halogen as a halide ion.

$RX + OH^- \rightarrow$ to $ROH + X^-$

b Hydrochloric acid already contains Cl^- ions.

2 a Isomers have the same molecular formula but a different arrangement of atoms in space.

b B carboxylic acid

C alcohol and ketone

C alcohol and aldehyde

c Add sodium carbonate solution and it will fizz.

Tests C and D with Benedict's or Fehling's solution. D will give a positive solution and C will not.

32.1

1 a a single peak

b Both carbon atoms are in exactly the same environment.

2 The upper one is propan-1-ol, as it has three peaks because all three carbon atoms are in different environments. The lower one is propan-2-ol, as it has two peaks because there are carbon atoms in just two different environments.

A brief theory of NMR

They are all odd numbers.

32.2

1 a Isomers have the same molecular formula but different arrangements of atoms in space.

b
propan-1-ol: $\overset{A}{CH_3}\overset{B}{CH_2}\overset{C}{CH_2}\overset{D}{OH}$

propan-2-ol: $\overset{A}{CH_3}\overset{B}{CH}\overset{C}{(OH)}\overset{A}{CH_3}$

c i 4 ii 3

d i A3; B2; C2; D1 ii A6; B1; C1

e i D>C>B>A

ii C>B>A

32.3

1 a A is ethanol, B is methoxymethane.

b A: 4.5 is R—O—H, 3.7 is $R—CH_2—O$-, 1.2 is $R—CH_3$

B 3.3 is $—O—CH_3$

2 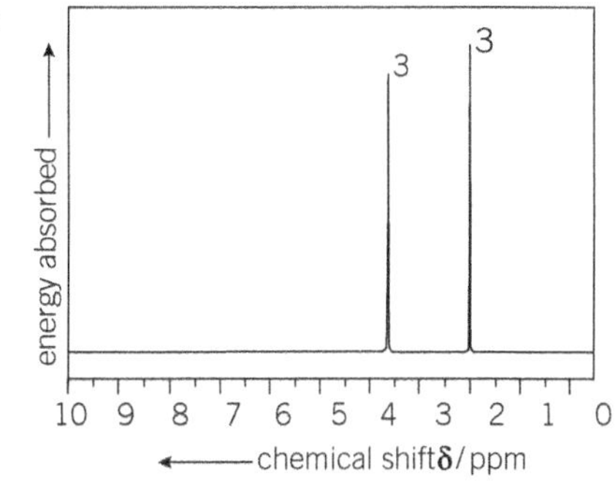

There is no spin-spin coupling because the two CH_3 groups are not on adjacent carbon atoms.

33.1

1 In column chromatography the eluent is a liquid. In gas–liquid chromatography it is a gas.

2 It can separate and help to identify minute traces of substances. For example, (among others) substances used in making explosives. [Many other suggestions are also acceptable.]

3 a A b C c A d C

Maths Section Answers

1 a Could be 1, 2, 3, 4 or 5 – you need to state the number

b 4

c 2

d 4

2 a It is also quadrupled

b $x \propto \frac{1}{y}$

c It reduces to one-third of its original volume

3 a $P = RT/V$

b $V = RT/P$

c $T = PV/R$

d $R = PV/T$

4 a $q = pr$

b $m = n/t$

c $h = fe/g$

d $r = se/p$

5 2.9 $cm^3\ s^{-1}$

6 0.23 $cm^3\ s^{-1}$

7 a 3

b −2

c −4

d 1.683

e −1.431

8 a 1000

b 0.01

c 1×10^{14}

d 1.584×10^{8}

e 2.344

Index